SOIL ECOLOGY

SOIL ECOLOGY

by

Patrick Lavelle

Université de Paris VI/ORSTOM

and

Alister V. Spain

CSIRO Land and Water

SPRINGER-SCIENCE+BUSINESS MEDIA, B.V.

Library of Congress Cataloging-in-Publication Data

ISBN 978-94-017-5281-7 ISBN 978-94-017-5279-4 (eBook)
DOI 10.1007/978-94-017-5279-4

Printed on acid-free paper

TABLE OF CONTENTS

ACKNOWLEDGEMENTS

We would like to acknowledge the assistance of the many friends and colleagues who have helped us in the preparation of this book. Because of its inter-disciplinary nature, a wide range of colleagues in particular fields have provided us with information and comment in the areas of their specialities and have criticised early drafts. We owe them a great debt of gratitude but accept the inevitable errors as our own.

We particularly thank the following:

Alain Albrecht, Jean André, Pierre Arpin, Jacques Balandreau, France Bernhard-Reversat, Nicolas Bernier, Jean-Marie Betsch, Eric Blanchart, Marcel Bouché, Alain Braumann, Keith Bristow, Michel Brossard, Georges Brown, Yves Cambefort, Remi Chaussod, Armand Chauvel, Claire Chenu, Marie-Madeleine Coûteaux, H.G. Diem, Yvon Dommergues, Vincent Eschenbrenner, Patricia Folgarait, Claude Girard, Benoît Jaillard, Gérard Kilbertus, Jan Lagerlöf, François Le Tacon, Ken Lee, Michel Lepage, Jean-Pierre Lumaret, Marc Montadert, Laurent Palka, Georges Pedro, Jean-François Ponge, Georges Reversat, Skule-Adam, Roger Schaefer, Mike Swift and the TSBF group, Michel Robert, François Toutain, Guy Vannier, Geneviève Villemin and Paul Woomer.

We thank library staff members, Norma Penny and Robyn Maconachie of the Davies Laboratory, CSIRO, Townsville and Emmanuelle Aldebert and her team of Centre IRD de Bondy for their considerable assistance. Their skills in helping to locate and verify sources of information are gratefully acknowledged. We also thank Christine Spain for her cover design.

Finally, we should like to acknowledge the debt we owe to our families in supporting us during the extended period over which this book was written and in forgiving us for spending the time involved that would otherwise have been devoted to them.

FOREWORD

A number of excellent textbooks on general ecology are currently available but, to date, none have been dedicated to the study of soil ecology. This is important because the soil, as the 'epidermis' of our planet, is the major component of the terrestrial biosphere. In the present age, it is difficult to understand how one could be interested in general ecology without having some knowledge of the soil and further, to study the soil without taking into account its biological components and ecological setting. It is this deficiency that the two authors, Patrick Lavelle and Alister Spain, have wished to address in writing their text. A reading of this work, entitled 'Soil Ecology', shows it to be very complete and extremely innovative in its conceptual plan. In addition, it follows straightforwardly through a development which unfolds over four substantial chapters.

Firstly, the authors consider the soil as a porous and finely divided medium of bio-organomineral origin, whose physical structure and organisation foster the development of a multitude of specifically adapted organisms (microbial communities, roots of higher plants, macro-invertebrates). In this context, and apart from discussion of the present work, the three elements necessary for consideration in all ecological approaches are:

- consideration of the soil as a matrix and a habitat for organisms, noting particularly its heterogeneity;
- a precise definition of the physical constraints (largely thermal and hydrological) that characterise the unique soil environment situated as it is at the interface between the atmosphere, the hydrosphere and the biogeosphere;
- an understanding and evaluation of the mineral and organic resources of the soil necessary for the existence and for the development of its populations of living organisms.

In the second chapter, P. Lavelle and A. Spain (as did B.B. Dokoutchaev in 1883) approach the soil as a *natural object,* whose scientific study comprises the science of Pedology. The importance of this chapter is that it demonstrates to the readers that while soil materials have a number of common characteristics, the soils of the planet differ substantially as a function of the geological and climatic conditions that occur across terrestrial environments. Consequently, a clear understanding of soil ecology, also demands a thorough knowledge of the laws governing soil formation, the different pedological processes that occur and a detailed appreciation of the major types of soils that occur on the surface of the globe. It is indeed the diversity of the pedological properties that, in a certain way, allows an understanding of soil biodiversity. Having worked much in this area during the course of my career, I am grateful to the authors who, as specialists in ecology, understand the value of the pedological approach in developing a deeper appreciation of soil ecology. It is this last point in particular that is developed over the course of the following two chapters.

In the third chapter, the work approaches the study of the principal living soil organisms including microbial communities, roots, microfauna, mesofauna and macro-invertebrates. However, this chapter is unique in the sense that, apart from purely taxonomic and biological information, it also considers the factors of spatial and temporal distribution and community structures of a diverse range of soil organisms. Consideration of such factors leads to the development of new concepts in the domains of faunal biodiversity; it also impinges on the related factors of adaptive strategies, the determination of species richness and on the description of community characteristics.

This work concludes with a chapter entitled 'Functioning of the soil system', which develops in a synthetic way the authors' ideas in the area of soil ecology. Finally, after having presented an hierarchical model of the biotic and abiotic factors governing such phenomena as the decomposition of organic components and recycling, this chapter successively examines the different sub-systems of the soil: litter, rhizosphere, drilosphere and termitosphere. This presents an entirely new perspective on eco-pedological problems.

In conclusion, it is clear that this voluminous work would undoubtedly have fascinated those of our forebears who have dedicated their thoughts in one way or other to these questions, for example, C. Darwin, E. Haeckel, P.P. Grassé, R. Dubos ... However, I am sure that it will particularly interest today's pedologists and ecologists. It is clear that such workers will find in this volume a summation of knowledge that has not been previously assembled. Moreover, in so doing, they will find a new view of soils which, in the future, will no doubt alter the fundamental conceptions of pedology and ecology.

Georges Pedro
Membre correspondant de l'Académie des Sciences

FOREWORD

There was recently a debate in the letter pages of Nature as to whether the term 'soil' should properly be applied to the particulate matrix at the surface of the planet Mars. The argument centred around uncertainty as to whether, in the absence of evidence of biological or organic influence, the concept of soil is applicable. The very fact that this was in question is indicative of the rather weak status that soil biology has within soil science, a viewpoint reinforced by the contents of many of the most popular textbooks.

Soil biology has indeed lagged far behind the sister disciplines of chemistry and physics; and also in comparison with the ecological study of above-ground biotic communities. Yet the biological diversity of soil is invariably high including as it does a wide range of bacteria, fungi, protists and representatives of the majority of the terrestrial invertebrate phyla. The greater fraction of these organisms participate, in one way or another, in the process of decomposition, which is the essential and fundamental counterpart in ecosystem function of photosynthesis. As with the organisms, so with the functions; the literature on decomposition is but a small fraction of that on primary production.

The reasons for this relative neglect are many but, as in all areas of science, growth is significantly related to method. Theory remains in a vacuum without the means to test it. The last decade or two have seen the significant lifting of a methodological hiatus that previously seemed to forbid progress. I believe that this has been based not so much on advances in techniques (although there have been many such) but more on a change of approach - to the use of function as a means of investigating the organisms rather than the converse which dominated soil biology for so long. The two authors of this book have been leaders in this process of change and it is most appropriate that they provide the first comprehensive synthesis of the recent work.

This book is not just a review of academic research results however, it also constitutes a contribution to the wider and more important revolution that is taking place at the interface of agriculture and ecology. The driving force for a more ecological approach to agricultural practice and to soil management in particular, has come from the sustainable development agenda in which central concern with the maintenance of agricultural yield is closely associated with the need to conserve natural resources for future generations. Emergence from the circumstances of nutritional deficiency and poverty which characterise the lives of many of the small scale farmers of the world can only be achieved where there are sufficient resources to raise food productivity. With very few exceptions this will only be possible by the provision of external sources of some nutrient elements. But many farmers in developing countries, in contrast with those of the industrialised world, still have only limited access to inorganic fertiliser. This skewed distribution, and the high cost of fertiliser in most parts of the world, emphasises the need for increasing the efficiency of fertiliser use. There is now abundant evidence that the highest and most persistent responses to nutrient additions come with combination of inorganic with organic sources. And management of organic inputs is a proxy for management of the soil biota.

The utilisation of organic inputs for soil improvement is as old as the history of agriculture itself, but the scientists' concept of organic matter management has failed in the past to reach a degree of sophistication comparable to the insights of farmers. I believe this was largely due to a failure to explore the biological basis of soil fertility regulation. Soil biologists have now however begun to provide predictive explanations for the responses of crops to different types of organic matter - a sign that the subject is maturing in a useful as well as an academic way. In this book the authors show time and again how the interactions of the food webs of soil organisms with their organic food sources is the basis for many of the phemonena of soil fertility.

Greater interest in soil fertility management through organic matter and the manipulation of the soil biota has served to refocus interests on many of the older traditions in agriculture and on indigenous sources of knowledge. But the relevance of soil biology is not confined to small scale or tropical agriculture. It has become increasingly apparent that many of the great gains in production made in the green revolution by use of high-yielding varieties with high inputs of inorganic fertiliser cannot be maintained indefinitely. Among the causes attributed to yield declines under long-term cultivation are changes in soil fertility associated with loss of organic matter, diminishment or disappearance of groups of the soil biota, and the accompanying decline in soil physical and chemical properties.

In this book the authors unravel, by numerous example, the intricate relationships between the different components of the soil biota, and their implications for energy flux and nutrient cycles in natural and agricultural ecosystems. This construct lays the essential scientific basis for an emerging paradigm for the biological management of soil fertility.

Mike Swift
Director Tropical Soil Biology and Fertility Programme, Nairobi

INTRODUCTION

Soils form a narrow interface between the atmosphere and the lithosphere and possess elements of both: water, a gaseous phase and mineral matter, together with a diverse range of organisms and materials of biological origin. They continually interact with the atmosphere above and the lithosphere beneath. Soils are that part of the earth's surface within which organic materials are broken down to form stable humic compounds thereby releasing their contained nutrient elements for uptake by organisms and dissipating their contained energy. This veneer of materials supports the growth of higher plants and therefore the primary production on which the human population directly depends. Soils provide other important services including the stabilisation of waste materials and part of the excess CO_2 released to the atmosphere by human activities.

The thin veneer of soil is readily-damaged or lost by misuse. Following such effects, it does not reform in any major sense within the time frames of human existence and must be considered a non-renewable resource. Many current systems of agricultural management are not sustainable in the longer term because of the pressures they place on the soil. Production levels may frequently be set on the basis of economic goals rather than the capacity of the soil to withstand particular stresses. Most cropping systems, for example, require substantial regular inputs of energy and nutrient elements and the sustainability of their use for this purpose is contingent upon continuing inputs. Similar principles apply in other situations. Continued over-grazing in pastoral environments situations soon leads to soil degradation and loss and stresses imposed by chemical contamination may eventually result in impaired functioning.

Faced with such pressures, soils are clearly a threatened resource. From this, one of the emerging challenges facing soil ecologists is the maintenance or amelioration of soil fertility in the development of long-term sustainable agricultural systems. This requires the integration of biological process knowledge into general models of soil functioning and the design of land management systems based on such models.

An understanding of soil functioning and the definition of appropriate management options demands a knowledge of the processes operating in both the above- and below-ground subsystems and identification and quantification of fluxes of energy and materials between and within them. The maintenance of the physical integrity and fertility of soils depends largely on these transfers.

The soil is a unique environment that combines solid, liquid and gaseous phases to form a three-dimensional matrix. The organisms that inhabit this porous, humid and amphibious environment face quite specific constraints due to: 1. the predominance of poor-quality food resources for species at the lower trophic levels; 2. spatial constraints in an environment where most organisms live within pores that differ broadly in size, shape and inter-connectedness; and 3. the rapid alternation, in both space and time, of air- and water-filled porosities. The adaptive strategies evolved under such conditions are certainly different from those of their above-ground counterparts and deserve more detailed study.

A major feature of the soil system is the intimate association of biological, inorganic and organo-mineral materials: their interactions are both the result and the source of soil processes operating at many spatial and temporal scales. In consequence, an understanding of soil functioning requires both the detailed 'reductionist' study of the individual structures and sub-processes involved and an 'holistic' or systems approach to energy and nutrient dynamics and the organisation of components within the soil matrix. Such approaches are few because of the traditional separation of soil science into relatively self-contained sub-disciplines and inequalities in the progress made within each sub-discipline. In particular, the extreme difficulty of integrating the effects of complex, interacting processes over the diverse spatial and temporal scales of importance in soil-related studies should not be overlooked.

Direct observation and accurate measurement of many soil phenomena is often difficult and much of our knowledge continues to be gained from indirect measurement and the partial to incomplete extraction of components and organisms. It seems likely that, due to the unique constraints faced by soil organisms, some major paradigms and models applying to above-ground ecological processes may be less relevant to below-ground populations and communities.

Definitions and roles of soils

Soil is an organo-mineral assemblage and, as such, its formation and properties depend largely on biological processes and influences. For the purposes of this book, soil is defined as an unconsolidated, stable, three-dimensional assemblage of organic, mineral and organo-mineral materials with a characteristic biota and located at the earth's surface (in part after Soil Science Society of America, 1997). Some pedological organisation (structure, horizonation) is apparent in almost all soils although, because they take considerable time to develop, it may be slight in incipient soils forming in recently stabilised materials. Concentrations of organic matter may range from negligibly-low in the latter soils to high in others, with carbon concentrations in the surface horizons of peat soils approaching those of plant materials.

As used here, the term soil encompasses the 'normal' soils that support a vegetation community comprising higher plants, including the familiar domesticated species. However, it also extends to the soils characteristic of extreme environments that may support only cryptogams, as in Antarctica, the high Arctic, the nival areas of the world and in certain saline environments. It also includes incipient soils and such soil-like materials as those forming on recently-stabilised mineral materials, notably geological materials exposed at the earth's surface by erosion, vulcanism and aeolian processes, or through the activities of man.

Soils have five principal and somewhat overlapping roles in ecosystems (partly after Buol, 1995):

(i) *Mechanical support.* The role of the soil in providing a mechanical support for plants is clear. However, to provide this support, the strength of the soil must fall within the range where plants are able to resist the pressures of wind and rain on their aerial parts without uprooting but not be so strong that root penetration is inhibited. Similarly, soil strength should not be so high that the excavation of habitat and feeding constructs by the larger invertebrates (mostly earthworms and social insects) is inhibited, nor so weak that the voids and other structures formed within the soil are unstable.

(ii) *Habitat provision.* The vast majority of soil organisms are too small and too weak to create their own habitats and these latter must exist within the soil to sustain the organisms essential to its functioning. Small organisms must live within existing voids formed through physical processes or through the activities of such larger organisms as plant roots and soil-dwelling invertebrates and vertebrates. In particular, decomposing roots comprise an important habitat for many soil organisms.

(iii) *Storage of organic matter.* The soil is an important store of dead organic materials present in all stages of decomposition from freshly-fallen litter and recently-dead plant roots to highly humified materials of great age and chemical complexity. Dead biological materials are important energy sources for many soil organisms and the more decomposed fractions interact with inorganic materials to form organo-mineral structures central to the organisation and stability of the soil matrix.

(iv) *Element release.* In addition to such elements as silicon and aluminium which normally dominate the soil mass, the soil also contains stores of elements that are of biological and pedological importance. Certain forms of such elements as iron, aluminium and silicon are important in both these respects. These and other elements may be held within the organic materials (living and dead) considered above but are also present in the soil solution, retained at and near the surfaces of the organic and inorganic soil colloids and, in less accessible forms, within the mineral soil particles. Decomposition of organic materials liberates the contained elements in inorganic forms (mineralisation) in a controlled or 'slow-release' way for uptake by plant roots and other soil organisms, or for involvement in pedological processes.

(v) *Water storage.* Soils possess a store of water that supports the growth of plants and other organisms. The magnitude of this store differs substantially between soils depending on soil depth, the size distribution and organisation of the soil particles and location in the landscape. Within soils, plants may have access to stores of different sizes, depending on their individual capacities to extract water from the smaller pores and on such factors as their rooting depths, mycorrhizal associations and salt tolerances. Other soil organisms also have their own characteristic physiological tolerance ranges beyond which they become inactive, or die.

Soil functioning is a concept used here to imply the continuing operation of the major soil processes (energy flow and its effects on the biological cycling of nutrient and other elements, the maintenance of structure) that determine the primary and secondary productivity of the soil mass and its stability. Within active soils, the biological units of soil structure are continually renewed thereby expediting gas exchange with the external environment and promoting the acceptance, storage and transmission of water at rates commensurate with the needs of their biotic components. Soils decompose their organic materials at rates dependent on climatic and other factors; the nutrient elements released are stored or supplied to other organisms at rates that support biological productivity.

Over longer time scales, soils undergo a pedological development appropriate to their parent materials and the broad environmental conditions to which they have been exposed. Soils that suffer from impaired activity may perform adequately or poorly with respect to particular physical and biological processes occurring within comparable unmodified soils but will show a syndrome of effects reflecting the physical, chemical and biological stresses to which they have been exposed.

Soil environment

The soil environment is dark, porous, still and semi-aquatic. Most energy inputs are concentrated at the soil-air interface and in the upper few centimetres of the soil matrix, where fine roots are largely distributed. This leads to a marked vertical gradient in the concentrations of soil organic matter, nutrient and other elements and in soil structure and biological activity. In conjunction with environmental variation, the unevenness of energy inputs also results in strong spatial variation at all scales of observation. Both aquatic and air-breathing organisms live together within the soil matrix although their movements are strongly constrained by its internal organisation: only a few of the larger organisms are sufficiently strong to create their own habitats within the soil. At the lowest trophic levels, the principal nutritional resources available within the soil for most heterotrophs occur as non-living, biologically-derived materials at all stages of decomposition; the nutritive quality of these materials is generally lower than that found in the aerial part of the ecosystem.

Approaches to the study of soil systems

Until relatively recently, soil biologists have been largely restricted to qualitative studies because of their limited capacity to observe soil processes and structures directly. Soil biologists are now increasingly able to evaluate quantitatively the roles played by organisms in the main soil processes. While such evaluations remain incomplete, it is clear that micro-organisms are always the ultimate mediators of soil chemical processes in both natural and intensively man-managed agroecosystems. Nonetheless, an increasing awareness has developed of the important regulatory roles played by macro-organisms (living roots, the larger invertebrates and a few vertebrates) in many soil systems. These organisms exert a major control over many soil processes through their influences on the decomposition of dead organic materials, nutrient cycling, modification and transport of soil materials and on the formation and maintenance of soil structure.

Some of the newer investigative tools available to soil scientists include improved analytical techniques for the quantitative characterisation of soil structural components, organisation and processes. Some major useful techniques include transmission and scanning electron microscopy, the use of radioactive elements and stable isotopes as tracers, the analysis of soil organic materials using nuclear magnetic resonance and other techniques and the microscopic observation of ultra-thin sections of soil. These and other techniques - notably recently developed statistical techniques - are providing new detailed insights into soil functioning over a range of observational scales. Molecular techniques have already been used successfully in several areas of soil ecology and promise much for future progress.

Advances in ecological theory have also contributed to our understanding of ecological systems. They have offered, for example, approaches to ecosystem development relevant to soil-based processes although they have been little tested in this environment. These advances have also allowed the characterisation of such important ecosystem parameters as community structure and hierarchical organisation, and assessment of the relative importance of biotic and abiotic factors. More fruitful approaches to the study of adaptive strategies have emerged at both species and community levels. Such concepts underpin descriptions of the functioning of soil systems.

Because of the lengthy periods required for full development, soil is essentially a non-renewable resource at human time scales. However management strategies aimed at developing sustainable soil use practices at time scales of decades to centuries are proving increasingly difficult to implement. This results from the substantial pressures for increased food and fibre production needed to satisfy the needs of a perhaps too-high and in some places rapidly-growing population (notably in the tropical world) and because of the frequent subordination of sustainability of use to economic and other social pressures. Excessive pressure on the land and accelerated erosion continue to lead to often-irreversible and widespread soil degradation, loss and desertification.

In many industrialised areas, inappropriate management has lead to the accumulation of toxic concentrations of metals and other pollutants in soils resulting in ecosystem stress and the contamination of ground and surface waters. The broad nature of the problems affecting soils is illustrated by the widespread soil acidification consequent on the release of nitrogenous and sulphurous atmospheric pollutants and the subsequent fall of acid rain. Increasing release of the greenhouse gases carbon dioxide, methane and the oxides of nitrogen from cultivated soils is a further matter of concern in climate change scenarios.

Many of the problems considered above are extreme and therefore readily perceived but less obvious and more slowly developing soil degradation is also widespread due to a diversity of widely-accepted land use practices. These include such factors as the over-application of fertilisers and biocides, over-cultivation, over-grazing and accelerated erosion due to the exposure of the unprotected soil surface to rainfall and wind. These may be considered as an 'over-use' syndrome, resulting often from socio-economic pressures, or ignorance.

Resolution of such broadly based problems requires the active involvement of the soil science community as part of interdisciplinary approaches. However, such studies often lack a common conceptual view of the nature of the soil system, its functioning and the limits to its exploitation. It is also clear that any approaches to problems of land management are likely to fail unless their socio-economic and political dimensions are resolved at the same time, in a global systems approach that combines biophysical and socio-economic processes.

Concepts and models

This book develops a comprehensive conceptual approach to the soil and its functioning. It views soils as integral parts of open ecological systems (natural and man-modified) set within the increasingly broader contexts of landscapes, regions, biogeographical domains, biomes and ultimately the terrestrial surface of the entire earth. In this, it provides a preliminary synthesis that conceptually integrates biological, physical and pedological structures and processes over the broadest range of spatial and temporal scales.

Chapter I starts with a description of the major soil components, examines the materials occurring in the different phase states and the important interactions that occur between them. Among the solid phase components, particular attention is given to the colloidal components since these comprise most of the reactive surfaces in soils. These include organic colloids and the clays whose properties govern much of soil functionality. Soil organic materials in all stages of decomposition largely fuel soil biological activity and contribute substantially to structural stability; the principal sources of these biological materials involved are here characterised. The spatial organisation of soil components - the soil 'structure'- is also discussed and the importances of both the inorganic and biological components are emphasised.

The major factors of the soil environment are also considered in Chapter I. These include consideration of the soil atmosphere and thermal regimes. Particular attention is given to the presence of water in the soil, the forms in which it occurs and the factors governing its availability to a range of organisms. A brief introduction to soil hydrological regimes is also presented.

The inorganic resources present within soils are considered in the latter part of Chapter I. The biological characteristics of the elements of major biological and pedological importance are presented. Soil organic resources are also categorised and the properties of the major types discussed and related to soil functioning. This chapter concludes with a brief introduction to soil heterogeneity.

Chapter II is concerned with the formation and internal organisation of soils. It initially examines the horizons typically present in many soils as the basis for a discussion of soil development. The major soil forming factors of climate, parent material, time, biota and topography are discussed in light of the definition of soil as a material owing its unique properties to biological influences and processes. Discussion of the dominance of the particular weathering processes (physical, chemical and biologically-mediated) in different regions leads on to the development of concepts relating to the formation and structure of soil organic reserves, and to the world-scale distributions of the major pedological and bio-pedological processes. The roles of the soil fauna in translocation, mixing and other soil developmental processes are also emphasised. The final part of this chapter introduces the differing bases of the various national and international soil classification systems, including some designed for man-made and heavily man-influenced soils.

Chapter III details the biology and ecology of the major groups of organisms living in soils: micro-organisms, roots and invertebrates. Basic information is provided on the organisation, life history and adaptive strategies and functional categories of each group. Quantitative data on population abundance and the major features of community structure are also considered.

At the end of the chapter, adaptive strategies and interspecific interactions are considered, both within and between guilds. Patterns specific to the soil environment are described with a strong emphasis on the mutualistic relationships between micro- and macro-organisms. A hierarchical classification of soil organisms is proposed based on these relationships. The question of biodiversity in soils, and its determinants is also considered.

Chapter IV describes the major processes that occur in soils, the decomposition of litter materials, soil organic matter formation and the maintenance of its physical structure. Their general properties are first defined. The main physical, chemical and biological determinants of these processes are then described and considered in a hierarchical way, at the level of functional domains dominated by certain macro-organisms. These domains include the following major systems: root systems and rhizospheres, arthropods and litter and pore systems, earthworms and drilospheres and termites and termitospheres. Particular emphasis is thereby set on the different scales at which such processes operate and provide tools for the interpretation of observations at scales that range upwards from the microaggregate to the ecosystem to higher levels of organisation

CHAPTER I

INTERNAL ENVIRONMENT, MICROCLIMATE AND RESOURCES

I.1 THE SOIL ENVIRONMENT

1.1 Soil components and structures - the major groupings

Soils form through the conjoint influences of lithological, topographic, climatic and biological factors that interact over time to condition the natures, directions and rates of soil forming processes. The varied and complex expression of these soil forming factors over the surface of the earth has lead to the development of an enormous diversity of soils with a correspondingly-large range of soil environments. Over time periods ranging up to the geological, soils develop from their parent materials (often but not necessarily rocks) and undergo a series of progressive changes to eventually become depleted in nutrient elements and, in the biological sense, exhausted.

On a global scale, the climatic regime is the dominant soil forming factor. In contrast, topography is regionally important in controlling the ways in which soils vary from site to site while, at a single site, considerable local spatial variability and vertical differentiation are features of the soil system. Finally, at the lowest level of pedologica organisation, the primary particles of the soil become associated to form aggregate which determine the micro-environments that occur within and between them.

At all spatial and temporal scales, the major influences defining the environments that exist within soils are the physical, chemical and surface properties of their components and their spatial dispositions. Further determinants include the soil microclimate, the nutritional resources available to the dominant biota and the presence of inhibiting compounds. An appreciation of these influences and the interactions that occur between them is fundamental to an understanding of the conditions under which the inorganic soil components, the biota and its by-products interact to mediate the unceasing changes that occur during soil development.

Soil is a complex and intimate mixture of materials distributed among the solid, liquid and gaseous phases and occurs in a wide variety of physical and chemical forms. The properties and functioning of the whole soil, both in natural and in applied situations, will largely be determined by the properties of its individual components and their relative proportions. The volume of the soil that may be occupied by the liquid or gaseous phases depends on the spatial arrangement of the solid phase materials. Because the properties and balance of these components largely determine how soils will behave in undisturbed and in agricultural and engineering environments, considerable effort has been expended in searching for useful ways to characterise soil components in these terms.

Soil components may be classified in many ways, depending on the intended purpose of the classification. Some common ways of classifying such materials are on their sizes, shapes and origins, the phases they belong to, their chemical or physical characteristics, their mineralogical compositions or on combinations of these. The classification of soil components employed here is arbitrary and hierarchical and attempts to characterise the individual components in ways that reflect their ecological interrelationships (Figure I.1).

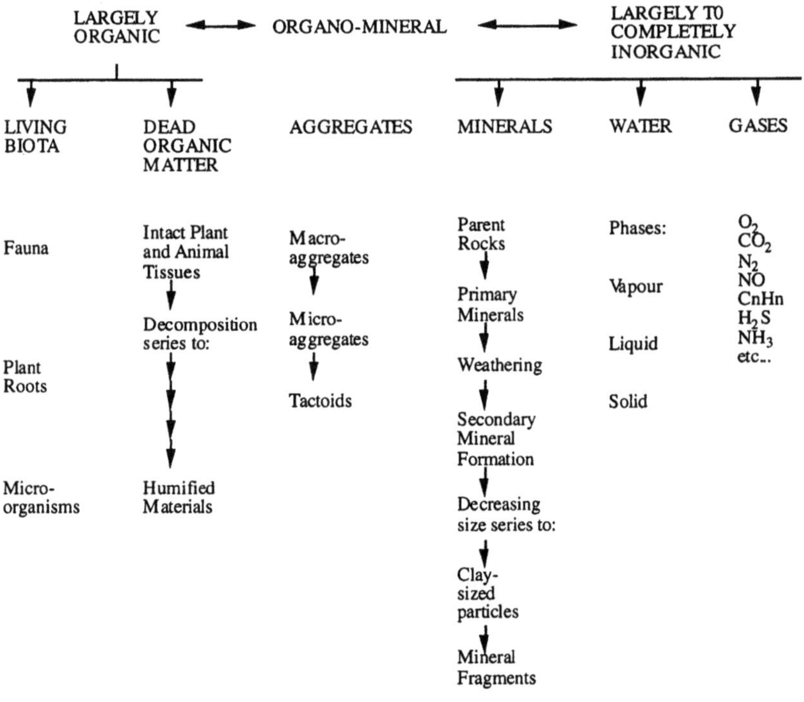

Figure I.1 An ecological classification of soil components.

The division of soil materials into separate components in no way implies that associations and interactions between the various components do not occur, or are unimportant. Indeed, nearly all solid and liquid phase entities that occur within soils include both organic and inorganic components, or are influenced by them. Further, as emphasised in several sections of this book, the formation and functioning of soil depends on the myriad interactions that occur between the organic and inorganic, and the living and non-living components of soils. An enormous variety of organic and inorganic components are available within soils to interact in the synthesis and development of the unique substance which is soil.

1.1.1 THE SOLID PHASE COMPONENTS

Apart from the organic horizons of peat soils, inorganic materials form the bulk of most soils both in terms of mass and, more importantly, surface area. This inorganic fraction of the soil may include mixtures of rock (or other parent materials) in all stages of weathering and physical disintegration, materials transported from elsewhere by wind, water, volcanism or gravity together with a variety of secondary components that have been formed by transformation *in situ,* or moved from other parts of the profile and elsewhere. These secondary components may be distributed throughout the profile or occur as local concentrations as nodules or concretions formed by reaction or precipitation from solution. Alternatively, they may form discrete layers or impermeable or slowly-permeable pans that modify soil drainage and other properties.

The average concentrations of the major elements in the surface of the earth's crust are presented in Table I.1 (Taylor and McClennan, 1985). It should be noted firstly that this is the average composition of the parent materials that weather to form soils (Chesworth, 1992) and, secondly, that the remaining, non-included elements of the periodic table comprise only a very small part of the total mass. The data presented in this Table are given both as oxides and as elemental percentages; these data are conventionally presented as oxides since oxygen is the major balancing ion (Birkeland, 1984). Two of the major elements critical to all biological organisms (nitrogen and sulphur) do not appear in Table I.1 and are ultimately derived from the air and minor minerals, respectively. As considered later in this book, shortages of these and other less-abundant elements commonly limit the growth of all organisms.

Table I.1 The average concentrations of the major elements in the surface of the earth's crust (Taylor and McClennan, 1985).

Element	**As Oxide**		**As Element**	
	Percentage	**Cumulative Percentage**	**Percentage**	**Cumulative Percentage**
SiO_2	67.0	67.0	31.3	31.3
Al_2O_3	15.2	82.2	8.0	39.3
FeO	4.0	86.2	3.1	42.4
CaO	3.6	89.8	2.6	45.0
Na_2O	3.8	93.6	2.8	47.8
K_2O	2.9	96.5	2.4	50.2
MgO	2.1	98.6	1.3	51.5
TiO_2	0.5	99.1	0.6	52.1
P_2O_5	0.1	99.2	0.04	52.14
MnO	0.08	99.28	0.06	52.20

Table I.2 presents the indicative total concentrations of selected major and minor elements in surface soils together with the ranges that may be expected to occur in areas where mineralisation has occurred or which have been polluted by human activities. Comparison of the concentrations of the major elements in Tables I.1 and I.2 shows that, on average, the less mobile elements (Al, Fe) occur in soils at similar or greater concentrations than their averages in the earth's crust. They may become even more concentrated in highly-weathered soils, due to the relatively greater losses of the more soluble elements over time. Further, certain elements required by the biota, such as Ca and P are also present in greater concentrations and C and N are important components of soil organic matter that generally occur in insignificant concentrations in the earth's crust.

Table I.2 Indicative concentrations (totals) and ranges of selected macro- and minor elements in surface soils. Data from various sources, approximate medians from Bowen (1979), Baker and Brooks (1989) and Kabata-Pendias and Pendias (1992).

i. Major elements (% unless otherwise indicated)

Element	Median	Range
Silicon	33	25-41
Aluminium	7.1	1.0-30.0
Iron	4.0	0.2-55.0
Carbon		
mineral soils	2.0	0.01-15
organic soils	-	20-40
Calcium	1.5	0.07-50.0
Potassium	1.4	0.008-3.7
Magnesium	0.5	0.04-0.9
Sodium	0.5	0.015-2.5
Titanium	0.5	0.01-2.5
Nitrogen		
mineral soils	0.2	0.02-0.8
organic soils	-	to *ca.* 2.0
Phosphorus ($\mu g\ g^{-1}$)	800	35-5300
Sulphur ($\mu g\ g^{-1}$)	700	10-1600

ii. Minor elements ($\mu g\ g^{-1}$), approximate medians and normal ranges where known, the approximate maximum concentrations known are also presented.

Element	Normal Soils		Contaminated soils and
	Median	Range	areas of mineralisation
Arsenic	6	1-70	2000
Boron	35	1-467	-
Cadmium	0.5	0.01-2.5	1780
Chromium	54	1-1500	2500
Cobalt	8	0.1-170	700
Copper	30	1-300	3700
Lead	25	3-70	18500
Manganese	437	7-10000	-
Molybdenum	1.2	0.01-40	-
Nickel	22	0.2-450	2500
Selenium	0.33	0.1-40	-
Thorium	9	3.6-96	20
Uranium	1	0.1-35	250
Vanadium	100	0.7-500	to 800
Zinc	64	1.5-770	180000

The most important of the inorganic components in terms of soil behaviour is undoubtedly the highly-diverse group of substances characterised generally as clays. The term clay may refer to three distinct entities. In the textural sense, clay refers to the important colloidal particles less than 0.002 mm (2 μm) in diameter. More loosely, it may also refer to a class of soils with a high proportion of such particles. It may also be applied to the phyllosilicate clay minerals discussed below. Here, the term clay will refer to clay-sized particles unless otherwise qualified. The phyllosilicate and other mineral particles in the clay size range have large surface areas relative to their masses and, in soils with appreciable clay contents, they control many reactions important to biological processes. Surface area is also closely correlated with a range of other properties that regulate the physical and chemical characteristics of the soil mass and influence plant growth (Farrar and Coleman, 1967).

Organic materials found in soils may be divided into the living organisms considered later in this book and non-living materials of biological origin. The latter comprise a diversity of materials including roots and other plant and animal remains in all stages of subdivision and decomposition. In addition, dead fungal hyphae, spores, bacteria and other larger constructs of microbial (*e.g.*, the sporocarps of mycorrhizal fungi) and faunal (*e.g.*, termitaria) origin are frequently present. During the later stages of decomposition of plant materials, less resistant materials are lost leaving only cell wall outlines and fragments which, in the terminal stages of decomposition, may appear amorphous under microscopic examination. Other directly-derived biological materials include such diverse products as plant root lysates and exudates, the faeces, excreta and secreta of animals, the cutaneous mucus secreted by earthworms and gels produced by micro-organisms.

In terms of chemical structure, an enormous variety of biologically-derived materials occur in soils. These range in complexity from simple aliphatic materials to polymers of high molecular weight and great complexity. Their temporal existence in soils ranges from the most ephemeral, with turnover times of a few hours or days, to highly degradation-resistant polymers whose age may be measured in tens of thousands of years.

Gases derived from the metabolism of micro-organisms and roots comprise an important part of the soil atmosphere and the state of aeration of the soil determines the pathways of many chemical and biological processes. These gases may be present as components of the soil atmosphere, partly dissolved in the soil solution and in some form of dynamic equilibrium with the soil atmosphere, or adsorbed onto other soil components. Such gases are of pedogenetic importance and may interact with roots to influence plant growth characteristics.

Through the form in which it is present, the tenacity with which it is held, and the pathways of its flow into, through and out of soils, water conditions the whole evolution and functioning of soils. Beyond its direct biological role in the hydration of plants and the other living inhabitants of the soil, it is the major agent of formation and change in soils. Apart from physical changes resulting from freezing and thawing, the effects of water are due largely to reaction (hydrolysis) and the dissolution and transport of a wide range of substances in solution and in suspension. This movement may take place vertically in either direction, laterally within the profile, or from one part of the landscape to others lower-lying. In regions where it has been active, glaciation has played an enormously important role over geological times in modifying landscapes, and in determining the characteristics and distributions of the soils present in these areas.

Because of its importance in pedogenetic processes and the biological functioning of soil, the physical processes associated with the status, entry and egress of water into the soil are treated in a separate section. However, the properties of the soil solution are arbitrarily considered in the section on the soil components.

The information presented here on soil components is of an introductory nature only; more detailed information may be obtained from the works of Birkeland (1984), Bonneau and Souchier (1982), Brady and Weil (1996), Duchaufour (1997, 1998) and Jenny (1980).

1.1.1.1 *The mineral components - soil texture*

A useful way of categorising soil inorganic material is by determining its texture which is defined as the proportions by mass of the mineral particles that fall into different size classes, the particle size distribution. Texture is an important characteristic of soil since it influences many aspects of its field behaviour including a number of properties important to the growth of higher plants. These include the amount of water held by the soil and its capacity to retain plant nutrient elements.

Texture directly determines water-holding capacity; soils with a high proportion of clay size particles (that is, of fine texture) may normally be expected to have lower infiltration rates but a greater capacity to store and supply water than those of coarse texture (that is, possess a high proportion of the larger particles). However, the moisture status of a soil also depends on its structure, location in the landscape, the nature of its vegetative cover, the seasonal distribution of rainfall and evapotranspiration, and the presence of impeding layers in the profile. Not all soils with high clay contents are poorly drained. As illustrated below (oxisol), the microaggregation of clay and silt particles into units (peds) equivalent to sand-sized particles may lead to certain clay-rich soils having the drainage characteristics of more coarsely-textured soils.

Texture also influences the distribution of soil organisms and, conversely, may be modified over time through the activities of soil animals. Finally, soil texture has a substantial influence on the ease of cultivation. Cohesion between soil particles is usually lower in coarse- than in fine-textured soils and tillage generally requires a greater expenditure of energy in the latter soils.

The proportions of the larger sized particles present in soils are conventionally determined by passing a dry and thoroughly-destructured sample of the soil through various meshes. Estimation of the finer particles depends on their settling rates in water, or under centrifugation. Recently, electronic counting devices have been developed and allow the characterisation of particle size distributions at a much more detailed level of resolution. The upper horizons of peat soils (histosols) are largely organic in nature although variable amounts of mineral matter may also be present and different methods of characterisation are required. Soil texture is normally determined in the laboratory although a broad characterisation of textural class may usefully be made in the field by wetting and manipulating a de-structured sample between the fingers (Birkeland, 1984).

The particles of less than 2.0 mm equivalent diameter comprise the **fine earth** fraction of the soil which, in most soils, predominates in terms of mass and, more importantly, surface area. It should be noted that the subdivisions of the fine earth fraction differ between classification systems, although clays are defined in both the International and United States systems as those particles less than 2 μm in diameter. Table I.3 presents the subdivisions of the fine earth and illustrates the very large increases in particle numbers and surface areas in descending from the largest to the smallest of the four particle diameter classes. As discussed below, the total surface areas per unit mass or volume of soil (the specific surface area) is correlated with a range of other soil properties.

Table I.3 Subdivisions of the fine earth fraction with the approximate numbers of particles and surface areas within each of the major groupings (Black, 1957).

Particle size class	Particle diameter (mm)	Approximate number of particles g^{-1}	Approximate surface area $m^2\ g^{-1}$
Coarse sand	0.2-2.0	5.4×10^2	0.21
Fine sand	0.02-0.2	5.4×10^5	2.1
Silt	0.002-0.02	5.4×10^8	21.0
Clay	0.000-0.002	7.2×10^{11}	230.0

The larger particles of greater than 2.0 mm are also of importance since coarse-textured soils are usually free-draining and thus tend to be more highly and deeply leached (Birkeland, 1984). The nature, placement and content of larger particles may substantially affect water infiltration and erosion, temperature fluxes and plant growth (Poesen and Lavee, 1994).

The inorganic soil particles belonging to different size classes may differ considerably in chemical composition. This is illustrated in Figures I.2a and b which present the variation of the total phosphorus and iron contents with size fraction and depth in the profile of a southeastern Australian alfisol (Chartres and Walker, 1988).

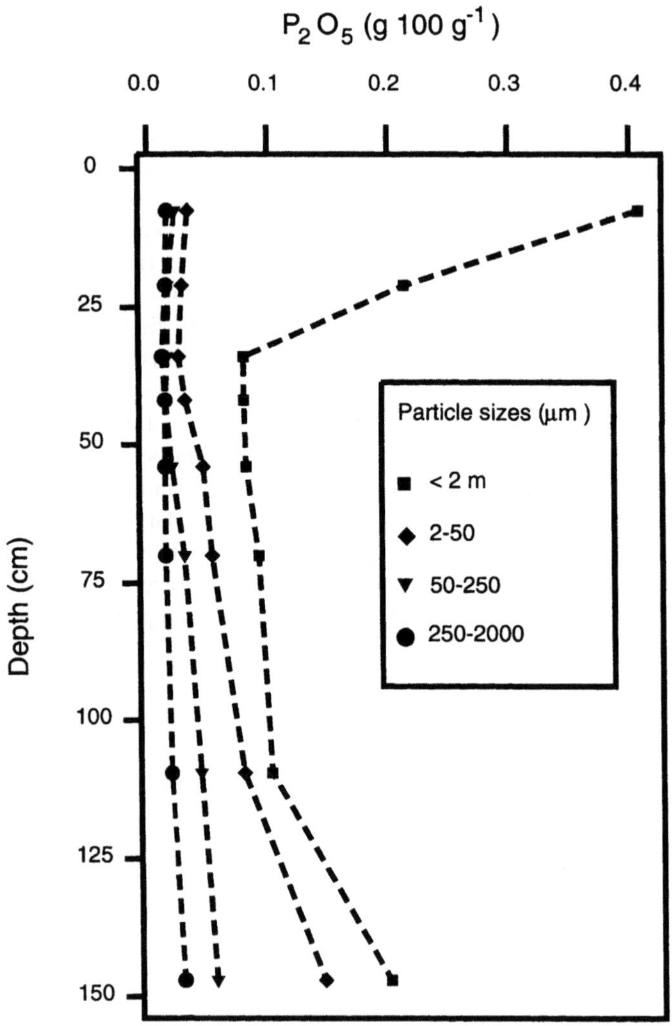

Figure I.2 Variation in total phosphorus (a) and iron (b) concentrations in an Australian alfisol with particle size fraction and depth in the profile (Chartres and Walker, 1988).

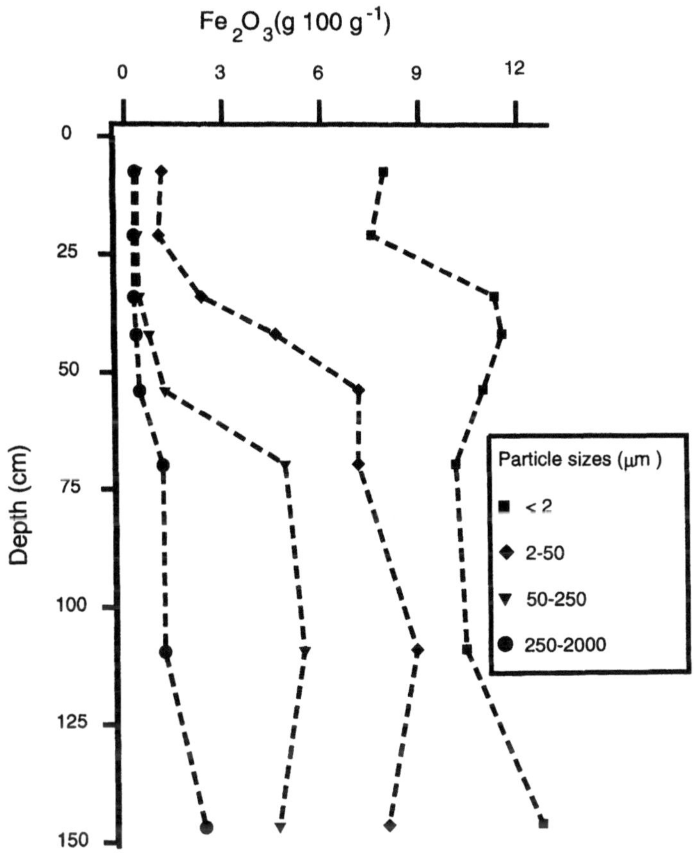

Figure I.2 Variation in total phosphorus (a) and iron (b) concentrations in an Australian alfisol with particle size fraction and depth in the profile (Chartres and Walker, 1988).

Information from each of the individual particle size classes may be integrated by plotting the data as cumulative curves or on triangular diagrams. Figure I.3a presents such cumulative plots for the surface horizons of four contrasting soils from Table I.4 (below). These curves allow a straightforward visual comparison of soil textures whether these are from different soils, different depths in a single profile or represent spatial variation, perhaps at different locations down a hillslope. Figure I.3b illustrates the changes in particle-size distribution that occur with increasing depth in the profile of the tropical alfisol described by Murtha (1982). The relative similarities of the particle size distributions of the A and E horizons and the sharp increase in clay content at the top of the B horizon are clearly illustrated.

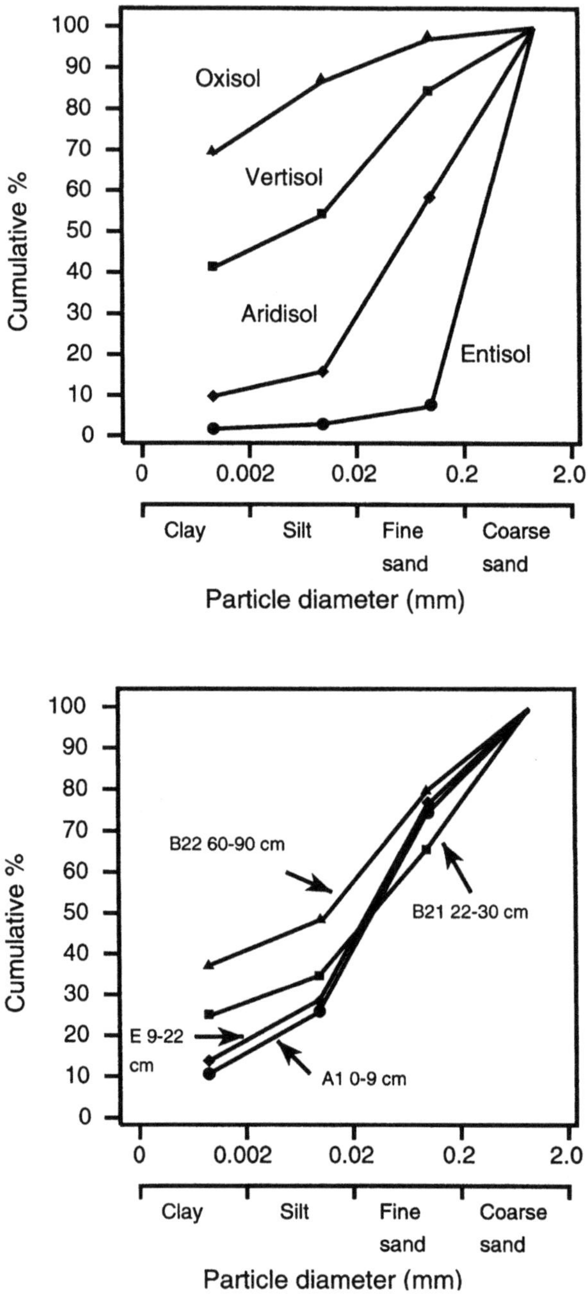

Figure I.3 Cumulative particle-size distributions for: (a) the surface horizons of four soils from Table I.4 and (b) four horizons from an aridisol (Murtha, 1982). Note that the horizontal scales are logarithmic and that the size classes are represented by their mid points.

Table I.4 Some textural and other properties of selected soils classified at the ordinal level of Soil Taxonomy.

Soil order	Depth (cm)	Particle size distribution %				% Carbon	Parent Material	Environment	Location	Reference
		Coarse sand	Fine sand	Silt	Clay					
1. Entisol	0-10	94	4	1	2	0.81	aeolian sand	Coastal	NE Queensland	Murtha, 1986
	20-30	97	2	0	1	0.10				
2. Entisol	0-1	100*		0	0	-	granodiorite	glacial morraine	Antarctica	Bockheim,1980a
	1-13	98*		2	0	-				
	13-42	97*		2	1	-				
3. Inceptisol	0-10	6	5	42	47	7.92	alluvium	tropical rainforest (poorly drained)	NE Queensland	Murtha, 1986
	20-30	4	13	35	48	3.52				
4. Andisol	0-5	13	47	35	5	9.4	volcanic ash	plantation	Papua New-Guinea	Bleeker and Sageman, 1990
	40-50	73	19	6	2	0.22				
5. Mollisol	0-13	-	-	43	13	6.0	dolerite	pasture	England	Avery, 1990
	13-28	-	-	31	22	4.0				
6. Alfisol	0-5	46	44	4	7	0.62	sand-stone	savanna	semi-arid Queensland	Coventry *et al.*, 1983
	20-30	41	51	5	3	0.025				
	65-75	42	29	4	25	<0.2				
7. Alfisol	0-9	25	48	14	12	0.8	granitic alluvium and colluvium	savanna	NE Queensland	Murtha, 1982
8. Ultisol	1-10	13	31	25	31	4.39	metamorphic rock	tropical rainforest	NE Queensland	Laffan, 1988
	20-30	12	30	28	30	1.90				

Table I.4 (cont.) Some textural and other properties of selected soils classified at the ordinal level of Soil Taxonomy.

Soil order	Depth (cm)	Particle size distribution %				% Carbon	Parent Material	Environment	Location	Reference
		Coarse sand	Fine sand	Silt	Clay					
9. Oxisol	0-10	2	11	17	70	4.80	basalt rock	tropical rainforest	NE Queensland	Spain *et al.*, 1989
	30-60	1	16	24	58	1.2				
10. Vertisol	0-7.5	8	33	15	44	3.8	shale	sub-tropical	Northern Queensland	Reeve *et al.*, 1963
	305-660	6	26	14	54	1.2				
11. Aridisol	0-5	42	43	6	9	0.5	sandstone	tropical savanna	Northern Queensland	Coventry *et al.*, 1983
	10-20	46	41	5	9	0.41				
12. Spodosol	0-8	59	28	8	5	12.0	sandstone	conifer forest	England	Avery, 1990
	8-18	62	28	7	3	0.9				
	18-30	58	30	5	7	2.9				
	30-40	59	29	3	9	1.2				
13. Histosol	0-25	-	-	-	-	31.0	plant material, rhyolite	coarse pasture	England	Avery, 1990
	25-33	39	21	31	9					

* Coarse plus fine sand

Triangular diagrams (Figure I.4) permit allocation of soils to broadly-defined textural classes and this can be of considerable value in predicting their handling and tillage properties. This may also be of use in biological characterisation since, as discussed below, moisture and organic matter relationships depend to a substantial degree on texture. Figure I.4 presents the uppermost mineral horizons of a subset of soils from Table I.4, classified according to International standards (McDonald *et al.*, 1990). It should be noted that because of the differing definition of silt in the British Standard System, the positions of the soils described in Avery (1990) are only approximate. Further, soil texture normally alters with depth in the profile and the alfisol described by Murtha (1982) (Table I.4, Figure I.3b) changes abruptly from a sandy loam to a sandy clay at the transition between the E and B horizons (*i.e.*, at 22 cm).

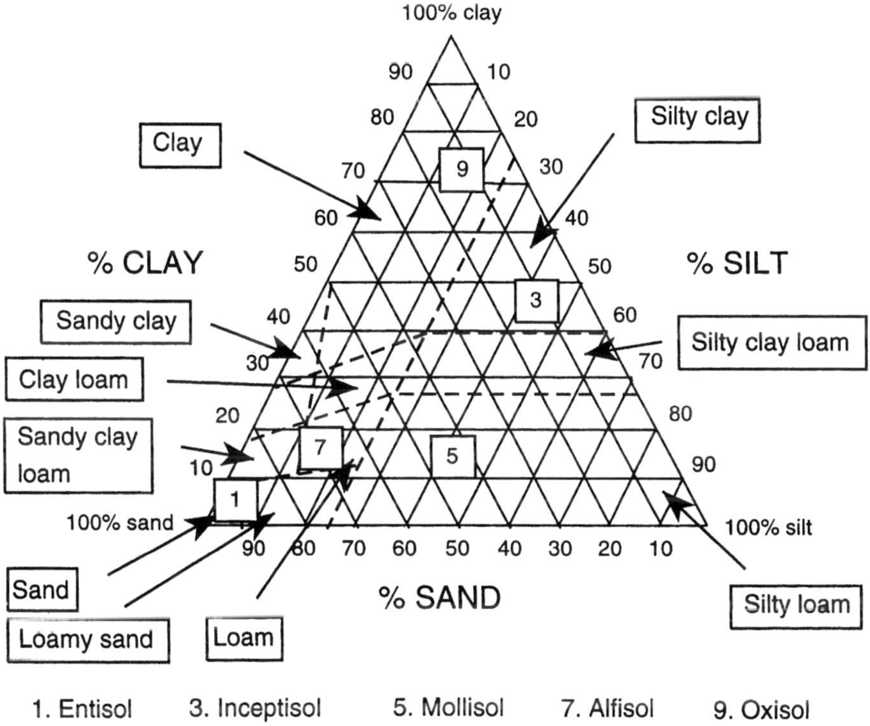

Figure I.4 Textural classes based on the International System (Marshall, 1947) showing the affinities of the upper mineral horizons of selected soils from Table I.4.

Textural differences between selected major soil groupings

The proportions of the primary size fractions of the fine earth (the particle-size distribution) often differ markedly between soils and horizons. Texture variation down the profile is a characteristic of certain orders in the American system of classification, Soil Taxonomy

(Chapter II) (Soil Survey Staff, 1999); detailed descriptions of each order are to be found in Wilding *et al.* (1983). Table I.4 presents examples of the particle size distributions, carbon contents, parent materials, environments and locations of a diverse range of soils classified at the level of Soil Order in Soil Taxonomy. It should also be recalled that the limits of the various size classes of particles differ between systems.

The **entisols** show little profile development. The first of these soils was formed on a beach ridge stabilised by tropical woodland vegetation and shows little horizon development beyond a slight organic matter accumulation near the surface (Murtha, 1986). In contrast, the second soil (Bockheim, 1980a) was formed on the terminal moraine of a glacier in the cold desert of Antarctica; it supports no macrophytic vegetation and has a negligible organic matter content. Despite its estimated age of 135,000 years, it has virtually no clay; this is a legacy of the slow rates of chemical weathering that occur in the Antarctic environment (Campbell and Claridge, 1987).

The **andisols** are a group of soils largely formed from volcanic ash parent materials and occur principally in regions with a history of recent vulcanism. These soils are found notably in countries of the Pacific rim but also occur in other continents and on certain volcanic islands such as Martinique (West Indies) and Réunion (Mascaraigne Islands). They are characterised by unusual physical properties such as low bulk densities and irreversible shrinkage on drying (Maeda *et al.*, 1977), high soil organic matter contents and the frequent presence of volcanic glass and the minerals allophane and imogolite. They are also rich in active aluminium and/or iron and have a substantial capacity to fix phosphorus in forms unavailable to plants (Wada, 1985). The andisol described by Bleeker and Sageman (1990) is from Papua New Guinea and is located on the lower footslope of an active volcano where it supports a vegetation of coconuts and cocoa. This soil is periodically rejuvenated by volcanic ash showers, the most recent of which took place during 1979.

The **alfisol** described by Coventry *et al.* (1983) is from a seasonally dry (*ca.* 600 mm annual average precipitation) tropical savanna environment in north-eastern Australia. It is a coarse-textured soil with little organic matter, consistent with its climate and sparse vegetation. High proportions of gravel occur throughout the profile; some gravel-sized particles have been transported from elsewhere or were formed secondarily by precipitation from solution and other processes, while further gravels at depth in the profile were inherited from the parent rock. The clay content of the fine earth increases at depth, probably due to the transport of clay-sized particles (illuviation) downward through the profile.

The **alfisol** described by Murtha (1982) from northern Australia was formed from granitic alluvial and colluvial parent materials in a seasonally-dry tropical climate, with an average annual rainfall of 1170 mm. It supports a tropical savanna vegetation with a tree stratum largely dominated by species of the genus *Eucalyptus* (Myrtaceae) and an herbaceous layer dominated by perennial grasses. As shown below, this soil is characterised by a coarse-textured A horizon that changes abruptly to a finer B horizon characterised by high concentrations of exchangeable sodium.

The **ultisol** described by Laffan (1988) was formed from metamorphic rock parent materials in a humid (approximately 2000 mm annual average rainfall) environment in tropical north-eastern Australia. It currently supports a montane rainforest through whose canopy the taller stems of *Eucalyptus* species emerge. It has an appreciable carbon content, consistent with its rainforest vegetation and elevation (620 m) (Spain, 1990).

The **mollisol** described in Avery (1990) was formed from doleritic parent materials in a humid temperate climate (England). The A horizon is thick (28 cm) and humic, as shown by the carbon figures presented (estimated from loss-on-ignition data using formulae in Ball, 1964). It has a well-developed fine granular structure and many large channels, a consequence of the substantial earthworm populations present.

The **oxisol** described in Spain *et al.* (1989a) was formed from basaltic parent materials and occurs in a humid (3609 mm annual average rainfall) tropical environment in northern Australia and supports a vegetation cover of lowland rainforest. Because of its considerable age, the nature of its parent material and the environment, it has weathered to produce a deep, permeable soil with a high proportion of clay-sized particles. The surface soil is rich in organic matter, consistent with its rainforest vegetation and high free iron (iron oxides soluble in dithionite) content (Spain, 1990). Despite its substantial clay contents it also transmits water readily because of micro-aggregation of the smaller particles into the larger compound units (peds) that comprise the smallest unit of structure. This is true of both infiltration and sub-surface lateral flow (Bonell *et al.*, 1983).

The parent material of the Australian **vertisol** described by Reeve *et al.* (1963) is a shale. The soil is alkaline in reaction with calcium carbonate nodules visible throughout much of the profile. It is fine-textured and shrinks and swells depending on its state of hydration, because of appreciable smectite among its clay minerals. Contraction on drying, leads to the formation of extensive patterns of deep surface cracking. Because of the smectite present, this soil has a higher organic matter concentration than most other soils occurring under similar rainfall and temperature regimes. This soil has formed under the influence of a sub-tropical climate with an average annual rainfall of 660 mm, most of which falls in the summer months. It supports a savanna vegetation dominated by *Acacia harpophylla* (Leguminosae) trees.

The English **spodosol** described in Avery (1990) was formed on a sandstone and supports a vegetative cover of *Pinus sylvestris* and *Larix decidua* with an herbaceous layer of grass and the fern *Pteridium esculentum*. It has a substantial litter layer of partly decomposed vegetation and a typically organic A horizon. The bleached E horizon is friable with particulate or single-grain structure. The B horizon shows characteristic accumulations of clay, soil organic matter and the oxides and oxyhydroxides of iron and aluminium.

Peats, or **histosols**, comprise a group of soils that possess a thick surface horizon (known as an H or O horizon) consisting largely of often little-modified plant remains overlying the mainly mineral soil. Peats usually form in situations where decomposition is inhibited by impeded drainage, low temperatures or a combination of the two. The histosol presented in Table I.4 was formed on rhyolitic parent materials in the north of England at an altitude of 677 m (Avery, 1990) where it developed under a cold and humid climate. It supports a vegetation of grasses, low shrubs and mosses and has a surface layer of fresh and partially-decomposed litter. The upper peaty horizon and the underlying mineral A horizon contain rhyolitic fragments, ascribed to the effects of cryoturbation (frost heave).

Faunal modification of soil texture
Soil texture may be altered by those soil animals that are most pedogenetically active, particularly termites, ants and earthworms (Chapter IV). Animals of these groups usually select the finer soil particles in building their surface constructions. These particles may be transported from a range of depths in the profile and used to build their surface mounds and temporary surface structures, or simply cast on or near the surface. Over time, erosion reduces the mounds and accumulated casts and the soil is re-distributed over the surface. On sloping sites, the finer particles may be transported away from the site by water, leaving the larger particles to accumulate as a coarser-textured surface horizon. This may result in a surface horizon of distinctly coarser texture overlying one of finer texture (see, for example, Williams, 1968; Wielemaker, 1984). Elsewhere, an accumulation of finer-textured surface soil results with an upper limit of particle size related to the capacity of the fauna to transport them.

1.1.1.2 *Clays and clay minerals*

The inorganic soil particles less than 2 μm in diameter are collectively termed clay. They are varied in nature, ubiquitous in soils and play a critical part, either directly or indirectly, in almost all soil processes and interactions. Many of the materials in this size range that are found in soils may be classed as minerals. That is, they are naturally-occurring homogeneous solids possessing a definite but not fixed chemical composition and an ordered atomic arrangement (Schulze, 1989). The occurrence of these minerals in soils results from three mechanisms, namely, inheritance from parent materials, formation *in situ* by crystallisation from solution or through the alteration of existing minerals into new forms (transformation) (Allen and Hajek, 1989) (see Chapter II).

These minerals are most usefully classified by their dominant anion or anionic group (Schulze, 1989) and the following are the groups of most importance in soils:
1. Sulphides
2. Oxides, oxyhydroxides and hydroxides
3. Halides
4. Carbonates
5. Phosphates
6. Sulphates
7. Silicates

Not all of these mineral groups are present in all soils and some only occur under particular environmental conditions. The simple mineral halite (NaCl) is one such example and is found naturally in arid areas or as an evaporite in certain tropical coastal saline soils. Other common simple minerals include calcite ($CaCO_3$), gypsum ($CaSO_4.2H_2O$) and dolomite ($CaMg(CO_3)_2$). In contrast, the silicate minerals are relatively stable, ubiquitous in soils and the types and proportions present are of the greatest importance in determining soil properties and behaviours in the face of agricultural and engineering usages.

Many soils retain a range of coarse fragmentation products derived from their parent rocks and therefore still possess some of the primary minerals derived from the parent rock and formed by high temperature or high pressure igneous and metamorphic processes (Chesworth, 1992). Quartz (a tectosilicate) is a major and common component of soils; it may form the major part of the sand and silt fractions and frequently also the coarser clay-sized components (Drees *et al.*, 1989). It may be either primary or secondary, that is, synthesised from Si within the soil rather than inherited from the parent material. Soils also contain a range of other secondary products notably clay minerals that have been re-synthesised from the weathering products of the parent rock and other materials, or inherited from the breakdown of sedimentary rocks. Inorganic soil particles of clay size therefore include a diversity of crystalline and non-crystalline materials that, in most soils, form associations with a variety of other substances, both inorganic and organic.

Because of their colloidal properties, the dominant phyllosilicate clay and other minerals must remain flocculated if the soils within which they occur are to have acceptable levels of stability. As colloids, they have large surface areas per unit mass or volume (specific surface areas) (Table I.5) and many of their properties derive from this (Petersen *et al.*, 1996). These include their capacities to retain both water and a range of plant nutrient and other elements at their surfaces. The large surface areas influence the rates and pathways of many soil chemical reactions and directly dictate the amount of water retained in the soil. Also, the amounts and types of the phyllosilicate clay minerals present influence the strengths and dispersabilities of the soils within which they occur. Clays are usually associated with other materials, both organic and inorganic, to form structural aggregates of all sizes that impart stability to the soil and control such important processes as water movement and aeration. Clays also play important roles in controlling the activities of micro-organisms and in reducing enzyme activity within soils (Burns, 1986) (see Chapter IV, section 4.1.3.2).

Classification of soil clay and other minerals

The principal materials of clay size that occur in soils are the crystalline phyllosilicate and oxide minerals, the 'amorphous' minerals often derived from materials of volcanic origin and the gels or metallo-hydrates. The crystalline phyllosilicate clays are the larger common clay minerals; as explained later in this chapter, the more reactive paracrystalline and amorphous materials are substantially smaller (Figure I.5).

The phyllosilicate clay minerals. The silicate minerals are based on two primary structures, the silicon tetrahedron and the aluminium octahedron. The terms tetrahedron and octahedron refer to the regular structures associated with co-ordination of the central cations; silicon is bonded to four O^{2-} ions arranged in a tetrahedron while the central aluminium ion of the octahedron is bonded to six O^{2-} ions to form a regular octahedron (Figure I.6) (Millot, 1979). By sharing certain of their oxygen ions with adjacent octahedra and tetrahedra, these structures may be arranged to form compound structures including chains, rings, three dimensional frameworks and sheets. Of these, the most widespread and important in soils are those that occur as sheets, the phyllosilicate clay minerals.

The phyllosilicate clay minerals are ubiquitous in soils and, except in the most coarsely-textured soils, may make up much of their volume. Their abundance in soils is a consequence of the dominance of their major constituents (Si, Al and O) in the earth's crust. These are also the best known of the clay minerals since a number have

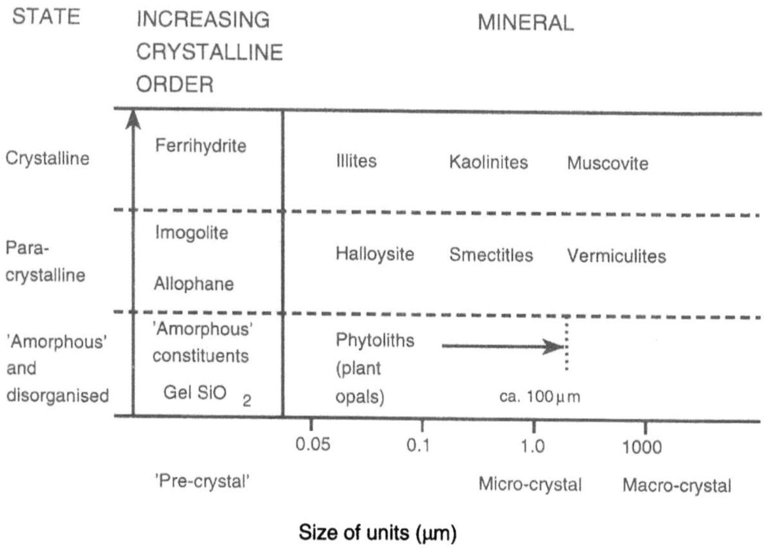

Figure I.5 Secondary soil constituents in relation to their size and crystallinity (from Pedro, 1983).

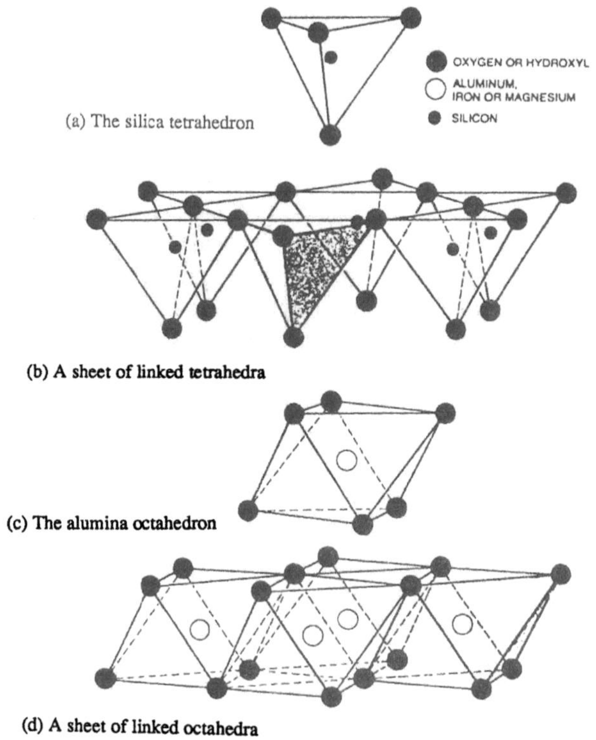

Figure I.6 The basic tetrahedral and octahedral units of the phyllosilicate clay minerals showing their respective inter-linkages to similar units to form sheets of uniform structure (Millot, 1979).

industrial uses. All are alumino-silicate compounds comprising layers of tetrahedrally-coordinated silicon atoms and layers of octahedrally-coordinated aluminium atoms.

These layers are arranged in various ways according to the clay mineral group involved; in some, the vertical repeating unit (the unit layer), consists of one layer of silica tetrahedra and one of aluminium octahedra, the so-called 1:1 minerals. The 2:1 minerals consist of two layers of silica tetrahedra with an intervening alumina layer. In the 1:1 clay minerals, the succeeding unit layers are tightly held together by hydrogen bonds and these minerals do not normally expand. In the 2:1 minerals, the bonds between the layers joining the repeating units are much weaker, consisting of Van der Waal's forces and the electrostatic attraction of the cations present. Water, other polar solvents and a range of other materials may penetrate these layers leading to variable spacings between the repeating layers, depending on the hydration state of the soil and the dominant cations present.

The central aluminium atoms of the octahedral layers and the silicon atoms of the tetrahedral layers may be replaced by other elements of appropriate size and charge, commonly magnesium and iron. The replacing ion is usually of lower valency and this process of isomorphous replacement leaves the sheet with a permanent net negative charge which is normally balanced by cations in the soil solution. A further source of charge arises from the dissociation of hydroxyl ions on the surfaces and particularly at the edges of the clay sheets; in these cases the charge is variable, being dependent on the pH and ionic concentration of the soil solution.

The common phyllosilicates fall into seven groups, based on the relative number of sheets in the unit layers, the types of ions that occur between the unit layers and the nature of the isomorphous replacement. These seven groups are (Millot, 1979): kaolin, illite, smectite, chlorite, vermiculite, mixed-layer and sepiolite-attapulgite, of which the first four are perhaps the most common in soils. Minerals in the sepiolite-attapulgite group are not organised as plate-like structures, but form fibrous, needle-like crystals.

The kaolin group is most commonly represented by kaolinite, a 1:1 clay mineral whose unit layer is made up of one tetrahedral and one octahedral sheet (Figure I.7) (Schulze, 1989). It has a relatively small surface area because its unit layers are held together tightly by electrostatic bonds and cannot be penetrated by polar solvents; its internal surfaces are thus unable to contribute to its often low overall apparent surface area (Table I.5).

Kaolinite normally has a low level of isomorphous substitution and thus little permanent negative charge. However, crystals of this mineral from tropical soils may be much smaller than those from kaolin deposits (*ca.* 0.1 μm in comparison to 1.0 μm) and possess high surface areas of 100-250 $m^2\ g^{-1}$. Schwertmann and Herbillon (1992) showed that the isomorphous substitution of iron within the kaolinite structure gave rise to smaller and more active crystals effecting substantial changes to such soil properties as phosphate adsorption. In addition to its permanent charge properties, this mineral has a pH-dependent charge based on the dissociation of hydroxyl ions. Kaolinite and halloysite, another member of this group, are common clay minerals in the highly-weathered and nutrient-poor soils of many tropical areas.

The illites are 1:1 clay minerals and are amongst the most common of the phyllosilicate clay minerals. The distance between the unit layers is fixed (1.0 nm) and they are bound together with a layer of potassium ions. Most illites appear to form through transformation of micas present in the soil parent materials (Brown, 1990).

The smectites (Figure I.8) are a group of 2:1 clay minerals in which water and other polar solvents are able to penetrate between the layers. The relative weakness of the interlayer bonds means that their distances apart are not fixed but vary with the humidity and the sizes of the hydrated and exchangeable cations present between the layers.

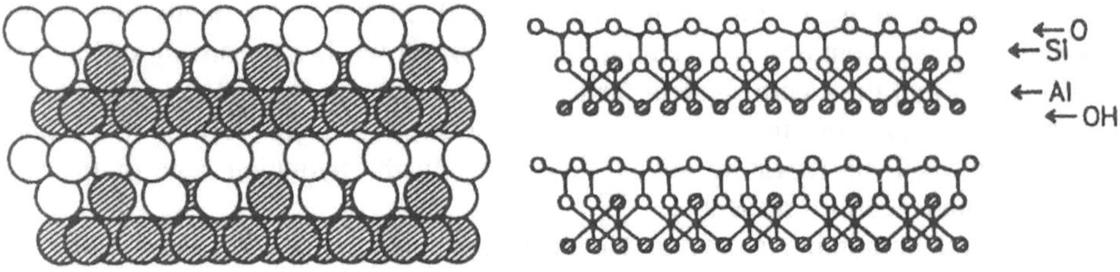

Figure I.7 Sheet and layer structure in the 1:1 phyllosilicate clay mineral kaolinite (from Schulze, 1989).

Interlayer spacing normally varies from 1.2 to 1.5 nm although it may be much larger at high humidities when sodium is the dominant interlayer cation present (Brown, 1990).

The surface areas of smectites are large since internal interlayer surfaces also contribute to their total surface areas. In the 33 smectite-rich soils described by de Kimpe *et al.* (1979) (Table I.5), the internal surfaces contributed an average of 82 % of the total surface area. Soils containing appreciable smectites (vertisols) shrink extensively and reversibly on drying to produce distinctive patterns of polygonal surface cracking.

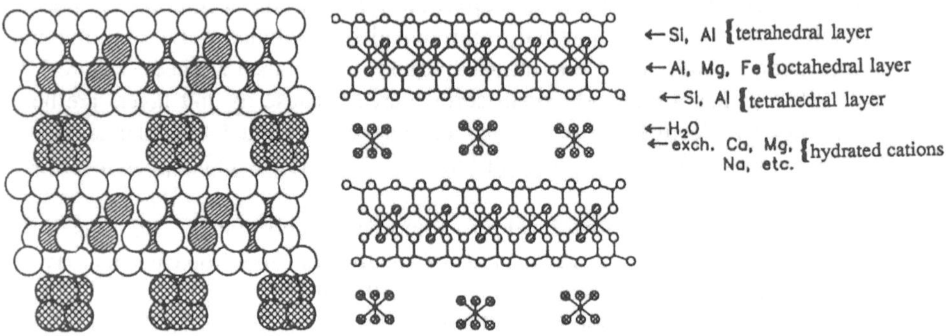

Figure I.8 Sheet and layer structure in the smectite group of 2:1 phyllosilicate clay minerals (from Schulze, 1989).

In the chlorite group, the 2:1 unit layers are bound together (at a distance of 1.4 nm) by layers of hydrated magnesium, aluminium or iron.

Not all clay minerals have the structure and behaviour typical of the major groups listed above. Mixed layer clays occur widely in soils and the inclusion of clays of

Table I.5 Specific surface areas, cation exchange capacities and particle sizes and shapes of selected minerals and soils (Cornell and Schwertmann, 1996; De Kimpe *et al.*, 1979; Schwertmann and Taylor, 1989; Uehara, 1982; Sumner *et al.*, 1992).

Mineral or dominant mineralogy	Particle diameter (μm)	Particle thickness (μm)	Specific surface (m^2 g^{-1})	Cation exchange capacity (cmol kg^{-1})
a. Minerals				
Smectite	0.03	0.001	600-800	110
Illite	0.3-1.0	0.02-0.07	65-200	20-50
Kaolinite	2	-	35	1.3-5.0
Halloysite (tubular)	0.07	0.04-1.0	21-43	-
Halloysite (spheroidal)	0.02-1	-	-	-
Allophane (hollow spheres)	0.003-0.005	-	1000	135-143
Imogolite (hollow, filiform)	0.002-0.003	1-3 (length)	1000	-
Haematite	0.02-0.05	0.01-0.02	34-45	-*
Goethite	0.02-0.10	0.05-0.1	14-200	-
Gibbsite	0.1	0.005	10-30	-
Ferrihydrite (spheroidal)	0.003-0.007	-	200-500	-
b. Soils				
Kaolinitic	-	-	15-49	3.7-21.8
Kaolinitic-oxidic	-	-	31-98	1.6-34.3
Kaolinitic-illitic	-	-	8-132	4.9-52.2
Smectitic whole soil	-	-	37-142	10.3-33.2
Clay component				
total	-	-	130-285	28.2-62.4
internal	-	-	107-234	-

*Charge on oxide minerals is dependent on pH and ionic strength of soil solution.

this type may alter soil properties considerably. In a soil derived from volcanic ash, the dominant clay mineral was the 1:1 clay mineral halloysite. However, samples also contained a substantial proportion of smectite resulting overall in a greater specific surface area, higher cation exchange capacity and other altered properties of agricultural importance (Delvaux *et al.*, 1990).

The crystalline oxides of iron and aluminium. Although other metallic elements (Ti, Mn, etc.) occur in soils and are sometimes important, iron and aluminium are among the most abundant metallic elements in the earth's crust. Due to their limited solubilities, both are particularly abundant in highly-weathered soils (notably of the tropics), the spodosols and the andisols where they may occur as discrete particles or form coatings on other materials. Their presence in significant quantity may result from the particular pedogenetic processes occurring within the soils involved, or the higher iron or aluminium contents of their parent materials. For example, soils derived from basaltic parent materials are usually richer in iron which, in reduced form, moves readily in soils and may become concentrated through reduction - transport - recrystallisation processes.

The oxides, hydroxides and oxy-hydroxides of both iron and aluminium occur in a range of chemical forms in soils, depending on pH, environmental conditions and such factors as organic matter concentration (Huang and Violante, 1986; Schwertmann *et al.*, 1986). Aluminium also routinely substitutes for iron in some iron oxide minerals commonly found in soils (Huang and Violante, 1986; Schwertmann and Herbillon, 1992).

The most common of the iron oxides in soil are goethite (α-FeOOH) and haematite (α-Fe_2O_3) although others may be prominent in specific environments. One such example is lepidocrocite (γ-FeO(OH)) which is characteristic of such poorly drained environments as swamps. Goethite may be the sole pedogenic iron oxide present in soils of the cool and temperate zones. In many of the warmer parts of the world, it occurs in association with haematite which imparts the vivid red colouring characteristic of many tropical soils (Schwertmann and Taylor, 1989).

Ferrihydrite ($Fe_{10}O_{15}.9H_2O$) is a poorly crystalline form of iron present in many soils; it is somewhat unstable and converts to goethite in cool temperate environments and haematite in warmer areas. Maghemite (γ-Fe_2O_3) is another iron oxide, probably largely formed by the conversion of other iron oxides through firing (Schwertmann, 1988).

Gibbsite ($Al(OH)_3$) is common in highly-weathered soils and in laterites. It occurs in highly crystalline form as small hexagonal plates or rods (Table I.5).

In contrast to the largely permanent charge of the phyllosilicate clay minerals, a variable charge is associated with the oxide minerals. Both the magnitudes and the signs of the net charges of the different oxide minerals vary with pH and, depending on the oxide, the sign may be negative, neutral or positive at normal soil levels. Both depend on the pH at which the net charge of the oxide is zero, *i.e.*, its point of **zero net charge**. Net charge is negative above this point and positive below it. The points of zero net charge of iron oxides occur between pH 6.5 and 8.0 and for aluminium oxides between pH 7.5 and 9.5. This has considerable importance in determining the retention of cations and anions (both inorganic and organic) in soils where these oxides are present in appreciable quantities. The implications for plant growth are important since nutrient cations may be readily lost from such soils, particularly in agricultural situations where organic matter is present in low concentrations. Exchangeable forms of aluminium in soils are mostly monomeric (Huang and Violante, 1986).

Apart from this, both iron and aluminium oxides are of particular agricultural importance in that they have high affinities for phosphates and some trace metals which they fix in non-exchangeable forms unavailable to plants. In addition, aluminium may occur in amounts increasingly toxic to higher plants in soils where the pH falls below *ca.* 5.5 since positive charges predominate and Al^{3+} ions are released in the soil solution.

Volcanic weathering products and other amorphous materials. As discussed above, the extremely-fine weathering products of volcanic glass are characteristic of the soils known variously as andisols or andosols.

The aluminosilicate materials that give these soils their characteristic properties (Maeda *et al.*, 1977; Wada, 1985; Table I.5) are the X-ray 'amorphous' and poorly crystalline minerals allophane and imogolite, although lesser quantities of both minerals also occur in podzols (spodosols) and other soils. The individual particles are very small and their surface charges are dependent on the pH and ionic strength of the soil solution. Allophane occurs in the form of small spheres and imogolite as small cylindrical or string-like particles; because of their small sizes, these minerals have very large specific surface areas (Table I.5).

In addition to that present in quartz and other crystalline and semi-crystalline forms, silica (SiO_2) may also occur in soils in an amorphous form (Drees *et al.*, 1989). Amorphous silica may either be inorganic in origin or derived from plants and sometimes animals. Biogenic silica occurs largely as structures called phytoliths or plant opals which are concentrations of hydrated silica and other materials that are precipitated within plant tissues. These are returned to the soil surface in litter or through root decay. They occur in a wide variety of shapes ranging up to *ca.* 100 μm in length although most are less than 50 μm. They differ considerably in solubility, depending on the presence of surface coatings of Al and other factors and the less soluble types may accumulate in soils to maxima of *ca.* 3 % or occasionally more (Drees *et al.*, 1989).

Annual Si return to the soil in dead plant parts has been estimated to range from 10-76 kg ha^{-1} y^{-1} (Duchaufour, 1997; Alexandre *et al.*, 1997). In a highly-weathered soil supporting an equatorial rainforest, Alexandre *et al.* (1997) estimated that the rate of silica release through phytolith dissolution was more than twice that due to silicate weathering. Because of its relatively high solubility, Si is readily taken up by plants (as monosilicic acid, $Si(OH)_4$) and is susceptible to loss in drainage waters; it is also a Si source for the formation of other minerals, including the phyllosilicate clay minerals (Duchaufour, 1997; Lucas *et al.*, 1993).

The evolution and distribution of clays

The clay minerals present in any soil depend on its parent materials, its weathering history and the amounts and differential solubilities of the elements present in the original parent materials. Millot (1979) states that clay mineral formation may be conceived of as a process involving the subtraction of elements in decreasing order of their solubilities (Na >K >Ca >Mg >Fe >Si >Al) followed by the formation of new minerals from those remaining.

The clay mineral composition of soils also varies qualitatively and quantitatively with depth and regionally with climate and the other soil forming factors (Chapter II).

Figure I.9 illustrates diagrammatically the distributions of the predominant terrestrial clay mineral groupings throughout the world (Millot, 1979). While the boundaries are considerably more complex than indicated, different environments clearly produce characteristic suites of clay minerals.

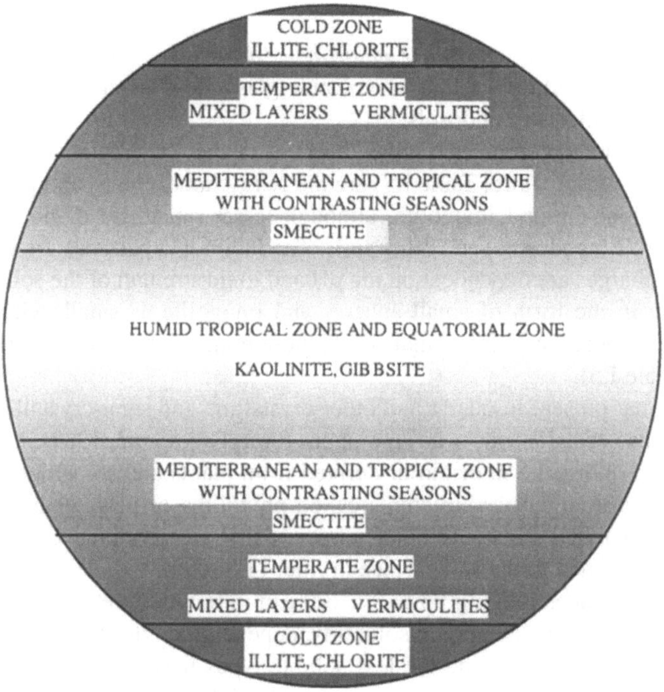

Figure I.9 Indicative distributions of the dominant clay mineral types throughout the world (Millot, 1979).

In Antarctica and other deserts, almost no leaching or hydrolysis occurs; little clay forms under these circumstances and nitrates and other soluble elements of aerosolic origin may accumulate (Campbell and Claridge, 1987). Under conditions of moderate leaching, mixed-layer minerals and vermiculites are typical. Where greater leaching and hydrolysis pertain, but where some Si, Fe and Mg still remain, smectites occur. Smectites also seem to form most readily where there is a clear alternation of wet and dry seasons. Where very strong leaching and hydrolysis occur, kaolinite and halloysite are typical and in the extreme, oxide minerals such as gibbsite ($Al(OH)_3$) may accumulate.

Even at more restricted geographic scales, climate is a dominant factor controlling surface soil clay mineral composition. Folkoff and Meentemeyer (1985) have shown that climate is the dominant factor determining the phyllosilicate clay assemblages of the A horizons of soils in continental United States of America. However, a number of soils were exceptions to this generality: the major group comprised older soils whose clay mineral suites had been determined under previous climatic regimes while another small group consisted of soils of different parent material composition. Clearly the balance of these effects must be expected to alter between locations. The soils developed on older geological surfaces may be expected to have clay mineral suites that reflect past rather than present climates. Further, when extreme parent materials such as volcanic ash and serpentine predominate, soil clay mineral assemblages are unlikely to reflect current climatic influences but will reflect parent material properties.

The cation exchange capacity of clays

Soil colloids (principally phyllosilicate clays and organic matter) have a mixture of positively and negatively charged sites on their surfaces to which are attracted a range of ions, both organic and inorganic. In most soils, the negative charges predominate to give a net negative charge to a soil or horizon; less commonly, a net positive charge may pertain in certain soils, particularly at low pH.

The permanent net negative charge on the phyllosilicate clays is of considerable importance in plant nutrition. It is the most important mechanism through which plant nutrient and other cations are held at the surfaces of the soil colloids. The major basic cations Ca^{2+}, Mg^{2+} K^{+} and Na^{+} (which raise the soil pH) and the acidic cations H^{+} and Al^{3+} (which lower it) are retained at negatively-charged sites on both organic and inorganic colloids within soils. The capacity to retain these positively charged ions is called the cation exchange capacity and is defined as the product of the specific surface area and the surface charge density.

An analogous anion exchange capacity, usually of lesser magnitude, also exists in soils and is responsible for the retention of such anions as PO_4^{3-} and SO_4^{2-} at colloid surfaces. It is associated with such positively-charged sites as the surfaces of variable charge oxide minerals, principally those of iron and aluminium. These elements are normally present at low concentrations in most soils but, because of their small particle sizes (Table I.5), have very large specific surface areas and thus reactivities.

A further source of charge, variable in nature, occurs on the surfaces and particularly at the edges of phyllosilicate clay mineral sheets due to the dissociation of hydroxyl groups. As considered in a later section, organic matter has a high CEC because of the dissociation of $COOH^-$ and OH^- groups. Its substantial net negative charge at most soil pH levels is entirely variable in nature. As will be seen, soil organic matter may be the only colloid able to retain appreciable quantities of the nutritionally-important cations in acid soils. This is particularly relevant in such highly-weathered soils as the tropical oxisols and ultisols (Uehara, 1982).

In soils with variable charge components, both the pH and ionic strength of the soil solution influence the net charge of the soil as illustrated in Figure I.10. This figure shows the relationships of these effects on the net charges of soils from two horizons of an iron and aluminium-rich oxisol (van Raij and Peech, 1972). The net charge becomes positive at pH values below the point of zero net charge (normally in the range 3.5-4.5). Such values are significantly lower than the points of zero net charge for the component soil oxides (pH 7-9) which will have net positive charges in most soils. In comparison, the points of zero net charge of many clay minerals lie in the pH range 4.5-6.0. From the substantial differences illustrated, it is clear that such agricultural practices as liming and fertilising will alter the net surface charge of the soil through their influence on both pH and soil solution properties.

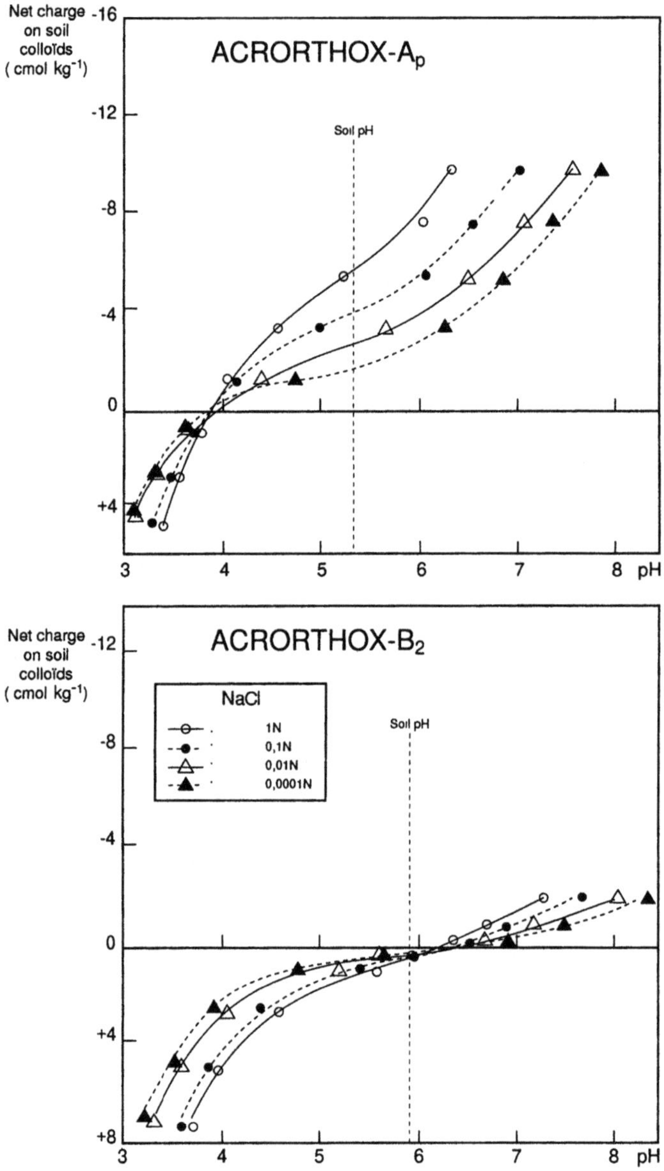

Figure I.10 Laboratory illustration of the effects of changing the pH and the soil solution concentration on the net surface charge of the colloids in two horizons of an oxisol rich in iron and aluminium (van Raij and Peech, 1972).

The exchangeable ions held at the surfaces of the phyllosilicate soil colloids are in a state of equilibrium with those of the soil solution and constant interchange occurs between them. Close to the particle surfaces, a layer of tightly bound cations occurs and this is succeeded by a more diffuse layer of cations further away. Cation concentrations are thus high close to the surfaces and decrease with increasing distance from the surface to a level similar to that of the soil solution. Anion concentrations may vary in an inverse manner.

Net exchanges occur when changes in ionic concentrations or pH occur in the soil solution. These may result from seasonal changes in soil moisture or through the addition of fertilisers or other soluble amendments. For example, the addition of KCl as a fertiliser may cause a displacement of calcium ions into the soil solution (Duchaufour, 1997):

$$Ca^{2+} \text{ (Colloid)} + 2KCl \text{ (Soil solution)} \leftrightarrow CaCl_2 \text{ (Soil solution)} + 2K^{+} \text{ (Colloid)}$$

As indicated in the above equation, such reactions are reversible; they also take place in terms of chemical equivalents. That is, one centimole of potassium ions will exchange for one centimole of calcium ions in the above exchange reaction.

Not all ions are adsorbed equally onto the colloid surfaces. With a few exceptions, the ions of higher valency are held more tightly than those of lower valency. Thus, in decreasing order, Al^{3+} is held more tightly than Ca^{2+} which is held more tightly than K^{+}. For ions of equal valency, the least hydrated ions are preferentially adsorbed. Ca^{2+} ions are preferentially adsorbed onto organic colloids and this ion therefore dominates in soils where organic matter concentrations are high.

Where exchangeable Na ions comprise 6-15 % or more of the cation exchange capacity (sodic soils), the large radius of the hydrated Na ion may move adjacent phyllosilicate colloid surfaces apart to the extent that deflocculation occurs. This leads to considerable instability in the soil and sodic soils are subject to severe erosion, surface sealing and the development of strong surface crusts together with other phenomena inimical to ecosystem function and productivity. The importance, morphology, structure and distribution of crusted soils are discussed in the volume edited by Sumner and Stewart (1992).

Not all reactions between ions in the soil solution and clay surfaces can be explained in terms of simple exchange phenomena. A range of cations, phosphate and other anions may all be subject to an initial adsorption process followed by slow diffusion into, particularly, iron oxides such as goethite (Barrow, 1989).

Table 1.6 presents examples of the cation exchange capacities of a range of soils classified at the ordinal level of Soil Taxonomy (Soil Survey Staff, 1999). A number of points are illustrated by the table. Firstly, considerable differences exist between the orders, with the vertisols having the highest overall values due to the expansive (smectitic) clay minerals present in these soils. With increasing depth in the soil, a clear reduction in cation exchange capacity occurs paralleling the substantial decline of organic matter concentrations with depth. Finally, considerable differences exist between estimates of cation exchange capacity depending on the measurement methodology. This is of particular importance in soils where variable charge components dominate, notably in such acid, highly-weathered soils as the ultisols, oxisols and exemplified here by the rainforested inceptisol (Cannon *et al.*, 1992) and the spodosol described by Mew and Lee (1981). Because of the variable charge components present, inflated estimates of cation exchange capacity result, particularly in acid soils where substantial differences

Table I.6 The cation exchange capacities of selected soils classified at the ordinal level of Soil Taxonomy (Soil Survey Staff, 1999)

Order	Horizon	Depth range (cm)	CEC (cmol kg^{-1}) at pH 7	CEC (cmol kg^{-1}) at soil pH	Vegetation	Author
Entisol	A1	0-10	-	1.0	*Casuarina*, *Acacia* woodland	Murtha, 1986
	B	20-30	-	0.4		
Inceptisol	A1	0-10	19.0	4.5	rainforest	Cannon *et al.*, 1992
	B2	55-85	9.0	2.5		
Inceptisol	A1	0-10	20.23	29.0	rainforest	Laffan, 1988
Andisol	A11	0-5	32.8	27.8	cultivated	Bleeker and Sageman, 1990
	A12	10-15	13.9	11.4		
	A13	20-30	12.0	7.7		
	C1	40-50	3.3	5.7		
Alfisol	A1	0-5	4.0	3.8	savanna	Cannon and Coventry, 1989
	B1	20-30	6.0	2.4		
	B2	40-60	2.4-3.3	2.9		
Ultisol	A11	0-10	20.0	2.9	rainforest	Cannon *et al.*, 1992
	B2	60-90	6.0	2.9		
Oxisol	A1	0-10	-	3.8	rainforest	Spain *et al.*, 1989
	B1	10-20	-	2.6		
	B2	20-30	-	0.6		
Oxisol	A1	0-2	31.9	15.4	rainforest	Bleeker and Sageman, 1990
	B2	22-42	3.0	0.5		
	B3	112-137	0.0	0.3		
Spodosol	A1	0-10	4.0	3.2	sclerophyll-*Casuarina* forest	Cannon *et al.*, 1992
	B2h	51-60	5.0	2.7		
	C1	120-140	<1.0	0.8		

Table 1.6 (cont.) The cation exchange capacities of selected soils classified at the ordinal level of Soil Taxonomy (Soil Survey Staff, 1999)

Order	Horizon	Depth range (cm)	CEC (cmol kg^{-1}) at pH 7	CEC (cmol kg^{-1}) at soil pH	Vegetation	Author
Spodosol	Ah	0-27	30.1	7.7	open swampy herbland	Mew and Lee, 1981
	E	45-78	7.6	6.8	and woodland	
	B2h	91-126	20.4	3.3		
Vertisol	Ap	-	27.2	-	cultivated	de Kimpe *et al.*, 1979
	Bg	-	22.2	-		
	Cg	-	22.0	-		
Mollisol	Ah	0-13	28.4	-	grassland	Avery, 1990
	Ah	13-28	22.7	-		
Histosol	O	0-40	154.0	-	peat marsh	Lévesque and Dinel, 1982

may exist between the soil pH and that at which the CEC was measured, often pH 7 or higher. In terms of understanding the field behaviour of soils, it seems appropriate to measure the cation exchange capacity under conditions approximating those pertaining in their environment.

1.1.2 THE SOIL SOLUTION

The water that occurs in the pore space of soils and moves through it contains a wide diversity of dissolved and suspended materials: organic, inorganic and organo-mineral. The soil solution is important in plant nutrition since the net flow of water to plant roots also moves a range of dissolved elements in the same direction. In pedogenetic terms, the soil solution mediates the vertical (in both directions) movement of dissolved and suspended materials both within the profile, and laterally from one part of the landscape to others downslope. Dissolved gases, including oxygen and carbon dioxide are also present; the latter gas occurs in the highest concentrations due to its greater solubility.

The soil solution can be sampled using lysimeters (with or without an applied suction) to collect the water draining through the soil. Alternatively, it may be displaced from soil samples by centrifugation, application of a suction or by infiltrating samples with fluids that displace the soil solution (Vedy and Bruckert, 1979). A problem with this methodology is the difficulty of deciding whether some of the material present in the sample is there because of the disturbance involved in installing the lysimeters, or because of the energy applied to extract the soil solution (McKeague *et al.*, 1986).

Nutrients dissolved in soil water can also be extracted using other methods (Arnold, 1958; Gibson *et al.*, 1985). Bags containing synthetic ion-exchange resins may be buried in the soil for periods; following collection, the adsorbed elements may be quantitatively leached to give satisfactory indexes of the concentrations of selected ions in the soil solution. Campbell *et al.* (1989) used an immiscible liquid to expel the soil solution from soils sampled in a largely agricultural environment. They found considerable seasonal variation in soil solution composition between different soil taxa, and between land use categories within such taxa.

1.1.2.1 *Biological nutrient and other elemental concentrations*

As shown below, the concentrations of many elements in soil solutions extracted under suction are normally higher than those draining freely into lysimeters. Lysimeter water is that held at high matric potentials in the larger pores while that extracted using suction, centrifugation or other methods is water that was previously held at lower matric potentials in the smaller pores. The soil solution extracted from the smaller pores has been in more intimate contact with the surrounding soil materials; residence times are also longer in these smaller pores.

Table I.7 presents the concentrations of a range of organic and inorganic solutes both displaced and freely-draining from the E horizon of an acid French spodosol supporting a pine forest (Bonne *et al.*, 1982). Several points emerge from this table. The first is the notably higher concentrations of many solutes in the displaced than in the lysimeter solutions, although this was not invariably so for Al, Si and N. Secondly, there is clear seasonal variation in the concentrations of the different solutes in both solutions although the maxima and minima did not coincide for all solutes. The more mobile monovalent ions have their maximum concentrations in winter, while the bivalent ions and manganese have their highest concentrations in spring and summer. Finally, the considerable concentrations of several different types of organic materials should also be noted since their movement and precipitation in the soil are important factors in the development of these spodosols, and of other soils.

Table I.7 Seasonal variation in solute concentrations (mmol dm^{-3}) in soil solutions collected from lysimeters (L) or displaced (D) from the eluvial horizon of a spodosol (Bonne *et al.*, 1982).

Solute	**Autumn**		**Winter**		**Spring**		**Summer**	
	L	**D**	**L**	**D**	**L**	**D**	**L**	**D**
Organic C	5.7	18.0	3.4	4.4	1.8	5.87	8.3	8.7
Phenol C	0.2	1.4	0.3	1.0	0.4	0.7	0.4	0.2
Carbohydrate	0.4	3.9	0.4	0.8	0.5	0.6	0.4	1.1
Organic nitrogen	0.61	0.87	0.14	0.14	0.12	0.30	0.51	0.74
α-amino N	0.18	0	0.13	0.14	0.12	0.09	0.17	0.11
NH_4^+ N	0.32	0.7	0.21	0.20	0.21	0.23	0.37	0.26
K	0.09	0.37	0.06	0.24	0.06	0.24	0.12	0.32
Ca	0.02	0.34	0.02	0.07	0.02	0.18	0.05	0.40
Mg	0.02	0.06	0.01	0.04	0.01	0.09	0.04	0.10
Mn	*	0.0005	*	0.0006	*	0.075	*	0.0004
Fe	0.0009	0.41	0.0007	0.170	0.01	0.127	0.014	0.17
Al	0.06	0.04	0.037	0.037	0.037	0.057	0.11	0.04
Si	0.07	0.002	*	0.01	*	0.045	0.07	0.07

* Less than detection limit.

1.1.2.2 *Pedogenetic importance*

In terms of soil processes, the transport of organic and inorganic materials of pedogenetic importance in the soil solution is well established (McKeague *et al.*, 1986). Table I.7 shows that the elements Fe, Al and Si are all mobilised into the soil solution and this is often a consequence of the complexes they form with organic materials of various types. Considerable variation in the properties of soil solutions occurs, depending on the pH, cation exchange capacity, redox potential, organic matter concentration, the microflora and vegetation present and whether fertilisers have been applied.

1.1.3 THE ORGANIC COMPONENTS

Organic matter accumulates in soil during its formation to reach abundances of 50 to more than 260 Mg ha^{-1} in mineral soils. It is mainly concentrated in the upper 10 to 20 cm where it comprises several *per cent.* of the total mass of the soil (Sánchez *et al.*, 1982; Post *et al.*, 1982).

Almost the whole range of naturally-occurring chemical substances may be found within soils, since the tissues of all living organisms are returned to the soil on their death. However, important differences exist in both the physical and chemical properties and the amounts of the organic materials returned to the soil in different environments and considerable efforts have been made to characterise soil organic matter in morphological, chemical and functional terms. Classifications resulting from these approaches are presented and the properties of the major soil organic components are described, including their energy density values and physical properties.

1.1.3.1 *Morphology and in situ localisation*

Soil organic matter occurs in the following forms:
(i) Living macro- and micro-organisms;
(ii) Decomposing animal, plant and micro-organism remains still retaining much of their original form;
(iii) Exudates and mucus produced by organisms;
(iv) Leachates from above-ground litter and vegetation;
(v) Humic materials.

The living macro-organisms (roots and macro-invertebrates) are readily separated from soil and are not normally considered to be part of the soil organic matter. Micro-organisms and the small invertebrates (the micro- and mesofauna) are included in soil samples and thus comprise that part of the soil organic matter known collectively as the microbial biomass. Micro-organisms can be readily observed in thin section preparations of the soil using microscopic and ultramicroscopic techniques. Two major components routinely noted in these preparations are fungal hyphae and bacteria. Fungal hyphae ramify through the soil pores often penetrating dead or living plant tissues and soil aggregates. Bacteria frequently occur as colonies of a few tens of individuals or less surrounded by polysaccharide capsules on the surfaces of which clay platelets may be adsorbed (see *e.g.*, Figure I.11 A and B).

The faeces of saprophagous invertebrates are a common feature of soil thin sections, particularly those of surface soil materials (see, for example, Bal, 1982 or Brewer and Sleeman, 1988). Serial ingestion and egestion of plant structural materials mediated by a succession of these animals progressively reduces the average size of the decomposing materials while removing some of the more readily-digested components.

Such decomposing plant materials may retain much of their original structure for long periods. As decomposition proceeds and the fragments of organic debris are reduced in size, cell contents are depleted and their walls collapse onto the vacuolar tannins which have resisted previous decomposition. At later stages, this structure becomes micro-divided and progressively lysed. Despite this, even in fragments less than 1 μm, the multi-lamellate structures of plant cell walls and cell membranes may be still recognisable (Foster, 1985).

Living organisms secrete copious metabolites (see Chapter III). The most important of these are:

(i) Root exudates and mucilages, which form a thick film over the actively-growing part of the roots;

(ii) Polysaccharide capsules and sheaths produced by bacteria and fungi;

(iii) Mucus and other secreta deposited by such invertebrates as earthworms and termites on the walls of their galleries, or admixed with ingested soil in order to facilitate digestive processes.

Finally, a considerable proportion of soil organic matter comprises small particles of uncertain origin falling in the micron to sub-micron size range. These particles are amorphous or granular in structure and include translocated and precipitated organic materials leached from the surface, humic substances synthesised *in situ* by microbial activity, or formed through spontaneous chemical reactions. These particles include much of what are chemically characterised as humic substances which may assume quite different shapes depending on the ionic environment, *i.e.*, pH and electrolyte concentration (Ghosh and Schnitzer, 1979).

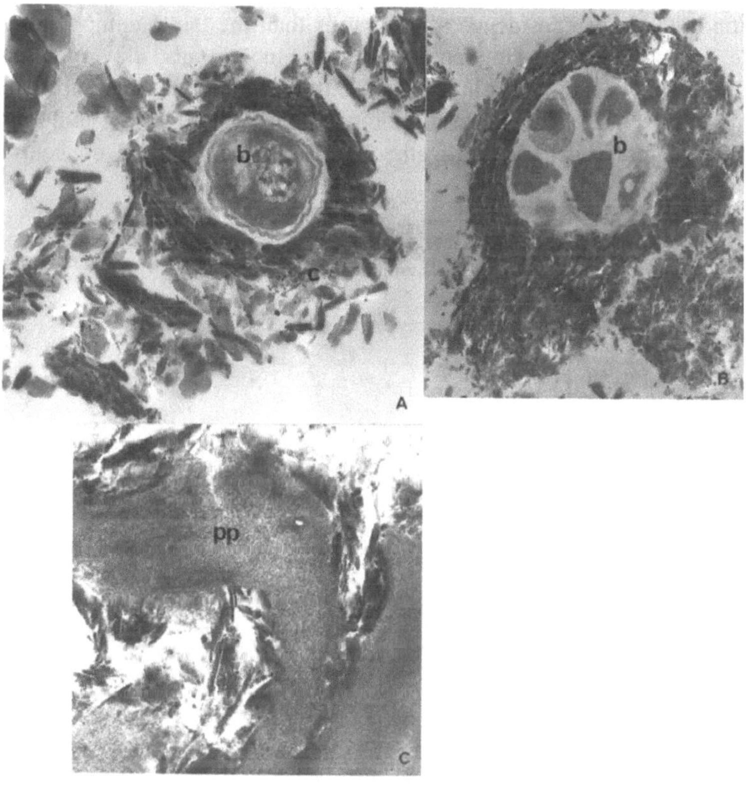

Figure I.11 (A) Clay platelets adhering to (b) a Gram negative bacterium (x22,000); (B) a polysaccharide capsule secreted around a bacterial colony (b) (x15,000); and (C) remains of leaf pectocellulosic cell wall (pp) (x22,000) (photographs by G. Villemin).

1.1.3.2 *Physical and chemical characterisation*

The chemical and other properties of soil organic matter differ considerably, depending on the relative proportions of the five major components defined above. It may be usefully fractionated into several components, defined by the methods used to separate them.

One fraction is defined as the **labile organic matter** and is composed of relatively simple organic compounds that may be rapidly assimilated by soil organisms; water-soluble polysaccharides leached from the litter layers, root exudates and earthworm mucus are some common examples. Such materials normally occur at low concentrations due to their rapid decomposition; in temperate soils, water-soluble carbon seldom exceeds 200 mg kg^{-1} (Davidson *et al.*, 1987). Similarly, fluxes of root exudates or earthworm mucus may be large although concentrations at any given moment are always low. The microbial biomass is considered to be part of the labile pool of organic matter although it is less labile than the above-mentioned materials. As discussed in Section I.3.1.2.1, microbial biomass carbon concentrations in most soils average between 1 and 5 % of soil organic carbon.

Decomposing root and leaf litters comprise the **light fraction** of the soil organic matter which may be separated by flotation on liquids in the density range 1.6 to 2.0 Mg m^{-3}. This fraction has a longer residence time in soils than the labile compounds and may accumulate over time, depending on soil climatic conditions, clay mineralogy and the presence of active decomposer invertebrates. It may comprise from 5 to 73 % of total soil carbon (Figure I.12). The chemical composition of this fraction is dominated by highly-condensed polysaccharides, together with high but variable concentrations of cellulose and lignin.

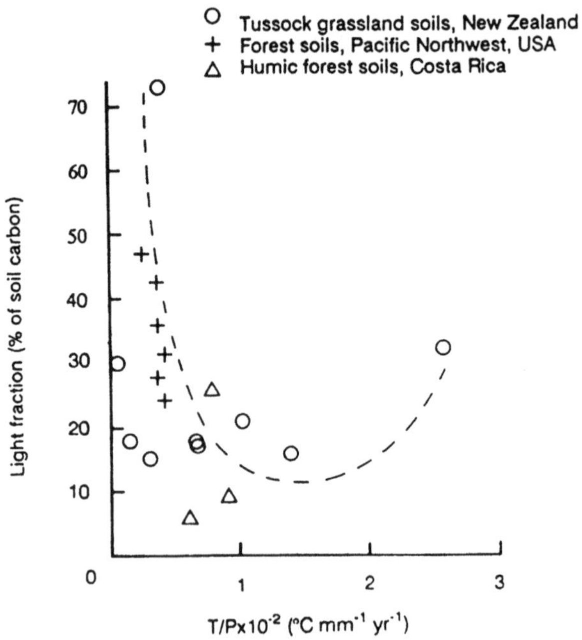

Figure I.12 Hypothetical relationship between the proportion of soil organic matter present in the light fraction and the temperature:precipitation ratio (Theng *et al.*, 1989).

The **heavy fraction** of soil organic matter is composed of humic substances which are linked to mineral constituents to form organo-mineral structures with bulk densities of more than 2.0 Mg m^{-3}. Humic substances are characterised by having a high proportion (*ca.* 50 % or more) of their carbon atoms incorporated within aromatic structures. At least part of the humic acid content may originate from such highly degradation-resistant phenolic materials as lignins and vacuolar polyphenols contained in the original plant material. Clay minerals may be the catalysts for such chemical condensation, perhaps by promoting the polymerisation of phenolic ring structures resistant to biological attack (Wang *et al.*, 1980).

Humic substances are characterised by their solubilities in certain extractants and the following three fractions are defined:

(i) humin, which is insoluble in NaOH or $Na_4P_2O_7$;

(ii) humic acids which are soluble in these extractants but are precipitated when the solution is acidified;

(iii) fulvic acids which remain in solution following acidification.

Fulvic acids have low molecular weights (1,000-30,000 Da) and variable shapes, depending on their bonds with other inorganic and organic soil constituents (Paul and Clark, 1989). Humic acids have higher molecular weights (10,000-100,000 Da) and are formed by the poly-condensation of elements containing a considerable proportion of their masses as aromatic rings, nitrogen in cyclic form and peptide chains. Humin is a mixture of humic and fulvic acids with some plant and microbial remains, all linked by clay minerals.

Variation in the chemical composition of humic acids has been related to annual rainfall, the nature of the land use and is known to differ between soils (Kumada, 1987; Almendros *et al.*, 1988; Arshad and Schnitzer, 1989). However, climatic differences (notably temperature), may not always be reflected in the composition of humic acids (Shoji *et al.*, 1987).

Recent characterisation of soil organic matter using solid-state ^{13}C NMR spectroscopy and other techniques is yielding useful information on the distributions of different groupings of chemical compounds in space, in depth and with changes in land use. Gregorich *et al.* (1996) compared the chemical compositions of surface-soil organic matter under a mixed-species hardwood forest and 90 years after conversion to maize (*Zea mays*) cropping. In the maize system, the surface soils were relatively depleted in C, N and in a number of high molecular weight compounds, particularly lipids and lignin. The compositions of the light fractions of both systems reflected those of the vegetation from which they were derived, apart from smaller amounts of carbohydrates and greater quantities of sterols, indicating that some decomposition had taken place. Compared with more decomposed materials greater than 53 μm, the light fraction had greater carbohydrate and aliphatic material but less aromatic material. Higher concentrations of lipids and lignin monomers and dimers were also present in the light fraction. No differences were apparent between the subsoils and the subsoil organic matter was almost free of carbohydrates, phenols and most lignins. The compounds present at depth were the most recalcitrant and included alkyl aromatics, complex nitrogen compounds, lignin residues and small quantities of lipids. Chemical changes following cultivation are largely associated with the light fraction (Skjemstad *et al.*, 1997).

1.1.3.3 *Particle size and aggregate fractionation*

Organic residues become increasingly fragmented as decomposition proceeds forming relatively-stable organo-mineral complexes with clay particles. Smaller particles may comprise the majority of soil carbon. Baldock *et al.* (1992) found that in five diverse soils including mollisols, andisols and oxisols, the amounts of organic carbon contained in particles less than 2 mm in size ranged from approximately 80 to 90 % of the total amounts present in these soils. Further, with diminishing particle size, an increasing proportion of this organic carbon is contained within the heavy fraction.

Consequently, smaller particles are often the most highly decomposed and frequently contain greater amounts of recalcitrant materials, including those derived from micro-organism activity. These are presumably more resistant to further decomposition than their larger counterparts. A knowledge of the particle size distribution of the organic fraction thus allows a broad assessment of resistance to decomposition although this is less clear in soils such as andisols and mollisols where a high degree of organic matter protection occurs through the formation of decomposition-resistant organo-mineral complexes (Duchaufour, 1997; Baldock *et al.*, 1992).

The proportion of total soil organic matter present between the aggregates may be contrasted with that protected from microbial degradation within the aggregates. Where aggregates are stable in water, these two organic matter fractions may be separated by sieving the whole soil in water. Total organic matter is assessed following dispersion of the soil into individual particles through the application of ultra-sound energy, or by shaking the soil for several hours in water with glass beads. The desired proportions are obtained by difference (Section I.3.2.4.1) (Feller, 1979; Brucker and Kilbertus, 1980; Elliott, 1986).

1.1.3.4 *Functional characterisation*

Organic materials have the following three complementary roles in soils. They:

1. Form a reserve of energy and nutrients;
2. Link mineral components within the soil matrix through their colloidal and charge properties;
3. Retain cations on their predominantly negatively charged surfaces.

Energy densities and nutrient contents

Litters and soil organic matter have relatively low energy densities. Values for several fresh litters from different litter systems in eastern France ranged between 18.7 and 20.5 kJ g^{-1}, depending on the tree species present (Loustau, 1984) (Table I.8). No significant changes were observed in the partly-decomposed litters of the F and H layers nor in the coarse (>50 μm) organic fractions of the soil. Fulvic and humic acids had lower energy densities (16.7 to 18.8 kJ g^{-1}) and humin even lower values (8.8 to 12.8), with the single exception (19.1) of samples from a poorly-drained soil.

Generally, 95 % or more of the nitrogen and sulphur, and between 20 and 75 % of the phosphorus of surface horizons is found in soil organic matter (Duxbury *et al.*, 1989). Ratios of C:N:S:P appear to be quite uniform with average values of 140:10:1.3:1.3

(Stevenson, 1986). Some differences occur between climates (*e.g.*, an increase of C: nutrient ratios in certain tropical Brazilian soils (Neptune *et al.*, 1975), and land use (*e.g.*, higher C: nutrient ratios in virgin than in cultivated soils (Stevenson, 1986).

About 30 to 40 % of N is included in amino acids or amino sugars (Stevenson, 1986) and the rest is present in unidentified forms. Organic phosphorus generally occurs as monoesters of orthophosphoric acids (Hawkes *et al.*, 1984) and up to 30 % of organic sulphur is contained in amino acids.

Table I.8 Energy densities (kJ g^{-1}) of leaf litter and soil organic matter in selected soils of eastern France (Loustau, 1984).

Fraction	Mean	Range	*n*
Litter	19.87	18.4-21.5	23
SOM fraction 50-2000 μm	18.62	17.5-21.4	14
SOM fraction <50 μm humic acids	18.10	16.7-18.8	18
humin	13.85	8.8-19.1	13
Glucose	15.6		
Protein	23.9		
Lipids	38.9		

Cation retention by soil organic matter

As stated in an earlier section, soil organic matter has a net negative charge at normal soil pH levels due to the dissociation of carboxyl (-COOH) and hydroxyl (-OH) groups and, to balance this charge, cations are retained at its surfaces. CEC's of purified humic acids range from 300 to 500 cmol kg^{-1} while those of purified fulvic acids vary between 500 and 750 cmol kg^{-1} (Burns, 1983). However, these substantial values are not expressed in soils because of the occlusion of many exchange sites by non-exchangeable iron and aluminium (Duxbury *et al.*, 1989) and the peat materials described from five histosols by Lévesque and Dinel (1982) had cation exchange capacities ranging from 122 to 171 cmol kg^{-1}.

Despite the occlusion considered above, organic matter does make a substantial contribution to soil specific surface area and to cation exchange capacity. Thompson *et al.* (1989) estimated that between 19 and 54 *per cent* of the total cation exchange capacity of two soils formed on loessial parent materials was due to variation in the organic carbon content. Soil organic matter may be the major source of negative charge in highly-weathered soils dominated by iron and aluminium oxides.

Functional pools

Soil organic matter may be classified into a number of pools that differ in their overall chemical compositions and in their degree of physical protection (see I.1.3.3, Figure I.13). Such protection of organic matter occurs when it is adsorbed onto the exterior surfaces of clay minerals, intercalated in their inter-layer spaces, within the inter-particle spaces of small aggregations of clay particles or within compacted soils. Separate pools of organic matter with different residence times in the soil have been defined on this basis. Jenkinson *et al.* (1987) and Parton *et al.* (1983) distinguish the following five pools which include two classes (1-2) of plant materials and three (3-5) of soil organic matter:

1. Readily decomposable "metabolic" plant material with a turnover time of 0.1 to 1 year;
2. Resistant structural plant material (2-5 year turnover time);
3. Active soil organic matter (*i.e.*, microbial biomass) (2-4 year turnover time);
4. Slow (or physically stabilised) soil organic matter (20-50 year turnover time);
5. Passive (or chemically protected) soil organic matter (800-1200 year turnover time).

An approximate correspondence has been found between soil organic matter fractions separated by various methods and the functional pools (Figure I.1.13)

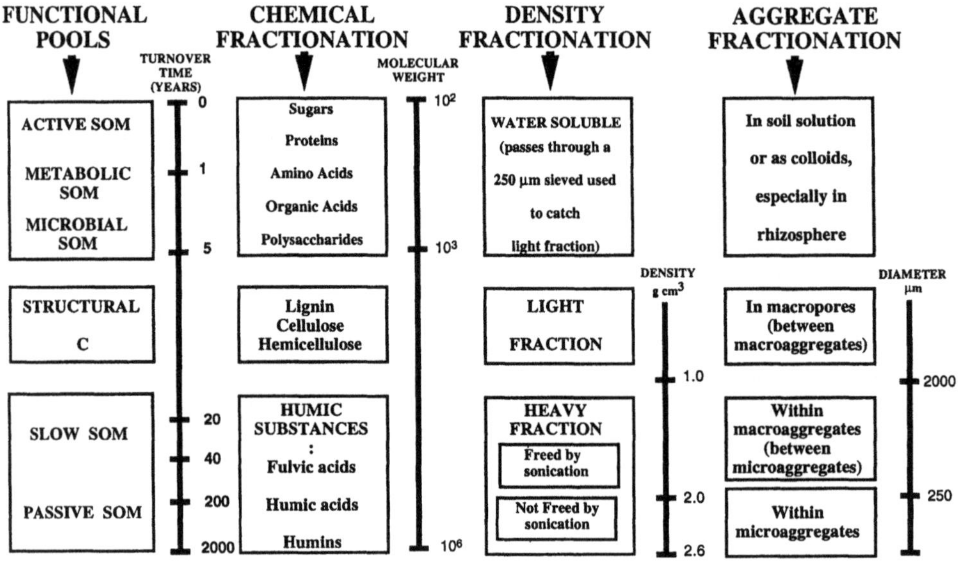

Figure I.13 Comparison of different soil organic matter fractionation schemes as related to functional pools, including current class limits (R.J. Scholes, M.C. Scholes and P.A. Sanchez, unpublished data).

1.2 The soil atmosphere

The mixture of gases and water vapour forming the atmosphere within the water-free pore space of the soil differs considerably from that above the surface. As discussed in a later section (Section I.2.2.2.2), the relative humidity of the soil air remains close to saturation until very low soil moisture potentials are attained. Carbon dioxide levels are also higher than those above the surface and a range of other gases may be present in substantial concentration, depending on physical conditions in the soil. The composition of the soil atmosphere is critical to the growth of plants and to the activities of the aerobic micro-organisms associated with them. Oxygen must be able to flow readily to both the roots and their microbial symbionts (notably the mycorrhizal fungi) to fulfil their respiratory requirements and carbon dioxide must also be able to diffuse away. A wide range of pedogenetically-important soil processes are also controlled by the nature of the soil atmosphere, including the general nature of decomposition processes, the activities of the aerobic members of the soil community and the direction of reduction: oxidation processes.

Gas movement within soils results largely from diffusion. However, barometric pressure changes, wind gusts over the surface, temperature gradients, the penetration of water during infiltration (which displaces the soil air ahead of the wetting front), fluctuations of shallow water tables and the extraction of water by plant roots may all induce significant mass flows into and out of surface soils (Patwardhan *et al.*, 1988). The diffusion rates of gases in soils depend on their diffusion coefficients in air, the concentration gradients present, the pore size distributions (particularly the proportion of air-filled pores), and the geometry of the pore system through which the gases move. Gas diffusion rates through soil water are too low to be of much biological consequence in terms of respiration. Nonetheless, the concentrations of dissolved gases certainly influence both biological and other pedogenetic processes.

Structures created by soil invertebrates, especially the large ecosystem engineers (notably earthworms, termites, ants and beetles, see II.2.4 and Chapter IV) may significantly influence gas diffusion. Kretzschmar and Monestiez (1992), for example, have demonstrated that earthworm burrows significantly increase gas diffusion rates from soils, particularly at high water potentials (ψ_m = -0.3 kPa) where only a small proportion of the pore space is filled with air.

I.2.1 THE COMPOSITION OF THE SOIL ATMOSPHERE

The concentrations of the major gases in the soil atmosphere are contrasted with those of the general atmosphere in Table I.9. In the atmosphere, water vapour may constitute <0.01 % to 3 % by volume of dry air, in addition to the gases. Concentrations in the soil are influenced by a range of soil physical, chemical and biological properties and by cultural practices. The most important properties include moisture status, texture, structure, soil organic matter concentrations and, at a smaller scale, density of and proximity to roots. Such cultural practices as the addition of organic materials, nitrogenous and other fertilisers may also affect the soil atmosphere.

The substantially lower diffusion rates of gases in soil water than in air mean that when soil moisture levels approach saturation, respiration by soil organisms will rapidly increase carbon dioxide concentrations, reduce oxygen levels and lead to some degree of anaerobiosis. Conversely, in equivalent dry soils, a greater proportion of the pore space will be filled with air and hence diffusion will be more rapid. Fine-textured soils have a greater capillary rise because of their higher percentages of smaller pores; thus, the effects of differences in texture on the composition of the soil atmosphere are clearly interlinked with those of structure and soil moisture.

Table I.9 Globally-averaged atmospheric concentrations (% by volume of dry air) of the major and some trace gases together with indicative concentrations of selected soil gases.

Gas	Atmospheric Concentrations	Range of Soil Concentrations	Comments
N_2	78.084	78.084	Varies little
O_2	20.948	18.0-20.5	Well aerated soil
		<2.0-10	Poorly aerated soil
Ar	0.934	-	-
CO_2	0.036	0.3-3.0	average range
		5->10	anaerobic soils
Ne	$1.818x10^{-3}$	up to $1.3x10^{-2}$	Magnusson, 1994
He	$5.24x10^{-4}$	-	-
CH_4	$1.5x10^{-4}$	up to $6.4x10^{-4}$	Magnusson, 1994
H_2	$5x10^{-5}$	$5-7x10^{-5}$	Conrad, 1994
N_2O	$3x10^{-5}$	$3.1x10^{-5}$ to $8x10^{-1}$	Amundson and Davidson, 1990
CO	$1.2x10^{-5}$	$7-17x10^{-6}$	Conrad, 1994
NH_3	$1x10^{-6}$		
NO_2	$1x10^{-7}$		
SO_2	$2x10^{-8}$		
H_2S	$2x10^{-8}$		

Soil structure (see Section I.1.3, below) has a major effect on soil aeration; a structure favourable for plant growth allows rapid infiltration and drainage and adequate aeration through the larger, air-filled pores while still permitting sufficient moisture to be retained for plant requirements. The clearly negative effects of excessive cultivation and compaction in reducing soil aeration are mediated through the modification of soil structure.

The spatial distributions of soil gases are notably heterogeneous. Steep vertical gradients occur towards the soil surface, although the nature of these gradients will depend on the properties of individual soils. Local concentrations of organic matter in the soil may, by providing a substrate for microbial respiration, result in locally-elevated carbon dioxide levels; declining concentration gradients also occur away from actively-respiring roots.

1.2.2 CARBON DIOXIDE

Carbon dioxide is a gas of major importance in soils and is produced by the respiration of plant roots, micro-organisms and, to a lesser extent, the soil fauna. This compound influences soil processes both as a gas and in solution; it controls soil pH in the mildly-acid to mildly-alkaline range (Bruckert and Rouiller, 1979), is an important leaching agent in solution as carbonic acid (Johnson *et al.*, 1977) and influences the uptake of plant nutrients. Carbon dioxide concentrations increase with depth in the profile and vary seasonally (see, *e.g.*, Buyanovsky and Wagner, 1983 and Amundson and Davidson, 1990); they are augmented at higher temperatures because of increased microbial metabolism and are also higher when the soil surface is covered by snow.

At a broad scale, soil carbon dioxide concentrations during the growing season vary with actual evapotranspiration (Figure I.14). Using this relationship, Brook and Box (1983) predicted growing season carbon dioxide concentrations of less than 0.1 % in deserts, 0.1-0.25 % in the conifer forests of North America and Eurasia, 0.6-1.6 % in North American deciduous summer green forests and 0.6-4.0 % for rainforests, monsoon and certain other tropical forests. Concentrations may thus be expected to be highest in those moist tropical environments where few moisture limitations exist and lowest in dry and semi-arid desert situations (Brook and Box, 1983). Vegetation change and other disturbances generally reduce stocks of soil organic matter and lead to lower soil carbon dioxide concentrations (Brook and Box, 1984).

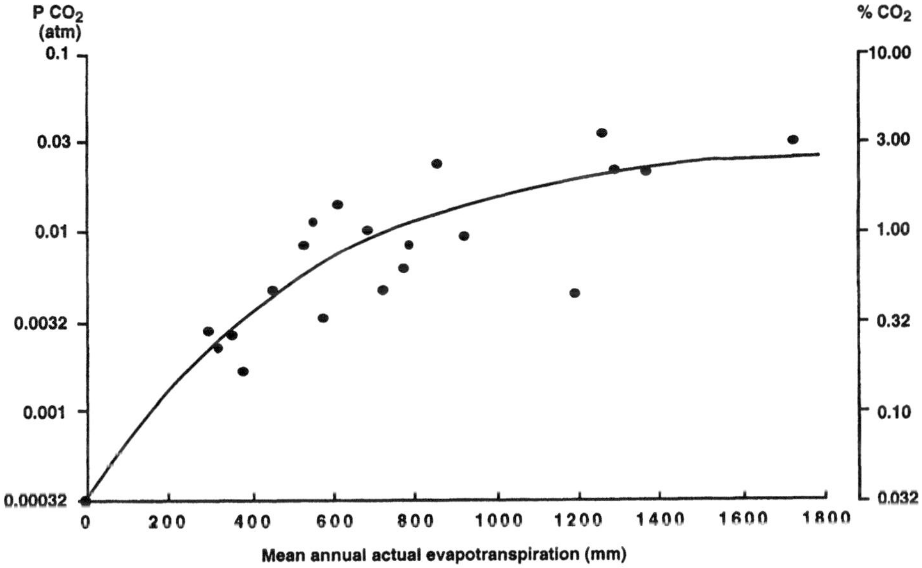

Figure I.14 Relationship between soil carbon dioxide concentrations during the growing season and mean annual actual evapotranspiration (Brook and Box, 1983).

1.2.3 ANAEROBIOSIS

In well-aerated soils, oxidative reactions dominate such processes as organic matter decomposition, the conversion of ammonium salts to nitrite and nitrate, of reduced forms of manganese and iron to oxidised and the oxidation of hydrogen and methane to other organic substances. Most of these processes are beneficial to plant growth. In contrast, where soils are flooded or waterlogged, partial or complete anaerobiosis soon supervenes because of continued microbial activity and the low diffusion rates of oxygen and carbon dixoide in water. The degree of anaerobiosis in the soil atmosphere is reflected in the concentrations of both gases and, as concentrations of both vary approximately inversely, either or both may be used as indicators (Stolzy *et al.*, 1981). It may also be measured as the redox potential (Eh) and this ranges from -300 mV (highly reducing conditions) to 900 mV (highly oxidising conditions).

Anaerobiosis may occur within aggregates even in well drained and structured soils where the inter-aggregate pore space is aerobic. Sextone *et al.* (1985) have shown the existence of strong oxygen gradients within aggregates from such soils; the oxygen concentrations at the centres of these aggregates were effectively zero (Figure I.15).

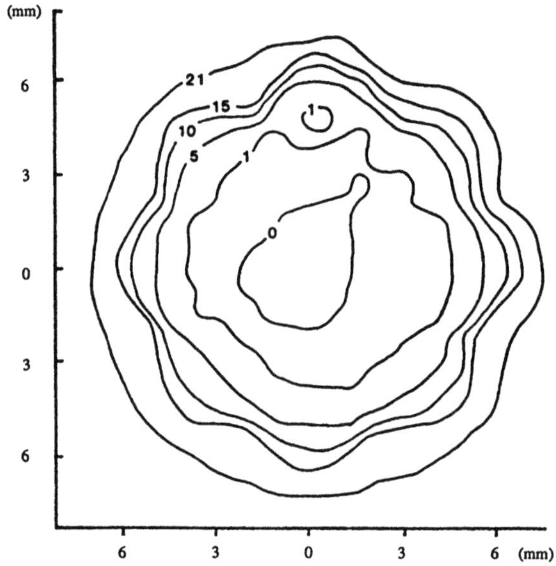

Figure I.15 Oxygen concentration gradient (%) within a single aggregate from a cultivated soil demonstrating an anoxic zone within an otherwise oxygen-rich environment (Sextone *et al.*, 1985).

As a consequence of microbial activity, a number of materials accumulate in anaerobic soils and in the anaerobic volumes of otherwise well-aerated soils. These include carbon dioxide, hydrogen sulphide, methane, ethane, ethylene, a range of higher hydrocarbons (van Cleemput and El-Sebaay, 1985) and reduced forms of such elements as manganese and iron. Denitrification gases such as NO and N_2O, may also be present although the actual gases present will depend on both the Eh and the nature of the microbiota present (Munch and Ottow, 1986).

As the Eh is progressively reduced, the oxygen concentration in the soil solution diminishes until at *ca.* 350mV, it disappears completely. Sequentially, other electron acceptors in the soil start to be utilised (Patrick and Jugsujinda, 1992):

at *ca.* 250 mV, NO_3^- starts to be denitrified to N gases;
at *ca.* 200 mV, Mn^{4+} starts to be reduced to Mn^{2+};
at *ca.* 100 mV, Fe^{3+} starts to be reduced to Fe^{2+}; and
at *ca.* -150 mV, SO_4^{2-} starts to be reduced to S^{2-}.

Anaerobic conditions have severe effects on unadapted higher plants. Small increases (1-2 %) in carbon dioxide concentrations may stimulate root growth but higher concentrations (>5 %) cause increasing impairment. Root function may also be impaired by a range of factors, including oxygen deficiency, that can lead to a shortage of the internal substrates needed to drive cell metabolism. In addition, toxins may accumulate in the root environment, or internal metabolites such as acetaldehyde and ethanol may build up and these, in turn, reduce photosynthesis, respiration and mineral uptake (Tiedje *et al.*, 1984). In agricultural situations, crop yields may be reduced by only short periods of soil inundation (Patwardhan *et al.*, 1988). The activities of such beneficial aerobic micro-organisms as mycorrhizal fungi and the aerobic nitrogen fixers are reduced. The decomposition rate of organic matter is depressed under anaerobic conditions (Tate, 1979) and the amount of this depression is related to the Eh (de Laune *et al.*, 1981) even below +350 mV, the level at which free oxygen is generally absent. In addition, much of the soil meso- and macrofauna may be eliminated where anaerobiosis is prolonged because of waterlogging. However, the activities of obligate anaerobic nitrogen-fixing bacteria such as the *Clostridium* spp. may be stimulated.

Inundation may range in degree from occasional to seasonal (as by the melting of snow or highly seasonal rainfall) to regular (as in tidal environments) and, in the extreme, to an occasional exposure to the atmosphere. Throughout the world, sites that are regularly saturated with water for varying periods of time on a diurnal, monthly or seasonal basis occur in areas known as wetlands. The plants that occur naturally in such waterlogged situations possess morphological and physiological adaptations that permit adequate root respiration to occur. These adaptations allow such plants to survive the deleterious effects of their variably anaerobic soil environments (Drew, 1983).

Among the most striking morphological adaptations to existence in anaerobic soils are the modified roots (pneumatophores) that occur in a range of mangrove species. These structures are specialised gas exchange organs and occur in several forms; they permit the species possessing them to succeed in the often anaerobic coastal and estuarine soils. Further adaptive air-conducting structures found in wetland plants include the internal aerenchyma (air conducting tissues) that occurs within the stems and roots of such plants as rice (*Oryza sativa*) and water lilies (*Nymphaea* spp.) and supplies oxygen to their roots and, in some species to the rhizosphere. Other species may respond to anaerobiosis by facultatively producing aerenchymatous adventitious roots, or increasing the proportion of gas-filled porosity in roots (van Noordwijk and Brouwer, 1993). A number of physiological adaptations also exist including the ability to avoid toxic concentrations of such metabolites as ethanol by promoting their leakage into the root environment, or the transpiration stream (Drew, 1983).

Figure I.16 summarises the results of a study of a 14-month study of soil redox potential carried out in a mangrove-forested tidal wetland in tropical northeastern Australia (Boto and Wellington, 1984). Nine sites were located along a transect laid out between two creeks; soils were sampled monthly at 5 cm intervals to a depth of 100 cm. Sites 1 and 2 were below mean sea level, sites 3 and 9 were below mean high water and site 6 was above the level of the mean high water spring tides.

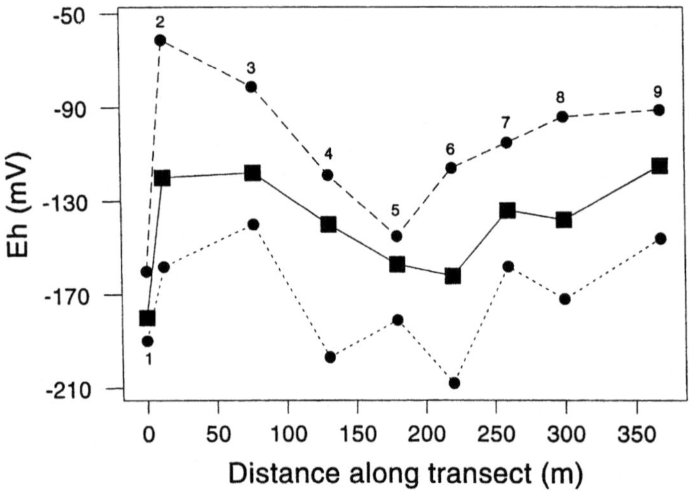

Figure I.16 Mean Eh (redox potential) values and their combined depth and sample ranges in nine soils sampled along a transect laid out between two tidal creeks in a tropical Australian mangrove forest (Boto and Wellington, 1984).

Since no significant depth or temporal variation was detected, the means and ranges presented are for all depths over the time course of the study. It may be seen from the Eh values that all the bulk soils remained consistently anaerobic over the sampling period. However some evidence of micro-scale increases in redox potential was apparent around the roots, suggesting that oxygen was being secreted into the mangrove rhizosphere.

1.3 Soil structure and micro-relief

At a number of locations and in particular soils, the surface may be organised into micro-relief features occurring at horizontal scales that range up to several metres or more. These features may occur as some combination of ridges, cracks, semi-regular to quasi-random patterns of holes, hollows and mounds. Micro-relief features may result from purely physical processes or they may be created by the biota, either directly or indirectly (Chapter IV) (Figure I.17). They are usually dynamic features of the landscapes and are constantly being degraded and recreated and have considerable implications for soil development (Chapter II.3.3.2). Such features may strongly affect the distribution and activity of the soil biota through their influences on water relationships, the distributions of soil organic matter and nutrient elements or by providing a favourable or sheltered habitat for other organisms.

Figure I.17 Surface cast of the endogeic earthworm *Millsonia anomala* and pores opening at the surface in an African savanna (size of cast 2 cm) (photograph by P. Lavelle).

Micro-relief features created by physical processes include the frost polygons characteristic of the soils of polar regions. These structures may form extensive semi-regular patterns of low mounds defined by systems of polygonal cracks. In the cold desert of Antarctica, snow preferentially entrapped in such cracks melts to provide a locally-favourable micro-habitat for Acari, Collembola and their supporting microbial food base (Wise and Spain, 1967). Gilgai formations commonly comprise patterns of alternating low hollows and mounds at scales of a few metres and are common in soils containing expansive clays. Biotic micro-relief structures include such features as coppice mounds formed through the trapping of wind-borne soil materials around the bases of shrubs in desert regions, the mounds that form at the bases of trees through bole expansion and the termite and ant mounds abundant in many savanna, tropical forest and other landscapes (Chapter IV.5.2.2.1).

At smaller scales than those of micro-relief, most soils possess some degree of structural development. Soil structure may be defined as the grouping of the primary soil particles into larger compound units (aggregates or peds) of different origins, sizes and shapes. These units are separated from those adjacent by pore spaces which permit water movement and gas exchange with the atmosphere. The pore space comprises voids of many types ranging from large planar cracks formed by the contraction of clay-rich soils during drying to irregular, partially-accommodating surfaces between adjacent units to pores with near cylindrical cross sections formed by root penetration and the burrowing activities of soil animals. Structure is not a static feature of soils, the component units are constantly being destroyed and renewed through the continuing and combined activities of physical forces and the biota; structural changes may be apparent even at a seasonal scale (Blackman, 1992). It is structure that largely defines the physical environment within the soil in terms of its degree of aeration and the amount of water it will absorb, retain and transmit. Following from this, it also determines the suitability of a soil as a habitat for small invertebrates, and for plant growth.

1.3.1 COMPONENTS OF SOIL STRUCTURE

The basic unit of structure in soils is the aggregate, or ped, which is defined as an association of soil particles that has a greater degree of internal, inter-particle cohesion than externally to the particles surrounding it. Clods, in contrast to aggregates, are larger structures (largest axis greater than approximately 25 mm) created by cutting and compressing the soil during tillage operations. Aggregates are considered to be formed through natural processes. Those created by physical processes are characterised by angular-blocky or prismatic forms. In contrast, faunal activities create aggregates with characteristically rounded shapes and many aggregates formed around a central core of plant material have elongate shapes (Oades, 1993). A variety of aggregate types will occur in most soils.

In most productive soils, aggregates occur in hierarchies of increasing size in which small aggregates are bound together to form larger compound aggregates. Aggregates of each size class in the hierarchy may possess properties distinct from those smaller and larger and will be bound together with characteristic suites of binding agents; these also differ between the size classes in the hierarchy.

Soil structure may also be considered from the viewpoint of the soil voids, or the spaces between the structural aggregates. This approach is advantageous when considering plant growth since the infiltration of water into the soil, its movement through the profile both laterally and vertically and the total store and availability of water to plants all depend on the presence and the size frequency distribution of the pores. Soil porosity may be divided into textural and structural components. Textural porosity is the minimal porosity resulting from the irregular distribution of the inorganic soil fragments; structural porosity is that component of porosity due to the generally-larger interconnected pores.

Pores are classified on the basis of their equivalent cylindrical diameters (ECD) although the boundaries of the size classes are somewhat arbitrary. Micropores are defined as those pores sufficiently small (less than approximately 30 μm in diameter) to retain water by capillarity and these contrast with the larger macropores which do not. Figure I.18 contrasts the pore size distributions of an Amazonian oxisol supporting respectively a primary forest and a pasture (Grimaldi *et al.*, 1993). Table I.10 presents examples of the typical size ranges of pores, biological and other structures commonly found in soils.

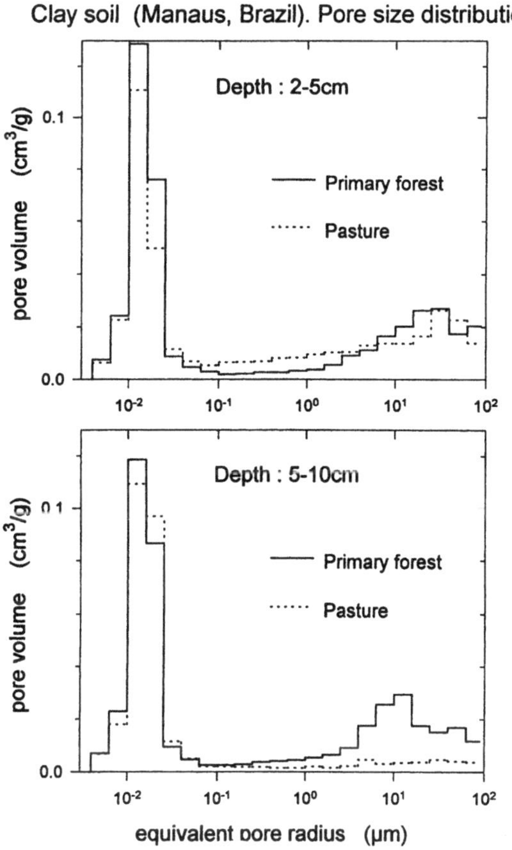

Figure I.18 Distributions of pore sizes in an oxisol supporting primary forest and a pasture near Manaus (Amazonia, Brasil). Note soil compaction at 5-10 cm leading to a reduction of macroporosity >10 μm. (after Grimaldi *et al.*, 1993).

Macropores are of particular importance in determining soil aeration and rapid water entry. In some soils (notably vertisols), large cracks form on drying and act in this regard at least during initial wetting. Large macropores created by the biota are known as biopores and functionally-effective, near-vertical pores are created in almost all soils by root growth and decay, earthworms, termites, ants and other burrowing invertebrates. Additionally, larger voids and horizontal galleries and chambers also occur that may not be continuous with the surface. In one particularly well-aggregated tropical inceptisol, Radulovich and Sollins (1991) showed that water flowed through the macropores at zero potential (and could be collected in zero potential lysimeters) even though the remaining soil was at less than field capacity (see definition in section I.2.2.2.2).

Table I.10 Indicative dimensions of selected soil organisms, seeds, galleries, nests and other biological structures.

Entity	Diameter or width	Comments	Author
Viruses	0.05-0.2 μm		Meetting, 1993
MICROBIAL ENTITIES AND CONSTRUCTS			
Soil bacteria:	0.6-0.75 μm		Kilbertus, 1980
Actino bacterial 'hyphae'	0.2-2.0 μm		Metting, 1993
Fungal hyphae:	0.5-2.5 μm		Griffin,1972
Vesicular-arbuscular mycorrhizal fungi:			Friese and Allen, 1991
Unspecialised hyphae:	2 - 10 μm	range of branching orders	
Germ tubes	2-5 μm		
Runner hyphae	10-15 μm		
***Fungal spores*:**			
Ectomycorrhizal fungi	4-30 μm		Brundrett *et al.*, 1996
VA mycorrhizal fungi	20-1000 μm		Brundrett *et al.*, 1996
***Sporocarps of mycorrhizal fungi*:**			
Vesicular-arbuscular fungi	5-20 mm		P. Reddell, personal communication
Ectomycorrhizal fungi	5-100 mm		
HIGHER PLANT STRUCTURES			
Pollen grains:	20-30 μm		Moore *et al.*, 1991
Phytoliths:	to 100 μm	most <50 μm	Drees *et al.*, 1989
Seeds (sizes):	from 50 μm	length	Arditti and Ghani,
	from 10 μm	diameter	2000
Seeds (masses, g):			
Orchids	$2*10^{-6}$		Harper *et al.*, 1970
Lodoicea maldivica	$2.7*10^{4}$		
Most temperate climate herbaceous angiosperms	$1*10^{-5}$ to $1*10^{-2}$		
Roots:			
Root hairs	80-1500 μm	length range	Hofer, 1996
	5-20 μm	diameter range	
Nodal roots of cereals	500-10000 μm		Hamblin, 1985
Seminal roots of cereals	100-1000 μm		
Lateral roots of cereals	50-100 μm		
1st and 2nd order laterals	20-50 μm		
Tap roots of herbaceous dicotyledons	300-10,000 μm		Hamblin, 1985
Tree roots:			
Final order lateral roots	>500 μm		
Major roots	to *ca.* 40 cm		

Table I.10 (cont.)

Entity	Diameter or width	Comments	Author
FAUNAL CONSTRUCTS			
Earthworm structures:			
Casts	1000 μm-20 cm		Lee, 1985
Burrows	500->11 000 μm		Lee, 1985
Termite structures:			
Microaggregates and faecal pellets	50-1500 μm		Eschenbrenner, 1986; Sleeman and Brewer, 1972; Humphreys, 1994
Galleries:			
Coptotermes acinaciformis	6.33-28.5 mm		Greaves, 1962
	1.58-3.2 mm	height	
	6.33-5.87 mm		Greaves, 1962
Pseudacanthatermes spiniger			Kooyman and Onck, 1987
Large galleries	10-20 mm		
Small galleries	1.2-1.6 mm		
Macrotermes michelseni			Darlington, 1982
Radial passages	30-70 mm		
Cross passages	10-25 mm		
Ascending passages	500-700 μm	circular	
Forage Storage Pits			
Macrotermes michaelseni	10 mm	breadth	Darlington, 1982
	30 mm	length	
	5 mm	depth	
Fungus chambers:			
Pseudacanthatermes spiniger	70-150 mm	sub-spheroidal	Kooyman and Onck, 1987
Subterranean nests:			
Range of species	5-400 mm	variable shapes	Noirot, 1970
Epigeal mounds (above ground parts):			
Amitermes laurensis			
i. Oriented wedge-shaped mound form		range of plot means	Spain *et al.*, 1983b
	1270-1560 mm	mean heights	
	1140-1640 mm	mean base lengths	
ii. Conical mound form		range of plot means	Spain *et al.*, 1983b
	390-610 mm	mean heights	
	370-553 mm	mean basal diameters	
Amitermes vitiosus		range of plot means	Spain *et al.*, 1983b
	250-840 mm	mean heights	
	200-850 mm	mean basal diameters	
Amitermes scopulus		range of plot means	Spain *et al.*, 1983b
	450-1280 mm	mean heights	
	170-330 mm	mean basal diameters	
Nasutitermes triodiae		range of plot means	Spain *et al.*, 1983b
	780-2440 mm	mean heights	
	1850-2460 mm	mean basal diameters	

Table I.10 (cont.)

Entity	Diameter or width	Comments	Author
Macrotermes mülleri	2000 mm	maximum height	Garnier-Sillam *et al.*, 1988a
	2500 mm	maximum basal diameter	
Cubitermes pretorianus	191 mm	mean maxium diameter	Ferrar, 1982
	234 mm	mean minimum diameter	
	176 mm	mean height	
High termitaria (Lamto, Côte d'Ivoire)			Observed maxima, Spain, unpublished
	to 22.1 m	length	
	to 20.5 m	breadth	
	to 2.0 m	height	
Ant structures:			
Galleries:			
Mycocepurus sp.	1-4 mm	diameter	Eschenbrenner, 1994
Aphaenogaster longiceps	5-10 mm	diameter	Humphreys, 1994
Iridomyrmex purpureus	20 mm	width	Ettershank, 1968
	15 mm	height	
	10 mm	diameter (verticals)	Humphreys, 1994
Lasius flavus, *L. niger*	2-50 mm		Green and Askew, 1965
Myrmecia gulosa	20-30 mm	diameter, (verticals)	Humphreys, 1994
Lasius neoniger	1.5-5.00 mm		Wang *et al.*, 1995
Nesting chambers:			
Atta vollenweideri	to >2m	height	Jonkman, 1980ab
	to >1m	diameter	
	to >8m	entire nest diameter	
Iridomyrmex purpureus	50 mm	diameter 1	Ettershank,1968
	70 mm	diameter 2	
	15 mm	height	
Myrmecia spp.	50 mm	width	Humphreys, 1994
	100 mm	length	
	30 mm	height	
Lasius neoniger	10-20 mm	width	Wang *et al.*,1995
	30-50 mm	length	
Fungus chambers:			
Mycocepurus sp.			Eschenbrenner, 1994
30-60 mm	diameter 1		
40-120 mm	diameter 2		
20-50 mm	height		
Aggregates:			
Aphaenogaster longiceps	0.5-1.5 mm	diameter	Humphreys, 1994
Camponotus consobrinus	2.0-3.0 mm	diameter	
Myrmecia spp.	3.0-6.0 mm	diameter	
Epigeal mounds:			
Lasius flavus	to 30 cm	height	Wells *et al.*, 1976
	to 60 cm	basal diameter	
Myrmecia brevinoda	to 70 cm	height	Higashi and Peeters, 1990
	to 80 cm	maximum basal diameter	

Table I.10 (cont.)

Entity	**Diameter or width**	**Comments**	**Author**
Subterranean nests:			
Lasius neoniger	15-70 cm	maximum depth	Wang *et al.*, 1995
Formica fusca	50 cm	maximum depth	Wiken *et al.*, 1976
Mammalian burrows, nesting and food storage chambers:			
Many species	few cm to *ca.* 1m.	diameter	Reichman and
	to many metres	length	Smith, 1990

As shown in Chapter IV, roots are important in the formation of both aggregates and biopores. During growth, they compress the surrounding soil to an approximate distance of one diameter (Dexter, 1991) and they cause the surrounding soil to shrink as they absorb water; they also act to bind small aggregates together mechanically. Living roots secrete a range of exudates which, in combination with those of the rhizosphere micro-organisms, cement soil particles together. Further, mycorrhizal fungi acting in conjunction with the roots are important in both physical binding and the cementation of soil particles. It has been found that monocotyledonous plants are more effective than dicotyledons at stabilising aggregates and that grasses are more efficient than cereals (Oades, 1993). Following root death and disintegration, a biopore remains in the soil providing a route for water entry and gas exchange and a preferential pathway for future root growth. Also, aggregates may form around nuclei of decaying root materials.

Soil animals, particularly earthworms, termites and ants strongly influence soil structure through the formation of both biopores and aggregates (Figure I.17) resulting from their different feeding and burrowing behaviours (Lee and Foster, 1991). Warner *et al.* (1989) reported the presence of faunally-created cavities at depths down to 560 cm in a North American mollisol supporting pasture; these were attributed to the activities of ants and earthworms although surface evidence of their activity was not necessarily visible. Structures created by soil invertebrates, especially the large "ecosystem engineers" (see II.2.4 and Chapter IV) may significantly affect gas diffusion. Kretzschmar and Monestiez (1992), for example, have demonstrated that earthworm burrows significantly increase relative diffusivity of gases, especially at high matric potentials (ψ_m = -0.03kPa) where a low proportion of porosity is filled with air.

These faunal structures may persist in the soil for long periods (years to centuries), long after the invertebrates that created them have disappeared (Wielemaker, 1984; Eschenbrenner, 1986; Blanchart *et al.*, 1997). The influences of these organisms are considered further in Chapter IV.

In agricultural terms, a well-developed soil structure implies the presence of porous, water-stable aggregates and is considered important both in protecting the soil surface against erosion and in promoting a satisfactory level of plant growth. The favourable effects of well developed structure on plant growth are achieved by facilitating aeration and water flow into and through soils. A high level of porosity is advantageous since, as considered in the next section, it is desirable that aerobic processes dominate within the pores and the aggregates. In terms of agricultural production, Dexter (1988) considers it advantageous to have most pores in the size range 0.2 to 30 μm to maximise water storage but for 10 % of the soil volume to be in pores larger than 30 μm, to promote aeration. In terms of aggregates, it is desirable that the whole range of size classes are developed and are stable to mechanical stresses. Further, it is particularly important that the aggregates exposed to raindrop impact and surface water flow are stable in water since their too-ready slaking and dispersion may lead to excessive erosion of surface soils, the blocking of soil pores and the formation of surface crusts.

1.3.2 CLASSIFICATION OF SOIL STRUCTURE

Soil structure has been classified in a variety of ways. Perhaps the best-known of the morphological classifications is that of the USDA (Soil Survey Staff, 1993). Under this system, the degree of structure in the soil, the **grade**, ranges from zero to strong on a four-point scale, depending on the proportion of the total soil that is occupied by aggregates. The peds are further classified on the basis of size, shape and degree of distinctness. Nonetheless, a classification that is based on the effects of the soil forming factors (a genetic classification), may be preferable and Bonneau and Levy (1982) have suggested the classification presented in Table I.11.

Such systems are useful for characterising the macroscopic physical and chemical structures found in soils and may imply other features of the soil environment. Few of these systems satisfactorily incorporate the complexities of biological and other small-scale structures found in soils.

1.3.3 AGGREGATE FORMATION AND STABILISATION

The soil components of colloidal size (clays and humic materials) must remain flocculated for the soil to maintain its stability. Floccules initially form relatively-open structures and undergo a process known as condensation to form more stable, overlapping arrangements of the component particles, usually clays. The formation of small, stable units of multiple clay particles (assemblages, quasi-crystals, domains or, more generally, tactoids) is followed by coalescence with organic matter and other cements to form increasingly-larger aggregates.

Aggregates may be formed by the fragmentation of larger particles such as clods or through agglomerative processes from smaller units. Aggregates may also be formed through the action of the soil fauna, either by passage through the gut or through their constructional activities.

1.3.3.1 *Genesis and size of aggregates*

The factors responsible for the initial formation of small aggregates include expansive and contractive forces associated with wetting and drying or freezing and thawing. Aggregate formation and stabilisation is therefore related to soil clay content and the nature of the clay minerals present. Oades (1993) considers that biotic influences are most important in coarse-textured soils but diminish with increasing clay content.

Table I.11 A genetic classification of macroscopic soil structure (Bonneau and Levy, 1982).

A. **Particulate Structure**

The soil is loose with no interlinking between particles, *e.g.*, unconsolidated sands.

B. **Massive Structure**

1. *Sintered Structure*

The soil is moderately cemented, it may be slightly porous or penetrated by plant roots or faunal channels, *e.g.*, earthworm burrows. May be divided artificially into coherent clods.

2. *Cemented Structure*

The soil is bound by hard precipitated cements (oxides, organic matter, carbonates, silica) which fix the primary units into a hard resistant mass. This mass may range from discontinuous (*e.g.* concretions) to continuous structures such as iron pans, calcareous hard pans, etc.

C. **Fragmentary Structure**

1. *Fine aggregate structures*

i. These include aggregates up to approximately 1 mm equivalent cylindrical diameter (ECD); they result from chemical linkages involving interactions between organic matter and clays, metal hydroxides and precipitated humic substances. They may include the excrements of small animals.

ii. Built-up (agglomerative) structures. These aggregates are formed through soil working, either through the agencies of soil animals or man:

a. Granular structures include subspherical aggregates ranging from 1 mm to 1 cm ECD and are often found in the surface layers of tilled soils;
b. crumb structures are irregularly-rounded aggregates ranging from 1 mm to 1 cm ECD;
c. nutty structures are irregularly-rounded aggregates greater than 1 cm ECD.

2. *Fissure structures*

i. Blocky structures are coherent, isodimensional aggregates with sharp edges; they range from a few millimetres to several centimetres in size.
ii. Prismatic structures are prism-shaped aggregates in which the vertical dimension is the greatest (to >20 cm). Columnar structures are prisms with rounded tops.
iii. Lamellate or platy structures have their greatest dimension in the horizontal plane, *e.g.*, fragipans.

Aggregate formation through abiotic processes becomes more important in soils with appreciable clay contents which crack along lines of weakness due to the tensile forces created during drying. Other factors include the pressures associated with drying as roots absorb water from the soil and the working of soil by man and soil invertebrates. Aggregates are bound together by cements which may be either organic or inorganic in nature.

Aggregates are classified into size groups that reflect their differing internal compositions. Particles up to approximately 250 μm equivalent cylindrical diameter (ECD) have been termed micro-aggregates and may be involved in dispersion: flocculation reactions dependent on surface properties (Oades, 1987). Macro-aggregates are defined as those greater than 250 μm ECD and are bound together by cements such as oxides of iron, aluminium, polysaccharides and humic polymers. At a larger scale of organisation, a different type of binding occurs through the action of fungal hyphae (notably those of mycorrhizal fungi, Miller and Jastrow, 1990; Chenu, 1993; Dorioz *et al.*, 1993; Foster, 1994) and small roots (Figure I.19).

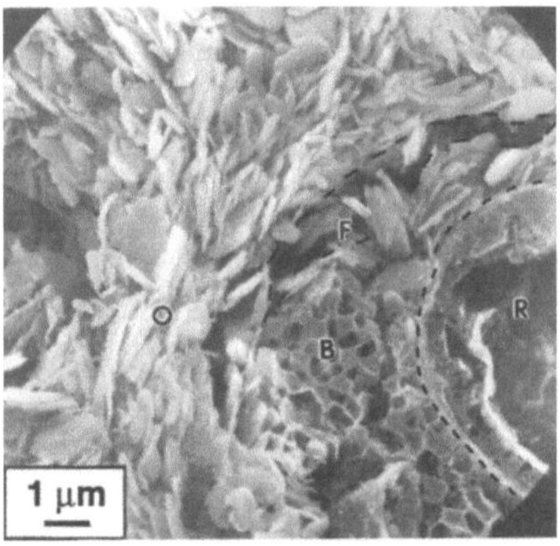

Figure I.19 SEM images of microaggregates at two different scales:
(a) Microstructures of clay-polysaccharide associations (kaolinite and xanthan),
(b) Microstructures formed in the rhizosphere of *Festuca rubra* grown in kaolinite at -0.01 MPa
(Chenu, 1993; Dorioz and Robert, 1987) (R = root hairs; B = bacterial colonies; O = oriented clay particles).

In well-structured soils, aggregates do not increase in a smooth gradation of sizes. Large aggregates comprise those of the smaller sizes held together in a hierarchical arrangement by characteristic suites of binding agents. Tisdall and Oades (1982) have described the nature of the binding agents operating at different scales within aggregates from an Australian red-brown earth (alfisol) (Figure I.20). While the processes involved may result in different size groupings of aggregates in particular soils, the principles are likely to apply to those soils in which organic matter is the main binding agent.

Water-stable aggregates less than 2 μm ECD include structures in all stages of condensation from relatively-open floccules to condensed tactoids. Some may be cemented with iron and aluminium oxides. Aggregates in this size range may be highly resistant to dispersion because of particle bonding by the stabilising aromatic and insoluble organic matter present. These aggregates are produced only slowly under pasture and may persist in cultivated soils for many years (Oades, 1987).

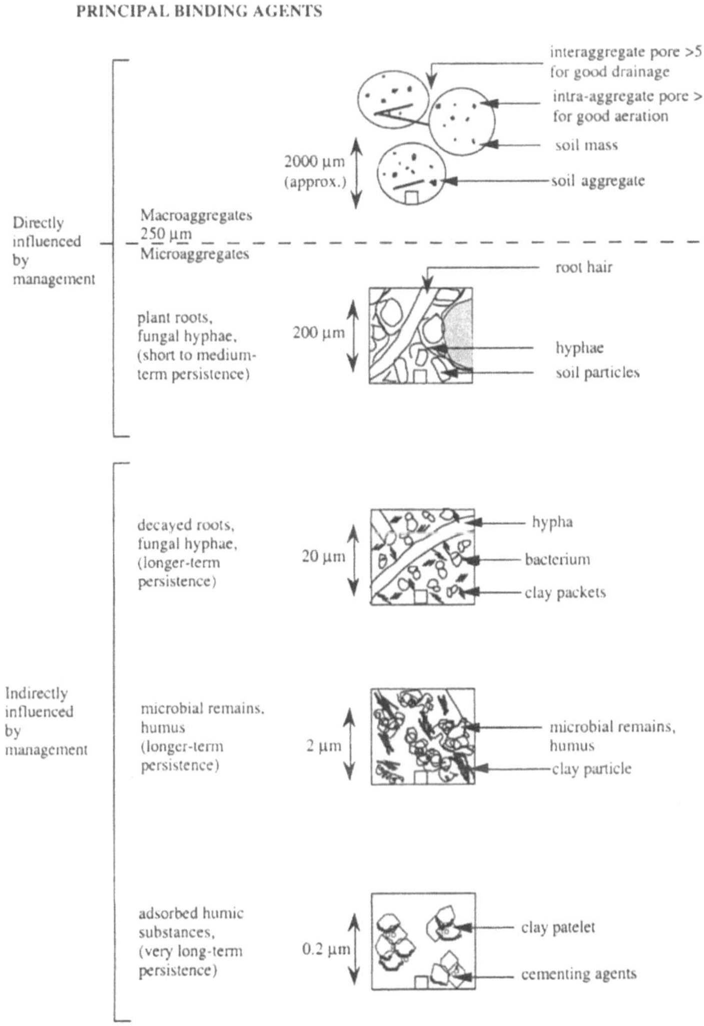

Figure I.20 Model of binding agents operating at different scales within aggregates (modified from Theng, 1987 and Tisdall and Oades, 1982).

Aggregates in the size range 2-20 μm ECD may be large floccules or tactoids. However, the majority probably comprise associations of clays and microbial remnants. Figure I.11a shows orientated clay particles associated with such microbial structures as bacteria, fungal hyphae and plant roots. The clay particles and tactoids are bound to the substantial quantities of polysaccharides produced by these organisms (Figure I.11b).

In the size range 20-250 μm ECD, aggregates are stable to rapid wetting but less so than those in the range 2-20 μm ECD. Materials bonding clay in particles of the size range 20-250 μm ECD include persistent organic matter, oxides of iron and aluminium and highly-disordered silicate clay minerals; the additive effects of these bonding agents produce strong, water-stable aggregates.

In both undisturbed and agricultural soils, water stable macro-aggregates in soils with adequate carbon concentrations are held together by a fine, three-dimensional network of small roots and hyphae of mycorrhizal and saprophytic fungi and other organisms. Miller and Jastrow (1990) consider that much of the effect of roots on aggregate formation is due to the indirect effects of associations with mycorrhizal fungi and that the strength and nature of this association also depend on the morphology of the root system. In contrast, ephemeral macro-aggregates bonded by transient binding agents are found in soils of lower organic matter content. Other binding agents, such as iron and aluminium oxides and highly disordered silicate clay minerals may also be present.

Microbial biomass carbon, nitrogen and phosphorus concentrations, and probably those of other elements, differ between aggregates of different size groups. In comparing aggregates from forest, savanna and cropped soils in a seasonally-dry tropical environment, Singh and Singh (1995) found that microbial biomass carbon concentrations were higher in macro-aggregates (more than 0.3 mm ECD) while micro-aggregates had greater concentrations of microbial biomass nitrogen and phosphorus.

The faeces of soil animals may be deposited within soils as void infillings, in remodelled form as linings to galleries and other structures, and as discrete aggregates. Such aggregates occur in a variety of shapes, sizes and mineral contents and may comprise a notable feature of the soil structure, particularly in undisturbed or little-disturbed soils. Aggregates produced by soil animals are therefore highly variable in composition since their genesis is dependent on the feeding habits of the species that produced them. The casts of soil-feeding earthworms are perhaps the most well-recognised form of faunal aggregate. Schemes exist for their classification in soil micromorphological studies (Bal, 1973).

1.3.3.2 *Organo-mineral bonding*

As indicated above, the intimate associations that occur between the various types of organic matter and the mineral fractions of the soil result from a range of chemical and physico-chemical bonding mechanisms (Emerson *et al.*, 1986; Theng, 1987). Because of their small sizes and elevated surface areas, interactions between clays and organic matter assume great significance in structural stability (Theng, 1987); important interactions also occur with larger particles (Turchenek and Oades, 1979; Chenu, 1993).

The most important organic materials in terms of soil aggregation are the polysaccharides and humic materials, although they act at different temporal and spatial scales. Monnier (1965) categorised the differing effects on soil structure of incorporating organic materials of varying degrees of humification into the soil. The incorporation of green materials leads to a flush of microbial activity and the production of such effective but ephemeral bonding agents as extra-cellular polysaccharides from the organisms involved (Chenu, 1993). If more highly decomposed materials are incorporated, the effect is lesser but more long-lasting. Finally, if highly decomposed materials rich in humic compounds are incorporated, an even smaller but very long-lasting stabilising effect is obtained. In the spatial sense, the larger aggregates are stabilised by organic matter derived from crop residues and other regularly-supplied plant materials both of which are constantly being degraded and renewed. Thus, changes in soil organic carbon concentrations mainly affect the larger aggregates; the organic matter that stabilises the smaller aggregates is incorporated in the spaces between the tactoids and may be highly protected and persistent.

1.3.4 DYNAMICS AND IMPAIRMENT OF SOIL STRUCTURE

As considered above, the types and spatial dispositions of the pores and aggregates that comprise the soil structure at any specific time are the result of a large number of physical, chemical and biological processes operating at different spatial and temporal scales. Thus, because of the dynamic nature of soil structure, a balance exists between the creation of aggregates and macropores and their destruction.

The physical forces leading to the shrinking and swelling of clays play an important role in the dynamics of structure by forcing particles into close contact. They may act differently between the seasons and years, depending on the climate and its variation. Inputs of photosynthetically-derived energy as root growth and expansion vary seasonally, as do inputs of the organic matter that provide a source of energy for faunal activity directed towards the creation of burrows and other biopores, casts and faecal pellets.

Aggregate stability is determined by the nature of the binding agents and will clearly differ between the various aggregate size classes and between soils. In soils that possess a hierarchy of aggregates, the larger aggregates may be disrupted by the decomposition of the fungal hyphae or the fine roots that bind them. The subsequent breakdown of the smaller more stable aggregates may result from disruption of their organic or inorganic glues.

The important role of the macrofauna in aggregate formation in most soils (Bal, 1982; van Breemen, 1993) is due to their promotion of primary binding through their mechanical activities and intestinal mixing (Barois *et al.*, 1993). They also produce macro-scale structures ranging from faecal pellets (250 μm to several centimetres) to large subterranean structures such as termite and ant nests. Earthworm faecal pellets may accumulate to such an extent that certain soils have been called 'vermisols' (Pop and Postolache, 1987). Probably the two most important faunal groups in aggregate production are earthworms (Hopp and Hopkins, 1946; Marinissen, 1995; Blanchart *et al.*, 1997) and termites (Eschenbrenner, 1986; Garnier-Sillam *et al.*, 1988a; Miklos, 1992) while such groups as enchytraeids (Didden, 1990; van Vliet *et al.*, 1993) are probably of lesser importance.

The stabilities of these structures largely depend on the strength of the primary bindings. Certain structures, such as the fresh casts of large endogeic earthworms, possess little structural stability although, on ageing, they become highly resistant to mechanical breakdown. For example, casts that have undergone a number of wetting and drying cycles (Blanchart *et al.*, 1993) may remain intact for several years. Production of such casts may exceed 1000 Mg ha^{-1} yearly and the large earthworms that produce them feed exclusively on small soil aggregates. As discussed further in Chapter IV, such earthworms do not re-ingest their own casts. Maintenance of an appropriate aggregate size distribution therefore depends on the activities of the smaller earthworm species, other invertebrates and non-biological processes to break down the large aggregates thereby restoring the pool of small aggregates (Blanchart *et al.*, 1997).

Other components of soil structure, especially the larger biopores, are also consistently renewed through biological activity. Consequently, in terms of both macropores and aggregates, soil structural maintenance should be considered to be largely under biological control in all but the finest-textured soils with high concentrations of expansive clays. Furthermore, such structures may persist for much longer periods than the organisms that created them or they can be destroyed by other organisms long before the cements and other materials binding them have lost their effects.

Apart from the biological destruction considered above, aggregates may be disrupted by internal or external forces. Disruption may result from the high air pressures associated with rapid wetting or through the dissolution of their internal "glues". External forces leading to aggregate breakdown include the compactive and shearing pressures induced by tillage implements and the trampling of domestic and other large animals. Aggregates may also be disrupted by the forces induced through drying and freezing and by the expansion of salt crystals in saline soils.

Soil structure may be readily degraded in agricultural and other situations through the application of external energy, through soil baring, excessive cultivation and mechanical compression. Soils differ widely in their susceptibility to structural disruption. In most soils, the exchange complex is dominated by Ca^{2+} and this results in strong inter-particle bonding leading to a stable structure resistant to the disruptive forces listed above. In contrast, in sodic soils (Section I.1.1.1.2), inappropriate management can quickly lead to the situation where rapid wetting by rainwater causes surface aggregates to slake and their clays to disperse (Sumner, 1995). Such processes can lead to the deposition of a layer of fine particles over the soil surface and the blocking of pores important in water transmission and aeration. This surface sealing process may lead to diminished plant growth, either directly because of the higher mechanical strength of the surface crust (which may reduce seedling emergence) or indirectly by restricting water entry and gas exchange.

Where its physical condition has been impaired by unsuitable land use practices, the soil may be rendered prone to accelerated erosion through the energy of wind or water, and its capacity to support satisfactory plant growth may be reduced. This impairment may be due to collapse of the surface soil structure following exposure to rain drop impact or excessive working during tillage operations, particularly of very wet or very dry soil. Agricultural operations (including the trampling and puddling activities of farm animals) may also cause severe soil compaction, particularly when the soil is wet, and this may cause part of the profile to become too strong for roots to penetrate.

Finally, oversimplification of soil fauna communities associated with conventional tillage or cropping may result in an overdominance of either "compacting" or "decompacting" species with negative effects on water infiltration or resistance to erosion. Severe compaction and significant losses of production have followed the establishment of large populations of the endogeic earthworm *Pontoscolex corethrurus* in sweet potato fields in New Guinea and in Amazonian pastures (Rose and Wood, 1980; Barros *et al.*, 1996). However, lower populations of this species normally have significant positive effects on plant growth (see Chapter IV.4.4).

I.2 THE SOIL MICROCLIMATE

The microclimate of the soil is defined in terms of its internal temperature and hydrological regimes and is therefore generally determined by the external climate. Since this varies broadly with latitude, elevation, rainfall distribution and, to a lesser degree, with vegetation type and cover, aspect and a range of other factors, it is clear that soil microclimates are likely to differ almost as widely as those of the surface, albeit increasingly buffered from rapid change with greater depth in the profile.

The soil properties that determine water entry and movement within the soil are clearly important in defining the microclimate. Soil temperature regimes and the factors governing their variation in time and space form the remaining part of its definition.

2.1 Soil temperatures

The temperature regimes that pertain in soils influence many processes that occur therein and play a part in controlling the rates and processes of soil development and the composition and activities of the biota. Agricultural productivity is frequently limited by either low or excessively high temperatures, although both effects are often related to moisture status. In the pedogenetic sense, water is a major agent of physical weathering through expansion: contraction processes and particularly the frequency with which the 0 °C boundary is crossed (Figure I.21a). The rates of many biological and chemical weathering processes are also temperature dependent.

All species have required minimum and maximum temperatures for growth and survival and, in the arthropods, these may differ between developmental stages. However, most species usually have a somewhat narrower preferred range.

The species present in extreme environments usually possess specific adaptations to permit their survival therein. However, a minimum requirement for the persistence of most species is that temperature and moisture regimes be regularly within a favourable range for a sufficient period to permit successful reproduction and development.

2.1.1 SURFACE ENERGY BALANCE

Soil temperatures are directly related to the energy balance at the surface. This may be stated as the following equality (Bonneau, 1979):

$$(1 - a) . R_g + R_a + C = E + R_t + Q$$

where a is the albedo, R_g is the incident solar radiation (direct + diffuse), R_a is the long wave atmospheric radiation, C is the energy gained by the condensation of water, E is the energy lost by evaporation, R_t is the energy radiated from the surface and Q is the energy gained or lost by the soil through conduction and convection of the air, and conduction by the soil. If Q is positive, the soil surface heats up and will transmit energy to the lower layers of the soil, if they are cooler. Where Q is negative, the soil surface cools and gains energy from the subsurface layers, providing they are warmer.

Transmission of heat to the lower soil layers depends on the thermal capacity and the thermal conductivity of the soil. Thermal conductivity is dependent on the particle size distribution, the water content, the bulk density and, in organic soils, the organic matter concentration. Due to the high thermal capacity of water in comparison with that of the dry soil, the thermal capacity of the soil is much higher when it is wet. Additionally, the high thermal conductivity of water and its intimate contact with the soil ensures that wet soils transmit heat readily. Because of rapid heat transmission to greater soil depths, wet and water-logged soils take much longer to warm in Spring, a matter of considerable importance to agriculture in colder climates. Soil water content thus plays a major part in controlling the absorption and transmission of heat.

From the foregoing, it is clear that both thermal capacity and conductivity will normally vary substantially down the profile and between soils. Additionally, regular temporal variation may be expected with seasonal and other changes in soil water status and with surface radiation.

2.1.2 DAILY AND SEASONAL VARIATION

Because of latitudinal differences, the radiation incident on the soil surface increases systematically from the poles to the equator, although similar patterns of daily and seasonal variation may be expected. Additional local variation results from changes in vegetation cover, aspect, slope and exposure.

Daily temperature variation at the soil surface follows an approximately sinusoidal pattern and produces a wave that is propagated into the soil. This quickly attenuates with depth and the amplitude of daily variation is slight below approximately fifty centimetres. The rate of penetration of the wave of heat energy through the soil is relatively slow leading to the situation where daily maxima at depth lag substantially behind surface temperatures.

The above points are illustrated with data from two very different locations. Figure I.21a presents the march of temperatures over a 24-hour period during summer at a site near the Shackleton Glacier, Antarctica (84°30'S, 174°W) (Wise and Shoup, 1971). This area is near the southern limit of animal existence and the site supports a sparse population of Acari based on a simple microbial food chain (Wise and Gressitt, 1965). Soils at the site remain frozen for much of the year but, during summer, thaw daily to depths of at least 15 cm for sufficient periods to permit the animals to complete their life cycles. The lag in temperature change at depth is illustrated by data from the different depths; the maximum occurred at the surface at 1400 h while that at 15 cm did not occur until 1600 h. Air temperature during the 24 hour period of observations ranged between -6 °C to 4.5 °C. Surface temperature varied more widely than that at depth; the rapid rise and fall of temperature at 0600 h was due to a brief period of insolation. The biota of this cold desert is strongly water-limited and soil animals are largely restricted to areas receiving water from melting snow and ice.

Data from the un-vegetated tropical alfisol studied by Bristow (1988) (Figure I.21b) illustrates a similar pattern of daily temperature variation, and the elevated temperatures that may be attained on bare tropical soils. This figure also confirms several of the points listed above. Soil temperatures show a much diminished diurnal variation with increasing depth and the magnitude of this is reduced to less than 1 °C at 50 cm. Also, the time lag between daily maxima at increasing soil depths is clear; the surface soil temperature attains its maximum at 1300 h while at 10 cm this does not occur until 1600 h. At 50 cm, the maximum temperature occurred at *ca.* 0500 h the subsequent day. Air temperatures during the 24 hour study period ranged from 20.9 °C to 32.8 °C.

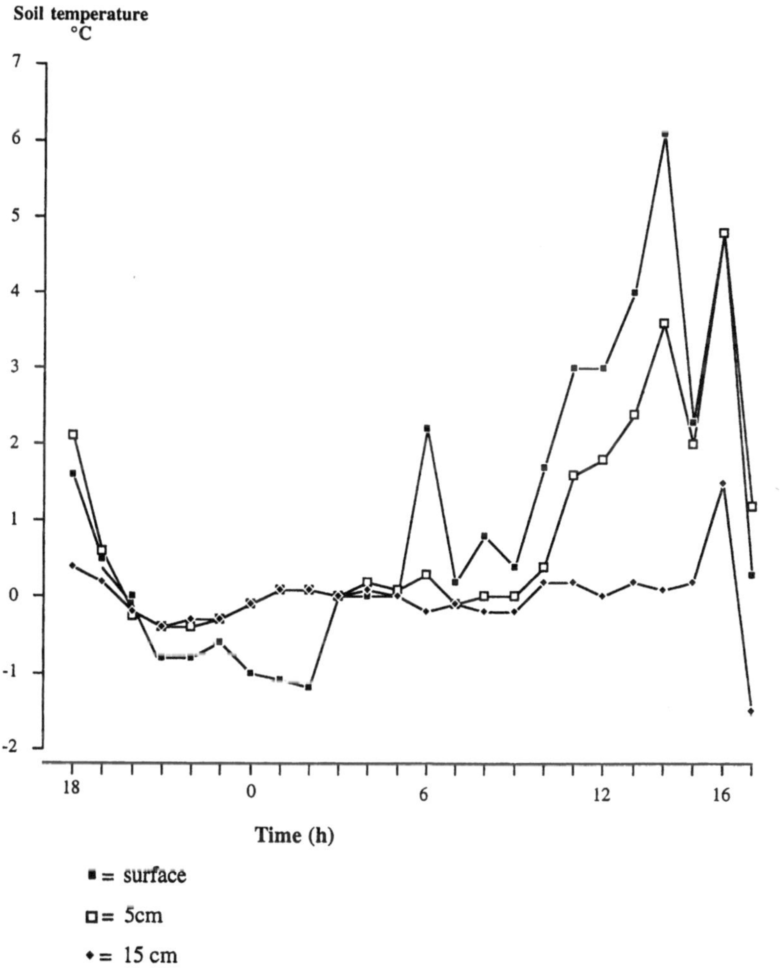

Figure I.21a Near-surface daily temperature variation at selected depths in two soils: (a) Near-surface soil temperatures during a 24 hour period at a site near the Shackleton Glacier, Antarctica (Wise and Shoup, 1971)

Finally, the rapid temperature decreases that occur between 1400 h and 2000 h and from 0400 to 0800 h can have important consequences for soil-dwelling animals since free water may condense in soil, given the constant saturation of the air-filled porosity. This is illustrated by the biomodal pattern of daily surface casting by two endogeic earthworms in African savannas at Lamto (Côte d'Ivoire): peaks in casting occurred at 1800 to 2100 h and again at 0600 h following two daily falls in temperature (Lavelle, 1978).

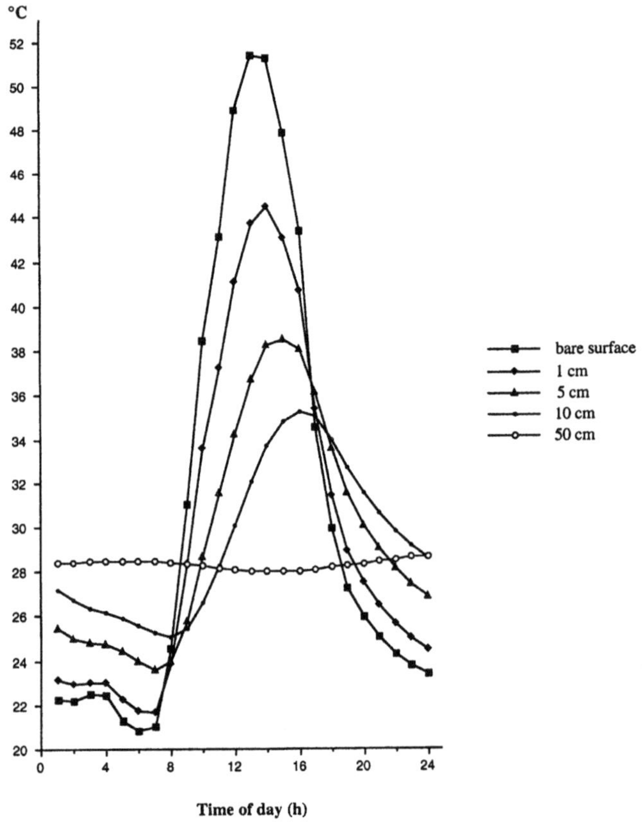

Figure I.21b Near-surface daily temperature variation at selected depths in two soils: (b) A tropical Australian alfisol (Bristow, 1988).

In a way analogous to diurnal temperature changes, seasonal variation in energy input to the soil surface strongly influences soil temperatures. However, seasonal depth changes penetrate to depths of some metres (Figure I.22) and lag behind those at the surface producing maxima (and minima) at depth later in the year than those at the surface. Further, the amplitude of seasonal temperature variation is progressively damped with increasing depth in the solum.

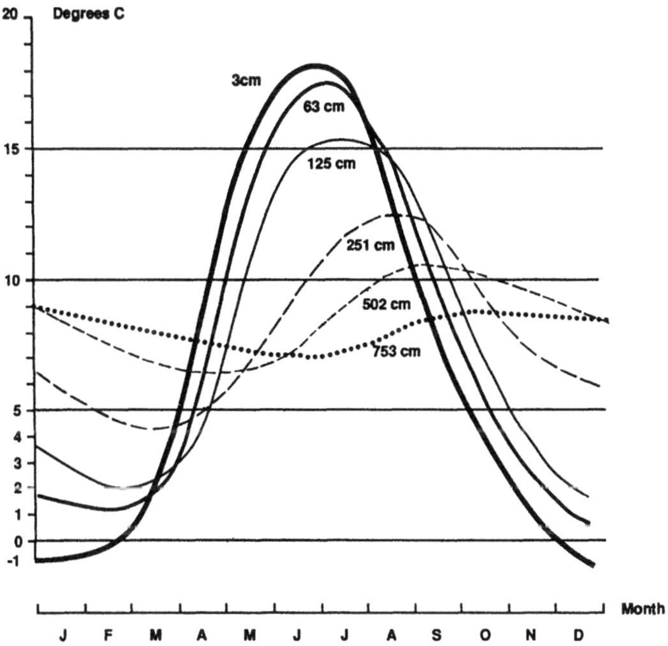

Figure I.22 Seasonal temperature changes at six depths in a soil at Konisberg, Lithuania (Bonneau, 1979).

2.1.3 EFFECTS OF VEGETATION

A vegetation cover dampens the extremes of soil temperature variation on both a diurnal and a seasonal basis; soils beneath a layer of vegetation are both cooler in summer and warmer in winter. This is of considerable agricultural importance since extremes of temperature are inimical to plant establishment and growth.

In the tropics, temperatures may become very high on unvegetated soil surfaces and in the upper part of the solum immediately beneath. Unduly high temperatures will reduce the germination and establishment of seedlings while low soil temperatures frequently limit growth rates in cool climates. Further, high soil temperatures are conducive to rapid organic matter oxidation and are thus undesirable from the viewpoint of conserving this resource.

Soil temperatures are effectively reduced by the presence of a living canopy or a layer of dead vegetation, either standing or as a surface mulch. This vegetation layer may be a forest, a pasture canopy, chemically-killed standing vegetation or a superficial layer of harvest residues or other organic materials. The effect of such a layer is to dampen the extremes of soil temperature by intercepting part of the incoming radiation and reducing convective exchange between the soil and the atmosphere.

Figure I.23a compares the daily march of the temperature of a bare soil surface with that beneath a surface mulch of standing dead foliage of the pasture legume *Stylosanthes hamata* in tropical northeastern Australia (Bristow, 1988). The mulch effected a reduction of 7.3 °C in the daily surface maximum at this date although this effect was much reduced when the soils were wet. The daily temperature regimes of bare soil surfaces may be even more severe than that presented above; Ross *et al.* (1985) recorded temperatures as high as 60 °C at the surfaces of bare soils while surface mulches reduced this by as much as 20 °C. In contrast, Figure I.23b compares the diurnal temperature changes at the surface and beneath a moss cushion at the Shackleton Glacier site presented in Figure I.21a. The amplitude of variation at the exposed soil surface is clearly greater, with lower temperatures pertaining during the hours when the sun was low in the sky or not incident on the site and higher and more variable temperatures when the surface was exposed to greater incident radiation.

The influences of vegetation may be just as important over longer time intervals. Bonneau (1979) considered that a forest canopy in a temperate climate may increase winter soil temperatures by 1-2 °C and reduce those during summer by as much as 7-8 °C, compared with soils outside the forest.

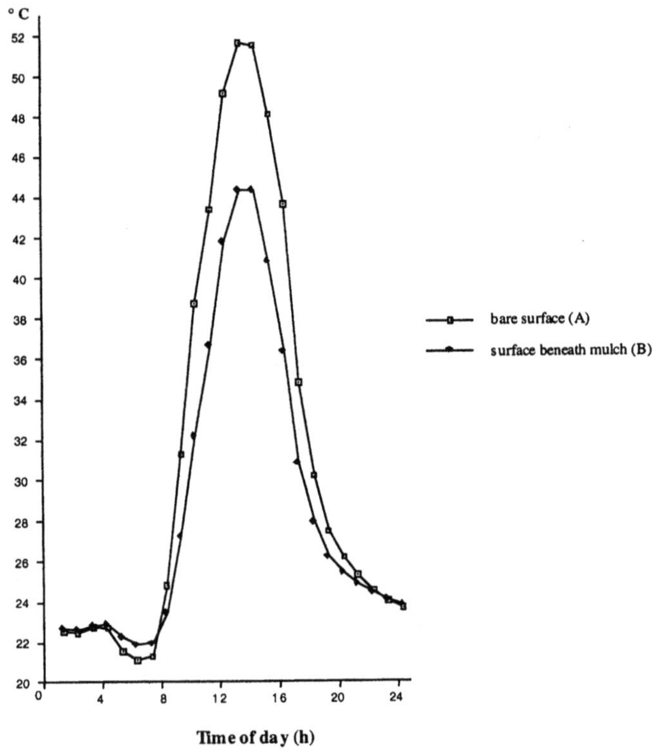

Figure I.23a Comparison of the effects of a vegetation cover on the daily march of surface soil temperatures at two contrasting locations. A bare surface and beneath a mulch of the dead foliage of *Stylosanthes hamata* (Leguminosae) at a tropical Australian site (Bristow, 1988).

Figure I.23b Comparison of the effects of a vegetation cover on the daily march of surface soil temperatures at two contrasting locations. A bare surface and beneath a moss cushion in Antarctica (Wise and Shoup, 1971).

2.2 Water in soils

The amounts of water in soils and the forms and tenacity with which it is held are the major determinants of rates and directions of soil processes at all spatial, temporal and functional scales. This includes soil biological activity and most of the chemical and physico-chemical processes occurring in the soil system.

The water molecule is not symmetric (Figure I.24). Two hydrogen ions with a fixed angle (104.5°) between them occur at one end of the molecule inducing a positive charge at this end of the molecule and a corresponding negative charge at its other end. This polarity allows the formation of electrostatic bonds with other water molecules, cations in the soil solution and the predominantly negatively-charged surfaces of clays and other surfaces. Water molecules do not act independently but are linked to surrounding water molecules to form short-lived, two and three dimensional polymer-like associations. They are also attracted to positively charged cations to form hydration shells around them (Figure I.24).

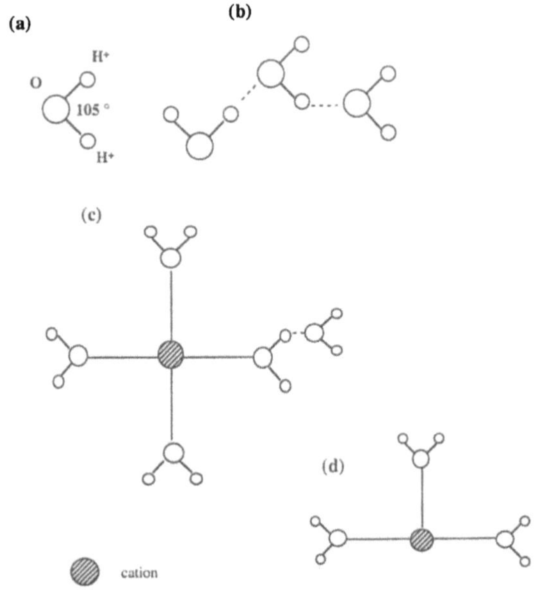

Figure I.24 The polar nature of water and some of the possible bonds that it may form with other molecules: (a) polarity of the water molecule, (b) water-to-water molecular bonds, (c) hydration shells around a cation, (d) cation adsorption with water bridges to a clay surface (partly after Fenwick and Knapp, 1982).

In the liquid state, water expends part of its potential energy in the physical transformation of parent materials and transport of solid particles, either at the surface (erosion) or within the soil profile (translocation). Because of its polarity, water adheres to solid particles, and this is of especial importance in the small pores and capillaries where solid, liquid and gaseous phases interact. As stated above, water is also a good conductor and therefore has an important role in heat transmission. As a reactant, water promotes hydrolysis and dissolution, two important processes in the weathering and transformation of both organic and inorganic components during soil formation (Chapter II).

Also, as considered later in this chapter (Section I.2.2.2), variation in soil water content exerts an important control over the nature and activity of the biota present and on the directions and rates of many soil reactions. Water-logging that lasts for more than short periods results in substantially-reduced soil oxygen concentrations, reducing conditions and the production of compounds inimical to the growth of most plants. The plants that inhabit wetlands normally possess special structures that facilitate gas exchange with the atmosphere, or physiological adaptations that allow them to cope with these conditions (Section I.2.2.3).

As the major component of living entities, water may be considered as a resource. Its availability in time and space, the competition that occurs for it between organisms and the energy necessary for plants and other organisms to absorb it from soil pores are characteristics of paramount importance in controlling the activities of the soil biota. The energy relationships of soil water are given particular emphasis in this section.

2.2.1 THE FORMS OF WATER IN SOILS

Water may be present in soils as a solid, a vapour or in its more usual liquid state. While water in the latter state is of the greatest interest in terms of its utilisation by plant and other soil organisms, the occurrence of water in other states assumes great pedological and biological importance at certain locations and times.

At high latitudes, soils may remain frozen for a substantial part of the year and only thaw in summer to shallow depths dependent on site insolation. In extreme climates such as Antarctica, soils may remain frozen virtually throughout the year; as considered above, free water may be present at the surface only briefly in the early afternoon for a short period during summer. Below the maximum depth of thawing lies the permafrost, a zone that remains permanently frozen. With decreasing latitude, the permafrost occurs progressively deeper in the profile and finally ceases to exist. However, surface soils in cool temperate climates and at high elevations may freeze for variable periods during the winter or only for periods each day, depending on the heat balance of the site.

The air within the larger pores of most well-drained soils is normally completely saturated with water vapour. The amount of water-vapour-saturated air in a given soil will depend on the total pore space and the pore size distribution. In addition, the amount of liquid water present has a bearing on this, filling the pore spaces progressively from small to large as the soil is wetted (Papendick and Campbell, 1981). Soil structure is a major factor in determining the volume of air in a soil and thus how much water will be present in the vapour phase. Water vapour movement in the soils of dry environments may be of considerable importance to the biota.

2.2.2 MECHANISMS OF WATER RETENTION

Because of their polarity, water molecules are attracted to the mineral and organic matter forming the walls of soil pores (adhesion or adsorption) and to other water molecules (cohesion). Water lenses occur at contact points between solid particles and surface films form due to capillary forces; the bonding energy of these molecules is directly related to their distances from the surface involved (Brady and Weil, 1996).

The soil pore space may be partly or completely filled with water which, depending on the pore-size distribution, will be retained within the soil with varying levels of tenacity. However, the pattern of water retention in soils is extremely complex since it depends on both the organisation of the pore space and the composition and arrangement of the solid particles.

2.2.2.1 *Soil water potential*

The water potential (ψ) is defined as the free or potential energy of a defined mass of water in the soil system relative to that of a reference body of pure free water (Papendick and Campbell, 1981). It is the sum of its component potentials:

$$\psi = \psi_g + \psi_m + \psi_\pi - \psi_p + \psi_\Omega$$

where ψ is the gravitational potential, ψ_m is the matric potential due to adsorptive and capillary forces, ψ_π is the osmotic potential due to the presence of solutes, ψ_p is the pressure potential as in applied air or hydraulic pressure and ψ_Ω is the overburden pressure resulting from the weight of the overlying material and is of importance in swelling soils. At the scale of a soil sample, only ψ_g and ψ_m are normally of importance, although ψ_π assumes great biological importance in salt-affected soils.

The water potential of a defined mass of soil water is related to the work required to extract it. Water potential may be expressed in any pressure unit (bars, centimetres of water, or the S.I. unit, the Pascal (Pa) or in pF units which are the logarithms (to base 10) of soil water potential expressed in centimetres of water (Table I.12). This table also presents the sizes of the pores that remain full of water at a range of specific potentials. Water potential in unsaturated soils is almost always negative although the numerically equivalent positive term 'water tension' is sometimes used for convenience. Water potential is formally defined as the equivalent pressure that must be applied to the soil water to bring it to hydraulic equilibrium through a porous membrane with a reference body of free water (Soil Science Society of America, 1997). Soil water potential values presented here will be expressed in megapascals (MPa).

Table I.12 Relationship between soil matric potential (MPa, pF), relative humidity of the soil air and the diameters of the pores remaining full of water at these potentials, at a constant temperature of 20 °C (from Papendick and Campbell, 1981).

Matric potential (MPa)	Matric potential (pF)	Relative humidity (%)	Effective pore diameter (μm)
-0.0001	0.01	100.0	2908.0
-0.0002	0.3	100.0	1454.0
-0.0005	0.7	100.0	582.0
-0.001	1.0	100.0	291.0
-0.002	1.3	100.0	145.0
-0.005	1.7	100.0	58.2
-0.010	2.0	99.99	29.1
-0.020	2.3	99.99	14.5
-0.050	2.7	99.96	5.82
-0.100	3.0	99.93	2.91
-0.200	3.3	99.85	1.45
-0.500	3.7	99.63	0.582
-1.000	4.0	99.26	0.291
-2.000	4.3	98.53	0.145
-5.000	4.7	96.37	0.058
-10.000	5.0	92.86	0.029
-20.000	5.3	86.24	*
-50.000	5.7	69.06	*
-100.000	6.0	47.69	*
-200.000	6.3	22.74	*
-500.000	6.7	2.47	*

2.2.2.2 *The energy status of water in soils*

Water falling on the soil surface enters the pores and soil aggregates under the influence of gravitational and capillary forces. Part of the applied water will run off over the surface if the rate of input per unit area of surface is greater than its infiltration capacity, the rate at which the soil surface will accept water.

The size-frequency distribution of the soil pores determines the maximum retentive capacity of soils which is defined as the maximum volume of water that may be retained in the absence of drainage. At this stage, water potential is close to zero. When water input ceases, part of the soil water drains to the water table under the influence of gravity and a new state of approximate equilibrium, field capacity, is reached. Field capacity is defined as the water content of a drying profile after drainage from it is arbitrarily judged to have become negligibly small. At this stage, capillaries less than a given diameter remain full of water and the forces retaining water in the soil approximately balance those of gravity. Matric potentials at field capacity depend on texture and may vary from -0.01 MPa (pF 2.0) for a uniform sandy soil to -0.05 MPa (pF 2.7) for one of fine texture. Because of this, the diameters of the capillaries remaining full of water at field capacity range from approximately 6 to 30 μm.

As the soil continues to dry, a further characteristic stage is attained below which many mesophytic crop plants can extract no further water from the soil: this is termed the permanent wilting point. The capillary water extractable by plants (*i.e.*, from pores greater than *ca.* 0.15 μm in diameter) is exhausted and the water potential is -1.5 MPa (pF 4.2). However, plants are capable of drying soils to potentials far below -1.5 MPa. Root-zone matric potentials of -4.5 MPa (pF 4.7) have been recorded under winter wheat crops (Papendick *et al.*, 1971) and -7.0 MPa (pF 4.9) from soils supporting the North American xerophyte, *Artemisia tridentata* (sagebrush) (Campbell and Harris, 1977). Some few extreme xerophytes ("resurrection plants") may survive drought almost independently of soil water status and can withstand the near-complete desiccation of their tissues (Gaff, 1981).

With further drying below -1.5 MPa, water only remains in progressively smaller pores (Table I.12) until effectively no liquid water remains, although the soil atmosphere still remains almost completely saturated with water vapour. The fall in the relative humidity of the soil air with increasing water potential is slow and it is not until potentials of -20 MPa are reached that it falls much below 90 *per cent*.

If, during the progressive drying of a wet soil, soil water potential is plotted against the equilibrium gravimetric water content over a wide range of potentials, the resulting characteristic draining curve describes how much water will be held in that section of the profile. The nature of the characteristic draining curve differs from soil to soil depending on soil texture, with fine textured soils retaining more water at any given potential than those of coarse texture (Figure I.25). Soil structure also plays a part but mainly at high water potentials. The shape of this curve differs when the soil is taking up as opposed to losing water, a phenomenon known as hysteresis. This results from several causes including the non-uniformity of pores, differential changes in structure on drying and wetting, differences in the radii of curvature of advancing versus retreating menisci in the pores, and the presence of entrapped air bubbles that may block pores during water absorption (Papendick and Campbell, 1981). Characteristic draining curves also vary with temperature; at any given potential, less gravimetric water will be present at higher temperatures than at low, irrespective of whether the soil is taking up or losing water.

2.2.3 RESPONSE OF THE SOIL BIOTA TO WATER STRESS

A large part of the adaptive strategies of the soil biota is defined by the ability of these organisms to live either in the water- or air-filled soil pores, and by their capacity to survive desiccation. This is addressed further in Chapter III.

2.2.3.1 *Microbial tolerance of water stress*

Micro-organisms, as with higher plants, are highly sensitive to changes in water potential and considerable variation in water-stress tolerance exists between and within groups. The movement of many propagules and of such unicellular organisms as bacteria is dependent on the presence of free water or water held at very high potentials: bacterial movement, for example, becomes negligible at matric potentials less than -0.01 MPa (*i.e.*, when pores of a diameter greater than *ca.* 30 µm are not completely filled with water).

As soils dry, the respiration of micro-organisms declines below certain ranges of matric or osmotic potential, characteristic for the various groups. In wet but not saturated soils, most organisms may be active although as soils progressively dry the bacteria rapidly cease to be active. Soil bacterial respiration, for example, may be minimal at water potentials less than -1.5 MPa. Soil fungi are much more tolerant of low matric potentials than soil bacteria and may remain active to quite low potentials. Papendick and Campbell (1981) state that some micro-organism activity may still be detectable where soil air relative humidity exceeds 70 *per cent.* (ψ_m *ca.* -50 MPa).

The growth rates of fungi *in vitro* are clearly related to the water potential (ψ), although the limitations to growth below certain potentials may be as much due to their inability to absorb nutrients as water (Jennings, 1995). Responses to matric and osmotic potentials have different forms with growth declining nearly linearly with ψ_m, but that with ψ_π showing an increase to an optimum followed by an approximately linear decline (Griffin, 1981a).

Griffin (1981a) defined five overlapping groups of micro-organisms defined by the osmotic potential ranges within which they grow. Overall, this range includes poorly-tolerant species that fail to grow at less than -0.2 MPa to the highly xerotolerant species that continue to grow at -40 MPa or less and which may fail to grow at high potentials. Soil micro-organisms belong to several groups but are largely included in the range of ψ_π values from -0.1 MPa to those that grow little at osmotic potentials less than -15 MPa.

The reactions of micro-organisms over their active ranges of water potentials are complicated by other factors than the matric and osmotic potentials. Antagonisms occur between and within groups and both fungi and bacteria may grow at much lower potentials in pure than in mixed cultures, or in the soil. In the natural environment, fungal growth may not be closely related to immediate water potentials due to the capacity of these organisms to translocate water and support growth into materials or across sites where little water is present at potentials suitable for growth. The situation is further complicated by additional interactions with temperature and nutrition. Griffin (1981b) considers that water-retention energy alone is insufficient as a predictor of microbial activity and that pore-size distribution, volumetric water content and other soil factors are also of importance.

2.2.3.2 *Faunal responses to water stress*

Because of their eco-physiological adaptations to a soil-dwelling existence, the activities of most soil animals are closely circumscribed by soil water status (Ghilarov, 1983). With respect to soil water, common adaptations include the possession of permeable integuments and a predominantly cutaneous respiration. Among the tracheate groups living in the soil, spiracles are frequently absent; where present, their possessors are often unable to regulate the rate of gas exchange through them.

Soil animals may be divided into two groups based on the ways they move through soils (see Chapter III, section 4), namely, those that are constrained to move through existing interstices and those that can progress through the soil either by ingesting it or transporting it out of their path. In terms of energy conservation, the reduced strength of moist soils must be advantageous to members of the latter group. Four major macrofaunal taxa belonging to this group are the earthworms, termites, ants and certain beetles. Earthworms require high soil water potentials for activity and this is reflected in their virtual exclusion from arid areas. In contrast, many species of termites and ants live in the protected environment of the soil and members of the former group can at least partially control the humidity of their nests by transporting water from deeper horizons. For this reason, a range of species has been able to colonise dry tropical savanna ecosystems and deserts successfully.

Despite the well-developed digging and burrowing capacities of a few groups, most soil animals are constrained to move through existing soil pores and the size distributions and inter-connectedness of these pores must constrain their mobilities. Many small soil animals such as nematodes and protists are effectively aquatic and also require water at high potentials for activity. For example, Wallace (1961) showed that movement of the larvae of the nematode *Ditylenchus dipsaci* was optimal at potentials between approximately -0.002 and -0.003 MPa, but also subject to satisfactory pore sizes being available.

As soils dry, the larger mesofaunal species are affected by water availability only below critical matric potentials specific to each species and beyond which they can no longer maintain their internal water balance. For example, atracheate Isotomid Collembola are affected at water potentials less than -1.5 MPa while the more desiccation-resistant Cryptostigmata (Acari) are only affected below -9.8 MPa. A few Cryptostigmata may sustain their internal water balances at water potentials down to -98 MPa, reflecting the considerable variability present within some groups (Vannier, 1987).

2.2.4 A BROAD CLASSIFICATION OF SOIL WATER

While the amount of water retained in a soil is a continuous function of the energy required for its extraction (Figure I.25), it is convenient, and meaningful in terms of the behaviour of certain crop plants, to consider three broad categories of soil water, based on their energies of retention within the soil matrix:

1. **Gravitational water** fills the larger pores (minimum diameter range 6 to >30 μm) and is retained in the soil with little energy. It drains rapidly (from pores greater than 50 μm diameter) or slowly (from pores of 10 to 50 μm diameter) towards the water table under the influence of gravity. This fraction of the soil water is that held at potentials of greater than approximately -0.03 MPa and is indicated as region A in Figure I.25.
2. **Capillary water** corresponds to the water held in soil pores or on surfaces with potentials in the range -0.03 to -3.1 MPa (region B in Figure I.25). It also corresponds to water held in pores between the approximate diameters of 0.1 and 6.0 μm. Below potentials of -3.1 MPa, little further capillary water remains.
3. **Hygroscopic water** forms a thin, tightly-adsorbed (matric potential less than -3.1 MPa) layer held on the surfaces of the soil particles, particularly the clay minerals and is represented as region C in Figure I.25. This category includes the inter-layer water bound to the surfaces of clay particles by covalent and hydrogen bonds (Figure I.24) and inter-particle water adsorbed within small tactoids formed from aggregated clays.

The mass of water held in a given volume of soil between the approximate potential levels of -0.03 MPa (field capacity) and -1.5 MPa (the permanent wilting point) corresponds to the **plant-available** water. This is the reserve (R) of water available to certain mesophytic agricultural crop plants and is expressed in millimetres of water using the formula:

$$R = (F - W) / 100.D.BD$$

where F and W are the percentages of water per unit mass retained at, respectively, field capacity and wilting point, D is the depth of soil and BD is the bulk density (Duchaufour, 1997).

At a profile scale, the total water reserve may be calculated by summing estimates made for each horizon. Such reserves may range from 50 to 60 mm for a shallow (*ca.* 30 cm deep) soil to more than 400 mm for the water available to strongly rooting plants in a deep silty soil (Mériaux, 1979). The store of "plant-available" water calculated in this way will underestimate that actually available because the roots of many plant species are infected with mycorrhizal fungi. Water held at potentials substantially lower than -1.5 MPa may be made available to their host plants through the medium of these fungi (Chapter IV.3.2.1.2). Further water may also be extracted from the weathering zone beneath the soil (the R layer) by deeper-rooted species.

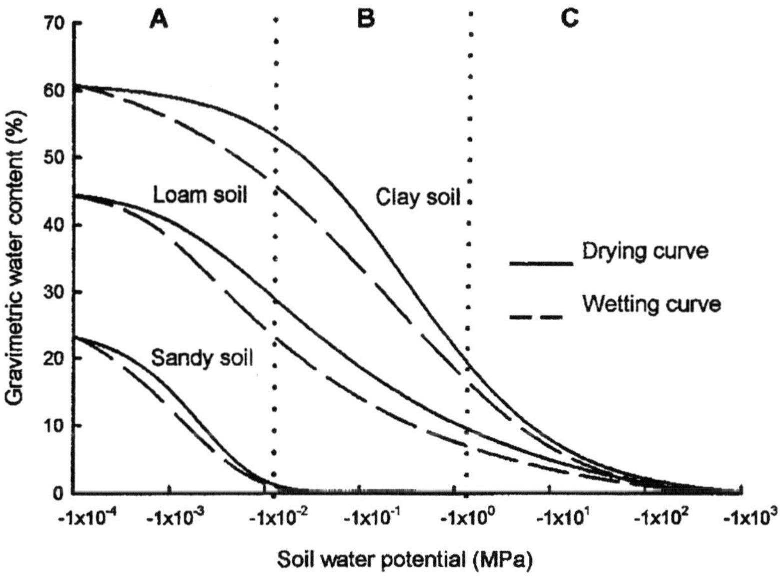

Figure I.25 Characteristic draining curves for three soils of contrasting texture including the division of the retained water into three classes (A gravimetric water, B capillary water, C hygroscopic water). Note the differences between the draining and wetting curves. (Redrawn from Yong and Warkentin, 1975).

2.2.4.1 *Water retention in different soils*

On either a gravimetric, or mass per unit volume basis, the equilibrium water content corresponding to any given water potential is strongly influenced by both soil texture (textural porosity) and structure (structural porosity). At a given potential, the gravimetric water content depends on the size-frequency distribution of the soil pores and, in particular, the total interconnected pore space, *i.e.*, that which can be filled with water. Since soils differ widely in pore space distributions, the total store of water in soils also varies accordingly.

At high potentials, much soil water is held in the larger pores and soil structure greatly influences the amount of water retained. The effect of structure is clear in strongly-structured soils, but only at high potentials since structure controls the volume of the large inter-aggregate pores that retain gravitational water. In a strongly-structured Hawaiian soil, Sharma and Uehara (1968) found that almost all the water stored between the aggregates had drained from the soil after drying to -0.029 MPa and that at lower potentials, water was largely derived from within the aggregates.

At lower potentials the lesser importance of structure means that the shape of the characteristic draining curve is largely dependent on texture and the specific surface area of the soil. Soils with high clay contents have large internal surface areas and a greater capacity to retain water at all potentials (Figure I.25). Coarse-textured soils retain less water overall and have less plant-available water than the finer textured loams and clays.

Certain soils may undergo volume changes in response to variation in moisture status. Vertisols shrink reversibly on drying, often presenting a surface pattern of deep polygonal cracking that disappears on rewetting. The B horizons of certain andisols may shrink on drying and beyond a certain level of drying this is largely irreversible (Maeda *et al.*, 1977).

2.2.4.2 *Water retention by dead plant material and soil organic matter*

Like clays, dead plant material and soil organic matter have very high water-absorbing capacities and may hold up to several times their own weight of water. Guano for example, may retain 273 % water at field capacity (Vannier, 1970). Papendick and Campbell (1981) report that a mulch of wheat-straw residues of 12 Mg ha^{-1} may absorb up to 2.5 mm of rainfall thereby preventing much of the water incident on a site from reaching the surface of the mineral soil.

The characteristic draining curve for plant material follows an exponential relationship (Papendick and Campbell, 1981) of the form:

$$\psi = a\theta^{-b}$$

where ψ is the matric potential, θ is the gravimetric water content and a and b are constants. Most of this water is held at relatively low water potentials and is of limited use to plants or micro-organisms except those, such as certain fungi, that are adapted to using it. Myrold *et al.* (1981) showed that at -1.5 MPa, chopped wheat straw residues still contained 45 % water and 26 % at -3.1 MPa while water contents at the same potentials in a mollisol of silt loam texture were 9 % and 7 %, respectively. These authors also showed that the osmotic potential of plant material may contribute appreciably to its total water potential.

As plant material decays, the characteristic draining curve changes such that water is held at a higher potential for any given level of gravimetric or volumetric moisture content. Dix (1985) showed that the leaves and wood of oak (*Quercus robur*) change with degree of decomposition such that water becomes progressively more available for the decomposing fungi (Figure I.26); this may have a substantial influence on the succession of organisms decomposing organic substrates.

2.3 Hydrological regimes

The fate of the water that falls on the landscape is a critical factor in all environments. The amount and intensity of the rainfall incident on the soil surface controls the severity of soil erosion, the quantity and quality of river water, recharge of subterranean aquifers and the supply of water for man's use. In the context of agriculture, the rainwater incident on a site and its subsequent movement have important implications for such cultural activities as the timing and nature of cultivation, the planting and harvesting of crops and the need for irrigation or drainage.

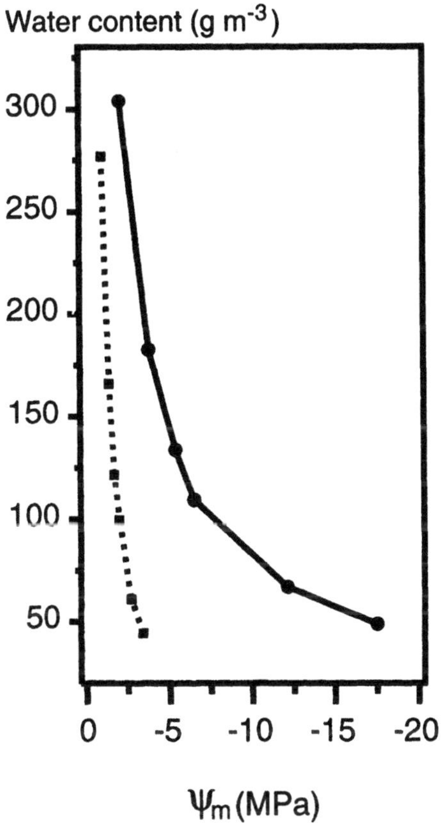

Figure I.26 Characteristic draining curves for undecayed leaves of *Quercus robur* (●) and after 11 weeks decomposition (■)(Dix, 1985).

Soil water regimes are dependent on both the amounts of water input to the soil surface and its form. While water is normally acquired as rainfall (and snowfall in the colder parts of the world) inputs as dew and fog may be significant to plants and O horizon processes in drier areas. Other factors affecting the amounts of water entering the soil are rainfall intensity in relation to the maximum rate at which the soil will accept water (the infiltration capacity) and the temporal distribution of the rainfall at both short term and seasonal time scales.

Soil factors affecting hydrological regimes include water holding capacity and profile depth, both of which define the maximum store of soil water that may be held in any given soil. Losses of water from the profile depend on soil hydraulic conductivities, site drainage and the thermal regime of the site. The latter factor determines the evapo-transpiration rate, which is the sum of evaporative losses from the soil surface and the water lost by plants to the air by evaporation and transpiration. Finally, surface micro-relief may slow water movement across the soil surface and thereby increase infiltration rates.

2.3.1 WATER BALANCE

The water balance at a site describes the partitioning and fates of the rain water that falls onto a defined area of ground over a specified period. It depends markedly on the type of vegetation present and the proportion of the soil surface that it covers. The factors of importance in a simple water balance model (Figure I.27) for a closed forest with a single canopy layer are the incident precipitation (P), the amount lost from the canopy by evaporation (Ec) of the intercepted precipitation stored by the canopy (Sc), that lost by transpiration (T) through the stomata of the leaves and the net amount of precipitation that reaches the forest floor by flowing down the stems (stemflow, SF) and by penetrating the canopy (throughfall, TF). In drier environments, substantial water may rise by capillary action (C) from deeper in the soil and enter the root zone. Where the capillary rise attains the surface, salts transported in this water may accumulate to salinate the surface horizons to such an extent that plant growth is inhibited.

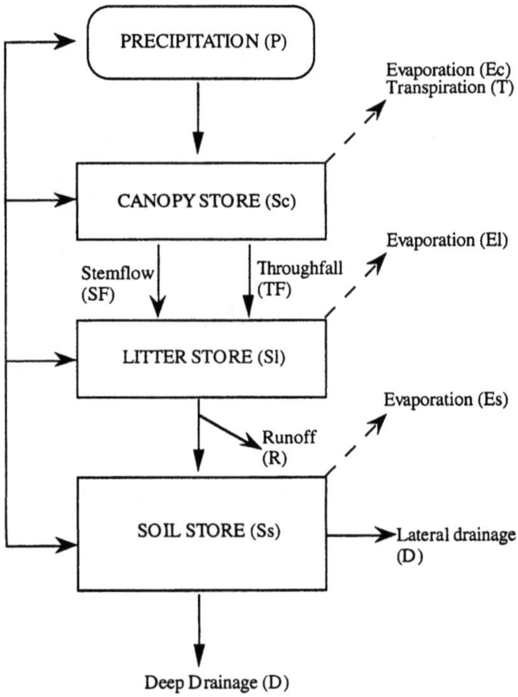

Figure I.27 Components of the hydrological cycle in forests and the transfers of water that occur between them and the general environment.

Once water reaches the forest floor, part will be involved in recharging the store of water contained in the litter (Sl), some will be lost by evaporation from the surface of the litter (El) and the soil (Es). A further quantity (runoff, R) may move by surface flow to sites lower in the landscape, and part will enter the soil. Some of the latter is retained to form the store of water available to the plants (Ss). That part of the water entering the profile that exceeds its water holding capacity will be lost by drainage (D), either laterally to adjacent sites or to deeper levels in the profile, beyond the range of plant roots. At sloping sites, surface runoff and sub-surface drainage waters from areas upslope of the site of interest also contribute to the local water balance.

Neglecting capillary rise, the water balance may be presented as follows:

$$P = TF + SF + Ec + Es + El + T + \delta Ss + R + D$$

where the symbols have the meaning expressed in the paragraphs above. The term evapotranspiration is often used to include the sum of the terms Ec, El, Es and T in the above equation.

In temperate France, Fardjah *et al.* (1980) compared the water balance of a bare soil with two vegetation types, a beech wood (*Fagus sylvatica*) and an area colonised by the grass *Calamagrostis epigeios*. The soil, a spodosol, was similar at all sites and the study area was horizontal; thus no runoff or run-on effects were present. Table I.13 presents the major fluxes of water in the study areas and their partitioning over a period of one year. The increased drainage component from the bare soil is due to the lack of evaporation and transpiration from the canopy and is typical of unvegetated areas; it is not compensated for by evaporation from the soil surface alone. The substantial interception losses from the forest are characteristic of this vegetation type and are due to the considerable turbulence that occurs at the top of the canopy.

Table I.13 Annual water balances (mm) under bare soil and two vegetation types at Fontainebleau, near Paris (Fardjah *et al.*, 1980) (See text for explanation of symbols).

Vegetation	**Component**					
	P	**Ss**	**Es+El**	**Ec**	**T**	**D**
Bare soil	481	209	310	-	-	171
Grassed surface	481	210	10	38	332	101
Forest (*Fagus sylvatica*)	481	206	10	99	316	56

2.3.2 VEGETATIONAL AND SEASONAL EFFECTS ON SOIL WATER

Figure I.28 presents the seasonal changes in the estimated reserves of plant-available water taking place over nine months for the three plots studied by Fardjah *et al.* (1980). The total, nominally 'plant available' soil reserve (*i.e.*, that held between water potentials of -0.03 and -1.5 MPa) was estimated as approximately 127 mm over all plots. Figure I.28 shows the effect of the predominantly winter rainfall in recharging the store of soil water depleted by evapotranspiration over the summer period.

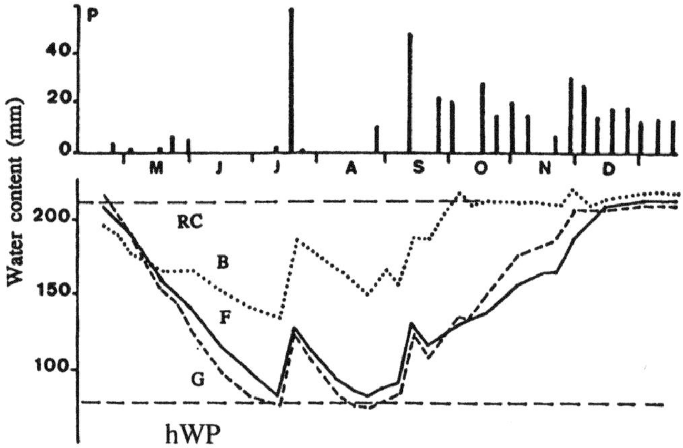

Figure I.28 Seasonal changes in daily rainfall (P) and 'plant-available' water (mm) in an unvegetated area (B), beneath a grassed surface (G) and a forest of *Fagus sylvatica* (F) at Fontainebleau, near Paris (RC is the water retention capacity of the profile, WP is the retention capacity at the 'wilting point') (from Fardjah *et al.*, 1980).

The marked effect of the vegetation present is clear from the greater soil water store in the unvegetated area and the longer period during which this reserve remained high. In the vegetated area, the available store of soil moisture was exhausted twice during the summer period. Precipitation after August restored the soil store by early October in the bare area but not until December for the vegetated sites. Differences between the two vegetation types are related to the effects of the forest canopy. As indicated above, increased turbulence at the top of the canopy meant that considerably less water reached the soil in the beech wood: net interception was estimated at 15.1 % in the beech wood and 8 % in the grassed area. Evaporation from the soil surface is similar in both vegetation types and, in the grassed area, the low evaporation rate was due to the substantial litter layer present.

2.3.3 EFFECTS OF TOPOGRAPHY ON SOIL WATER

Except for those in the highest parts of the landscape, soils at a particular site are usually influenced by those surrounding. Normally, they acquire materials and water from sites upslope and supply them, often in a modified form, to those downslope; such suites of related soils are known as toposequences or catenas (Chapter II.4.5). The water that flows to sites lower in the landscape by either surface runoff or subsurface lateral flow profoundly influences their hydrological regimes.

Figure I.29 presents seasonal changes in moisture status at three such sites located sequentially down a hillslope in a humid (1276 mm annual average precipitation) savanna in the Côte d'Ivoire (Lavelle, 1978). The highest plateau site is dominated by perennial grasses with a few woody shrubs, the sloping site has a similar cover with more woody vegetation and the next lowest site has a near-complete herbaceous cover with some palms. The soil of this latter site is described as hydric. That is, a soil whose development has been characterised by the frequent presence of excess water and which is commonly saturated. At the lowest site, a gallery forest lines the bank of a stream. The soil moisture data are partitioned into three matric potential ranges (> -0.03 MPa, -0.03 to -1.5 MPa, < -1.5 MPa) corresponding to saturation, "plant-available" water and water held at potentials below an arbitrarily-designated "wilting point".

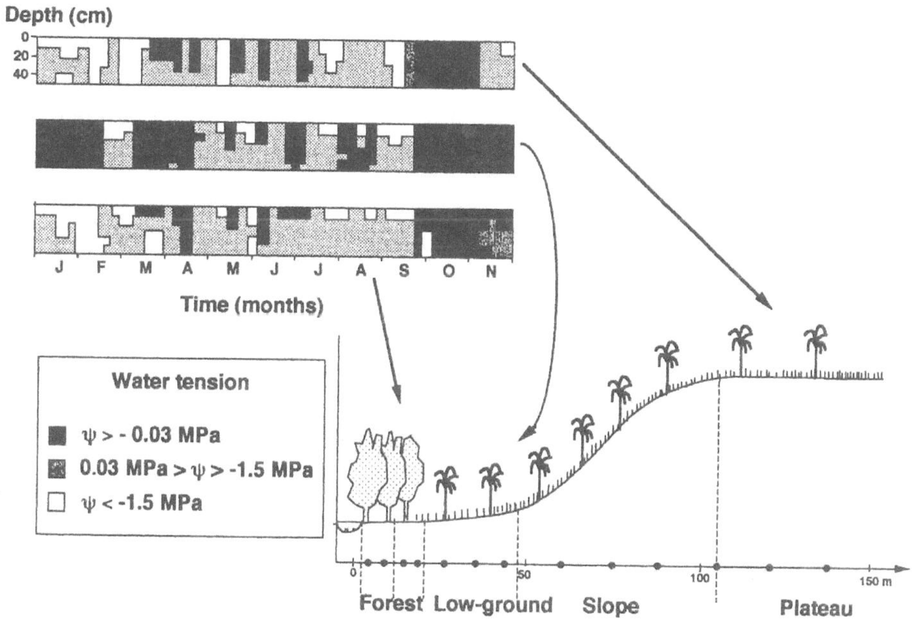

Figure I.29 Seasonal changes during 1969 of water held at three potential ranges in a toposequence of soils in a humid savanna at Lamto, Côte d'Ivoire (Lavelle, 1978).

From the figure, the effects of topography and vegetation are clear. Sites become wetter for longer periods with progression downslope. While all sites were saturated for considerable periods during the wetter months, the surface (0 10 cm) of the hydric savanna soil was saturated for 230 days during 1969, a dry year for the area. In contrast, the surface layer of the plateau soil was only saturated for 130 days. In the gallery forest, the surface soil was saturated for just 119 days during the same year, illustrating the substantially larger water usage of the rainforest canopy and the higher rates of internal drainage in this soil.

I.3 SOIL RESOURCES

A wide range of resources are available to the organisms that grow in, or live within or on the surface of the soil. These include mineral nutrients and a diverse range of organic materials, both of differing degrees of availability.

Organic resources consist predominantly of dead materials that accumulate on the soil surface (leaf, wood and other litter, dead micro-organisms or animals) or deeper within it, localised at specific places in the soil profile (dead roots and other organisms). In addition to these dead materials, soils offer a considerable variety of other living resources (roots and soil organisms) and their secretory products (root exudates, microbial mucilages and invertebrate mucus) to the organisms that are mechanically and physiologically equipped to utilise them.

Most mineral nutrient elements are derived ultimately from soil parent materials through weathering processes, although constant inputs of particular elements may also be derived from atmospheric sources. Some of these nutrients accumulate in the living biomass and are recycled through decomposition processes. A further part is retained in the soil in mineral form, adsorbed more or less tightly to clay particles or organic materials by electrostatic forces, or mechanically trapped within them.

3.1 Inorganic Nutrients

The major elements that combine to form organisms are, in order of importance: hydrogen, oxygen, carbon, nitrogen, phosphorus and sulphur. These elements possess distinctive chemical properties which allow them to contribute to the continuing creation and maintenance of life (Taube, 1992). To be capable of this, they must be abundant, stable and capable of reacting together to form the complex molecules that comprise structural elements, mediate the internal and external exchanges of materials and energy, transmit their genetic information and form the semi-permeable membranes responsible for the necessary isolation of organisms from their environments.

In addition to the six elements considered above, only a small proportion of the total of 91 naturally-occurring elements are known to be required for the normal development of organisms. The nutrient elements defined as essential for an organism are those without which it is not able to complete its life cycle (*i.e.*, produce viable propagules), or that form part of a molecule which is itself known to be essential (Epstein, 1972). The designation of an element as an essential nutrient implies that it has an indispensable structural or biochemical role. However, not all the roles of the essential elements are known in detail and most nutrient elements have multiple roles. Numerous complex interactions occur between the nutrient and non-nutrient elements. For some essential elements, partial substitution may occur by another not known to be so and certain elements may interact to inhibit the uptake of others (antagonism), or to promote it (synergism). Beneficial elements are mineral elements which will stimulate growth but are not essential or are essential only for certain plant species or under certain conditions.

A number of elements taken up by the roots and subsequently translocated throughout the plant serve no known structural or biochemical function within the plant. Among these are included such elements as selenium which are required by animals and constitute the major source of supply to the higher members of biological communities. These and other elements may also be taken up, even though some are actively inimical to the health of living organisms, either through their radioactivity or because of their capacity to inhibit natural processes at the functional level of the cell, organ or the whole organism.

A range of elements may also be absorbed through the leaf surfaces. Elements deposited on these surfaces diffuse through the cuticle and may be actively absorbed by the cells and transported elsewhere in the plant. This capability is used to alleviate deficiencies of certain trace and other elements in some tree and other crops; however, it also provides a pathway for pollutants such as lead to enter the leaves.

In an ecological sense, the nutrition of plants (and the animals dependent upon them) is concerned with rather more than the events taking place at the surfaces where nutrients are absorbed. Three separate and complex stages occur, namely, release of the nutrients from their sources, transport to the roots and uptake by the plant (Chapin, 1988). Each step potentially involves interactions between the inorganic components of the soil and organic processes or products. Also, plants are not passive absorbers of nutrient and other elements and may actively modify their root distributions in response to those of certain elements or alter the environments around their roots in ways that modify element uptake. Further, most higher plants enter into symbiotic associations with representatives of one or more groups of micro-organisms that aid or effect nutrient uptake (Chapter IV.3.1.2.2).

Plants form the productive nutritional base on which animals ultimately depend, whatever their trophic level. Through its effects on plant productivity, the uptake of nutrient and other elements by plants is of prime importance since it eventually affects the entire biological community, including the higher order predators and the saprovores.

3.1.1 CLASSIFICATION OF THE NUTRIENT ELEMENTS

In addition to the nutrient and non-essential elements, the major components of higher plant tissues are carbon and oxygen, each of which comprise *ca.* 45 % on a dry weight basis. Table I.14 lists the elements known to be required by higher plants and animals, the ionic forms in which they are most commonly taken up from the soil by plants together with some 'typical' concentrations in the tissues of unspecialised plants. In addition, for the elements needed by plants, the functional groups that they belong to (see below) are also indicated (Mengel and Kirkby, 1987). The concentrations presented in Table I.14 are indicative only and macronutrient tissue concentrations required for highest production in crop plants are generally greater than those presented (Marschner, 1995). Table I.15 presents the ranges of concentrations for six macronutrient elements found in plant tissues sampled from a variety of natural environments.

Table I.14 The elements known to be required for the nutrition of unspecialised higher plants (Salisbury and Ross, 1991) and animals. Beneficial elements are required by some plants or may confer such benefits as insect pest resistance. The ionic forms in which the elements are most commonly taken up by higher plants and some typical plant tissue concentrations (macroelements %, microelements $\mu g\ g^{-1}$) are indicated (partly after Salisbury and Ross, 1991).

Element	Group	Forms commonly taken up by plants	Typical concentrations in dried plant tissues
i. Macro-nutrient elements required by animals and most plants (%):			
N	I	NH_4^+, NO_3^-	1.5
Ca	III	Ca^{2+}	0.5
Mg	III	Mg^{2+}	0.2
K	III	K^+	1.0
P	II	$H_2PO_4^-$, HPO_4^{2-}	0.2
S	I	SO_4^{2-}	0.1
ii. Elements required by animals in micro quantities but either not needed by plants, or needed in very small amounts (%):			
Cl	-	Cl^-	0.01
*Na	-	Na^+	0.08
iii. Micro-nutrient elements required by plants and animals ($\mu g\ g^{-1}$):			
Cl	III	Cl^-	100
Cu	IV	Cu^{2+}, Cu^+	6
Fe	IV	Fe^{2+}, Fe^{3+}	100
Mn	III	Mn^{2+}	100
+Mo	IV	MoO_4^{2-}	0.1
Ni	IV	Ni^{2+}	1
Zn	IV	Zn^{2+}	40
iv. Elements beneficial to, or possibly beneficial to higher plants ($\mu g\ g^{-1}$):			
+Co	IV	Co^{2+}	0.5
F	-	F^-	10
I	-	I^-	3
++Na	III	Na^+	800
Se	-	SeO_4^{2-}, SeO_3^{2-}	25
Si	II	$Si(OH)_4$	3000
V	-	VO^{2+}	2

Table I.14 (cont.)

	Element	Group	Forms commonly taken up by plants	Typical concentrations in dried plant tissues
v.	**Micro-nutrient elements required by animals but not by plants ($\mu g\ g^{-1}$):**			
	As	-	As^{6+}, As^{3+}	1
	Cr	-	Cr^{6+}, Cr^{3+}, CrO_4^{2-}	0.4
vi.	**Micro-nutrients required by plants but not by animals ($\mu g\ g^{-1}$):**			
	B	II	$B(OH)_3$	20

* Beneficial or possibly beneficial element for higher plants (Marschner, 1995).

+ Required by nitrogen-fixing organisms.

++ Some C4 pathway plants only.

Certain elements may stimulate growth in higher plants even though they are not essential or may only be essential for particular species. These are known as beneficial elements and include Na, Si, Co, Se, Al and possibly I, Ti, V, La and Ce (Marschner, 1995). For example, rice is known to accumulate Si which interacts with other nutrient elements to confer resistance or tolerance to insect pests and diseases and to such abiotic stresses as Al toxicity and salt injury (Savant *et al.*, 1997). Na is essential for a number of species that use the C4 photosynthetic pathway such as the halophyte *Atriplex vesicaria* (Chenopodiaceae) but not for others such as sugarcane (*Saccharum* spp.).

While it is as yet unknown whether silicon and selenium are essential for higher plants, certain prokaryotes require these elements, cobalt and vanadium (Salisbury and Ross, 1991). Our knowledge of the trace element requirements of micro-organisms is poor and it is likely that further elements will be demonstrated as essential for certain organisms, but only in very minute amounts.

Certain micro-organisms may accumulate very high concentrations of essential elements such as copper and zinc and non-essential elements such as lead, mercury and cadmium (Lepp, 1992; Gadd, 1993). The human health consequences of ingesting contaminated wild mushrooms can be severe although large inter-specific differences in metal uptake occur. Further, eight to nine years after widespread contamination associated with the Chernobyl reactor accident, and possibly also nuclear weapons testing (Gadd, 1993), wild mushrooms in affected parts of the Russian Federation were shown to have substantially higher activity concentrations of radiocaesium (^{134}Cs, ^{137}Cs) than other human foods. In these contaminated areas, human radiocaesium body loads were highly correlated with mushroom intake (Skuterad *et al.*, 1997); other food chain effects are also likely.

Table I.15 Indicative ranges of concentrations of six major elements (%) in the tissues of higher plants and in forest litter layers (data from various sources, including: Rodin and Bazilevitch, 1967; Tanner, 1977; Bevege, 1978; Grubb and Edwards, 1982; Edwards and Grubb, 1982; Vercoe, 1987; Cuevas and Medina, 1988; Spain *et al.*, 1989b).

Material	N	P	K	Ca	Mg	S
i. Needles/Leaves						
Conifer forests	0.84-2.32	0.07-0.28	0.30-1.40	0.05-1.40	0.07-0.40	0.03-0.19
Deciduous forests	1.36-2.90	0.10-0.60	0.73-3.00	0.35-3.00	0.11-0.67	0.01-0.42
Tropical forests	0.5-2.29	0.03-0.20	0.25-1.90	0.17-3.20	0.10-0.82	0.18
Subtropical forests	1.08-3.96	0.05-0.78	0.60-2.45	0.97-5.10	0.29-0.77	0.01-0.42
Australian *Eucalyptus* forests	0.57-1.6	0.04-0.18	0.12-1.13	0.23-1.04	0.22-0.49	0.06-0.20
Australian *Acacia* spp.	1.42-2.30	0.05-0.11	0.46-1.17	0.30-3.52	0.21-0.37	0.12-1.13
ii. Stemwood						
Conifer forests	0.09-0.47	0.004-0.05	0.02-0.10	0.01-0.09	0.01-0.02	<0.01-0.02
Deciduous forests	0.14-0.36	0.01-0.07	0.06-0.62	0.05-0.65	0.01-0.11	<0.003-0.04
Tropical forests	0.06-0.34	0.002-0.08	0.06-1.2	0.03-1.1	0.008-0.19	0.07
Subtropical forests	0.25	0.05	0.17	0.3	0.06	0.05
iii. Fine roots						
Conifer forests	0.36-1.22	0.06-0.14	0.15-0.26	0.33-0.44	0.07-0.17	0.14
Deciduous forests	0.76-1.46	0.10-0.13	0.17-0.66	0.40-1.06	0.11-0.17	0.07-0.16
Subtropical forests	-	0.03	0.01	0.12-0.13	0.05-0.07	0.01
Tropical forests	0.74-2.30	0.03-1.13	0.01-1.22	0.09-0.73	0.09-0.61	-
iv. Litter layers						
Conifer forests	0.69-1.45	0.01-0.28	0.09-0.44	0.30-3.98	0.08-0.40	0.03-0.27
Deciduous forests	0.72-2.16	0.07-0.17	0.07-0.55	0.84-2.48	0.10-0.62	0.04-0.22
Tropical forests	0.93-2.17	0.01-1.44	0.09-0.70	0.88-4.40	0.13-0.36	0.16
Australian *Eucalyptus* forests	0.33-1.55	0.02-0.08	0.09-1.02	0.42-1.45	0.15-0.41	0.07-0.41
v. Grasses						
World	0.48-4.8	0.03-0.41	0.9-3.3	0.13-9.2	0.07-0.80	0.07-0.80

As shown above, plants may absorb elements for which they have no nutritional requirement or absorb nutrient elements to concentrations far in excess of nutritional needs. Many of these elements may be taken up in substantial quantities. Halophytes, for example, may accumulate substantial quantities of sodium and chloride. The summed mean concentrations of these elements in the leaves of a variety of mangrove species ranged up to 5.06 % of dry biomass (Walsh, 1974) while those of arid-climate *Atriplex* species (Chenopodiaceae) are known to occur in concentrations in excess of 10 % of dry weight (Sharma *et al*, 1972).

Plants may be classified as excluders, indicators or accumulators, depending whether their tissue concentrations of particular elements increase more slowly, in parallel with, or more rapidly than those in the surrounding medium. Glycophytes (non-halophytic species) may also accumulate substantial quantities of certain elements, some of which may cause problems in animal nutrition through food chain effects. For example, in certain accumulator plants growing in seleniferous soils, selenium may occur at concentrations that make them toxic to grazing animals (Marschner, 1995).

Certain plants (extreme metallophytes) may accumulate extraordinarily large amounts of metallic elements. Baker and Walker (1990) report the following concentrations of metals in the foliage of various species of vascular plants: Cu 1.2 %, Co 1.0 %, Mn 5.1 %, Ni 4.7 %, Pb 0.8 % and Zn 3.9 % of dry biomass. Accumulation of metals by certain plants and sometimes animals (indicator species) has been used by economic geologists to indicate the presence of subterranean orebodies (Brooks, 1983); such accumulations may also be indicative of contaminated sites.

Of the elements needed by higher organisms, some, the **macro-nutrients**, are required in relatively large quantities. Macro-nutrients usually have relatively well-defined roles in plant biology; for example, calcium is a major structural component of the cell walls of higher plants where it occurs in conjunction with pectin, while magnesium is a central element in the structure of chlorophyll, a magnesium porphyrin. In contrast, the **micro-nutrients** are needed in only very small amounts although, as indicated above, they too may accumulate in plant tissues in concentrations much greater than dictated by physiological requirements. Not all higher plants require all the recognised nutrient elements. Sodium, for example, is a recognised micronutrient only for those plants that possess C4 or crassulacean acid metabolism photosynthetic pathways, and some cyanobacteria (Brownell, 1979).

The nutrient elements may also be classified into four major groups based on the principal roles that they fulfil in plant tissues (Table I.16) (Mengel and Kirkby, 1987). The group I elements, including carbon, hydrogen and oxygen are the principal structural components of all living tissues and play important roles in enzymes and other functional groups. They are taken up as liquids (H_2O), gases (O_2, SO_2, CO_2, N_2) or as ions from the soil solution (for example, HCO_3^-, NO_3^-, NH_4^+, SO_4^{2-})

Members of group II are taken up from the soil solution as the oxyanions - phosphate borate or boric acid and silicate. The phosphate anion is particularly critical to energy transfer mechanisms within the plant while boron is important in membrane function. The group III elements are taken up from the soil solution and play osmotic and ion balance roles within plant tissues. Other roles include more specific functions in enzyme conformation and catalysis. Clarkson and Hanson (1980) state that members of this group occur largely as free or reversibly-bound ions rather than as metallo-proteins. Members of group IV occur in plants as structural chelates or metallo-proteins; Fe, Cu and Mo (and Mn) participate in redox reactions. Uptake occurs in the form of ions or as chelates from the soil solution. Table I.16 presents the forms assumed in plant tissues by a range of nutri ent elements and, for those important in forming salts and complexes, some major roles.

While many nutrient elements may accumulate in plant tissues at concentrations above physiological requirements, most become toxic when present in considerable excess. Although complicated by interactions with other elements and growth-related factors, individual concentrations may be considered to lie within a spectrum of levels which run from deficient to suboptimal to optimal to supraoptimal to toxic (Smith, 1962), a concept also pertinent to animal nutrition.

Table I.16 A functional classification of plant nutrient elements (Clarkson and Hanson, 1980).

Covalent		Salts and Complexes		
Reduced	**Oxidised**	**Enzyme Conformation**	**Redox**	**Uncertain**
N	P	K	Fe	B
S	S	Na	Cu	Cl
		Mg	Mo	
		Ca	Mn	
		Mn		
		Zn		

In field conditions, the elements Cu, Ni and Zn are the micronutrients most commonly implicated in toxicity. In glasshouse experiments, yields of maize (*Zea mays*) grown in certain soils formed from ultramafic parent materials (oxisols) in New Caledonia were negatively correlated with shoot Ni concentrations ($r = 0.78$). In addition, several crop plants grown on lowland soils under reducing conditions had higher Ni concentrations than those in well drained control soils (L'Huillier and Edighoffer, 1996).

In addition to growth impairment, plants exposed to either extreme may exhibit such characteristic symptoms as leaf deformation, discolouration or necrosis. Manganese is an example of an essential element that is deficient in some soils although, under reducing conditions, it may also be taken up in sufficient amounts to produce concentrations toxic to certain plants. In manganese-deficient plants, a chlorosis or yellowing occurs between the veins of the leaves while toxic concentrations lead to an uneven distribution of chlorophyll in the leaves and with brown spots mostly comprising precipitated manganese oxides (Mengel and Kirkby, 1987).

Because of the basic biochemical similarities of all living organisms, most elements required for the nutrition of higher plants are also needed by the lower plants and the animals. There are, however, some exceptions to this. Selenium is required by animals but not by plants and, conversely, boron is essential to the nutrition of higher plants but is not known to be required by animals. Cobalt is needed by animals, the bacteria and actinomycetes involved in nitrogen fixation and some fungi but perhaps not by higher plants. Nutrient supply is rarely optimal in nature and one or more of the nutrients listed above probably limit the growth and productivity of organisms in most situations.

Positive correlations occur between the concentrations of nutrient elements at the levels of both the tissues of higher plants and the microbial biomass. Garten (1976) has shown that, in terrestrial plants at least, foliar concentrations of nitrogen and phosphorus are highly correlated, as are those of magnesium and calcium. Singh *et al.* (1991) report correlations exceeding 0.85 among carbon, nitrogen and phosphorus concentrations in the microbial biomass of savanna soils. For the above reasons, correlations may also be expected to occur between a wider range of essential elements.

3.1.2 SPECIFIC ELEMENTS

3.1.2.1 *Carbon*

While not a nutrient element, carbon is a major component of the tissues of all organisms and is one of the elements most closely associated with life. Indicative concentrations (dry weight basis) in a range of organisms are (Matsumoto, 1976; Bowen, 1979; Anderson and Domsch, 1980; Spain, unpublished):

bacteria 43 %;
fungi 32-38 %;
marine algae 28-35 %;
terrestrial higher plants *ca.* 43 %;
Insecta (whole termites) 17-69 % (equivalent to 45-66 % on an ash free basis);
Insecta (whole ants) 45-60 %;
fish 47 %;
man: muscle 67 %, bone 36 %.

Carbon is the vehicle for biological energy transfer within the biosphere at landscape and ecosystem scales, and within organisms. Large stores of this element occur in the lithosphere (including soils), the biosphere, the atmosphere and the hydrosphere. Mass transfers of carbon occur between the lithosphere, the biosphere, the seas and other water bodies and the atmosphere as part of the global carbon cycle (Schlesinger, 1997).

The world's carbon stocks are distributed between four reservoirs (Table I.17) three of which are in dynamic equilibrium with each other. Apart from a small proportion extractable as fossil fuels (estimated mass 4000 Pg), the large geological reservoir ($65.6*10^6$ Pg) may be considered as inert (Eswaran *et al.*, 1993). The oceans comprise *ca.* 85 % of the dynamic stocks while the atmosphere contains less than 2 %. However, it should be noted that fossil fuel carbon stocks are more than five times the size of the atmospheric pool.

Table I.17 Estimated stocks in the world's dynamic carbon and nitrogen reservoirs (Schlesinger, 1997; Eswaran *et al.*, 1993; Davidson, 1994; Batjes, 1996) (Tg = g 10^{12}, Pg = g 10^{15}).

Reservoir		Carbon Stocks (Pg)	Nitrogen Stocks
1.	Atmosphere	750	3.866×10^{21} g
2.	Oceanic	38 000	6000 Tg
3.	Surface and near-surface terrestrial		
	Vegetation biomass	560	10 Pg
	Microbial biomass	10	2 Pg
	Fossil fuels	4000	-
	Soil organic matter (0-100 cm depth)	1184-2293	95-140 Pg
	Soil organic matter (0-200 cm depth)	2376-2456	-
	Soil carbonate carbon (0-100 cm depth)	695-748	-

The carbon in the world's vegetation biomass (Post *et al.*, 1992) is estimated to be about 1.2 % of the total dynamic pool, approximately 37 % of the soil reservoir and slightly less than that of the atmosphere. Organic carbon in the top 100 cm of the world's soils appears to be close to 1500 Pg (Eswaran *et al.*, 1993; Batjes, 1996) while an estimated further 900 Pg may occur between 100 and 200 cm (Batjes, 1996). Estimates for the carbonate carbon of the top 100 cm of the soil range from 720-930 Pg (Batjes, 1996). Total soil carbon stocks therefore represent approximately 6.9 % of the carbon contained in all except the largely-inert geological pool.

Figure I.30 presents the major compartments and flows occurring at an ecosystem scale. The major carbon stores are the above-ground plant structures, the litter layer and the soil store, including plant roots. Within the soil, the biomass comprises the living roots, the microbial biomass and the meso- and macro-fauna. The soil organic matter component consists of dead materials of a range of ages and degrees of decomposition varying from recognisable plant, microbial and animal remains to the highly-stable humic compounds so important for soil stability and functioning. Carbon is transferred onto the soil surface and into the mineral soil horizons through litter fall, root death and decomposition and in leachates and exudates of various types. Litter fall includes dead leaves, bark, stems and reproductive structures originating from the higher plants; it also includes faecal materials, secretory products and the dead bodies of the above-ground fauna. Carbon losses include the carbon dioxide respired by all sections of the biota and by abiotic oxidative processes (Warneke *et al.*, 1999), that in dissolved and particulate organic carbon exported off-site in deep drainage, overland or subsurface lateral flows.

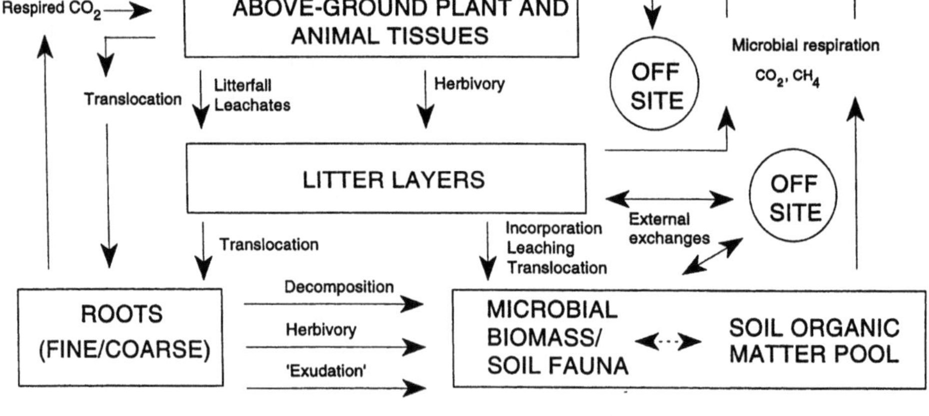

Figure I.30 Simplified short-term ecosystem-level model of the terrestrial part of the carbon cycle.

Soil carbon stocks are reduced by 20 to 40 % following clearing and cultivation and much of this is lost in the first two to five years, depending on climate (Davidson and Ackerman, 1993). The stocks that remained after cultivation formed a constant fraction of the pre-clearing amounts in the largely North American studies collated by these authors. Major off-site exports also occur through biomass harvesting in cropped agricultural and forest plantation systems.

Carbon is present within soils in all three phases and in the biota; its chemical composition, distributions, concentrations and movements within the soil are of great pedogenetic and applied importance. Carbon in soils occurs in an enormous variety of compounds ranging upwards in complexity from carbonates to simple gases to aliphatic acids to large structured macromolecules. It also forms part of a range of poorly-defined humic polymers, often colloidal in nature.

The carbon content is a major factor influencing such factors as the bulk density of the soil mass, as illustrated by Walker and Adams (1958) for 22 New Zealand soils (Figure I.31). However, as shown by its generally favourable effects in promoting soil structural development, this is much more than the simple dilution effect of higher organic matter concentrations. The carbon concentration strongly influences nutrient supply to plants and is important in determining soil stability.

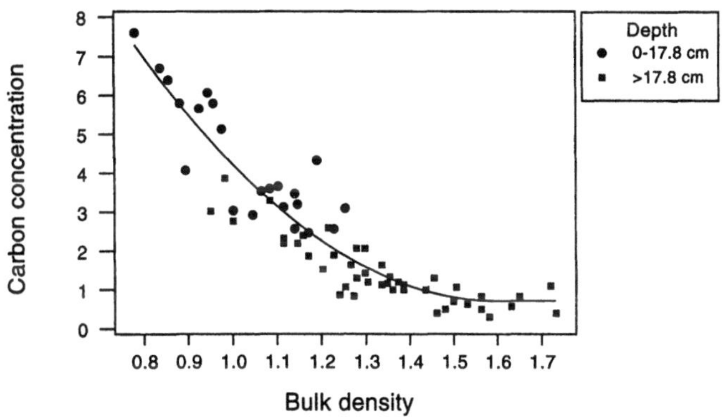

Figure I.31 Relationship between carbon concentration (%) and bulk density (Mg m^{-3}) in selected New Zealand soils (Walker and Adams, 1958).

In the solid phase, carbon occurs predominantly in the living biomass and dead organic matter and in the soils of drier areas as calcium, magnesium and other carbonates precipitated *in situ* and sometimes forming discrete layers or horizons. Soils formed from calcareous parent materials such as certain limestones may also contain substantial carbonates.

Considerable carbon may occur as charcoal in the upper profiles of soils whose vegetation cover has been subject to burning, or be present as bands in soils subjected to periodic ash showers. In addition to its presence as discrete bands in certain soils, finely-divided charcoal occurs widely in almost all Australian soils (Skjemstad *et al.*, 1997) and may be expected to be so in many tropical environments, and wherever biomass burning has been common. Skjemstad *et al.* (1996) estimated that finely-divided charcoal represented up to 0.83 % of the soil mass in three of four Australian soils (two mollisols, an oxisol and a vertisol) and formed a substantial proportion of total soil carbon. Charcoal consists of condensed or cross-linked aromatic rings resistant to microbial and probably chemical decomposition (Skjemstad *et al.*, 1996). It comprises a substantial proportion of the aromatic material less than 53 μm in size and may be confused with partially-decomposed lignin and humic substances. Charcoal is measured as organic carbon in conventional wet chemical estimations and this has considerable implications for studies of carbon turnover in soils.

Carbon comprises *ca.* 43 % of the dry biomass of fresh plant residues although concentrations decline as decomposition proceeds. The terminal products of organic matter decomposition include humic and fulvic acids, a range of carbon-rich organic polymers with high cation exchange capacities and affinities for trace metals. In addition, large quantities of polysaccharides are produced in the rhizosphere although these may have relatively short turnover times. Considerable amounts of carbon are included in the living biomass of the soil external to the roots; estimates of microbial biomass are in the order of 1 to 5 *per cent.* of organic carbon (see, for example, Brookes *et al.*, 1984; Sparling, 1992). However, this excludes the biomass carbon of the mesofauna and the macrofauna, particularly the earthworms, termites and ants, which may represent 10-20 % of the microbial biomass.

As considered earlier, carbon gases, notably carbon dioxide and methane, may occur in the soil environment at concentrations considerably higher than atmospheric. In the liquid phase, dissolved gases and other soluble carbon compounds play important biological and pedogenetic roles in chelation and complexation reactions and thus the transport of metals. Carbonate weathering and leaching will lead to the movement of bicarbonate ions in the groundwaters of environments where these minerals occur in the soil.

The chemical form of the dissolved organic carbon in soil interstitial waters is poorly known. Fulvic acids are generally present in greater concentrations than humic acids while a large proportion of the dissolved organic carbon consists of low-molecular-weight organic acids. Some compositional differences occur between drainage classes and the humic and fulvic acids of the interstitial waters differ from those of bulk soil. Current estimates of the concentrations of dissolved organic carbon in soil interstitial waters lie in the range 5-1000 mg C dm^{-3} (Malcolm, 1993).

Carbon concentrations in soils differ widely reflecting climatic conditions, local ecosystem productivity, the types and amounts of the clay and other minerals present (notably carbonates), internal and external environmental conditions and cultural practices. The organic horizons of peat soils (histosols) may have carbon concentrations similar to that of plant tissues (up to *ca.* 43 %) while those of mineral soils are much lower. Figure I.32a presents the carbon concentrations of the A horizons of 1975 soils from throughout Australia. It has been estimated that more than 75 % of the Australian land surface is covered with soils having less than 1 % organic carbon while soils having high levels of organic carbon are very limited in extent (Spain *et al.*, 1983a). Other less-arid continents have greater proportions of their soils and landscapes in higher concentration classes.

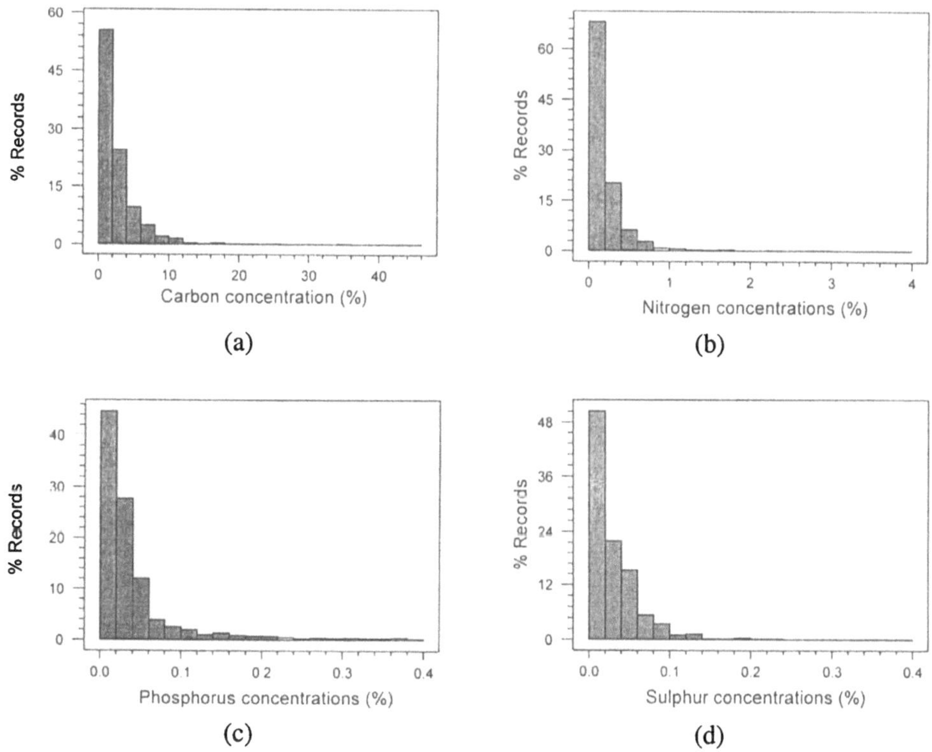

Figure I.32 Elemental concentrations in the A horizons of Australian soils. (a) Carbon, $n = 1975$, (b) Nitrogen, $n = 2099$, (c) Phosphorus, $n = 1788$, (d) Sulphur, $n = 572$.

Because of ecosystem transfer processes, carbon is normally most concentrated near the soil surface and, at a smaller scale, in close proximity to the roots. Figure I.33 presents the depth distributions of carbon in six contrasting soils. That in Figure I.33a is an alfisol formed from sandstone parent materials and supporting a savanna woodland dominated by *Eucalyptus* species (Isbell and Smith, 1976). The low carbon concentrations reflect the semi-arid environmental conditions and the constraints of a reduced nutrient status resulting partly from the considerable ages of these landscapes. The data in Figure I.33b are from a spodosol formed on a beach ridge in the humid (average annual rainfall >3000 mm) tropics of north-eastern Australia (Murtha, 1986). It supports a sclerophyllous vegetation (species of *Acacia, Casuarina* and *Melaleuca*) and illustrates the bleached E horizon and illuvial increase in humic materials and sesquioxides characteristic of the B horizons of these soils. Figure I.33c illustrates the distribution of carbon in an oxisol supporting a tall, species-rich, closed-canopy rainforest in humid (*ca.* 3000 mm) tropical northeastern Queensland (Spain *et al.*, 1989a). The substantial carbon stocks (234 Mg ha^{-1} to 120 cm) of this permeable, strongly-structured soil are related to the high secondary productivity of the rainforest (fine litterfall 8.6 Mg ha^{-1} y^{-1}, 3 year mean) and the iron and aluminium oxide-rich soils. The depth distribution of carbon presented in Figure I.33d is that of a very young andisol from the footslope of a recently-active volcano in Papua New Guinea (Bleeker and Sageman, 1990). The site receives *ca.* 3500 mm of rainfall annually and currently supports a cover of coconuts (*Cocos nucifera*) and cocoa (*Theobroma cacao*). The C:N ratios suggest ready decomposability (Chapter IV.1.3.3). However, the mineralogy of the soil is dominated by vermiculite, volcanic glass, allophane and imogolite which implies that some protection of the organic matter may occur (see Chapter IV.1.3.2). Figure I.33e presents data from a poorly-drained acid histosol supporting a forest cover of Southern Beech (*Nothofagus* spp.) and podocarps (*Podocarpus* spp.) located on the humid (2020 mm annual rainfall) western coast of South Island New Zealand at *ca.* 42°S (Mew and Lee, 1981). The thick (44 cm) peat O horizons (depths conventionally indicated by negative numbers) are underlain by loess and alluvium (largely greywacke and granite). Carbon contents in the upper peat horizons are equivalent to those of fresh plant material but diminish with increasing depth and undergo an abrupt reduction at the transition with the predominantly-mineral horizons. Appreciable carbon still occurs in the upper gleyed mineral horizons and diminishes with depth in the solum, apart from a slight increase in the upper B horizon. C:N ratios (see below) are typically high, up to more than 80 in the O horizon and greater than 53 in all mineral horizons. Figure I.33f presents the depth distribution of carbon concentrations in a vertisol from warm temperate eastern Australia (Ross, 1986). This soil developed on alluvial, slightly-calcareous parent materials beneath an open forest vegetation of *Acacias* with perennial grasses. Ratios are typical of other Australian vertisols and vary little with depth.

Close associations of carbon with nitrogen, sulphur, phosphorus and other nutrient elements occur in dead plant and animal tissues. As discussed later in this book, these tissues must generally be broken down by the soil biota to allow their contained nutrient elements to be recycled.

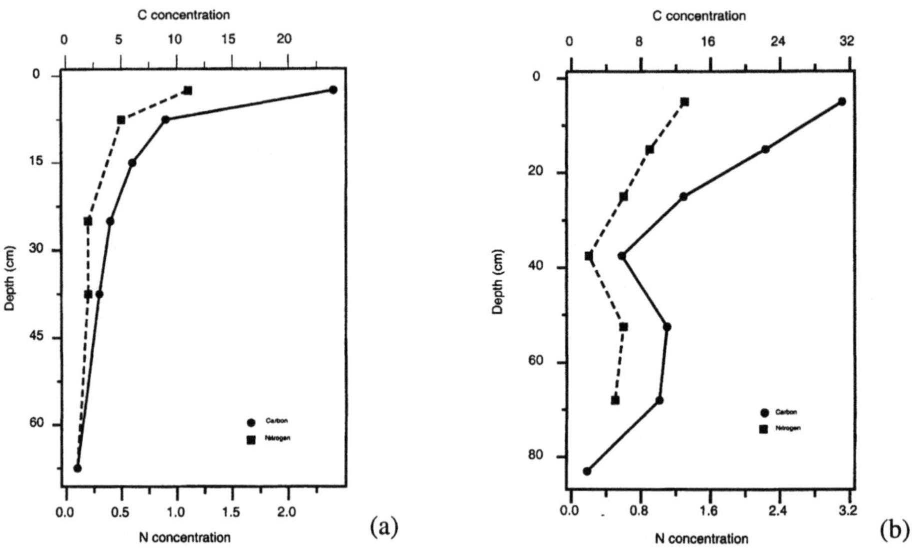

Figure I.33 Depth distributions of carbon (●) and nitrogen (■) concentrations (gkg^{-1})in selected soils: (a) an Australian alfisol, (b) an Australian spodosol.

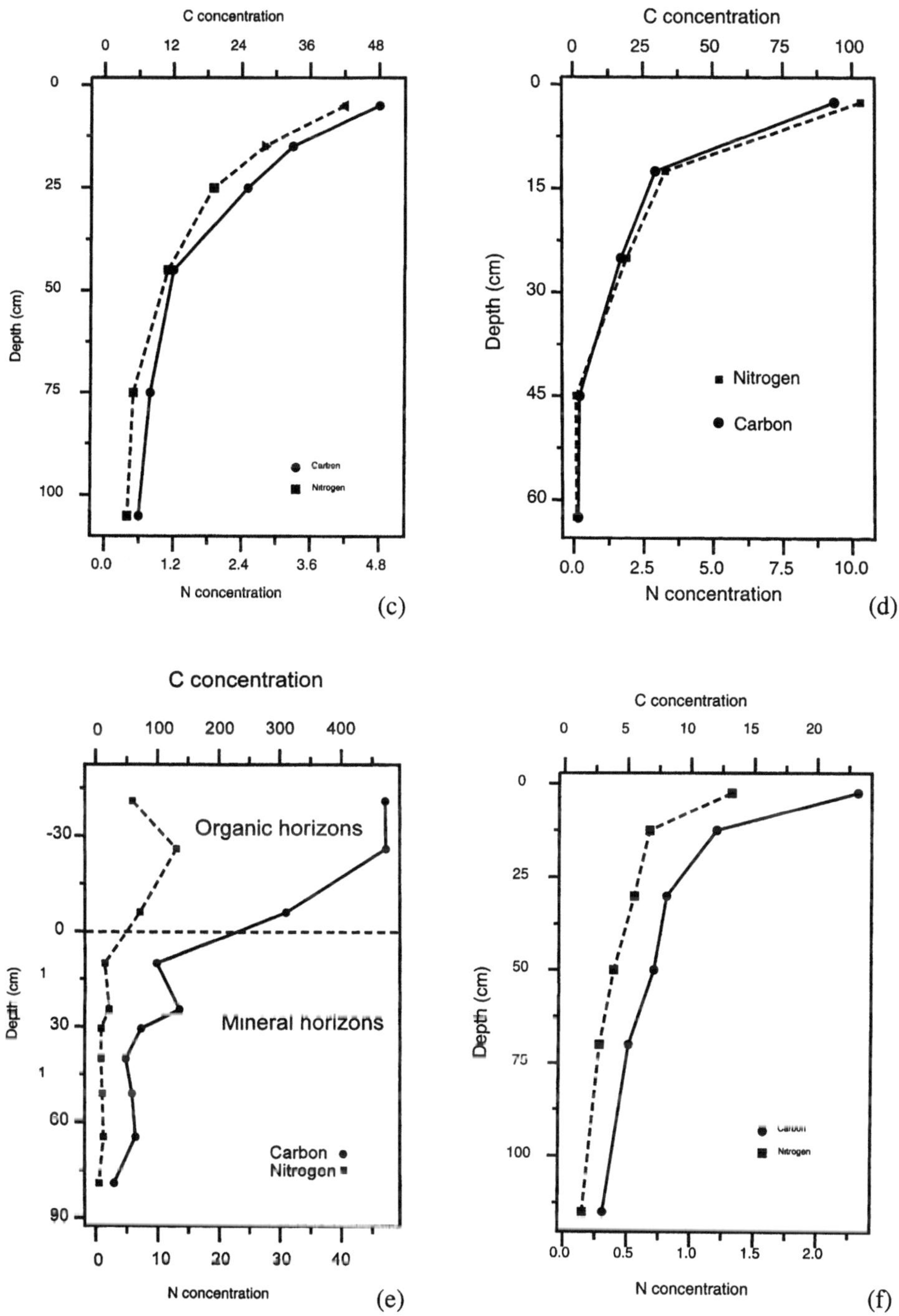

Figure I.33 (cont.) Depth distributions of carbon (●) and nitrogen (■) concentrations in selected soils: (c) an Australian oxisol, (d) an andisol from Papua New Guinea, (e) a histosol from New Zealand, (f) an Australian vertisol.

3.1.2.2 *Nitrogen*

Nitrogen is an element of major biological importance in all terrestrial and aquatic ecosystems. It is the dominant element in the atmosphere (Table I.9) where it occurs principally as dinitrogen gas (N_2) and in low concentrations as the various gaseous oxides and as ammonia. Large reservoirs of these gases also occur in dissolved form (and in suspension), often at saturation concentrations, in the oceans and other water bodies and in living organisms. Substantial amounts of nitrogen also occur in the lithosphere, in coal, shales and in liquid and gaseous hydrocarbons. Table I.17 contrasts the estimated sizes of the world's major pools of carbon and nitrogen. The overall carbon: nitrogen ratios in the vegetation biomass, the top 100 cm of the soil and the microbial biomass are thus 55, 8-13 and 5, respectively.

As part of the global nitrogen cycle, constant exchanges occur between the atmosphere, the biosphere and the oceans and other water bodies. Ecosystem level cycles occur in all terrestrial environments and internal cycles also occur in such large water bodies as the oceans (Schlesinger, 1997).

Substantial modification to the global nitrogen cycle has occurred in recent years through mans' activities. These include a greater than doubling of the rate of nitrogen transfer from the atmosphere to biologically-available terrestrial pools (Vitousek *et al.*, 1997) and a decoupling of the close linkages between the carbon and nitrogen cycles (Asner *et al.*, 1997). Very high rates of nitrogen deposition in certain North American and European environments (Holland and Lamarque, 1997) have contributed to soil acidification, nutrient element impoverishment, ecosystem degradation and damage to downstream environments (Vitousek *et al.*, 1997).

Nitrogen is critical to all forms of life. It is an essential component of such basic biological building blocks as amino acids and thus their polymers: peptides and proteins. In the micro-organisms, nitrogen-containing materials such as chitin (polymers of amino sugars) are major cell wall constituents of the fungi and the bacteria. This polymer is also a major component of the endo- and exo-skeleton of the arthropods. Nitrogen is a component of the nucleic acids DNA and RNA and is thus of importance in reproduction and evolution. As an important constituent of many biologically-active compounds, nitrogen plays a part in the integration of life at levels from the cell to the population. There is no major category of biological function in which nitrogen does not play a major part.

The productivities of micro-organisms, higher plants and animals are commonly limited by low nitrogen concentrations (Vitousek and Howarth, 1991). Nitrogen is found in soil mainly within the organic matter fraction where it occurs largely in humic compounds but also in plant roots, the microbial biomass and in decomposing organic materials. While low concentrations of inorganic nitrogen (nitrates, ammonium) are found in most soils, their turnover rates are usually high and seasonally-related changes follow from variation in environmental conditions and thus micro-organism activity. A variety of other nitrogenous compounds are present in soils and include dinitrogen gas, the gaseous oxides of nitrogen, ammonia and other excretory products.

The nitrogen concentrations of higher plants range from less than 0.10 % of dry weight in woody tissues to more than 4 % in actively-growing tissues (Table I.15) and even higher in seeds. Strong ontogenetic changes occur with much reduced nitrogen concentrations characteristic of older tissues and minimal concentrations in litter. In most environments, tissue nitrogen concentrations are higher in those species that have nitrogen-fixing symbionts associated with their roots than in those not so endowed. As estimated from laboratory cultures, mean tissue nitrogen concentrations of ten species of soil bacteria were 7.7 % while those of fourteen species of soil fungi ranged from 4.4 to 4.75 %, all on a dry weight basis (Anderson and Domsch, 1980). Values in nature may be much lower and Foster (1994) considers that most soil micro-organisms are in a constant state of near starvation. N concentrations in the fruiting bodies of a number of field-collected fungi ranged from 1.5 to 7 % of dry weight while mycelial values as low as 0.68 have been reported from wood-rotting Basidiomycetes (Cowling and Merrill, 1966). Whole termites have N concentrations of the order of 3.5-12.3 % of dry biomass (Matsumoto, 1976; Spain, unpublished) and ants 9.2-12.7 % of dry biomass (Spain, unpublished). In man, nitrogen concentrations in soft tissues are approximately 7.2 % while bone contains 4.3 %; hair and nail tissues are somewhat higher at 15-16 %, again on a dry weight basis (Bowen, 1979).

Considerable energy is required to break the N≡N bond in dinitrogen gas, whether this is carried out by organisms, or in the production of nitrogenous fertilisers. Thus, the large pool of atmospheric nitrogen is largely unavailable directly to higher plants and can only be obtained through fixation by lightning and the ability of certain symbiotic, associative and free-living prokaryotes to fix it into useable forms. It is also unavailable directly to most animals. However, N-fixing organisms exist in the intestinal tracts of many termites (Slaytor and Chappell, 1994) (Chapter III.4.3.2.1) and certain other insects and some evidence exists for N-fixation in the gut of certain earthworms (Barois *et al.*, 1987). The advantages of these symbioses to the host animal are clear given the very high C:N ratios of the diets of such animals as termites.

Despite the fixation of atmospheric nitrogen by prokaryote organisms, most of the nitrogen required for plant growth in terrestrial ecosystems must be supplied through faunally, microbially - and abiotically - mediated decomposition processes (Amundson and Davidson, 1990), or through fertiliser application. The nitrogen in plant tissues occurs principally as protein. On tissue death, these proteins are initially broken down through proteolysis and mineralised to yield ammonium ions. These ions may be adsorbed onto the internal or external surfaces of clays, assimilated by plants and micro-organisms or volatilised after conversion to ammonia gas. The rest largely undergoes nitrification by aerobic chemautotrophs (or, under certain conditions, by heterotrophs) to produce nitrite (NO_2^-) ions which may be reduced to nitrous oxide (N_2O) or oxidised further to nitrate ions (NO_3^-). The nitrate thus formed may move out of the soil system, be taken up by plants or micro organisms or broken down through anaerobic denitrification processes (biotic but also abiotic in acid environments) to give gaseous oxides of nitrogen and dinitrogen gas (Figure I.34).

Higher plants and many micro-organisms may take up inorganic nitrogen either as NH_4^+ or as NO_3^- although micro-organisms preferentially take up inorganic nitrogen as NH_4^+. Certain groups of higher plants may favour a specific form, while others appear to be indifferent. There is evidence that amino acids may form an important source of nitrogen for plants which may therefore compete with soil micro-organisms (Chapin *et al.*, 1993; Jones and Darrah, 1994). Micro-organisms may also take up nitrogen in organic form. For example, all fungi appear able to utilise some organic nitrogen sources while certain plant pathogenic fungi may be unable to utilise inorganic forms (nitrate or ammonium) (Jennings, 1989). Ericoid, some ectomycorrhizal and other fungi (notably wood-rotting species) produce diffusible enzymes around the roots and assimilate the lower molecular weight breakdown products of protein degradation (Read *et al.*, 1989) to meet their nitrogen (and sulphur) requirements (Chapter IV.3.1.2.2).

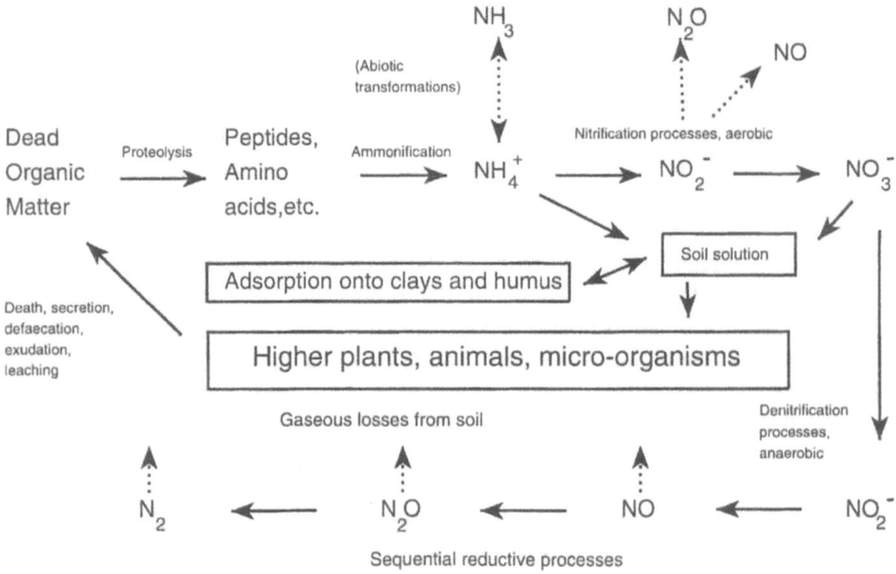

Figure I.34 Major biological nitrogen transformations associated with decomposing organic matter.

Nitrogen transformations (and other plant assimilation processes) have clearly different effects on soil pH. Assimilation of nitrogen as NH_4^+ (and other positively charged ions) by both plants and micro-organisms leads to proton production and soil acidification, while assimilation as NO_3^- results in hydroxyl or bicarbonate ion production. European examples of the former group (acidophiles) are pines (*Pinus* spp.) and spruces (*Picea* spp.); these species are well adapted to acid soils and take up their nitrogen as NH_4^+. In contrast, neutrophiles are species which are adapted to less acid soils and take up nitrogen as nitrate; they include the elms (*Ulmus* spp.), birches (*Fraxinus* spp.) and the maples (*Acer* spp.). The oaks (*Quercus* spp.) and the beeches (*Fagus* spp.) appear less exigent in terms of soil pH and may take up both forms of nitrogen (Duchaufour, 1997). Overall, nitrification and denitrification processes also have, respectively, acidifying and neutralising effects (Sprent, 1987).

The mobilities of nitrate and ammonium ions in soils differ substantially. While ammonium ions are held at the negatively charged surfaces of clays and organic colloids, nitrate does not interact appreciably with the exchange complex and moves readily in the soil solution. It is thus susceptible to loss off site through deep leaching or in overland and lateral subsurface flows. In soils subject to excessive nitrogenous fertiliser application, nitrates may leach through the soil to depths below the range of the roots and contaminate local aquifers. Nitrate is toxic in excess and health problems may ensue when contaminated waters are consumed by humans or domestic stock (Spalding and Exner, 1993) (see also IV.2.6.2).

Substantial quantities of nitrogen may be lost from ecosystems in other forms and losses of NH_3 through volatilisation and from living plants, and of nitric oxide (NO) and nitrous oxide (N_2O) during nitrification may all be significant. Denitrification occurs in anaerobic soils and in anaerobic microsites within aggregates, even within otherwise well-aerated soils (Figure I.15) leading to the loss of denitrification gases (NO, N_2O, N_2). Such 'greenhouse' gases are of importance in global climate change scenarios.

Ammonia volatilisation occurs through the hydrolysis of ammonium compounds and may be a source of major nitrogen losses where such compounds are applied as fertilisers to agricultural systems, particularly in alkaline soils. More than 50 *per cent.* of the nitrogen applied may commonly be lost, depending on the chemical form of the fertiliser and the depth of its application, the temperature and the buffer capacity of the soil (Peoples *et al.*, 1995). Nitrogen losses from animal secreta and from alkaline areas where nitrogenous fertilisers have been applied can be particularly severe in certain grazing systems (Jenkinson, 1990). In environments where biomass burning is frequent, nitrogen losses through volatilisation are also likely to be substantial (Vitousek and Howarth, 1991).

Since *ca.* 90 % of soil nitrogen reserves are organic (Sowden *et al.*, 1977), the distributions of nitrogen and carbon are usually closely correlated. This is apparent in the similarities of the frequency distributions of Figures 1.32a and 1.32b which compare the frequency distributions between carbon and nitrogen in the A horizons of 2099 Australian surface soils. Figures 1.33a-f present the related depth distributions of carbon and nitrogen in the six selected soils discussed in the previous section. In most of these, the general nature of the concentration decline with depth is similar although the C:N ratio clearly alters with depth in the alfisol and the spodosol. In contrast, the histosol has a characteristically higher C:N ratio throughout although it varies somewhat within both the peaty O and the mineral-dominated horizons.

Soil organic nitrogen comprises *ca.* 40 % proteinaceous materials (proteins, peptides, amino acids), 5-6 % amino sugars, *ca.* 35 % heterocyclic compounds (including purines and pyrimidines) and *ca.* 19 % NH_4-N (Schulten and Schnitzer, 1998). Some geographic variation has been detected with, for example, amino acid and amino sugar N forming a greater proportion of total N in soils from cooler areas (Sowden *et al.*, 1977). Much soil organic N appears to be protected in associations with humic materials, clay or the hydrous oxide minerals.

3.1.2.3 *Phosphorus*

Phosphorus is a further major nutrient element required by all forms of life. It plays an important role in energy metabolism (through its involvement in ATP and ADP) and in a range of metabolic pathways. It is a structural component of coenzymes, phosphoproteins and phospholipids; as a component of nucleic acids, it forms a part of the genetic information system. In higher animals it plays an important additional structural role in skeletal tissues and *ca.* 80 *per cent.* of body phosphorus occurs in teeth and bones (McDonald *et al.*, 1988). Deficiencies in phosphorus limit the productivities of many biological systems and phosphatic fertilisers are widely applied to agricultural systems to compensate for this.

Concentrations of phosphorus in the tissues of higher plants range up to more than 1 % of tissue dry weight (Bowen, 1979) although typical values are generally much lower than this (Tables I.14, I.15). Phosphorus concentrations differ substantially between species, between tissues within species and with organ development. Because phosphorus is relatively mobile within the plant, it is most concentrated in the biologically-active tissues to which it is actively translocated from senescing leaves. Plant tissue concentrations also depend on the readily plant-available phosphorus present in the soil.

Within the lower plants, uncontaminated bryophyte tissues contain 0.07-0.2 *per cent.* phosphorus while bacteria may contain much higher concentrations, up to 7.7 *per cent.* of dry weight in laboratory cultures (Anderson and Domsch, 1980). Field collected fungi, principally Basidiomycetales, typically have lower concentrations ranging from 0.10 to 1.4 *per cent.* of dry weight (Beever and Burns, 1980).

In the higher animals, phosphorus concentrations differ notably between hard and soft tissues. Bone concentrations in man range from 6 to 7 *per cent.* of dry weight while those of a range of organs vary from 0.3 to 0.9 *per cent.* (Bowen, 1979).

While phosphorus has no gaseous phase in its biogeochemical cycle, dry and wet deposition from the atmosphere may be in the range 0.07 to 1.7 kg ha^{-1} y^{-1} (Newman, 1995). Thus, within an ecosystem, phosphorus must be supplied largely through organic matter decomposition, turnover of the microbial biomass, weathering of the parent material or, in agro-ecosystems, as fertilisers. Therefore, the efficient cycling of phosphorus assumes great importance in natural ecosystems, particularly in older landscapes where soils have been weathered to depths below the zone occupied by the roots. Phosphorus plays a particularly important part in pedogenesis (Walker and Syers, 1976).

A wide range of phosphorus concentrations occurs in soils, depending principally on the parent materials (Walker and Adams, 1958) and the age of the landscape (Walker and Syers, 1976) or the degree of soil development (Cross and Schlesinger, 1995). For example, soils formed from basic igneous rock parent materials usually contain higher concentrations of phosphorus (Norrish and Rosser, 1983). A factor in long term phosphorus retention in soils and the unavailability to plants of much of that present, is the occurrence of other elements with which phosphorus forms insoluble associations, notably iron, aluminium, manganese and calcium and other carbonates. The forms in which inorganic phosphorus occurs in soils and their different availabilities depend much on pH (see Figure I.35, and below) and the other minerals present in the soil; Brady and Weil (1996) discusses this from an agronomic perspective. Plant-available forms are principally $H_2PO_4^-$ and HPO_4^{2-} (Table I.14).

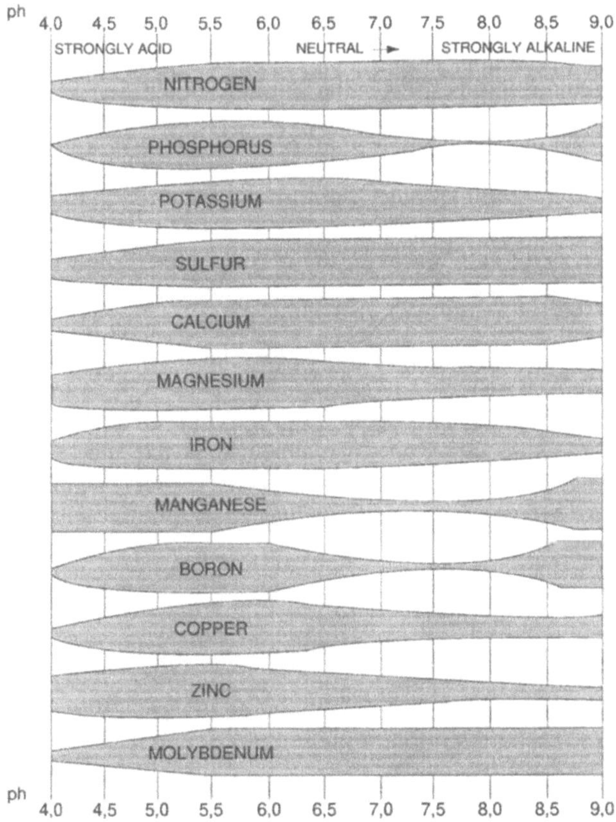

Figure I.35 Relationships between soil pH and the availabilities of selected nutrient elements in organic soils (Lucas and Davies, 1961).

Figure I.32c presents the concentrations of total phosphorus in the A1 horizons of 1788 Australian soils. It can be seen that most of these soils have very low concentrations; continents with younger landscapes and wider distributions of basic igneous rocks are likely to have a greater proportion of their soils in higher concentration classes.

Much of the phosphorus in non-agricultural soils is organic and this form of phosphorus may assume very high proportions of total phosphorus in highly-weathered soils. Singh *et al.* (1991) reported that organic phosphorus comprised 98 *per cent.* of total phosphorus in the highly weathered soils of an Indian savanna environment. In a contrasting climate, Walker and Adams (1958) found that, over 22 different New Zealand soils and at depths to 53 cm, organic phosphorus was highly correlated with carbon ($r = 0.79$) and comprised a large part (median 67 %, interquartile range 52-80 %) of total (organic plus inorganic) phosphorus. Not surprisingly, the lowest ratios were found in a few soils containing phosphorus-rich limestone. Further, the amount of organic phosphorus declined from a median value of 78 *per cent.* in the surface 18 cm to 59 *per cent.*

at a depth of 36-53 cm. Over all soils and depths in the Walker and Adams data set, organic phosphorus was highly correlated with sulphur ($r = 0.86$) and nitrogen ($r = 0.86$). However, there also appeared to be separate relationships between organic carbon and phosphorus at different depths in the profile.

The individual components of the organic phosphorus fraction of soils are poorly known; most (10-50 *per cent.*) occurs as inositol phosphates, probably of microbial origin while phospholipids and nucleic acids account for 1-5 and 0.2-2.5 *per cent.*, respectively. As considered earlier, from 2-20 *per cent.* of the organic phosphorus in soils may be contained within the microbial biomass (Brookes *et al.*, 1984).

As indicated above, much of the inorganic phosphorus in soils is occluded in minerals or occurs in insoluble forms unavailable to higher plants. The phosphorus available to these plants appears to be largely taken up in inorganic form from the soil solution, although a small proportion is also taken up in low molecular weight organic forms. The soil solution inorganic phosphorus is maintained by the mineralisation of organic phosphorus from soil organic matter and decaying plant residues and through the turnover of soil microbial populations. This may be augmented by the slow solubilisation of some inorganic forms (McBride, 1994). However, in most soils, much of the inorganic phosphorus added as fertiliser quickly becomes unavailable through fixation processes.

Phosphorus fixation refers to its removal from solution through chemisorption and precipitation processes (McBride, 1994) and its subsequent inclusion in insoluble inorganic forms, very largely unavailable to plants. The forms in which phosphorus is fixed differ with pH, although the forms present at any pH value may overlap those present at adjacent values. At very acid values, fixation mainly occurs through chemical reaction with soluble Fe, Al and Mn while at slightly higher pH, fixation occurs on the surfaces of the hydrous oxides of these elements. Solubility, and thus availability to organisms, is greatest at slightly acid to neutral pH values while at alkaline values, phosphorus is increasingly fixed as calcium phosphates (Brady and Weil, 1996).

Certain forms of soil phosphorus may be preferentially available to some soil micro-organisms over the higher plants. Fungi can grow towards sources of phosphorus and can also utilise a wider range of forms including condensed phosphates, insoluble phosphates and organic forms. Some fungi, together with certain bacteria and actinomycetes are able to solubilise otherwise insoluble phosphates through the secretion of organic acids and phosphatase enzymes (Jennings, 1989, 1995).

The phosphorus in soils that is available to plants is particularly difficult to measure in a direct way. A range of methods has been developed which involve leaching the soil with various chemical extractants. While some of these may provide agronomically-useful indexes of availability under restricted circumstances, the variety of chemical forms of soil phosphorus, the different plant rooting strategies and the complexities involved in defining the interactions with mycorrhizal and associated organisms suggest that a simple universal chemical assay is unlikely to be found. Because of this, approaches that determine foliar phosphorus concentrations (and sometimes those of other elements) are often used, particularly in tree cropping situations.

3.1.2.4 *Sulphur*

Sulphur is an essential major element for all living organisms and plays important roles in a number of metabolic pathways. It is a component of the essential amino acids cystine, cysteine and methionine and thus forms a structural part of tissue proteins and enzymes important in photosynthesis and nitrogen fixation. In higher animals, it also occurs in certain vitamins and in the hormone insulin (McDonald *et al.*, 1988).

In soils, sulphur occurs in inorganic form as sulphates, sulphides, elemental sulphur, sulphur-containing gases and a variety of other forms, depending on the climate, soil pH, the degree of soil aeration and biological activity. Concentrations depend on the parent materials and are generally higher in soils formed from basaltic than other parent materials. Total soil sulphur concentrations in surface soils are usually in the range 10-1600 $\mu g\ g^{-1}$ (Table I.2). Figure I.32d presents the frequency distribution of the concentrations of total S in the surface horizons of 572 Australian soils while Table I.15 presents typical concentrations in plant materials from a range of natural and semi-natural environments.

In most soils, sulphur occurs largely in organic form and this may be subdivided between that associated with the soil organic matter and that with decaying plant residues. As with the other essential nutrient elements, a further small labile pool occurs in the microbial biomass which may be either a source or a sink of sulphur. In fourteen temperate-climate acid soils this pool ranged between 0.7 and 1.9 *per cent.* of microbial biomass carbon with an average of 1.2 *per cent.* (Wu *et al.*, 1994); Banerjee *et al.* (1993) found that this pool comprised slightly higher values, *ca.* 3 *per cent.* of microbial biomass carbon. Organic sulphur is mineralised to sulphate which is the form normally taken up by plants. The sulphate ion is relatively mobile in soils and may be subject to leaching losses. Sulphur dioxide (SO_2), a common atmospheric pollutant, is absorbed through the leaf surfaces and while it may contribute to plant nutrition at low concentrations, it quickly becomes toxic with increasing concentration.

In arid and semi-arid areas, sulphate sulphur may accumulate in soils as gypsum ($CaSO_4.2H_2O$), epsomite ($MgSO_4.7H_2O$) or as thenardite (Na_2SO_4) and other minerals. Toward the drier end of the climatic spectrum, gypsum may form cemented horizons at varying depths from the surface. More complex sulphur salts such as the jarosites (*e.g.*, $KFe_3(SO_4)_2(OH)_6$) may occur in certain environments depending on the pH and other factors.

In contrast, in wetland soils and those that are submerged frequently enough to become effectively anaerobic, inorganic sulphur may occur largely in reduced form as sulphides such as pyrite (FeS_2) and in other reduced forms. Hydrogen sulphide (H_2S) is one of the characteristic end products of the mineralisation of organic sulphur under such conditions and is largely responsible for the distinctive odour associated with such wetland environments as mangroves communities. This gas is also produced in aerobic environments although it is quickly oxidised to elemental sulphur. Other sulphur containing gases such as CS_2 and COS are also produced in soils through microbial activity.

Unlike that of phosphorus but like nitrogen, the biogeochemical cycle of sulphur has a significant gaseous component. The importance of this component has recently been emphasised through the effects on the atmosphere of fossil fuel consumption and the emissions from certain industrial processes. These have lead to increased atmospheric concentrations of sulphur (and nitrogen) and increased deposition of acid forms of these elements in both particulate form and in solution, particularly in areas associated with industrialisation. In such areas, a major consequence has been the reduction of rainfall pH to low levels, the 'acid rain' phenomenon, although this is also due to inputs of nitrogen-containing acids.

When sulphur-rich wetland environments are drained during the processes of land reclamation, the sulphur present in reduced form may be oxidised to sulphuric acid leading to a drastic reduction in pH and the solubilisation and mobilisation of metals (Dent and Pons, 1995). The effects of sulphur oxidation may be severe. The acid-sulphate soils that result from such acidification are particularly difficult to manage and environmental damage, including fish kills, may be incurred in down stream aquatic ecosystems. A similar situation frequently follows exposure at the surface of sulphide-containing overburden and processing waste materials associated with the mining of many minerals of commercial importance (see, for example, Ripley *et al.*, 1996).

While much of the sulphur needed by agricultural crops is provided through the mineralisation of organic matter, low sulphur concentrations commonly limit the growth of crops and pastures. Sulphur adequate for the needs of many crops is supplied through the application of such phosphatic fertilisers as superphosphate. In situations of intensive production, however, regular sulphur additions may be required to maintain satisfactory levels of productivity. The application of inorganic sulphur has an acidifying effect on soils; elemental sulphur may sometimes be applied to alkaline soils to reduce their pH.

3.1.2.5 *Calcium and magnesium*

Calcium and magnesium are bivalent elements belonging to the second group of the periodic table, the alkaline earth elements; a group whose hydroxides form strong bases. They are amongst the few most concentrated elements in both the earth's crust (Table I.1) and in soils (Table I.2).

Calcium is one of the soil macro-elements and median concentrations are *ca.* 1.5 % in non-calcareous soils. In contrast, airborne calcium inputs may lead to the formation of discrete horizons of calcium and magnesium carbonates or gypsum ($CaSO_4 . 2H_2O$) in the calcareous soils of semi-arid and arid areas (Birkeland, 1984). Surface efflorescences or crusts dominated by calcium may form in depressions, in areas where subsurface lateral flows emerge to the surface or through the capillary rise of saline waters. Shallow calcareous soils also form from such calcium-rich parent materials as limestone and chalk.

Both elements have essential roles in the nutrition of living organisms and, through a strong association with soil pH status, calcium concentrations (in particular) strongly affect the availability of other elements to plants. Both elements may strongly influence soil structural stability.

Roles of calcium in plants and animals
Calcium occurs at a median level of *ca.* 0.5 % by dry weight in higher plant tissues while maximum concentrations in excess of 9 % are known in some calcicole species (Table I.15). Monocotyledonous plants require lesser quantities than dicotyledons. Most of the calcium in higher plant tissues occurs in exchangeable form associated with the cell walls or the plasma membrane; intra-cellular calcium is most concentrated in the mitochondria and leaf vacuoles (Marschner, 1995).

As a Group III element, calcium functions to regulate ion balance and osmotic potential in plants. In addition to this, it has a major role in stabilising membranes and maintaining cellular integrity. Calcium also has an important role in stabilising cell walls and is essential to root elongation. It also has a minor role in activating a number of enzymes.

Calcium is poorly mobile within the phloem, in cell-to-cell movement and between tissues. Because of its low mobility in the phloem, the calcium needs of roots must be largely supplied from the soil solution surrounding the root tip (Marschner, 1995); this requirement may limit the penetration of the subsoil by plant roots in acid soils. Little calcium is withdrawn from leaf tissues prior to abscission and forest and other litters may often have higher concentrations of this element than the foliage from which they were derived.

In higher animals, calcium comprises a major proportion of the structural tissues (bone, teeth) and occurs in notable concentrations in the blood plasma. It plays important enzymic roles in the transmission of nerve impulses and in muscle contraction. It is also important in the coagulation of blood. In some soil animals (*e.g.*, Isopoda, Amphipoda, millipedes), calcium is an important structural component of the exoskeleton; calcium concentrations strongly influence the distribution of earthworms (Lee, 1985) and other soil organisms.

Roles of magnesium in plants and animals
Magnesium normally occurs in lower concentrations than calcium in the earth's crust, soils, higher plants and other terrestrial organisms. Typical concentrations in the earth's crust are about 1.3 % while non-calcareous soils contain *ca.* 0.5 %; concentrations in higher plants are *ca.* 0.2 % with maximum values up to approximately 1 % of dried weight.

In addition to its important role in controlling intra cellular osmotic potential, magnesium is the central atom of the chlorophyll molecule and is therefore critical to photosynthesis, and in protein synthesis. It is also responsible for activating a wide range of enzymes and, together with calcium, has a structural role in stabilising the cell wall (Marschner, 1995). In contrast to calcium, magnesium is highly mobile in the phloem.

In higher animals, magnesium is associated with calcium and phosphorus and *ca.* 70 % of the magnesium in the body is found in the skeleton (McDonald *et al.*, 1988). As in plants, magnesium plays an important role in activating a number of enzymes.

Roles of calcium and magnesium in soils
In contrast to its immobility in plant tissues, calcium is relatively mobile in soils within which it moves largely by mass flow. Encrustations around plant roots testify to its movement in the mass flow of water and its contained elements caused by the transpiration stream.

Calcium dominates the exchange complex of most well-structured soils where it exists in an equilibrium state with the soil solution. Site losses occur through erosion, illuviation, leaching, subsurface lateral flows, and plant uptake. It is the dominant metallic element in areas dominated by silicate rock parent materials (Schlesinger, 1997).

Soil pH is strongly influenced by calcium concentration and it therefore indirectly affects the supply of a number of nutrient elements. At high pH, for example, the availability of many of nutrient elements is severely reduced. In particular, phosphorus availability is substantially reduced at pH values above 7, largely by fixation on inorganic calcium minerals. Calcium promotes the flocculation of soil colloids and in agricultural and other environments where it dominates the exchange complex, a well-developed aggregate structure favourable to aeration and water infiltration normally occurs. In contrast, magnesium (together with sodium) may have a dispersive effect.

Applied aspects

Because of the acid-producing effects of nutrient uptake, the removal of plant parts in crops and the losses of calcium through erosion, leaching and the use of acidic fertilisers, intensively-used agricultural lands tend to become progressively acidified. The application of calcium as agricultural lime to raise soil pH is therefore a widespread and regularly-required feature of intensive agricultural practice. The primary purpose of this is to raise the soil pH to the range in which most nutrient elements are readily available (Figure I.35) although the calcium supplied is also available as a nutrient element.

Calcium may be applied to the soil as agricultural lime in carbonate, oxide or hydroxide form. Crushed limestone (largely $CaCO_3$) is the most frequently used medium; limestones may also contain a variable proportion of dolomite ($CaMg(CO_3)_2$), depending on their sources. Calcium is also applied to soils as gypsum ($CaSO_4.2H_2O$) to alleviate the high dispersivities and often crusted states characteristic of sodic soils. The dispersive nature of these soils is reduced by replacing a proportion of the sodium ions with Ca^{2+} leading to improved flocculation of the soil colloids, lessened dispersivities and a reduced tendency to form surface crusts.

3.1.2.6 *Potassium*

Potassium is required by all living organisms, normally in relatively large amounts. Its major roles in plants are in pH stabilisation, osmoregulation, enzyme activation and membrane transport processes; these impact on protein synthesis, photosynthesis and many other basic processes at both the cellular and physiological levels (Marschner, 1995). Potassium is very mobile in plants at the level of the cells, the tissues and at the longer distance transport level of the phloem and the xylem; it is thus readily redistributed in response to changing metabolic requirements. Leaves usually have higher concentrations than the other plant parts (Table I.15) and potassium is, after calcium, the most abundant metallic element in the cytoplasm of the leaves. Fertilised crop plants in particular may take up substantially more than their metabolic requirements, that is, so-called 'luxury' consumption. Potassium is not an important component of the structural tissues.

In animals, potassium similarly plays important roles in osmoregulation and in maintaining the acid:base balance. It is also necessary for nervous conduction, muscle excitability and carbohydrate metabolism. It is the major cation within cells in contrast to sodium, which is the major cation at the physiological level (McDonald *et al.*, 1988).

In soils, potassium enters the soil solution largely through the weathering of such primary minerals as micas, feldspars, etc. It is slowly released into the soil solution where it equilibrates with potassium on the exchange complex or is taken up by plants. It may also be fixed on to 2:1 clays in non-exchangeable forms. Potassium is continuously leached from the aerial parts of plants, mineralised biological materials and soil organic matter. As with other nutrient elements, it is taken up and released by micro-organisms under constraints imposed by environmental conditions.

Likens *et al.* (1994) consider that potassium reserves in natural soils are not large but are maintained by weathering and conservative cycling mechanisms. Plant uptake may account for considerable proportions of the potassium available in the soil, particularly in high-producing agricultural crops. To maintain crop production in these situations, it must be replaced by appropriate fertiliser inputs (Brady and Weil, 1996). Because potassium is highly mobile in soils, particularly those of coarse texture, the potassium cycle is sensitive to such disturbances as the removal of a forest cover. Such practices can lead to large site losses that may continue over many years (Likens *et al.*, 1994).

3.1.3 GENERAL NUTRIENT SOURCES

The nutrient elements absorbed by plants are derived from geological, biological and, to a lesser extent, atmospheric and anthropogenic sources. Chemical weathering of saprolite below the soil horizons, fragments of the parent rock, or of other parent materials resident in the soil may all release nutrient and other elements into the soil solution (Chapter II.3.1). Mineral nutrient elements continue to be released from this source until, over geological periods of time, all the weatherable minerals in the rooting zone have been broken down. Because of this, many highly-weathered tropical soils have little capacity to supply plant nutrient elements from this source.

Within the soil, nutrient elements are available through displacement from exchange sites on soil colloids. These may include both phyllosilicate and other clays and organic matter. Nutrient elements may also become available as a consequence of organic matter decomposition, through either the consumer or decomposer food webs (Chapter IV), or from turnover of the microbial biomass. In particular, organic matter decomposition supplies much of the nitrogen, phosphorus and sulphur available in unfertilised soils.

It has become clear from work carried out over the past few years that the microbial biomass is an important component of plant nutrition. The microbial biomass comprises a small pool of nutrient elements with high turnover rates and it may act as either a sink or source of these elements (Singh *et al.*, 1989). The phosphorus in microbial biomass comprises 3-24 *per cent.* of total organic phosphorus in soils, although most values lie in the lower part of this range; equivalent values for nitrogen are 1.7-6 *per cent.* (Singh *et al.*, 1991), and for sulphur are 1-3 *per cent.* (Banerjee *et al.*, 1993; Wu *et al.*, 1994). Microbial biomass and its contained nutrients vary with climate, the type of vegetation present, land use and cultural treatments (Wardle, 1992) (see also Chapter III.2.2.2).

Turnover of the microbial biomass appears capable of supplying a substantial part of the nutrient requirements of the higher plant community in both semi-natural and cropped ecosystems (Wardle, 1992).

There may be considerable temporal variation in nutrient supply due to seasonal variation in such biological factors as litterfall and root litter inputs, and of climatically-associated soil conditions including soil water and temperature status. In many non-agricultural situations, nutrients such as nitrogen and phosphorus may be partly or largely supplied to plants through the activities of symbiotic micro-organisms associated with the roots (Chapter IV.3.2.1).

Atmospheric inputs may play a significant role in the accession of nutrient and other elements. Volcanic eruptions may provide nutrient inputs ranging in extent from minor to substantial or may blanket an area with a thick layer of ash or other ejecta from which new soils will form. The soils of large areas of the North Island of New Zealand (Molloy, 1988) and other regions bordering the Pacific Ocean owe their existence to the volcanic activity that has showered their landscapes with ash. Appreciable annual inputs of nitrogen may occur through physical fixation processes occurring in the atmosphere. Substantial quantities of both sulphur and nitrogen are distributed widely from sources of industrial and other pollutants.

In the arid areas of the world, aeolian processes are responsible for considerable surface erosion. Substantial masses of fine materials become detached from exposed surfaces and entrained in air currents. These materials may subsequently be transported for very long distances from their source areas, sometimes entraining further materials in their paths (Pye, 1987). The amounts and sizes of the particles transported vary with distance from the source. Accessions of fine particles over time may be large; Khalaf and Al-Hashash (1983) reported an average annual depositional rate of *ca.* 23 Mg ha^{-1} in Kuwait, representing a layer of 1 mm thick. This particularly high value may represent conditions relatively close to the source. Studies in western Africa (Stoorvogel *et al.*, 1997) showed that deposition rates from the seasonal Harmattan winds fell off rapidly with distance from the source area falling from *ca.* 920 kg ha^{-1} y^{-1} at 1000 km from the source to *ca.* 60 kg ha^{-1} y^{-1} at 2700 km. The particle size distribution also changed commensurately: materials 1000 km from the source contained small proportions of materials greater than 100 μm in diameter while at 2700 km only materials less than 20 μm were present.

On deposition, wind-transported materials may form superficial mantles of parent materials (silt, loess, parna) from which soils have subsequently formed (Butler and Churchwood, 1983). Lesser effects include modification of the textures of the surface horizons of existing soils (Simonson, 1995).

Smaller amounts of materials and airborne salts are continually deposited by such processes all over the earth's surface. In the highly-weathered, impoverished soils of Amazonia and Africa, nutrients in atmospheric inputs may be sufficient to support the continued functioning of certain high-biomass rainforest ecosystems which survive because of their efficient nutrient conservation mechanisms (Jordan, 1982; Stoorvogel *et al.*, 1993). Airborne salts may provide substantial inputs of sodium, chloride and associated elements particularly to near-coastal areas; site differences are often related to their distances from the sea and the prevailing winds (Schlesinger, 1997).

Biological fixation by higher plants and some micro-organisms provide inputs of atmospheric carbon to ecosystems. Nitrogen, despite its predominance in the atmosphere, is largely unavailable to higher plants except through fixation by prokaryotes. These organisms may be symbiotic, associative or free living, although species of intermediate habits also occur. Of great agricultural and silvicultural importance are the bacteria of the genera *Rhizobium*, *Bradyrhizobium* and *Azorhizobium* that form symbioses with a wide range of tree, shrub, pasture and cropped legumes; however, not all legumes form such associations. A number of other groups also fix nitrogen with varying degrees of effectiveness. Actinobacteria of the genus *Frankia* form nitrogen-fixing symbioses with a range of species (mainly trees) belonging to several families (Quispel *et al.*, 1993). Other symbiotic systems include the ancient associations of the lichens and the cycads with cyanobacteria (Grobbelaar, 1993). Sugarcane and some other tropical grasses, have been shown to fix nitrogen in association with bacteria of the genus *Azospirillum* (Sprent and Sprent, 1990). The cyanobacterium *Anabaena azollae* forms a symbiosis with aquatic ferns of the genus *Azolla* which is important in the nitrogen economy of rice crops in some areas. A range of free-living bacteria, both aerobes and such anaerobes as *Azotobacter* spp. and *Clostridium* spp., are widespread in soils and fix lesser but still significant quantities of nitrogen.

Lastly, in many agricultural situations, fertilisers are frequently the major source of a number of plant nutrient elements for crop plants and the simplified ecosystems of which they form the autotrophic base.

3.1.4 MECHANISMS OF STORAGE AND RELEASE

3.1.4.1 *Nutrient stocks (standing crops)*

As indicated above, nutrient elements are derived from several different sources and possess a wide range of mobilities and turnover times within the soil. The stock of a particular element at a given site is defined as the total mass per unit area of that element lying within a defined soil depth range, often the root zone. It may also be defined in terms of the masses of particular fractions (often chemically defined) of such elements. Such fractionations frequently aim to define a 'plant-available' fraction and this may have restricted value in the context of crop growth within limited ranges of soil and environmental conditions. Despite the above, fractionations based on functionally-realistic criteria have proved very insightful, as discussed above in the context of phosphorus fractionation (Cross and Schlesinger, 1995). As discussed below, fractionations on various bases have also proved useful in studies of soil organic matter (see Figure I.13) and in estimation of the nutrient element composition of the microbial biomass.

Nonetheless, the magnitudes and depth distributions of nutrient stocks in soils provide an estimate of the total amounts of particular elements contained within a given soil mass, or volume. Such estimates may reflect soil age, the pedological processes occurring within them, the nature of their vegetative cover and environmental conditions.

Nutrient elements are not distributed evenly within soils: through biological and pedological processes, they become concentrated at particular locations and remain substantially lower throughout the remaining soil volume. Because of ecosystem recycling processes, greater reserves of most nutrient elements occur near the soil surface and, at a smaller scale, close to the roots. Nutrient reserves are high near roots because of exudation and their continual death and replacement. Close to the surface, such reserves are a consequence of the greater inputs of litter and are associated with the higher organic matter status of the A horizon. Figure I.36 shows the similar and decreasing patterns with depth in the concentrations of carbon, exchangeable calcium, magnesium and potassium in a tropical oxisol (Spain *et al.*, 1989a). In contrast to the other elements presented, the concentration of total phosphorus varies little down the profile, probably due to fixation on the considerable amounts of iron- (haematite, goethite) and aluminium-containing (gibbsite) minerals present in these soils (Isbell *et al.*, 1976). However, as considered below (see also section I.3.1.2.3), a substantial proportion of this element occurs in organic form in most soils.

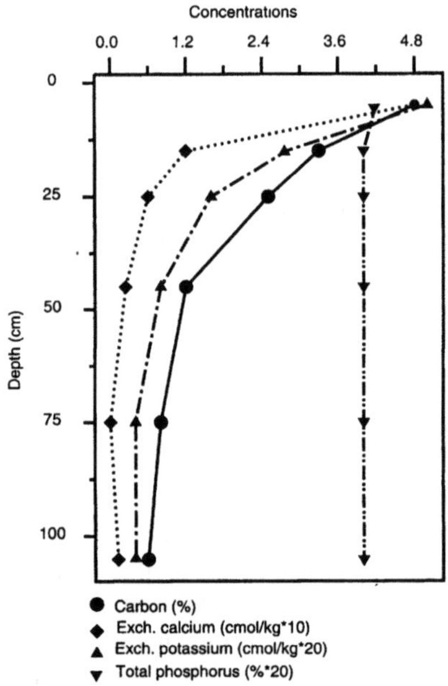

Figure I.36 Depth distributions of the concentrations of selected elements in an Australian oxisol supporting a rainforest vegetation.

Where the rate of accession of organic matter by the soil exceeds its rate of decomposition or where decomposition rates remain depressed, as occurs with waterlogging, little-modified plant remains may accumulate as a litter or peat layer (O or H horizons). The presence of higher organic matter concentrations near the surface of the mineral horizons has a number of consequences. Because of the high cation exchange capacity of soil organic matter (see below), the cation exchange capacity of the whole soil will normally be greatest near the surface and in locations where the roots are most densely distributed. Such concentrations of organic matter are normally associated with higher microbial and faunal activities and thus a more rapid turnover of the contained nutrient elements.

Stocks of nutrient and other elements may be calculated for specific areas by summing the products of their concentrations and the bulk densities for successive horizons or layers sampled within the soil. These values provide an estimate of the total amounts of the elements in this soil and, for the nutrient elements, an estimate of overall reserves. However, they provide little insight into biogeochemical processes, as considered below.

For many elements in many environments, nutrient element stocks included in the vegetation biomass are small in relation to those held within the rooting depth of the plant community. However, this may not be true of highly-weathered soils. This is illustrated in Table I.18 which presents the distributions of four selected nutrient elements in a Amazonian rainforest ecosystem formed on an extremely acid (pH 3.5) oxisol (Klinge in Salati and Vose, 1984). The stocks of total nitrogen and phosphorus held in the phytomass are equivalent to approximately 37 and 47 *per cent.* of those present in the soil. In contrast, soil calcium concentrations were below detection limits and the amount of magnesium in this highly-weathered soil comprised approximately eight *per cent.* of total ecosystem stocks. Therefore, the stocks of these last two elements, together with those of potassium and sulphur, occur predominantly in the phytomass.

Table I.18 The distributions of total N, P, Ca and Mg in the phytomass and soil of an Amazonian rainforest ecosystem growing on an oxisol. The data for each element are presented as kg ha^{-1} for the phytomass and for the soil to 1m. The data for stocks of individual elements in the phytomass are also presented as percentages of those in the soil (Klinge, 1975 in Salati and Vose, 1984).

Ecosystem component	Biomass	Total N		Total P		Total Ca		Total Mg	
	as OM (t ha^{-1})	kg ha^{-1}	% mass	kg ha^{-1}	% mass	kg ha^{-1}	% mass	kg ha^{-1}	% mass
Phytomass									
Tree stratum	504	2428	19.9	59.0	27.3	424	80.3	202	67.8
Roots	45.9	561	4.6	6.9	3.2	82.9	15.7	55.1	18.5
Dead biomass	21.2	293	2.4	3.0	1.4	21.1	4.0	17.9	6.0
Total	571	3282	26.9	68.9	31.9	528	100	275	92.3
Soil									
0-30cm	113	4263	34.9	71	32.8	0	0	17	5.7
30-100cm	120	4661	38.2	76	35.2	0	0	6	2.0
Total	233	8924	73.1	147	68.0	0	0	23	7.7
Ecosystem store	804	12,201		216		528		298	
Phytomass: soil store (%)	246	37		47		*ca.* 100		92	

Nutrient stocks diminish with depth, often markedly in highly-weathered soils paralleling the decline in organic matter concentrations. This is illustrated in the context of organic matter studies in Table I.19 which lists the stocks of carbon and nitrogen at selected depths in the oxisol profile presented in Figure I.33c. Other factors affecting the magnitudes of useful soil reserves include the depth of the soil profile and nutrient availability.

Table I.19 Depth distributions of carbon and nitrogen stocks (Mg ha^{-1}) in the profile of an oxisol supporting a rainforest cover in humid tropical Australia (Spain *et al.*, 1989a).

Depth range (cm)	Bulk density (Mg m^{-3})	Mass of C (Mg ha^{-1})	% of C in interval	Mass of N (Mg ha^{-1})	% of N in interval
0-10	0.80	57.8	24.7	4.44	27.4
10-20	0.89	43.3	18.5	3.51	21.7
20-30	0.97	33.7	14.4	2.59	16.0
30-60	1.13	51.0	21.8	3.06	18.9
60-90	1.23	26.0	11.1	1.47	9.1
90-120	1.26	22.5	9.6	1.13	7.0
Total		234.3		16.2	

The 'plant-available' amount of a nutrient element in a soil is considered to be the fraction of it that could potentially be taken up by plant roots and their associated mycorrhizal fungi, at any given time. It must therefore be defined in relation to the particular plant community or crop present and in relation to plant mycorrhizal status and other factors affecting the ability of the root system to extract nutrients from the soil. To be available, nutrient elements must be physically accessible, that is, not sequestered within unweathered primary minerals or undecomposed organic matter. Further, nutrients in parts of some profiles may not be accessible to roots because of genetic constraints. For example, roots may not be able to grow to sufficient depths to exploit deeper reserves. Inaccessibility may also result from the presence of impermeable pans or a layer of permafrost, partial waterlogging leading to anoxia or undue soil dryness. In some agricultural soils, past poor soil management practices may restrict the roots to the soil volume above a compacted zone.

In non-calcareous soils, much of the store of readily plant-available basic cations (Ca, Mg, and K) is retained at colloid surfaces and may be held in a state of equilibrium with cations in the soil solution. The colloids involved are largely clay particles or organic matter and comprise the major part of the whole-soil cation exchange capacity.

Nutrient and other elements may occur in soils in chemical forms that are insoluble or can not be taken up by plants. This may occur because they are poorly soluble at existing soil pH levels, as in the case of manganese considered above (Figure I.35), or are occluded through reaction with other soil components. Low phosphorus concentrations commonly limit plant productivity in many soils and phosphorus may be of particularly low availability in highly-weathered soils due to absorption by iron and aluminium. Thus, measures of the total amounts of many elements in defined parts of a profile may be of little value in predicting their availability to the biota although they may have some use as indicators of the overall reserves present.

3.1.4.2 *Release from dead plant materials and soil organic matter*

The rates of nutrient release from decomposing plant remains depend on environmental conditions and the nature of the materials under consideration. Where leaves are shed naturally, a substantial proportion of the more mobile nutrient elements contained in the living tissues are normally withdrawn prior to abscission. This contrasts with the situation in such crops as sugarcane which, because they are harvested when the leaves are still actively photosynthesising, leafy residues may still contain somewhat greater quantities of elements. Such processes are analogous to those that occur through herbivory. In pastures and other systems, the grazing of plant materials that possess a relatively high nutrient status results in a direct transfer of a large proportion of the included nutrients from the above ground biomass to the soil through their deposition in excreta and secreta (Swift *et al.*, 1979).

Table I.20 presents changes in the standing crop biomasses of selected nutrient elements in sugarcane harvest residues decomposing as a surface mulch in the humid tropical environment of northeastern Australia (Spain and Hodgen, 1994). The biomass of dead plant material initially declines steeply and then less rapidly; such changes are characteristic of many decomposing plant residues although such materials differ enormously, depending on the chemical and physical characteristics of the plant materials concerned and environmental conditions. Percolating rainwater leaches away a portion of the nutrient elements and other materials solubilised through microbial activity.

Table I.20 Changes in the standing crop biomasses (g m^{-2}, dry weight) of selected elements in decomposing sugarcane harvest residues from tropical northeastern Australia (Spain and Hodgen, 1994).

Period of decomposition (days)	Cumulative rainfall (mm)	Litter biomass (g m^{-2})	% Remaining	Elements (g m^{-2})				
				N	P	K	Ca	Mg
0	0	1583	100	4.16	0.58	9.65	1.25	0.70
21	85	1424	90	3.13	0.36	5.23	0.65	0.39
42	87	994	63	1.94	0.25	3.03	0.50	0.24
75	265	834	53	1.52	0.20	1.61	0.38	0.21
148	1300	834	53	2.47	0.27	0.52	0.58	0.18
204	2561	446	28	1.69	0.18	0.37	0.55	0.05
282	2972	443	28	2.54	0.21	0.51	0.98	0.13
338	3027	297	19	1.57	0.12	0.39	0.52	0.08

Nutrient release during the initial phase may be rapid; during the first 21 days of decomposition, 25 % of the nitrogen, 38 % of the phosphorus and 46 % of the more mobile potassium was lost, compared with only 10 % of the biomass. The rate of release per unit of biomass decomposed is slower during the later stages of decomposition and its measurement is complicated by other factors, including admixture with soil materials through earthworm casting into the remaining residues, the activities of other soil animals and nitrogen fixation (Hill and Patriquin, 1990). Decomposition rates and the dynamics of nutrient release from decomposing resources are highly variable since they depend on a suite of hierarchically-organised factors (Chapter IV.1.2).

3.1.4.3 *Ratios of carbon to nitrogen and other elements (see also Chapter IV.1.3.3)*

The ratios of carbon to other nutrient elements in decomposing tissues have a controlling influence on their breakdown and on the recycling of their nutrient elements. In particular, the ratio of carbon to nitrogen (C:N) (Table I.21) has been widely used as an index of tissue decomposability and of the capacity of various materials to supply nitrogen to higher plants and to micro-organisms. However, other tissue properties, including the lignin content, the ratio of structural tissues to cytoplasm and the presence of secondary plant compounds also strongly influence the pattern of breakdown.

Heterotrophic micro-organisms decomposing plant tissues in terrestrial environments normally have to cope with materials with higher C:N ratios than those of their own tissues. In such environments, micro-organisms in increasing their numbers will absorb the available inorganic nitrogen to incorporate into their own biomass. Since micro-organisms preferentially absorb nitrogen as NH_4^+ (or as low molecular weight organic compounds), they may have a competitive advantage in this and immobilise much nitrogen in their tissues that could have otherwise been available to higher plants.

During decomposition, the C:N ratios of litters and harvest residues vary in similar ways although the time scales differ widely due to both intrinsic and environmental factors. In the harvest residues of sugarcane (*Saccharum* spp.), initially high values (*ca.* 170) increase slightly which may result from the loss of the most assailable materials; this is followed by a lengthy decline and perhaps a plateau at a C:N of *ca.* 70 (Figure I.37) (Spain and Hodgen, 1994). Description of the changes occurring during the later phases of decomposition is complicated by N fixation in the decomposing residues (Hill and Patriquin, 1990) and by their admixture with earthworm casts. In soils, lower C:N ratios generally pertain in the heavy fraction of the soil organic matter and, in both the light and the heavy fractions, the C:N ratio generally declines with decreasing particle size (Baldock *et al.*, 1992).

As considered above, nitrogen deficiencies may limit the productivities of both micro-organisms and plants, depending on the nature of the decomposing materials and their stage of decomposition. Stevenson (1986) considered that net mineralisation leading to an increase in inorganic nitrogen will occur below a C:N ratio of 20, an approximate equilibrium state will pertain between 20 and 30 and that, over 30, net immobilisation will take place constraining the supply of nitrogen to plants, as described above.

Table I.21 Indicative C:N ratios of selected organism: organic materials, soils and some soil fractions (Lee and Wood, 1971a; Anderson and Domsch, 1980; Matsumoto, 1976; Spain *et al.*, 1983a; Spycher *et al.*, 1983; Sprent, 1987; Sarathchandra *et al.*, 1984; Spain *et al.*, 1989a; Singh *et al.*, 1991; Srivastava, 1992; Attiwill and Adams, 1993; Spain and Hodgen, 1994; Garnier-Sillam and Harry, 1995; Batjes, 1996; Hasegawa and Takeda, 1996; Spain and Reddell, 1996).

Materials	C:N ratio
Predominantly Organic Materials	
Amino acids, amides	1.7-7.7
Nucleic acids	3
Protein	3.1
Bacteria	3.7-13.0
Cyanobacteria	7-26
Fungi	7.6-8.25
Lichens	12-23
Microbial biomass	6.4-16.0
Root exudates	2.5-13
Insecta:	
i. termites	3.9-7.8
ii. ants	4.0-5.2
Leaves:	
i. living	5->40
ii. senescent	18->170
Woody tissues	50-1000
Decomposing litters:	
i. tropical rainforest	33-53
ii. eucalypt forest	70
iii. conifer	40-160
Sugarcane harvest residues	72-240
Termite mounds (carton):	
i. wood feeders	37-107
ii. grass and litter feeders	21-25
iii. soil feeders	11-15
Soils and predominantly earthen materials	
Andisol	13
Aridisols	10
Vertisols*	12
Other vertisols	14
Oxisols	14
Mollisols	14
Spodosols	29
Histosols	to >70
Inceptisol:	
i. whole soil	16-26
ii. light fraction	31-50
iii. heavy fraction	13-22
Termite mounds (earthen):	
i. grass and litter feeders	9-27
ii. wood feeders	24-47

* Note that these soils support a forest cover dominated by the leguminous tree *Acacia harpophylla*.

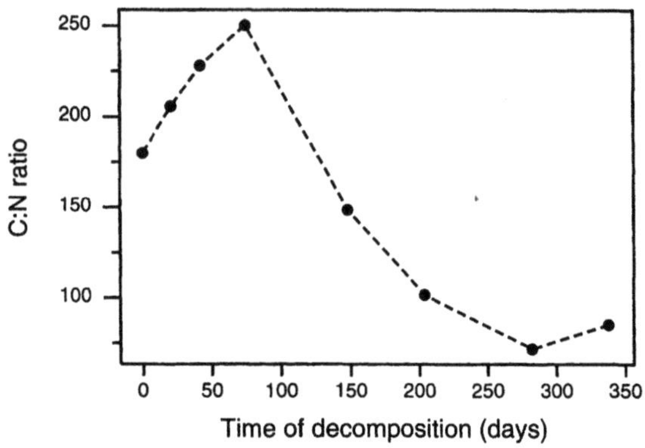

Figure I.37 Temporal changes in the carbon to nitrogen ratio of decomposing sugarcane harvest residues (Spain and Hodgen, 1994).

In undisturbed, fully-developed soils and those at approximate equilibrium with their management regimes, C:N ratios are relatively constant, although clear differences exist between environments and soil taxonomic groupings at a range of scales (Batjes, 1996). Spain *et al.* (1983a) analysed the C:N ratios of the A1 horizons of more than three thousand Australian soils and the data for three selected areas are presented in Table I.22. As shown, few consistent differences were apparent between temperate and tropical climate soils. In general, however, soil C:N ratios tend to be lower in arid regions because of accumulation of inorganic N in the soil while in areas with higher rainfall, they are more dependent on decomposition rates. Thus, soils of hot humid climates tend to have lower C:N ratios than those of cold areas (Post *et al.*, 1985). Differences between vegetation types also exist and cultivated lands tend to have lower C:N ratios (*ca.* 10) than forests (*ca.* 14-30) (Attiwill and Adams, 1993).

Table I.22 C:N ratios of the A1 horizons of soils from three regions of Australia (Spain *et al.*, 1983a).

Area	Median	Inter-quartile range	Sample size
Tropical areas	16	13-20	783
Tasmania (cool temperate)	16	13-22	142
West Australia (temperate)	21	16-29	162

Clear differences were also apparent between some soil taxonomic categories. Spodosols and histosols, rarer in Australia than in many cool temperate and boreal regions, tended to have the highest C:N ratios (up to more than 70). Slight decreases occur with depth in many soils reflecting the greater humification and greater age of the organic carbon stored deeper in the profile; however, this does not occur in vertisols because of the profile churning that occurs (Batjes, 1996). In some soils, C:N ratios may decline with increasing soil depth because of the greater amounts of fixed NH_4^+ at depth.

The situation discussed above may be generalised to other elements. Stevenson (1986) states that, while some local variation occurs, the ratio C 140:N 10: S: 1.3:P 1.3 remains relatively constant across a wide range of environments. Where C:organic P and C:S ratios are less than 200 and 400, respectively, net mineralisation will occur. Net immobilisation of P and S will occur at C:P and C:S ratios above 300 and 400, respectively.

3.1.4.4 *Roles of symbiotic micro-organisms*

As considered above, certain micro-organisms (for example, *Rhizobium* and related species, *Frankia* spp., Actinobacteria) enter into symbioses and looser associations with many legumes and some non-legumes to fix atmospheric N a variable portion of which may become available for use by the host plants. The biological N fixation process underpins the agricultural economies of many areas and offers the species involved a competitive advantage in N-deficient environments. The other micro-organisms important in mediating nutrient acquisition for their hosts are the diverse groups of fungi that form mycorrhizal relations with many higher plants.

For most plant groups in most environments, it is not likely to be profitable to consider their nutrition independently of the actual or potential activities of these organisms. Their roles are addressed in greater detail in Chapter IV.3.2.1.2 and IV.3.2.1.3.

3.2 Organic resources

In soils, energy flows either through the consumer or decomposer foodwebs (see definitions in Chapter III). The decomposer foodwebs are the most complex in that decomposing organic resources are progressively transformed through sequences of ingestion-egestion processes which form what Swift *et al.* (1979) have aptly called **cascade structures**. Consequently, organic resources are extremely structurally and chemically diverse and, of the enormous range of compounds known, almost all are likely to be found in soils. These resources are here characterised by their quality, availability in time and space and, where possible, quantified.

Organic resources present in soils are distributed among the following four major classes of materials:

1. Dead leaves, woody litter and reproductive structures which fall to and may accumulate on the surface of the mineral horizons;
2. Living and dead roots;
3. Organic matter accumulated in the soil profile; and
4. Soluble organic compounds, including bark, foliar and litter leachates, root exudates, polysaccharides and other materials produced by both micro-organisms and invertebrates.

3.2.1 RESOURCE CLASSIFICATION

In ecological systems, the quality and availability of resources in time and space largely determine patterns of consumer usage and the organisation of the consumer communities.

The relationships between consumer organisms (competition, commensalism or mutualism) will also depend on resource properties: quality (Swift *et al.*, 1979; Lavelle, 1985), distribution in time (Price, 1984) and space (Southwood, 1977) and, eventually, interactions (Tilman, 1982).

The quality of a resource is defined by its chemical composition (which determines its ease of digestion) and such physical properties as its size, shape and hardness which govern its susceptibility to ingestion by consumers. Resources may be of high, medium or low quality, depending on the degree of adaptation required for their exploitation (Lavelle, 1985) (Table I.23).

Table I.23 Classification of resources based on quality.

Quality	Adaptation necessary to use the resource	Examples
HIGH	none specific	glucose or amino-acids to heterotrophic organisms
MEDIUM	individual genetic or ethologic	cellulose to organisms secreting their own cellulases
LOW	mutualistic relationships with other organisms	Cellulose to lower termites with cellulolytic protists in their hindgut

Resource availability over time is defined as the temporal pattern of its presence in relation to the generation time of the organisms using it. Price (1984) recognises the following five classes of resources (Figure 1.38):

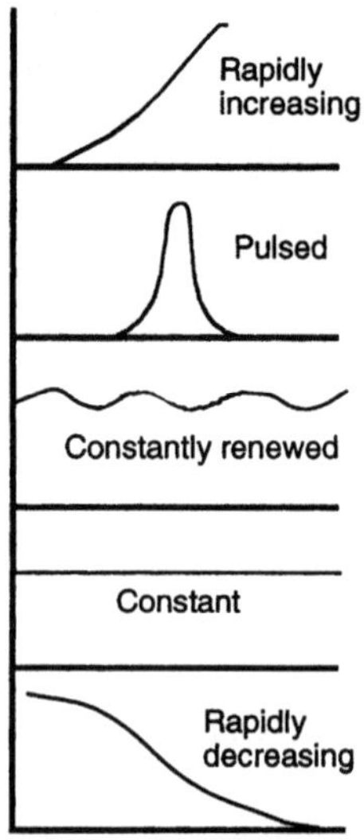

Figure I.38 A classification of resources based on their occurrence in time (Price, 1984).

1. Rapidly increasing (example: foliage of deciduous trees in relation to herbivores in temperate climate spring);
2. Rapidly decreasing (example: rapidly decomposing litter in tropical rainforests in relation to primary decomposers);
3. Constantly renewed (example: soil organic matter in relation to geophages);
4. Constant (example: bird nesting sites, soil pores at certain scales of time in relation to micro-organisms);
5. Pulsed (example: root exudates, litter leachates in relation to the soil microflora).

Variation in resource availability over time is of considerable importance since interactions between consumers competing for the same resource will depend on both the timing of its availability and the capacities of consumers to respond to it by increasing their populations. To illustrate, the exploitation of a rapidly-increasing resource is unlikely to lead to competition among consumers since their populations will increase too slowly to create density-dependent effects. In contrast, such competition is more likely where resources are regularly-renewed or rapidly-decreasing. However, even in these circumstances the likelihood of competition taking place may be reduced by the spatial distribution patterns of the resource (Table I.24; Price, 1984).

Table I.24 Probabilities of competition between species using the same resources in relation to their spatial-temporal distributions and population responses to change.

	Spatial distribution	Uniform		As mosaic	
	Population response	**Fast**	**Slow**	**Fast**	**Slow**
Resource characteristics	Rapidly growing	+	-	-	-
	Pulsed	(+)	-	-	-
	Regularly renewed	+	-	-	-
	Constant	+	+	+/-	-
	Rapidly decreasing	+	+	+/-	-

The ways in which consumers exploit a resource and the nature of the relationships they develop with other consumers of that resource is dependent on the **spatial distribution of the resource**. The spatial distribution of a resource may be evaluated by relating resource patch size to the home and migratory ranges of its consumers (Southwood, 1977).

The following three classes of resource spatial distribution may therefore be defined (Figure I.39):

1. Evenly distributed and available to all consumers in the area;
2. Occurring in large, isolated patches located further apart than the normal migratory ranges of the consumer species; and

3. Occurring in small but closely-spaced patches falling within the home ranges of the consumer species.

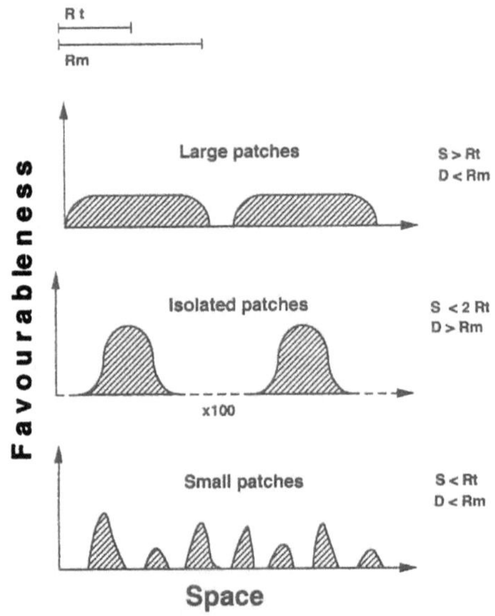

Figure I.39 A classification of resources based on their spatial distributions and their relationships with the ranges of the species that exploit them (Southwood, 1977) (S = size of patch, D = distance between favourable patches, Rt = trivial or home range, Rm = migratory range).

Resource interactions: the classification proposed by Tilman (1982) is based on the extent to which resources are necessary for plant growth and how they interact. In considering the relative importances of two pairs of resources (R1 and R2) and their influences on individual plant growth, eight different outcomes may result (Figure I.40):

Resources may be *perfectly substitutable*: replacement of a given quantity of R1 by a same amount of R2 does not change plant growth.

Resources are *complementary* when a given amount of the mixture of R1 and R2 gives a better growth than the same amount of R1 or R2 alone.

Resources are *antagonistic* when a mixture of the two resources gives lower growth than the same amount of any single resource. Two resources that are totally antagonistic are said to be switching.

Resources may also be *essential.* They are said to be interactive essential if an optimal combination of both resources allows growth maintenance with a lower overall consumption. If one of these resources may be replaced by the other, it is said to be hemi-essential. Finally, they are *inhibitory* if an increase in each resource beyond a given threshold diminishes growth. This classification does not consider only the feeding value of resources, interactions described by Tilman integrate the costs associated in obtaining and exploiting the resources together with their distribution patterns. The drawback of this approach is that it limits the analysis to single pairs of resources.

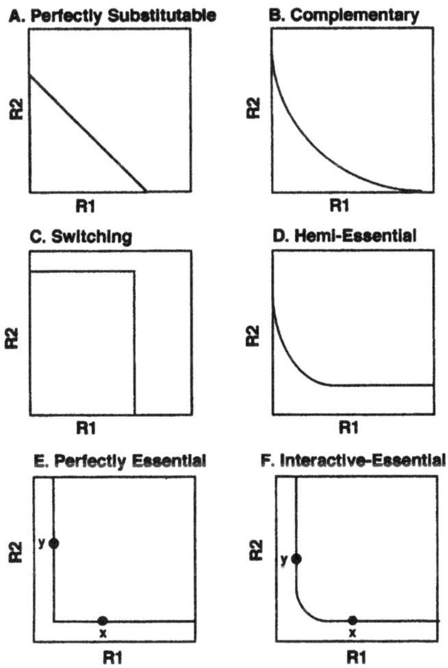

Figure I.40 Classification of nutrient resources for plants (Tilman, 1982). The solid curve is the zero net growth isocline of a population. A zero net growth isocline shows the availability of resources 1 and 2 for which growth and reproduction exactly balance all sources of loss. Note that R1 and R2 are resource levels, *i.e.*, the measureable concentration of the usable form of each resource in the habitat..

3.2.2 SURFACE LITTER

3.2.2.1 *Quality*

The dead leaves, wood and reproductive structures which form the major part of the surface litter differ strongly in quality both within and between species. Quality also depends on soil nutrient status and changes markedly with time, as decomposition proceeds.

Composition of fresh litter

Leaf litter may be broadly classified as being of low to moderate accessibility. Its quality is positively related to its concentrations of soluble carbohydrates, hemicellulose, nitrogen and other nutrient elements and negatively to its concentrations of lignin, cellulose and secondary plant compounds, including polyphenols and other toxic modifiers

Fresh leaf materials usually possess high concentrations of soluble carbohydrates, mainly reducing sugars, *e.g.*, 15 to 18 % in savanna grasses (Lavelle *et al.*, 1983a), 7.2 % in oak leaves (Waksman, 1952) and 29.6-33.8 % in the leaves of three *Salix* species (Slapokas, 1991). However, concentrations of these substances may decrease sharply during senescence and prior to leaf abscission, partly through their translocation to other parts of the plant.

The tanning of cytoplasmic proteins by polyphenols released from vacuoles at cell death is a further process of particular importance for decomposition (see Chapter IV). This results in the formation of highly decomposition-resistant tannin-protein complexes (Handley, 1954; Toutain, 1987a). The accessibility of remaining water-soluble sugars may be further reduced by the presence of such soluble toxic components as polyphenols, waxes and resins (Mangenot and Toutain, 1980).

Litters may thus be classified into the three following broad groups:

1. Litters with high concentrations of soluble carbohydrate whose exploitation is not limited by extreme physical or chemical characteristics. Examples include the litters of hornbeam (*Carpinus* spp.) and birch (*Betula* spp.) in temperate areas and those of such tropical legumes as *Sesbania* or *Erythrina*;
2. Litters with high concentrations of soluble carbohydrates, waxes and/or polyphenols including those of conifers, European oaks (*Quercus* spp.), chestnut (*Castanea sativa*) and the tropical leguminous shrub *Inga edulis* (Palm and Sánchez, 1991);
3. Litters with low concentrations of soluble carbohydrate such as European beech (*Fagus sylvatica*).

The chemical quality of fresh leaf litter thus depends on the compositions of the leaves of the individual plant species present and their relative contributions to the overall mass of litter present.

Many plants accumulate quite specific sugars and a wide range of waxy materials, terpenes and alkaloids, especially those of tropical areas (Waterman and McKey, 1989; Baas, 1989). The chemical variability of leaf litter in tropical forests may well exceed that of temperate forests because of the greater species diversity and complexity of tropical plant material. However, in a tropical Bornean rainforest, only 4 of 306 species present contributed more than 5 % to litterfall, the maximum proportion being 10 % (Burghouts, 1993). Similarly, at a single location at the foot of an individual tree in a French Guyanese rainforest, only 16 to 21 % of the litter sampled came from the tree above and more than 20 different species contributed significantly to the litter collected at this location throughout a yearly cycle (Leroy, unpublished data).

The quality of leaf material also changes with its position on the tree ('light' as opposed to 'shade' leaves) and the nutrient status of the soil (Heath *et al.*, 1966; Dickinson and Pugh, 1974; Ellsworth and Reisch, 1993).

Woody litter

Wood has more structural supporting tissue implying higher cellulose and lignin and substantially-lower nutrient element concentrations than foliage. The accessibility of woody litter is reduced by its hardness and the low surface to volume ratio it offers to potential consumers; its accessibility is therefore substantially lower than that of leaf litter (Table I.25).

Table I.25 The major organic components of selected decomposer resources % dry weight (from sources in Swift *et al.*, 1979).

	Lipid	Water-soluble	Hemi-cellulose	Cellulose	Lignin	Protein	Ash
Deciduous leaf - young *Quercus* sp.	8	22	13	16	21	9	6
Decidous leaf- old *Quercus* sp.	4	15	16	18	30	3	5
Conifer needle-old *Pinus* sp.	24	7	19	16	23	2	2
Grass leaf *Deschampsia flexuosa*	2	13	24	33	14	2	-
Grass stem *Zea mays*	2	15	18	30	11	1	8
Tree litters:							
Carpinus betulus	6.84	18.95	26.32	17.37	28.95	-	1.58
Fagus sylvatica	5.76	10.1	26.26	15.66	40.60	-	1.62
Quercus robur	6	14	27.8	16	34.8	-	1.4
Deciduous wood, range	2-6	1-2	19-24	45-48	17-26	-	0.3-1.1
Conifer wood, range	3-10	2-8	13-17	48-55	23-30	-	0.2-0.5
Invertebrate faeces, *Glomeris marginata*	-	2	-	38	-	11	8
Vertebrate faeces, Horse	2	5	24	28	14	7	9
Bacteria (range)	10-35	5-30	4-32	-	0	50-60	5-15
Fungi (range)	1-42	8-60 (chitin)	2-15	-	0	14-52	5-12
Earthworms, *Lumbricus terrestris*	2-17	11-17	-	-	-	54-72	9-23
Arthropods, various	13-26 (glycerogen)	-	5-9	14-31 (chitin)	-	38-50	-

Changes in chemical composition with decomposition

Once deposited at the soil surface, the chemical composition of litter changes continuously as decomposition proceeds. Water-soluble carbohydrates and hemicellulose decompose steadily while cellulose and lignin decompose slowly; phenols and waxes are resistant to decomposition. Similarly, Table I.20 illustrates the decline in concentrations of five biological nutrient elements and thus the reduction in quality that occurs during the decomposition of sugarcane harvest residues. Such changes imply a clear overall diminution in the chemical quality of decomposing materials associated with the increasing proportion of resistant compounds (Minderman, 1968 and Chapter IV.1.3.3).

Where temperature and moisture conditions do not limit biological activity, mineralisation is rapid and little resistant material remains after periods of a few months. This situation is typical of leaf litter dynamics in wet tropical lowland forest situations. However, when drought, water-logging or low temperatures restrict decomposition, highly-polymerised chemical compounds (*e.g.*, fulvic and humic acids) may accumulate and impart a very low chemical quality to the decomposing material. This is a common situation in cold and temperate-climate regions.

In some circumstances, resource quality may increase over time, following the removal of strong chemical or physical constraints to decomposition (Heal and Dighton, 1985) (Figure 1.41 and see Chapter IV.1).

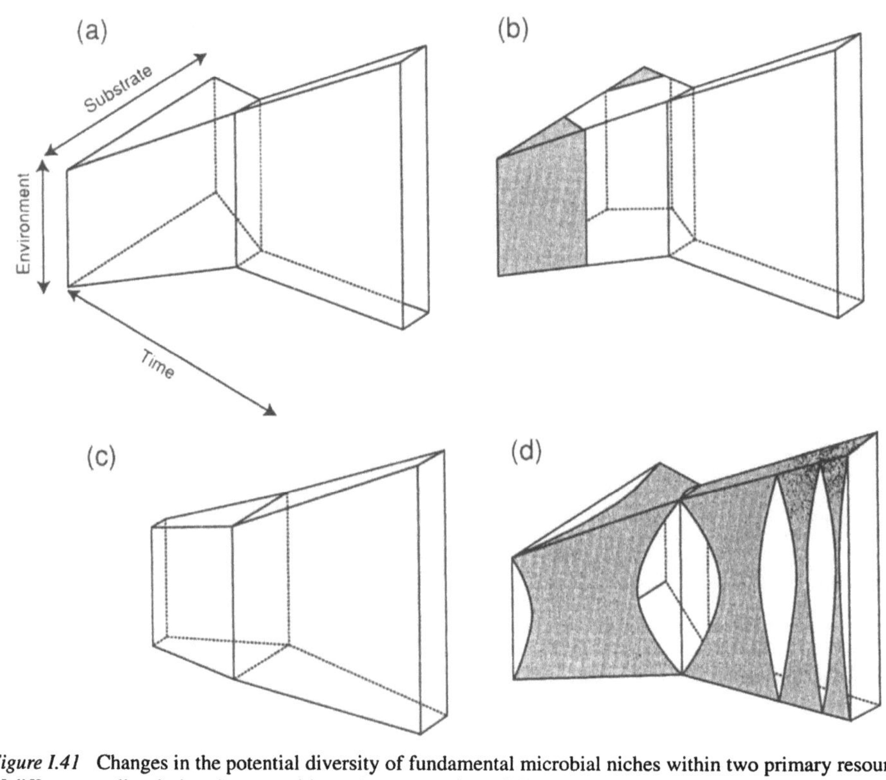

Figure I.41 Changes in the potential diversity of fundamental microbial niches within two primary resources of different quality during decomposition. A cross section of the volume at right angles to any point in time describes the available niche space in relation to the major determinants, environment and source. The time scale is simplified to represent two stages in decomposition: the initial rapid exploitation phase characteristic of primary resources and the much slower interaction phase of the terminal sources. (a) general model representing high resource quality, (b) initial substrate availability masked by the presence of inhibitory chemicals, (c) low quality of source, *e.g.*, wood with a limited range of substrates, (d) availability of niches limited by periods of severe climate, *e.g.*, drought or cold.

3.2.2.2 *Availability in time*

Fresh litter inputs are normally seasonal, even in humid tropical forests (see, for example, Spain, 1984) (Chapter IV.2.6). The general pattern of availability is a rapid accumulation of fresh material during the dry or cold season followed by a progressive decrease in the standing crop biomass up to the start of the next dry or cold season.

Where decomposition is rapid, as in some temperate mull systems and many lowland tropical rainforests, most of the annual litter input may be decomposed or exported in as little as six months. Under these circumstances, litter is defined as a rapidly-diminishing resource,.according to Price (1984).

Conversely, in environments where decomposition is slow, the O horizon may comprise several layers corresponding to the accumulated remains of successive yearly inputs. Under such circumstances, litter is considered a regularly-renewed resource, regardless of the chemical differences between successive layers. This situation is illustrated in Figure I.42 (Garay *et al.*, 1986a).

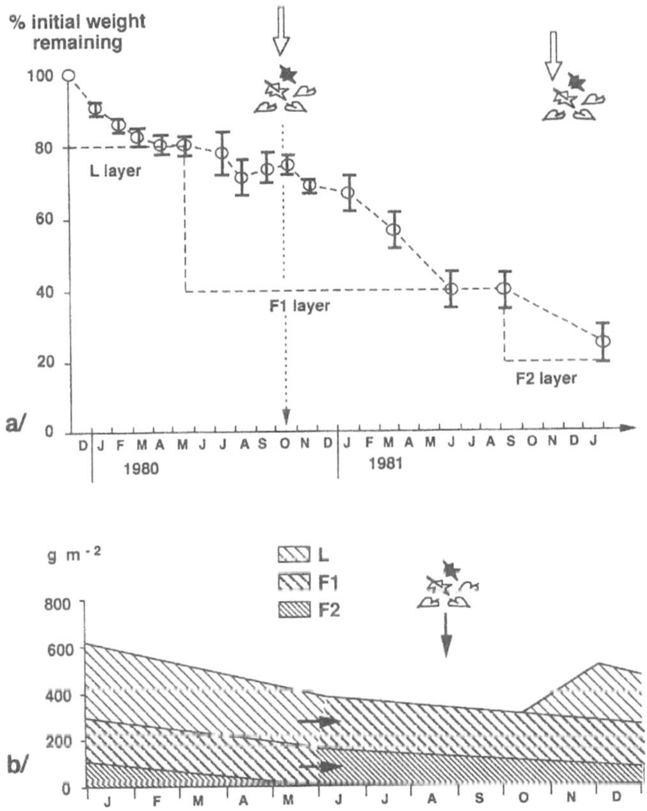

Figure I.42 Seasonal changes in the accumulated masses of successive litter layers in a temperate forest (Foljuif, France) (Garay *et al.*, 1986a)(Bars indicate 2 se intervals).
a) decrease of litter weight in litterbags;
b) localisation in layers of 0 horizon.

3.2.2.3 *Spatial distribution*

Horizontal distribution
Initial impressions often suggest that surface litter is evenly distributed in many ecosystems. However, interactions between wind, microtopography, gravity and the distribution patterns of such entities as trees and efficient litter consumers like termites and white-rot fungi lead to the formation of accumulation and depletion zones in which decomposition processes may proceed differently (see, *e.g.*, Garay, 1989; Hafidi, 1989).

As shown above for tropical forests, further spatial variability may result from inter-specific differences in litter quality. This heterogeneity is likely to increase with the number of species present and may therefore be lower in grasslands than in forests, and in temperate or cold regions than in tropical regions.

Vertical distribution
In many forest environments, litter accumulates at the soil surface and forms relatively compact layers of accumulated material in intimate contact with the underlying soil. However, in certain grasslands the pattern may be quite different with dead leaves sometimes remaining attached to the plant for long periods leading to the formation of a more diffuse 'duff' with poor soil contact.

Finally, in humid tropical environments, significant amounts of litter are intercepted by ferns, orchids, Bromeliad and other epiphytes, large branches, or *Pandanus* and palm-tree leaf axils. These form characteristic suspended litter systems with no soil contact (Lavelle and Kohlmann, 1984).

3.2.2.4 *Annual inputs and standing crop*

As with primary production, leaf litter production is determined by a suite of hierarchically-organised factors. Climate is the predominating influence and is expressed through temperature and rainfall conditions (Figure I.43). Meentemeyer *et al.* (1982) used the close relationship between litter production and actual evapo-transpiration to estimate world total fine litterfall (excluding large wood) and leaf litterfall production.

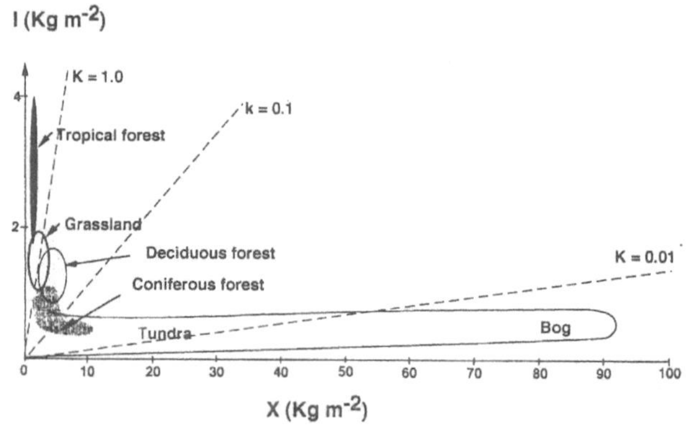

Figure I.43 Annual litterfall inputs (I) and standing crops of litter (X) in different terrestrial ecosystems (Heal *et al.*, 1981).

Total biomass production was estimated at 54.8 Pg y^{-1} (equivalent to *ca.* 23.6 Pg of carbon per year) of which 64-70 % was leaf material. Across a temperature and latitudinal gradient of biome and ecosystem types, litter accumulates in increasing quantities from tropical forests to the tundra, depending on climatic conditions. Nonetheless, soil characteristics and vegetation type are important secondary factors, as discussed in Chapter IV.2.

At Butte Montceau, near Fontainebleau (France), the outcropping of superimposed sand and limestone sediments creates downslope a pH gradient in the A1 horizon with values increasing regularly from 3.8 to 8.0 over a distance of 500 m; climatic conditions and the composition of the tree cover remain identical over this distance. In the upper part of this gradient, acidity limits earthworm populations and depresses decomposition rates leading to the accumulation of substantial litter layers containing large arthropod populations (Lavelle and Faille, unpublished data) (Figure I.44). With rising pH, earthworms become more abundant and increasingly dominate the litter system (see the drilosphere system, Chapter IV.4.1). Rapid earthworm burial of litter means that it increasingly decomposes within the soil, rather than on the soil surface.

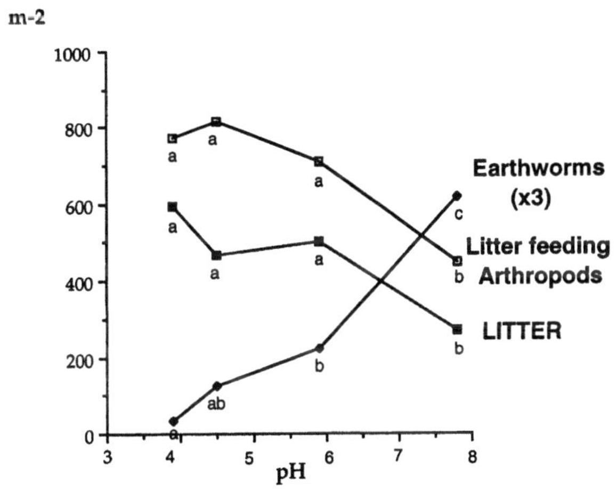

Figure I.44 Correlated changes in pH, litter mass (g m^{-2}), the population densities of macroarthropods and earthworms (individuals m^{-2}) at Butte Monceau (Fontainebleau, France) (P. Lavelle and A. Faille, unpublished data) (data with different letters on the same curve are significantly different ($P < 0.05$).

3.2.3 ROOTS

Root biomass may be extremely high in soils and often constitutes the major resource for soil organisms. This is especially true of ecosystems in which surface litter is regularly removed by cropping (agro-ecosystems) or burned (sugarcane crops, many savannas). Relative to surface litters, root material is still a rather poorly-known organic pool since assessment of root abundance and decomposition patterns has long been limited by the lack of an appropriate methodology. Root material may be classified as being of slight to moderate accessibility.

3.2.3.1 *Chemical quality*

For the same reasons, much less information is available on root than on leaf litter chemical composition and even fewer studies differentiate between coarse and fine roots. Apart from their mineral nutrient contents, little attention has been paid to the chemical composition of roots. Nutrient concentrations are often higher in roots than in leaves: in the strongly nutrient-limited soils of the Amazonian *tierra firme* forests, fine roots have much higher concentrations of all nutrient elements except calcium (Table I.26) (Cuevas and Medina, 1988).

Table I.26 Comparative nutrient contents (mg g^{-1} dry weight) of fresh leaves and roots in an Amazonian *tierra firme* forest (Cuevas and Medina, 1988).

Element	Leaves	All fine roots (mean, se)
N	11.2-17.1	23.0 (0.8)
P	0.48-0.87	1.13 (0.10)
K	3.4-5.8	6.7 (0.6)
Ca	1.7-3.2	2.1 (0.2)
Mg	1.0-1.5	2.6 (0.1)

In a temperate-climate European deciduous forest, coarse roots generally had lower concentrations of most nutrient elements than fine roots, except for potassium and sodium (Table I.27). Dead roots of all sizes had greater concentrations of the nutrient elements calcium, magnesium, sulphur and aluminium (not a nutrient) with values approximately twice as high as those of living roots) but only slight differences in potassium, nitrogen and phosphorus. Finally, comparison of nutrient element concentrations in the fine roots of Amazonian and temperate climate forests (Tables I.26, I.27, see also Table I.15) highlights some substantial differences: in the tropical forest, higher concentrations of potassium (x 15-29), magnesium (x 1.8-2.4) and nitrogen (x 2-2.6), slightly higher concentrations of phosphorus (x 1.2-1.8), and much lower concentrations of calcium (x 0.18-0.22) were apparent.

In humid African savannas, roots of the dominant perennial grasses contain more lignin and cellulose than leaves and are therefore much harder materials. Water-soluble compounds are present in lower concentrations, although most are only available to consumers when lipids and waxes have been removed; some appear to contain toxic compounds (Lavelle *et al.*, 1989). Consequently, the overall quality of these roots is much lower than that of the leaves. This may be interpreted as a defence against soil herbivores, comparable to those developed against folivores (Table I.28).

Table I.27 Nutrient and other elemental contents (mg kg^{-1}) in root materials of Beech (*Fagus sylvatica*) and Spruce (*Picea abies*) in European temperate climate forests (van Praag *et al.*, 1988)(nd not determined).

Species	Element	Fine roots (diameter <1 mm)		Coarse roots (diameter 1-5 mm)	
		Live roots	Dead roots	Live roots	Dead roots
Beech	Ca	9.42 (1.68)	18.43 (2.38)	4.10 (1.20)	5.92 (nd)
	Mg	1.14 (0.30)	2.03 (0.26)	0.72 (0.39)	0.72 (nd)
	K	0.44 (0.07)	0.54 (0.09)	0.55 (0.13)	0.51 (nd)
	Na	0.73 (0.14)	1.20 (0.25)	0.81 (0.21)	1.79 (nd)
	N	9.99 (1.06)	16.40 (1.02)	5.84 (0.51)	7.70 (nd)
	P	0.63 (0.06)	0.98 (0.06)	0.38 (0.03)	0.47 (nd)
Spruce	Ca	11.51 (2.02)	24.09 (3.09)	4.16 (0.47)	8.01 (3.31)
	Mg	1.10 (0.17)	2.12 (0.31)	0.50 (0.08)	1.11 (0.53)
	K	0.23 (0.04)	0.28 (0.06)	0.44 (0.12)	0.35 (0.12)
	Na	0.52 (0.05)	0.76 (0.07)	0.38 (0.06)	2.13 (1.80)
	N	11.31 (0.30)	12.24 (0.19)	7.01 (0.70)	8.94 (1.17)
	P	0.97 (0.07)	1.01 (0.06)	0.63 (0.06)	0.91 (0.29)

Table I.28 Compositions of the mature leaves and roots of the perennial grass *Loudetia simplex* in a Guinean savanna (Lamto, Côte d'Ivoire) (reducing power measured by Hagedorn and Jensen's 1923 micromethod).

		Water-soluble I	Lipids	Water-soluble II	Hemi-cellulose	Cellulose	Lignin	Ash	N%
LEAF	weight %*	15.3	12.7	2.1	40.10	12.1	26.5	3.9	0.59
	Energy value (mg eq glucose g^{-1})	24.6	6.3	4.0	24.3	21.3			
ROOT	weight %*	12.2	10.8	22.4	15.60	10.8	33.6	4.8	0.39
	Energy value (mg eq glucose g^{-1})	9.8	5.4	4.5	22.1	12.1			

* except for lipids: mg 100 g^{-1}

In the same soil, the nitrogen concentrations of roots decrease with depth (Abbadie, 1983). As decomposition proceeds, the chemical composition of the roots changes. In the early stages, the quality of this material (as assessed by the growth rates of invertebrates feeding on it) increases and this is probably related to the loss of soluble toxic compounds.

3.2.3.2 *Availability in time*

As a first approximation, roots may be considered a constantly, although irregularly-renewed resource. Most studies of seasonal changes in root biomass in grasslands or forests show no clear patterns. In humid African savannas, there appear to be phases of intense growth or mortality directly related to immediate conditions (Cesar, 1971; Abbadie, 1983). In climates with more marked seasonality, root growth may have clear patterns with pulses of fine root production in the growing season (van Praag *et al.*, 1988).

However, roots comprise at least two distinct resources, living and dead (see Chapter III.3). Living roots may be classified as constantly-renewed resources whereas dead roots may vary from constantly-renewed to a pulsed pattern of production, following sudden or unpredictable mortality. The scales at which decomposers operate substantially affect the ways in which such changes are perceived.

Coarse perennial roots are a constantly renewed resource whereas fine roots are ephemeral, living for only a few weeks to months. While fine roots are constantly present, pulses of growth and mortality occur regularly. At certain scales of time and space, they may thus be classified as rapidly decreasing resources (or increasing, if dead fine roots are considered) (see, *e.g.*, Fairley and Alexander, 1982; Kalela, 1957; Persson, 1978 and 1980; Vogt *et al.*, 1980 and 1981).

3.2.3.3 *Spatial distributions*

Root distributions differ greatly from those of litter materials in that roots may be dispersed throughout the entire upper soil volume. Also, they are normally in intimate contact with the soil matrix over their entire surfaces. Roots may occur at any depth although they are particularly concentrated in the upper part of the profile and diminish sharply with depth. This pattern is clearer in grassland than in forest, in situations where the plant-nutrient status of the soil is low or where the water table frequently comes close to the soil surface (Chapter III.3.1). Tree root systems are more deeply and evenly distributed with depth than those of grasses. Tree roots are also clearly differentiated into coarse perennial supporting structures and finer, short-lived absorbing roots. In contrast, grasses have fasciculated, or bundled systems of functionally-similar roots.

A further factor in the differing "rooting strategies" that occur between plants groups is the variation that occurs in their patterns of horizontal distribution.

3.2.3.4 *Annual inputs and standing crop*

Root biomass may vary from a few hundreds of kilograms to more than 300 Mg ha^{-1}, depending on the dominance of grasses or tree components and soil characteristics. As developed further in Chapter III.1, forest ecosystems normally have a higher root biomass (especially coarse roots) than grasslands and, for a given type of ecosystem, the ratio of above-ground to below-ground biomass tends to be higher in tropical than temperate regions.

3.2.4 SOIL ORGANIC MATTER

3.2.4.1 *Quality*

In mineral soils, organic matter comprises only a small proportion of the total soil volume. In terms of its utilisation, a major factor is the size of the organisms exploiting it. Soil organic matter has a markedly discontinuous distribution at the micro-organism scale while, at the larger scale of geophagous invertebrates, it is effectively continuous, although of low energy density. A sample from the upper 10 cm of a savanna soil with *ca.* 1 % organic matter in West Africa had an energy value of 235 J g^{-1} (Lavelle, 1978) while, in an English beech (*Fagus sylvatica*) wood, a soil with 18 % organic matter had an energy value of 4,500 J.g^{-1} (Bolton and Phillipson, 1976).

The quality of soil organic matter, when treated as a resource, depends on its:

(i) chemical properties (as defined in section I.3.2.2.1);

(ii) the shape and proportional representation of different particle size classes (the particle-size distribution); and

(iii) accessibility in the soil matrix (which, in turn, depends on its inclusion within aggregates and whether it is combined with minerals to form organo-mineral complexes).

Particle size distribution

Soil organic matter is a mixture of plant and animal residues normally present in all stages of degradation and comprises particles of a wide range of size classes. Part comprises amorphous colloids flocculated following the precipitation of soluble compounds leached from the surface litter. The larger particles (*i.e.*, those greater than 50 μm) are assumed to be more susceptible to decomposition than those of smaller size, partly because the latter have already undergone a degree of microbial degradation. These smaller particles contain high concentrations of such complex molecules as lignin and humic materials and are therefore more resistant to microbial attack; such particles are considered to be chemically protected. Other forms of chemical protection include the formation of clay:humus and clay:metal ion:humus complexes. Such effects are greatest with allophane and progressively lower in smectites, illite and kaolinite (van Breemen and Feijtel, 1990).

Tiessen *et al.* (1984) have confirmed the relationship between soil organic matter quality and particle size (Table I.29). Carbon and nitrogen concentrations were shown to increase with decreasing particle size down to 0.2 μm, while the C:N and C:P ratios decreased indicating a greater degree of humification of the organic material. This was associated with increased concentrations of highly aromatic humic acids which further contribute to their recalcitrance to decomposition. Notably, the finest fraction (<0.2 μm) differs from the other fine fractions in that it has higher concentrations of fulvic acids (predominantly aliphatic) and hydrolysable nitrogen (Anderson *et al.*, 1981; Tiessen *et al.*, 1984; Catroux and Schnitzer, 1987).

Table I.29 Chemical compositions of different organo-mineral size fractions from a prairie soil (a black chernozemic soil from Blaine Lake, Saskatchewan, Canada) (Tiessen *et al.*, 1984).

	C	N	P	Extractable P**	$\delta^{15}N$*	C:N	C:P
	mg g^{-1} fraction				‰		
Whole soil	50	4.0	0.58	0.22	11	12.5	87
Sand (>50 μm)***	27	1.9	0.19	0.06	7	14.0	141
Coarse silt (50-5 μm)	41	3.6	0.47	0.17	9	11.2	86
Fine silt (5-2 μm)	85	7.9	1.17	0.43	10	10.8	73
Coarse clay (2-0.2 μm)	82	8.7	1.67	0.56	12	9.5	51
Fine clay (<0.2 μm)	63	8.0	1.16	0.37	17	7.9	54

* $\delta^{15}N$ indicates the ratio of ^{15}N to ^{14}N relative to that of the air as parts per thousand excess

** Organic P extractable in sodium bicarbonate and hydroxide

*** The sand fraction also includes coarse, floatable (<1.00 g cm^{-3}) organic matter

The particle size distribution of soil organic matter varies considerably between soils, and with depth in the profile (Figure I.45). The results presented below were obtained in regions of comparable climate (average annual rainfall: 1200-1300 mm; mean temperature 26-27 °C) and suggest that:

(i) Soils with high concentrations of active clays such as vertisols accumulate much larger amounts of organic matter than sandy soils which have a relative dominance of coarse (>50 μm) and fine (<5 μm) particles;

(ii) Tropical forest soils have a larger proportion of coarse organic particles (39.5 % as compared to 23.4 % in the adjacent grass savanna), and a larger overall abundance of organic matter;

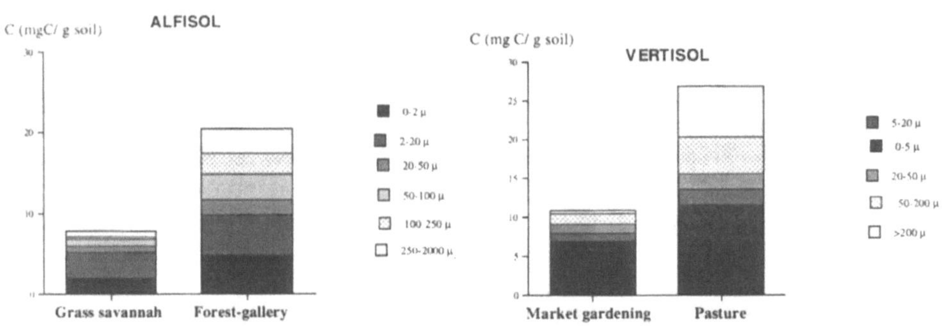

Figure I.45 Particle size distribution of organic matter of alfisols in forest and savanna (Lamto, Côte d'Ivoire) (Martin *et al.*, 1990) and in a vertisol (Ste Anne, Martinique) (Feller *et al.*, 1995).

The particle size distribution of soil organic matter is strongly affected by land use practices. In Martinique, ten years of intensive vegetable production in a vertisol resulted in the loss of two thirds of the organic matter, particularly from the coarse (>50 μm) (loss of 78 %) and the finest (<2 μm) fractions (Martin, 1990; Albrecht *et al.*, 1986)

Distribution within aggregates

Organic particles may be enclosed within organo-mineral aggregates and thus physically protected from microbial digestion. This protection appears to be most effective when oxide minerals are abundant; their significant amounts of positive charge attract the negatively-charged humic molecules. At the scale of larger aggregates (*i.e.*, 50 to 2000 μm and more), simple dilution of organic particles in the often compact mineral matrix makes them less susceptible to microbial attack. For example, mineralisation of soil organic matter in the casts of geophagous earthworms is much slower than in smaller aggregates (less than 2 mm) formed in the same soil. These casts occur as large aggregates (>5 mm diameter) with low internal porosity which limits microbial activity (Martin, 1989)(Chapter IV.4).

From the above, the distribution of soil organic matter particles between the different size classes of aggregates may affect their accessibility to decomposers. The sharp decrease in soil organic matter concentrations following cropping, especially in the tropics, is largely correlated with a reduction in the degree of aggregation (Feller *et al.*, 1995).

In most soils, the difference in particle size distributions after (a) gentle separation of aggregates (by dry-sieving or gentle shaking in water) and after (b) total disruption by sonication (use of ultrasound energy) provides a measure of the effective strength of aggregates. Protection of soil organic matter within aggregates depends on both the degree of aggregation and the resistance of these aggregates to dispersion by water or more powerful forces.

Scheu *et al.* (1996) found considerable increases in microbial respiration following removal of protection by the disruption of aggregates collected from three temperate-climate forested soils. The effect was most marked in the larger aggregates and in the most clay-rich of the three soils sampled.

Temporal variation of soil organic matter quality

When ecosystems are in an approximate state of equilibrium, the amounts and qualities of soil organic matter vary little with time. However, in situations of strong seasonality and rapid soil organic matter turnover, the quality of the soil organic matter may differ seasonally. This occurs, for example, in sandy alfisols in the moist savannas of the Côte d'Ivoire where the nutritional value of the soil for geophagous earthworms is greatest at the end of the rainy season and lowest a few weeks after the onset of rains (Lavelle and Meyer, 1983). Microbial respiration follows a similar temporal pattern in these soils (Abbadie, personal communication).

At longer temporal scales, changes in the vegetation cover are reflected in different soil organic matter characteristics. Table I.30 illustrates the changes measured in a West African humid savanna due to an increased dominance by trees following protection from fire (Martin *et al.*, 1990). While only slight differences in carbon concentrations are apparent between the grassy and the shrub savanna, substantially-increased surface soil carbon concentration result from fire protection. As shown by changes in the stable carbon isotope compositions (as $\delta^{13}C$ values, see section I.3.2.4.2 below) of the sites studied, the effects on organic matter are most apparent in the surface soils. Increased $\delta^{13}C$ values also occur at depth in the fire-protected soils although only in the surface soils do these values approach those of the gallery forest.

In addition to large reductions in organic matter concentrations, substantial changes in its quality follow clearing and cultivation, as shown in Sections I.1.1.3.2. and I.1.1.3.3. Such changes appear to be largely confined to the light fraction (Skjemstad *et al.*, 1997) although fertiliser nitrogen applied to the cultivated soil may also be stabilised into a number of heterocyclic forms (Gregorich *et al.*, 1996).

Table I.30 Depth distributions of carbon concentrations (mg C g^{-1} soil) and ^{13}C natural abundance ($\delta^{13}C$ ‰) in soils from adjacent savanna and forest sites on a sandy alfisol (Lamto, Côte d'Ivoire) (Martin *et al.*, 1990).

Depth (cm)	Grass savanna		Shrub savanna		Gallery forest		Fire-protected savanna	
	$\delta^{13}C$	C	$\delta^{13}C$	C	$\delta^{13}C$	C	$\delta^{13}C$	C
0-10	-12.8	7.9	-13.5	8.2	-27.5	20.4	-23.1	12.6
10-25	-12.8	6.2	-12.7	5.7	-24.8	6.1	-16.0	4.9
25-40	-12.4	4.2	-12.7	4.2	-24.3	5.2	-16.1	4.2
40-80	-13.4	1.7	-14.1	3.4	-23.6	3.4	-17.7	3.0
80-120	-16.1	0.4	-16.1	2.7	-23.5	2.7	-19.7	1.5

3.2.4.2 *Isotope natural abundance*

The ratio of the two stable isotopes of carbon (^{13}C and ^{12}C) in biological and other materials may be expressed as $\delta^{13}C$ values which relate this ratio to an international standard. $\delta^{13}C$ values vary considerably between plants depending whether they use the C3 or the C4 photosynthetic pathways and, to a lesser extent on a number of other environmental factors. C3 plants incorporate the carbon of atmospheric CO_2 into C3 compounds and have $\delta^{13}C$ values of -22 to -38‰ (mean *ca.* -27 ‰) while C4 plants incorporate their carbon into C4 compounds and have $\delta^{13}C$ values of -9 to -21‰ (mean *ca.* -13 ‰); CAM plants are intermediate with average values of *ca.* -17‰ (Tieszen, 1991).

The $\delta^{13}C$ values of soil organic matter reflect the history of the organic materials that have been input to the soil system, with only small changes. There is no evidence that differences in isotopic composition affect the nutritive value of soil organic matter for soil organisms and this property of soil organic matter is a useful label, especially when rapid changes of vegetation or organic inputs of a contrasting nature have taken place (Smith and Epstein, 1971; Cerri *et al.*, 1985; Balesdent *et al.*, 1987; Martin *et al.*, 1990).

Most plants in cold and temperate regions belong to the C3 group while, in the tropics, grasses usually belong to the C4 group and trees and shrubs to the C3 group. Organic matter formed from decomposed plant residues retains the $\delta^{13}C$ value of its parent plant with only small changes (Table I.30) and these values may thus be used to determine its origin. Continuous cultivation of corn (a C4 plant) for 23 years lead to significant changes in the $\delta^{13}C$ value of soil organic matter in a soil which had previously received inputs of solely C3 organic matter (Balesdent *et al.*, 1987). The mean $\delta^{13}C$ value had increased significantly, with the most marked changes taking place in the coarser (200-2,000 µm and surface litter) and in the finest (0-0.2 µm) fractions (Figure I.46). These differences have been used to assess the turnover times of organic matter (Cerri *et al.*, 1985; Martin *et al.*, 1990) and to investigate the origin of organic matter assimilated or transported by invertebrates (de Niro and Epstein, 1978; Spain *et al.*, 1990; Martin *et al.*, 1991; Spain and Reddell, 1996) and are illustrated further in Chapters III.4.3.2.1 and IV.5.3.2.4. $\delta^{13}C$ values increase by 0 to 1‰ per trophic level along food chains (Petersen and Fry, 1987).

Particle size (µm)	C in % of soil C	C/N	X%
0-30 cm Depth			
Litter	ND	ND	100
200 - 2000	10	23	61
50 - 200	17	14	31
20 - 50	3	15	17
2 - 20	37	16	12
0.2 - 2	29	10.5	18
0 - 0.2	4	8.5	26
30-40 cm Depth			
50 - 200	7	16	16
20 - 50	3	ND	8
2 - 20	47	14	6
0.2 - 2	31	12.5	13
0 - 0.2	9	7.5	9

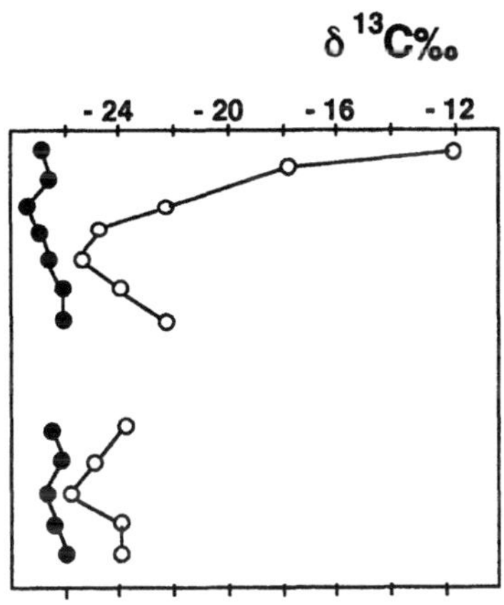

X % of organic C derived from maize
• forest soil
o maize crop

Figure I.46 Organic carbon contents, C:N ratios and $\delta^{13}C$ values of different soil organic matter particle size fractions after 23 years of continuous corn cultivation (Balesdent *et al.*, 1988).

Stable isotope ratios of other elements may also be useful as source tracers and process indicators in the same way as those of carbon. The ratio of ^{15}N to ^{14}N (adjusted to relate the ratios of these isotopes to the air as a standard, $\delta^{15}N$ values) has been used in this way and has potential uses in soil-based and related studies as an indicator of trophic level and other processes (Petersen and Fry, 1987).

$\delta^{15}N$ values of the tissues of plants possessing nitrogen-fixing symbionts may be used to indicate the proportional contributions of N fixation to total plant nitrogen supply, provided that certain other information is available, or may be assumed (Shearer and Kohl, 1989). In forests, $\delta^{15}N$ values of decomposing materials increase with depth in the litter layers and with increasing depth in the mineral soil, probably due to the cumulative effects of a number of small fractionations (Nadelhoffer and Fry, 1994). Addition of nitrogenous fertilisers may lead to substantial changes in the $\delta^{15}N$ values of the organic matter breakdown products.

Sulphur isotopes fractionate very little between trophic levels and are therefore efficient tracers (Peterson and Fry, 1987). $\delta^{34}S$ measurements have recently been used to plot the change in plant sulphur source that occurs during the growth of the North American desert xerophyte *Tamarix aphylla* (Yang *et al.*, 1996). $\delta^{34}S$ values of the rings of older trees were shown to change systematically from the centre of the stem (*ca.* 13‰) to the exterior (*ca.* 20‰). Yang *et al.* (1996) used a three stage model and a knowledge of the $\delta^{34}S$ values of the groundwater (17-20‰) and the atmospherically-derived sulphur of the near-surface unsaturated zone (*ca.* 7‰) to explain the distribution of sulphur stable isotopes. During the first stage of early growth, $\delta^{34}S$ values of *ca.* 13‰ indicated that plant sulphur originated from atmospherically-derived sources in the upper soil horizons but, with ongoing root development (stage 2), these values progressively rise reflecting the increasing utilisation of phreatic water sources. In the third stage, the tissues have stable $\delta^{34}S$ values of *ca.* 17‰, similar to those of the dominant phreatic sulphur source.

3.2.4.3 *Spatial distribution patterns*

Vertical distribution

Soil organic matter concentrations and overall stocks generally decline from the surface to deeper horizons (see, for example, Tables I.19, I.30, Figure I.36) reflecting the localisation of inputs, *i.e.*, litter deposited at the surface and roots largely concentrated in the upper horizons. Further, because of their deeper root systems, higher organic matter concentrations may occur at depth in forest than in grassland soils.

This is illustrated in Table I.19 which presents the distributions of C and N in the upper 120 cm of the profile of the undisturbed rainforested oxisol described in Figure I.33c. Both C and N masses decline with depth and more than 57 % and 65 %, respectively, of the masses of these elements occur in the top 30 cm of the soil. As shown, substantial masses of carbon still occur at depth in this and in similar soils elsewhere in the region. At this site, from 4.7 to more than 12 Mg ha^{-1} of C were present at depths of 150-180 cm in the profiles.

Some grasslands may have carbon stocks equivalent to those illustrated above. Fisher *et al.* (1994) reported the large carbon mass of 269 Mg ha^{-1} in the top 100 cm of a soil supporting improved pastures in humid tropical Colombia; more than 40 % of this amount was present below a depth of 40 cm.

As considered in Section 1.1.3.2, the quality of the organic matter resource changes notably with increasing depth. Only the most recalcitrant materials remain in the subsoils.

Horizontal distribution

Soil organic matter concentrations vary greatly in the horizontal dimension. This is a consequence of small to medium-scale variation in the quantity and quality of inputs and of such environmental conditions as moisture, temperature and soil depth.

While not all plants affect soils this way, organic matter may vary systematically at the scale of single perennial plants to produce surrounding islands of enhanced fertility with organic matter concentrations declining with increasing distance from the plant (see, for example, Figure I.47, Jackson and Caldwell, 1993). At a larger scale, Figure I.48 presents a contour plot of the distribution of carbon in the A horizon (0-5 cm) of a soil (inceptisol) derived from basaltic parent materials in semi-arid (535 mm annual average rainfall) tropical northeastern Australia (McIvor *et al.*, 1991). The site (*ca.* 40 ha) supports a savanna vegetation, is nearly level and was chosen for its apparent uniformity. The contours were generated from samples taken at the intersections of a square 40m grid.

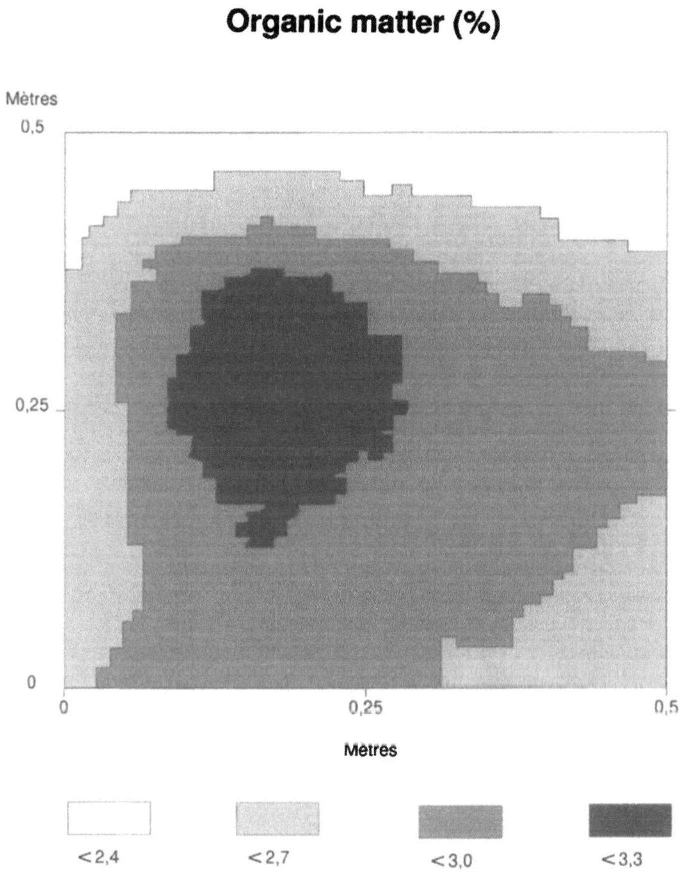

Figure I.47 Soil organic matter concentrations (g $100g^{-1}$), around a single tussock of the grass *Pseudoroegnaria spicata*; the border of the tussock corresponds with the area of highest organic matter concentration (redrawn from Jackson and Caldwell, 1993).

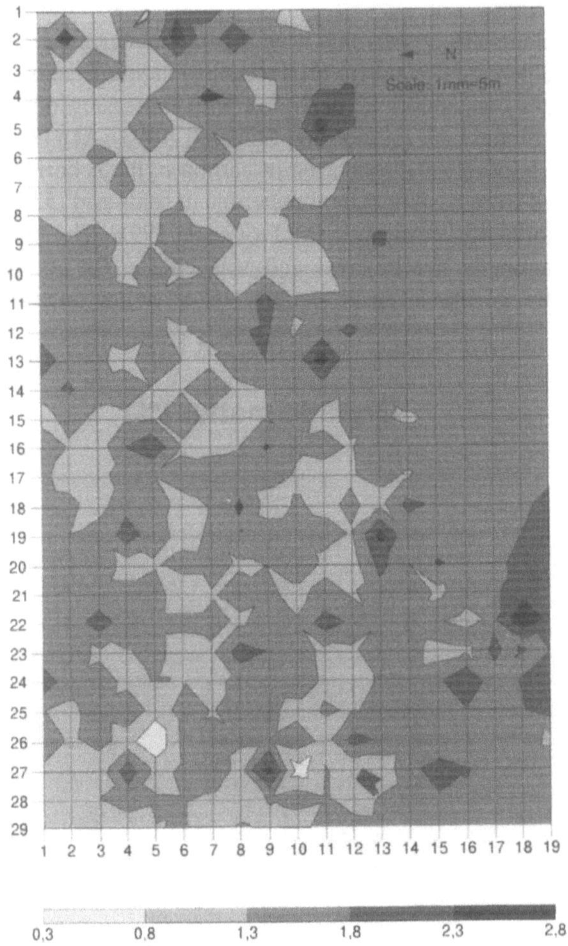

Figure I.48 The horizontal distribution of C (g 100 g^{-1}) in the A horizon of a tropical Australian inceptisol (McIvor *et al.*, 1991).

3.2.4.4 *Quantity*

Organic matter accumulation in soils depends on three processes, *viz.*:

(i) Plant production, which provides most of the inputs;

(ii) The capacity of soil to stabilise and store organic matter, which depends on soil depth, the amount and types of the clay and other minerals available to form stable organo-mineral compounds; and

(iii) The rate of mineralisation through biological oxidation which is, in turn, determined by a suite of hierarchically-organised factors, namely, climate, certain soil properties (clay type and amount, pH) and biological activities (especially of the larger organisms) (Lavelle *et al.*, 1993).

Because of variation in the above, substantially different amounts of organic matter accumulate in soils. This may be considered at the following different scales:

(i) Large biomes;
(ii) Climatic zones (as defined by Holdridge, 1947); and
(iii) Regional and locational differences due to changes in the vegetation cover.

Recent estimations of the stocks in the world's soil and other carbon pools were presented in Table I.17. Soil carbon stocks differ widely between soil taxa and with climate. They generally rise with increasing annual average rainfall and, for a given average rainfall, increase as temperatures fall (Post *et al.*, 1982) (Figure I.49). Carbon stocks also increase with soil clay content and are influenced by litter quality (Schimel *et al.*, 1994).

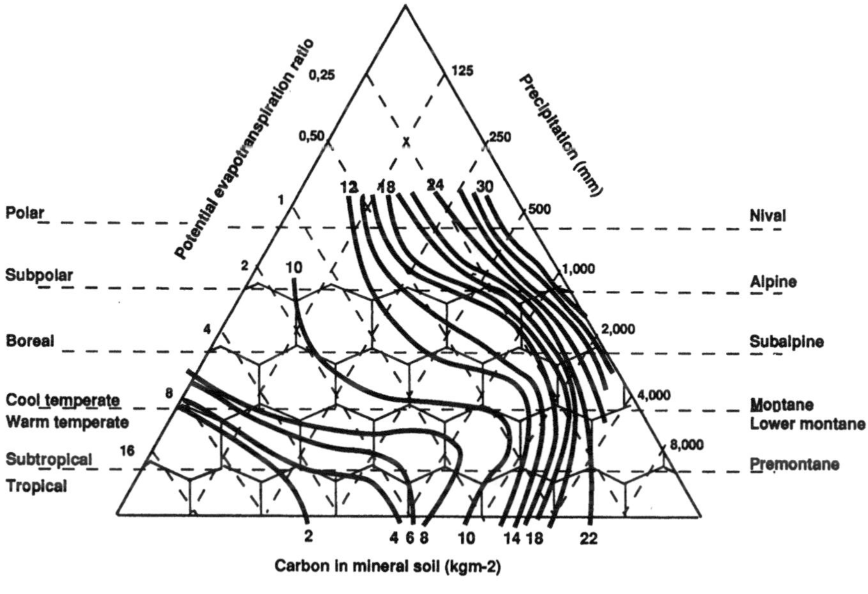

Figure I.49 Contours of soil carbon density plotted on the Holdridge schematic for world life-zone classification. Values of biotemperature and annual average rainfall uniquely determine a life zone and its associated vegetation (Post *et al.*, 1982).

The lowest carbon stocks recorded (Batjes, 1996) occur in certain warm temperate deserts (1.4 kg m^{-2}) and subtropical dry forests (2.6) while, apart from the peats, the highest values in mineral soils are found in rain tundra (36.6), boreal rainforest (25.6), wet tundra (22.2) and in tropical wet forest (21.0). Further, large amounts of soil organic carbon are stored below 100 cm in a number of deep tropical soils (Batjes, 1996). Thus, the belief that tropical rainforests have much lower organic reserves than temperate forests (see, *e.g.*, MacNeil, 1964; Bohn, 1976) is seen to be incorrect (Greenland *et al.*, 1992).

Sánchez *et al.* (1982) have shown that no significant differences in organic matter reserves exist between soils of temperate and tropical areas, provided that they belong to the same order of Soil Taxonomy (Soil Survey Staff, 1999). In tropical areas, the oxisols discussed by Sanchez *et al.* (1979) have significantly higher overall organic reserves (11.3 kg C m^{-2}) than the alfisols (5.4 kg C m^{-2}) and ultisols (4.9 kg C m^{-2}). In temperate regions, mollisols were found to have higher C reserves than alfisols (5.0) and ultisols (3.8).

In addition to the above, large differences exist among soils, even within small areas, depending on variation in the local soil forming factors. At the regional scale of tropical northeastern Queensland, rainforested soils derived from basaltic materials have substantially greater organic matter levels than those formed from granitic or acid-volcanic parent materials (Spain, 1990). In the same study, both carbon and nitrogen concentrations (Figure I.50) were negatively correlated with temperature and precipitation; topography was also shown to be of importance through its control of drainage and soil depth.

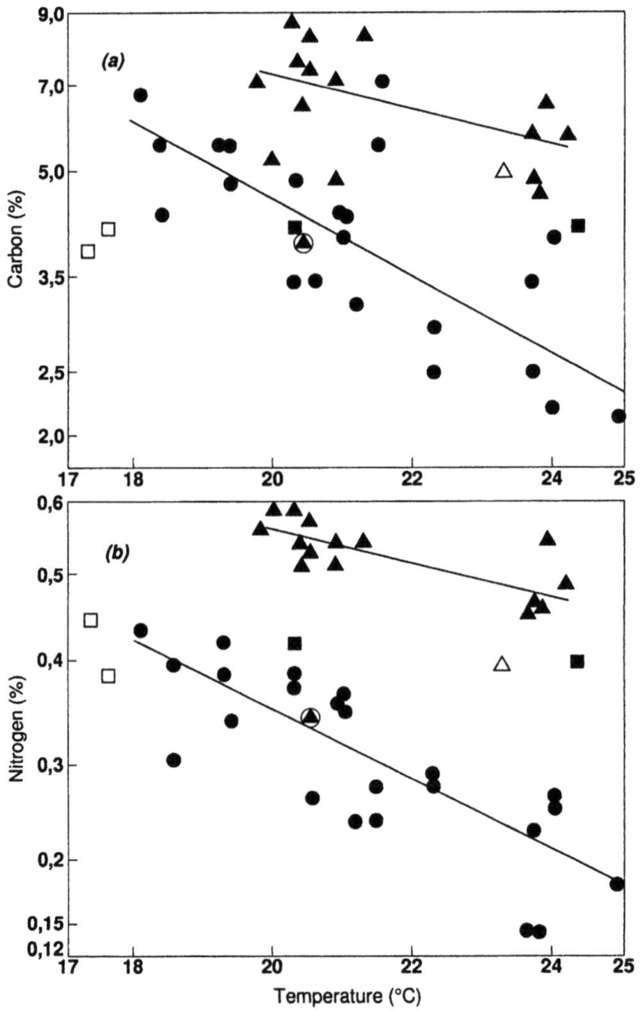

Figure I.50 Relations between (a) C and (b) N concentrations and annual average temperature in the surface horizons of rainforested soils formed from basaltic (●) and combined granitic and acid volcanic (▲) parent materials (Spain, 1990).

Important differences also exist between vegetation types. In tropical areas, forest soils have greater organic reserves than savannas in the upper 15 cm, but may have equivalent amounts in the soil profile down to 100 cm. In contrast, temperate climate grasslands have significantly more organic matter than forest soils.

3.2.4.5 *Conclusions*

The low quality of the soil organic matter resource is partly due to its dilution in large volumes of inorganic soil materials but it may be due to its chemical or physical protection against decomposer digestion through humification and soil aggregation processes, or through associations with particular minerals (Chapter IV.1). It is a regularly-renewed resource, with inputs concentrated at the soil surface and in the root zone. Soil organic matter concentrations decline rapidly with depth in most soils and commensurate changes in quality also occur. It has a relatively homogeneous lateral distribution at the scale of decimetres to metres and this contrasts with its discontinuous spatial distribution at microscopic scales.

3.2.5 SOLUBLE RESOURCES

A range of water-soluble resources are produced in soils by roots (exudates *sensu lato*), earthworms (cutaneous or intestinal mucus) and other invertebrates; soluble organic matter may also originate from the leaching of foliage, bark and litter. These assimilable materials are pulsed resources as defined by Price (1984) since their availability to the soil biota is highly dispersed in time and space, and largely unpredictable. They are characterised by large fluxes but low stocks and are known as "ecological mediators" for their important role in activating the microbiota at selected scales of time and space (Lavelle *et al.*, 1994a) (see Chapter IV.1.4).

Estimates of the rates of production of this relatively-diverse suite of soluble and colloidal resources are still poor since they degrade rapidly and techniques are not yet available to measure their fluxes under field conditions. Production of root exudates is probably in the order of several Mg ha^{-1} y^{-1} in most ecosystems (Chapter III.3.2.3).

Certain invertebrates also produce such readily-assimilable organic resources as mucus. Endogeic earthworms may add from 6 to 40 % of the dry weight of the ingested soil as intestinal mucus, a mixture of a low molecular weight glycoprotein, sugars and amino-acids. This mucus is released in the anterior part of the gut, and may be largely reabsorbed in the mid and posterior gut, after part of it has been used by the ingested soil micro-organisms to increase their activity. Although similar products have not been identified and quantified in other soil invertebrates, it is likely that they exist.

The production rates of foliar and litter leachates have rarely been quantified. Their occurrence in ecologically-meaningful amounts is clearly demonstrated by the changes that occur in the redox potential (Eh) of the mineral soil underlying the litter layers. These are related to the temporary accumulation of water-soluble organic compounds, mainly sugars with high reducing potentials (Toutain, 1974) (Figure I.51). In one of the very few measurements of the production of this resource, Carlisle *et al.* (1966) measured an annual flux of 350 kg ha^{-1} of water soluble organic matter in a English *Quercus petraea* forest.

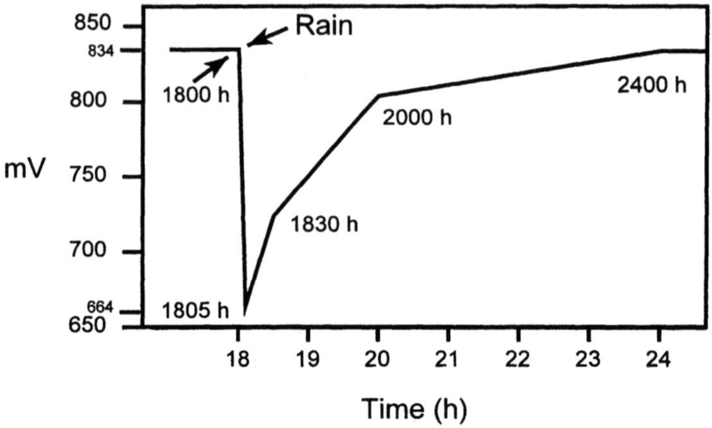

Figure I.51 Changes in Eh after a 19 mm rainfall event in a temperate climate forest soil (Vosges, France) following leaching of water-soluble organic compounds from a moder litter type (Toutain, 1974).

I.4 SOIL HETEROGENEITY

The complexity of the soil environment results from the interactions of four factors. These are the types of energy input to the soil system, spatial and temporal patterns and functional structures.

4.1 Diversity of energy types

Three different forms of energy dissipate within the soil. *Physical energy* linked to gravity and the electrostatic forces of capillarity and attraction are responsible for the movement of materials, in solution or suspension, which play such a predominant role in soil formation. Another form of physical energy is the important release of mechanical energy during mineral crystallisation (Pedro, 1964). *Chemical energy* may also play an important role through abiotic oxidation and reduction reactions. Scharpenseel *et al.* (1984) have demonstrated that up to 20 % of the decomposition of humic acids in sandy and silty soils may result from the abiotic oxidation of these compounds. *Photosynthetic energy* becomes available to soil organisms as calorific energy contained in the organic matter deposited on the soil surface and incorporated into the mineral soil through such diverse biological activities as oxidation (respiration), hydrolysis (digestion) and mechanical mixing (burrowing, transportation, construction).

The dissipation of these energies has quite separate effects. Chemical and physical energies induce diffusive effects whereas the energy produced through photosynthesis promotes the cycling of nutrient and other elements, and modification of the soil environment by a specific group of invertebrates called "ecological engineers" (II.2.4 and Chapter IV).

4.2 Heterogeneity

Heterogeneity is a normal feature of ecosystem structure. It is not an expression of disorder but originates from the diverse and multiple interactions between ecosystem components. In soil, heterogeneity is expressed in the spatial and temporal distributions of biotic and abiotic elements, and through their inclusion into different functional entities.

4.2.1 SPATIAL HETEROGENEITY

Spatial heterogeneity in soils occurs at widely different scales. At continental and regional scales, the units shown in pedological maps reflect the effects of the soil forming factors (largely climate, topography and the distributions of parent materials discussed in Chapter II) on their distributions. At the scale of several hectares, the toposequences (or catenas) that form down slopes and the vertical succession of horizons that occur within soil profiles are pre-eminent features. Micro-relief structures occur at a smaller scale. Finally, within specific soil horizons, macro-aggregates are subdivided into a diminishing hierarchy of smaller macro- and micro-aggregates which comprise the subsequent levels of the aggregate structure. Common soil properties such as clay and organic matter concentrations often have clearly-defined spatial patterns that vary depending on the scales at which they are considered (Figures 1.47-1.49).

4.2.2 TEMPORAL HETEROGENEITY

Chemical and physical soil transformations may occur over a broad range of time scales. In terms of profile formation processes, these range from the near-geological times of the *ca.* 1×10^6 years required for the formation of a fully-developed oxisol profile to less than 100 years which is the approximate time needed for the formation of a fully-developed A horizon (Birkeland, 1984). Biological processes may operate on time scales that are effective over the hundreds to thousands of years needed for succession in certain plant communities to the annual and seasonal scales, to those of a day or an hour, the generation time of fully active bacteria.

4.2.3 FUNCTIONAL HETEROGENEITY

Soils comprise a wide variety of components and structures varying in nature from biotic to abiotic and intermediate (the necromass). Any one of these units, from the simplest inorganic molecule to an aggregate may form part of larger composite functional entities and, therefore, potentially have many different effects. For example, soil nutrients sequestered within a pool of resistant organic matter with a slow turnover rate will have a much different impact on plant growth than the same nutrients incorporated into labile compounds with rapid rates of turnover. As indicated above (section 1.1.3), soil organic matter may be divided into pools of differing qualities, turnover times and abundances. Similarly, the effects of the broad biological systems of regulation (for example, the rhizosphere and the drilosphere, Chapter IV) and the histories of the soil components within which they are included (*e.g.*, successional processes taking place in the rhizosphere, ageing process within invertebrate faecal pellets) influence the present and future of these molecules.

Some of the difficulties associated with the study of soil processes arise from the variety of factors involved. Each phenomenon results from the multiple interactions of a diversity of components and sub-processes operating at different scales of time and space. It is necessary to understand the structure of this heterogeneity in order to incorporate the systems of interaction into a coherent intellectual framework. An hierarchical approach has been developed in Chapter IV to provide a more integrated approach to the study of soil processes (Mesarovic *et al.*, 1970; Allen and Starr, 1982).

CHAPTER II

SOIL FORMATION

The rocks that outcrop at the surface and other surficial parent materials are generally not in equilibrium with the atmosphere and the biosphere and are slowly transformed into soil, an intimate mixture of fine mineral particles and organic matter. As sub-components of ecosystems, soils are open thermodynamic systems through which energy flows and dissipates and this is reflected in the diversity of biological and physical structures present within them.

Soils are the products of complex interactions between the lithosphere, atmosphere and biosphere, and reflect the specific influences of each. While mainly comprising mineral materials from the lithosphere and organic matter, the pore spaces of soils possess distinctive atmospheres and hydrological environments. Water moves under the influence of gravity and capillarity, transporting solutes, colloidal and larger particles, while gases and vapours move by diffusion and mass flow. Water and gases react with the mineral matter in both physical and chemical interactions to release chemical energy through hydrolysis and oxidation.

Plants inject energy into the soil system partly through the growth and penetration of roots. More importantly, organisms also add energy to the soil as the partially-degraded products of their metabolic and digestive processes inluding such metabolites as root exudates, leachates, microbial polysaccharides, earthworm mucus and, eventually, their own dead biomasses.

Soil formation processes are driven by the expenditure of energy derived from predominantly biological sources and its dissipation through a series of inter-related physical, chemical and biological processes. Soils are dynamic entities, highly reactive, constantly-evolving and subject to the marked daily and seasonal rhythms of their environments but buffered from their extremes. They are also subject to longer-term variation associated particularly with climate change; the time scales of these range from a few years to those measured at the scales of geological processes.

Considerable spatial variation occurs in soils due to the lack of uniformity in the way internal processes act. Soil differentiation proceeds downwards from the surface leading to the formation of characteristic horizons and layers representing zones in the soil dominated by specific processes. The pore space is the site of the major interactions that occur between soil components (minerals, gases, water, living organisms and organic matter); in the confined environment of the soil, such interactions are a major feature of soil development and function.

Soil formation is a slow process and the formation of fully-developed soils occurs over time scales of tens to hundreds of thousands of years. It results from a balance between such aggradational processes as weathering and translocation, and degradational processes, principally erosion and leaching. The nature and properties of the soils that eventually form depend on the relative influences of the five soil forming factors discussed below and their expression through the medium of soil processes.

This chapter describes the differentiation of soils into horizons or layers and examines the factors important in soil formation. It discusses the major weathering processes that alter parent and soil materials and the pedogenetic processes that are the proximal agencies of soil formation and modification. It briefly describes the geographical distributions of the major weathering and pedogenetic processes, the principal soils that result and discusses approaches to their classification under various national and international systems.

II.1 SOIL HORIZONS AND THEIR NOMENCLATURE

Because of the influences of soil processes acting over time, the profile usually becomes vertically differentiated into a variable number of master horizons or layers that reflect the dominance of particular processes (Figure II.1). The genetic horizons and layers may differ in morphology, texture, the types of clay and other minerals, organic and nutrient reserves. The term 'genetic' implies that the horizon concerned has developed under the influence of the soil-forming processes considered later in this chapter. One horizon (the C) and one layer (the R) may occur beneath the more superficial horizons. The C horizon has been less influenced by pedogenetic processes and the R layer only slightly or not at all. In some soils, morphological and textural differentiation is slight with only gradual and somtimes continuous changes occurring with increasing depth.

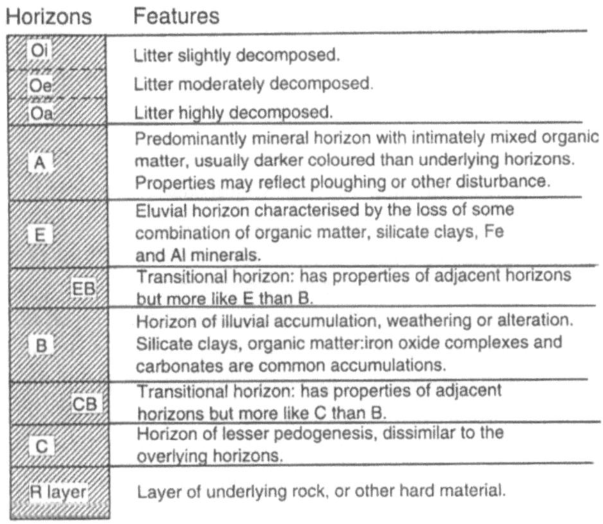

Figure II.1 Sequence of horizons in a hypothetic profile of a well-drained soil in the humid temperate climate region.

Various schemes exist for naming the master horizons. However, those used in the two principal international soil classification schemes, Soil Taxonomy (Soil Survey Staff, 1999) the FAO-UNESCO scheme (FAO-UNESCO, 1990) and its potential successor (The World Reference Base For Soil Resources / ISSS Working Group RB, 1998) are, with the exception of the H horizon, consistent at the master horizon level and will be used here. Full definitions of the master horizons recognised in the above systems are available in FAO-UNESCO (1990) and Guthrie and Witty (1982).

It should also be noted that not all of the master horizons considered below occur in every soil.

1.1 Predominantly organic horizons

The O or surface organic horizon is dominated by organic materials usually increasingly decomposed with depth. The FAO-UNESCO (1990) scheme requires that O horizons should contain more than 20 % organic carbon by weight. Under this system, O and H horizons are differentiated on the basis that the former are saturated with water for no more than a few days per year while H horizons undergo extended saturation. Due to faunal mixing, mineral matter increases in concentration towards the bases of many O horizons.

The H horizon
The H horizon comprises accumulations of organic matter that remain saturated for long periods each year. This horizon is characteristic of peats, or the histosols of the two classification schemes mentioned above. In Soil Taxonomy, the term H horizon is not used and such materials are considered to be O horizons.

The O horizon
A range of classification schemes exists for O horizons and most depend on the degree of decomposition of the accumulated organic materials. Because O horizons are best developed under forests, some older classification systems have been narrowly based on the layers pertaining in particular forest types. One such example is the classification of the litter layers of temperate-climate conifer forests in which three distinct layers may be recognised namely, the L (litter) layer comprising freshly-fallen needles, the F (fermentation) layer comprising needles in an advanced stage of decomposition but still recognisable as such and the H (humus) layer in which the original form of the needles is no longer recognisable. However, grasslands may also develop substantial O horizons, particularly where anecic earthworm populations are not sufficiently active to bury the surface litter. These may be quite different morphologically from those described above and sometimes comprise a diffuse layer with poor ground contact.

Other systems of classifying the O horizons divide them into layers depending whether the original form of most of the vegetation is still recognisable to the naked eye. One other system (Guthrie and Witty, 1982) bases its subdivisions on the reduction of the fibre content with state of decomposition after the material has been rubbed between the fingers.

Table II.1 presents the approximate equivalents of the divisions found in O horizons. The dotted line represents the approximate division between materials whose origin remains apparent to the unaided eye and those that are not.

Table II.1 Approximately-equivalent nomenclature for the morphological sub-divisions of the litter layers (O horizon) commonly reported in the literature. The dotted line indicates the approximate division between materials in which plant remains can be clearly identified and those in which it no longer can.

Degree of decomposition	Guthrie and Witty, 1982	Hesselman, 1926	Soil Survey Staff, 1951	Soil Science Society of America, 1965
slight	Oi	L		
			A00	O1
	Oe	F		
			A0	O2
advanced	Oa	H		

1.2 Predominantly mineral horizons

The A horizon

The A horizon normally occurs at the surface or beneath an O or H horizon, except in the less common situations where such profiles are buried, for example, beneath volcanic ash or alluvial deposits. A horizons are characterised by an intimate admixture of humified organic matter and the predominantly mineral soil. They usually appear darker than those below due to the organic matter present, some of which may form coatings on the mineral particles. The organic matter of the A horizon is usually derived directly from plant and animal remains and is incorporated into it by biological activity, rather than through physical translocation processes.

The E horizon

The E horizon usually underlies an H, O or A horizon and shows a concentration of sand and silt-sized fractions high in resistant materials due to the loss of such relatively mobile components as organic matter, silicate clays, iron and aluminium minerals. The absence of organic coatings on the resistant grains remaining frequently gives this horizon a lighter and sometimes an ashy colour.

The B horizon

This horizon is characterised by the complete or almost complete loss of original sedimentary or rock structure. B horizons possess one or more of the following:

(i) Accumulation of illuvial materials, such as clays or organic matter translocated downwards from overlying parts of the profile;

(ii) The presence of residual accumulations of materials such as sesquioxides (the oxides and oxy-hydroxides of iron and aluminium); or

(iii) Evidence of alteration of the soil material from its original state.

The C horizon

This horizon comprises unconsolidated material in various stages of weathering and from which the soil is considered to have formed. Roots are often active within this horizon and play important roles in weathering. Illuvial accumulations of organic matter, clays and larger particles, calcium carbonate and silica are often present in fractures and other voids within this horizon (Graham *et al.*, 1994).

The R layer

This is the hard layer of underlying rock which may include quartz, basalt, sandstone and other parent rock types. It supports no significant root activity.

1.3 Transitional and sub-horizons

The master horizon designations may be modified by the addition of qualifying suffixes to permit differentiation between vertical subdivisions within horizons and to identify the horizons in a general sense across different soils. One example is the need to reflect different textural properties within the B horizon and such sub-horizons may be denoted as B1, B2, B3, etc. Similarly, the designation Ap denotes a sub-horizon disturbed by tillage and the designations Bh, Bt, Bk and Bs denote B horizons with, respectively, accumulations of organic matter, translocated clay, pedogenic carbonate and sesquioxides combined with organic matter. Indurated horizons are indicated by m. More than one suffix may be used and an horizon that is largely or completely cemented with deposits of pedogenetic calcium carbonate is denoted by the term Bkm. It should be noted that the qualifying suffixes do not always have the same connotation in different systems of classification.

The boundaries between adjacent master horizons may be discrete or transitional horizons may occur with some of the properties of both. Such horizons are named from their adjacent master horizons, that most ressembling the transitional horizon taking precedence. Thus, a transitional horizon lying between the A and B horizons but most ressembling the B would be designated BA. Alternatively, horizon boundaries can be convoluted and the horizons may interdigitate into each other; occasionally small parts of one horizon may become completely surrounded by the materials of that adjacent.

While many soils acquire dust and other inputs over long periods, certain soils may have a layer of some other material, such as volcanic ash, loess or alluvium deposited on their surfaces and these then become the parent materials of new soils. Over time, such parent materials differentiate into new horizons and, where such additions recur over time, sequences of buried horizons may be apparent in such exposures as road cuttings, notably in the volcanically active areas of the world.

II.2 THE SOIL FORMING FACTORS

During the latter part of the nineteenth century, the Russian pedologist Dokuchaev developed the integrating concept that soils form under the influence of the five soil forming factors: climate, parent material, time, the biota and topography. In the intervening years, considerable efforts have been made to further develop this theory of soil formation and to quantify the effects of the individual factors (Jenny, 1941, 1980). However, these have only been partially successful. Jenny's ideas of independence between the soil forming factors and his limited appreciation of the biological influence (notably omission of the role of biomechanical processes) on soil development much constrained its further development. In addition, no account was taken of the capacities of soils to develop without corresponding changes in the soil forming factors. Despite these drawbacks, development of this theory has provided a stable conceptual basis for soil formation; its substantial contributions to soil and related sciences are undoubted (Johnson and Hole, 1994).

With the exception of time, the soil forming factors are now considered as dependent and mutually interacting with multiple feedback effects occurring between them (see, for example, Chesworth, 1992). Furthermore, an hierarchy exists with climate playing a dominant role over parent materials and topography. The biotic factor is now considered to encompass the effects of all groups of the biota that impinge on soils either by living within them, on their surfaces or contributing organic matter to them. Time is a different sort of factor but of great significance because of the differing and extended periods over which soil development occurs. Hole (1961) considered that a sixth factor, space, should also be included in soil formation theory because of the importance of lateral differentiation.

The soil forming factors operate wherever soil and soil-forming materials occur, although their relative influences differ between soils and with location on the earth's surface. The soil forming factors do not directly influence the soil but are considered to act through the medium of soil processes. These processes act in potentially different combinations in each environment and their myriad combinations and degrees of expression are reflected in the wide diversity of soils found on the surface of the earth.

2.1 Climate

The regional climate is a dominant factor controlling the formation of all soils (Birkeland, 1984), since it circumscribes the forms and rates of local weathering, the translocation of weathering products and other pedogenetic processes. It interacts with, and conditions the effects of the other factors in determining the biota that can survive in particular environments and the seasonality of its activities. Through its control of temperature and moisture regimes, climate determines the phase of the soil water and the intensity of water fluxes. It thereby controls the transport of solid particles and dissolved materials within developing soils, over their surfaces and laterally in the landscape.

2.2 The parent material

This is the basic inorganic material from which the soil will be formed and, depending on its physical, chemical and mineralogical composition, strongly influences the composition and texture of the resulting soil.

The chemical and mineralogical composition and the physical structure of the parent material set the initial conditions of the incipient soil. Thus, soils formed from basaltic rock parent materials usually contain higher concentrations of phosphorus, alkaline earth elements and iron than those formed from other parent materials because of the higher concentrations of such elements in these compared with most other rock types. The forms in which they occur and the concentrations of these elements affect the ability of the soil to supply biological nutrient and other elements and thereby condition the way in which the profile develops.

The grain size of the parent rock is important in that coarse grained igneous rocks usually weather more rapidly than do those of finer grain size (Birkeland, 1984). It also sets the size of the units that are initially detached from the parent rock by weathering processes. Thus, soils derived from fine-grained basaltic parent materials have a finer texture than those derived from coarse-grained granitic parent materials.

The type of the parent rock affects soil formation particularly through (Brewer, 1954):

(i) The amount of clay that may be potentially formed by *in situ* weathering. This varies with the nature of the parent rock and intensity of water fluxes. It may vary from quite small amounts to large: for example, 12 % kaolinite from a sandstone to 33-55 % from biotite granite to 99 % kaolinite from an amphibolite-pyroxine schist (Macias and Chesworth, 1992);

(ii) The amounts of alkali (notably Na, K), alkaline-earth (largely Ca, Mg) metals and iron that can be released by weathering;

(iii) The ease of release of the above metals;

(iv) The permeability of the parent rock.

While rocks are the most common parent materials, soils also develop on a diverse range of materials transported by wind, water, gravity and vulcanism, and on the truncated profiles of previously eroded soils. Agriculturists profoundly modify soil chemical and physical properties through such practices as drainage, flooding and the addition of organic and inorganic ameliorants; these are also new parent materials for further soil development.

In addition to the parent materials considered above, a range of incipient soils are currently developing on parent materials provided through man's activities. These include such materials as urban organic and inorganic wastes and materials dredged from wetland environments. A variety of overburden materials are deposited at the surface from mining operations and these, together with waste materials from initial mineral processing, manufacturing and power-generating industries also provide a diversity of unusual parent materials, some with properties that may be inimical to soil development processes. For this reason, composite soils are often formed by topdressing with a layer of fully-developed soil transported from elsewhere to provide a favourable base for the growth of stabilising vegetation.

2.3 Time

Soils undergo extended and complex series of reactions and processes during formation from their parent materials. The net effects of these eventually lead to the differentiation of fully-developed profiles and, at a landscape scale, of suites of related soils. Depending largely on parent materials and environment, the development of fully-differentiated profiles occurs at widely variable rates. Chesworth (1992) considered that in humid temperate climates, the formation of a metre of temperate soil from aluminosilicate parent materials is likely to take approximately 20,000 years and half that on carbonate-rich parent materials. The formation of a similar depth of ferrallitic soil (oxisol) in the Côte d'Ivoire may take up to 77,000 years (Leneuf and Aubert, 1960). In contrast, an Andisol with a distinct AC horizon structure may form in Japanese volcanic ash in as little as 500 years although more than 1000 years is required for a fully differentiated ABC profile to develop (Yamada, 1968 in Wada, 1985).

Duchaufour (1982, 1997) divides soils into those forming over short and long cycles. The short-cycle soils develop over periods ranging from less than 10^3 to 10^4 years. Some soils in this group owe many of their properties to those of their organic matter inputs; they include such groups as the peats (histosols) and the spodosols. Other soils in this category include those whose pedogenetic differentiation has progressed little (entisols, inceptisols), those forming on rapidly weathering parent materials (for example, certain mollisols forming on carbonate-rich parent materials) and those whose pedogenetic development requires relatively small changes to the parent materials (vertisols). The long-cycle soils require periods of 10^5 to 10^6 years for development and are typified by the highly-weathered soils of the humid tropics, the ultisols and oxisols.

Over long periods of time, soils continue to change with on-going inputs and weathering while climatic change and mechanical disruption may influence profile differentiation. In the absence of disturbance, most changes that occur are smooth (Bockheim, 1980b) although exceeding a threshold in some critical process may invoke relatively rapid change in the direction of development, as illustrated in Section II.4.1 (Birkeland, 1992).

As discussed later in this chapter (Section II.4.4.3), extended weathering leads eventually to soils that are highly depleted in plant nutrient elements and possess characteristic suites of the simpler clay minerals, typically with low and pH-related cation exchange capacities and therefore very limited capacities for nutrient retention. Chesworth (1973a) points out that differences between soils formed on various parent materials are greatest in younger soils and diminish but rarely disappear completely with time. The chemical compositions of old, highly-weathered soils become remarkably uniform with chemical compositions increasingly dominated by the elements Al, Fe and Si, irrespective of their original parent materials (Chesworth, 1973b).

Table II.2 illustrates the approximate minimum times required for fully-developed profiles to differentiate from their parent materials in selected orders of Soil Taxonomy (Birkeland, 1984). However, full development within these orders may take considerably longer. Further, a number of soils may develop sufficiently over time to require classification into other orders. As pointed out by Birkeland (1984), categorisation of the landscape distributions of soils at the level of order provides information on the relative and approximate actual ages of the different parts.

Table II.2 The approximate minimum times required for development of soils recognisable as belonging to selected orders of Soil Taxonomy (Duchaufour, 1982; Birkeland, 1984).

Order	**Approximate time for development (years)**
Entisols	100
Andisols	500-1000
Histosols	<1000
Vertisols	<1000
Inceptisols	1000
Spodosols	few centuries to >2000
Mollisols	1000-10 000
Aridisols	1000-10 000
Alfisols	>10 000
Ultisols	100 000
Oxisols	1 000 000

2.4 Biota

As developed further in later chapters, the biota is the major factor in soil formation in that its effects define the difference between soils and 'not soils'. The biota has continuing inputs into soil genesis from its earliest stages and in soil maintenance and long-term development. In the past, the effects of this factor have been wholly or largely attributed to the effects of higher plants (see, for example, Jenny, 1941; Birkeland, 1984). However, it is clear that the important roles of other sections of the biota must not be neglected if further progress is to be made in understanding soil formation (see, for example, Johnson and Hole, 1994). The biotic factor is now considered to include the trophic and non-trophic effects (Jones *et al.*, 1997) of all living organisms that impinge on the soil, both directly and through the products of their metabolism and decay.

The initial colonisation of a new parent material by organisms marks the start of its stabilisation and of soil formation. Links between initial ecosystem establishment and development processes and those of soil formation are so complete that it is inappropriate to attempt to separate them, except perhaps in the case of the most incipient soils, some man-made and heavily man-influenced soil systems.

The initiation of simple energy flow and basic material cycling pathways occurs at the earliest stages of ecosystem development and soil formation and these trophically-based systems continue to develop in degree and complexity over time (Odum, 1983). Of major importance in ecosystem and soil development and maintenance are those species that directly or indirectly control the availability of resources to other species, the ecosystem engineers (Jones *et al.*, 1997). Through their non-trophic activities, these organisms physically modify, maintain or create new habitats for other organisms. One effect of such organisms is to create greater habitat diversity which may in turn lead to increased species diversity.

Physical ecological engineers may be divided into two groups. The first comprises the autogenic physical engineers that modify their environments through endogenous processes such as tree growth that alter the structure of the engineer which then remains as part of the environment. Allogenic engineers alter their environments by transforming living or non-living materials from one physical state to another, as occurs in the creation of large voids in the soil through root decay (Jones *et al.*, 1997). Trees also act as geochemical engineers in that deep roots take up soluble silica released through litter decomposition processes in the surface horizons, thereby ensuring the stability of clay minerals (Lucas *et al.*, 1993).

A number of organisms may also be designated ecosystem chemical engineers for the roles they play in mediating changes to inorganic and organic substrates through secretory and excretory activities. This group includes the micro-organisms because of their roles in such diverse processes as rock weathering, organic matter breakdown, nutrient cycling, nitrogen fixation and the acquisition of nutrients through mycorrhizal associations. Other organisms including higher plants and animals also play substantial roles in this respect. Many examples of both major types of engineers appear throughout this book.

Close interactions between the biota and the other soil forming factors are also clearly apparent. Climate, parent materials and topography control the continually-changing chemical and physical environment within which the biota acts. In turn, the biota acts to progressively modify topography, the parent materials and the developing soil.

2.5 Topography

The importance of topography lies in its control of site hydrological regimes and it therefore exercises an important influence on soil development and maintenance. In addition, through its influence on the distribution of erosional products, topography modifies the effects of such pedogenetic processes as creep and frost heave thereby affecting the landscape distributions of such parent materials as colluvium. The lateral and downhill transport processes that occur at and below the soil surface in all landscapes are largely controlled by topography and lead to the formation of the suites of spatially-related soils known as toposequences or catenas, discussed later in this chapter.

II.3 THE STAGES OF SOIL FORMATION AND DEVELOPMENT

The formation and evolution of soils involve a series of physical, chemical and biological processes which act progressively over time, are controlled by climatic variables and are greatly influenced by topography. Simonson (1978) divides such processes into additions from without the soil system, removals or losses, translocations (or transfers) within the system and transformations of contained materials.

The original parent material is transformed by *in situ* weathering into a mixture of stable mineral components which blend intimately with organic materials to form the soil. The parent material is first broken down into its primary minerals whose decomposition products may be partially transformed into secondary or neoformed minerals.

From this early stage, the nutrients necessary for plant production and such other essential components as Al and Fe accumulate progressively in the upper parts of the incipient soils. Clay fractions are formed firstly through alteroplasmation *i.e.*, transformation of primary minerals into clays with no subsequent modification of rock structure. Pedoplasmation is a subsequent transformation whereby clay minerals acquire a pedological structure and such specific properties as swelling and shrinkage.

The initiation of biological activities within the developing substrate leads to the accumulation of organic matter. This organic matter mixes with the weathering mineral components to form an A horizon that becomes an active source of further physico-chemical changes in the underlying parent material to develop a C horizon. CO_2 evolved from decomposing organic matter also participates in the process. With further development, weathering and downward transport of materials progressively modify the deeper strata of the parent material and, depending on the processes operating, E and B horizons may form. At this stage, translocation, biological transport and erosion become the dominant processes in the evolution and differentiation of the soil profile.

Nutrient and other elements (*e.g.*, Si and Al, Cornu *et al.*, 1997) and organic materials are continually lost in solution and suspension through erosion and by transport in sub-surface water flows. Such losses are expected to be greatest in incipient soils with juvenile ecosystems and to diminish with ecosystem and soil development (Odum, 1983).

3.1 Weathering

Weathering is the sum of the processes involved in the alteration of materials at and near the surface through complex interactions between the lithosphere, the atmosphere, the hydrosphere and the biosphere that occur over time. It may extend far below the surface and includes all the physical and chemical processes responsible for rock fragmentation and the production of dissolving ions. Weathering in the upper part of the parent materials has been considered separately as pedochemical weathering because of the considerable influence of the biomass in the production of, *inter alia*, complexing agents, substances that attack clay minerals or form organo-mineral complexes. However, biological weathering influences may extend well into the underlying regolith (Graham *et al.*, 1994), although increasingly attenuated with depth.

While physical weathering is often considered separately from chemical weathering, in reality they operate together, often in a synergistic way (see, for example, Ugolini, 1986b). Weathering involves the simultaneous activities of a range of processes including physical fragmentation, inorganic chemical processes (hydrolysis, oxidation, hydration and dissolution) and biologically-mediated processes (*e.g.*, acidolysis and acido-complexolysis). The weathering processes predominating at a site are determined by climatic, biological and lithological factors and the degree of evolution of the soil (Jenny, 1980).

In all weathering systems, water plays dominant physical and chemical roles. In the solid phase it is a major agent of landscape sculpting and transport while as a liquid, it is also an important agent for the diffusion and transport of materials. It is a potent medium of physical disruption through volume change, both as a consequence of phase change and through involvement in hydration and related reactions. Chemically, it is an effective solvent, a component of many reactions and of neoformed products, and an important buffering agent (Chesworth, 1992). Little chemical weathering occurs in very dry and frozen environments.

3.1.1 PHYSICAL FRAGMENTATION

Physical fragmentation of the parent rock is linked to the development of stresses inside the rock or individual minerals. These initially lead to the development of micro-fissures, then to systems of fissures and extensive fracturing (Robert and Tessier, 1992) thereby opening the materials to chemical weathering processes. Fragmentation may result from a range of mechanisms although the relative contributions of chemical and physical processes are still a matter of some debate (Ugolini, 1986b).

Water plays a major role in physical weathering both directly through abrasion and in transporting the products of breakdown. Past glaciation, in particular, has been broadly effective in modifying landscapes and producing till which is a common parent material in some parts of the world. In such cold, arid environments as Antarctica, the efficiency of fluid water based physical weathering is substantially limited by the scarcity of water and the brief period during the year when it is available for transport and other processes (Campbell and Claridge, 1987).

On freezing, water increases in volume by approximately 9 % and when it penetrates rocks and subsequently freezes, sufficient pressure may be exerted to fracture rocks along lines of weakness. The effects of freezing and thawing differ between rock types but are most effective where this boundary is crossed frequently and in rocks that contain considerable water. In certain rocks, such effects result in the production of considerable silt but also clay-sized materials (Ugolini, 1986b). Legros (1992) reported that the freeze: thaw boundary is crossed *ca.* 165 times per year at an elevation of 2500 m in the French Alps and leads to the fragmentation of exposed rocks and scree formation.

Insolation and fire are other temperature-related factors involved in rock fragmentation through the differential expansion of rock constituents and layers. Disruptive pressures may also build up through volume increases consequent on the weathering of such minerals as biotite and plagioclase to clays (Birkeland, 1984). Volume increases in rock constituents may also result from such salt weathering effects as crystallisation, changes in the level of hydration and thermal expansion, and through chemical reactions such as the formation of iron oxides (see, for example, Robert and Delmas, 1984).

Wind erosion is highly effective in desert environments through both particle transport and abrasive detachment, that is a 'sand-blasting' effect. In the northern polar desert and the cold desert of Antarctica, this is evidenced by the preferential erosion of the sides of boulders facing the prevailing winds (ventifacts) and the patterns of transport and sorting of surrounding finer particles (Campbell and Claridge, 1987). It is also evidenced by the widespread formation in these areas of lagged surfaces (desert pavements) in which a layer of small hard rocks forms a protective cover over the finer underlying materials. Measured rates of wind erosion on exposed rocks are of the order of 1 mm per year (Ugolini, 1986b).

Fungal hyphae are capable of penetrating intact rock surfaces thereby accelerating micro-division (Robert and Berthelin, 1986). Lichens are common inhabitants of rock surfaces in many environments and secrete chelating and other agents which affect rock surfaces. In a study of the lichen:rock surface interface of a volcanic rock, Adamo and Violante (1991) recorded the extensive physical disintegration of the rock surface with separation of surface particles and widespread etching of the mineral grains. They also found a mixture of rock-forming minerals, clay minerals and poorly ordered alumino-silicates thus demonstrating the close interactions between physical and chemical weathering. More complex communities of bacteria, algae and fungi exist in lithic communities (Vincent, 1988) although their roles in rock weathering remain to be defined.

Roots and burrowing soil animals may penetrate the matrix of the weaker and chemically-weathered rocks and indurated layers. As an example of the latter situation, Johnson *et al.* (1987) reported the widespread local disruption and surface deposition of a dense 0.5m-thick calcrete (Bkm) horizon and the underlying C horizon through the burrowing activities of wombats (Marsupalia: Vombatidae) in South Australia. In stronger rocks, roots may penetrate and enlarge cracks and fissures thereby exposing enlarged areas of reactive mineral surface to the actions of the major chemical weathering agents CO_2, O_2 and water (Graham *et al.*, 1994).

3.1.2 CHEMICAL PROCESSES

Chemical weathering results from the reactions that take place between primary and secondary minerals and a weathering solution (Figure II.2). The primary minerals are degraded, secondary minerals including clays and clay-humus complexes are formed, some of the weathering solution is retained and a leachate is produced. Silica is exported and new crystalline minerals may be precipitated from materials dissolved in both the retained solution and the leachate through the process of neoformation (Pédro, 1979):

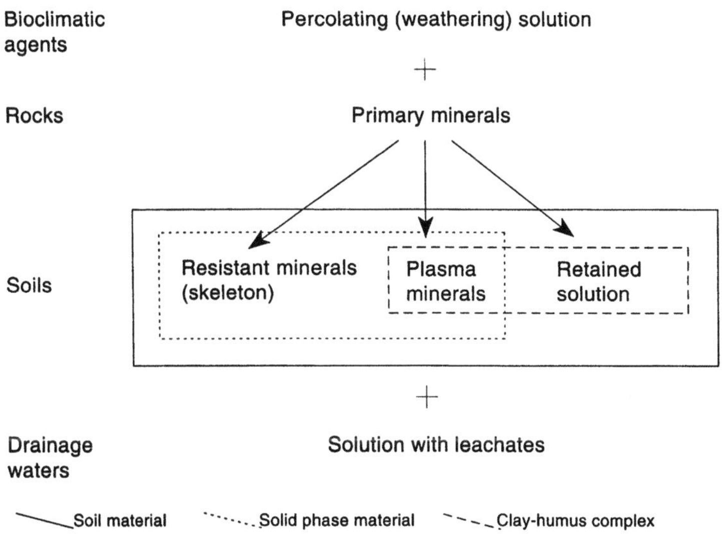

Figure II.2 A generalised scheme of chemical weathering (redrawn from Pédro, 1989).

Depending on the pH and the composition of the leaching solution, different chemical weathering and a wide range of processes may ensue (Table II.3). Chesworth (1992) classifies such processes into three major weathering trends (acid, alkaline and reducing).

Table II.3 The dominant processes in the chemical weathering of aluminosilicate minerals in solutions of differing pH and salinity (Pédro and Sieffermann, 1979).

Salinity	**pH ranges**		
	<5	**5-9.6**	**>9.6**
<0.001 N	ACIDOLYSIS	HYDROLYSIS	
>0.001 N		SALINOLYSIS	ALKALINOLYSIS

Dissolution

Dissolution is the movement of materials into solution and occurs in the presence of solvents; it may be either rapid or slow, depending on the solubilities of the materials involved. Most common minerals are poorly soluble and dissolution often complements other processes by removing their more soluble products in solution.

Neutral hydrolysis
Hydrolysis is the degradation of aluminosilicate minerals by the free H^+ or OH^- ions of water in weak saline solutions and at pH values between 5 and 9.6; it is accelerated by the presence of dissolved CO_2. As an example, the hydrolysis of feldspars may be written as follows (Pédro and Sieffermann, 1979):

$$(Si_3Al)O_8K \xrightarrow{H_2O} 3\ Si(OH)_4 + Al(OH)_3 + K^+, OH^-$$

where the $Al(OH)_3$ occurs in the form of the crystalline solid gibbsite.

Acidolysis
Acidolysis occurs in the presence of such acid solutes as the organic acids released from biological materials; the reaction is:

$$(Si_3Al)O_8\ K \xrightarrow[HCl]{H^+} 3\ Si(OH)_4 + Al^{3+}, 3Cl^- + K^+, Cl^-$$

where all components are soluble.

Through these reactions, the primary silicates are broken down, dissolved and their constituents transformed into secondary minerals (silicate clays, oxides, hydroxides). Pédro (1964) showed in laboratory weathering experiments simulating tropical conditions, that a variety of parent rock materials (basalt, granite and andesitic lava) are rapidly hydrolysed to form secondary sesquioxides (*i.e.*, Fe and Al oxides and hydroxides) within a few months. In contrast, crushed granite placed in lysimeters under simulated temperate climate conditions showed a significant microdivision of the rock, but no mineralogical modification. Hydrolysis therefore appears to be the dominant process occurring under conditions of high environmental temperatures and available free water.

Partial acidolysis may lead to aluminization, a transformation in which the 2:1 clay sheet structure is preserved leading to aluminized clays (Pédro, 1989).

Acido-complexolysis
Acido-complexolysis occurs when organic acid components (simple aliphatic or fulvic acids) form mobile soluble complexes with the Fe and Al of the primary clay minerals and impede the formation of secondary minerals. This is the characteristic mode of weathering in the spodosols.

Salinolysis
In highly saline solutions, hydrolysis is replaced by salinolysis in the pH range of 5 to 9.6 (chlorides or sulphates) and alkalinolysis (solonisation) where the pH is greater than 9.6 (carbonates).

Salinolysis of feldspars leads to the following reaction:

$$(Si_3Al)O_8K \xrightarrow{H_2O\ -\ NaCl} 3\ Si(OH)_4 + Al(OH)_3 + K^+ + Cl^-$$

Alkalinolysis may be written:

$$2\ (Si_3Al)O_8K \xrightarrow{OH^-,\ (Na_2CO_3)} 6SiO_3^{2-}, 2Na^+ + 2AlO_2^-, Na^+ + 2K^+ + CO_3^{2-}$$

Hydration and oxidation

Hydration and oxidation may supplement the above processes. Hydration transforms the original minerals through the addition of water molecules. A few specific hydration reactions are relevant to soil formation *e.g.*, formation of the iron hydroxide (α-FeOOH. goethite) from iron oxide which constitutes the cement of some sandstones and may lead to their subsequent degradation. Oxidation is often mediated by oxygen dissolved in water. When associated with hydrolysis, it particularly affects reduced iron which may be retransformed into ferric oxide (Section I.1.2).

Neoformation

As weathering progresses, primary rock-forming minerals break down through hydrolysis and other processes. Secondary minerals may form through the degradation of the primary minerals or new (neoformed) minerals may crystallise from the ions released into solution. These new minerals are more stable in the near-surface environment.

The types and amounts of neoformed minerals that develop depend on the weathering environment and the age of the landscape. The distributions of the dominant chemical weathering processes throughout the world and the typical products of neoformation are discussed below in relation to soil-forming processes (Section II.4).

3.1.3 BIOLOGICALLY-MEDIATED PROCESSES

The four chemical weathering processes described above may involve purely abiotic interactions and can occur in the absence of biological influences. However, they may often be promoted directly by biological activity, or indirectly by the activities of simple metabolites or organic compounds. Nutrient element uptake by plants is offset by the release of equivalent quantities of protons to maintain a net electrochemical balance (Grimaldi and Pédro, 1996). It has been well documented that simple organic and humic acids from decomposing litter can destroy silicates through acidolysis (Lossaint, 1959; Juste, 1965). Carbon dioxide produced by respiration and other metabolites may also promote acid hydrolysis, alkalinisation, oxido-reduction or acido-complexolysis (*i.e.*, acidolysis by acid organic complexes) (Berthelin, 1976).

Direct substrate alteration results from:

(i) Microbial activity in the rhizosphere (Leyval and Berthelin, 1982; Stucki *et al.*, 1992);

(ii) Enzymatic reduction of iron (Berthelin and Kogblevi, 1974);

(iii) The direct activity of roots which may absorb and concentrate such minerals as calcite in their cells (Jaillard, 1983), or at their surfaces.

Roots may produce some secondary silicates (Callot *et al.*, 1992), and this over time may influence much of the soil volume through exploration by the root system. Silica is precipitated in substantial quantities in plant tissues and is returned to the soil in litter as phytoliths (Section I.1.1.1.2). Also, clay minerals may be altered during transit through the earthworm gut and by termites through manipulation in the buccal cavity (Boyer, 1973a).

Indirect alteration is the modification of mineral structures by soluble organic compounds which have the capacity to attack minerals and thereby promote rapid weathering and leaching. However, products of this attack may be insolubilised through precipitation or flocculation-coagulation processes in the presence of active minerals (clays, active Fe cations) (see *e.g.*, Toutain, 1974) or metabolised when microbial activity is intense (Berthelin *et al.*, 1979). In temperate regions, weathering and subsequent leaching are also intense although aggressive organic compounds may be neutralised rapidly through insolubilisation. Soil invertebrates often play a cardinal role in mediating these processes.

Soil invertebrates produce two kinds of faeces within which contrasting reactions occur:

(i) The casts of anecic earthworms (those that ingest a mixture of soil and surface litter and inhabit galleries) are macro-aggregates of 2.5 to 10 mm in diameter and form what are known as 'macroaggregate closed-systems', since they are a mixture of organic and mineral elements supporting intense microbial activity. This leads to the rapid flocculation of soluble organic compounds and largely restricts mineral weathering.

(ii) Arthropods that live in the surface litter and holorganic horizons produce small (<100 μm) and unstable faecal pellets. Such aggregates are composed of partly comminuted, but largely-untransformed organic matter with low concentrations of associated minerals. Microbial activity in these aggregates is generally low. These aggregates are subject to intense leaching and the aggressive organic compounds released promote intense weathering and the loss of minerals. They form the basis of what is known as 'aggressive leaching systems'.

Therefore, under comparable conditions of climate, mineralogy and cation concentration, the structure of invertebrate communities can substantially influence prevailing weathering patterns. Equivalents of both these systems may also be found in the rhizosphere and their activities will depend on the chemical compositions of the root exudates and root litter present.

3.1.4 THE MAJOR WEATHERING PATTERNS

Weathering transforms the original bedrock into a mixture of large, slightly-altered mineral fragments and other materials, the soil skeleton; such materials, except through pedoturbation, are little subject to movement during soil formation. Smaller secondary minerals and soluble materials (the soil plasma) result from weathering of the primary minerals (alteroplasmation), or secondary crystallisation following the precipitation or flocculation of solutes and colloids (pedoplasmation) (Flach *et al.*, 1968). This plasma mainly comprises clays and other minerals of colloidal size which may subsequently be redistributed during soil formation.

Two main forms of weathering have been defined (Pédro, 1983):

3.1.4.1 *Aplasmogenic weathering*

Aplasmogenic, or congruent weathering occurs in situations where no plasma is formed because of the complete dissolution of the alterable minerals and the absence of secondary minerals from the weathered substrate. This type of weathering pattern is characteristic of soluble salt minerals such as halite and gypsum which dissolve into their component ions. Calcite dissolves in the same way in the reversible reaction (Birkeland, 1984):

$$CaCO_3 + CO_2 + H_2O \leftrightarrow Ca_2^{2+} + 2HCO_3^-$$

Further examples include the acidolysis and acido-complexolysis that occurs in podzols, alkalinolysis by sodium carbonate in certain alkaline soils (solonetz, Duchaufour, 1997), or attacks on minerals by acidic cations at the interface between dry and flooded parts of the soil, ferrolysis. This last process is a special case of salinolysis by ferrous cations in soils affected by intermittent waterlogging; it leads to the loss of bases, acidification and clay mineral destruction, as occurs in planosols (FAO-UNESCO, 1990) or argialbolls (Soil Survey Staff, 1999).

3.1.4.2 *Plasmogenic weathering*

Plasmogenic, or incongruent weathering is characterised by the formation of secondary minerals which are precipitated to comprise the plasma of the weathering materials. Two forms may be distinguished, depending whether weathering is complete or partial.

If weathering is complete, no primary minerals remain and the plasma may be composed entirely of secondary minerals. This process is particularly favoured in humid, moist conditions where hydrolysis is complete, and the plasma comprises neoformed clay minerals, ferric and aluminium oxides and oxy-hydroxides (oxisols, Soil Survey Staff, 1999). Three separate stages may be defined, depending on the intensity of hydrolysis.

Bisiallitisation

Bisiallitisation occurs where clay minerals of the 2:1 type (Section I.1.1.1.2) are formed secondarily. This occurs, for example, in vertisols where smectites form under conditions of impeded drainage. In these soils, silica is retained although partial dealkalinisation (*i.e.*, removal of alkaline cations associated with the clay sheets) occurs.

Monosiallitisation
Monosiallitisation is the process which leads to the formation and accumulation of secondary 1:1 clay minerals, most commonly kaolinite. In this situation, the alkaline cations from clay sheets and part of the silica have been lost a situation typical of ferrallitic soils (oxisols and ultisols; Soil Survey Staff, 1999).

Allitisation
Allitisation is an extreme stage of hydrolysis, rarely achieved, in which all the silica and alkaline cations have been lost and only gibbsite ($Al(OH)_3$) derived from the original clay minerals and hydrated ferric oxides remain.

Apparent bisiallitisation
Where weathering is limited to slow and incomplete hydrolysis, as in cold or dry climates, primary minerals remain largely unmodified for long periods. Under these circumstances, the plasma comprises primary 2:1 minerals (micas) inherited from the parent material and secondary minerals resulting from their transformation. The formation of secondary minerals is limited to hydrated iron oxides and $CaCO_3$. Clay minerals of the 2:1 type are dominant and weathering involves a partial hydrolysis of the primary clays through replacement of the interlayer K^+ with hydrated Ca^{2+} and Na^+ ions from the weathering solution. This leads to the formation of vermiculite and then smectite (Pédro and Sieffermann, 1979) in an apparent bisiallitisation. Examples of this are found in the brown soils (udalfs; Soil Survey Staff, 1999) of temperate regions and fersiallitic soils (ustalfs; Soil Survey Staff, 1999) of semi-arid regions.

3.1.5 EFFECTS OF CLIMATE ON THE WORLD DISTRIBUTION OF THE MAJOR WEATHERING PROCESSES

Climate is the major determinant of the weathering process through its control of temperature and moisture, water flux and the decomposition rates of organic materials. In extremely cold climates, weathering is dominated by physical processes while in cool and temperate climates a limited biochemical weathering dominates. As the climate becomes increasingly warm, a more intense geochemical weathering gradually supervenes to become completely dominant in humid equatorial soil environments (Duchaufour, 1997).

The coldest soils of the world occur in the cold desert of Antarctica and perhaps in northern Greenland (Campbell and Claridge, 1992); only slightly less cold are those of the subadjacent northern polar desert (Ugolini, 1986a). The dominant weathering processes in these largely unvegetated environments are physical although a limited chemical weathering of rocks and soils is evidenced by the red and brown staining that occurs on rocks, the oxidised horizons that occur in the soils and the presence of soluble salts derived from rock weathering. Some hydration and transformation of the 2:1 clays present has also been reported (Campbell and Claridge, 1992). The major chemical weathering processes in these environments are salination (and sometimes alkalinisation), calcification and oxidation (Ugolini, 1986ab).

Bockheim (1979) found several well-defined relationships between soil chemical properties and age in Antarctic soils. For example, there was a linear relationship between water-soluble salt content (as electrical conductivity, EC) in the horizon of maximum salt content and age for soils ranging from 18 000 to 135 000 years:

$$EC\ (dS\ m^{-1}) = -36.6 + (3.8\ \log_e.\text{age})$$

The tundra occurs to the south of the polar desert zone and is characterised by a closed vegetation (Ugolini, 1986a); organic components thus have a much greater organic influence on soil development. Most soils in the tundra are formed in heavy-textured alluvium in low-lying environments where the permafrost is close to the surface. These soils occupy 90-99 % of the landscape and are largely saturated with meltwater during the summer thaw. Soil processes are therefore dominated by reduction, gleying and low rates of mineral weathering. The low rates of organic matter decomposition lead to peat formation. At well drained sites, the weathering processes are transitional with the southern tundra soils showing some of the acidolysis, acido-complexolysis and translocation typical of spodosols. The northern tundra soils evidence little of this with an A horizon in which the soil organic matter is stabilised in humus-Fe^{3+}-clay complexes and a B horizon that shows little evidence of illuviated materials, apart from some silt (Ugolini, 1986ab).

In cool climates with vegetation, decomposition is slow to intermediate and water-soluble acid solutions are very active in promoting hydrolysis, acidolysis or acido-complexolysis. Biological activity in temperate regions is normally high, and may lead to a rapid immobilisation or insolubilisation of organic acids; partial hydrolysis becomes the main weathering process. However, where the parent material is of low nutrient status (for example, 'acid' materials high in silica) or the vegetation produces 'low quality' litters resistant to decomposition, active biological communities may not develop which leads to weathering conditions dominated by acidolysis or acido-complexolysis.

In contrast, weathering in hot and humid climates is considered to be dominated by geochemical weathering processes that extend over very long periods (Duchaufour, 1997). In these climates, decomposition is so rapid that intermediate organic compounds may have little effect beyond the surface horizons. Hydrolysis becomes massive because of high temperatures and high carbon dioxide concentrations in the soil solution, particularly at depth. CO_2 concentrations normally increase with soil depth, sometimes markedly (Section I.1.2.2). Concentrations found at depth in an Amazonian oxisol (Lucas and Chauvel, pers. comm.) have been up to forty four times greater those of the atmosphere and may accelerate the alteration of primary minerals by lowering the pH (by up to 0.5 units).

Monosiallitisation and allitisation are the dominant processes of humid tropical climates, unless drainage is impeded when neoformed smectitic 2:1 clay minerals may accumulate. Where longer dry seasons occur, weathering is limited to mono- and bisiallitisation. In hot deserts, physical weathering processes predominate although important chemical weathering processes may still occur (Watson, 1992).

Table II.4 summarises the major chemical weathering processes. The minerals that form in hot and humid tropical environments are typically kaolinite and goethite and, at the wetter extremes, gibbsite. In warm areas with alternating wet and dry seasons, smectites are typically the dominant clay minerals. Degradation rather than neoformation is typical of temperate climates and incompletely altered minerals such as exfoliated illite and chlorite, irregular mixed layers, vermiculite and degraded smectites are characteristic. In the cold areas of the high latitudes and altitudes and in the arid areas of the world, hydrolysis is limited by the lack of water. Few necformed clays are formed and the clay sized materials present consist largely of the minerals also present in the parent rocks. The broad scale patterns given above are modified by local conditions of relief, poor drainage, salt accumulation and other factors.

Table II.4 The major weathering processes, their characteristics and major determinants (after Pédro and Sieffermann, 1979).

	PROCESS	SOLUTION	CHARACTERISTIC MINERALS
APLASMOGENIC (congruent)	Ferrolysis	Alternating redox state with perched water table; acid weathering of clays	Skeleton of resistant minerals
	Acidolysis and acido-complexolysis	Water-soluble organic acids	(quartz,...) No plasma
PLASMOGENIC (incongruent)	Partial hydrolysis	Weak acid (CO_2) solutions	Transformed and inherited 2:1 clay minerals
	Total hydrolysis		Neoformed minerals
	Bisiallitisation	Weak acid (CO_2) solutions	2:1 clays (smectites)
	Monosiallitisation	Weak acid (CO_2) solutions	1:1 clays (kaolinite, halloysite), Fe and Al hydroxides
	Allitisation	Weak acid (CO_2) solutions	Al and Fe hydroxides

Figure II.3 presents a highly generalised map of the dominant weathering processes occurring in different parts of the world. This is based on the four major weathering types recognised by Pédro and Sieffermann (1979) and also indicates the largely ice-covered continents of Antarctica and Greenland where appreciable chemical weathering is limited to the non ice covered areas.

The dominant weathering processes in the acidolytic area are acidolysis and acido-complexolysis and are associated largely with the cold humid environments of North

America and Asia. The hydrolytic environments cover the largest areas of the terrestrial surface and the major processes dominating in these areas are, in decreasing order of importance, hydrolysis, acido-complexolysis, salinolysis and alkalinolysis. The areas of arid climate have lower rates of chemical weathering and are dominated by salinolysis and alkalinolysis; little chemical weathering occurs in the extreme desert environments.

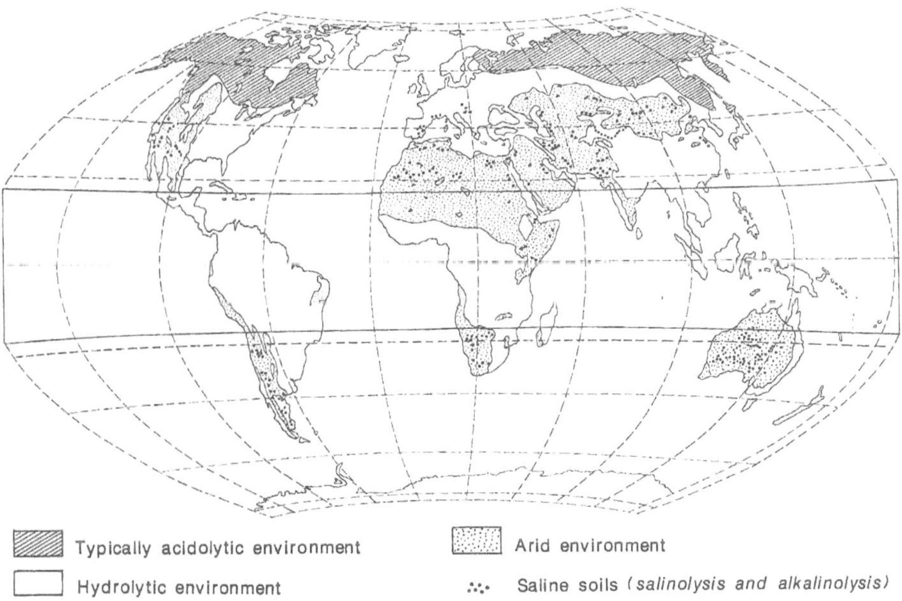

Figure II.3 Distribution of major chemical weathering processes on a world scale (redrawn from Pédro and Sieffermann, 1979).

3.2 Formation and structure of organic reserves

3.2.1 DYNAMICS OF ACCUMULATION

Living organisms colonise soils from the very early stages of their formation. Photolithotrophic micro-organisms and plants fix atmospheric carbon and nutrients from the weathering parent materials and incorporate them into their tissues.

While they live, plants release such metabolites as root exudates and foliar and bark leachates into the soil. These are supplemented with microbial by-products and animal mucus and polysaccharides to form part of the metabolic organic matter defined by Parton *et al.* (1983) (Figure II.4).

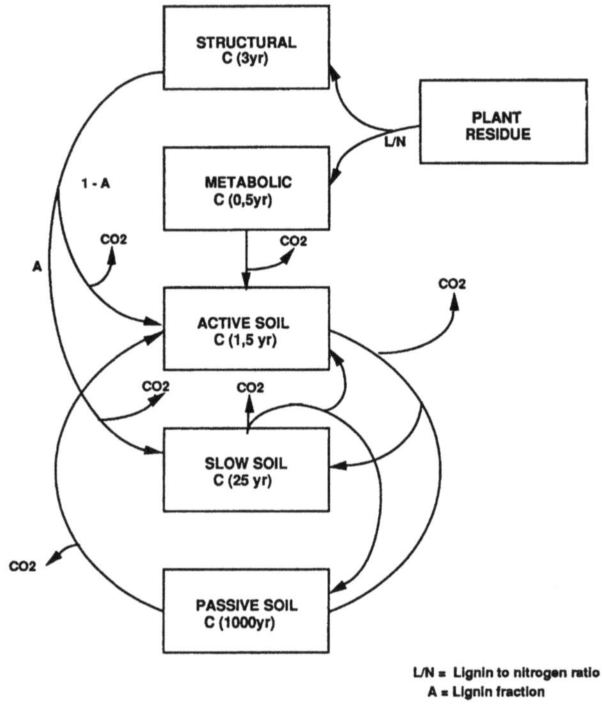

Figure II.4 Flow diagram for C-dynamics in terrestrial ecosystems showing the five major compartments (Parton *et al.*, 1983).

Dead organisms quickly come in contact with the soil surface, or die within it, and immediately start to decompose. The biomass of these organisms is arbitrarily divided into decomposable and resistant materials (Jenkinson and Rayner, 1977). Metabolites and decomposable organic matter are considered to degrade rapidly and are incorporated in the 'soil biomass' compartment, which is the sum of the biomasses of all living organisms, including the micro-organisms and soil fauna. Because of physical or chemical protection from decomposer attack, a considerable proportion of the original organic matter accumulates in forms resistant to further decomposition and undergoes protracted decomposition.

Thus, as a result of this extended decomposition, organic matter gradually accumulates in pools with increasing levels of resistance to decomposition (Jenkinson and Rayner, 1977) (Table II.5). Balesdent *et al.* (1987) estimated the origin of soil carbon by using natural isotopic labelling in temperate soils that had been cropped first to C3 forages and then to corn (*Zea mays*) (a C4 plant) for more than 30 years. They showed the remarkable stability of the older organic matter from the original grassland (57 to 75 % of total C) and a gradual replacement of C3 carbon by C4 organic matter produced by the corn crop. After 35 years, less than 20 % of C3 carbon remained and an equivalent amount of C4 carbon from the maize had accumulated (Figure II.5).

Table II.5 Size and turnover time of the five pools of organic matter in agricultural plots at Rothamstead, England (Jenkinson and Rayner, 1977)

Fraction	Half life (years)	Pool size (Mg ha^{-1})
Decomposable plant material	0.156	0.01
Resistant plant material	2.31	0.47
Soil biomass	1.69	0.28
Physically stabilised organic matter	49.5	11.3
Chemically stabilised organic matter	1980	12.2

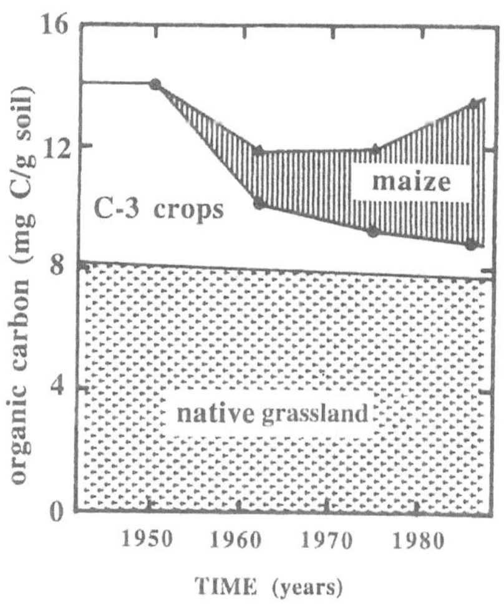

Figure II.5 Changes in the amounts and origins of soil organic C (0-20cm depth) accompanying long-term cultivation of a fertilized soil supporting corn (*Zea mays*) but previously cultivated to C-3 forages for 61 y (Balesdent *et al.*, 1987).

Natural radiocarbon (^{14}C) isotope dating has proved to be a useful tool for investigating the residence times of carbon in soils. The half life of ^{14}C (5730 years) provides a tool for aging the old carbon in soils. In contrast, the recent injection of ^{14}C into the atmosphere by nuclear tests in the 1960s ('bomb radiocarbon') labelled all photosynthesised material and provides a tracer of younger organic carbon. Figure II.6 shows a radiocarbon profile of a temperate-climate, forested alfisol (Elzein and Balesdent, 1995). Positive $\delta^{14}C$ values in the upper layers denote the incorporation of bomb radiocarbon in the soil. The L layer had a radiocarbon content equal to atmospheric CO_2 in 1993. In the F layer, a five-year mean carbon residence time was calculated. Higher values in the upper 5 cm of the A horizon correspond to the peak of bomb-radiocarbon in the 1960s. The authors simulated the carbon and radiocarbon profiles (lines) with a model that couples the decomposition of carbon compounds with downward movement. In applying the model to this soil, diffusion due to bioturbation and soil homogenisation by the soil fauna accounted for most of the movement. Since burial occurs more slowly than decomposition, the rapidly decaying components are confined in the upper part of the soil. In contrast, old organic matter is more homogeneously distributed throughout the entire profile.

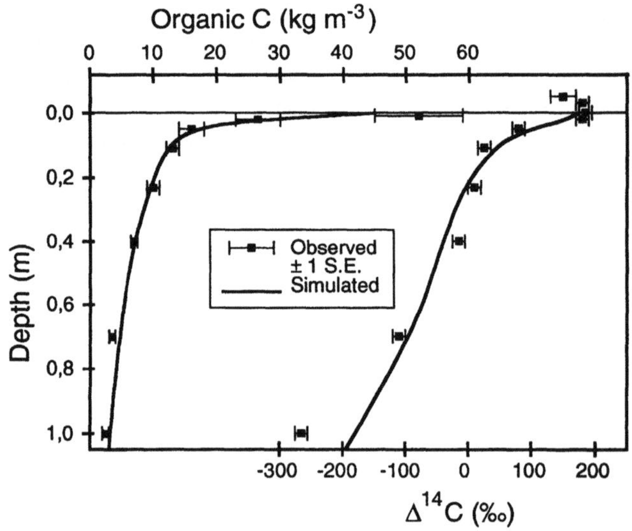

Figure II.6 Profiles of organic carbon (left) and natural radiocarbon content (right) in a temperate forest soil (typic hapludalf with a mull-moder humus) (adapted from Elzein and Balesdent, 1995).

3.2.2 FACTORS REGULATING ABUNDANCE

The total amount of organic matter accumulated in any soil varies with the rate of supply of biological materials and the nature of its timing. It also depends on soil age, whether the soil environment is conducive to rapid organic matter decomposition, the chemical composition of the material to be decomposed and the capacity of soil to retain organic matter. This latter capacity is largely dependent on the types and concentrations of the soil clays.

Time

Balesdent *et al.* (1987) considered that organic stocks in recent volcanic soils (Andisols) in Martinique comprised two pools of different ages. The first had a mean age of 38 years and corresponds to the decomposable plant material, soil biomass and physically-stabilised pools of Jenkinson and Rayner (1977). The mean age of the second pool was estimated at 1,400 years and included the chemically-stabilised organic matter. Accumulation of organic matter in the second pool was considered to cease after *ca.*

2,000 years. However, from studies carried out in several other environments, much older organic matter (from 3,000 to 12,000 years) has been found to occur in the resistant pool (Stout *et al.*, 1981).

Climate

Decomposition of organic matter is a combination of two contrasting processes. The first is mineralisation, which is associated with the disappearance of organic matter and its reduction to mineral components, and humification which leads to the formation of highly-polymerised and decomposition-resistant compounds. Jenny *et al.* (1949) found a satisfactory relationship between temperature and moisture conditions and both C and N contents in soils of the Great Plains of the USA. It has been conventionally considered that the high temperatures and favourable moisture conditions that prevail in the wetter tropics would favour mineralisation over humification and therefore result in high rates of organic matter breakdown. Since decomposition rates increase faster with temperature than primary production, it was also considered that, despite their higher organic inputs, soils of the tropical regions would have lower organic reserves than those of colder areas (see *e.g.*, Beck, 1971; Scharpenseel, 1988). However, extrapolations of Jenny's model systematically under-estimated soil organic contents (Laudelout, 1990) and the reality is much more complex (Greenland *et al.*, 1992).

Sánchez *et al.* (1982) compared the total carbon contents of 61 randomly-chosen tropical profiles with 45 similarly-chosen profiles from the temperate region, classified as belonging to equivalent orders of the USDA classification (Soil Survey Staff, 1999). These authors could find no significant difference between these two groups. Thus, it appears that the increased inputs and decomposition rates in the tropics do not necessarily lead to significant differences in accumulated carbon stocks. Coefficients of variation within each group, however, ranged from 32 to 79 %, which indicates that many local variables influence the dynamics of organic matter accumulation.

Local factors

Among the local factors that influence the accumulation of organic matter are the types and quantities of the different clay minerals present. Some clays may form very stable organo-mineral associations with soil organic matter. Organic colloids may be physically protected from enzyme attack by their insertion between the clay platelets, or by adsorption onto their surfaces.

The nature and amount of inputs also helps to determine soil organic matter dynamics and further stabilisation (Swift *et al.*, 1979).

3.3 Translocation

Translocation may be defined as any movement of soil components within a profile that affects soil development. Such movements may either promote or act to inhibit or partially reverse (pedoturbation) profile differentiation (Hole, 1961). Translocation may result from both physical and chemical processes that involve the movement of soil materials by such agencies as water and air flows and gravity; it may also occur through volume changes consequent on freezing:thawing or hydration:dehydration, as in the mechanical pedoturbation (churning) that occurs in vertisols.

Biological processes may also effect a translocation of soil materials (bioturbation) and can result from the activities of soil animals (faunal pedoturbation) and through such processes as windthrow by plants (floral turbation) (Hole, 1961). Bioturbation largely takes place within the upper 30 to 60 cm of the soil whereas physico-chemical processes may operate throughout the whole soil profile, and within the underlying regolith (Brabant, 1991).

One consequence of continued bioturbation has been the widespread formation of biomantles. These are defined as differentiated zones in the upper parts of soils produced largely through bioturbation but aided by other subsidiary processes (Johnson, 1990). Such biomantles may be single to multilayered depending on the combination of agents that gave rise to them. Over time, continued erosional redistribution of bioturbated materials may give rise to surface layers characteristically-different from the underlying materials in particle size distribution and in structural and other properties.

3.3.1 PHYSICAL AND CHEMICAL PROCESSES

Some soils are formed by weathering in place. They have a Bw (w = weathering) horizon in which the accumulated clay minerals (colloids) result from *in situ* weathering of the parent material. In other soils, the differentiation of a Bt (translocated) horizon results from the movement or redistribution of certain components of the weathered materials (particularly clays) down the soil profile by two important processes: leaching and eluviation.

Leaching has been defined previously as the movement of material through the soil profile both in solution and suspension. However, it seems desirable to restrict the word leaching to the transport of materials in solution and to define eluviation (or pervection) as the transport of solid particles in suspension (Duchaufour, 1982). Leaching occurs when there is substantial movement of water through the profile and this may occur either downwards or laterally beneath the surface. It affects the more soluble salts such as potassium or sodium rather than the less soluble alkaline earth metals, calcium and magnesium. Under some circumstances, upward movement of salts may occur through capillaries. As shown in Chapter I (Section I.1.1.1.2), even poorly-soluble compounds such as silica, particularly where present as phytoliths, may be broken down to silicic acid and either leached from the system or taken up by plant roots (Lucas *et al.*, 1996).

Translocation in soils may occur through the following five major processes differing in their intensities (Duchaufour, 1997) (Figure II.7).

Salination

Salination typically occurs in arid or irrigated soils when soluble potassium and sodium salts present in the profile accumulate close to the surface. These salts move to the surface in capillary water as a consequence of high evaporative demand, or because of high saline water tables. In soils where the deeper horizons contain appreciable salt, salination may result from tree clearing since the water not transpired or evaporated can leach sub-surface salts which may appear as saline surface seeps in lower-lying parts of the landscape.

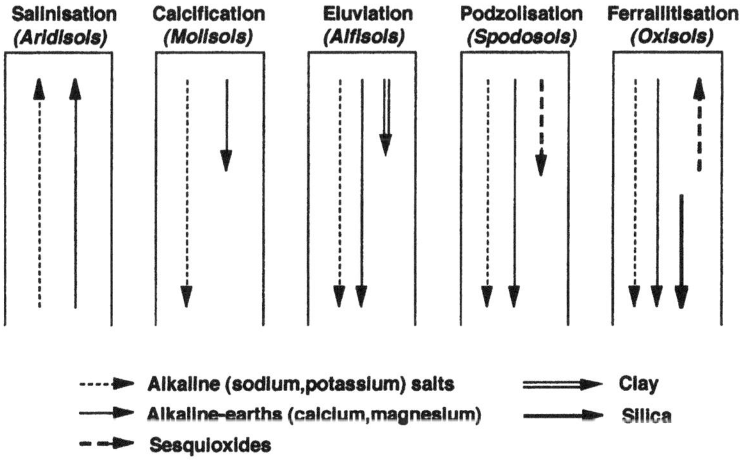

Figure II.7 Soil translocation processes (Fenwick and Knapp, 1982).

Calcification

Calcification is characterised by the downward movement of Ca salts through the profile either by leaching and precipitation, or by the eluviation of small particles from surface horizons (see Milnes, 1992). This process is typical of arid and semi-arid environments subject to alternating periods of wetness and dryness. Leaching is most effective for the highly soluble salts but less so for those of lower solubility such as the salts of the alkaline-earth elements, notably Ca and Mg. These may precipitate as calcite or dolomite and accumulate in the soil profile to form Bk horizons or, following complete cementation, dense petrocalcic horizons known as caliche or calcrete (Bkm horizons).

Eluviation

Eluviation is a stronger translocation process and describes the removal in suspension of soil materials from particular layers and is most likely to occur when the soil particles are not strongly aggregated and in soils where large macropores are present. Under these conditions, clays, sesquioxides, silica (notably as phytoliths) and humic compounds may be translocated downwards depending on the intensity of eluviation and climatic conditions. Clays and other substances may be displaced down the profile to accumulate in the B and sometimes in the C horizon. This process is characteristic of alfisols and ultisols (*e.g.*, leached brown soils, fersiallitic or ferruginous tropical soils).

Podzolisation

Podzolisation is a comparable process in which a pronounced downwards translocation of iron, aluminium and organic matter takes place. Clays in the upper horizons are decomposed under acid conditions into sesquioxides and silica by soluble organic compounds. The breakdown products are then translocated downwards to precipitate and reform in the B horizon. Thus, an eluvial E horizon typically overlies the illuvial accumulation of these materials in the B horizon.

Podzolisation is often associated with the presence of a fluctuating perched watertable created by the blockage of pores in the deeper horizons by translocated clay minerals (Guillet *et al.*, 1979).

Andosolisation
This process is characteristic of soils forming on eruptive materials containing volcanic glass or readily-weatherable fine-grained crystalline materials. A major feature is the presence of active aluminium, either amorphous or associated with allophane. In this process, the precursors of humic compounds are rapidly insolubilised by active aluminium giving substantial protection against mineralisation. In turn, this protection reduces the production of crystalline clay minerals and only small amounts of halloysite are normally produced. Andosolisation is accompanied by a continuing loss of basic cations and silica (Duchaufour, 1982, 1997).

Vertisolisation
Vertisols are soils with high clay contents that form under conditions of contrasting seasonal climates, notably in the tropics and sub-tropics and characteristically in base-rich locations with restricted drainage. The clays typical of these soils are largely neoformed smectites (Section I.1.1.1.2) which shrink markedly and deeply on drying to form polygonal patterns of cracks and swell on rehydration. The consequence of this dehydration:rehydration process is a churning or a turnover of the upper profile leading to its homogenisation. This occurs through materials falling into the cracks that form during dry periods; on rehydration, the centres of the polygons formed are lifted upwards to accommodate the extra materials deposited at depth. This churning gives rise to characteristic structures (slickensides) in the soils formed where blocks of soil slide against each other.

Ferrallitisation
Ferrallitisation is characterised by the contrasting apparent downward movement of silica and the accumulation of sesquioxides. Such processes are characteristic of conditions of active tropical soil weathering. As occurs during podzolisation, the original clays are broken down and poorly crystallised kaolinite or the aluminous mineral gibbsite are recrystallised. Active alteration and translocation result in a significant lowering of the soil surface associated with the loss of materials (Chauvel, 1976; Lucas *et al.*, 1996).

Additional processes
Three additional processes are recognised which may occasionally superimpose their effects on those listed above:
(i) Gleying, movements of iron and manganese and their accumulation in reduced forms under conditions of impeded drainage and low redox potential.
(ii) Rubefication, the *in situ* weathering of iron oxides and their transformation into deep red, poorly-hydrated ferric oxides (α-Fe_2O_3, haematite) under climates with marked seasonal contrasts.
(iii) Solonisation (or alkalinisation), the leaching and illuvial accumulation of clays and sesquioxides in the B horizons of alkaline soils (Duchaufour, 1997).

3.3.2 BIOLOGICAL TRANSLOCATION: BIOTURBATION

The translocations considered above result from the expenditure of physical energy (gravity, capillarity, osmosis) within the soil. Movements of soil materials also result from the expenditure of biological (photosynthetic) energy by roots and soil animals, largely invertebrates but sometimes mediated by the burrowing and other soil-disturbing effects of vertebrates (see authors in Meadows and Meadows, 1991; Butler, 1995). Such effects may be direct (effective transportation) or indirect, for example, the translocation of soil materials through galleries and other channels (biopores) formed by soil invertebrates.

Faunal pedoturbation is most apparent when it involves the deposition of soil materials at the surface. However, much faunal pedoturbation is less apparent since it involves the vertical and horizontal movement of ingested and transported materials within the soil. The materials displaced may simply be deposited behind the animal as it moves through the soil, as with certain earthworms or packed into pre-existing voids, as occurs with certain termites, ants and some burrowing vertebrates.

Because they are unable to move through existing cracks and crevices in the soil, the larger invertebrates and vertebrates must rearrange the soil to accommodate their general life patterns and movements. Johnson *et al.* (1987) recorded the contrasting effects of two burrowing rodents in California. During burrow construction, the Ground Squirrel *Ostospermophylus beecheyi* causes considerable profile homogenisation extending to the C horizon. However, it deposits much of the soil it excavates into abandoned burrows. In contrast, the Pocket Gopher *Thomomys bottae* deposits most of the materials excavated during burrow construction on the surface and therefore acts to promote profile differentiation. Soil materials deposited at the surface become available for erosional detachment and transport and may contribute substantially to hillslope processes.

Some of the larger invertebrates simply eat their way through the soil, thereby transporting soil upwards or downwards, depending on the net direction of their movements. Most selectively ingest the finer soil particles which are thereby concentrated in their faeces. The main agents of such biological translocations are the earthworms and certain termites. Both termites and ants also transport fine particles of an upper size limit related to that of their mouthparts. Members of these three groups frequently effect a net transport of finer particles to the surface during the excavation and construction of nests, burrows and other structures.

Humphreys (1994) quantitatively described a number of micro-morphological features of a southeastern Australian ultisol heavily faunally-turbated by ants, termites and other animals (Figure II.8). It is clear that most activity occurs in the coarse-textured A and E horizons with about 8 % of the soil volume being represented by faunal constructs. However, the surface of the B horizon was being mined for clay and biologically formed structures (pedotubules) were distributed throughout the C horizon, although in diminished amounts within the lower Cr horizon. Of particular note is the large proportion (80-98 %) of the A and E horizons comprising macular colour since this represents the remnants of reworked pedotubules. Faunal deposition of soil materials on the surface exposes them to rainwash with the finer particles being transported away in runoff waters leaving the coarser particles to form the surface soil horizons. It is clear that the A and E horizons of this soil owe much of their structure to faunal activity and therefore comprise a biomantle.

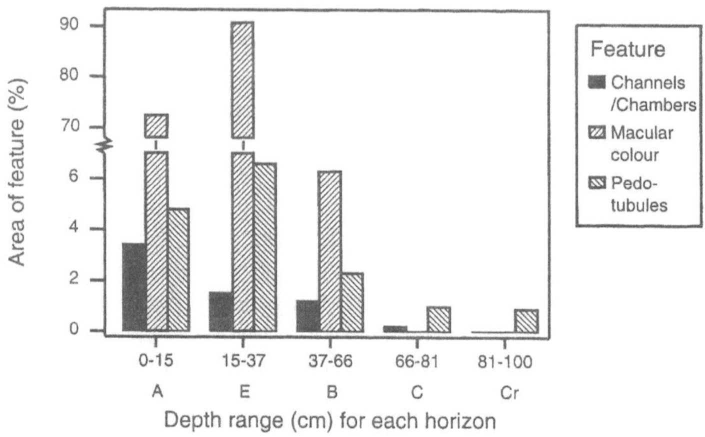

Figure II.8 Faunal features in each horizon of the profile of a southeast Australian ultisol (Humphreys, 1994).

Biomantles may also be defined in terms of the activities of the three major groups discussed below and in particular locations to the effects of certain surface and soil-dwelling vertebrates. Other groups such as the Crustacea are also responsible for considerable bioturbation in a few terrestrial environments (Butler, 1995) and their effects are particularly clear in many tropical mangrove forest soils.

Earthworms

In the temperate grasslands of Europe, earthworms may mix and displace up to 500 Mg of soil ha^{-1} $year^{-1}$ on a dry weight basis (Bouché, 1975) although only part of this soil is deposited as casts on the surface. Darwin's (1881) estimates from English pastures showed that casts deposited on the soil surface annually represented the approximate equivalent of 0.25-0.50 mm of sub-surface soil and reflect the significant inversion and mixing of the upper profile caused by these animals.

In the humid grass savannas of Africa, the estimated annual transit of soil through the intestinal tracts of earthworms may be as high as 1,000 to 1,200 Mg of soil $ha^{-1}year^{-1}$, also on a dry weight basis. Again, only a small amount (25 to 30 Mg or *ca.* 2.5 %) of the total is deposited on the soil surface as casts. These amounts are equivalent to layers of, respectively, 8-10 cm overall and 1-2 mm of fine particles deposited annually on the soil surface (Lavelle, 1978).

Since earthworms have marked diurnal and seasonal vertical movements, soils from different horizons may be mixed and in some cases, this may reduce or completely offset the effects of leaching and eluviation. However, the activities of most earthworm species are essentially concentrated within the A horizon and very little material is normally brought to the surface from the deeper horizons.

Despite the above, the impacts of soil mixing by earthworms on soil function and structure may be large. Three great groups of mollisols (vermustolls, vermudolls and vermiborolls) are defined as having more than 50 % of the volume of their A horizons (below any Ap horizon present) occupied by such earthworm constructs as galleries, casts and infilled voids (Soil Survey Staff, 1999).

Termites

In the tropical areas of the world where they are dominant, termites can be potent agents of pedogenetic change and their activities may extend throughout the whole soil profile and well into the underlying regolith. The mass of soil brought to the surface annually to be built into mounds and temporary foraging shelters has been estimated at up to 1.8 Mg ha^{-1}, in African savannas (Nye, 1955c; Lepage, 1974; Roose and Lelong, 1976). Roose (1980), however, calculated that, in the savannas of Burkina Faso, 4 to 33 % of this clay-enriched soil was washed away by sheet erosion, resulting in the formation of a poor, sandy top soil (Josens, 1983).

To a limited degree, termites select the sizes of the particles that they ingest or transport. Most select the finer particles, although this varies with the soil and species involved (Lee and Wood, 1971a). For example, Kemp (1955) found that the mounds of *Cubitermes sankurensis* have a small proportion of particles between 100 and 500 μm in diameter but greater proportions less than 100 μm and greater than 500 μm. This results from the transport of fine particles in their crops while larger particles are carried in their mandibles.

Wielemaker (1984) reported the upwards and downwards transport of soil materials by termites in Kenyan soils and the illuviation of clays through galleries created by these animals.

The effects of termites on soils are further considered in Chapter IV.

Ants

Although perhaps of less pedogenetic importance than termites, ants are ubiquitous, especially in the tropics and may effect substantial changes to the soil (see also Chapter III). At a local scale, ant bioturbation (Lévieux, 1976; Cowan *et al.*, 1985; Lockaby and Adams 1985) involves continuing profile modification with differences between species largely related to their nesting habits. Such effects also depend on whether they are mound-building species or nest underground and whether the ground is sloping or not so that bioturbated materials deposited on the surface may eventually be transported downslope. Within the soil, their effects form part of continuing cycles of void formation and infilling and they contribute to processes influencing infiltration, gas exchange, organic matter incorporation and nutrient flows in the ecosystems of which they form a part. Information available on their roles in soil processes is limited to effects recorded from a small range of sites and species and, beyond acknowledging their undoubted importance, only limited generalisation appears possible at our current state of knowledge.

Many ants build sub-surface nests and their contributions to soil mixing and transport have been considered as largely restricted to the surface deposition of soil excavated from their subterranean nests on the surface. The amounts may be considerable. Lévieux (1976) estimated that 0.3 Mg ha^{-1} $year^{-1}$ of B horizon material was deposited on the soil surface by the ant *Paltothyreus tarsatus* in the humid savannas of the Côte d'Ivoire. Lyford (1963) reported the surface deposition of some 0.6 Mg ha^{-1} over the course of one year in a North American oak forest. In a Nearctic desert environment, Whitford *et al.* (1986) estimated that 0.8 Mg ha^{-1} was moved to the surface during the same time period, a similar amount to that reported by Humphreys (1981) from an open forest environment in southeastern Australia.

The leaf-cutter ants of the tribe Attini (Chapter III) occur throughout much of the Neotropical and the southern part of the Nearctic regions. Certain members of this group excavate large quantities of soil materials during nest construction, much of which may be deposited on the surface. The pedogenetic effects of these ants are indisputable. Alvarado *et al.* (1981) reported an amount equivalent to 460 Mg ha^{-1} of AB and B horizon subsoil spread over the soil surface and Boulet *et al.* (1995) estimated at 0.21 to 0.23 mm the thickness of the layer annually deposited at the soil surface.

Other ants construct epigeal nests on the soil surface and these may become abundant in certain environments. Population densities of the mounds of *Lasius flavus* recorded by Wells *et al.* (1976) from British grasslands were locally equivalent to 5000 mounds per hectare and their bases occupied 15.6 % of the site surface area. The selective nature of their contribution to vertical translocation is evidenced by the finer texture of the soils comprising the mounds.

Apart from these single nests, compound structures built by ants may achieve several metres in diameter and rise well above the local topography; these have entirely different soil characteristics from those surrounding (Cox *et al.*, 1992). Construction of these large mounds occurs through generations of ants building nests in the same location, starting from a local topographic high point and gradually extending through a combination of vertical and lateral transport of materials selected from the underlying and adjacent soils.

In addition to translocating soil materials to the surface, ground-dwelling ants return organic foodstuffs to their nests. In particular, the downward translocation of organic materials into the nests of certain leaf-cutting ants may be substantial; reported consumption estimates range up to the equivalent of several tonnes per hectare per year of dry plant materials (Fowler *et al.*, 1986). Further, ants can have considerable effects on profile morphology. Eschenbrenner (1994ab) reported the presence of chambers that had been back-filled with loosely-packed mineral soil in attine ant nests in Martinquan Andisols while Alvarado *et al.* (1981) found similar structures infilled with organic matter and mineral soil in Costa Rican inceptisols. The latter authors also reported large modifications to the number and type of the horizons present because of the large amounts of soil translocated. Underground galleries and chambers may be infilled by illuviation and other transport processes after their abandonment.

Because of their abundances, particularly in the inter-tropics, ants indirectly influence soil processes through predation and other trophic interactions with the rest of the biota. Non-trophic interactions (ecological engineering; Jones *et al.*, 1994), include habitat creation for other animals. Humphreys (1994) reported constructs formed by small mesoarthropods active within infilled ant galleries excavated in soils formed from quartzose sandstones in southeastern Australia.

Faunal induction of translocation and creep

Where they occur together, termites and earthworms may have complementary effects. In forested areas of Nigeria, Nye (1954, 1955abc) described a topographic sequence of soils with two primary horizons: a surface horizon of soil material moving slowly downhill under the influence of soil creep (Cr) and an underlying sedentary horizon (S). The Cr horizon is subdivided into three separate layers namely, a CrW, formed from earthworm cast material, a CrT formed by material transported by termites, and a CrG of accumulated gravel. CrT varies from 10 to 60 cm depending on the degree of maturity; it is underlain by a 1.25-2 cm CrW horizon with a high humus content. Nye considered that the CrT derived from the CrG and S, and CrW derived from CrT. The gravel-free horizon formed by the fauna promotes soil creep down-slope and plants constantly acquire minerals from the continuously decomposing parent material. The total amount of soil brought to the surface by termites and accumulated in the CrT was estimated to be 30 cm in 12,000 years (*i.e.*, 0.025 mm $year^{-1}$ and up to 0.10 mm in some situations). This soil would clearly qualify as a biomantle (Johnson, 1990).

In another tropical forest of Côte d'Ivoire, Nooren *et al.* (1995) hypothesised that selective erosion of earthworm casts rich in clay was responsible for the creation of sandy surface soils.

3.4 Erosion

The removal of solid particles from surface soils through erosion is a normal geomorphological process and a dominant factor in landform evolution through the movement of soil materials to lower parts of the landscape. In contrast, accelerated erosion occurring as a consequence of poor land use practices is a major problem in the management of a wide range of agricultural soils and in the urban environment.

The factors controlling soil loss through rainfall erosion are given by the revised universal soil-loss equation (Renard *et al.*, 1994):

A = RKLSCP where A, the predicted soil loss, is the product of
R = climatic erosivity (rainfall and runoff);
K = soil erodibility;
L = slope length;
S = slope gradient or steepness;
C = cover and management;
P = erosion control practices.

Wind may be just as important an agent of erosion as water, notably in drier areas. Predicted losses due to wind are described by the following interacting factors (Brady and Weil, 1996):

$$E = f(ICKLV)$$

where:
E = the predicted wind erosion;
I = soil erodibility factor, including soil properties and slope;
C = the climatic factor which includes wind velocity, soil temperature and precipitation (through its effect on soil moisture);
K = the soil-ridge-roughness factor which includes the size distribution of the soil surface unit (clods in cultivated soils), the vegetation cover and whether ridges are present;
L = is the width of the field in the down-wind direction; and
V = is the vegetation cover factor which describes the characteristics of the vegetation cover present.

Erosion is normally a slow process, generally in equilibrium with the creation of soil through pedogenesis. For example, Leneuf (1959) estimated natural erosion in forested areas of the Côte d'Ivoire to be 0.15 Mg $ha^{-1}year^{-1}$, *i.e.*, 1 m in 100,000 years. The period for complete alteration of 1m of granite in the same area was considered to be between 20,000 to 100,000 years. In natural humid tropical ecosystems, soil loss estimates of 0.05 to *ca.* 6 Mg ha^{-1} $year^{-1}$ have been reported (Smith and Stamey, 1965; Roose and Lelong, 1976; Capelin and Prove, 1983). Lombardi Neto and Bertoni (1975) consider that values as high as 4.2 to 15 Mg ha^{-1} $year^{-1}$ are the limits of acceptable soil loss tolerance within which the functioning of these soils would not be significantly impaired.

In contrast, extremely large annual soil losses have been reported from sugarcane agro-ecosystems in northern Queensland, due to inappropriate land use practices. Average values under a cultural regime of burning the crop prior to harvest and any post-harvest residues and cultivating the soil was 148 Mg ha^{-1} $year^{-1}$ with extreme values exceeding 500 Mg ha^{-1} $year^{-1}$. The mean soil losses under regimes involving no-tillage, no pre-harvest burning of the crops or of post-harvest residues reduced this to approximately 21 Mg ha^{-1} $year^{-1}$. In the United States, rates of total erosional soil losses considered tolerable in terms of maintaining long-term agricultural production range from 5 to 11 Mg ha^{-1} $year^{-1}$ (Brady and Weil, 1996).

Depending on its intensity, natural erosion slowly alters the nature of the soil profile (Roose, 1980; Lal, 1984). Under conditions of low slope gradients and thick vegetation cover, sheet erosion and sub-surface lateral translocation may remove fine particles from the upper horizons. The soil fauna may partially offset such losses by depositing fine materials derived from deeper horizons on the soil surface.

When the rates of soil loss by sheet or creep erosion become greater than those of the formation of humified surface soil, the soil profile may be altered to form either an AC or a truncated (BC) profile. In the first case, the soil, although shallow, may be relatively fertile due to the constant input of nutrients from newly-weathered parent material. Conversely, in the second case, extended soil weathering may lead to the formation of biologically-exhausted infertile horizons. Eroded material accumulates in low-lying areas as footslope and valley floor deposits and a continuing flux of new material from upper parts of the slope or the watershed may be an important feature of the evolution of landscapes and soils.

II.4 MAJOR PROCESSES OF SOIL FORMATION

Soil formation is an enormously complex series of reactions and processes that may continue to operate over long time periods. It encompasses weathering but also involves the range of reorganisational and translocational processes discussed above, together with the build up of soil organic matter and all that this entails in terms of biological processes and bio-mineral interactions (see, for example, Ugolini, 1986a).

It is the interactions of the soil-forming factors (particularly climate) that determine the general direction of weathering and pedogenesis in any given area and, as time passes, the soil will develop towards a idealised transient steady state, the pedoclimax. However, because of continuing weathering losses and changing environments, the directions of pedogenesis will also change. The differing climatic conditions experienced during the development of soils in the older landscapes will usually be reflected in their morphological, chemical or physical properties.

Local factors may be important determinants of pedogenesis which normally alters gradually along continua, except where abrupt changes of parent material, vegetation or drainage occur (Ugolini, 1986a). These may lead to the formation of very different soils within the same environment.

4.1 The pedoclimax concept and its limitations

Climate is the dominant factor in pedogenesis and early systems of soil classification defined 'zonal soils' as the idealised pedoclimaxes of individual climatic zones. However, Duchaufour (1982) considers that this concept should be restricted to processes related to humification. Different major trends occur in pedogenesis, depending on the prevailing climate: ferrallitisation in humid tropical soils and brunification in temperate regions. Temperature and precipitation are the two major aspects of climate. Temperature is important in that it determines the rates of chemical reactions and the metabolic rates of most organisms. Precipitation is the major determinant of moisture regimes and water fluxes, and hence of chemical weathering, pedogenesis and water supply to organisms.

As discussed earlier in this chapter (Section II.2.3), extended periods of time may be required for soil development (see also Table II.2) and the changes that occur are mostly gradual. However, the evolution of soils over time is not always continuous. Climatic or other changes that affect ecosystem stability may lead to further phases of pedogenesis in which an existing soil is replaced by one that differs in its mineral composition and fabric (Lucas and Chauvel, 1992) (Figure II.9). Transformation fronts appear which may progress laterally. In Central Amazonia for example, the lateritic cover developed on Tertiary sediments is being replaced by spodosols (Lucas and Chauvel, 1992). These transformations may be determined by rapid changes in plant cover following deforestation. On the plateau, elements released by surface degradation are transported deep into the profile. Al is concentrated as kaolinite and gibbsite ($(Al(OH)_3)$ and Si is leached. On slopes, lateral water flow is more intense and greater amounts of Al are leached resulting in a relative concentration of SiO_2 as quartz sand. Podzolisation occurs in these sandy areas, starting from low-lying areas and progressing upslope as the slope develops. This transformation is related to percolating water and starts at the plateau edges from where it progresses towards the centre, further contributing to slope development. In French Guiana, contrasting systems occur in which podzolisation progresses from the centre to the edges. In the latter case, podzolisation is initiated by the formation of a perched water table following clay eluviation and deposition lower in the profile (Boulet *et al.*, 1984). The properties of very old soils (palaeosols) are discussed in Section II.4.4.3.

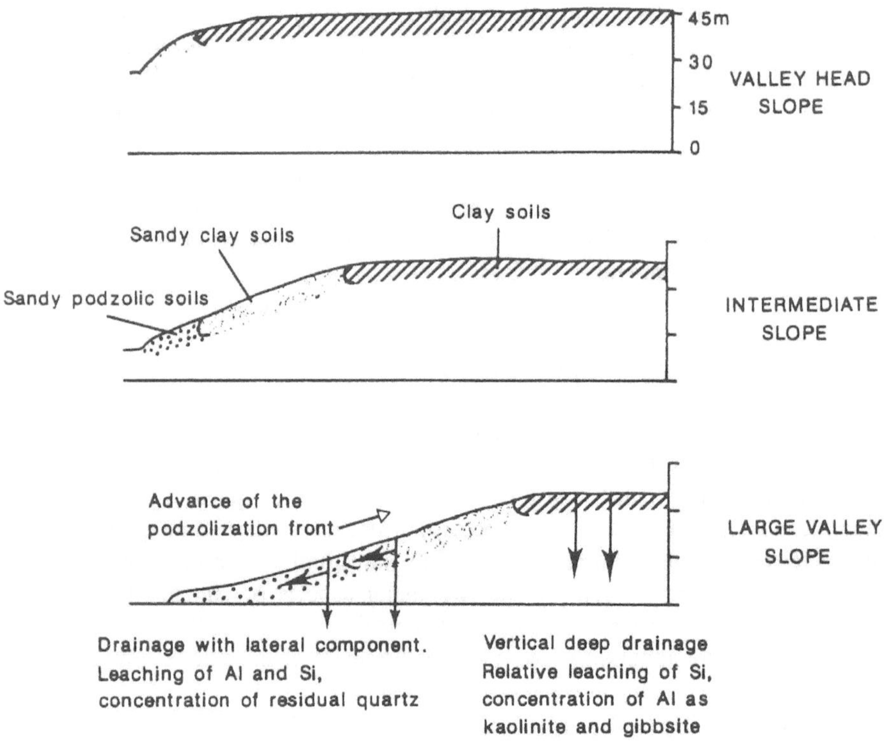

Figure II.9 Transformation of soils on sediments in Central Amazonia (Lucas and Chauvel, 1992).

Local differences in parent materials (Section II.2.2), drainage, topography and the biota may lead to large differences between soils over short distances. Pedogenetic processes may be conditioned by the characteristics of the original parent material. As discussed earlier in this chapter, clear differences may result between soils formed on different parent materials although these diminish with soil age (Chesworth, 1973a).

The contrasting soils that form on calcareous and acid rock parent materials are discussed below.

Both climatic factors and the nature of the parent materials generally determine the broad compositions of biological communities and pedogenesis in many areas may largely result from the activities of organisms and the reactions induced by their metabolites. This will exclude the areas of hot and cold deserts and those soils subject to deep geochemical weathering where the biological influence on soil weathering is much reduced. Whether occurring naturally or through man's activities, major structural changes in the plant cover, as from forest to grassland, across sites with similar soils usually lead to large changes in the soil environment. These include different litter and sometimes mycorrhizal types and thus altered chemical weathering processes and microclimatic regimes in the upper soil and the area immediately above it.

A major factor in pedogenesis across all zones is drainage status since soils differing in this regard also differ strongly in chemical, physical and biological regimes. Where drainage is inhibited, soils often have low redox status and hydromorphic properties characterised by reduction and segregation of Fe (Section I.1.2.3) and a reduced or inhibited rate of organic matter decomposition. A wide range of these soils exists - permanently to intermittently saturated associated with poorly-drained coastal and inland areas and such water bodies as rivers, ponds and lakes and areas flooded for rice production. Apart from the coastal hydromorphic soils and the areas cultivated for rice, such soils are distributed from sea level to the highest mountain ranges.

Whole regions may be influenced by inhibited drainage. In the tundra environment, for example, seasonal melting of the upper permafrost layer leads to saturation of the upper profile resulting in acid reducing conditions, reduced decomposition rates and peat formation. Across the tropical world, peats (histosols) occur in poorly-drained environments with the greatest areas being in South East Asia, Papua New Guinea and Africa (Andriese, 1988).

Table II.6 summarises the dominant weathering and pedogenetic processes operating across the earth's surface and the characteristic soils that result. It also lists the principal processes involved in the transformation and neoformation of clays. Areas included in the table do not include hot deserts and ice-covered areas where minimal chemical weathering occurs or areas of carbonate-rich or tephritic parent materials.

4.2 Pedogenetic processes in tropical climates

Ferrallitisation is the dominant process in the hot, moist climates of the equatorial regions covered by rainforests. Under such conditions, the original bedrock becomes deeply and highly-weathered, irrespective of its initial composition. Hydrolysis is the dominant process because of the acid soil solutions resulting from elevated concentrations of dissolved carbon dioxide. Reactions with soil organic matter are considered to be of lesser importance than in other environments and geochemical weathering is considered to dominate in the freely-drained areas of the tropical regions (Duchaufour, 1997).

Table II.6 A summary of the dominant weathering processes, types of clay transformation processes and the production of neoformed clays together with characteristic soils and soil forming processes in a range of terrestrial environments defined largely by climate (Birkeland, 1984; Ugolini, 1986ab; Duchaufour, 1997; Campbell and Claridge, 1992).

Climate	**Dominant weathering processes**	**Clay formation**		**Characteristic soils**	**Characteristic or dominant soil-forming process**
		Transformation	**Neoformation**		
Cold desert	oxidation, salinolysis, calcification	inheritance	None	Entisols, Gelisols	Wind erosion, cryoturbation, salt accumulation
Polar and sub-polar desert	oxidation, salinolysis, calcification	inheritance	None	Entisols, Inceptisols, Mollisols	Wind erosion, cryoturbation, salt accumulation, carbonatation:decarbonatation
Tundra (wet)	reduction, organic matter accumulation	inheritance	None	Histosols	Peat formation, gleying
Tundra (dry)					
northern part	Acidolysis	inheritance, slight transformation	None	Entisols, Inceptisols	Brunification, eluviation
southern part	Acido-complexolysis, oxidation	inheritance, minimal transformation	None	Spodosols	Podzolisation
Taiga	Acido-complexolysis	Degradation, solubilisation	None	Spodosols	Podzolisation
Temperate environments (mor)	Acido-complexolysis	Degradation, solubilisation	None	Spodosols	Podzolisation

Table II.6 (Cont.)

Climate	Dominant weathering processes	Clay formation		Characteristic soils	Characteristic or dominant soil-forming process
		Transformation	**Neoformation**		
Temperate environments (mull)	Acidolysis (moderate degradation)	Transformation	Weak	Alfisols (*Sol brun*)	Brunification
Mediterranean	Neutral hydrolysis	Inheritance, limited transformation	Medium (bisiallitisation)	Alfisol (Fersiallitic soils)	Fersiallitisation
Tropical (well drained)	Total hydrolysis	Limited inheritance, transformation	Medium-strong (monosiallitisation)	Alfisols, ultisols (Ferruginous tropical soils)	Ferrugination
Tropical (impeded drained)	Total hydrolysis	Limited inheritance, transformation	Strong (bisiallitisation)	Vertisols	Vertisolisation
Equatorial	Total hydrolysis	Inheritance, no transformation	Strong, monosiallitisation, allitisation	Oxisols (Ferrallitic soils)	Ferralitisation

In tropical environments, weathering is intense and as much as 57 % of the original material, mainly comprising bases and silica, may be leached or eluviated out compared with 24 % under equivalent temperate-climate conditions (Lelong and Souchier, 1972). Primary minerals disappear while sesquioxides accumulate and low activity clays such as kaolinite are formed. These clays may be further degraded in the upper horizons and nutrients leached out. Depending on their final nutrient status, these soils are defined as unsaturated ferrallitic soils (ultisols and oxisols; Soil Survey Staff, 1999).

Under drier conditions, weathering is less complete and a proportion of the primary minerals may persist leading to the formation of ferruginous soils (alfisols; Soil Survey Staff, 1999) which have a higher nutrient status than the ferrallitic soils.

As discussed above (Section II.3.3.1), vertisols are predominantly formed in areas of low relief and from basic parent materials in climates with strongly differentiated wet and dry seasons (Millot, 1979). Where unexploited, most have appreciable nutrient reserves. Such soils are widespread in Australia, Africa, India and in parts of Central and South America.

4.3 Pedogenetic processes in cold and temperate climates

Soil processes differ very strongly below and above the Arctic treeline and in the ice-free areas of Antarctica. The latter two areas are underlain by frozen ground and northwards of the tundra are subject to an increasing aridity, also characteristic of the Antarctic climate.

4.3.1 COLD CLIMATE PROCESSES

In going northward from the Arctic treeline, the tundra supervenes with its continuous cover of vegetation (Ugolini, 1986a) and this progressively gives way to the sub-polar desert, the polar desert and finally the cold desert. This latter zone may occur in northern Greenland but its major representation is in the ice-free areas of Antarctica (Campbell and Claridge, 1992). North of the tundra zone, the environment becomes increasingly arid and the vegetation becomes increasingly discontinuous with a commensurate reduction in its influence on soil processes. Finally, in the cold desert, the vegetation is restricted to a few scattered mosses and lichens in favourable locations. To the north of the Arctic treeline, the soils are largely frozen or underlain by permafrost. These are all subject to the substantial physical effects of wind erosion and cryoturbation including the patterned ground common to the polar and cold desert soils (Ugolini, 1986b).

Because of the ice melting during the summer thaw, most soils in the tundra are very wet. They typically have a surface horizon of decaying vegetation up to 15 cm thick overlying 10-30 cm of light olive to olive-gray very wet silt loam overlying very dark grey permanently frozen silt loam. Physical processes associated with freezing and thawing dominate the landscape while reduction, gleying and the very slow decomposition of organic matter dominate the pedogenetic processes (Campbell and Claridge, 1992). In the lesser areas of dry tundra, acid Arctic Brown soils (inceptisols) are typical with humic A horizons and dark brown to yellow brown colours in the B horizon. Weathering is minimal in these youthful soils and most clays are inherited from the parent materials (Ugolini, 1986a).

North of the tundra zone the soils are marked by decreasing biological inputs and increasing dryness. Desert pavements become more apparent and A horizon development is minimal with a brownish-yellow Bw horizon often present below the desert pavement (Ugolini, 1986a). The soils are structureless, coarse-textured, saline and pH ranges from neutral to alkaline (Campbell and Claridge, 1992).

Soils of the Antarctic cold desert are typically coarse-textured, with desert pavements at the surface and an horizon of salt accumulation. Salinity is high and salt efflorescences and calcite crusts are common. The salts present include nitrates and sulphates of aerosolic origin inland and those derived from rock weathering; marine aerosols deposit salts close to the sea. Soil reaction ranges from mildly acid to alkaline. A permafrost typically occurs within 100 cm of the surface. Very limited chemical weathering occurs although older soils may have an oxidised horizon comprising thin oxide coatings over the mineral grains (Campbell and Claridge, 1992). While the soils are generally of coarse texture, up to *ca.* 5 % of clay sized materials occur within the older soils, often largely comprising finely subdivided parent materials. However, limited transformation of micas and illites has been noted and smectites have been recorded from alkaline environments (Campbell and Claridge, 1992).

4.3.2 TEMPERATE CLIMATE PROCESSES

In temperate climates chemical weathering is greater than in the cold regions but is less intense than in tropical areas. Consequently, for a specific landscape age, the parent material influences are greater and the roles of plants and soil organisms and their metabolites are of particular importance.

Pedogenesis follows very different pathways depending whether the parent materials are calcareous (limestones) or acid (*e.g.*, granite, sandstone or gneiss).

4.3.2.1 *Calcareous bedrock*

The main weathering processes occurring in the softer calcareous rock parent materials are the dissolution of carbonates by carbon dioxide (decarbonation), organic acids and nitrates dissolved in the soil solution and accumulation of the remaining silicate clay minerals. However, Al and Fe may originate from sources other than bedrock residues including aeolian deposits (Cornu, 1995). Dissolved minerals are leached to subterranean waters or accumulate in low-lying areas; such monovalent ions as Na^+ and K^+ are leached much more rapidly than their divalent counterparts (Ca^{2+}, Mg^{2+}). Biological activity is intense and mineral element losses are reduced by the intimate mixing of mineral and organic components which promotes the formation of stable organo-mineral complexes. Under such conditions, clay mineral alteration is reduced and the neoformation or the conservation of clays with high silica contents is favoured.

The soils that form on hard indurated limestones are of a very different character and are often old polycyclic soils that have formed on heterogeneous parent materials and undergone additions of other materials during development. In contrast to soils formed from the softer limestones, the amounts of active carbonate are small and the profiles may be rich in sesquioxides (Duchaufour, 1982).

4.3.2.2 *Acid rocks*

Brunification is the dominant pedogenetic process occurring on all well-drained, non-calcic parent materials although they must be capable of supplying adequate clays and iron through weathering. This process also requires a mull humus type and a pH in the range 5-6. It is characterised by the formation of silicate clay-ferric (Fe^{3+}) ion-humus complexes. In the upper organic horizons, the ferric cations act to bridge the humic acid molecules and clay minerals to form stable complexes that protect the organic materials from decomposition and the clays from eluviation. In the predominantly mineral horizons, iron also forms films of hydrated amorphous oxides surrounding the clay minerals (Duchaufour, 1982).

Chemical weathering rates in these parent materials are lower than in tropical conditions. Active earthworm populations lead to the formation of biological closed-aggregate structures through their casting activity and this protects minerals from the aggressive hydrolytic activity of soluble organic acids. Consequently, such soils may contain variable proportions of largely-unweathered minerals. Their clay contents are generally low and minerals, mainly smectite and illite, are largely derived from neoformation processes. Most of the elements released by the dominant carbonic acid weathering are soluble and are progressively translocated down the profile leading to the formation of leached brown soils (alfisols, Soil Survey Staff, 1999).

With more acid substrates, and particularly where biological activity is restricted by low quality or toxic litter (*e.g.*, that of most conifers) and seasonal waterlogging, biological aggressive leaching microstructures dominate. They accelerate the destruction of existing clay minerals while neoformation rates are very slow. This leads to the intense leaching and eluviation of clays, bases, sesquioxides and organic matter from the surface horizons to form an E or eluvial horizon. The leached and eluviated materials accumulate (podzolisation) in the B horizon to form the spodosols characteristic of, but not restricted to, the areas where conifers dominate.

4.4 Chronosequences

Soils gradually pass through a series of intermediate stages during profile development. Under some circumstances, the profile may be constantly rejuvenated by creep or erosion and the formation of a fully-developed profile may be much delayed. Thus, under defined conditions of climate, topography, parent material and the biota, chronosequences (*i.e.*, age sequences) of soils occur that correspond to the successive stages of increased weathering, translocation, and other transformations. However, Birkeland (1992) points out that earth-surface processes have not been constant over time and that soils older than the Holocene have been subject to variable soil forming influences due to changes in the climate and therefore in other soil forming factors.

4.4.1 TEMPERATE CLIMATES

On acid parent materials in the temperate regions, fine particles derived from weathering mix with newly-accumulated organic matter to form a young 'sol brun' (brown soil) with an AC profile (Figure II.10). In cold regions, most soils are too youthful to have had sufficient time to have evolved further and organic matter may accumulate continuously, due to the prevailing slow decomposition and mineralisation rates.

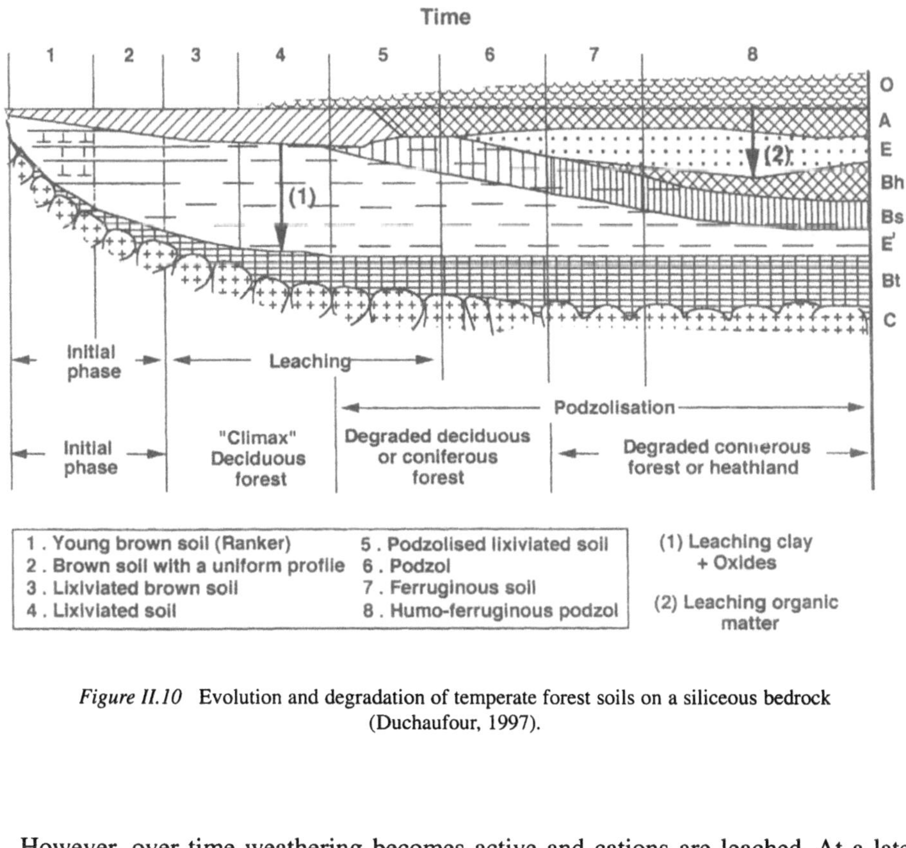

Figure II.10 Evolution and degradation of temperate forest soils on a siliceous bedrock (Duchaufour, 1997).

However, over time weathering becomes active and cations are leached. At a later stage, clays, iron oxides and oxyhydroxides are translocated and accumulate in a Bt horizon (possessing translocated illuviated clays) to form a 'sol brun lessivé' (acid leached brown soil, or alfisol).

Under even more acid conditions, biological activity gradually changes and closed-macroaggregate systems are progressively replaced by aggressive-leaching systems. Epigeic litter-dwelling arthropods become dominant and replace the anecic earthworms which normally drag litter into their galleries and mix it with the soil. Under such conditions, Fe and Al complexes are translocated down the profile to form a Bs horizon that is; possessing high concentrations of illuvial materials apparently amorphous under X-ray examination, comprising carbon, the oxides of Al and perhaps Fe. The A horizon is reduced and a bleached E (eluvial) horizon is formed. This is the beginning of the podzolisation process, which is completed by the translocation to depth of humic material which accumulates in a Bh (rich in dispersible, amorphous humus) horizon (Duchaufour, 1982). As vegetation changes, pedogenesis proceeds; the pioneer grassland vegetation gradually evolves toward a deciduous forest which is, in turn, progressively replaced by coniferous forest or heath, depending on the climate.

Where the parent material is calcareous, its dissolution rapidly leads to the formation of a rendzina (rendoll, mollisol; Soil Survey Staff, 1999). These soils have thick and active A horizons, in which earthworms are generally abundant and within which casts act as stable closed-aggregate systems. As dissolution proceeds, carbonates are progressively eliminated leaving a residuum of clay and silt. At a further stage, leaching and eluviation translocate cations and clay minerals further down the profile to accumulate in a Bt horizon. The fully developed soil that forms under such conditions is a *'sol brun lessivé* ' (leached brown soil or alfisol; Soil Survey Staff, 1999) characterised by an intense biological activity in the A horizon due to the neutral to alkaline pH (Figure II.11).

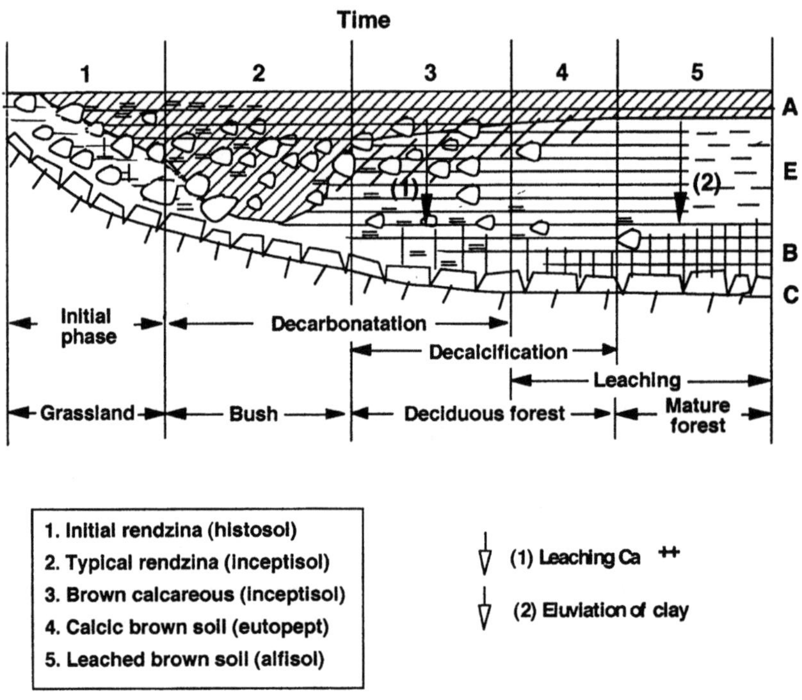

Figure II.11 Evolution and degradation of temperate forest soils on a calcareous bedrock (Duchaufour, 1997).

The development of spodosols on an age series of aeolian beach ridges forming on dominantly-quartzitic parent materials in southern New South Wales provides an example of a chronosequence that has formed during the Holocene (Bowman, 1987). The developing soils were accurately aged from the carbon contained within shell fragments. Figure II.12 presents a cross section of the site indicating the short time required for initial horizon formation (less than 1000 years) and its rapid development with age in these permeable materials. A number of other soil properties were found to be related to soil age and this is illustrated in Figure II.13a which shows the exponential increase in organic matter content of the B horizon and, in Figure II.13b, the linear decline in C horizon pH with time before the present.

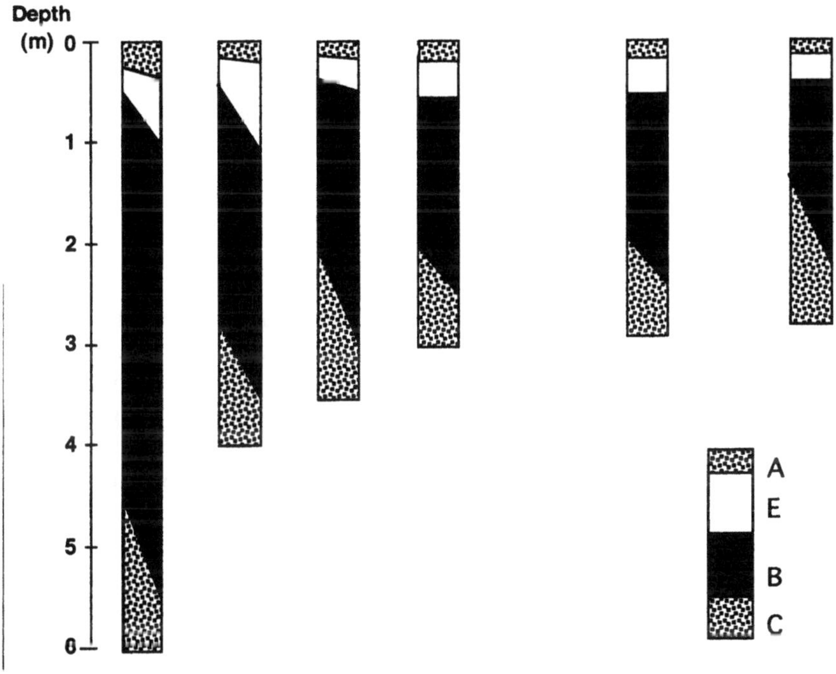

Figure II.12 Cross section and profile development with age in a series of soils (spodosols) formed on beach ridges in south eastern Australia (Bowman, 1987).

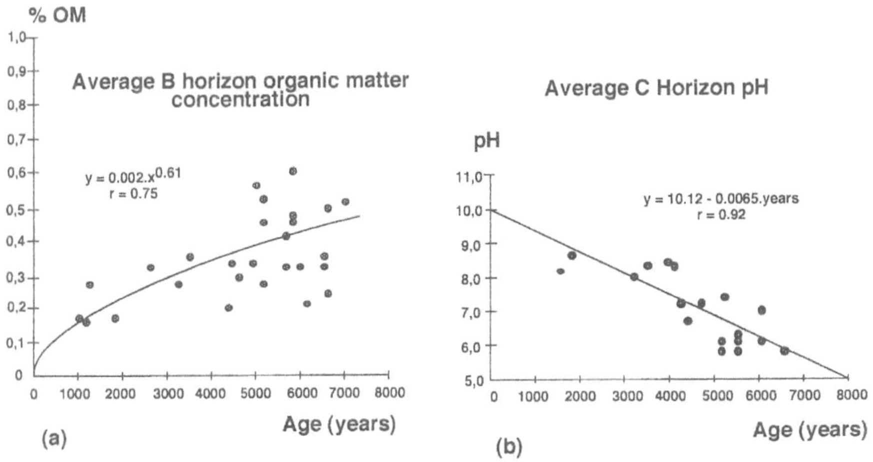

Figure II.13 Changes in a spodosol with age: (a) average concentration of organic matter in the B horizon, and (b) average C horizon pH (Bowman, 1987).

4.4.2 TROPICAL CLIMATES

In the humid tropics, the intense weathering is considered to result largely from hydrolysis, rather than the action of aggressive organic compounds. These latter appear to have a limited influence on the weathering zone since, with exception of certain deeply-weathered rainforested soils (Section I.3.2.4.3), organic matter generally accumulates close to the surface. Ferrallitisation may be achieved through three successive stages leading to a series of progressively more weathered soils (Duchaufour, 1991):

Fersiallitisation

Fersiallitic soils are those in which neoformed or inherited silica-rich 2:1 clay minerals persist in the profile and the exchange complex remains saturated, or nearly so. They occur in environments with strong seasonal contrasts. Clay eluviation leads to the gradual formation of an argillic Bt horizon and its red colour betrays the common presence of haematite (fersiallitic soils, Duchaufour, 1982; largely alfisols, some ultisols; Soil Survey Staff, 1999).

Ferrugination

In ferruginous soils much of the original clay has been destroyed and some desilicication has occurred. The primary minerals have been largely replaced by neoformed kaolinite although some 2:1 clays are still present. Base status is variable depending on the length and intensity of the dry season. These soils are known as ferruginous tropical soils (Duchaufour, 1982; some alfisols, mostly ultisols, Soil Survey Staff, 1999).

Ferrallitisation

Ferrallitisation is a later stage during which all of the primary minerals except quartz have been weathered. Neoformed kaolinites are the dominant or sole clays and because of their resistance to movement by water, eluviation has effectively ceased. Most of the silica has been lost and the oxide minerals gibbsite, haematite and goethite are characteristically present. Finally, further bases are lost, leading to the formation of acid, base-unsaturated ferrallitic soils (Duchaufour, 1982) (Figure II.14) (oxisols; Soil Survey Staff, 1999).

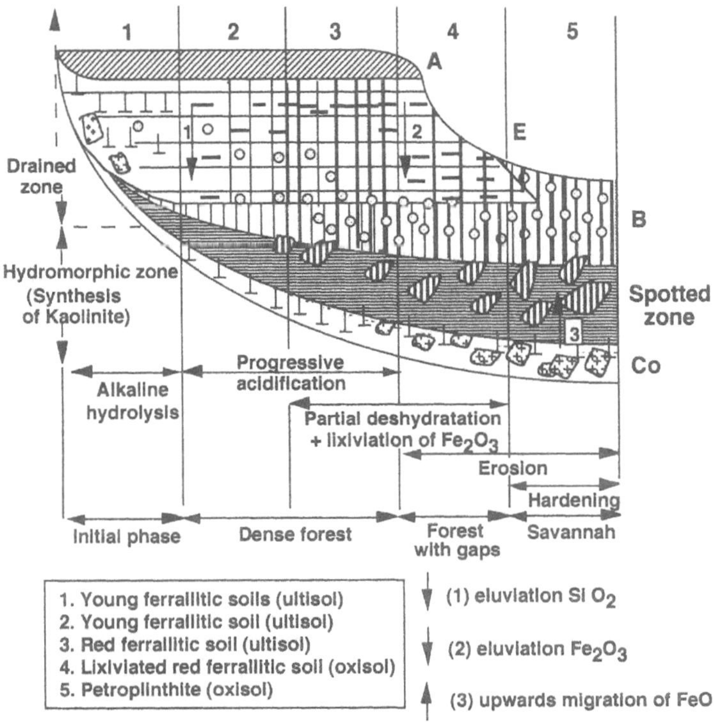

Figure II.14 Evolution of tropical soils (Duchaufour, 1997).

4.4.3 PALAEOSOLS

Weathering and leaching processes continue to affect soils even after the formation of fully-developed profiles and lead to gradual changes over time. Some changes are reversible and involve variably-lagged adjustments that follow those of the environment. Soil organic matter status may alter in this way and such adjustments to environmental conditions occur over relatively-short time periods. However, most are irreversible and include progressive changes to the minerals present in the soil and losses of nutrient and other elements that inevitably accompany such weathering processes as leaching, eluviation and runoff.

The soils that result from extended periods of weathering are known as palaeosols. These soils possess characteristic suites of properties that depend on the properties of their parent materials, the weathering environments to which they have been subsequently exposed and any additions that have been made in the form of, for example, dust or volcanic ash. Within their profiles, structures such as hardened layers or horizons of silica (silcretes) or iron (ferricretes) may reflect pedogenetic processes that occurred under climatic conditions that no longer exist. Two major ecological features of these soils are their diminished capacity for biological production and their susceptibility to the loss of a large proportion of their often-limited nutrient capital following disturbance.

Perhaps the world's most fertile soils are those associated with the geologically active areas of the world which include such areas of recent vulcanism as the Pacific 'Rim of Fire', recently uplifted and glaciated areas and locations where recent deposition has occurred. In contrast, highly-weathered and poorly-fertile soils are abundant in stable continental areas of low relief, remote from active plate boundaries (Fyfe *et al.*, 1983). Such areas are widespread in the old landscapes of the world's tropics, notably Australia, Africa and South America.

Rejuvenation of these landscapes is dependent on large-scale depositional or tectonic events to provide a new parent material base for weathering into soil. A further source of rejuvenation is large-scale erosion which may expose underlying, unweathered parent materials.

Highly-weathered soils tend to be rich in neoformed clays because of the gradual replacement of rock-forming minerals by those more stable in the near-surface environment. The dominant phyllosilicate clays present in these soils are those with low cation exchange capacities and the mineralogy of many older tropical soils is dominated by such 1:1 clays as kaolinite and halloysite together with the oxides of iron and aluminium (Section I.1.1.1.2). Because of the oxide minerals and organic matter present, the cation exchange complexes of these soils often possess an appreciable pH-dependent charge component.

Concentrations of the biologically-important elements in such highly-weathered soils are characteristically low. Phosphorus is a key element in both biological productivity and in pedological development and is supplied very largely by the parent material (Walker and Syers, 1976). However, Newman (1995) reports annual aerial deposition rates of 0.07 to 1.7 kg m^{-2} $year^{-1}$ and some sites may be in approximate balance with weathering losses. Nonetheless, in highly-weathered soils, concentrations are low and a large proportion of that present occurs in organic form; depending on the pH, much of the remainder may be occluded and therefore largely unavailable to organisms (Section I.3.1.2.3). Deficiencies of phosphorus and other essential elements commonly limit productivity in palaeosols.

Because of the low nutrient status of highly-weathered soils, aerial inputs of elements in rainfall and dust (Section I.3.1.3) may become significant components of site nutrient balances in areas where such soils occur (Artaxo *et al.*, 1988). Jordan (1982) considered that, in tropical Amazonia, the maintenance of the large evergreen rainforest biomass is dependent on such inputs.

The biological implications of extended soil weathering are substantial. Since palaeosols have a poor capacity to supply and retain plant nutrient elements, they naturally support highly mature ecosystems whose continued function is dependent on the efficient recycling of nutrients and the fixation of nitrogen, consistent with Odum's (1983) hypothesis of the diminished 'leakiness' of mature ecosystems.

Stark (1978) presented a theoretical model of the biological consequences of the evolution of soil properties over time. After an initial period where shallow soils support only cryptogams, a vegetation develops that is largely dependent on indirect nutrient cycling, that is the take up of inorganic nutrient elements directly from the soil. As the soil becomes progressively impoverished in nutrient elements over time, the biological community adapts to the diminished nutrient supply by developing such nutrient conservation mechanisms as efficient mycorrhizal networks which cycle nutrients directly from the biomass to the host plant. However, nutrient stocks continue to diminish with continued weathering leading over time to a series of vegetation communities dominated by, sequentially, small trees then shrubs then cryptogams. Finally, a terminal exhausted state was considered to supervene wherein effectively no nutrients are available.

Because much of the ecosystem stock of certain nutrients may be tied up in the biomass or is associated with litter and surface-soil organic matter, these systems are highly susceptible to degradation. Such disturbances as clearing, high-temperature firing and severe erosion may deplete the biomass and soil organic matter stocks and lead to further substantial reductions in the remaining nutrient stocks. This is consistent with Odum's (1985) hypothesis that disturbance tends to return ecosystems to earlier 'leakier' developmental stages. Following disturbance of some very highly and deeply weathered soils, the previous ecosystem trajectory may not necessarily ensue.

Successful agricultural development of highly-weathered soils may therefore be difficult. Clearing normally involves the destruction of most of the above ground biomass and considerable disturbance to at least the surface soils. The above ground biomass is usually disposed of by firing and this contributes directly to site nutrient losses through volatilisation and leaching of the ash by subsequent rainfall (Stark, 1978). Because soil organic matter may be the major source of negative charge in these soils, its depletion following forest clearing can lead to greater proportional reductions in cation exchange capacity and nutrient retention capacity than in less highly-weathered environments (Allen, 1985). After clearing, loss of soil structure may ensue and continuing soil disturbance through cultivation often leads to an even lower organic matter equilibrium in the surface soils and an even more diminished capacity to retain nutrient elements. Overstocking is of frequent occurrence in pastoral situations and may lead to a reduced vegetation cover, structural degradation and accelerated erosion.

4.5 Toposequences

Topography is an important factor in soil formation at micro, local and greater scales and erosional transport of materials downslope may substantially influence landscape morphology and soil development processes (Millot, 1977). Soils in different parts of the landscape do not operate independently from others around them; they contribute water, solutes, suspended and larger materials to those downslope and are similarly influenced by those further upslope. Thus, under defined conditions of climate, parent material and soil age, suites of related soils (toposequences or catenas) will develop across landscape assemblages of similar topography in which the soils will vary in predictable ways between different landscape locations. This also implies that at specific landscape locations within comparable toposequences, similarities in soil profile morphology, the suites of clay minerals present and soil nutrient status may be expected.

The differences between soils within toposequences are therefore attributed to the effects of gravity, drainage and relief. It should be noted that within toposequences, biotic effects may not necessarily act equally on all member soils. This occurs because of the mutual interdependence of the activities of the biota in soil formation and particular soil properties. Thus, between the members within a particular toposequence, different biotic communities may be expected, particularly where drainage extremes occur.

Toposequences form under general erosional influences and any processes (biotic or abiotic) that lead to the detachment and transport of soil materials will contribute to their formation. With downhill movement, coarser colluvial materials tend to be deposited near the bases of slopes while finer materials are deposited further away. Of particular importance are the movements of soluble soil materials and fine particles entrained in lateral and downslope water flows, both below and above the soil surface. In addition, local modification of Eh conditions often plays a significant part in their differentiation (Duchaufour, 1997).

A great variety of toposequences have been formed due to variation in the balance of the soil forming factors. Figure II.15 presents a toposequence of five related soils comprising four spodosols and a peat (histosol) formed on greywacke (fine sandstone) parent materials and supporting a vegetation dominated by *Nothofagus cliffortioides* in an alpine, ice-cut, environment in South Island, New Zealand (Molloy, 1988). Processes of downslope movement, leaching, eluviation:illuviation, creep and differential movement have lead to a suite of related soils with many properties of the downslope members depending on those further upslope.

Drainage is free in the upper slope positions but poor at the base of the slope leading to some gleying and a reduced rate of organic matter decomposition. The predominantly mineral soil is covered by an O horizon of decomposing litter comprising (with increasing depth) the following three sub-horizons: Oi (slightly decayed), Oe (moderately decayed) and Oa (highly decayed). The O horizon increases in thickness downslope and the sequence culminates at the toeslope position in a series of deep peat sub-horizons overlying the C horizon.

The A horizon only appears at the lower footslope position while the characteristic E horizon lies immediately below the O in the upper three profiles. The E is subject to reducing conditions at the lower footslope position and is absent in the histosol. The B horizon has an upper thin (Bhs) sub-horizon enriched in illuvial humus and the oxides and hydroxides of iron and aluminium (sesquioxides) in the upper three profiles and, in the lower footslope position, this horizon also contains some illuviated phyllosillicate clays (Bhst). At the summit and backslope positions, a weathered zone (Bw) overlies an R layer of the fine sandstone parent material. At the footslope position, the lower Bg horizon has some gleying indicating reducing conditions and overlies a modified C horizon of parent rock weathering *in situ*. At the lower footslope position, weathered parent rock is evident with some illuvial accumulation of sesquioxides and amorphous organic matter. Finally, at the toeslope location, the organic horizons overlie an horizon of weathering parent rock.

The changes in profile morphology illustrated and described above are paralleled by similar differences in the suites of clay minerals present; allophane and sometimes gibbsite are features of the Bw and Bs horizons, particularly. All subsoils are acid, deficient in a number of nutrient elements and have limited cation exchange capacities.

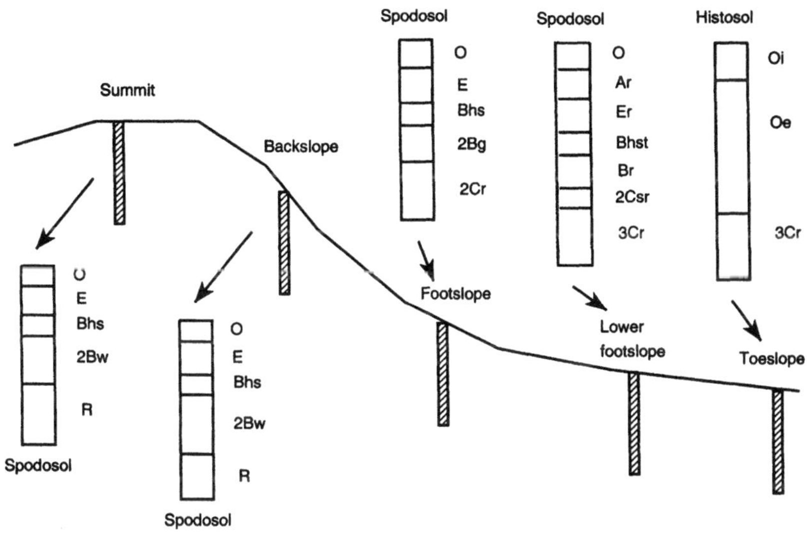

Figure II.15 Stylised profile changes along a toposequence of soils from alpine South Island, New Zealand (redrawn from Molloy, 1988). Horizon suffixes: hst an illuvial accumulation of amorphous and dispersible organic matter as a complex with significant quantities of sesquioxides and some illuviated clay; w an horizon with the development of structure or a redder hue relative to the C horizon but with no evidence of illuvial accumulation of materials; g strong gleying indicating a reducing environment; r weathered or soft bedrock, s an illuvial accumulation of sesquioxides.

4.6 Classification of soils

4.6.1 INTRODUCTION

As with the study of all natural phenomena, there is a need to group or classify soils into meaningful categories. That is, into groups that possess an internal homogeneity with respect to those not so included, for some defined combination of characteristics. This need arises from the enormous variety and complexity of natural materials and environmental conditions which is expressed in the wide diversity of soils that exist on the surface of the earth: natural, influenced or formed by humankind. Appropriate soil classifications are indispensable to the acquisition, organisation and dissemination of knowledge about soils and in defining their capabilities and limitations in many aspects of their use and function.

Soil classification has its roots in antiquity. Systems of classifying soils on the basis of their suitabilities for growing food plants existed in China and Rome approximately 4000 and 2000 years ago, respectively (Simonson, 1984). This has remained a common influence on many soil classifications up to the present. However, there are many reasons for wishing to classify soils, commensurate with the diversity of uses to which they are put. Classifications exist in which soils are rated in terms of their suitabilities for various categories of land use within agriculture or forestry, although there is a developing need for classifications that are more environmentally oriented. Soils are also commonly classified (and mapped) on their capabilities for particular engineering uses, including road making and architecture. Indeed, Kellogg (1974) reported that *ca.* 50 % of the benefits of soil survey in the U.S.A. were derived from urban planning, *ca.* 25 % from planning the locations of engineering structures and the remaining 25 % of the benefits were derived from planning in relation to agriculture, forestry and recreation.

The above are known as technical classifications and aim to characterise soils in terms of their capabilities for single or restricted purposes. They may therefore have little utility beyond the purposes for which they were created or beyond a narrow range of soils. Technical classifications contrast with the more general pedological classifications which do not aim at classifying soils for any specific end land use or other function but treat them as natural bodies. However, the pedological classifications are usually effective in broadly predicting many soil functional and technical properties.

In contrast with the technical classifications, pedological classifications are relatively recent in conception. Initially, these were based largely on geological characteristics. However, the Russian pedologist Dokuchaev is credited with creating, during the latter part of the nineteenth century, the first system of soil classification based on an appreciation of soil formation as the product of the five soil forming factors discussed above (Simonson, 1984).

As discussed below, a range of modern pedological soil classification systems exist and many countries have developed their own national classifications. Some of these include, either implicitly or explicitly, concepts related to soil genesis although most recent systems rely more on directly-observable soil characteristics.

4.6.2 PROBLEMS OF SOIL CLASSIFICATION

In classifying soils, decisions need to be made as to the size and properties of the smallest unit needed for adequate description and accurate representation. In contrast to biological classification where the limits of non-colonial individuals are generally clear, the lateral boundaries of soil units are often relatively diffuse and the problem of subdividing a continuum emerges in delineating units.

The profile is the basic unit of study in many systems and is defined as a two dimensional section extending through the vertical depth of the soil and into the C horizon, A three dimensional extension of this, the pedon, has been defined as the equivalent unit of primary study in Soil Taxonomy (Soil Survey Staff, 1999). Whatever the basic unit used, it must be deep enough to cover the full depth of the soil and of sufficient lateral extent to include small-scale cyclic variation. In practice and depending on the intensity of the sampling program, mapping individual soils is often carried out from a knowledge of landscape relationships and verified using limited profile sampling at locations considered representative for the unit under study. Under Soil Taxonomy, the individual soil is considered to be a polypedon, an association of similar and closely-associated pedons. Practically, the profile remains the individual study unit with a very limited lateral dimension related to sampling. Other decisions to be made include the extent of variation acceptable within any one taxonomic unit, at all levels.

Other units have been advanced as the appropriate basic unit of classification systems. Fitzpatrick (1967) considered that the soil horizon was the appropriate unit in this regard and, more recently, the *Référentiel Pédologique Français* (discussed below) takes a similar approach. Holmgren (1988) presented a scheme for the point recognition of soil properties which avoids some of the conceptual ambiguities of the current pedon definition.

A particular problem in soil classification is that of deciding the relative importances of different characteristics. This becomes especially subjective in trying to interpret imperfectly-known characteristics related to soil genesis. Of similar importance are the large-scale, systematic geographical differences that occurs in the distribution of soils across the face of the earth. Because of the varied environments, surficial materials, landscape ages and biotas that occur world wide, most countries possess only a small portion of the spectrum of world soils; this has lead to differing conceptions and emphases in studies of soil genesis and relationships. In consequence, no agreed, world-wide system of soil classification exists and this continues to pose real problems of communication between soil scientists and those concerned with information transfer associated with soil management.

4.6.3 TYPES OF SOIL CLASSIFICATIONS

Two pedological systems of soil classification dominate the world soil science literature, namely, 'Soil Taxonomy', the American system of classification (Soil Survey Staff, 1999) and the 'Revised legend to the Soil Map of the World' published by FAO/ UNESCO (1990). More recently the 'World Reference Base for Soil Resources' (WRB) has been established (ISSS Working Group RB, 1998a) to build on and extend the FAO/UNESCO sytem through a merger with one of the reference based systems, IRB. Other systems exist but have not received widespread recognition.

A range of national systems of soil classification has been developed. Some examples include those of Australia (Isbell, 1996), Brazil (Camargo *et al.*, 1986), Canada (Canada Soil Survey Committee, 1978), China (Chinese Soil Taxonomic Classification Research Group, 1990), France (*Association Française pour l'Etude du Sol*, 1998), Germany (Muckenhausen, 1965), New Zealand (Hewitt, 1992), South Africa (Soil Classification Working Group, 1991) and the United Kingdom (Avery, 1980). However, most have little currency beyond areas within which they were produced. In many countries a succession of systems has occurred, new ones replacing the previous when their utility is perceived to have diminished. In addition to the national classifications, lower-level, local systems have also proved necessary (see, *e.g.*, Isbell, 1990)

Taxonomic groupings may not necessarily be correlated between the different soil classification systems because different criteria are used in defining the various categories.

However, partial correlations may sometimes be made for certain groupings, particularly at the higher levels of classification. The New Zealand system, for example, attempts to link the national system with 'Soil Taxonomy' by making explicit correlations at certain hierarchical levels (Hewitt, 1992).

Duchaufour (1997) divides the systems of soil classification into hierarchical and reference-based and, within both of these categories, into those based on combinations of soil characteristics and those with a genetic base. The hierarchical systems are those in which soil individuals are divided into a ranked series of categories within which each member is grouped with others of similar properties into a common member at a higher level. Information is most specific at the lower levels and become increasingly generalised with progress up the hierarchy.

Reference-based systems, for example, the *'Référentiel Pédologique Français'* (*Association Française pour l'Etude du Sol*, 1998), take a more flexible approach, reducing or eliminating the hierarchical structure of taxonomic units and defining soil individuals in relationship to a series of well-described soils that serve as bases for comparison.

Soil Taxonomy, the American system of classification (Soil Survey Staff, 1999), is an hierarchical system with six levels. The highest level of classification is the order of which there are currently twelve. The soil orders are defined by their degree of development, by the possession of specific diagnostic horizons (either at the surface or forming part of the B horizon), environmental factors, the degree of cracking, the possession of highly organic surface horizons or, for the andisols, by their largely pyroclastic parent materials. Recent keys to the orders and subordinate categories of this evolving system are presented in Soil Survey Staff (1999). Table II.7 lists the orders and summarises some major properties (Fanning and Fanning, 1989); Wilding *et al.* (1983) provide detailed discussions of the ordinal concepts of Soil Taxonomy.

In contrast, the WRB as a potential successor to the FAO/UNESCO (1990) classification, has two levels of classification with 30 highest level groups and a likelihood that 200-250 second level groupings will eventually be defined (ISSS Working group RB6, 1998). This system concentrates on creating major soil groupings at a global scale. At its upper level of classification, soils are divided on the basis of major pedogenic processes with some groups being differentiated on the basis of specific parent materials. Its emphasis is on morphological rather than analytical properties and it has borrowed extensively from a number of national systems.

Table II.7 The twelve orders of Soil Taxonomy (Soil Survey Staff, 1999) (partly after Fanning and Fanning, 1989).

Order	**Selected Properties**
i. Soils with little or no pedogenic horizon development except an A or an Ap (an horizon created by cultivation)	
Entisol	Incipient soils with, perhaps, only an A horizon over a C horizon.
Inceptisol	Juvenile soils. Normally possess an A horizon over a weakly-developed B horizon.
Vertisol	Soils that, because of their expansive clays, expand with hydration and contract on drying to form regular patterns of sometimes deep cracking. Little horizonation develops because of pedoturbation.
Gelisols	Soils underlain by permafrost and showing evidence of cryoturbation or ice segregation.
ii. Generally possess distinct pedogenic horizons	
Andisol	Immature soils often derived from volcanic ash parent materials. Frequently with thick, humic A horizons, low bulk densities and a tendency for their physical properties to alter irreversibly on drying. Their mineralogy is dominated by amorphous minerals.
Aridisol	Soils of arid environments; many are saline.
Mollisol	Soils with a dark-coloured, well-structured surface soil mollic epipedon (diagnostic surface horizon) and high base saturation throughout the soil.
Spodosols	Acid soils with an accumulation of clay and often organic matter in the B horizon, often with a thin iron pan. Commonly possess a bleached eluvial horizon.
Alfisols	Soils with a moderate to relatively-high base status and an accumulation of clay in the B horizon.
Ultisols	Leached soils with a low base status and an accumulation of clays in the B horizon indicative of a high degree of weathering, commonly kaolinite dominant.
Oxisols	Dominantly soils of the tropics, these are highly-weathered soils whose mineralogy is dominated by 1:1 silicate clays and the oxides and hydrous oxides of Fe and Al.
Histosols	These soils have highly-organic surface horizons overlying mineral soil horizons at variable depths. They occur in poorly-drained areas.

4.6.4 MAN-MADE AND MAN-INFLUENCED SOILS

In most parts of the world, soils increasingly occur that have been strongly influenced or directly created by humankind. This varied group includes agricultural soils whose original properties have been substantially modified by such practices as deep mixing, cultivation carried out over long periods, flooding, drainage, re-contouring, large and continued additions of organic matter and other ameliorants, and other cultural practices. Further groups of soils include the so-called urban soils, those created from domestic and industrial refuse, dredged materials and land-fill materials of various types.

Another class of materials is those derived from various geological strata that are exposed on the surface as a consequence of mining operations. Depending on the type of mining and the strata present, these may range from highly inimical materials often extremely acid or extremely alkaline as are common in the metalliferous and coal mining industries. Some materials, such as those derived from the mining of bauxitic laterite, are relatively benign in terms of their propensities for supporting the growth of higher plants and require little more than appropriate fertilisation and seeding (Reddell *et al.*, 1992). Waste products such as the tailings created by initial ore processing may be similarly variable.

Some of these materials scarcely qualify as soils. However, in many countries, particularly in the case of mining spoils and wastes, there is a social or legislative requirement to stabilise them with a vegetative cover or return them to some form of agricultural production. Where rehabilitation procedures are successful, these materials may develop some of the properties of incipient soils in relatively short periods (see, for example, Reddell *et al.*, 1992).

A number of classifications include categories for many of these man-influenced and man-made soils, often at the highest level of organisation (see, for example, ISSS Working Group RB, 1998; Hewitt, 1992; Isbell, 1996). In these systems they may be designated as anthrosols or anthropomorphic soils. Other systems include these soils at lower levels within existing groups; in Soil Taxonomy, for example, they may be included among the Entisols. Because of the continuing widespread creation of these soils, their incipient nature and their often unusual properties, these is a clear need to develop a more effective approach to their classification.

CHAPTER III

SOIL ORGANISMS

III.1 GENERAL FEATURES

Soil is a dark, still, porous, semi-aquatic medium within which temperature and moisture conditions are highly buffered. Soils were among the first terrestrial environments to be colonised since they possess environmental conditions intermediate between aquatic and aerial media. As such, soils have played a key role in evolution and the presence of many primitive organisms in soils is considered a testimony of this phase of biological evolution (see, *e.g.*, Ghilarov, 1983; Vannier, 1985).

The composition and structure of soil communities reflect both the spatial organisation of soil and the major role this environment plays in decomposition processes. Since decomposition is the main role of soil, many soil organisms are adapted to feeding on decomposing organic materials which are relatively poor quality resources. Soil organisms are subject to severe spatial constraints since they occupy a semi-discontinuous network of pores filled with air and/or water, or water films which cover solid particles.

Soil organisms have responded to these constraints through a variety of adaptive strategies of which body size and respiratory patterns are the major characteristics (Table III.1).

- **the hydrobionts** are aquatic organisms which live and feed in the free water and soil. They belong mainly to the **microflora** (Kingdoms Fungi and Bacteria, including the Actinobacteria or Actinomycetes), and **microfauna** (protists and nematodes). Their size rarely exceeds 10 to 50 µm and most are only a few micrometres long. They are generally well adapted to occasional desiccation and/or shortage of food in their micro-environment. The spatial scale at which they operate is that of a few millimetres and their generation times are of the order of a few hours to a few days.
- **hygrobionts** are invertebrates which possess an aerial system of respiration but still depend on high moisture levels and often require free water in their environment. Depending on their size, two main categories are defined. The **mesofauna** include the microarthropods and the Oligochaeta (Enchytraeidae); these groups have body widths of 0.1 to 2 mm, which allows them to move freely through the large pore network within the surface litter and soil. They operate at scales of space and time of, respectively, centimetres and weeks to months. The larger invertebrates, the **macrofauna,** are much wider than most soil pores; they live in the surface

litter (*e.g.*, most Arthropods) or dig galleries in the soil (most earthworms, termites or endogeic Coleoptera larvae). The average scales of time and space at which they operate are, respectively, months to years and decimetres to metres.

Table III.1 Main parameters of the adaptive strategies of organisms in soils.

FUNCTIONAL GROUP	MICROFLORA	MICROFAUNA	MESOFAUNA	MACROFAUNA
BODY WIDTH	0.3-20 μm	<0.2 mm	0.2-10 mm	>10 mm
TAXA	BACTERIA FUNGI	PROTISTS NEMATODES	MICROARTHROPODS ENCHYTRAEIDAE	TERMITES EARTHWORMS MYRIAPODA ANTS
Water relationships	HYDROBIONT	HYDROBIONT	HYGROBIONT	HYGROBIONT
Interactions with micro-organisms	ANTIBIOSIS + OTHERS	PREDATION	PREDATION	MUTUALISM (external rumen, facultative or obligate internal mutualism)
Ability to change the physical environment	NONE	NONE	LIMITED (faecal pellets)	HIGH (galleries, burrows, macro-aggregates)
Resistance to environmental stresses	HIGH (cysts, spores...)	HIGH (cysts, spores..)	INTERMEDIATE	LOW (with possibility of behavioural compensation)
Intrinsic digestive capabilities	HIGH	INTERMEDIATE	LOW	LOW

Plant roots are considered as hygrobiont organisms. Fine roots operate at meso- space and time scales. Large roots operate at macro scales since they have long life-spans and may ramify for distances up to 30 metres from the trunk.

III.2 MICROBIAL COMMUNITIES

2.1 Taxonomic composition

The soil microflora comprises five main groups: Viruses, Bacteria, Actinobacteria, Fungi and Algae (Table III.2).

Table III.2 General classification of micro-organisms.

GROUP	KINGDOM	ORGANISATION	SIZE
Viruses*	None	Acellular	20-300 nm
Bacteria	Bacteria	Prokaryotic cell	0.1-10 μm
Fungi	Fungi	Eukaryotic cell	μm to m
Algae	Protoctista	Eukaryotic cell	μm to cm

* viruses are inanimate and therefore not considered as part of any kingdom

2.1.1 VIRUSES

Viruses are the smallest micro-organisms, arguably living or inanimate (Margulis and Schwartz, 1998); they are acellular and comprise a central nucleic acid core (DNA or RNA) with a protein covering. They are specific internal parasites of plants, animals, bacteria and occasionally fungi. Their size generally ranges from 0.01 to 1 μm and their role in controlling other members of the soil microflora is still poorly understood.

Some viruses are specifically associated with soils. Many viruses released from decomposing infested debris do not survive a dry period in the soil. Nonetheless, a few species such as the tobacco mosaic virus retain their infectivities for long periods and survive heated-steam sterilisation treatments. These may be considered as true soil-borne viruses.

They are usually transmitted in soil, by either nematodes or fungi. When transmitted by nematodes, viruses occur in patches and their spread is slow. Dissemination by fungi is much faster, probably due to the aerial dispersion of infected spores by wind or animal vectors.

2.1.2 BACTERIA

Bacteria are unicellular prokaryotes, *i.e.*, their nuclear material is not separated from the cytoplasm by a nuclear membrane. They range from 0.3 to more than 3 µm in size, and possess great specific and functional diversity (Figure III.1). The higher classification of the prokaryotes is currently somewhat unstable. It is considered to comprise two sub-kingdoms, the Eubacteria and the Archaea, a group of methanogenic, sulphate-reducing and extremely halophilic and thermophilic organisms (Woese, 1992). Most groups of concern to soil functioning belong to the Eubacteria, using the classification of Margulis and Schwartz, (1998). They will generally be referred to here simply as bacteria.

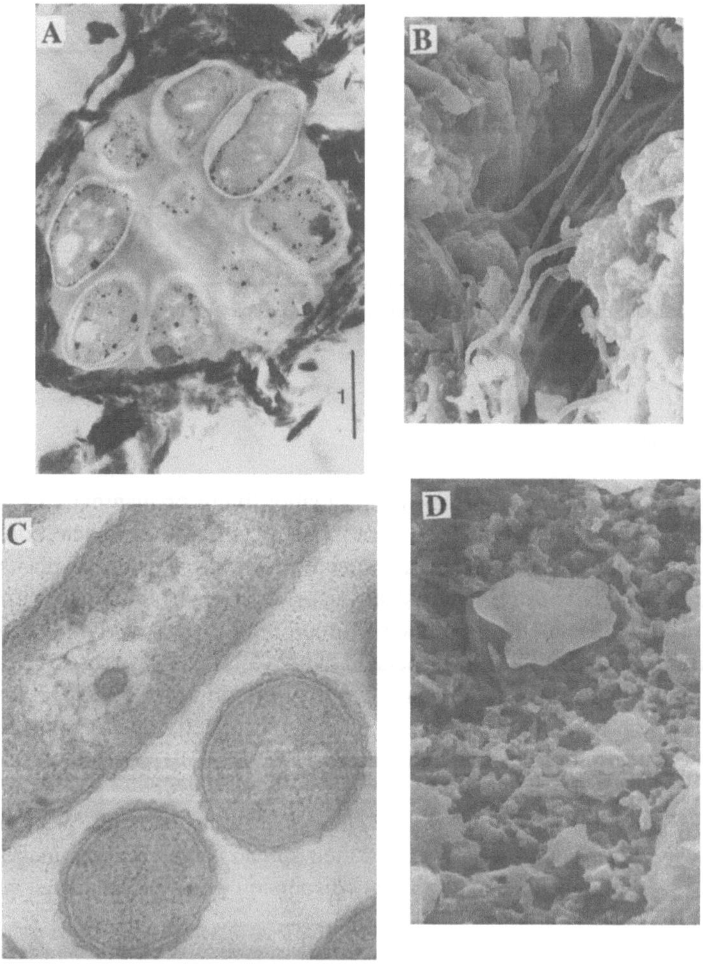

Figure III.1 Soil micro-organisms *in situ*. A: bacterial colony protected by a cover of mucopolysaccharides and an external layer of adsorbed clay platelets (scale is 1 µm). B: fungal hyphae in pore space outside micro-aggregates (x 3450); C: Gram-negative bacteria (x 130,000); D: Actinobacteria fixed to a soil microaggregate (x 11,400) (G. Kilbertus, Université Nancy I-Microbiologie ESSTIB and M. Robert, INRA).

The species concept is difficult to apply to bacteria because of their genetic plasticity (Tiedje, 1995). Recent methodological advances have allowed the establishment of a molecular basis for the definition of genera and species of bacteria. Characterisation of bacterial 16S rRNA, allows the separation of bacteria at the genus level; DNA/DNA and DNA/rRNA hybridisation techniques separate species and, finally, RFLP (Restriction Fragment Length Polymorphism) techniques allow an evaluation of genetic polymorphism among strains within a single species (Murray *et al.*, 1990). The concept of bacterial species has been recently redefined on the basis of experimental results obtained using these techniques. Two bacterial strains belong to the same species if the percentage of DNA reassociating at 60 °C, using the S1 nuclease technique, is greater than 70 % (Grimont *et al.*, 1980; Wayne *et al.*, 1987). These authors suggest that higher levels of classification such as classes and families, which cannot be defined by experimental criterion, should be abandoned (Heulin and Berge, 1992).

The classification of bacteria has conventionally been based on phenotypic characters including *e.g.*, cell morphology (rods, cocci, bacilli, ...), cell wall structure (separation into Gram-positive and Gram-negative through specific staining), the presence of endospores, the mobility of cells and the shape and position of flagella, where present. Physiological characteristics usually complemented such descriptions as recognisable morphological and anatomical characters are often few and variable in bacteria. Physiological or "functional" classifications are often preferred as they are more informative. Four categories are first defined depending on the energy source that they use (light or energy from redox reactions) and the nature of the electron donor (organic or mineral):

(i) photolithotrophic bacteria which acquire energy from light and use oxidisable mineral substrates (essentially sulphides) as electron donors;

(ii) photo-organotrophic bacteria are photosynthetic organisms that use oxidisable organic substrates (*e.g.*, organic acids) as electron donors;

(iii) chemo-lithotrophic bacteria use the energy produced by redox reactions and four kinds of mineral substrates as electron donors (*i.e.*, reduced nitrogen, sulphur or iron compounds and hydrogen);

(iv) chemo-organotrophic bacteria are typical heterotrophic organisms which take their energy from redox reactions and use organic compounds as electron donors.

Finally, "functional categories" are defined by the chemical transformations bacteria perform as geochemical agents in laboratory cultures. Some of the most common examples are the bacteria that mediate the reactions of the carbon cycle (*e.g.*, aerobic or anaerobic cellulolytic, chitinolytic...) or the nitrogen cycle (*e.g.*, nitrifiers, N-fixers, denitrifiers, ...).

2.1.3 ACTINOBACTERIA

The Actinobacteria (formerly the Actinomycetes) are filamentous bacteria. They are Grampositive, except for members of the family Sphaerophoraceae. They possess a ramified pseudomycelium whose diameter (0.5 to 1 μm) is much smaller than that of fungi. They are generally poorly tolerant of acid pH environments and most are unable to grow in media more acid than pH 5 (see *e.g.*, Waksmann, 1952; Dommergues and Mangenot, 1970). Their classification is based on the structure of the vegetative apparatus (rods or mycelia), reproductive patterns (fragmentation, formation of vegetative conidia or sexual sporangiospores) and their tolerance to acid pH levels.

Actinobacteria are usually set apart because of their unusual metabolic capabilities, *e.g.*, the production of pigments and antibiotics and the assimilation of acetylene. Species of the genus *Frankia* form symbiotic, nitrogen-fixing associations with Casuarinales and a few other higher plants.

2.1.4 FUNGI

Fungi are eukaryotes with a filamentous vegetative structure called a mycelium. Most are *Eumycota*; they have chitinous membranes and, in the most common soil fungi, their reproductive organs do not possess flagellae. Four main groups commonly occur in soils, *viz.*, Zygomycota, Ascomycota, Basidiomycota and Deuteromycota (*Fungi imperfecti*). They differ in the structure of their mycelium and reproductive organs. Zygomycota have a non-segmented mycelium of which only the growing extremes are living. Ascomycota and Basidiomycota have segmented mycelia with pores in the transverse septa which allow successive segments to communicate and live cytoplasm to be translocated towards the growing ends. Sexual reproduction leads to the formation of spores produced by several different kinds of organs: sporangia, ascini or basidia (Figure III.2).

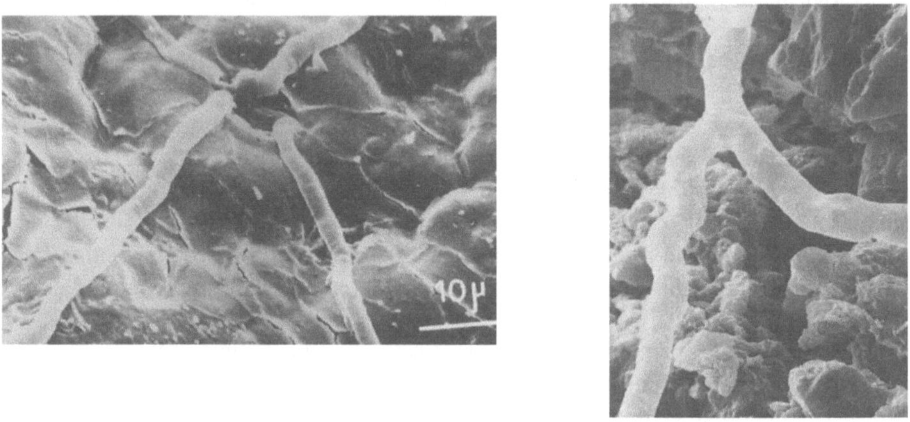

Figure III.2 Soil fungi. LEFT: Mycelium penetrating the stomata of a *Fagus sylvatica* leaf (G. Villemin and F. Watteau, CPB Nancy). RIGHT: mycelium on the outside of a soil aggregate (G. Kilbertus, Université Nancy I-Microbiologie ESSTIB) (1 cm = 2 μm).

Another important group in soil is the Deuteromycota (= *fungi imperfecti*), which comprises species with no known sexual stages.

Finally, yeasts (Ascomycota) are unicellular fungi that are normally poorly represented in the soil system, although they may occasionally play important roles.

2.1.5 ALGAE

The soil algae are a rather heterogenous assemblage of uni- and multicellular, motile and immotile eukaryote organisms. They comprise several hundreds of different taxa. They appear to become important only in extreme conditions such as dry deserts or extremely nutrient poor systems. Nonetheless, the roles they play in, *e.g.*, humid and temperate ecosystems, may well have been underestimated since they have been little studied to date. On the other hand, they are highly sensitive to all the relevant parameters of soil fertility and pollutants *e.g.*, pH, nutrients and pesticides (Pipe and Shubert, 1984).

2.2 Quantitative estimates

The satisfactory estimation of microbial abundance in soils has long been limited by the lack of reliable techniques. Improved methods are presently available and a better appreciation of the distribution patterns and biology of micro-organisms is helping to define appropriate sampling designs.

Bacteria and fungi are the major components of microbial communities and generally account for most of their biomass. Actinobacteria are often included with the bacteria and may occasionally comprise 30 to 50 % of overall abundance.

2.2.1 TECHNIQUES FOR ESTIMATING ABUNDANCE

Micro-organism abundances may be assessed by direct methods such as counting cells or analysing soils for specific components of their cells. Indirect or "physiological" methods may be preferred since they are less time-consuming to perform and provide a valuable index assessment of microbial functions in soils (see reviews of Jenkinson and Ladd, 1981; Nicolardot *et al.*, 1982; Parkinson, 1982).

(a) Indirect counting by the soil dilution plate method has long been the most commonly used method. Dilution series are prepared from a soil suspension and inoculated on to an appropriate solid or liquid culture media. The number of colonies that develop gives an estimate of the number of colony forming units (CFU). Despite the use of a variety of culture media, this method generally results in underestimations of 2 to 4 orders of magnitude and its use should be restricted to semi-quantitative estimates in comparative studies (Brierley *et al.*, 1928 and detailed procedures in Pochon and Tardieux, 1962). The major drawbacks of this method are (i) that culture media are selective towards particular groups of micro-organisms and are exaggeratedly nutrient-rich in comparison to soil conditions and (ii) that a single fungal CFU may be derived from either a spore or from a large piece of mycelium, which are not comparable in either biomass or associated activity.

(b) Direct counts of bacterial cells and measurements of fungal hyphae may be conducted on thin soil sections or, rather, thin agar films prepared from a soil suspension (Jones and Mollison, 1948 modified by Thomas *et al.*, 1965) or through filtration of determined amounts of the suspension and direct observation of the filtrate (Sundman and Sivelä, 1978; Bingle and Paul, 1986).

The soil suspension may be obtained by a gentle trituration in water, or a more intense dispersion using a blender or ultrasound. The latter methods which disintegrate microaggregates generally give higher estimates (see, *e.g.*, Babiuk and Paul, 1970).

Micro-organisms are observed in phase contrast or fluorescence microscopy, using appropriate stains such as aniline blue and aceto-orcein for fungal hyphae and acridine orange for bacteria. Stains may be used in combination as proposed by Anderson and Westmoreland (1971) and fluorescent stains such as fluorescein isothiocyanate (FITC) may also be used (Jenkinson and Ladd, 1981).

(c) Chemical methods include estimation of the soil concentrations of ATP (Ausmus, 1973), DNA (Torsvik and Goksoyr, 1978), phospholipids (White, 1988; Zelles *et al.* 1995), nucleic bases (Cortez and Schnitzer, 1979) or chitin, an important component of hyphal membranes, and hexosamines (Swift, 1973).

(d) Physiological methods assess microbial enzymatic activity or respiratory activity under different conditions.

Relating the activities of soil enzymes to microbial abundance has proved difficult because a number of other soil characteristics also determine enzyme activity. An unknown proportion of this activity may be attributed to extracellular enzymes fixed on clay minerals (Burns, 1986). Schnürer and Rosswall (1982) proposed that the hydrolysis of fluorescein diacetate (FDA) into fluorescein is a useful index of microbial activity. Most soil micro-organisms are able to effect this transformation and the method is therefore considered valuable.

(e) Respirometric methods are widely used. Direct respirometry is widely used for comparative studies (see, *e.g.*, Petersen's method as modified by Rashid and Schaefer, 1985; Verdier, 1983). Quantitative estimates may be derived by measuring the increased respiration that occurs after adding glucose to the soil (Anderson and Domsch, 1978). The use of selective respiratory inhibitors allows the separation of bacterial and fungal respiration. However, their utilisation may be difficult.

(g) The biocidal method (Jenkinson and Powlson, 1976) is currently the commonest technique used to estimate microbial biomass. The soil sample is fumigated with chloroform to kill the microflora and is then incubated under controlled conditions. Microbial biomass is estimated from the magnitude of the "flush" of carbon dioxide produced from the decomposition of dead micro-organisms by surviving or newly-inoculated ones. Microbial biomass estimated by this technique includes such soil microfaunal components as protists and nematodes.

The biocidal method has been further developed and improved by a number of authors. Specific methodologies have been designed to measure the amount of nitrogen contained in biomass (*e.g.*, Brookes *et al.*, 1985; Amato and Ladd, 1988; Alef *et al.*, 1988). A simplification of the method consists of an extraction and titration of microbial organic C rendered extractable by 0.5 M K_2SO_4 (Vance *et al.*, 1987). A similar method has been proposed by Davidson *et al.* (1989) for measuring microbial biomass nitrogen. Finally, Hendricks and Pascoe (1988) have demonstrated that micro-organisms may be killed by microwave irradiation instead of chloroform with equivalent results.

(f) Comparison of methods. Several authors have compared the results obtained using the different methods available.

Culture techniques produce the most variable results. Variability decreases with direct counts and chemical methods and is least when physiological methods are used

(Domsch *et al.*, 1979). Bae *et al.* (1972) calculated that plating techniques can only detect 0.1 to 1 % of the organisms present.

Biomass estimates obtained from the chloroform fumigation technique, direct counts and ATP analyses are often comparable (see compilation in Jenkinson and Ladd, 1981 and Jenkinson, 1988). However, the fumigation technique is less efficient at acid pH levels less than 4.5 (Anderson and Domsch, 1978; Amato and Ladd, 1994) and the conversion of ATP content or FDA activity into biomass estimates may be of dubious value (Ross *et al.*, 1980). Finally, Jenkinson and Ladd (1981) recommend using more than a single method and Nicolardot *et al.* (1982) make a clear distinction between methods which estimate biomass directly (direct counts, chloroform fumigation) and index methods which asses the activity of microflora (ATP content, FDA activity, respirometry). The cross calibration of the two sets of methods is potentially difficult.

2.2.2 DENSITY, BIOMASS AND PRODUCTIVITY OF THE MICROFLORA

There have been relatively few estimates of the population densities of micro-organisms. The use of methods with unequal efficiencies makes comparisons and comprehensive approaches difficult. However, the studies that use direct counting methods associated with specific staining are considered to give realistic assessments. Densities in soils may be as high as 5.77 10^{10} g^{-1} bacteria (in Japanese paddy fields, Hasebe *et al.*, 1984) and 18.1 km g^{-1} of fungal hyphae (in the F layer of a *Pinus* stand in Sweden, Söderström, 1979).

In twenty soils from temperate regions, bacterial populations ranged from 1.1 to 19.7 10^9 ind.g^{-1} dry wt of soil, with a mean value of 4.64 (Table III.3). Hyphal length ranged from 0.14 to 2.03 kmg^{-1} dry wt with an average of 0.88 km. Differences have been found between agricultural treatments with bacterial densities decreasing from grasslands to fallows while cereals and other crops have intermediate densities. Hyphal length showed a different pattern of variation, with cereals having the maximum values, three times greater than that of fallows.

Table III.3 Densities of bacteria and fungi in 18 soils from temperate regions (Shields *et al.*, 1973; Jenkinson and Powlson, 1976; Sundman and Sivelä,1978; Domsch *et al.*, 1979; Bååth and Södeström, 1982; Nannipieri *et al.*, 1978; Ingham and Klein, 1984; Schnürer *et al.*, 1986 (derived from Hansson *et al.*, 1990)).

Land use	Bacteria ($10^9$$g^{-1}$ dry wt. soil)			Total hyphal length (10^3 mg^{-1} dry wt soil)		
	range	mean	n	range	mean	n
arable	1.1-5.5	3.1	4	0.14-1.77	0.80	6
cereals	3.5-6.6	5.1	6	1.03-1.79	1.32	6
grassland	1.0-19.7	7.1	5	0.04-2.3	0.90	6
fallow	1.4-3.0	2.2	2	0.19-0.68	0.44	2

Microbial biomass, obtained either from direct counts or estimated by the fumigation technique, has been estimated for a wide range of soils. Values of 300 to 2240 kg dry weight ha^{-1} have been measured in a variety of temperate and tropical soils (Jenkinson and Ladd, 1981). This usually represents *ca.* 1 to 5 % of the total soil carbon and 2 to 6 % of nitrogen (Figure III.3) (Insam; 1990)

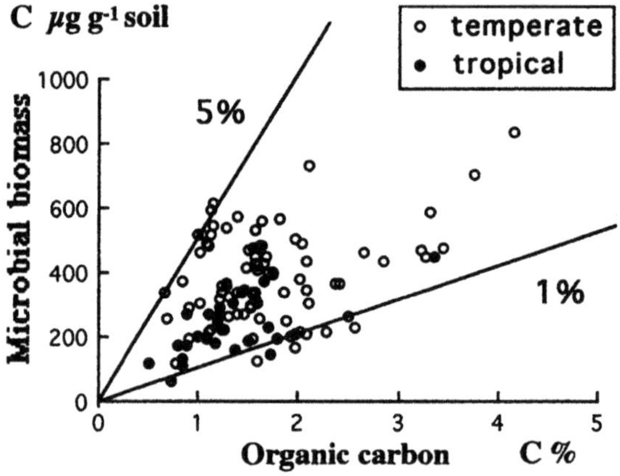

Figure III.3 Relationship between microbial biomass and soil organic carbon for a range of temperate and tropical soils (after Theng *et al.*, 1989 and Insam, 1990).

Values of up to 8.7 % carbon have been measured by Hasebe *et al.* (1985) in rice paddy soils of Japan. However, these authors considered that the fumigation technique overestimates the biomass since decomposition rate of the non-microbial, easily-decomposable organic matter is enhanced. Microbial biomass is generally high where substantial carbon inputs occur. Several authors have observed increased biomasses where organic fertilisers were added to the soil (see, *e.g.*, Kaczmarek, 1984; Hasebe *et al.*, 1985; Schnürer *et al.*, 1985).

Microbial biomass is also influenced by soil texture and nutrient content (*e.g.*, Chaussod *et al.*, 1986), and by roots and faunal activities (Bakonyi, 1989; Edwards and Bohlen, 1996; Fyles *et al.*, 1988; Helal and Sauerbeck, 1986; Holt, 1996 and Chapter IV).

Fungi and bacteria are the main components of the microbial biomass. Their relative importance varies substantially with organic matter content and such other soil characteristics as texture and pH. In seventeen agricultural and forest soils selected for their wide range of carbon contents, the relative contributions of bacteria and fungi to microbial biomass averaged, respectively, 25 % (10-40 %) and 75 % (60-90) (Anderson and Domsch, 1980; Christensen and Funck-Jensen, 1989). Similar proportions have been indicated by Clark and Paul (1970) in Canadian grasslands. Fungi are usually dominant in forest ecosystems. Nonetheless, bacteria may represent 78 to 98.2 % of biomass in paddy soils although organic fertilisation increases the fungal biomass and thus their relative importance (Hasebe *et al.*, 1985). In some cultivated Polish soils, bacteria represented 72 to 85.6 % of biomass with the lowest values occurring in plots receiving organic fertilisers (Kaczmarek, 1984).

These differences appear to be largely attributable to differences in the overall composition of decomposing resources and to pH (see Chapter II Resources): fungi are mainly primary decomposers feeding on carbohydrates whereas bacteria are generally more capable of digesting substrates with higher protein contents and low C:N ratios. Fungi tend to dominate at low pH values since bacteria are favoured by more neutral pH conditions.

2.3 Biology of Soil Micro-organisms

The biology of micro-organisms determines their roles in soils, either directly or in association with other soil organisms. It is also important to identify the major constraints facing these organisms when living in soil and their adaptations. The other important task for soil ecology is that of integrating the multitude of chemical transformations carried out by micro-organisms and the results of their interactions with biotic and abiotic soil components. The gap between overall estimates of abundance or activity (*e.g.*, data presented in section III 2.2) and the detailed knowledge accumulated on individual species under laboratory conditions is still too large. Its reduction is a tremendous challenge and probably constitutes one of the major limitations to the progress of soil ecology.

2.3.1 GENERAL CHARACTERISTICS

2.3.1.1 *Bacteria*

Water relations

Bacteria are aquatic organisms which live in the free capillary or gravitational water of the soil. As such, their activities are directly dependent on relatively high soil water contents and optimal activity occurs at water potentials between -0.01 and -0.03 MPa (*i.e.*, within pores larger than 2-4 μm filled with water) (Figure III.4).

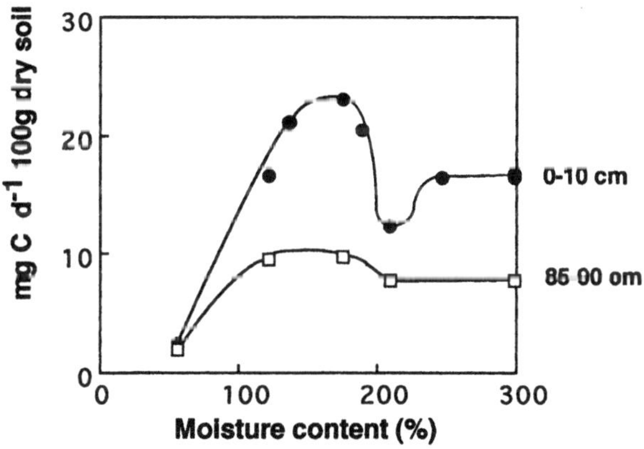

Figure III.4 Respiratory activity of an andisol at different soil moisture concentrations. Note the bimodal shape of the upper curve due to the activity of micro-aerophilic bacteria at high soil moisture concentrations (Legay and Schaefer, 1980).

As soils dry to matric potentials less than -0.05 to -0.3 MPa, most bacterial activities (movement, fungal lysis, chemical transformations, respiration) are substantially reduced. Below potentials of -0.3 to -0.6 MPa, bacterial respiration declines rapidly and becomes negligible at -2.0 MPa (Griffin, 1981b). Bacterial activity generally ceases at potentials below -1.5 MPa. Nonetheless, the lower tolerance potentials of some bacteria may be much more negative (*e.g.*, -10 MPa in *Clostridium*) (Paul and Clark, 1989), and significant activity has been recorded at potentials of -15 to -30 MPa.

The ability of bacteria to survive in dry conditions is probably due to the selective colonisation of favourable microhabitats, and differences in the vertical and horizontal distributions of soil moisture.

Mobility

Bacteria are normally associated with the surfaces of mineral or organic particles and assemblages. Most are unable to move and even those which have flagellae (mostly Gram-negative species) have very limited ranges of movement. Consequently their dispersion is dependent on water movement or root and faunal activity.

Digestion

As with all micro-organisms, bacteria have a system of external digestion mediated through the production of extracellular enzymes, especially in the Gram-positive bacteria (Hattori, 1973).

Part of the metabolites released by extracellular digestion may be used by other organisms thus creating a trophic stimulus for opportunistic or cooperating micro-organisms (Figure III.5). Intracellular enzymes may also be released by dead cells which can contribute to the nutrition of the colony. When active, bacteria have high levels of metabolic activity of 50-1200 µl O_2 mg^{-1} dry cells hr^{-1}.

Functional Diversity

Bacteria are able to perform an extremely wide range of chemical transformations. They are, however, only active over a very narrow range of environmental conditions. For example, non-symbiotic fixation of atmospheric nitrogen by *Azotobacter* is only possible if easily metabolisable carbon substrates are available and N-compounds such as ammonia or nitrates are either absent or present at extremely low concentrations. Fixation occurs between pH levels of 6 and 9 and is at a maximum between 7.2 and 7.6. Growth factors, vitamins and high levels of phosphorus are also required. Soil water potentials of greater than -2.4 to -0.4 MPa, temperatures between 10 and 37 °C (with optimal fixation at 25-30 °C) and low oxygen tensions are other necessary conditions. *Azotobacter* is inhibited by a large range of toxic mineral and organic compounds but may tolerate relatively high salinity. Colony growth and N-fixation is enhanced in the presence of clays. Finally, *Azotobacter* develops synergistic relationships with a number of other bacteria (particularly cellulolytic bacteria) which provide assimilable-carbon (Dommergues and Mangenot, 1970).

Recent development of techniques using fluorescent rRNA-targeted oligonucleotide probes has enabled the detection of bacteria of different phylogenetic levels, from the subspecies to the kingdom level (Stahl and Amann, 1991).

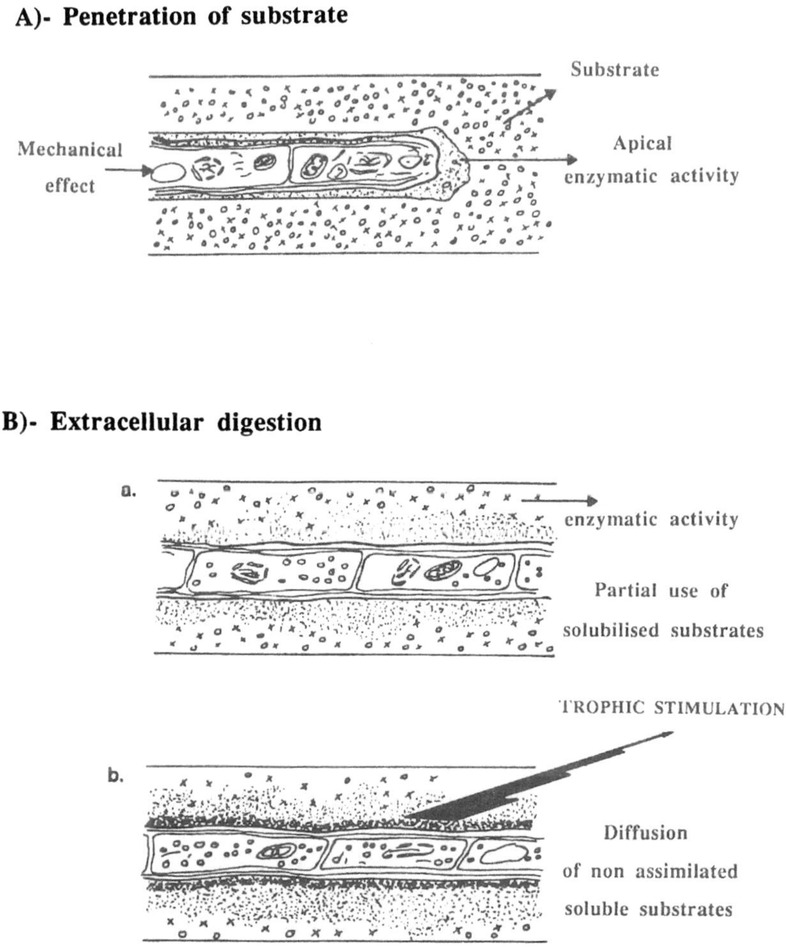

Figure III.5 External digestion in micro-organisms (Reisinger and Kilbertus, 1980).

2.3.1.2 *Fungi*

Fungi are primarily heterotrophic decomposer organisms although it is clear that they have a more diverse metabolic capacity than generally recognised and may mediate a range of chemical transformations in soil (Wainwright, 1988). Many species are pathogenic on the tissues of higher plants while others form symbiotic mycorrhizal associations with their roots. Because of their importance to the growth of their host plants, the fungi involved and the associations that they form, mycorrhizal fungi are considered in more detail in Section IV.3.1.

Saprophytic fungi derive their carbon and energy requirements largely from the products of cellulolysis, cell wall materials, cellulose, hemicellulose and lignin. White-rot Basidiomycota and some other fungi are among the few organisms able to degrade lignin and the brown phenol-protein complexes of leaf and root litter. They are much less dependent on water than other micro-organisms and while there is a considerable range of tolerances to low moisture potentials, some xerophytic fungi such as *Aspergillus* spp. survive in pure culture at potentials as low as -40 MPa (Parr *et al.*, 1981). Facultative anaerobiosis is also widespread in fungal species. However, in the field, tolerance is effectively reduced due to interactions with other micro-organisms. Temperature, solute concentrations in the soil solution and the nutritive resources available will also influence the relationships between fungal activity and soil moisture levels (Griffin, 1972). Their oxidative capacity, which is in the range of 10 to 12 $\mu l\ O_2\ mg^{-1}$ dry cells hr^{-1} is substantially inferior to that of the bacteria.

Digestion is external. In contrast to bacteria which attack organic substrates from the outside by direct contact, some fungi produce mycelial extensions which breach the cell-walls (Figure III.2a). These then penetrate the living or dead tissues allowing a rapid digestion of cell contents and a further breakdown of cell walls. Digestive capacities and nutritional requirements are diverse and Garrett (1951) has classified the fungi into five trophic groups: saprotrophic sugar fungi which live on simple sugars and the products of cellulolysis, lignin-decomposing, coprophilous, root associated (mutualistic or antagonistic) and predaceous fungi that feed on nematodes. The activity of fungi starts in the living leaves and roots, is at a maximum in decomposing leaves and wood and tends to diminish in the later stages of decomposition.

It is known that the fungi active in the soil environment may simultaneously use a wide range of carbon sources including gases and volatiles. Attempts to classify fungi into exclusive ecological categories, have proved difficult due to the wide versatility of their metabolic capacities and the diverse patterns of growth, reproduction and dispersal (Swift 1976; Pugh, 1980; Wainwright, 1988).

Fungi have the ability to move by growing new mycelium and may translocate living protoplasm towards the growing ends (Dowding, 1976; Schnürer, 1985). This leads to an accumulation of large masses of dead mycelium which may exceed the biomass of living mycelium by several times. In six European temperate forest soils, live mycelium comprised from 2.4 to 82 % of the total. Average maximum values ranged from 50 to 70 % in the L and F layers and minimum values of approximately 17 % were found in the H layer and A horizon (Kjöller and Struwe, 1982).

Five main growth patterns have been distinguished (Burges, 1960):

(i) In the *Penicillium* type, the fungi densely colonise organic substrates but mycelia do not extend into the surrounding soil;

(ii) In contrast, in the *Mucor ramannianus* type, the fungal mycelia grow into the surrounding soil and produce numerous spores following colonisation and exploitation of an organic substrate;

(iii) Fungi of the *Zygorrynchus* type produce isolated hyphae in the soil although these have no obvious relationship with organic substrates;

(iv) In the Basidiomycota, growth from a colonised substrate occurs by the production of mycelial strands and rhizomorphs;

(v) In the "fairy ring" type, fungi form an expanding annulus with a dense network of mycelium unrelated to substrate accumulation. Associated changes in soil chemical properties make such zones conspicuous, especially in grasslands.

Despite the relative mobility due to their growth patterns, many fungi are dependent on invertebrates, especially earthworms (Reddell and Spain, 1991a), microarthropods and Diptera larvae, for their dispersal through the soil sometimes involving selective grazing and spread (Arpin *et al.*, 1980; Swift and Boddy, 1984; Hanlon and Anderson, 1980).

2.3.2 SURVIVAL

Bacterial cells and colonies and fungi may live for more than 50 years in vegetative form, or as spores (Hattori, 1973). Sneath (1962) surveyed the organisms present on plant roots in herbaria and observed that while few micro-organisms survived more than 50 years, it would take 1000 years to sterilise the roots. Micro-organisms spend most of their lives just surviving. Estimates of the turnover time of microbial biomass in soils are 1,000 to 10,000 times less than the values measured in optimal laboratory conditions (*ca.* one day). This means that, on average, bacteria may be inactive for several successive months or even years after a period of activity **and may thus be fully active only every 3 to 30 years.** Environmental conditions seldom meet their often-specific requirements. Unable to move and search for suitable conditions, they possess adaptations to permit them to survive long periods of inactivity.

Four main mechanisms are involved in this survival:
(i) cannibalism;
(ii) cell maintenance with minimal energy expenditure;
(iii) physical protection in favourable micro-environments; and
(iv) the production of resistant spores.

When nutritional conditions decline and micro-organisms start to die, part of the colony may survive by ingesting the products of the lysis of dead cells. This cannibalism may be efficient and 13 dead cells were found to be sufficient to allow the doubling of one living cell (Nioh and Furusaka, 1968 quoted in Hattori, 1973). The liberation of endoenzymes after lysis of dead cells may also allow the production of assimilable substances in the remaining live cells. Short-term survival of colonies through these processes is obviously dependent on their initial size.

The maintenance energy of dormant cells is in the range of 1.6-4.3 10^{-6} mg glucose mg^{-1} biomass dry wt h^{-1}. Equivalent values for active cells are in the range 0.004-0.17, *i.e.*, three orders of magnitude smaller than those measured for growing cells in laboratory cultures (Anderson and Domsch, 1985a, b). Jenkinson and Ladd (1981) estimated that the average maintenance energy for inactive microbial biomass in a cropped soil at Rothamsted (England) was 21 % of biomass per year. Such low values can only be explained by assuming that starving cells have a different metabolism and may use endocellular rather than exocellular substrates for their respiration.

Long-term survival involves either endospore production or physical protection in suitable micro-environments. When environmental conditions are no longer suitable for growth, some individuals produce endospores, for example, the Gram-positive bacteria which live outside microaggregates (Hattori, 1973). However, Kilbertus *et al.* (1977), observed that, in laboratory cultures, few individuals were able to produce these endospores because sporulation requires specific chemical compounds which may be absent.

Physical protection in suitable sites within microaggregates allows many fragile Gram-negative bacteria to survive unfavourable conditions. The simplest situation is where clay particles are adsorbed onto the cell surface: under appropriate conditions of pH, bacteria and clay particles may have opposing electrical charges which favour their association (Dorioz and Robert, 1987). Adhesion may also be effected by a simple attraction effect (Van der Waal's forces) or, more commonly, by the production of a polysaccharide cover (Figure III.1).

Kilbertus and Mangenot (1981) consider that adsorption of clay particles would constitute a protection under natural conditions. This may be more efficient than the production of endospores because of the specific nutrients required for cells to produce spores.

Another efficient protection system is that afforded to the Gram-negative bacteria that live in the micropores that occur within microaggregates. These micropores are generally filled with water and offer suitable physical conditions for microbial life. Hydrological conditions are generally favourable but oxygen diffusion may be slow and promote anaerobic conditions. This protection is very efficient and Gram-negative bacteria may survive drastic soil treatments (Gray and Postgate, 1976). For example, most bacteria surviving the chloroform fumigation treatment of soils are found in narrow (<1 μm) pores of microaggegates or enclosed in thick layers (>2 μm) of their own extracellular polysaccharides (Foster, 1988).

The survival of fungal populations is mainly achieved by the translocation of living cytoplasm towards the growing mycelial ends, and the production of resistant spores.

2.4 Community structure

The structure of microbial communities varies considerably in time and space at all scales: micro-, meso- and macro. Such variation is related to their population characteristics, their ability to use specific substrates, their colonisation strategies, and to external factors including the characteristics of the physical environment and available resources.

Spatial and temporal distributions of microbial communities reflect the combined effects of these four factors. Successions at different stages in the decomposition of a resource especially reaffirm the major roles of resource quality and competition in determining the balance of different groups between microbial populations at all stages (Swift, 1982).

2.4.1 SPATIAL DISTRIBUTION

Variation on a geographic or **macro scale** has been little studied. Mishustin (1966 quoted in Hattori, 1973) in comparing different soils from the USSR, identified a clear North-South gradient of microbial abundance with total numbers increasing towards the South but the relative importance of other bacteria decreasing in favour of the Actinobacteria.

Swift *et al.* (1979) confirm this pattern with hyphal lengths increasing from tundra (*ca.* 1000 m g^{-1} dry weight) to boreal forests (4000 m g^{-1}) temperate grasslands and forests (3000 m g^{-1}) and tropical forest (6000 m g^{-1}). However, these are only broad approximations since regional variation may be very high: British and Scandinavian tundra soils, for example, have densities 3 to 100 times higher than equivalent North American ecosystems (Kjöller and Struwe, 1982). Furthermore, microbial biomass seems to be roughly proportional to the amount of organic matter (Figure III.3). Since there is no clear geographical pattern for organic matter accumulation in the soils of vegetated regions (Sánchez *et al.*, 1982), regional and local factors are more likely to determine microbial abundance than factors operating at larger scales.

On a regional scale, vegetation, soil characteristics and types of land-use influence the abundance and composition of the microflora by modifying water and temperature regimes and oxygen and energy supply. In Swedish agricultural soils, the density of bacteria increased with soil organic matter concentration in cultivated fields (Figure III.6) (Schnürer *et al.*, 1985). The effect of soil type has also been observed by Kaczmarek (1984) in Polish cultivated fields with mean densities of 5 10^9 and 18 10^9 g dry soil respectively in sandy and muck soils.

The **meso-scale** may be considered as that represented by vertical heterogeneity down a soil profile or horizontal heterogeneity created by plant distribution or microrelief.

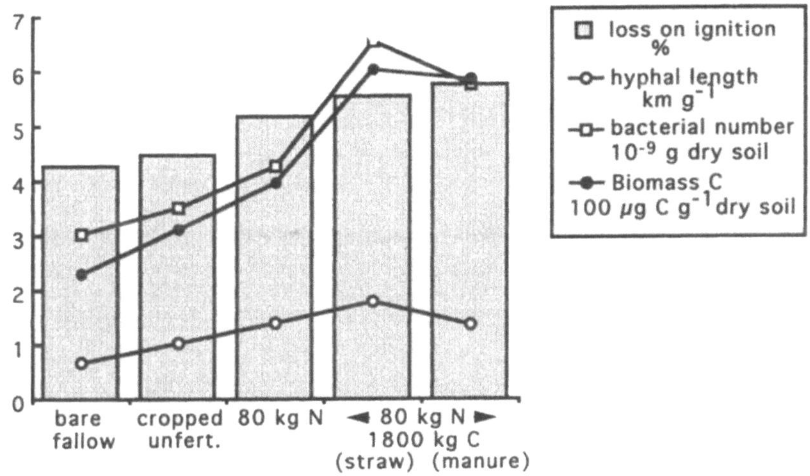

Figure III.6 Microbial abundances in cultivated soils with five fertiliser treatments and different organic matter contents (Schnürer *et al.*, 1985).

2.4.1.1 *Vertical distribution*

Bacterial population densities diminish with depth in parallel with organic matter contents. However, other factors such as the water-logging of deep horizons or differences in the physical structure of successive horizons may confound this pattern leading to different patterns of decrease with depth (Dommergues and Mangenot, 1970).

Population densities of most micro-organisms decrease regularly with depth (type I): maximum density occurs close to the surface and the rate of decrease may vary depending on the species. However, some micro-organisms attain their maximum densities at particular depths (type II). This may be due to the leaching of toxic substances from the surface litter *e.g.*, the distribution of *Azotobacter* in forested rendzinas (mollisols). Micro-organisms may also have a concave type of distribution. This has been observed in soils where chemical compounds toxic to micro-organisms, have accumulated in intermediate soil layers thus depleting the microbial community. The deeper horizons of higher pH in calcareous soils are favourable for the greater development of populations of micro-organisms.

The depth distributions of bacteria also change seasonally. In the upper 8 cm of a forested rendzina (mollisol), Proth (1978) observed three patterns of distribution, depending on the season (Figure III.7). Kjöller and Vestberg (1985) made similar observations on seasonal variation in the distribution of bacteria in the holorganic (L + F + H) layers of an alder forest soil waterlogged for most of the year. At their study site, variation in environmental conditions during decomposition was responsible for these changes and differences in the representation of particular functional groups of micro-organisms.

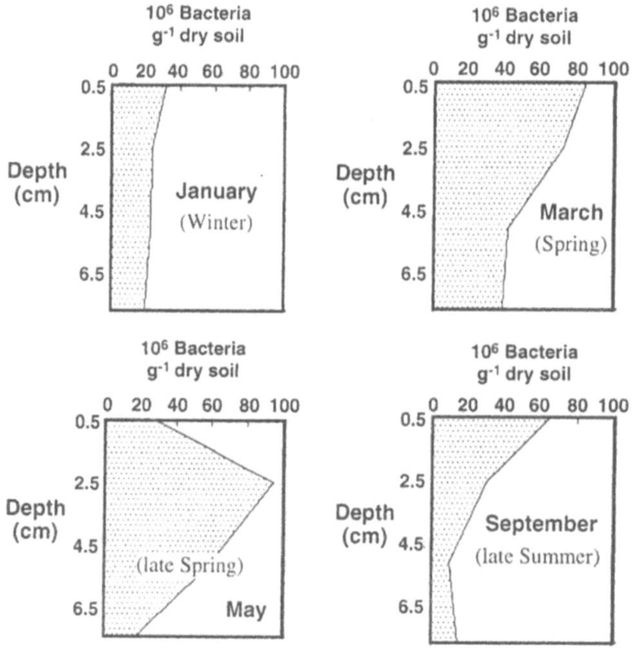

Figure III.7 Seasonal variation in the vertical distribution of bacterial colony forming units in the soil of a temperate climate deciduous forest (Proth, 1978).

The abundance of fungi generally decreases with depth, especially in forest soils where they preferentially colonise the litter and humus layers (*e.g.*, reviews by Kjöller and Struwe, 1982; Bissett and Parkinson, 1979; Rodriguez *et al.*, 1990). Within the O horizon, abundance increases from the surface to the F or H layer and then decreases sharply. In grasslands, the decrease with depth may be far less clear and Harris (1971) has reported greater abundances at 10-30 cm than in the upper ten centimetres.

2.4.1.2 *Horizontal distribution*

The horizontal distribution of micro-organisms on a mesoscale is determined by geological and pedological factors. Plant distribution also greatly influences the distribution of soil micro-organisms on the soil surface by the litter effect, and within the predominantly inorganic soil, by the rhizosphere effect. Large invertebrates such as termites and earthworms may also significantly affect the distribution of micro-organisms in their respective functional domains (see Chapter IV). The development of bacterial populations in leaf litter is limited and they are most abundant within the upper mineral soil horizon where their horizontal distribution is influenced by the availability of energetic resources. They are most concentrated in the rhizosphere where they live on readily assimilable root exudates and dead root material. Other sites in the soil with concentrations of organic matter may be preferentially colonised by bacteria and include structures built by soil invertebrates, such as the linings of earthworm burrows or the walls of termite galleries and fungal chambers (see Chapter IV).

In leaf litter, fungal populations often show patchy distribution patterns. This is especially clear for white rot fungi (Basidiomycota), essential decomposers of the litter system, which form patches of a few square centimetres (Bartolozzi, 1984 in Garay, 1989).

The **microscale distribution** of bacteria is that at the microaggregates, *i.e.*, small structural units of a few tens of micrometres in size. Bacteria may live inside microaggregates or outside, on their surfaces or in the pores (>10 μm) between them (Hattori and Hattori, 1976; Balkwill *et al.*, 1977).

Outside the aggregates, the flux of water is rapid which facilitates the supply of energy, nutrients and oxygen, although conditions may quickly become limiting when soils dry out. The microflora is mainly composed of fungi, spore-forming Gram-positive bacteria and Actinobacteria (Figure III.8). Members of these groups are opportunistic organisms whose populations can vary rapidly depending on microclimatic conditions and their ability to resist adverse conditions.

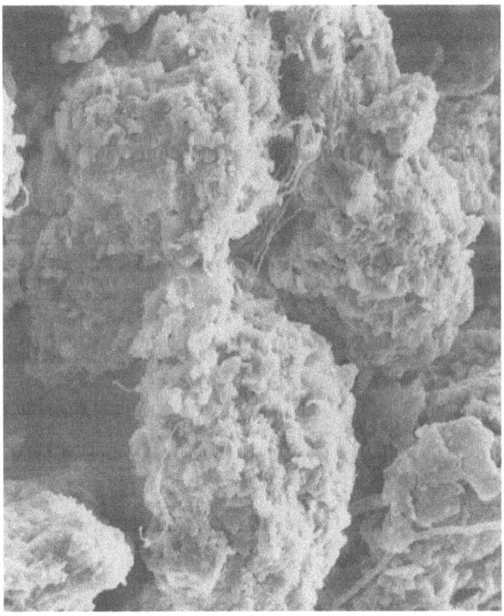

Figure III.8 Soil microaggregates showing an external network of fungal hyphae (Photo: Michel Robert) 1 cm = 7.5 μm.

Within the aggregates, the microflora essentially consists of largely dormant Actinobacteria and Gram-negative bacteria which exist there under buffered moisture conditions, at times limited by oxygen supply (see Figure III.1 A). They occupy micropores and may be released into the 'external' medium when microaggregates are disrupted under conditions of high moisture or assimilable energy supply. In six different temperate climate soil-vegetation associations, the average diameter of micropores occupied by bacteria inside microaggregates, was 1.43 to 2.24 μm (Kilbertus, 1980; Schwartz,1981). The ratio d/D of mean bacterial diameter (d) to the mean diameter of the pores they colonise (D) was 0.33 on average, with remarkably little variation among the six soils investigated (Table III.4).

Table III.4 Relationship between bacterial diameter (d) and the pores that they colonize (D) in six different soil-vegetation associations (Kilbertus, 1980; Arpin *et al.*, 1986; Schwartz, 1981)

Soil or humus type and Vegetation	Diameter of colonized pores (D μm)	Diameter of bacteria (d μm)	d/D
Dystrophic brown soil, mixed forest	2.24±1.40	0.75±0.44	0.33
Rendzina, deciduous forest	1.74±0.49	0.60±0.08	0.34
Chernozem	1.81±1.33	0.60±0.21	0.33
Mull-moder, mixed forest	2.07±0.95	0.65±0.21	0.31
Mull-moder, deciduous forest	2.03±0.81	0.63±0.22	0.31
Moder, coniferous forest	1.43±0.48	0.51±0.11	0.35

Postma and van Veen (1990) have proposed the classification of soil porosity into three categories based on the ability of pores to host bacterial cells of the genus *Rhizobium* and protect them from predators: accessible pore space that is large enough to let micro-organisms enter, habitable pore space that comprises pores larger than 0.8 μm and protective space (>0.8 μm but <3 μm) that is inaccessible to bacterial predators.

Bacteria rarely occur as isolated individuals. In a forested rendzina soil (mollisol), 90 % of individuals were grouped in colonies of 6 to 300 individuals, most colonies having *ca.* 10 individuals on average (Proth, 1978). Inside the microaggregates, an average number of 0.8 bacteria was found per μm^3 of pores colonised. Bacteria represented 0.29 % of the volume of the aggregate and the overall volume of the pores they colonised was 6.21 %.

The above results are important in that they clearly demonstrate that potentially-assimilable organic matter may be protected from microbial attack because of its location in non-colonised micropores, or within non-porous microsites within aggregates (Foster, 1985).

They also explain why, at low moisture potentials, non-spore forming cells may survive in moist microsites. Finally, they illustrate how opposing processes (*e.g.*, nitrification and denitrification) may occur at any time in closely adjacent microsites (see Fig. I.15, p40), as emphasised by Berg (1986).

Different types of micro-aggregation, and therefore soil structure, may influence the composition of the overall microbial community. Some soils will have more opportunistic micro-organisms located in non-protected microsites whereas others with a well-developed micro-aggregate structure will have large amounts of Gram-negative specific bacteria protected in micropores (see Section III 2.4.3).

2.4.2 TEMPORAL VARIATION

Microbial populations may change rapidly. In a few hours, or days, rapid changes in environmental conditions, *e.g.*, a sudden input of assimilable organic matter or moistening of a dry soil, may trigger a flush of microbial activity and rapid cell division.

On a seasonal scale, progressive changes in climatic or energy supply conditions may be reflected in changes to the densities and biomasses of microbial populations. Over longer time-scales, organic matter transformations during decomposition may be followed by changes in micro-decomposer populations resulting in very different microbial successions.

2.4.2.1 *Short-term variation*

Dormant micro-organisms may rapidly reactivate when conditions again become favourable. Bacteria which are protected inside microaggregates may leave this shelter 24 h after a sucrose solution has been added to the soil (El Bahlki *et al.*, 1978) and it takes a similar time for fungal or bacterial spores located outside the aggregates to germinate. Consequently, bacteria can increase their numbers dramatically within a few days, although little variation occurs within 24 h (see, *e.g.*, Zaitzeva and Zviagintzev, 1978; Clarholm and Rosswall, 1980).

In the humus layer of a Swedish conifer forest, Clarholm and Rosswall (1980) observed increases in population density and biomass of, respectively, 2 to 6.10^{10} ind g^{-1} and 4.2 to 17.6 mg dwt g^{-1} dry wt soil over three days following rainfall. An analysis of the structure of bacterial populations showed that, under such conditions, 15 to 30 % of the formerly-dormant population grew actively with a generation time of 14 to 17 hours. After a maximum has been reached, both density and biomass decrease following intensive grazing by protists and nematodes (Figure III.9).The initial rise in biomass resulted from an increase in the size of individual cells, as shown by the significant rise in the number of large rods observed in the population. Numbers subsequently declined, probably because they had started to divide into small coccoid cells.

Bottner (1985), simulated the alternation of 8 to 10 day dry and 15 to 20 day moist periods in a Mediterranean soil. Drying killed 25 to 33 % of the biomass (bacteria + fungi) although during the wet periods, biomass returned again to the same level present before drying.

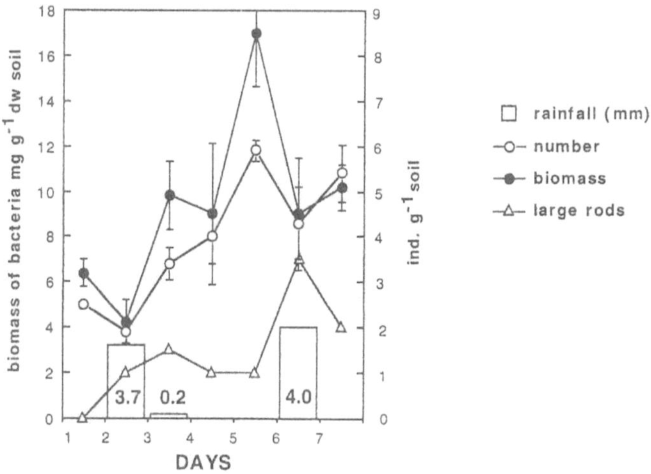

Figure III.9 Fluctuations in bacterial numbers and biomasses in the humus layer of a Swedish forest over a five day period (after Clarhom and Rosswall, 1980). Vertical bars indicate S.E.

2.4.2.2 *Seasonal variation*

In some environments, monthly sampling conducted in several soil and vegetation types has revealed clear patterns of variation, in the numbers, biomass and activity of bacteria. For example, in the Swedish coniferous forest soil studied by Clarholm and Rosswall (1980), bacterial biomass underwent a clear seasonal pattern of variation with a minimum of 5 g dry wt m^{-2} in June and a maximum in September, nearly 10 times higher (Figure

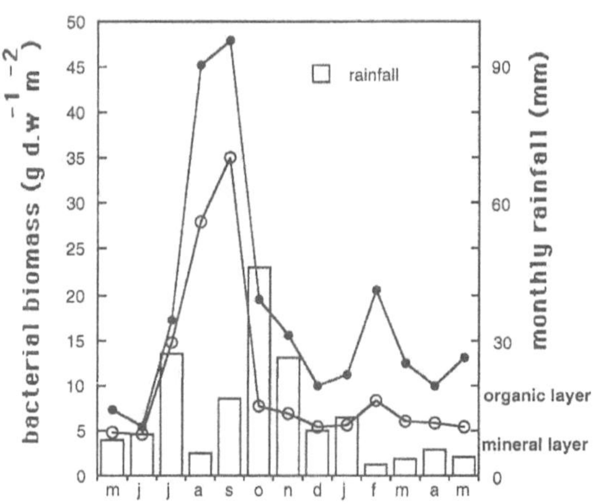

Figure III.10 Monthly estimates of bacterial biomass in the top 10 cm of a Swedish pine forest growing on a podzolic soil (Clarholm and Rosswall, 1980).

III.10). The seasonal variation observed in a Japanese grassland by Higashida and Takao (1985) had a lesser amplitude; the maximum density occurred in August which was twice as high as the May minimum. In cultivated soils at Kjettslinge (Sweden), seasonal variation of total biomass was much smaller with a difference of 25 % between the minimum value in June (*ca.* 400 μg g^{-1} d.w) and the August maximum (*ca.* 500 μg) (Schnürer *et al.*, 1985).

Kilbertus *et al.* (1982) observed quite large variation in the bacterial population density of the top few centimetres of the A horizon of a conifer forest: the maximum density, in June, was nearly 100 times greater than the minimum in December. The proportion of bacteria protected in microaggregates (released by sonication of the soil samples) varied greatly over the year, with minimum values in the middle of the year and maximum in winter (Figure III.11). A similar pattern was observed in an adjacent deciduous forest with a lesser proportion of protected micro-organisms (40 % on aver-

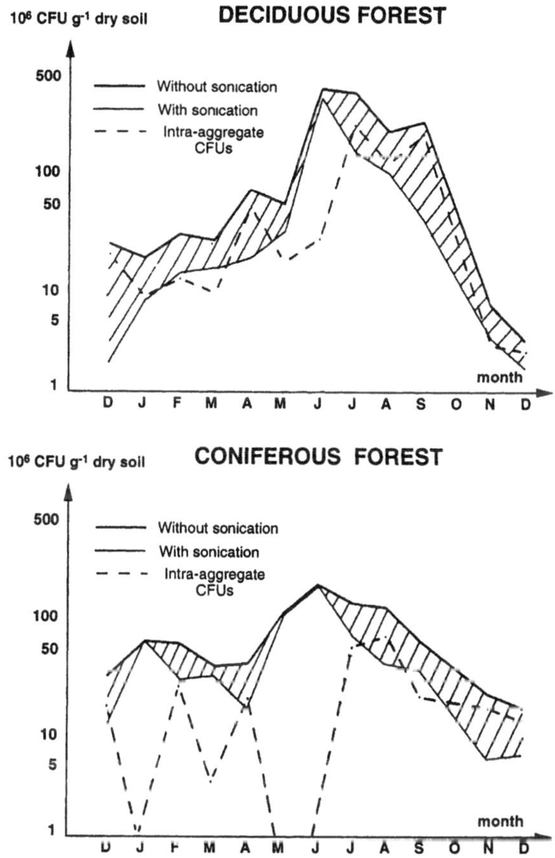

Figure III.11 Seasonal variation in numbers of colony-forming units in the A horizon of a forest soil with and without a previous sonication.

age compared with 58 % in the coniferous site).

Several studies of fungal populations in temperate forests have revealed some significant seasonal variation in the relative abundances of species in microfungal communities (*e.g.*, Bissett and Parkinson, 1979; Widden, 1986). In volcanic soils on slopes of

Mt Popocatepetl (Mexico), clear seasonal variation has been observed in the composition and abundance of fungal communities, especially at high altitudes (3660 m) (Bettucci *et al.*, 1990). In temperate forest ecosystems, fungal activity, as measured by the growth of mycelium on buried nylon gauze, was at a maximum in Spring, early Summer and Autumn although no significant change in the amount of live hyphae was observed (Nagel de Boois 1971; Cancela da Fonseca, 1975; Tôth and Hammer, 1977). Nonetheless, significant differences have been observed in the growth of mycelia in litter of contrasting quality, *i.e.*, *Pinus sylvestris* and *Carpinus betulus* (Garay, 1989).

2.4.3 TURNOVER OF MICROBIAL BIOMASS: THE SLEEPING BEAUTY PARADOX

The turnover time of microbial biomass has been estimated in a limited number of soils by comparing the average microbial biomass to plant-C input to the system, assuming that the production of microbial biomass is equivalent to 50 % of the organic inputs to

Table III.5 Microbial biomass and substrate input in three temperate agroecosystems (Hansson *et al.*, 1990).

	Kjettslinge (Sweden) Andren *et al.*, 1987	Rothamsted (England) Voroney, 1983	Saskatchewan (Canada) Voroney, 1983
Cropping system	Barley	Wheat	Wheat-fallow
Sampling depth (cm)	0-27	0-23	0-24
Soil-C %	2.6	1.2	2.4
Plant-C input (µg C g^{-1} dry wt yr^{-1})	700	500	600
Microbial biomass-C (µg C g^{-1} dry wt yr^{-1})	400	300	600
Microbial generations (per year)	0.9	0.8	0.5

the soil (Table III.5).

Jenkinson and Ladd (1981) used a mathematical model to calculate generation intervals of 912 and 456 days in, respectively, English and Australian soils. Chaussod *et al.*, (1988) gave similar estimates of 522-659 days in cultivated fields near Paris, although, under laboratory conditions (28 °C at field capacity over 56 days), the turnover time decreased to 133-179 days. Turnover time may however be significantly greater in soils that experience thermic or hydric stresses, especially those with a sandy texture. For example, turnover time of the microbial biomass was estimated at 1600 days in Saskatoon (Canada), 570 days at Rothamsted (England) and 460 days in Brazil (Paul, 1984).

These estimates are 1,000 to 10,000 times longer than those obtained for isolated microbial colonies under optimal laboratory conditions and imply that soil micro-organisms are only active during short periods of time in a limited number of microsites. Further, cytochemical tests indicate that very few cells are physiologically active (Foster, 1988). Microbial communities thus appear as 'a huge, largely dormant population, with an enormous richness of species and an ability to survive hard times' (Jenkinson and Ladd, 1981). The first characteristic is considered a response to the heterogeneity of soils and the second, a response to their inability to search for food. This striking contrast between the potential for an extremely fast turnover and field reality has been called the 'Sleeping Beauty Paradox' (Lavelle *et al.*, 1994a). Macro-organisms (*i.e.,* roots and invertebrates) that have the ability to move the soil and change environmental conditions at the scale of micro-organisms can interrupt this dormancy (acting as 'Prince Charming') and hence appear to be major regulators of microbial activities (see Chapter IV). Interactions among micro-organisms with high capacity to digest almost all organic substrates and 'macro-organisms' that have potential for mechanical activities, are the basis of the biological systems of regulation that determine soil function (see Chapter IV).

These results also emphasise the need to assess microbial processes at much finer scales than a hectare or a year. Identifying the appropriate scales requires an in-depth knowledge of the biology and population dynamics of these organisms.

2.4.4 COMMUNITY STRUCTURE

2.4.4.1 *Species richness*

The total estimated number of microbial species is very high. Fungi are the most diverse group with an estimated number of 1,500,000 species whereas algae and bacteria may comprise, respectively, 60,000 and 30,000 species (Hawksworth and Mound, 1991). Studies of microbial DNA extracted from a Norwegian forest soil showed the existence of more than 4000 independent genomes of bacteria in a gram of soil (Torsvik *et al.*, 1990). Based on the assumption that individuals with more than 70 % DNA homology pertain to the same species (Wayne *et al.*, 1987), these authors suggest that there may be as many as 20,000 to 40,000 bacterial species per gram of soil.

The soil microflora is commonly said to be poorly specific to any one environment as many species of bacteria and fungi are found in a great variety of soil types and microclimates; some species have world-wide distributions. Comparison of three fungal communities from forests in southern Quebec (Widden, 1986), heathland from northern England (Widden, 1987), and tropical montane forests of Mexico (slopes of Mt Popocatepetl, Rodriguez *et al.*, 1990) showed that four species are common to these very different environments *i.e., ca.* 15 % of the species represented at each site. This assessment however, is a matter of controversy as the concepts of species and ecological niches are difficult to apply to micro-organisms (especially bacteria and actinobacteria) because of their high genetic variability and ecological versatility (see *e.g.,* Szábo, 1974; Swift 1976). Further, morphological criteria are few and often difficult to observe. Indeed, such criteria may be entirely misleading since colonies with very different morphologies have been shown to have the same DNA and, thus belong to the same species (Tiedje, 1995). Finally, different isolation techniques and sampling designs may lead to large differences in the estimated species richness of a single site.

The number of fungal species found in the litter of temperate forests may be as high as 161 species although numbers generally decrease with depth towards the H layer and A horizon. In contrast, the Actinobacteria tend to have greater species richness in the soil than in the litter (Table III.6).

Table III.6 The numbers of species of micro-organisms isolated from the litter layers and A horizons of forest soils. Differing methodologies do not permit vertical comparisons within the table (after Swift, 1976).

Organisms	Dominant litter types	Litter Oi	Humus Oa	Soil A	Reference
Fungi					
	Eucalyptus maculata	33	25	75	Eicker, 1973
	Pinus strobus, Acer saccharum, Ulmus americana	59	30	30	Widden and Parkinson, 1973
	Fraxinus pennsylvanica, Quercus bicolor	161	96	nd	Novack and Whittingham, 1968
	Mixed deciduous	27	nd	21	Proth, 1978
Actinobacteria					
	Quercus robur, Fraxinus excelsior	91	126	130	Williams *et al.*, 1969

Twenty seven species of fungi were identified in the mixed deciduous litter. However, the number of fungal species found on individual leaves was very low on fresh litters (0-3 depending on the tree species) but increased to 1-6 after four months of incubation. In the presence of Collembola, 1 to 7 species had colonised the fresh litter after 2 months incubation (Proth, 1978). Species richnesses of up to 161 species for a single mixed deciduous temperate forest have been reported (Novack and Whittingham, 1968).

These results demonstrate the effect of resource quality on the species richness of microbial communities. The rapid changes in quality that occur during decomposition, are followed by similar changes in the microbial community, as shown by the marked successional patterns that occur.

2.4.4.2 *Community structure and dynamics*

Large scale patterns: microbial successions on decomposing substrates

Primary resources often form a spatial mosaic comprising such structures as the decomposing twigs, branches, leaves or inflorescences of the individual species present. As a result, microbial communities which colonise these substrates have a hierarchical structure (Swift, 1984). On each fragment of the mosaic, a community develops, generally consisting of a limited number of species.

Swift (1976) found that small pieces of branches of beech-wood, 10 cm long and 2 to 5 cm in diameter were colonised in the early stages by up to nine species with a modal value of four. Fifty six species were identified among the 32 branch samples; most of these species were found on a small number of branches and only five species were found on more than five branches. The efficiency of the different 'unit-communities', expressed in terms of decomposition rate, was rather similar.

As predicted by the islands theory of MacArthur and Wilson (1967), the number of species is proportional to the size of the substrate. According to Swift (1976) two main processes determine the dynamics of such unit-communities: stress and disturbance (Grime, 1979; Pugh, 1980; Boddy *et al.*, 1988). Stress is defined by Grime as the external constraints which limit the production of plants. Disturbances are unpredictable events which limit the plant biomass by destroying it, partly or entirely.

On fresh substrates, stress is first represented by the initial quality of the resource (*e.g.*, high C:N ratio, lignin and polyphenol contents). At later stages, shortage of substrate, unfavourable microclimatic conditions or antibiosis are common stresses that micro-organisms are subject to. Disturbances are by definition unpredictable events. This may be the fall of a dead branch which may occur after it has lost 20 to 80 % of its weight. Another frequent example of disturbance is the ingestion of the substrate by a soil invertebrate or the grazing of growing fungi. Microclimatic events such as drought or frost are other possible disturbances.

Stress leads to successions of communities over time. Species disappear when they can no longer adapt to the changed conditions of the substrate and new colonisers replace them. The general pattern is an increase in the number of species to a maximum value and a decrease when the substrate becomes exhausted or unsuitable for micro-organisms (Figure III.12). Disturbance generally breaks the dynamic equilibrium between species by accelerating the extinction of species and the immigration of new colonisers.

Micro-organisms may adapt through two supra-strategies: the 'occupation' and the 'window' strategies. Micro-organisms of the first group are adapted to persist on a changing resource. In the second group, the strategy is one of rapid colonisation when conditions are suitable and dispersion when they have changed. Within each of these two types, *r* and *K* strategies occur, as defined by Pianka (1970) or 'ruderal',' stress-tolerant' and 'competitive' as defined by Grime (1979) for higher plants.

In such a context, the formation and evolution of unit-communities appears to be a highly stochastic process. Important differences are expected to occur however, depending on the nature and size of the substrate; communities certainly behave differently on large substrates such as decomposing logs, in comparison to those on leaves or small inflorescences.

In the subsoil, the patterns of community structure are not known. Outside the rhizosphere, soil organic matter is highly dispersed in the soil matrix and bacteria, the most important micro-organisms, have a higher potential metabolism but, on average, depend more on specific environmental conditions. Their strategies are thus likely to be more of the 'window' category and the process of assemblage of their species in communities more stochastic than observed in fungal communities of the surface litter.

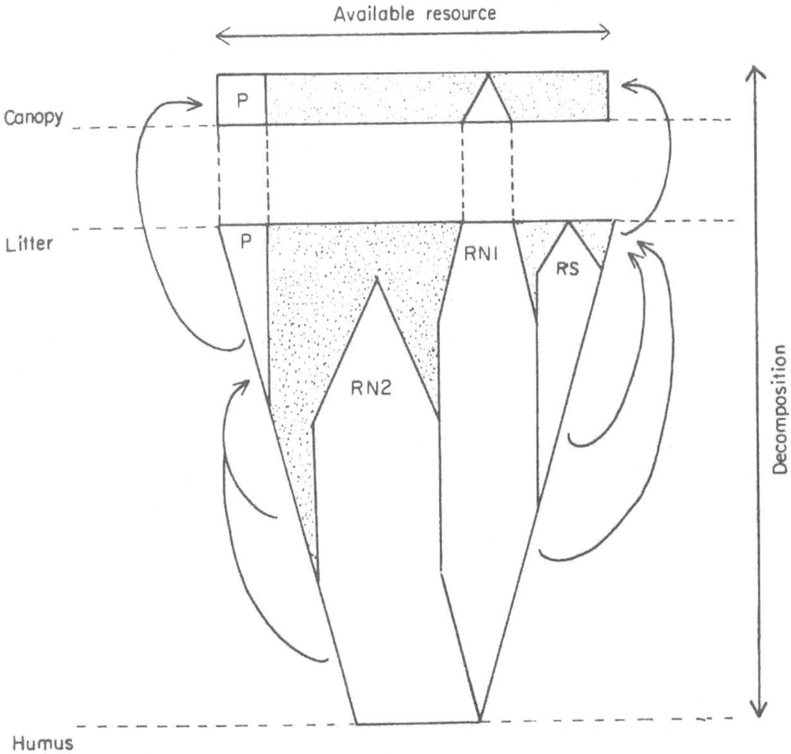

Figure III.12 Diagrammatic representation of the cyclical changes in fungal community structure on an'average' resource unit. Unstippled areas represent the extent of occupation of available resource by fungi, the arrows represent the dispersal of fungal propagules. P = parasitic species; RS = resource-specific saprophytes; RN1 = primary resource-nonspecific species; RN2 = secondary resource-nonspecific species. This 'succession' may be diverted if the resource is consumed by soil animals causing subsequent changes in the patterns of colonisation, chemical composition of the resource and particle size distribution (Swift, 1976).

Successions on decomposing substrates

Decomposition processes generally start while the plant material is still alive. In the early stages, the yeasts and phyllosphere bacteria present on the surfaces of leaves and fungi may attack both leaves and roots, especially when they have been damaged by frost or insect attack. After death of the plant organs, the cytoplasm is rapidly transformed with tannin-protein compounds being important components of the decomposing material (see Chapter IV.1.3.3).

Decomposition rates subsequently increase. In the initial stages, fungi predominate, since they are more capable than bacteria and actinomycetes of degrading the sugars and polysaccharides of the primary resources. Fungi colonise fresh litter, dead roots and logs. Unlike bacteria, they may tolerate relatively unfavourable environmental conditions in order to penetrate cell walls and metabolise the condensed cytoplasm and cell wall constituents.

Primary resources often have a mosaic pattern of distribution and communities of a few species may develop on each part of this. Early colonisers may be 'sugar-fungi' as defined by Garrett (1963). Although polysaccharolytic fungi soon colonise the substrate, Swift (1976) points out that even though simple sugars decompose more rapidly than cellulose, more cellulose than sugar disappears at the beginning of decomposition because of its much greater abundance. Beside sugar-fungi which colonise the leaf before its death, resource-specific polysaccharolytic fungi develop. The metabolites released by their external digestion create a trophic attraction ('*appel trophique*') and commensal secondary sugar-fungi, also defined as 'resource non-specific', may develop (Garrett, 1963; Reisinger and Kilbertus, 1980). Under some circumstances, the presence of secondary sugar fungi may benefit the polysaccharolytic organism by preventing the catabolic suppression of enzyme production that can follow the accumulation of simple carbohydrates (Figure III.12).

During the later stages of this original phase of decomposition, fungal communities include increasing numbers of resource non-specific organisms, species diversity decreases and unit-communities tend to be more homogenous (Hogg and Hudson, 1966; Swift, 1976). As the substrate changes, specific resources become exhausted leading to competition. The most persistent fungal flora includes components that have the highest capacity to produce antibiotics (*Penicillium* spp.) and the widest range of enzymatic capacities (Basidiomycota) (Webster, 1970; Swift, 1976).

Finally, fungi produce spores and die. New colonisers spread which are better adapted to the changed environmental and trophic conditions of the substrate.

The decomposing substrate changes and its relative nitrogen content increases as the proportion of mycelial material increases. After three months of decomposition in temperate forests, 50 to 80 % of biomass of decomposing hornbeam (*Carpinus betulus*) foliage is composed of fungi mycelium. Two months later, in early spring, the proportion increases up to 90 to 95 % (Proth, 1978). Environmental conditions change as this material is progressively incorporated into the soil. Bacteria and actinobacteria then become predominant. They are more capable than fungi of using substrates with relatively high nitrogen contents and many of them possess chitinases to decompose mycelial cell-walls.

III.3 ROOTS

Roots are the principal biological components of the soil system since they comprise the largest part of the living soil biomass and exercise a major control over such processes as pedogenesis, soil organic matter, nutrient and water dynamics. In terms of their function in the soil system, they act as true heterotrophic organisms since they obtain their energy and carbon from the above-ground part of the plant and their nutrients and water from the soil. Roots supply energy and return absorbed nutrients to the soil through the production of below-ground root litter and, while living, through the production of exudates, a mixture of readily-assimilable compounds and modifiers (*i.e.*, growth factors, hormones and allelopathic substances).

Their importance as pedogenetic agents is widely recognised and they have major roles in soil reinforcement and in the maintenance of structure (see Chapter I.1.3.3). The spatial and temporal patterns of their growth through soils are particularly important to the ways in which they utilise soil water and nutrient resources.

Finally, roots have close relationships with the free-living microflora and fauna, and in a wide range of natural and agricultural ecosystems, frequently form associations with such symbionts as nitrogen-fixing bacteria, fungi (mycorrhizae) or actinobacteria (actinorhizae).

3.1 The morphology of the root system.

3.1.1 DEFINITIONS

Root topology influences both the exploration and transport characteristics of the root system. Differences in the efficiencies of seven theoretical systems, all with 16 tips, but with different branching systems are illustrated in Figure III.13. System B is most effective at exploiting a restricted soil volume and is most common in nutrient rich soils; system A is less efficient in exploiting zones of high nutrient concentration, but better adapted to exploration (Fitter, 1985). Systems C to F are intermediate in nature.

Root systems may initially be described in terms of their vertical branching patterns (see *e.g.*, Jenik, 1978; Kahn, 1984). The order and importance of their ramifications may also be considered. Basically, two main types of root systems may be defined:

(i) A sinker or tap rooting system in which secondary roots originate from a primary root which grows vertically into the soil;

(ii) A lateral, or fasciculated rooting system in which roots expand from the base of the stem in all directions. Many intermediate patterns occur between and within species and represent modifications of a genetically-determined design imposed by local environmental conditions. The root system of the grape vine (*Vitis vinifera*), *e.g.*, comprises three types of roots: (i) large tap roots which anchor the plant to the soil and convey the phloem sap, nutrients and water; (ii) thin lateral roots, which ramify mainly close to the surface and in a nutrient-rich layer of marl deposited at 10-30 cm depth; and (iii) deep lateral roots which extract water from the lower horizons, at 60-120 cm depth. (Figure III.14)

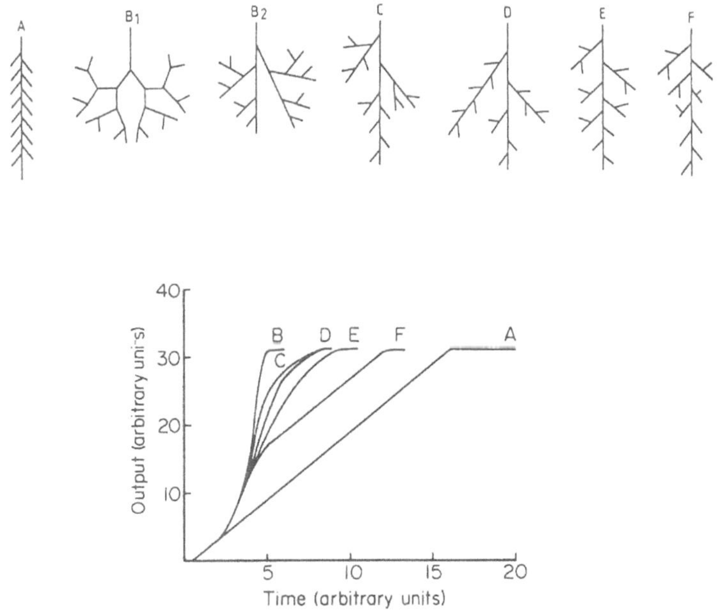

Figure III.13 Effect of topology of seven root systems comprising sixteen root tips (upper part) on the efficiency of water and nutrient transport (lower part) (Fitter, 1985).

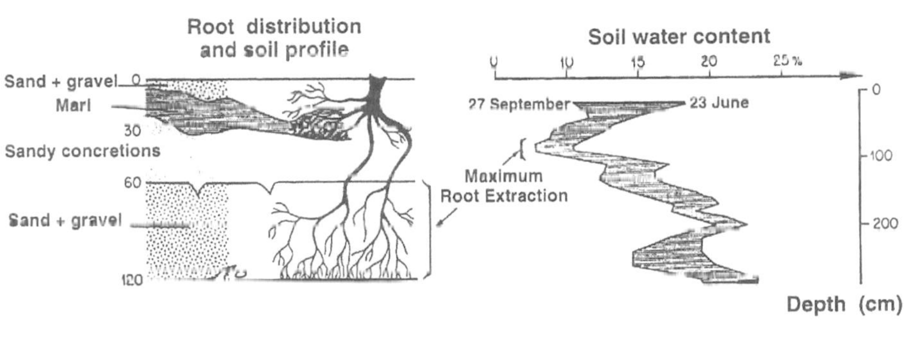

Figure III.14 Vertical distribution of the root system of vine (*Vitis vinifera*) and soil water before and after dry season in gravelly soils in Gironde (France) (note that different scales are used on each part of the figure).

Root diameter is also an important parameter. Thick suberised roots have a clearly different function than fine, unsuberised pale-coloured roots. An important step towards a functional interpretation of root morphology was made when coarse and fine roots were distinguished, depending whether their diameter was more or less than an arbitrary limit of 2 mm. Jenik (1978) uses the terms 'proximal' and 'distal' roots to define coarse and fine roots while Lyford and Wilson (1964) proposed the terms 'macrorrhizae' and 'brachyrrizhae' to qualify the two above-mentioned categories.

Coarse roots are generally highly suberised and lignified. They support the mass of the plant and conduct water and nutrients from the soil to its aerial parts, and soluble carbohydrates formed in the leaves to the root extremities and out into the soil. Coarse roots are perennial and form the framework on which new root growth occurs (Bowen, 1984). These roots are longeval and may persist throughout the life of the plant.

Fine roots are specialised for the assimilation of water and nutrients. They are neither lignified nor suberised and are generally pale-coloured and covered with hairs, which increase their absorptive surface area. Their life-spans are generally of the order of months but may vary from a few days for non-mycorrhizal roots to several years when they are infected with mycorrhizae. A number of other biotic and abiotic factors may determine root lifespan such as soil fertility, root herbivory, climate and competition with other plants (Eissenstat and Yanai, 1997). Fine roots will therefore be replaced many times during the life of a perennial plant and may be considered the endogeic analogues of leaves. Annual or ephemeral plants produce only fine roots and, in perennial grasses, it is difficult to differentiate between coarse and fine roots. Root hairs that grow on fine roots have even shorter life spans, of the order of a few days (see, *e.g.*, Fusseder, 1987)

Specific root length (SRL), the ratio of total root length to weight, is a useful index for estimating the relative importance of fine and coarse roots.

3.1.2 ROOTING 'STRATEGIES'

The importance of the soil-root contact may be evaluated by the overall surface area of roots (Callot *et al.*, 1982). This surface is relatively small, and part of it may actually not be in contact with the soil solution when roots follow old termite or earthworm galleries or cross soil macropores. The root surface of a mature wheat plant is about 1 m^2 whereas the overall surface of soil particles in the volume of soil exploited by these roots may be close to 10,000 m^2. On average, the ratio of the total surface of soil particles to roots is approximately 10^6, or 10^4 including the surface area of the absorbing hairs. In contrast, the diffusion of water and nutrients in soil is relatively slow and may be quite limited in amount [see Chapter I]. In consequence, the total mass of roots produced, and their horizontal and vertical distributions within the soil, are particularly important factors in plant fitness.

Rooting strategies may be defined as the way roots exploit soil nutrient and water resources. This, together with the energy that plants allocate to growing roots, may be described in terms of root topology, biomass, root:shoot ratio (above-ground to below-ground biomass), specific root length, horizontal and vertical distribution and concentration in fertile patches of the soil. Several factors may influence these parameters, including nutrient and water supply and soil physical characteristics.

3.1.2.1 *Root:shoot ratios and specific root length*

As indicated by (Bray, 1963), root:shoot ratios are characteristic of species and communities although they are sensitive to environmental influences and may vary markedly over the life of a single plant. They generally increase with increasing nutrient, water or cold stress and decrease with the amount of light or increasing biomass. High phosphorus availability favours greater root production (Callot *et al.* 1982; Strong and La Roi, 1985). Grazing by nematodes also tends to increase the root:shoot ratio (Rovira, 1978). Comparable patterns have been observed at the community level (Scholes, unp. data) (Figure III.15) and temperate ecosystems tend to have much higher root:shoot ratios than tropical.

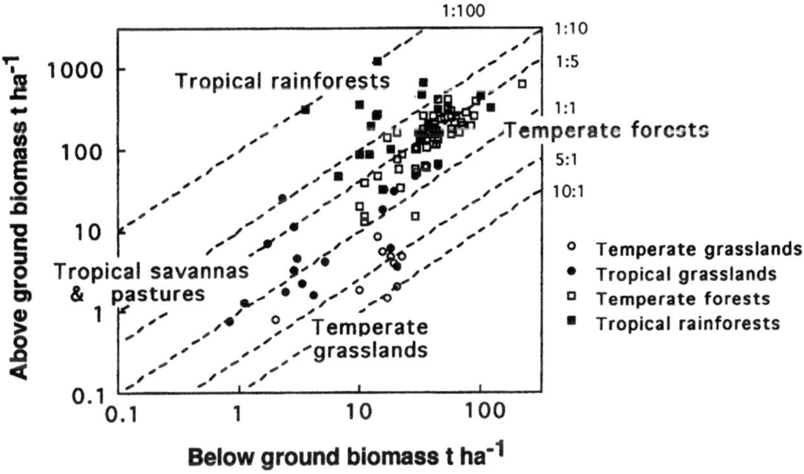

Figure III.15 Root to shoot ratios for major types of ecosystems (Scholes, 1986; pers. comm.).

Grasslands tend to have much higher root:shoot ratios than forests. Measured values range from 1:5 to 10:1 in temperate ecosystems, and 1:100 to 1:2 in forests. Root:shoot ratios tend to be lower in tropical ecosystems than in temperate or boreal situations since above-ground biomass increases much more rapidly than below-ground. The root:shoot ratio is five times greater on average in temperate than in tropical ecosystems. Nonetheless, considerable variation has been noted, reflecting the local and regional effects of different soil types, nutrient status and soil water regimes, together with variations among plant taxa.

At a smaller regional or local scale, root development may be severely curtailed by high soil strength and, where this occurs, roots may follow large communicating pores or channels where these are present. Low porosity often has inimical effects on root development, particularly when associated with high bulk density, fine texture, high clay contents or compact structure since the size, distribution and continuity of macropores directly influences oxygen supply to roots (van Noordwijk *et al.*, 1993). When grown on clay, loam and alluvial sand, the total root lengths of mature wheat plants were, respectively, 2.5, 10.3 and 17.5 m (Callot *et al.*, 1982). Aeration is also important for root growth and anaerobiosis often limits elongation while the frequency of pathogenic lesions is increased. In poorly structured and compacted soils, distances between macropores and cracks may be large and roots tend to concentrate in cracks (Ringrose-Voase, 1987). Aluminium toxicity may also limit root extension in acid tropical soils of low nutrient status (Alva *et al.*, 1986).

Specific root length varies with a range of factors (Fitter, 1985) including:

(i) Interspecific variation (Figure III.16) ;

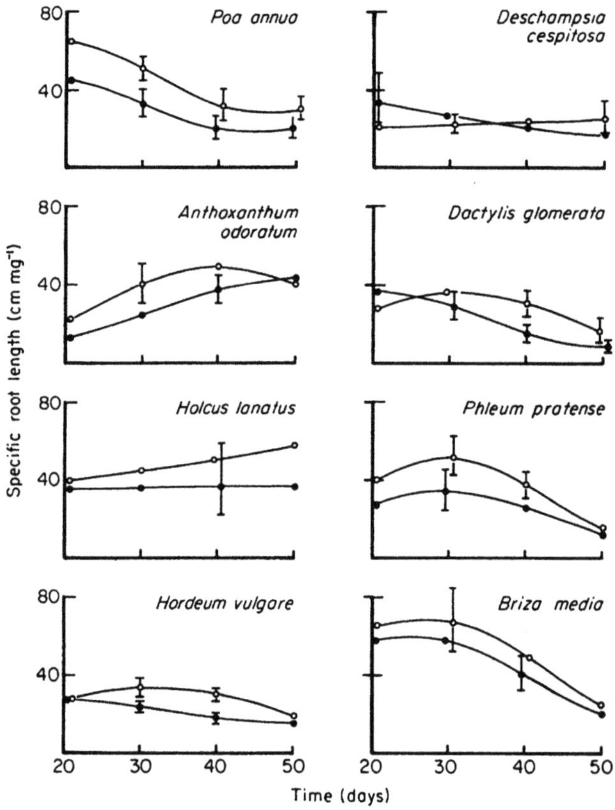

Figure III.16 Specific root lengths of eight temperate climate grasses over a period of 50 days, at two levels of soil fertility (● high amount of fertiliser (167.4 mg NPK per pot), O low rate (18.6 mg)), vertical bars are 95 % confidence limits (Fitter, 1985).

(ii) Genotype. Variation from 11.8 to 21.8 has been observed in different maize genotypes;

(iii) Age. Higher values are generally found in young plants, simply because coarse roots have not completed their growth;

(iv) Soil fertility. Higher values usually occur in poor soils, as shown in Figure III.16 for eight grass species;

(v) Temperature. High specific root lengths are associated with low temperatures (Veresoglou and Fitter, 1984) whereas moisture responses appear to differ with the species;

(vi) Grazing by rhizophagous herbivores. Root consumption by nematodes has been reported to affect the specific root length of wheat plants which is doubled in their absence (Rovira, 1978) (Table III.7);
(vii) Depth. The specific root length of wheat plants increases markedly from the upper few centimetres (1.4 cm mg^{-1}) to 1 m depth (50 cm mg^{-1}) (Drew and Saker, 1980). This may be interpreted in terms of the age of the roots which decreases with depth.

Table III.7 Specific root lengths (cm mg^{-1}) of wheat plants in relation to nematode infestation and fertilizer application (after Rovira, 1978).

		Phosphorus added	
		-	+
Nematicide	-	6.9	8.7
applied	+	14.3	15.1

3.1.2.2 *Horizontal and vertical distribution of roots in the soil profile*

Plants develop different strategies in the lateral exploitation of soils. In tropical forests, Kahn (1984) observed four phases in the horizontal development of the root system: i)- formation of a tap root axis and production of a few lateral rootlets; ii)- production of large lateral roots from the initial tap root axis; iii)- development of these secondary axes which moves the exploitation zone of distal roots away from the axis; and iv)- production of new large lateral roots from the tap root (or from the trunk, or the bases of the major lateral axes); these roots will exploit the soil close to the axis which is no longer exploited. During the last two phases, the tap root system develops through the elongation of the primary axis and the formation of new axes with positive geotropism from the large lateral roots (Figure III.17).

In most plants the root system seems to develop at random. Nonetheless, roots tend to maintain a minimum distance from each other in order to avoid crossing and to optimise the extraction of water and nutrients available in the volume of soil that they exploit. This is especially true for perennial species in arid environments (*e.g.*, Fernandez *et al.*, 1988) (Figure III.18). In a maize plantation, the half mean distance between roots was at a minimum in the upper 0-6 cm of soil (0.5 cm) and increased towards deeper horizons to a maximum value of 2.86 cm (Tardieu, 1988). In compacted areas, much higher values of up to 6.3 cm were observed in the deepest (71-78.5 cm) horizon. In arid environments, the average distances of fine roots of perennial shrubs ranged from 1.62 to 2.34 cm with higher values in surface than in moister deep horizons (Figure III.19).

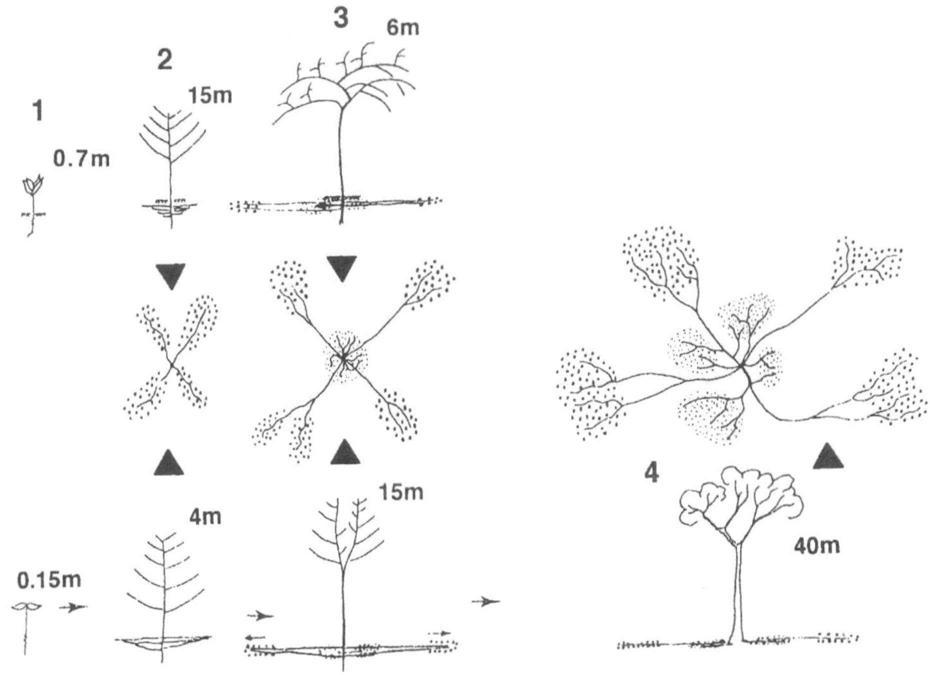

Figure III.17 Horizontal development of root system of trees in a tropical rainforest (Kahn, 1984): 1. Seedling with a single tap root; 2. Shrub with coarse roots and fine roots concentrated near the base of the tree; 3 and 4. Large coarse roots grow and zones of nutrient exploitation become more distant from the trunk; 4: new roots grow to exploit nutrients close to the trunk.

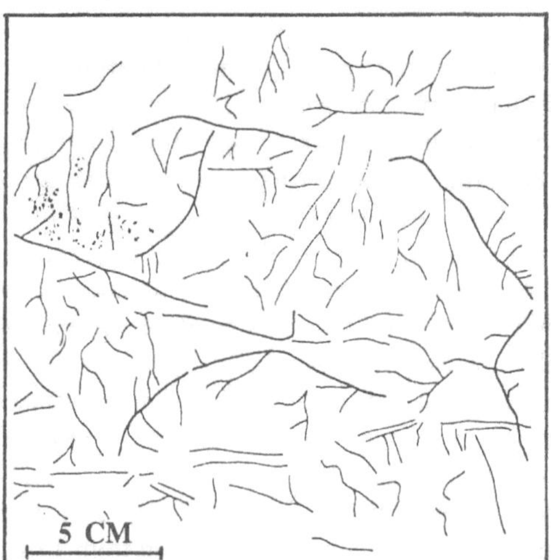

Figure III.18 Distribution of roots of shrub *Geoffroea decorticans* from semi-arid soils of Utah (USA) on a subterranean glass window. Note distances between roots; the dotted area indicates a zone of preferential absorption (Fernández *et al.* 1988).

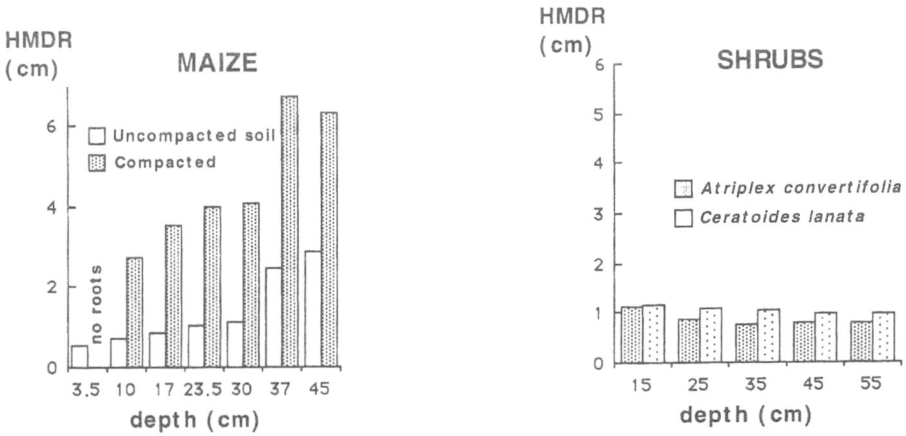

Figure III.19 Half mean distance between neighbouring roots (HMDR) of two shrub species in a semi- arid area of Utah (USA) (Fernández *et al.*, 1988) and a maize crop in France (Tardieu, 1988).

Depth distributions depend on the rooting strategies of individual species, soil physical conditions and nutrient status. However, the basic determinants are the vertical gradients of nutrient availability and the simple fact that roots generally start growing from close to the surface. As a result, root biomass is generally highest in the upper part of the profile and diminishes with depth (Figure III.20). Departures from this general pattern are due to the effects of four main factors *viz.*, strategies of the individual plants, nutrient availability, depth of the water table and the horizontal distributions of nutrient elements.

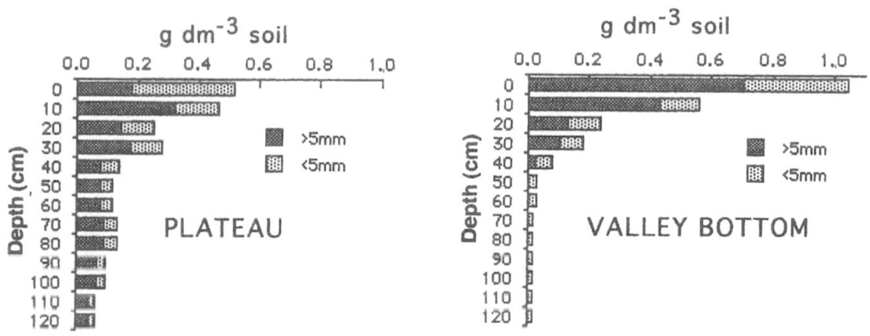

Figure III.20 Depth distributions of roots of two size classes at two landscape locations in a tropical rainforest in the Côte d'Ivoire (Huttel, 1975).

In the Banco tropical forest, coarse roots are evenly distributed in depth on the plateau whereas they are concentrated in the upper 40 cm in the humid low lying areas of the valley bottoms. Fine roots are more abundant close to the surface in both situations.

In a temperate forest at Ardennes (Belgium), fine roots have grown deeper than coarse roots and the general pattern of rooting varies with tree species and the age of the stand (van Praag *et al.*, 1988) (Figure III.21).

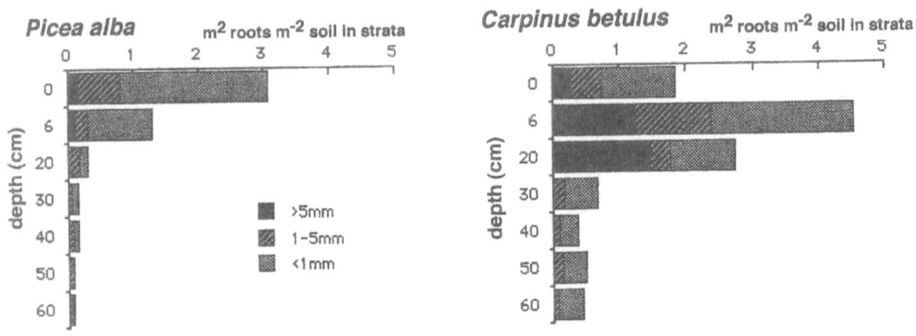

Figure III.21 Depth distributions of roots in a 35-year old spruce (*Picea abies*) stand and a 120-year old Beech (*Fagus sylvatica*) stand in the Ardennes (Belgium) (van Praag *et al.*, 1988).

Individual strategies

Trees usually have much deeper root systems than grasses (see, *e.g.*, Singh and Srivatsava, 1984; Fournier, 1991). Their root systems are often differentiated into a lateral-superficial system specialised for absorbing nutrients from decomposing litter and the upper organic soil and tap roots that penetrate deeply into the soil searching for water (Figure III.14). In grasslands, most roots are concentrated near the surface. In the humid savannas of Lamto (Côte d'Ivoire) for example, César (1971) collected 78 % of root biomass in the first 20 cm. In shrub savannas, the root distribution showed a bimodal pattern with the first mode corresponding to grass roots close to the surface (65 % in the upper 20 cm) and shrub roots comprising a second mode at a 30-40 cm depth.

The maximum depths at which roots have been found exceed 61 m, for *Juniperus monosperma* roots from underground mine workings. The maximum radial extent of tree roots is more than 50m from the stem and 30m has been recorded for a number of species (Stone and Kalisz, 1991).

Nutrient availability

In highly nutrient-limited soils the superficial roots may form mats up to 40 cm thick which grow into the litter layers. Mycorrhizal fungi link the roots and the litter layers ensuring direct nutrient reabsorption from the decomposing leaves (Stark and Spratt, 1977). This root mat has proved to be highly efficient in the interception of nutrients and, in a South American rainforest, 99 % of a radioactively-labelled solution of phosphate sprayed on its surface was taken up by the plants (Stark and Jordan, 1978).

Dynamics of water in the soil profile

The distribution of deep roots appears to be closely related to soil water status and particularly the depth to the water-table. In two oxisols of the equatorial forest, Humbel (1978) showed that root penetration was markedly reduced where water tables occurred at shallow depths; in well-drained soils, roots normally explore much deeper soil strata. In the Banco tropical rain forest (Figure III.20), the presence of a water-table in the low-lying areas of the valley bottom limits the extension of coarse roots to 40 cm in depth whereas on the plateau, they grow down to the bedrock, at a depth of 130 cm. In soils restored after open-cast mining, Fairley (1985) showed that drainage tends to reduce the root biomass and specific root length in Perennial Ryegrass (*Lolium perenne*). Conversely, roots may also extend deep into the soil to seek water from the water table or moister deep soil horizons. As an example, (Seguin, 1970) showed that the deeper roots of the grape plants were concentrated at 60-120 cm and that maximum water extraction occurred at that depth (Figure III.14). In the Australian dry tropics, the roots of *Ficus* spp. growing on karst landscapes have been observed to extend many metres underground through faults and caves to reach the water table (Clarke, 1968 in Bowen, 1984).

Soil horizontal variability

The distributions of absorbing roots and their associated mycorrhizae are greatly affected by the horizontal distributions of soil resources. St. John *et al.* (1983) have shown that tree roots in the Amazonian forest grow at random but, when a nutrient-rich site is encountered, branching is increased and favours a more intensive exploitation of the site. Similar results have been found by Robinson and Rorison (1983) in 'split-root' experiments. In such experiments, parts of the root systems supplied with high levels of nitrogen show large increases in specific root length. The other part of the root system may also show a similar response, but with a significant time-lag (Figure III.22). Such a growth pattern may explain the negative binomial distribution of roots and their associated mycorrhizae observed in this forest.

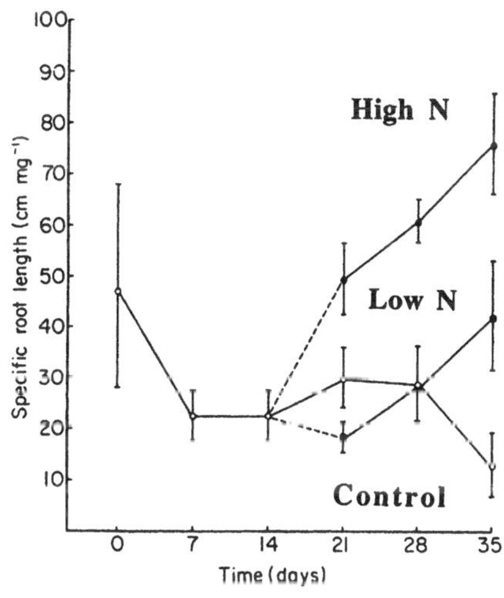

Figure III.22 Specific root length of *Lolium perenne* in a split-root experiment. That part of the root system supplied with high nitrogen shows an immediate increase in SRL, while that in low N does so after a time-lag. Controls exhibit a further decline in SRL (from Robinson and Rorison, 1983).

3.1.2.3 *Root biomass and production*

Root biomass is always very high, with average values ranging from 0.9 Mg dry wt ha^{-1} in dry savannas to maximum values of up to 100 Mg ha^{-1} in temperate and tropical forests (Figure III.15).

Roots generally account for 40 to 85 % of primary production in temperate ecosystems (Fogel, 1985) and 16 to 68 % in tropical savannas and grasslands (Scholes, unp. data) (Table III.8). In high-input annual crops, root production is relatively low (*e.g.*, 104 g m^{-2} and 16 % in a fertilised plot of Barley as compared to 150 g m^{-2} and 21.5 % in an unfertilised adjacent plot in southern Sweden (Hansson *et al.*, 1987). Perennial grass leys have proportionally higher production of respectively 460 g m^{-2} and 31.4 %, and 480 and 30.4 % in a grass and a lucerne ley in the same locality (Andren *et al.*, 1990).

Table III.8 Contributions of roots (% total) to total net primary production in different ecosystems (Fogel, 1985; Scholes, 1986, unp. tab.).

ECOSYSTEM	TOTAL PRODUCTION g dry weight m^{-2}	CONTRIBUTION %	REFERENCE
Coniferous forest			
Douglas fir (55y)		73	Fogel, 1983
Pacific silver fir (23y)		60	Grier *et al.*, 1981
Pacific silver fir (180y)		71	Grier *et al.*, 1981
Scots pine (14y)		60	Agren *et al.*, 1980
Lodgepole pine (74y)			
mesic water regime	11.8-12.0	20-27	Comeau and Kimmins, 1989
xeric water regime	7.9-9.5	38-46	Comeau and Kimmins, 1989
Deciduous forest			
Yellow poplar		40	Harris *et al.*, 1980
Shrub			
Atriplex steppe		65	Caldwell and Camp, 1974
Ceratoides steppe		66	Caldwell and Camp, 1974
Herbaceous			
Tall grass prairie		50	Kucera *et al.*, 1967
Short grass prairie		85	Sims and Singh, 1971
Crops			
Lucerne ley	0.96-2.26	27-47	Pettersson *et al.*, 1986
Fertilized barley	1.04	16	Hansson *et al.*, 1987
Unfertilized barley	0.56	23	Hansson *et al.*, 1987
Perennial grass ley (*Festuca pratensis*)	1.46	31	Andren *et al.*, 1990

Production is generally calculated by summing the positive increments of root biomass observed at regular time intervals. Such a method certainly underestimates production because it does not take into account the roots which may have died and decomposed between adjacent sampling occasions. Methods taking into account root decomposition or using direct measurements based on rhizotrons give estimates up to 180 % higher than simple core harvest techniques (see, *e.g.*, Hansson and Steen, 1984; van Praag *et al.*, 1988). Root production, particularly of fine roots, may be highly seasonal and heterogenous in space. Root systems seem to develop in an opportunistic way, taking advantage of favourable conditions to develop new fine roots; when conditions become difficult, a large part of these rootlets may die. Consequently, root biomass may vary greatly with season and because of particular climatic episodes. Methodologies which do not accurately track these variations are likely to give highly-biased estimates of biomass and production.

Estimates of fine-root production vary from 0.6 to 11 Mg ha^{-1} over a range of temperate climate ecosystems (Fogel, 1985; Aber *et al.*, 1985; van Praagh *et al.*, 1988). Turnover of fine roots is rapid and the ratio of fine root production to standing crop may be as high as two in a young stand of *Pinus sylvestris* in Sweden, and 0.9 in an adjacent 120-year old stand (Persson, 1985). Similarly, the proportion of fine roots dying annually ranged from 40 to 92 % of total weight in the series of deciduous and coniferous temperate forests listed by Fogel (1985). In a coniferous forest, estimates of the annual input to the soil of dead fine roots and associated mycorrhizae ranged from 14.6 to 18.8 Mg ha^{-1}. These values are two to five times greater than the contributions of leaf litter and branches to soil organic matter. Forests have the highest production levels, whether they are coniferous (with production of fine roots of 1.2 to 11 Mg ha^{-1} $year^{-1}$) or deciduous (2 to 9 Mg ha^{-1} $year^{-1}$). In a tall-grass prairie, production was 5.1 Mg ha^{-1}. In crops, estimates range between 0.6 and 4.2 Mg on the same areal basis, depending on the type of crop and soil conditions.

In tropical grasslands and savannas, root production increases over that of cooler climates and, within the tropics, increases with rainfall. Root production estimates as high as *ca.* 20 Mg ha^{-1} y^{-1} have been measured in moist African savannas at Lamto (César, 1971). Nonetheless, in many situations, waterlogging or nutrient deficiencies may limit root growth (see *e.g.*, Cuevas and Medina, 1988). In South American savannas, Fisher *et al.* (1994) found that conversion of savannas into highly productive pastures might account for the sequestration of 100-507 Mt carbon per year. Too few measurements of root production in tropical forests have been made to identify general trends.

Root production is a highly seasonal process. For example, in temperate areas fine root production by apple (*Malus pumila*) trees showed two marked annual peaks in spring (April to June) and late summer (September) respectively (Figure III.23). In African savannas, root production by perennial grasses seems to be more opportunistic since rapid increases of root biomass may occur at any time of the year, provided that the soil moisture status is favourable (César, 1971; Fournier, 1991).

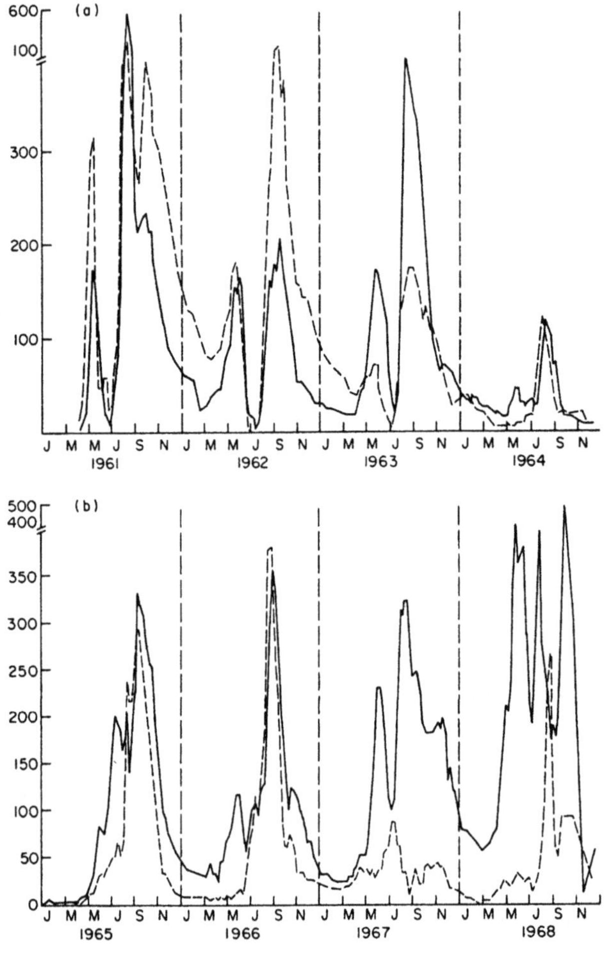

Figure III.23 Periodicity in the production of fine root (cm per 3 windows of a rhizotron) by two apple trees (a) and (b), as observed in rhizotrons (Atkinson, 1985).

3.2 Biology of Roots

The major factors in root biology are the uptake of water and nutrients, resistance to water stress and their relationships with microbial symbionts.

3.2.1 UPTAKE OF WATER AND NUTRIENTS

3.2.1.1 *Plant 'feeding regimes'*

A wide range of chemical elements have been found in plants including eleven major or macro-elements and fourteen trace or oligo-elements (Callot *et al.*, 1982). The four macro-elements, carbon, hydrogen, oxygen and nitrogen comprise most of the biomass. Sulphur, phosphorus, potassium, calcium, magnesium and occasionally silicon are other important components and chloride and sodium, to a lesser extent. Their concentrations are of the order of mg g^{-1}; that of the minor or trace elements is of µg g^{-1} (Chapter I.3.1).

Plants absorb water and nutrients selectively and their mineral compositions are thus different from those of the soil solutions from which they derive their nutrients. Each plant species has its own specific composition in terms of its concentrations and balance of nutrient elements (see, *e.g.*, Callot *et al.*, 1982).

Thus, it may be possible to define for plants the botanical equivalent of the animal dietary regimes, listing the nutrients used and the amounts of each needed to sustain growth. A further characterisation of nutrient resources may also be envisaged using Tillman's (1982) system of classification for pairs of resources. Depending on the shape of the nutritional isoclines observed, resources may be recognised as substitutable, complementary, antagonistic or essential (see Chapter I, Figure I.30). However, the data necessary to define such hypothetical nutritional regimes are still largely lacking.

3.2.1.2 Internal and external factors of nutrient assimilation.

Nutrient absorption by roots depends on:
(i) The availability of nutrients in solution in the surrounding soil;
(ii) The relative mobility of ions, and,
(iii) The capacity of the rooting system to absorb these elements.

Availability of nutrients in soil solutions

Callot *et al.*, (1982) calculated that 200 kg of potassium per ha, (*i.e.*, about five times the stock of exchangeable K in a current temperate brown soil, inceptisol) in solution would only be in contact with 45 % of solid particles of an arable soil. As nutrient solutions are generally much less concentrated and discontinuous, an adequate nutrient supply implies appropriate levels of nutrient transport through diffusion or mass flow. Nutrient supply to the plant actually depends on the mobility of ions and root density (Figure III.24) and Bowen (1984) states that 'the limiting factor in uptake of ions moving to the root by diffusion is not usually the absorption ability of the root but the movement of ions to the root'. This is especially true for non-mobile ions such as phosphorus, whereas the absorption of mobile ions such as nitrates may be limited in some circumstances by the ability of the root to assimilate this element (Scholes *et al.*, 1993).

Thus, all the factors which affect diffusion and mass flow such as texture, structure and soil moisture will determine the accessibility of nutrients to roots. Drying especially affects diffusion and massflow by increasing the tortuosity of the water film. At potentials less than -0.2 MPa, soil water retreats to voids less than 1.5 µm in diameter which are inaccessible to root hairs (Clarke and Barley, 1968).

In some cases, roots may dissolve specific nutrients (especially nitrogen) through interactions with micro-organisms that have been triggered by the input of high-energy exudates (see Section IV.3.2.1.1) (Elliott *et al.*, 1979; Clarholm, 1983; Sallih and Bottner, 1988; Billès *et al.*, 1990).

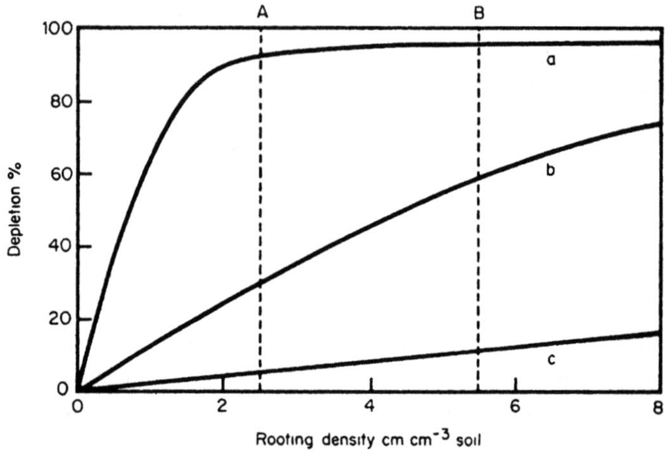

Figure III.24 The influence of rooting density on the depletion of labile pools of nutrients. Curves a, b and c represent the depletion curves for, respectively, highly mobile ions (*e.g.*, NO_3, SO_4^{2-}), poorly mobile ions (*e.g.*, K^+, NH_4^+, $H_2PO_4^-$) and immobile ions (*e.g.*, Cu^{2+}) in a soil at field capacity (Barley, 1970 in Bowen, 1984).

Specific allelopathic compounds in soil solutions (such as phenols) may directly inhibit nutrient absorption by roots or impede the transformation of ammonium into nitrate. In nutrient poor soils, two mechanisms seem to limit the production of nitrates *i.e.*, i - the low rate of ammonification due to low amounts of assimilable carbon to trigger the reaction, and ii- inhibition of nitrification by water soluble substances produced by roots (*e.g.*, Meiklejohn, 1962; Munro, 1966; Lensi *et al.*, 1992). Conservation or immobilisation of NH_4-N instead of nitrification are considered efficient ways to conserve N (Marrs *et al.*, 1991) (see also Chapter IV.3.2).

Absorption of nutrients by roots

Absorption of water and nutrients by plants may be active (*i.e.*, against electrochemical or water potential gradients) or passive (*i.e.*, with the gradient)

Passive absorption is due to the accumulation of ions in a 'free' space, which is equivalent to 20-30 % of the root volume. This 'free space' is the sum of a diffusion space, *i.e.* cell spaces from which ions may diffuse in or out freely and the 'Donnan space' in which ions are retained by fixation and exchange due to the negative charges borne by cell wall constituents and cytoplasmic proteins (Callot *et al.*, 1982). The maximum concentrations attainable by passive absorption may exceed those in the soil solution by this exchange process as the exchange occurs on the basis of equality of chemical potentials rather than concentrations. The total amount of cations that roots are able to absorb through this 'Donnan equilibrium' is called the cation exchange capacity of roots. Low cation exchange of roots, as for example, in the Gramineae, is considered an adaptive advantage as anion diffusion occurs more readily. If the concentration gradients are reversed, ions which have been passively absorbed are desorbed and move towards the external solution.

Active absorption refers to the selective accumulation of ions and is generally more important than passive absorption. It is an energy-consuming process which may be limited when energy reserves are depleted. Active absorption is only possible after negative potential and pH gradients have been created by the hydrolysis of ATP by ATPases. Some ions, such as NH_4^+, are absorbed by passive absorption while others, such as NO_3^- are only absorbed by active processes. The selectivity of absorption processes is still poorly understood.

3.2.1.3 *Dynamics of absorption*

Plant growth is more likely to be limited by the availability of solutes rather than by the intrinsic absorption capacity of roots. In general, as little as one fifth of the root system immersed in a suitable nutrient solution may provide sufficient root area for adequate nutrition. For example, maize plants 45-75 days old have a root biomass of *ca.* 500 g although only 30 to 35 g of this is necessary to adequately supply the plant (Nye and Tinker, 1978; Callot *et al.*, 1982).

Under natural conditions, low nutrient availability and metabolic limitations in the utilization of absorbed nutrients necessitate the existence of a well-developed root system and its further extension with mycorrhizal mycelia.

In laboratory experiments, all parts of the root system of maize plants appear to have equal absorptive capacities, even those parts called coronary roots specialised for anchoring. Nonetheless, absorption is usually highest near the apex of the young unsuberised roots and by mycorrhizae (Bowen and Theodorou, 1973). When the plant is immersed in a nutrient solution, branching does not appear to affect absorption (Callot *et al.*, 1982). Similarly, age also has little influence although the relative positions of absorbing roots are important; when two roots are left, absorption is greater if they are diametrically opposed rather than adjacent. Genotypic variation may also be important. In seven genotypes of corn, nitrate uptake by six-day-old root systems varied from 44 to 86 μmol NO_3^- g^{-1} f wt during an 8 h period; differences in lateral root proliferation and root elongation were considered to explain these differences (Pan *et al.*, 1985).

Finally, the importance of the volume of the above ground part of the plant influences the rate of uptake by roots through a 'sink' effect. As a result, excised roots have lower rates of uptake than similar amounts of roots with their above-ground parts still attached (Maertens and Clauzel, 1980).

3.2.2 RESISTANCE TO WATER STRESS

The uptake of water by roots is closely linked to that of ions. The major mechanism is evapotranspiration of leaves and shoots which creates a net flow of water and ions from the soil solution to the roots.

As with most soil organisms, roots are very sensitive to water stress and have developed both short and medium-term mechanisms of coping with it.

3.2.2.1 *Short-term mechanisms*

Under conditions of moderate water-stress, root elongation may initially be favoured (Sharp and Davies, 1985) and root production is enhanced (Comeau and Kimmins, 1989). In this situation, a continuous proliferation of roots into unexplored regions of the soil profile may help to avoid the effect of rapid water depletion at the soil-root interface (Newman, 1969). With persistent water stress, short-term resistance results from the accumulation of amino acids (particularly proline) and soluble reducing sugars (*e.g.*, sucrose, glucose) due to limitations in the allocation of translocated photosynthates to growth and in the hydrolysis of starch. As a result, osmotic pressure is increased which helps to prolong turgor and delays wilting (Vartanian, 1977). In roots of *Sinapis alba* submitted to water stress, the total carbohydrate contents of the roots did not change although the ratio soluble sugars:polysaccharides increased up to six times. Similar observations have been made on *Carex* spp. (sedges) and *Gossypium* sp. (cotton) plants (Hubac, 1978) although, after some time, the accumulation of soluble sugars and amino acids may inhibit enzymatic activities and impede root recovery.

3.2.2.2 *Medium-term mechanisms*

When transpiration starts to decline due to water stress, some plants develop short tuberous roots which neither grow in length nor produce root hairs. Cortical cells swell and accumulate starch reserves. When moisture conditions again become favourable, these roots extend rapidly and produce active root hairs (Sabatier and Vartanian, 1983; Vartanian and Chauveau, 1986).

3.2.3 EXUDATION AND RHIZODEPOSITION

3.2.3.1 *Definition*

Part of the carbon fixed in the leaves as photosynthates and translocated to the roots via the phloem is transferred to the soil as a mixture of soluble and insoluble substances, together with sloughed cells, through the process of rhizodeposition. Rhizodeposition is important because of the huge flux of carbon and nutrients involved (up to 10-25 % of carbon fixed by photosynthesis and 30-40 % of the photosynthates translocated to the roots). Through this process, a wide variety of chemical compounds is added to the soil and this is of critical importance to the organisms close to the roots.

Rhizodeposition gives the soil-root interface its characteristically high metabolic activity and roots are therefore the major regulators of biological activities and nutrient cycling in their sphere of immediate influence, the 'rhizosphere' (see Chapter IV.3).

Several constituents may be distinguished (Rovira *et al.*, 1979; Hale *et al.*, 1981) fig III.25:

1. Exudates *sensu stricto* are water-soluble, low molecular weight compounds leached from the roots without metabolic control by the plant;
2. Secretions include compounds of low molecular weight released by metabolic processes;
3. Plant mucilages are insoluble organic compounds of four different origins:

a) mucilage secreted by Golgi organelles in the root cap cells;
b) hydrolysates of the polysaccharides of the primary cell wall between the epidermal cells of the primary wall and sloughed root cap cells;
c) mucilage secreted by epidermal cells and root hairs. The development of a secondary wall of epidermal cells proximal to the root tip stops the production of secretions and mucilages;
d) mucilage produced by bacterial degradation of dead epidermal cells;

4. Mucigel is the gelatinous material at the surface of roots grown in non-sterile soil. It includes plant mucilages, bacterial cells and their metabolites, together with colloidal mineral and organic matter from the soil;
5. Lysates and sloughed cells from the epidermis and cortex.

Hale *et al.* (1981) and Curl and Truelove (1986) also consider leakages as compounds of low molecular weight which diffuse into the apoplast (cell walls) and, via the apoplast, move to the root surface or leak directly from the epidermal or cortical cells. Leakages include a number of volatile compounds that have a wider sphere of influence on soil micro-organisms, beyond the immediate environs of the root surface.

The term exudates is often used to designate the mixture of exudates *sensu stricto* and secretions because of the difficulty of identifying the origin of these products.

3.2.3.2 *Sites and mechanisms of exudation*

The meristematic region behind the root cap is an area of active exudation, although it may also occur along the whole length of the roots of herbaceous species, including parts covered with root hairs (Figure III.25). Two principal features of the exudation occurring in the active, non-woody roots of herbaceous and woody plants are (Smith, 1970; Dart, 1971; Greaves and Darbyshire, 1972; Bowles and Northcote, 1974; Hale *et al.*, 1981):

(i) An intense release of sloughed cells and mucilage from the apices of lateral roots;
(ii) A diffuse exudation of soluble compounds all along the root axis, with a maximum occurring immediately behind the root tip (Pearson and Parkinson, 1961)

Exudation involves both passive leakage and active secretion processes. The photosynthates produced in leaves are translocated to the roots via the phloem by a 'source-to-sink' effect. They accumulate in a soluble pool and are transformed into other compounds. They may be leached out by passive diffusion (*i.e.*, exudation *sensu stricto*) along a concentration gradient from the root to the surrounding medium. A pool of soluble compounds has to be formed in the root prior to any release into the soil. McDougall (1970) observed that fixed ^{14}C was translocated to roots as sucrose and that after 10 hours, 50 % of the sucrose had been transformed into other molecules (mainly other sugars and amino acids); 33 % of the ^{14}C appeared as exudates after 24 hours.

Secretion implies the expenditure of metabolic energy to transport molecules across cell membranes to the soil, occasionally against electrochemical or chemical potential gradients. Morré *et al.* (1967) and Burke *et al.* (1974) have shown that the Golgi apparatus of the outer root-cap cells of wheat and maize secretes polysaccharides which are transported out of the cells and form viscous droplets at the root tips.

Exfoliation of sloughed epidermal and cortical cells is part of the normal growth processes. These cells have a life span of a few weeks (Holden, 1975) and after their death, cell walls and cytoplasm mix with the root and microbial mucilages adhering to the extremity of the root to form the mucigel.

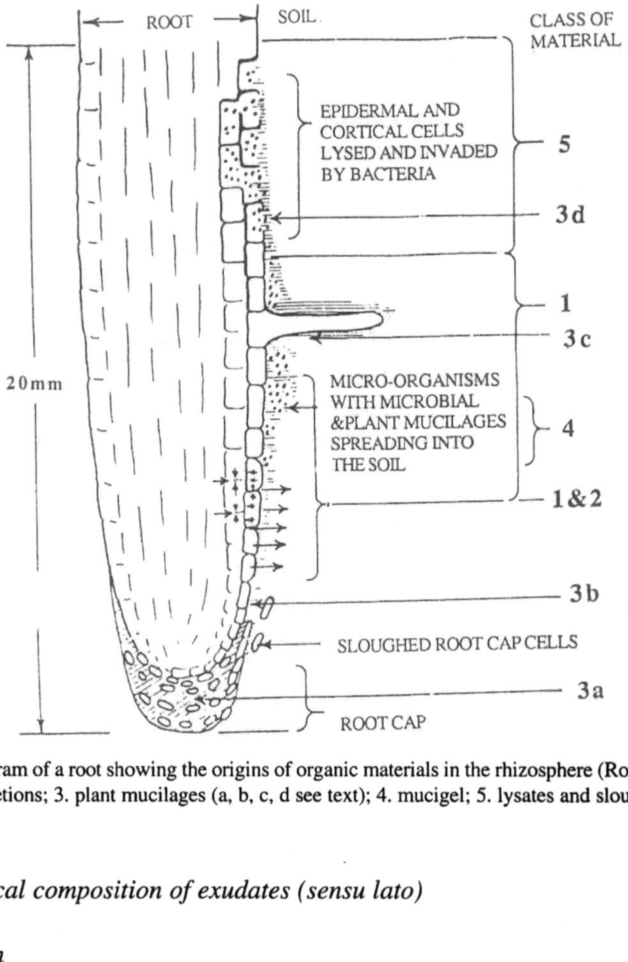

Figure III.25 Diagram of a root showing the origins of organic materials in the rhizosphere (Rovira et al., 1979): 1. exudates; 2. secretions; 3. plant mucilages (a, b, c, d see text); 4. mucigel; 5. lysates and sloughed cells.

3.2.3.3 *Chemical composition of exudates (sensu lato)*

General pattern

Root exudates are basically a mixture of carbohydrates (simple sugars and polysaccharides), amino compounds, organic acids, nucleotides, flavones, enzymes and growth factors (see, *e.g.*, Smith, 1970; Rovira and Davey, 1974; Chaboud and Rougier, 1981; Curl and Truelove, 1986) (Table III.9). The chemical diversity of exudates is comparable to that of living cells since most cytoplasmic components have the capacity to be leached. For example, Bowen and Theodorou (1973) identified 10 carbohydrates, 14 organic acids and 21 amino acids from the seedlings of ten forest species. Exudates of seven species of cultivated plants (wheat, oats, maize, peas, lucerne, rice and peanut) included 21 amino acids, 9 organic acids and 10 carbohydrates (Lespinat and Berlier, 1975). In both cases, significant differences among species were observed.

Table III.9 Substances detected in plant root exudates (Curl and Truelove, 1986).

Kind of compound	Exudate components	Plants most studied
Sugars	Glucose, fructose, sucrose, maltose, galactose, rhamnose, ribose, xylose, arabinose, raffinose, oligosaccharides	*Triticum aestivum*, *Hordeum vulgare*, *Phaseolus vulgaris*, *Pinus* spp.
Amino compounds	Asparagine, α-alanine, glutamine, aspartic acid, leucine/isoleucine, serine, -aminobutyric acid, glycine, cystine/cysteine, methionine, phenylalanine, tryosine, threonine, lysine, proline, tryptophane, β-alanine, arginine, homoserine, cystathionine	*Triticum aestivum*, *Zea mays*, *Avena sativa*, *Pisum sativum*, *Phalaris* spp., *Trifolium* spp., *Oryza sativa*, *Gossypium barbadense*, *Lycopersicon esculentum*, *Pinus* spp., *Robinia pseudo-acacia*, *Bouteloua gracilis*
Organic acids	Tartaric, oxalic, citric, malic, acetic, propionic, butyric, succinic, fumaric, glycolic, valeric, malonic	*Triticum aestivum*, *Zea mays*, *Phaseolus vulgaris*, *Lycopersicon esculentum*, *Brassica* spp., *Pinus* spp., *Robinia pseudo-acacia*
Fatty acids and sterols	Palmitic, stearic, oleic, linoleic, linolenic acids; cholesterol, campesterol, stigmasterol, sitosterol	*Phaseolus vulgaris*, *Arachis hypogaea*
Growth factors	Biotin, thiamine, niacin, pantothenate, choline, inositol, pyridoxine, *p*-amino benzoic acid, n-methyl nicotinic acid	*Triticum aestivum*, *Phalaris* spp., *Phaseolus vulgaris*, *Pisum sativum*, *Trifolium* spp., *Medicago* spp., *Gossypium barbadense*
Nucleotides, flavonones and enzymes	Flavonone, adenine, guanine, uridine/cytidine, phosphatase, invertase, amylase, protease, polygalacturonase	*Triticum aestivum*, *Zea mays*, *Pisum sativum*, *Trifolium* spp.
Miscellaneous compounds	Auxins, scopoletin, fluorescent substances, hydrocyanic acid, glycosides, saponin (glucosides), organic phosphorus compounds, nematode cyst or egg-hatching factors, nematode attractants, fungal mycelium growth stimulants, mycelium-growth inhibitors, zoospore attractants, spore and sclerotium germination stimulants and inhibitors, bacterial stimulants and inhibitors, parasitic weed germination stimulators	*Avena sativa*, *Medicago* spp., *Trifolium* spp., *Pisum sativum*, *Lycopersicon esculentum*, *Lactuca* spp., *Fragaria vesca*, *Musa paradisiaca*, *Zea mays*

Growth factors may be present together with a range of chemically-undefined substances that possess demonstrated stimulatory effects on micro-organisms, including plant pathogens and such parasites as nematodes. However, not all exudates are stimulatory in nature as evidenced by the presence of hydrocyanic acid and saponins. No relationships have been observed between total sugar content of root exudates and the degree of VA mycorrhizal infection of annual plants (Azcon and Ocampo, 1984).

Allelochemicals (principally phenolic acids) have also been identified (see, *e.g.*, Hoagland and Williams, 1985; Pope *et al.,* 1985). Ions are generally absent from exudates with the exception of phosphorus which has been found in the exudates of wheat seedlings. Volatile organic compounds may also be produced. Most compounds found in exudates have low molecular weights and in axenic cultures of wheat, 98 % of exudates had molecular weights below 25,000 Da and 90 % of that fraction was below 800. An important proportion of these compounds is water soluble: 17-25 % for three-week old wheat plants (Barber and Martin, 1976), *ca.* 30 % for maize (*Zea mays*) (Guckert and Balandreau, 1981) and 44 % for axenic cultures of oats (*Avena sativa*) (Trofymow *et al.*, 1987).

Most exudates have a high C:N ratio (*ca.* 30) due to a predominance of polysaccharides (80 %) and low protein content (1-5 %) (Lynch, 1982). As excreted into the soil, they are mixed with substantial amounts of water and this is of great importance in relation to the reaction of the rhizosphere communities.

Factors influencing the composition of exudates

The composition of exudates varies depending on the species, age of the plant and general environmental conditions. Exudate production is sensitive to a variety of external factors such as light intensity, temperature, aeration, nutrient availability and soil characteristics (Lespinat *et al.*, 1975). Defoliation, as with any adverse effect on plant metabolism (pesticides, insect attacks), may also affect exudation.

Qualitative and quantitative variations

Seeds also produce exudates that differ from those produced by seedlings. Seedlings, in turn, often produce quite different exudates than the unsuberised root-tips of mature plants; exudates from the unsuberised root tips of mature sugar maple (*Acer saccharum*) trees had less abundant and diversified carbohydrates than seedlings. Nonetheless, mature trees produce more abundant and diversified amino acids than seedlings, and six times greater amounts of organic acids (Smith, 1970). These results indicate the difficulties in extrapolating results from seedlings to mature plants.

Lighting, through its control of photosynthesis, is logically a major factor in the amount of root exudates produced and Rovira (1959) observed that shading significantly modified the composition of tomato root exudates. Exudation from maize plants still occurs during the night although certain pigments are only produced during light periods (Richter *et al.*, 1968).

Stress generally leads to increased exudate production (Lespinat *et al.*, 1975). Low temperatures (Vancura, 1967), high carbon dioxide concentrations in the rhizosphere (Rittenhouse and Hale, 1971), nutrient deficiencies (Shay and Hale, 1973) and the alternation of dry and moist periods have all been shown to increase exudation (Vancura and Garcia, 1969). However, the response to progressive drying is not simple and exudation increases from potentials of 0 to -0.19 MPa, decreases between -0.19 and -0.55 MPa and increases again below the latter figure. Finally, accidental defoliation (Smith, 1972) also leads to increased exudation of specific compounds (*i.e.*, fructose, cystine, glutamine, lysine, phenylalanine and tyrosine).

Interactions with soil organisms are a particular feature of root exudation since plants tend to produce more exudates in the presence of the soil microflora than under axenic conditions. Heulin *et al.* (1987), *e.g.*, have observed a 16 to 37 % increase of exudation following the inoculation of axenic rice seedlings with strains of *Azospirillum*. The soil microflora acts as a sink for root exudates and may stimulate production by up to 100 % (Barber and Martin, 1976; Prikryl and Vancura, 1980). Thus, extrapolating results from plants grown in axenic solutions to soil conditions may well be misleading.

Exudation decreases following attacks by parasites such as nematodes and by pathogenic fungi (Jalali and Suryanaryana, 1971; Claudius and Mehrotra, 1973)

3.2.3.4 *Quantitative estimates*

The quantitative production of root exudates has been measured in a number of situations. The variety of the methods used (plants grown in axenic nutritive solutions or the tracking of ^{14}C-labelled photosynthates), the units employed (percentage of biomass or percentage of photosynthates fixed by the leaves) and plant species considered (trees or cultivated herbaceous plants) make comparisons difficult.

Overall, this work indicates that a large part of plant production, *ca.* 10 % of net production and 33 % of the assimilates translocated to roots may be exuded into the rhizosphere annually. This is an overall input of several Mg ha^{-1} $year^{-1}$, a vitally-important flux of readily-assimilable resources for the soil organisms (Table III.10).

Insoluble mucilages and cell debris are the main constituents of exudates. Their production has been estimated at 80 % of the total carbon released into the soil by wheat roots (Guckert, 1985). Maize mucilage contains 94.4 % of polysaccharides, including 21.9 % uronic acids and 3.6 % total protein (Guckert *et al.*, 1975).

The production of sloughed cells is an important component of the release of organic materials into soils since the functional life of cells from the root cap on the plant may be as short as 22 to 33 hours (Clowes, 1976). For example, Griffin *et al.* (1976) have calculated values of 1.5 mg g^{-1} dry root per week, *i.e.*, 0.15 % of the root biomass for groundnuts (*Arachis hypogaea*). In some cases, the production of cell debris may exceed that of secretions (Holden, 1975; Martin, 1977).

Table III.10 Rhizodeposition in the rhizosphere of cultivated cereals.

Crop species	% total photosynthates	% assimilates translocated to roots	Mg ha^{-1}	References
MAIZE	-	25	-	Haller and Stolp, 1985
	-	40	-	Leclère-Sonrier,1977
	10-20	-	-	Breisch, 1974
WHEAT	5-10	29-44	-	Martin, 1977
	-	-	*ca.* 5	Sauerbeck and Johnen, 1977
	15 (including sloughed cells)	-	-	Milchunas *et al.*, 1985
OATS	5-25	-	-	Trofymow *et al.*, 1987
RICE	24-34	-	-	Heulin *et al.*, 1987

III.4 INVERTEBRATE COMMUNITIES

The biology and ecology of the major groups of soil invertebrates are also considered in this chapter. Emphasis is given to the way in which these organisms adapt to the constraints imposed by living in soil, the factors affecting their abundances and distributions and the patterns of the community structures.

Soil invertebrates have been classified on the basis of size and water relations (see Table III.1). They may also be classified in terms of the nutritional resources they use, their microhabitats and possibly, related morphological and demographic characteristics. Special attention is paid in this chapter to ecological classifications based on such criteria. They are useful in that they permit preliminary evaluation of the assessment of the impact of invertebrate communities without the need to consider each individual population.

4.1 MICROFAUNA

4.1.1 PROTOCTISTA (FORMERLY PROTOZOA)

4.1.1.1 *General biology*

Taxonomic classification

Soil protists belong to three phyla within the Kingdom Protoctista: Rhizopoda (or the former Sarcodina) which include the naked Amoebae and the Testacea (amoebae protected by a resistant test or outer covering), the Zoomastigina (formerly Mastigophora, or flagellates) and the Ciliophora (or ciliates) (Dommergues and Mangenot, 1970) (Figure III.26). The parasitic Apicomplexa (formerly the Sporozoa) also occur in soils as many infective stages must pass through the soil (Stout *et al.*, 1982).

For convenience, it is proposed to refer to the above soil dwelling Protoctista in general as protists.

Microhabitat

Soil-dwelling protists occupy the pore space outside the microaggregates and their growth rates have been reported to decrease when pore space has been reduced by compaction (Coûteaux, 1985, Vargas and Hattori, 1986, Coûteaux *et al.*, 1988). However, this is not a general rule as those protists that possess fluid body forms such as the pseudopods of the amoeba *Acanthamoeba castellanii*, may penetrate into microaggregates to ingest bacteria colonising this microhabitat (Vargas and Hattori, 1986).

Size, metabolic rate and growth patterns

Soil protists are generally much smaller than those of aquatic habitats and have a size range from a few micrometres to less than 100 micrometres. The larger species are restricted to the litter layers and the average size of individuals tends to decrease with depth (Stout and Heal, 1967). Testacea tend to have simple and flattened tests with a reduced pseudostome and a vestibule (Bonnet, 1975). Species living at the soil surface tend to have the most flattened tests (Schönborn, 1968, Coûteaux, 1972).

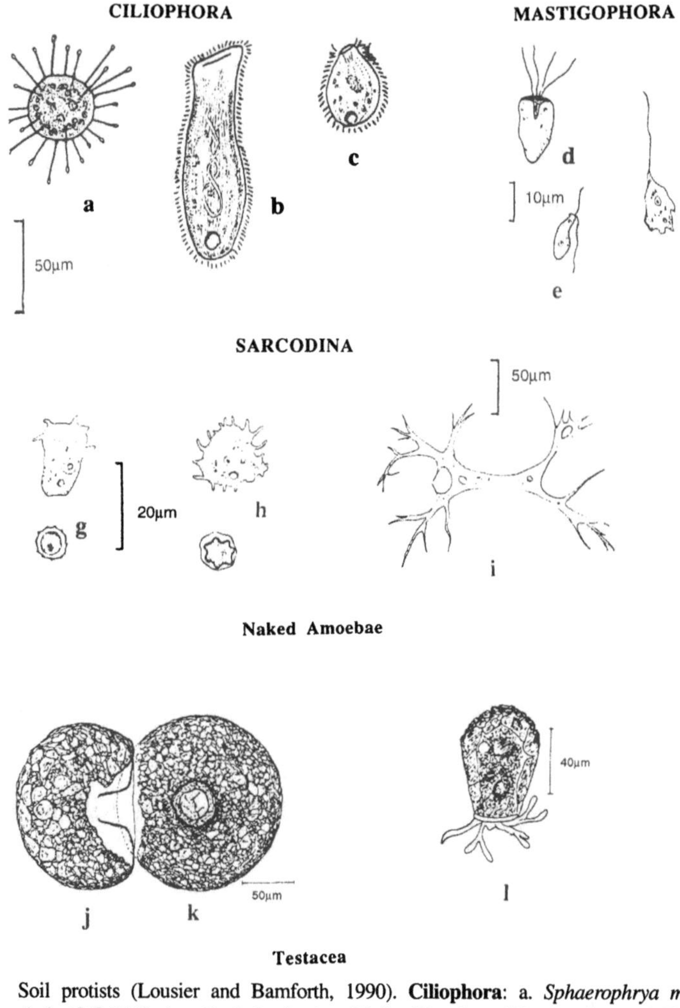

Figure III.26 Soil protists (Lousier and Bamforth, 1990). **Ciliophora**: a. *Sphaerophrya magna* Maupas; b. *Spathidium spathula* (O.F. Müller); c. *Enchelys farcinen* Müller-Ehrenberg. **Mastigophora**: d. *Tretamitus rostratus* (Perty); e. *Spiromonas angusta* (Dujardin) f. *Mastigella simplex* (Kent) **Sarcodina: Naked Amoebae**: g. *Stachyamoeba lipophora* (Page); h *Acanthamoeba castellanii* (Douglas); i. *Biomyxa vagans* Leidy. **Testacea**. j. *Cyclopyxis puteus* Thomas, lateral view; k. same, ventral view; l. *Heleopera rosa* Penard, lateral view.

The main factor limiting the size of protists is considered to be the pore-size distribution (Stout and Heal, 1967). Pores must be water filled to allow the development of protist trophozoïtes (active growing cells). In order that pores with an average diameter of 30 μm remain filled, the matric potential must be around *ca.* -0.01 MPa (pF 2)

(Darbyshire, 1976). Alabouvette *et al.* (1981) estimated that *ca.* -0.15 MPa (pF 3.2), which corresponds to the filling of all pores of < 2 µm is the theoretical lower limit for water potential which allows protist activity. That is, small ciliates may swim freely in the small micropores, and testate amoebae can immerse their pseudopods in the water film covering solid particles.

Growth may be rapid, under favourable conditions generation time varies from two to 48 hours for most species. The Testacea, however, have slower growth rates (of days to weeks) and they always appear last in community successions after, respectively, flagellates, naked amoebae and ciliates (Stout and Heal, 1967; Coûteaux and Ogden, 1988).

Feeding habits

Many flagellates are considered osmotrophic (Stout and Heal, 1967). However, most protists are mainly bacterial feeders and, of 258 species examined by Biczok (1965), 76 % were at least partly bacteriophagous and 40 % could not use another food. Bacteriophages are not generally considered to be specific in their diet (Coûteaux and Pussard, 1983). They may however distinguish edible from inedible species, which are determined by the presence of specific pigments (review in Stout and Heal, 1967). Under some circumstances, the presence of edible bacteria has been reported to induce a rapid emergence from cysts in certain protist populations (Kunicki-Goldfinger *et al.*, 1967). Other food resources may be used, either as an obligate food or as an alternative source of nutrition when edible bacteria are exhausted. They include fungi, algae, yeasts, protists, small Metazoa and also humic substances and cellulose (Lousier and Bamforth, 1990).

Water relations - Survival

Soil water potential is a major determinant of protist life. When the soil dries, the osmotic potential in free water decreases and protists may cease activity if salinity becomes too high. However, most tolerate salinities of up to 25-45 ‰.

Protists are unable to move more than a few centimetres per day and may encyst when conditions become unsuitable for their activity. True resistant cysts may withstand extreme conditions of exposure to drought, temperature or acidity for many years. Soil desiccation, increased salinity and the exhaustion of nutritive resources are commonly considered as the main factors determining encystment. However, Coûteaux *et al.* (1988), consider that cyst formation requires favourable nutritional conditions. Following Singh (1941), Corliss and Esser (1974) and Coûteaux (1976) who found greater numbers of cysts when feeding resources were not limited, they consider that encystment starts as soon as cells have accumulated sufficient reserves. Conversely, where nutritional resources are low, encystment is limited and many cells die if the soil dries out.

Community densities and biomasses

Most estimates of community density fall in the range 10 to 1000 10^6 individuals m^{-2} and biomass between 50 and 3000 mg f wt m^{-2} (Petersen and Luxton, 1982). Testate amoebae are dominant in forest soils whereas naked amoebae replace them in cultivated soils. Densities are highest in soils with elevated organic matter levels. Thus, amoebae may be more numerous in woodlands with moder rather than mull-type humus (Volz, 1964). This trend was confirmed by Schönborn (1982), although production was higher in mull than in moder. Amoebae also have high densities in tundra soils (450-900 10^6 individuals m^{-2} at Signy Island (Heal, 1965), and minimal densities in the equatorial dry forests and savannas of Zaïre (0.5 to 4.5 10^6 m^{-2}) and savannas of Ivory Coast (2.4 10^6 m^{-2}) (Coûteaux, 1978). There is some indication that protist abundance and diversity may be greater in environments with relatively high levels of environmental stress.

Exceptions to this pattern are however known. Persson *et al.* (1980) presented data for Swedish coniferous forests which clearly suggest that regional and local environmental variability is at least as important as inter-regional differences in determining the abundance of protist populations.

4.1.1.2 *Community structure*

Species richness

Protist communities of temperate and subantarctic communities may comprise several tens of species in each of the three classes. Many species are cosmopolitan due to the efficient dissemination of viable cysts and genetic stability (Stout and Heal, 1967). Consequently, the overall world fauna is unlikely to be greater than a few hundred species.

In New Zealand soils, Stout and Heal (1967) reported species richnesses of 14 to 47 for rhizopods and 21 to 59 for ciliates, with values increasing from grasslands to forests and subantarctic peats. In England, Stout observed an increase in testacean species from calcareous mull to acid mor soils (10 to 20) while ciliates exhibited a reverse pattern (35 to 21). In African savannas, at Lamto (Ivory Coast) 49 species of Testacea have been found (Coûteaux, 1978) and comparable species richnesses (63) have been recorded in a variety of forests and savannas in French Guyana (Coûteaux and Chardez, 1981).

Spatial distribution

Because of the nature of their feeding regimes and their water dependence, protists are particularly concentrated in the upper organic horizons of soils. They are thus generally more abundant in the litter layers than in the mineral horizons.

Generally, the spatial distribution of protists is similar to that of the micro-organisms they feed on. This explains why they are such an important component of the rhizosphere; populations may reach high densities and levels of activity while feeding on the abundant microbial populations present (see *e.g.*, Clarholm, 1985, Coûteaux *et al.*, 1988).

Temporal distribution

In temperate soils, protist populations sometimes show clear seasonal patterns of variation. Daily counts of populations in English cultivated soils revealed highest population densities in autumn and lowest in summer (Cutler *et al.*, 1922). Similar patterns have been reported by other authors for temperate soils (Stout and Heal, 1967). Short-term variation in population numbers is of similar amplitude to that occurring at seasonal scales. In a pot experiment with a dry arable soil, Clarholm (1981) induced a sudden increase of bacterial biomass by watering. Populations of bacterial-feeding naked amoebae had peak biomasses 2 to 3 days after those of the bacteria, and their population density was increased 30-fold (Figure III.27).

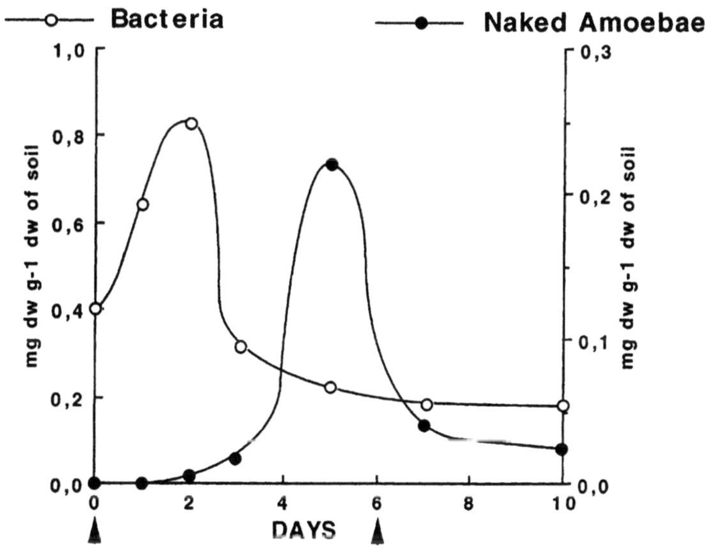

Figure III.27 Development of the biomasses of bacteria and naked amoebae in a small container experiment with a wheat plant grown in an arable soil. Watering is indicated by arrows; each point is the mean of three replicates (after Clarholm, 1981).

As illustrated above, any sudden change in soil conditions which induces a burst of microbial activity will affect protists, after a time lag of a few days. Among the events which may trigger a burst of bacterial development, and subsequent protist abundance are watering of a dry soil, rhizosphere activity (especially close to root-tips) and activation of microflora due to invertebrate activity, particularly that of earthworms.

4.1.2 NEMATODES

4.1.2.1 *General Biology*

Nematodes are filiform, acoelomate Metazoa, with neither circulatory nor respiratory systems, although they have a complete digestive apparatus. In soil environments, they may be free-living in soil water films, or phytoparasitic and live at the surface or within living roots.

Free-living, soil-inhabiting species are small (0.15 to 5 mm long and 2 to 100 μm wide with individual weights of 20-60 ng) with the exception of the Mermithidae, a group of giant nematodes up to 20 cm long which are a constant component of some humid tropical soil communities (Goodey, 1963). They are hydrobionts and live in the water which fills soil capillaries and covers roots and solid particles, or within plants as parasites.

Respiration occurs by simple oxygen diffusion through the cuticle, which is permeable to water. Respiratory rate is high (*ca.* 100 ml O_2 mg^{-1} f wt h^{-1} according to Nielsen, 1949). A general relationship between respiration rate (R) and individual weight (m) is given by the formula $R = 1.40\ m^{0.72}$ (Klekowski *et al.*, 1972). Respiratory rates of phytoparasitic nematodes decrease sharply when they are deprived of food (Reversat, 1981a).

Reproduction is either amphimictic or parthenogenetic. Nematodes produce eggs and most larvae reach the adult stage in *ca.* 20 days (5 to 50), after four stages separated by moulting of the cuticle (Kühnelt, 1961). However, significant differences occur between free-living nematodes which become adult in *ca.* eight days, and phytoparasitic species which have much longer cycles of *ca.* 30 days. The length of the development period may be significantly influenced by temperature conditions (Norton, 1978).

Nematodes progress by undulatory movements and crawl in thin films or swim in thicker ones. Unable to move soil particles, they are restricted to pores and capillaries of a diameter larger than their own. The presence of solid particles may be essential to their growth although they can sometimes be bred in agar or liquid media (Cheng *et al.*, 1979).

Basically, all nematodes feed on living protoplasm, including the cell contents of plants, bacteria, microscropic algae, protists, small Metazoa or plant sap. Their simple digestive systems do not permit assimilation of decomposing material that is not readily assimilable. They often have chitinous stylets, spears or buccal armatures that they use to pierce living cells; cell contents are ingested by a muscular pharynx, which acts as a pump (Figure III.28).

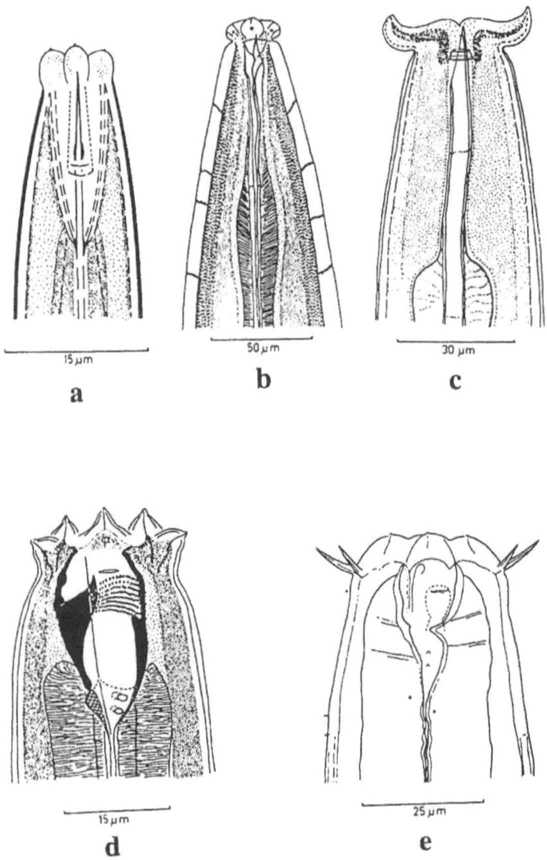

Figure III.28 Anterior parts of nematodes showing stylets and spears. a. *Seinura tenuicaudata* (de Man); b. *Sectonema rotondicaudata* Goodey; c. *Discolaimus major* (Thorne) and teeth d. *Mylolonchus brachyuris* (Butschli); e. *Tobrilus gracilis* (Bastian) (Arpin *et al.*, 1980 after several authors).

Bacterial feeders may ingest up to 5000 cells minute^{-1} or 6.5 times their own weight per day. Forty to 60 % of cells are assimilated which results in a growth efficiency of 4-13 % of the consumed material. Overall consumption may be as much as 800 kg bacteria ha^{-1}year^{-1} and the amount of N turned over in the range of 20-130 kg (review by Coleman *et al.*, 1984a).

Resistance to water-stress
When the soil starts to dry, nematodes may initially migrate. Since they have a limited capacity for movement, they are unable to avoid desiccation and part of their population will die. For example, nematodes were virtually eliminated from a fallow soil dried out to < -6.0 MPa (pF 5.5) (Simons, 1973). However, in most cases, a proportion of the population will survive since nematodes, as all soil hydrobionts, have developed a considerable resistance to desiccation (Demeure and Freckman, 1981).

Resistance varies among species and, within populations, among the successive stages. In some cases, eggs have an unexpectedly low resistance. Adults generally tolerate desiccation better than juveniles; in a 22 week experiment with alternating dry and moist periods, 84 % of adults, but only 24 % of juvenile *Helicotylenchus* sp. survived. With *Tylenchorynchus* sp., survival rates dropped to, respectively, 25 and 16 %, and 10 and 0 % with *Paratylenchus* sp. (Simons, 1973). On the other hand, juveniles of the phytoparasite *Meloidogyne javanica* resist starvation better than the adults. In a laboratory experiment, starved juveniles consumed their body food reserves 7.5 times more slowly than adults and their motility and infectivity remained unchanged during the 10 weeks of the experiment (Reversat, 1981b).

In the first stages of desiccation, nematodes may cease activity and coil, thus limiting subsequent water losses (Demeure *et al.*, 1979). Further resistance is obtained by anhydrobiosis, encystment or protection in galls, *i.e.*, specific tissues elaborated by plants in response to the infection. In the genus *Heterodera* cysts are the bodies of dead females whose cuticle has been tanned to form a resistant sac containing the eggs.

Ecological categories
Ecological classifications of nematodes are based on their feeding regimes and they are divided into the following five categories (Banage, 1966; Wasilewska, 1971): phytophages, bacterivores, fungivores, predators, and miscellaneous feeders.

Movements and dissemination
Most nematode genera, and even species, have cosmopolitan distributions which indicates a capacity to live in diverse environments and high vagility. For example, Maggenti (1961) records *Plectus parietinus* from Hawaii, continental USA, Antarctica, Australia, Canada, England, Ireland and the Netherlands.

Cysts and active nematodes are disseminated by wind and circulating waters. At the scale of a soil profile, nematodes may be attracted by decomposing organic residues (Griffiths and Caul, 1993), roots (Prot, 1980) and bacteria (Andrew and Nicholas, 1976). This attraction is mediated by specific chemicals such as CO_2, ammonium, unidentified components in root exudates, heat and light (Bilgrami *et al.*, 1985 and diverse authors quoted by Yeates, 1981). Another efficient medium-scale dissemination process is transport on growing root-tips. Selective reproduction in suitable environments is an additional mechanism which favours the colonisation of soils (review by Yeates, 1981).

Density and biomass of populations

Nematodes have a remarkable ability to resist environmental stress. As a result, certain community parameters, *e.g.*, species diversity, density and biomass tend to increase along a gradient of increasing environmental stress. This trend slows and later reverses in environments with particularly high stress levels (Procter, 1990).

Most estimates of nematode populations are in the range of 10^6-10^7 individuals m^{-2} with a corresponding fresh biomass of 1-10 g (Sohlenius, 1980; Petersen and Luxton, 1982). Densities up to 30-50 ind. 10^6 m^{-2} have been recorded in such different environments as temperate grassland in France (Ricou, 1978), a meadow-steppe in Russia (Ghilarov and Chernov, 1974) and an oak-ash forest with a mull-type soil in Gemany (Volz, 1951).

Relatively high densities have been recorded in such extreme environments as the high arctic tundra at Devon Island (4 10^6 ind. m^{-2}) and the Tajmyr Peninsula in the North of Siberia (10^6 m^{-2}). In sub-desertic areas of the USA (1.2 10^6 m^{-2} in Nevada) and USSR (1.1 to 6 10^6 m^{-2}) intermediate values have been reported (Chernov, 1975; Freckman *et al.*, 1975; Procter, 1977; Ryan, 1977).

Population densities less than 10^6 ind. m^{-2} have only been recorded in desert soils (Steinberger *et al.*, 1988), extreme tundra soils (Point Barrow, Alaska) and some tropical forests of Puerto Rico, Kenya and Uganda (Coleman, 1970; Kitazawa, 1971). In the latter case however, the method used (centrifugation-flotation) is suspected to have given underestimates and the densities of 1.5-1.9 10^6 m^{-2} reported for tropical forests of the Congo Basin using classical methods (modified Baerman technique) (Maldague, 1961) are considered more representative of that type of ecosystem (Petersen and Luxton, 1982). In New Caledonian tropical forests with ultrabasic soils, Yeates (1991) reported densities of 95,000 to 265,000 m^{-2} and suggested that nematode populations in tropical forests typically occur at the scale of hundreds of thousands m^{-2} compared with the several millions m^{-2} found in temperate deciduous forests and coniferous forests. In the tropical montane grasslands of Montagne Pelée (Martinique), Cadet and Van den Berg (1992) reported population densities of 1.8 10^6 m^{-2}.

The main variables influencing abundance are the soil hydrological regimes and the relative accumulation of organic matter on the surface. Populations are more numerous in ‘mesic’ than wet soils with the lowest densities being found in ‘dry’ soils. Abundances, and particularly biomasses, are greater in grasslands than in forests (due to an increase of phytoparasitic populations), and in mineral mull-type soils than in moders or acid mors with large surficial accumulations of organic matter (Petersen and Luxton, 1982).

Finally, in grasslands in the USA, Yeates and Coleman (1982) observed positive correlations between the abundance of nematodes and root biomass and above-ground production. Such a relationship appears to be general in grasslands whereas, in forest soils, contrasting observations, have been made.

4.1.2.2 *Community Structure*

Species Richness

Nematode communities may comprise up to 70-100 species in such different ecosystems as an English pastures (71), a Danish beech-wood (75), a prairie site in Dakota (80) a Slovakian mixed deciduous forest (92) or wheat-soybean fields in Tennessee (100) (as reviewed by Petersen and Luxton, 1982 and Bernard, 1992). An absolute maximum of 431 species has been recorded in a tropical forest in Cameroon (Bloemers *et al.*, 1997). Intermediate richnesses of 30-40 species have been recorded in temperate forests (Arpin *et al.*, 1986) and minima of 1-16 species are found in tundra soils (Signy Island) (Petersen and Luxton, 1982).

Functional structure: interbiome comparison

The distribution of trophic groups within nematode populations (*i.e.*, the functional structure) has been assessed in a variety of environments including tundra, moorlands and derived grasslands, temperate coniferous, deciduous and mixed forests. There is no clear pattern of variation in functional structure between these biomes as variations across biomes and between sites are often less marked than intra-site variation (see, *e.g.*, Schmitt and Norton, 1972). Some trends, however, may be derived from the existing data:

- bacterial feeding populations generally represent 30 to 50 % of the total community (range 12.9 to 72 %) with higher proportions in coniferous forests and crops (*ca.* 50 %) than in grasslands and deciduous forests (30-40 %);
- fungus-feeding nematodes commonly comprise 20-30 % of total density (range 3.1 to 50 %) with maximum abundances in forests. This may reflect the composition of microbial communities with, typically, a dominance of bacteria in many grasslands and crops; in forests, especially with mor or moder-type humus, fungus-feeders are more abundant (Arpin *et al.*, 1986), commensurate with the fungal dominance common in microbial communities in those environments.
- root-feeding parasitic nematodes are of little importance in forest soils (range 0 to 14 %), but represent 25-40 % on average of grassland communities and up to 83 % in moorlands. In crops, they comprise 17-33 % of the populations;
- predators comprise a few percent of the community; they are nearly absent from moorlands and coniferous forests; highest values (10 %) have been noted in a short-grass prairie in Colorado (USA);
- miscellaneous feeders have abundances of *ca.* 5-25 % with no clear interbiome differences.

Nematode communities vary at a local scale with changes in the vegetation type, with drainage and earthworm activities (Dash *et al.*, 1980; Yeates, 1981) probably through their effects on available food resources, bacteria, fungi, algae and fine roots. Tillage also affects nematode abundance and community trophic structure. In a rye (*Secale cereale*) field in Georgia (USA), the mean populations of bacterivorous, fungivorous and total nematodes were greater in a conventional tillage system; in the summer, plant parasitic and fungivorous nematodes were more abundant in the no-tillage system (Parmerlee and Alston, 1986).

Spatial distribution

At a mesoscale level of approach, that of an experimental plot, the horizontal distribution of nematode populations is very uneven. At sites in nine prairies in Iowa, Schmitt and Norton (1972) found more heterogeneity within than between sites. In a soy-bean field in Michigan, Robertson and Freckman (in Roberston, 1994) found nematodes to be distributed in large patches *ca.* 200 m in diameter with population density variations spanning two orders of magnitude. In tundra soils, nematode communities reflect the mosaic of microsites with densities a hundred times greater in grass stands than in bare soil (Kuzmin, 1976). Similar differences have also been reported for soil within and between rows in cultivated soils (Ferris and McHenry, 1976) and in a sugarcane field in Martinique (Rossi *et al.*, 1996, Figure III.29).

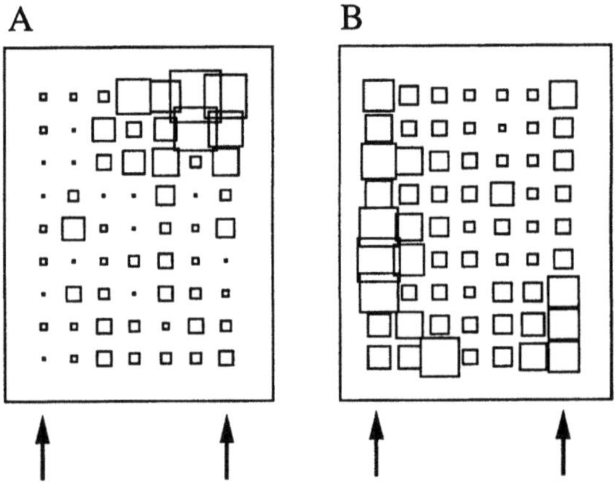

Figure III.29 Horizontal distribution of phytoparasitic nematodes in a sugarcane plantation of Martinique (Rossi *et al.*, 1996). A:*Criconemella anoensis*, B: sum of *Helicotylenchus erythrinae*, *Hernicriconemoides cocophilus* and *Pratylenchus zeae*. Plot size is 1.6 m x 1.2 m. Arrows indicate location of plant rows.

Such factors as total-C, assimilable P, pH, Ca, aeration and moisture may influence the micro-distribution of nematodes (Yeates, 1981). Interactions with other organisms, especially roots (which attract them) or earthworms (whose activities depress nematode populations) also modify distribution patterns (Yeates, 1981). In consequence, the distribution of phytoparasitic nematodes in cultivated fields largely reflects that of the plants (Francl, 1986).

Vertical distribution

Nematode populations are generally concentrated in the upper 5 to 10 cm of soil although they may also occur at depths of 50 cm and more (Sohlenius and Sandor, 1987). Their distribution follows those of site organic resources and thus forest communities have more superficial patterns of vertical distribution than grassland communities (see, *e.g.*, Yeates and Coleman, 1982; Hota *et al.*, 1988). Vertical migration is limited, with most movement being due to transport on growing root-tips (Yeates, 1981). Significant differences exist between populations of different species and ecological categories

(Sohlenius and Sandor, 1987)

Temporal variation
Samples taken at monthly intervals generally show some degree of seasonal variation with different patterns emerging with changes in climate, vegetation (Arpin, 1983) and with the trophic group under consideration (Seastedt *et al.*, 1987). Large inter-annual variation has also been observed (see, *e.g.*, Arpin *et al.*, 1985).

In a *Brachiaria humidicola* tropical grassland, Smith and Wallace (1976) found centres of permanently high and low populations of *Helicotylenchus dihystera* which had similar patterns of seasonal variation, but which fluctuated asynchronously at shorter scales of time. Short term fluctuations may be expected to be frequent and intense since the generation times of nematodes vary from 5 to 50 days. Further, in desert soils, Freckman and Womersley (1983) have observed that anhydrobiotic microbivorous nematodes may resume activity 15 minutes after rainfall has started and lay eggs immediately.

The species composition of nematode communities may also change with time. In sugarcane plantations in the Côte d'Ivoire, Cadet and Debouzie (1990) showed that population changes appear to depend more on changes of the environment (soil microclimate, plant activity) than on interspecific interactions.

Nematode communities are highly sensitive to environmental change. The addition of nitrogenous fertilisers results in decreases of the density of predaceous and omnivorous nematodes whereas bacterivorous and phytoparasitic populations increase (Wasilewska, 1989).

4.2 Mesofauna

MICROARTHROPODS

Microarthropods are constant inhabitants of leaf litter and soil interstices that, as a group, function mainly as epigeic decomposers. Apterygote Collembola and Acari are generally dominant by far, both numerically and in terms of biomass. However, other Apterygota (Protura, Diplura and Thysanura), Micromyriapoda (Symphyla, Pauropoda and Polyxenidae) or small Diptera larvae and Coleoptera may be locally important. All respond, although differentially, to the dry funnel ('Berlèse') extraction technique.

4.2.1 COLLEMBOLA

4.2.1.1 *General Biology*

The Collembola or springtails, are a very diverse taxon with 21 families and twenty thousand currently-described species. However, the total number of existing species is expected to be close to 300,000 (Schaller, 1970; Massoud, 1976). They are apterous hexapods close to the true insects, small and elongate with a characteristic saltatory organ that, when fully developed as in surface-dwelling species, allows rapid jumping movements (Figure III.30). A further typical structure is the ventral tube whose many functions include sensing the moisture status of the substrate, absorbing liquids, ions and gases and attachment to the substrate (Sedlag, 1951). Their body lengths range from a few tenths of a mm to 1 or 2 cm with individual biomasses of 1-20 μg dry weight (*ca.* 3-60 μg fresh weight).

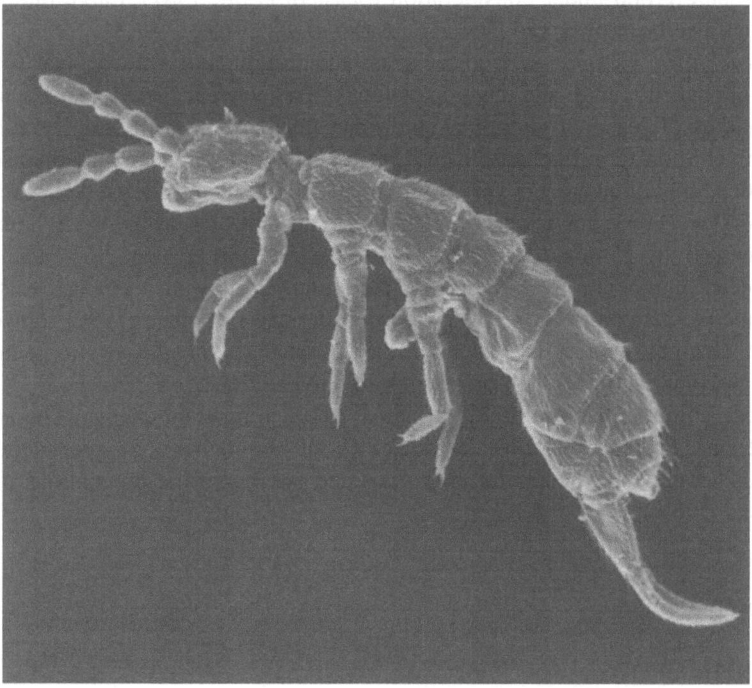

Figure III.30 *Isotoma violacea* (Isotomidae), an epiedaphic species of Collembola from a Swedish pasture soil (Photo, Jan Lagerlöf).

They live in the litter or in the pore space of the upper 10 to 15 cm of soil and are saprophagous, feeding mainly on fungi, bacteria or algae growing on decomposing plant litter (Christiansen, 1964; Anderson and Healey, 1972; Kilbertus and Vannier, 1979, 1981; Ponge, 1991). There is, however, considerable variation between species, especially between litter and soil dwellers. Saur and Ponge (1988) showed that the soil dwelling *Paratullbergia callipygos* ingests a substantial amount of soil (47.6 %) and can digest cellulose as well as starch, tannins, fungal cytoplasm and chitin. Zinkler (1983) presented evidence of digestion of hemicellulose by Collembola although it was assumed that it may have been due to the activity of micro-organisms in the gut rather than the activity of the animals' own enzymes. Consumption and assimilation by such litter-feeding species as *Onychiurus zschokkei* increases on the soil surface (Sadaka and Poinsot Balaguer, 1987) due to a decreasing concentration of polyphenols and an increase in nitrogen concentration of food resources.

Growth is continuous with 4 to 50 stadia, according to species (Hale, 1965a, b). Large species generally have more stadia; nonetheless, for a given size, species with tracheae have fewer stadia than others. Development through the reproductive instars generally takes 40 to 400 days and moulting is continuous throughout life (Massoud, 1971).

Reproduction may be parthenogenetic or bisexual. In the latter case, copulation does not take place; males deposit spermatophores and females fertilise themselves by rubbing their genital aperture with these spermatophores. A number of different behaviours may facilitate this process, *e.g.*, primitive or specialised courtships and the production of pheromones (Schaller, 1953; Betsch-Pinot, 1977; Blanquaert and Mertens, 1977) although Petersen (1980) considers this system to be inefficient. Parthenogenesis is more frequent in species living in the soil interstices than in litter dwellers. This situation may have been selected to obviate the problems faced by females in locating spermatophores in the restricted environment of the soil pore space. Each female may lay 100 to 600 eggs during her adult life time, which rarely exceeds one year.

Resistance to desiccation and low temperatures has been developed in some species. *Isotoma saltans* is a common inhabitant of the surface of alpine glaciers and *Anurophorus subpolaris* has been reported to survive temperatures of -50 °C in Antarctica (Wise, 1965; Block, 1983). The physiological state of the individuals concerned partly determines their ability to tolerate low temperatures (Vernon and Vannier, 1990).

A few cases of anhydrobiosis have been reported in Collembola (Poinsot, 1968). Desiccation tolerance is generally limited to a matric potential of -5.0 MPa (pF 4.7) (Vannier, 1970), although it varies greatly among species (Verhoef and van Selm, 1983). The amount of water lost before activity ceased varied from 16.2 to 33.8 % of body fresh weight in six litter-dwelling species from woodlands in the Netherlands (Vegter and Huyer-Brugman, 1983). Vannier (1978) defined three types of transpiration regime depending on the efficiency of body water control. Hydrophilic species exert no control, mesophilic a limited control and xerophilic a strong control. Again this pattern separates soil-dwelling from the litter-dwelling species which have a much greater drought resistance (Figure III.31). Tolerance may change during life as juveniles are usually much less desiccation tolerant than older animals (Betsch and Vannier, 1977). In *Allacma fusca*, the difference was due to the absence of tracheae in the early stadia. In some species, populations persist because of the survival of eggs whose drought tolerance is greater than that of adults (Wallace, 1968).

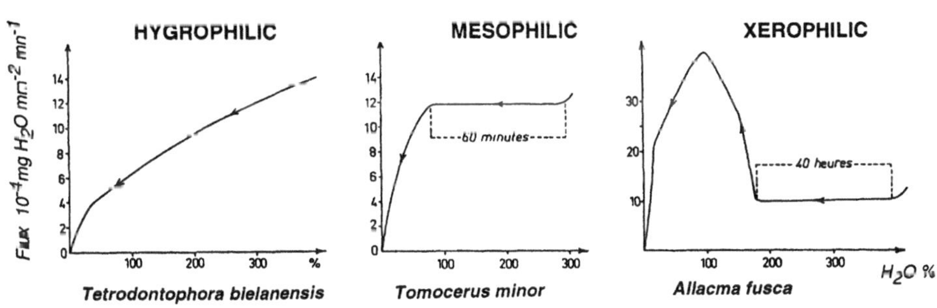

Figure III.31 Desiccation patterns in the three main ecological types of Collembola (Vannier, 1978).

Ecological categories

Different life-forms ('Lebensformen') were recognised early by Gisin (1943) on the basis of morphological characters while Petersen (1980) and Vannier (1978) have supplemented these with ecological and physiological characters. Three main groups may be distinguished: epigeic, hemiedaphic and euedaphic (Table III.11).

Table III. 11 Principal morphological and ecological characteristics of the main ecological types of Collembola (after Gisin, 1943; Petersen, 1980; Vannier, 1978); hemiedaphic types are intermediate.

	EPIEDAPHIC	EUEDAPHIC
Vertical Distribution	Surface + L layer	H + A
Size	large	small
Morphology	long antennae, legs and furcae, pigmented, 16 ommatidia.	reduced appendages, furcae absent, unpigmented, ommatidia absent or reduced
Reproduction	bisexual, small progeny, many eggs, seasonal	parthenogenetic, large progeny, few eggs, throughout the year
Metabolic activity	high but intermittent	low but constant
Food	high quality	low quality
Foraging activity	must move to acquire food, feed intermittently	feed in the immediate vicinity, feed constantly

Epiedaphic Collembola (*i.e.*, living in the 'Atmobios') live on the plant surfaces and in the upper litter (L) layer. They are generally large species with long legs, antennae and furcae; the eyes have numerous ommatidia and they have a dark pigmentation. Because they live in an unpredictable environment, these species have developed a resistance to desiccation. These Collembola use high quality nutritional resources which allow them to sustain a high metabolic rate when changing conditions require increased locomotory activity. In consequence, they have definite seasonal patterns of activity and reproduction and their demographic profiles are often of the '*r*' type.

Euedaphic species are adapted to life in the pore space of the H and A horizons where movement is limited and food is of relatively low quality, although environmental conditions are more stable. These species are small and unpigmented and their appendices and ocelli are reduced. Their demographic profiles appear to be of the '*K*' type.

The relationships between demographic strategy and ecological categories of Collembola are still unclear. Certain epigeic species have a high mortality rate due to active predation and epidemics (*e.g.*, *Parisotoma notabilis*); as a result, they are highly fecund. Others, *e.g.*, *Orchesella* spp. escape more readily from predators and have a lower oviposition rate. In the soil, Collembola may be better protected against predators although some specialised predators (*e.g.*, Chilopoda, Geophilidae) have developed adaptations for moving within the pore space. As soil-dwelling Collembola cannot readily escape predators, some species have developed mechanisms for reducing predation, *e.g.*, the production of toxic secretions in the subfamily *Onychiurinae.*

Density and biomass of communities

As is normal in studies of microarthropods, Collembola are generally extracted by the dry funnel technique, or an improved variant of this apparatus such as the 'high gradient' canister extractor (Macfadyen, 1961). Extraction is never complete due to the variable reactions of different species and stages to the extraction stimuli. The efficiency of extraction may be as high as 85-89 % when using the best techniques (Marshall, 1972); extraction using simple dry funnel methods may be 2 to 10 times lower than with high gradient designs (Macfadyen, 1961). A further source of error may arise from an inefficient separation and counting of the extracted material.

Population densities reported from more than 200 sites varied from 100 to 670,000 individuals m^{-2} (Petersen and Luxton, 1982) with most data falling in the range 10,000-100,000. There is a strong thermo-latitudinal effect, with low values in colder tundra sites (Devon Island, Canada) but maximum values in cold temperate grasslands (670,000 in an ornithogenic soil of Signy Island) and cold temperate forests (145-244,000 in a Norwegian spruce forest). Densities in temperate forests range from 40,000 to 70,000. In tropical forests, abundances may vary from a few thousands (Betsch and Betsch-Pinot, 1983) to 20,000-80,000 depending on soil type and drainage conditions. Tropical savannas have significantly lower abundances (1100 to 3500 at Lamto, Ivory Coast) and finally, a site in an American desert had the lowest densities (100 m^{-2} in a Californian desert) (Figure III.32).

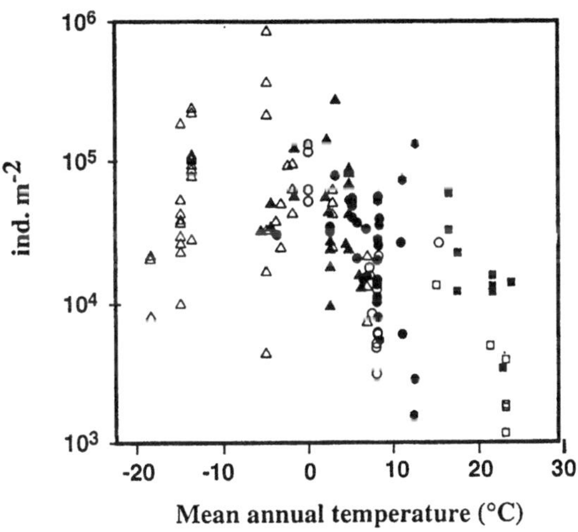

Figure III.32 Relationship between mean annual air temperature and population densities of Collembola across a wide range of ecosystems from the tundra to tropical rainforest (Petersen and Luxton, 1982).

In most areas, no significant differences in abundance have been observed between wooded and non-wooded areas. This may be explained by the feeding habits of Collembola: they do not directly feed on litter, but rather on the soil microflora which is of comparable abundance in pastures, crops and forest soils. The effects of land management are not yet clearly understood. In southern Sweden, Lagerlöf and Andrén (1991) found densities of 12,430 to 30,410 ind. m^{-2} in four arable crops (fertilised and unfertilised barley (*Hordeum* sp.), grass and lucerne (*Medicago sativa*) leys) while Persson and Lohm (1977) recorded much higher values (66,000 ind. m^{-2} in summer, 194,500 in winter) in a grassland soil. Low soil water potentials may also limit populations.

Biomass generally varies from 80 to 200 mg dry weight m^{-2} (*i.e.*, 240-600 mg fresh wt). The highest values have been recorded in cold temperate ecosystems, notably forests, where individual mean weights are high. In tropical forests individuals are small and biomass may be as low as 10-20 mg m^{-2}.

4.2.1.2 *Community structure*

Species structure

The species richness (α diversity) at single sites may be as high as 60 to 80 species in temperate deciduous forests although large variations occur among environments, with a clear thermo-latitudinal effect (Petersen and Luxton, 1982). Tundra environments have low species richnesses with 1-3 species at Signy Island (Antarctica), 7-13 at Devon Island (Canada), and larger numbers at Barrow (Alaska): 20, Moor House (England): 13-25 and Hardangervida (Norway): 21-29. In temperate deciduous forests, species richness is in the range 27-60. In a mixed forest near Paris, 27 species were found over an annual cycle; nonetheless, only 12 to 19 species were collected at each sampling time (Massoud *et al.*, 1984). A maximum of 85 species has been recorded across a range of biotopes (ß diversity) in mixed forest at Senart in France (Ponge, 1980). In temperate grasslands, communities may comprise 16-32 species, comparable to the arable land sampled in Southern Sweden (32 species in cropped plots at Spiboke, Lagerlöf and Andrén, 1991). No complete list of species seems to be available for tropical environments.

An elevated species richness has also been noted in a number of other soil arthropods and has been discussed by several authors (see discussion in Section III.4.4.2). High levels of predation on litter arthropods (up to 70 % mortality in adults, Ernsting and Joose, 1974) and a high mortality due to environmental factors maintain populations at low densities, with limited, if any, competition (Ponge in Vannier, 1985; Persson, 1980).

Spatial distribution

Collembolan populations are largely confined to the surface holorganic horizons (Petersen and Luxton, 1982). This illustrates the general function of Collembola as epigeic decomposers and, in many sites, 75 to 99 % of the population is concentrated in the upper 2.5 cm. However, in such ecosystems as the boreal coniferous forest (McLean in Petersen and Luxton, 1982) or moist African savanna (Athias, 1975), there may also be a significant proportion of euedaphic populations living below 10 cm depth, down to maximum depths of 30 or 40 cm. Persson and Lohm (1977) have also observed seasonal variation in vertical distribution with a marked migration to depth taking place in winter (Figure III.33).

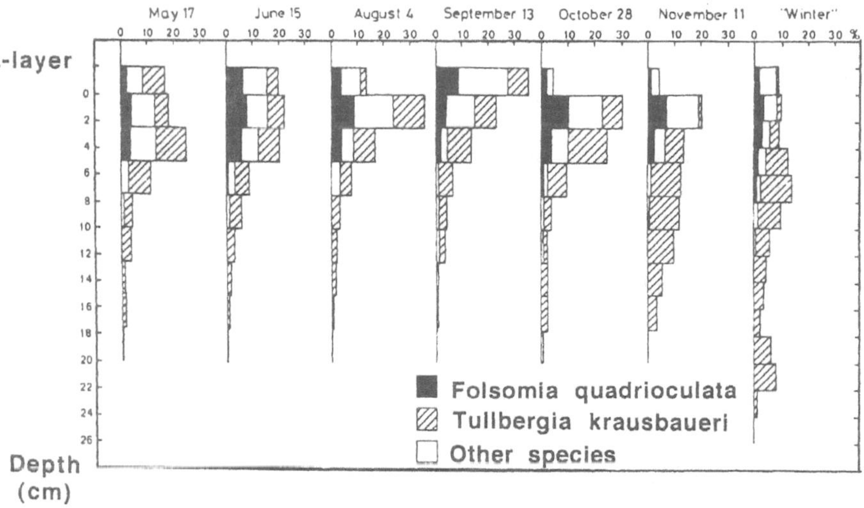

Figure III.33 Seasonal variation in the vertical distribution of Collembola in a Swedish grassland soil (Persson and Lohm, 1977).

In woodlands in Alberta (Canada), Hassall *et al.* (1986) have related vertical migrations in populations of *Onychiurus subtenuis* to moisture conditions in the litter layers. Populations that were concentrated in the deeper litter horizons during dry periods returned to the surface litter layers within a few hours after summer rain storms and remained there until the litter dried out again. These movements were instigated by the development of populations of highly palatable micro-organisms in the surface litter, including yeasts and the fungus *Cladiosporum*.

The horizontal distribution of Collembola is often highly aggregated at small spatial scales, as some species may form colonies of 8 to 30 cm diameter (Kühnelt, 1961). There is some evidence that this aggregation may be caused by the production of chemical signals (pheromones) which attract the animals to the most suitable micro-environments (Joose, 1970; Verhoef and Nagelkenke, 1977). At a meso-scale, local variation in litter abundance and quality and/or microclimate may affect population distribution. Where cushion-like vegetation forms or tussocks occur (see *e.g.*, Stanton and Tepedino, 1977; Garay, 1981b; Athias, 1976), Collembola generally accumulate in these structures. At a regional scale such factors as land use practices (see, *e.g.*, Massot and Cancela da Fonseca, 1986; Lagerlöf, 1987), humus type (Ponge, 1983) and local differences in soil moisture status, depth, light and soil type (Ponge, 1980) influence the abundance and composition of collembolan communities.

Seasonal variation

Seasonal variation has been observed on many occasions and in Swedish grasslands at Spiboke, density was at a maximum in winter (194,500 ind. m^{-2}) and a minimum in summer (65,800) (Persson and Lohm, 1977). However, most euedaphic and hemiedaphic species may reproduce during any favourable period throughout the year. Consequently, in soils where these groups are dominant (*e.g.*, the arable land studied by Lagerlöf, 1987 or the moist African savannas of Lamto, Ivory Coast (Athias, 1976), abundance may be highly variable with rapid week-to-week variation and a slight but clear seasonal pattern.

Successions

Collembola are clear indicators of ecosystem change in successional situations because of their large specific and functional diversity (see, *e.g.*, Huhta *et al.*, 1969; Bonnet *et al.*, 1976; Betsch and Betsch-Pinot, 1983). Recolonisation may be rapid although euedaphic species tend to recolonise more slowly than epiedaphic species (Dunger, 1975). Recolonisation of reclaimed areas does not appear to proceed evenly; population aggregates appear at the edges of the colonised area and further split into other such aggregates which progressively invade the area (Petersen, 1995).

4.2.2 ACARI

4.2.2.1 *General Biology*

Acari are small Arachnida with an unsegmented posterior parts and, except in the Prostigmata, no constriction between the anterior and posterior part of the body. The Acari are subdivided into five main sub-orders based on the location of their stigmata or respiratory organs: Mesostigmata (= Gamasida), Prostigmata (= Actinedida), Astigmata (= Acaridida), Cryptostigmata (= Oribatida) and the parasitic Metastigmata (= Ixodida). In all, they comprise 1200 families and probably as many as 500,000 species of which many are soil and litter inhabitants. They have chelicerae, primitively forceps-shaped although in some predatory groups, they have been transformed into claws, cogged tips (Prostigmata Trombidiforma) or harpoons (Metastigmata) (Kethley, 1990; Krantz and Ainscough, 1990; Norton, 1990; Philips, 1990) (Figure III.34).

Their morphology is highly variable. Predators generally have long legs and elongated bodies whereas saprophages are frequently globular with a thick chitinous carapace; some species have lateral extensions that can close hermetically over the legs, as in the family Phthiracaridae (Cryptostigmata). Some species have ommatidia (1-5) and all have a variety of sensory organs whose functions are still poorly understood (Bachelier, 1978). Excluding the parasitic Metastigmata, their sizes range from 0.1-0.2 mm to 1.5-2 mm and fresh body weight from 0.25 to 300 μg (Petersen and Luxton, 1982) equivalent to 0.1-120 μg in dry weight.

As with most mesofauna, soil-dwelling Acari are chiefly epigeic and live in the leaf-litter and the pore space of the upper 10-15 cm of soil. They have highly diverse feeding habits and exploit all the resources available in litter and soil using a wide range of ecological strategies (Vannier, 1985). Three major groups may be distinguished (Luxton, 1972):

(i) macrophytophages, that mainly occur in the Cryptostigmata and Mesostigmata Uropodida and which feed on leaf litter;

(ii) microphytophages which feed on bacteria, fungi and algae and may be found in all the major taxonomic groupings;

(iii) predators, *e.g.*, Mesostigmata or Prostigmata, which feed on small invertebrates, mainly microarthropods and Enchytraeidae. A number of species prey

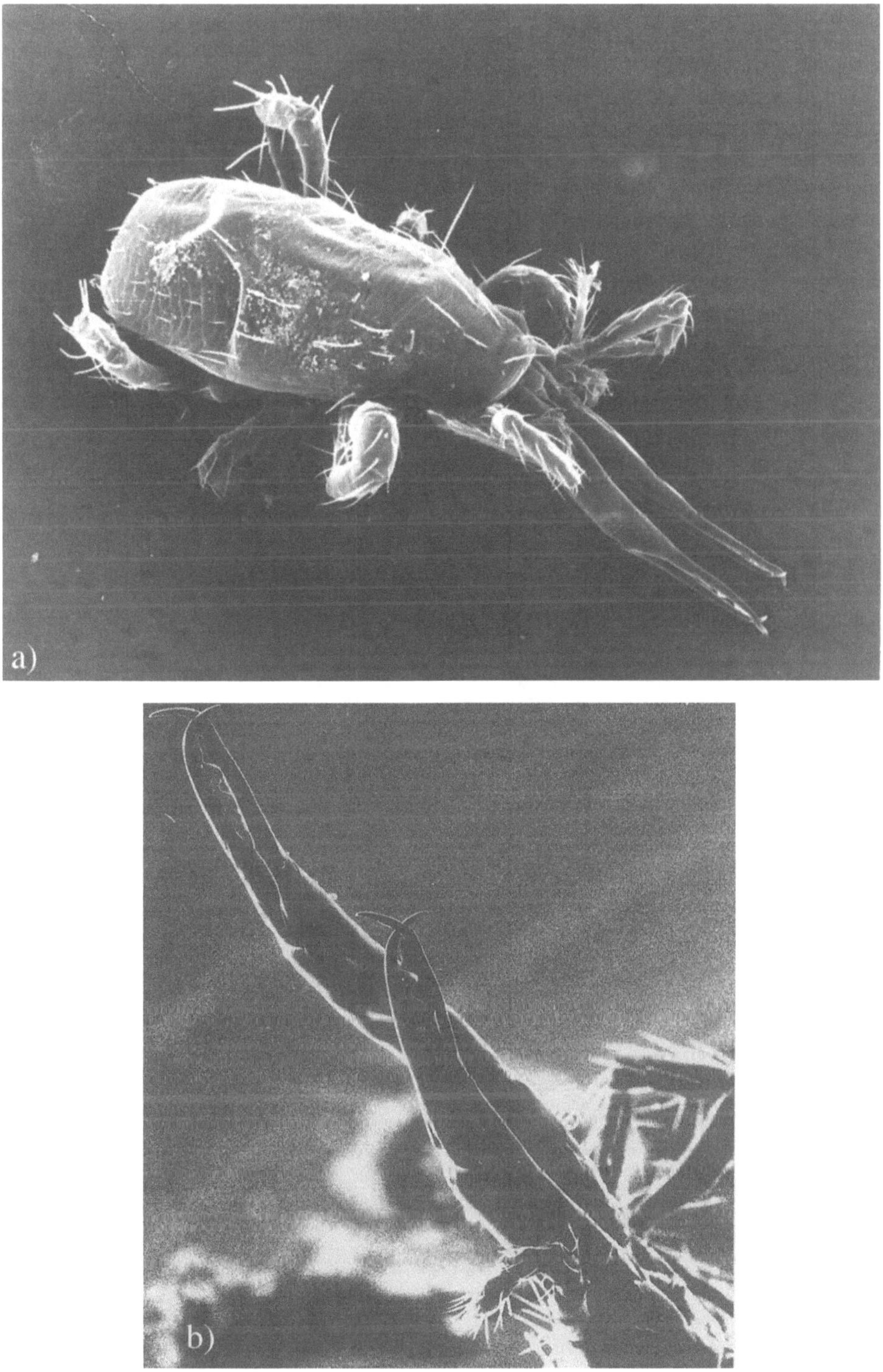

Figure III.34 a). *Veigaia nemorensis* (Gamasina, Mesostigmata), a soil-dwelling predatory mite from Sweden (photo Jan Lagerlöf) (length: 0.8 mm); b). close up of chelicerae.

on vertebrates (including man) during the immature stages; some transmit various pathogens and are thus of public health importance.

A number of species, however, have mixed feeding habits some, as in the Astigmata, are panphytophagous (both micro- and macrophytophages), others (*e.g.*, some Prostigmata seem to be able to feed on plant roots and certain species of this order are parasitic on the leaves of higher plants) are microphytophagous and predatory. Anderson (1975) found that a simple trophic classification of mites was inadequate because of marked seasonal changes in the feeding habits. Nonetheless, gut content analysis of Cryptostigmata carried out over a twenty-month period showed a relationship between size and feeding regime (Anderson, 1977). The largest species, *Phthiracarus* sp. and *Steganacarus magnus* which, respectively, measured 0.8 and 1.0 mm, tended to feed mainly on leaf litter. As body size decreases, the proportion of fungal material ingested increases, *Eniochthonius minutissimus* (0.55 mm), for example, has a purely mycophagous regime. The smallest species tend to ingest large amounts of amorphous material; *Oppia* sp., the smallest one (0.25 mm), ingests small amounts of fungal and no higher plant material (Figure III.35).

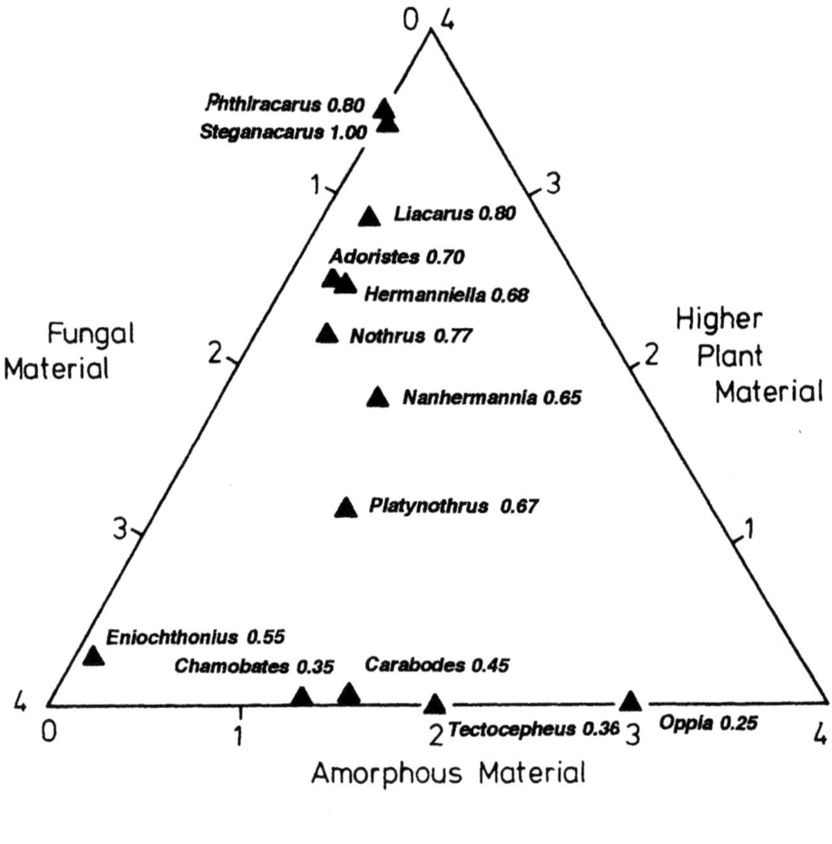

Figure III.35 Cumulative mean proportions of major food items selected by cryptostigmatid mites over a 20 month period in a temperate forest, in relation to body size. Values beside the species names refer to mean body length in millimetres (Anderson, 1975).

Litter assimilation by saprophages is of low efficiency. It was estimated at *ca.* 25 % for a community of cryptostigmatid mites in an old field and only 5 % of the assimilated energy was allocated to production. Estimations of assimilation made by a number of authors are in the range 24-52 % and the percentage allocated to production in the range 3-28 % (see review by Luxton, 1982).

Post-embryonic development may last for several months. The parasitic Mesostigmata have only two immature stages, a hexapod larva and a quadriped nymph before moulting to the adult form. Acari of the three other non-parasitic groups pass through six development stages, *i.e.*, prelarva, hexapod larva, protonymph, deutonymph, tritonymph and adult. Pupation occurs between the larval and nymphal stages and again between nymphal and adult stage. As a result, inactive forms may be relatively common in acarine populations: Cryptostigmata, *e.g.*, may spend nearly 30 % of the annual cycle in moulting or resting stages (Luxton, 1982). Most Cryptostigmata have one generation per year. Nevertheless, larger species, or those from boreal and arctic environments may take 2-3 years to complete their life cycles (Luxton, 1972; Mitchell, 1977). Similar figures are reported for Mesostigmata Uropodida (Athias-Binche, 1985).

Reproduction is generally bisexual but certain species may be parthenogenetic (*e.g.*, Oliver, 1971; Athias-Binche, 1985); as in the Collembola, males deposit spermatophores and females fertilise themselves. Cryptostigmata females lay only 1 to 6 eggs on average (maximum 16 or more) which hatch from 1 to 6 weeks later, depending on temperature. In other groups, *e.g.*, Prostigmata, the number of eggs produced varies between 10 to 100, depending on the family and species involved (Kethley, 1990).

Resistance of Acari to water and temperature stress appears to be greater than that of the Collembola. Cryptostigmata can withstand desiccation up -6.0 MPa (pF 5) before moving to wetter areas (Vannier, 1970). As a consequence and in contrast to most soil invertebrates, their population densities may be highest during the dry and hot seasons. This is the case, *e.g.*, in the savannas of Lamto (Côte d'Ivoire) where maximum densities were observed at water potentials equivalent to -1.6 to -5.0 MPa (pF 4.2-4.7) and soil temperatures of 33-42 °C (Athias, 1976). In tundra soils at Point Barrow (Alaska), an inverse correlation is observed between the density of Acari and soil moisture in certain habitats (McLean, 1975). They are regularly found in such extreme environments as littoral sands of the Mediterranean basin and in the fine sand dunes of the Namib desert (Coineau and Seely, 1983). Application of water to a dry soil of the Chihuahuan desert in New Mexico did not significantly increase their density (Kamill *et al.*, 1985).

Acari also exhibit remarkable resistance to low temperatures which allows them to colonise polar regions and they are the macroscopic organisms with the most southerly populations. Wise and Gressitt (1965) reported the prostigmatid mite *Nanorchestes antarcticus* to be the most southerly terrestrial animal species. *Alaskozetes antarcticus* and *Cryptopygus antarcticus,* two species from Signy island (Antarctica) had supercooling points of -25 to -30 °C (Block *et al.*, 1978). Resistance to low temperatures was greatest in starved individuals and nymphs were more resistant than adults.

Ecological categories

In most ecological studies, Acari are simply divided into the four suborders Cryptostigmata, Astigmata, Mesostigmata and Prostigmata which exhibit some homogeneity in their general ecology (Table III.12). The Metastigmata which are parasitic on vertebrates are excluded from this discussion.

However, the ecological classification used for Collembola (into epi-, hemi- and euedaphic) may also be used for the Acari (see, *e.g.*, Wallwork, 1967; Cancela da Fonseca, 1987). The mean size of individuals tends to decrease with depth as the mean pore size diameter becomes a limiting factor. Epiedaphic Acari tolerate desiccation better than hemi- and euedaphic forms.

Each of these categories may be further divided into hygrophilous, mesophilous and xerophilous. This concept of life forms is much more difficult to apply to Acari than to Collembola. This is because most Acari would be classified as hemiedaphic and, as such, face rather limited environmental constraints on their morphology.

A classification of Acari among 12 life history groupings has been proposed by Siepel (1994). It is mainly based on the occurrence of phoresy (obligate, facultative or absent), feeding habits, demographic parameters (life duration, the type of resting stage (obligate diapause or quiescence) and type of reproduction (sexual or parthenogenetic). The relative frequency of each type varies depending on soil and vegetation types. There are also clear differences in the species present at the various stages of organic matter decomposition.

Table III.12 Main ecological characters of the four non-parasitic sub-orders of the Acari (after Wallwork, 1967; Petersen and Luxton, 1982).

	Vertical Distribution	Relative abundance (%)	Feeding regimes	Individual biomass µg dw
CRYPTOSTIGMATA	soil	30-70 in most ecosystems	macrophytophagous, panphytophagous, microphytophagous	*ca.* 5
PROSTIGMATA	litter and soil	0.2-83*	microphytophagous, predaceous, rhizophagous	0.1-1.0
MESOSTIGMATA	litter and soil	0-20, more abundant in woodland	macrophytophagous, microphytophagous, predaceous	3-270
ASTIGMATA	deep organic horizons	very low	macrophytophagous, microphytophagous	0.7-5.8

* highly variable due to uneven sampling efficiency

Density and biomass

Acari are often the most numerous group of soil arthropods. Their resistance to drought and extreme temperature has allowed them to colonise almost all soils, from the southernmost south polar cold desert to hot tropical deserts. In the 180 sites reviewed by Petersen and Luxton (1982), population density varied from a few thousands in desert soils and some arctic habitats, to a mean annual maximum of 792,000 ind. m^{-2} in a Swedish pine forest (Persson *et al.*, 1980) and an absolute maximum of 1,783,000 m^{-2} in a late autumn sample from a beech (*Fagus sylvatica*) forest site at Solling in Germany (Schauermann, 1973). Most data fall into the range 20,000-200,000 m^{-2}.

There is a clear thermo-latitudinal gradient of population density with low abundances in most tundra sites (2,000-80,000), high densities in subarctic tundra and temperate coniferous forests (20,000-600,000), similar values in temperate deciduous forests (25,000-250,000) and grasslands (10,000-200,000) and decreased densities in tropical soils (20,000-70,000). Some known exceptions to this pattern include populations densities of 213,000-242,000 ind. m^{-2} recorded from a rainforest in Zaïre and 187,000-520,000 in a South African montane forest (Figure III.36).

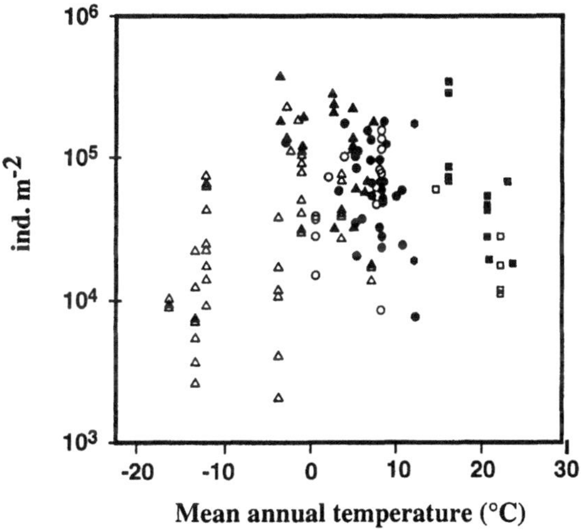

Figure III.36 Relationship between mean annual temperature and acarine population densities across a wide range of ecosytem types (Petersen and Luxton, 1982).

At a regional scale, Acari are generally most abundant in soils with well-developed O horizons with a mor type of humus, low pH and 'mesic' rather than dry or moist water regimes. They are more numerous in wooded than non-wooded areas. Finally, abundance varies in different cultural systems. Generally, both abundance and diversity are lowest in cultivated annual crops and highest in pastures and crops with reduced tillage (see, *e.g.*, Winter *et al.*, 1990; Hansson *et al.*, 1990).

Biomass ranges from 10 to 1500 mg dry weight m^{-2} with most values ranging between 100 and 1000 mg. The mean weight of individuals tends to decrease from tundra sites (10 μg in a number of sites) to the tropics (0.7 μg in savannas and forests of Zaïre).

This is particularly due to an increase of the proportion of Prostigmata whose average weight is often *ca.* 0.10 µg. In the soil of a Swedish coniferous forest site, a high proportion of Prostigmata and juvenile Cryptostigmata Brachythioniidae depressed the average individual weight to a low 0.16 µg. As a result, biomass estimates do not show such a clear pattern of thermo-latitudinal variation as population density. It is still not clear, however, whether these differences reflect existing patterns or simply depend on the efficiency of the separation of the smallest individuals (Garay, pers.com.).

Tundra communities contain populations whose biomasses range from the lowest (100 mg m^{-2} f wt) to the highest values, *i.e.*, 1-1.5 g m^{-2}. Temperate-climate coniferous forests are in the range 400-1000 mg whereas deciduous forests are split between mor and moder soils with biomasses greater than 1000 mg and mull soils with lower biomasses of 100 to 400 mg. In temperate grasslands and in most tropical environments and deserts, biomasses are low, most often less than 130 mg m^{-2}.

4.2.2.2 *Community structure*

Species Richness

The species richness of acarine communities is often very high. More than a hundred species have been identified from an Irish grassland (Whelan, 1986) and 72 in a Swedish grassland (Persson and Lohm, 1977). In temperate forests, species richness may be of similar order and as many as 52 species of Cryptostigmata have been recorded in a mixed wood at Foljuif near Paris (Garay, 1981a) and 90 species in a chestnut (*Castanea* sp.) wood (Anderson, 1975). In some tundra soils, high species richnesses have been recorded, *e.g.*, 37 species of Acari at Point Barrow Alaska (McLean, 1980). In more extreme environments, such as the maritime antarctic island of Signy, however, only one or two species, with outstanding resistance or tolerance to frost were noted (Block, 1983).

In deserts, water-stress also limits the number of species. Nonetheless, 30 species have been recorded from the Chihuahuan desert of New Mexico (Kamill *et al.*, 1985). In sclerophyllous Mediterranean vegetation associations, Asikidis and Stamou (1991) found 28 species of Cryptostigmata mites.

Taxonomic structure

Cryptostigmata are generally the most abundant group in soils but their relative abundance varies greatly within single vegetation types. They represent more than 90 % of Acari in blanket bogs and ferns (Petersen and Luxton, 1982), 17 to 71 % in the tundra soils of Prudhoe Bay (Alaska), 49-75 % in temperate forests (Wallwork, 1967; Huhta *et al.*, 1986; Garay, 1981a), 26 to 45 % in temperate and tropical grasslands (Persson and Lohm, 1977; Athias, 1976), and 16 to 51 % in hot or alpine (Venezuelan paramo) deserts (Petersen and Luxton, 1982; Garay, 1981b; Kamill *et al.*, 1985).

Mesostigmata are the second most important group with relative abundances ranging up to a maximum of 20 %. They are more abundant in wooded than non-wooded ecosystems. The density of Astigmata rarely exceeds 1000 m^{-2} although they may achieve some importance where a deep organic horizon is present. Prostigmata have generally low, but highly variable abundances with a maximum relative value of 55 % in soils of the Negev desert (Steinberg and Wallwork, 1985). However, their densities may often have been underestimated, due to their small size.

Niche partitioning

Communities of Acari tend to form 'loosely-defined species assemblages' (Anderson and Hall, 1977). Despite the apparent lack of a simple association with environmental factors (see *e.g.*, Ghilarov, 1977), a few general rules have been observed (Vannier, 1985):

(i) some clear differences exist in feeding regimes, mainly related to size (Luxton, 1972; Anderson,1977) (Figure III.35) ;

(ii) some populations are distributed in aggregations that have little spatial overlap: in samples of 20 cm^2, containing at least one out of three potentially competing Oribatid species, only 8 % had three and 14-19 % had two (Cancela da Fonseca, 1987). Similarly, Stanton and Tepedino (1977) found that small patches of litter (100 g dry weight) deposited on the surface of temperate and tropical soils had, after three weeks, been colonised by a similar number of species (11 to 14) forming assemblages with equivalent diversities irrespective of the overall α diversity which was greater in the tropical environment. In a *Pinetum-Vaccinietum myrtilli typicum* association, Blaszak *et al.*, (1977) observed a clear difference of species assemblages in patches with the lowest vegetation stratum dominated by either mosses or *Vaccinium*. Topographic effects were also reported;

(iii) in patches of equal size, species diversity is directly related to those of microhabitats and food resources present (Anderson and Hall, 1977) (Figure III.37).

In some cases, associations of species with specific environmental conditions have been identified. However, many species are ubiquitous and generalisation of such findings is difficult (Wallwork, 1967).

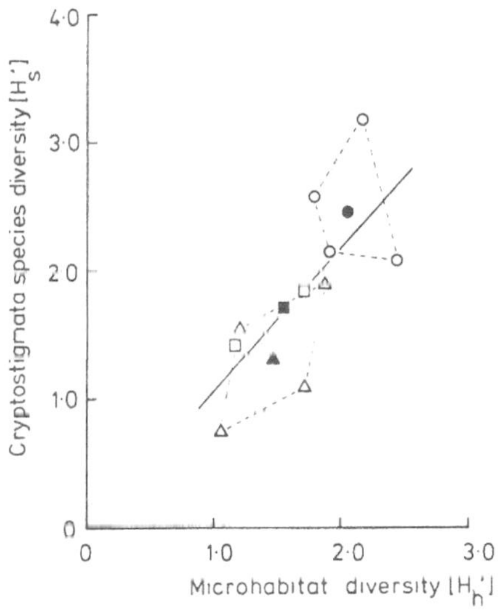

Figure III.37 Relationship between the species diversity of Cryptostigmata Acari and soil microhabitat diversity in four woodland soils with different humus types. Microhabitat diversity is based on the occurrence and abundance of 23 different types of physical (*e.g.*, cavities of a given size) or biological structures (*e.g.*, roots, faeces of a given group of soil invertebrates). △ = L-subhorizon, ○ = F-subhorizon, □ = H-subhorizon ($r = 0.70$, $p < 0.05$). Solid symbols indicate mean values.

Horizontal distribution

At a broad geographical scale, many species appear to be widely distributed (Swift *et al.* 1979). Cryptostigmata such as *Tectocepheus velatus* (Mich.) and *Oppia nova* (Oudms) have a world wide distribution across a wide diversity of habitats. New Zealand has a largely cosmopolitan fauna and only the extreme habitats of Antarctica have a higher degree of specific endemism (Anderson, 1977).

At a mesoscale, the distribution of acarine populations is influenced by (i) a frequent tendency to population aggregation and (ii) the heterogeneity of vegetation and litter inputs. In the moist African savannas of Lamto (Côte d'Ivoire) four groups of Acari had highly aggregated distributions, seven were aggregated and four loosely aggregated. Sixty six % of the individuals (in plots submitted to fire) to 81.8 % (in protected plots) were found in soil under tussocks. In the latter case, however, the proportion varied from 36 to 88 % depending on the group. Five taxonomic groups out of 13 were significantly associated with tussocks whereas two were significantly more numerous in bare soil (Figure III.38).

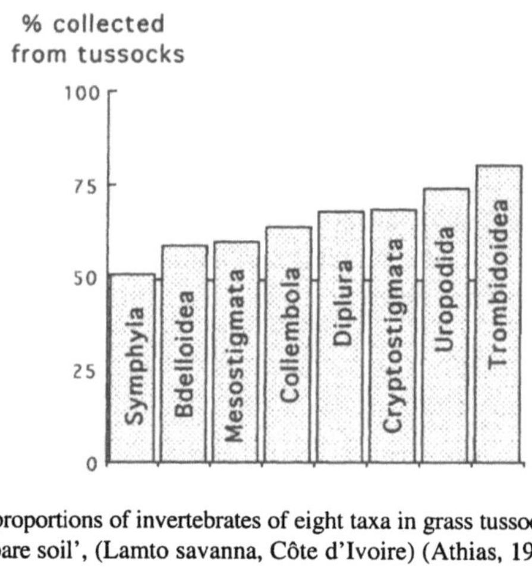

Figure III.38 Relative proportions of invertebrates of eight taxa in grass tussocks compared with those in 'bare soil', (Lamto savanna, Côte d'Ivoire) (Athias, 1976).

Comparable effects of plant distribution and microtopography have been observed in coastal Arctic tundra at Point Barrow (Alaska) (McLean, 1980) with an increasing gradient of density from 'wet' to 'dry' areas. In general, mesoscale variations in temperature, moisture, litter accumulation, or variation ascribed to litter quality (Garay 1981a; Asikidis and Stamou, 1991) often influence the distributions of acarine populations. Nonetheless, a few examples of relatively homogenous distribution patterns in field populations at the scale of a few square metres have been reported (*e.g.*, Cancela da Fonseca *et al.*, 1967).

Vertical distribution

In temperate climate forests, Acari typically inhabit the organic surface layers (L, F and H) and the upper few centimetres of the mineral soil (see, *e.g.*, Anderson, 1971). However, they may also penetrate much deeper into the soil (down to 40 and even 120-180 cm) in tropical and desert environments where holorganic layers are often lacking and the temperature is very high close to the soil surface (Athias, 1976).

Similar overall patterns occur over a variety of situations with large interspecific, seasonal and even daily variations, as observed in leaf litter of North American desert ecosystems (McKay *et al.,* 1987).

In relatively bare soils such as savannas after fire and in cultivated soils, populations are distributed much more deeper than in soils with a surface covering of vegetation and litter (Lagerlöf and Andrén, 1989; Athias, 1974).

Seasonal variation

Acarine communities comprise a large number of species with diverse life cycles. Their generation times average about one year and populations occasionally show clear seasonal abundance patterns (*e.g.*, Asikidis and Stamou, 1991; Lagerlöf and Andrén, 1989). Quite often, no clear pattern of seasonal variation may be detected, probably due to the overwhelming influence of spatial variability which tends to mask temporal variation (see *e.g.*, Athias, 1976 for a discussion of variability in savanna ecosystems).

4.2.3 ENCHYTRAEIDAE

4.2.3.1 *General Biology*

Enchytraeidae are small white-coloured Oligochaeta, 1 to 50 mm long and possessing fasciculate setae. They live in aquatic and, particularly, terrestrial environments. Respiration is cutaneous and these animals require a constantly wet integument for efficient gas exchange. Enchytraeidae have limited ability to move and work the soil and thus they live in the litter and organic horizons of the upper few centimetres of the soil. Some species have been reported to make burrows in the sand and transport material as faeces to a depth of 4 cm. Others move vertically and tunnel through earthworm casts, or conifer needles (Standen, 1984). In agroecosystems in the Netherlands, 21 to 35 % of the enchytraeid population had mineral grains in the gut (Didden, 1990) although they have rather limited consequences for the soil physical structure (Springett *et al.*, 1970). In contrast to earthworms, they have no gizzard, which implies an inefficient comminution and mixing of decomposing material (Bal, 1982).

They are rather homogenous in their anatomy and are divided into 21 genera and *ca.* 600 species (Dash, 1990) (Figure III.39).

Enchytraeidae do not have an adequate suite of enzymes to digest complex polysaccharides (Dash *et al.*, 1981) and may therefore rely on the 'external rumen' type of digestion. As a result, they are principally saprophagous and first order decomposers. They preferentially ingest parenchymatous tissues rather than the veins of decomposing leaves, fungal mycelia and faecal pellets of Collembola or earthworm casts, but do not feed on purely microbial material. They are attracted by plant debris with high N but low tannin contents, low C:N ratios (Latter and Howson, 1978; Dash, 1990; Wolters, 1988) and lignin contents. Consequently, they grow better when ingesting bleached leaves which have undergone some preliminary decomposition by white-rot Basidiomycota and are clearly selective in the plant materials that they ingest. The feeding regime of individual species may vary greatly around this general pattern with some species ingesting more fungal or mineral material than others (Springett and Latter, 1977).

Figure III.39 Morphology of Enchytraeidae and their effects on leaf litter (Albrecht, 1984).

Enchytraeidae are hermaphroditic although reproduction is bisexual in most species. However parthenogenesis is not uncommon and asexual reproduction by fragmentation occurs in a few (but abundant) species. In species with sexual reproduction, adults lay cocoons which are mucilaginous bags containing 1 to 48 eggs, depending on the species (O'Connor, 1967). Some species cover their cocoons with organic and mineral particles to protect them from drought. Estimates of hatching rates are generally high and vary from 19 to 97 %. In British meadows, Enchytraeidae are reported to produce 4 to 10 immatures adult^{-1} year^{-1} with a developmental period of 4 to 12 months (Brockmeyer *et al.*, 1990). Under optimal conditions, total life cycle from cocoon hatching to maturity may range from 60 to 120 days.

Asexual reproduction occurs through the fragmentation of individuals in a few species. In *Enchytraeus fragmentosus,* individuals break into 3 to 14 fragments of about five segments, each of which develops into a complete worm in 10 days (Dash, 1990). In *Cognettia sphagnetorum*, a dominant species in acid soils of northern Europe, worms start fragmenting when they have more than 42 segments. They divide into 2 or 3 fragments and the time taken to regenerate varies from 8 to 26 days. Minimum regeneration times occur for head fragments and are proportional to temperature (15 °C rather than 10 °C or 5 °C) (Standen, 1984). In natural populations, fragmentation occurs throughout the entire year with obvious maxima in spring or early summer, especially when relatively dry periods occur (Standen, 1973; Lundkvist, 1981), or in the middle of winter (Ellenberg *et al.*, 1986).

Drought is a major limiting factor for the Enchytraeidae since they cannot survive water potentials lower than -1.0 MPa (pF 4) (Abrahamsen, 1971; Bell, 1959). Populations may survive through some combination of colonising favourable microsites and laying drought-resistant cocoons. Fragmentation permits rapid recolonisation and conserves energy and nutrients (Standen, 1973). Resistance to low temperature appears rather limited and the temperature range over which reproduction occurs in temperate species ranges from 1.5 ° to 25.5 °C for several species, with optimal values between 5 ° and 15 °C (Standen, 1973). Nonetheless, Enchytraeidae have been found in Arctic tundra soils (Ryan, 1977; McLean, 1980) and in the high altitude South American paramo (Sturm, 1983)

Ecological categories have not been defined for Enchytraeidae probably because of the relatively homogeneous way in which they exploit soil and litter resources. Most Enchytraeidae seem to be hemiedaphic or euedaphic and thus relate to the polyhumic category defined for earthworms (see Chapter III.4.3.1.2).

Density and biomass of communities

The abundance of Enchytraeidae is clearly determined by soil temperatures and moisture regimes. While they are virtually absent from 'extreme' tundra environments in Signy Island (maritime Antarctica), they have mean densities of 46,000 ind. m^{-2} (from 11,400 in 'raised polygons' to 93,600 in 'polygon troughs') in the milder tundra environment at Point Barrow (Alaska) (Dash, 1990). The highest mean density, 200,000 ind. m^{-2} (82,000-289,000 ind. m^{-2}) was observed at Moor House (England) in a *Juncus* moor (Peachey, 1963; McLean, 1980).

Even higher densities of 90,000 to 150,000 ind. m^{-2} were found in moist temperate coniferous or deciduous forests, *e.g.*, Douglas fir (*Pseudotsuga menziesii*) plantations in Wales (O'Connor, 1967), beech (*Fagus sylvatica*) forest at Solling (Germany) (Schauermann, 1977) and in Japanese subalpine forests (Kitazawa, 1977). However, low temperatures and moisture constraints strongly limit populations of Enchytraeidae. Much lower densities have been recorded from forests in Scandinavia, USSR and Alaska: 27,500 to 90,000 ind. m^{-2} in a range of Finnish forests (Huhta *et al.*, 1986), 10,000 to 25,000 ind. m^{-2} in Swedish pine forests (Lundkvist, 1981) and Hungarian deciduous forests (Dozsa-Farkas, 1987) and only 3,800 to 6,800 ind. m^{-2} in coniferous and deciduous woodlands in northern Sweden (Lohm *et al.*, 1977). In temperate climate grasslands, relatively low densities have been recorded: for example, 18,000-35,000 ind. m^{-2} at Spiboke (Sweden) (Persson and Lohm, 1977), 10,000-25,000 ind. m^{-2} in a grassland on limestone at Moor House (England) and 5,000-10,000 ind. m^{-2} in grass and lucerne leys at Kjettslinge (Sweden) (Axelsson *et al.*, 1984). In tropical soils, Enchytraeidae are either absent or limited to a few hundred individuals per square metre in grasslands (Athias *et al.*, 1974) to a few thousand in forests (Chiba *et al.*, 1975).

The abundance of Enchytraeidae is also related to the organic matter regime with higher concentrations in mor than mull-type humus. These relationships, however, may be minor in relation to the major temperature and moisture effects.

Most biomass estimates range from 1.3 to 13 g f wt m^{-2} (Petersen and Luxton, 1982). The ratio of fresh to dry weight being 6.7 on average (Persson and Lohm, 1977). A maximum value of 53 g f wt was recorded from a *Juncus* moor at Moor House in England (Cragg, 1961). The highest biomasses (>1 g dry wt, *i.e., ca.* 6.6 g f wt on average) have generally been recorded in 'mild' tundra soils where individuals are generally larger than in temperate and tropical soils. In temperate soils, biomass generally ranges between 1 and 10 g f wt m^{-2} (Petersen and Luxton, 1982) with exceptionally high values at Solling, in a German beech forest on a moder soil (Schauerman, 1977).

4.2.3.2 *Community Structure*

Species richness

Enchytraeid communities are generally not highly speciose. A maximum species richness of 22 was recorded from meadows in the North of England (Standen, 1984), 21 in mull-type soils in Norway (Abrahamsen, 1971) and 20 in hornbeam (*Carpinus* sp.) forest in Hungary (Dozsa-Farkas, 1987). However, 17 species were found in an apple orchard in Bohemia (Chalupsky and Leps, 1985) and 10 in northern deciduous woodlands in Sweden (Axelsson *et al.*, 1984). Nonetheless, most communities include 3 to 7 species (Persson and Lohm, 1977; Petersen and Luxton, 1982; Lagerlöf, 1987) and in some Scandinavian coniferous forests, *Cognettia sphagnetorum*, an asexually reproducing species may represent 90 to 100 % of the community (Lundkvist, 1981).

Spatial Distribution

Most Enchytraeid species have quite wide geographical distributions and some may colonise a wide range of different habitats (*e.g.*, water courses, dry sandy soil or raw humus) (Dash, 1990).

At a mesoscale level, Enchytraeidae have strongly-aggregated horizontal distributions in all types of ecosystems (*e.g.*, O'Connor, 1967; Albrecht, 1984; Chalupsky and Leps, 1985; Römbke, 1991). They form multispecies aggregates, 30 to 50 cm in diameter, very similar from one site to another (Albrecht, 1984).

These aggregations may be caused by (i) - restricted dispersal of individuals after hatching from cocoons or the fragmentation of asexual individuals; (ii) - alternatively, individuals may favour microsites with the most attractive abiotic conditions and highest food availability. The latter effect may well predominate as adults appear slightly more aggregated than juveniles and asexually reproducing species show similar patterns; and (iii) - gregariousness that may be interpreted as a semi-social behaviour to increase efficiency of the 'external-rumen' type of digestion (see Chapter III.4.4.1.2).

Most Enchytraeidae are concentrated in the upper 6 to 10 cm with some differences among species (Healy, 1980). Most species are epigeic (hemi-rather than epiedaphic) but some polyhumic endogeics (*i.e.*, euedaphic) populations may descend down to 20-40 cm in the soil. (see *e.g.*, Persson and Lohm, 1977; Lagerlöf, 1987). Seasonal changes in depth distribution may occur as Enchytraeidae descend deeper into the soil to escape drought or frost at the soil surface.

Temporal variation

Enchytraeid populations generally show distinct seasonal changes with highest densities occurring during winter (Figure III.40).

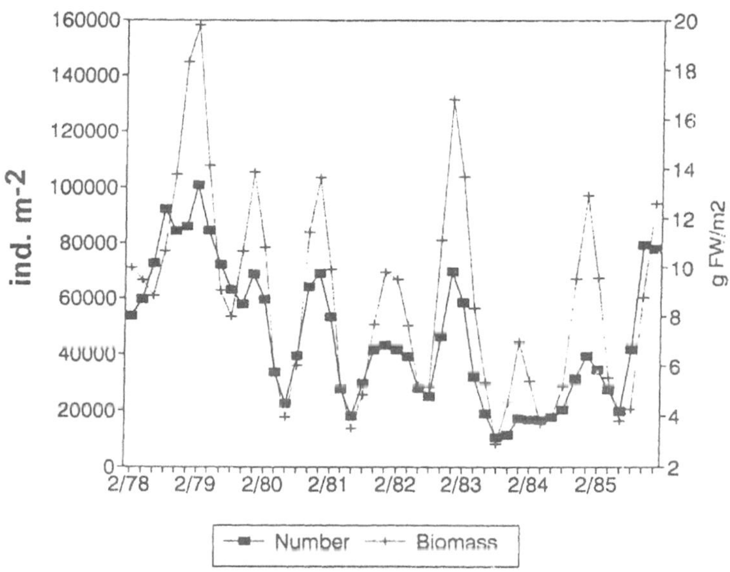

Figure III.40 Seasonal fluctuations in abundance (ind m^{-2}) and biomass (g f wt m^{-2}) of Enchytraeidae in a Beech (*Fagus sylvatica)* forest in Germany (Black Forest) (Römbke, 1991).

4.2.4 OTHER SOIL MESOFAUNA

Acari and Collembola are generally the major taxa in mesofauna communities. Nonetheless, four minor groups, Protura, Diplura, Pauropoda and Symphyla may be locally important because of their relative abundances and the unique roles they play in soils.

Protura, Diplura

Protura and Diplura are Apterygota Insecta which are found consistently in soils, although at low densities.

Protura are minute insects characterised by the absence of wings, antennae and eyes which live in humid organic environments. They are panphytophages with highly aggregated patterns of distribution. The highest density, 28,000-50,900 ind. m^{-2} was recorded in a spruce mor with a moss cover in Norway. Figures of a few hundred to few thousands of individuals m^{-2} are commonly cited for a variety of ecosystems namely, temperate forests (Nosek, 1973), temperate grasslands and crops (Persson and Lohm, 1977; Axelsson *et al.*, 1984; Lagerlöf, 1987) and African moist savannas (Athias, 1974). As these insects are very small (0.4 to 8 μg individual dry weight) biomass does not exceed a few mg m^{-2} in most environments.

Diplura are another consistent though minor component of the soil mesofauna. Their lengths range from a few tenths to several millimetres. They may be panphytophages (*Campodeidae*) or predators of other microarthropods (Fox, 1957). They are concentrated in the upper part of the profile and have highly aggregated distributions. Density estimates range from *nil* to several thousands with highest values reported from such different habitats as a *Liriodendron* forest in Tennessee (790 ind. m^{-2}) (McBrayer *et al.*, 1977) or the moist African savannas at Lamto in the Côte d'Ivoire (700-1,000 ind. m^{-2}) (Athias, 1974). Maximum biomass estimates do not exceed a few mg dw m^{-2}.

Micromyriapoda

Three diverse groups of the Myriapoda may be classified as mesofauna *i.e.*, Pauropoda, Symphyla and Polyxenida. Pauropoda are small (less than 1 mm), blind and pale panphytophagous invertebrates which live in decomposing litter and organic horizons and feed on fungi and decomposing leaves (McBrayer *et al.*, 1977). They have patchy distributions and their densities generally range from a few hundred to 2000 ind. m^{-2}; biomasses are limited to a few mg dw in both temperate and tropical soils. The largest populations recorded are of 6,200 ind. m^{-2} in a West African pastures (Belfield, 1956) and 17,900 ind. m^{-2} in a Californian *Pinus ponderosa* forest (Price, 1973). They do not seem to occur in extreme tundra and desert environments.

Symphyla are unique in that their small size and slender body shape allows them to move readily through the larger soil pores and colonise soils to one metre in depth (Belfield, 1956). They are polyhumic endogeics (euedaphic), and feed mainly on dead and living roots and on micro-organisms. Some of them may become serious pests of a wide range of plants (Anglade, 1967). Densities are generally a few hundreds ind. m^{-2} although they may exceptionally reach maximum values of 1.2 to $2.4.10^3$ ind. m^{-2} in cultivated temperate soils and 10,800 ind. m^{-2} in West African pastures (Belfield, 1956).

Biomass is normally only a few mg dry weight and is unlikely to ever exceed 50 mg m^{-2} (Petersen and Luxton, 1982).

4.3 Macrofauna

In terms of their abundances and their biological and pedogenetic roles, earthworms, termites and ants are the most important macrofaunal components of soils. Indeed, the importance of their activities has caused them to be designated 'ecosystem engineers' (Stork and Eggleton, 1992; Jones *et al.*, 1997). Macroarthropods and Mollusca are constant inhabitants of litter and, to a lesser extent, of soils, but they have generally more specific ecological roles.

4.3.1 EARTHWORMS

4.3.1.1 *General Biology*

Earthworms belong to the class Oligochaeta (Annelida: Clitellata) which started colonising terrestrial environments 600 millions years ago. Since that time, they have become the predominant component of the soil fauna in all but the dry and cold environments. From their aquatic origins, they have retained the primitive cutaneous respiration and a narrow dependence on hygrophilous soil moisture regimes. They comprise 20 families, 693 genera and an estimated 6000 species of which 3700 have been so far described (Reynolds and Cook, 1976; Fragoso *et al.*, 1997). Most species are relatively restricted in their distributions although a few peregrine species, generally with parthenogenetic reproduction, have world-wide distributions and a clear preference for soils affected by man (*e.g.*, Omodeo, 1952; Gates, 1970; Ljungström, 1972; Lee, 1985; Fragoso and Lavelle, 1992; Fragoso *et al.*, 1997).

Earthworms are segmented coelomate Metazoa in which the coelom functions as an hydrostatic skeleton. Their circulatory system is closed and they have a complete digestive apparatus (Avel, 1959; Edwards and Bohlen, 1996). The body may be roughly divided into two parts: an anterior part with segments containing cephalic ganglions, reproductive organs, gizzards, calciferous glands and hearts, and a posterior part rearward of the hindgut comprising a series of rather similar segments.

In length, earthworms generally range from a few centimetres to 2-3 m with most species falling within the range of 5 to 15 cm. Size varies considerably within single species populations and the largest adults may be more than 100 times those of newly hatched individuals.

Most earthworm species are hermaphroditic and cross-fertilisation is predominant. However, parthenogenesis is frequent in peregrine species which represent *ca.* 3 % of the total number of species (Lee, 1985). Bouché (pers. comm.), however, indicates that the lack of spermathecae is not an absolute indication of pathenogenesis since a significant number of species produce spermatophores. The reproductive apparatus is relatively complex due to the presence of reproductive organs of both sexes and delayed external fertilisation. In the family Lumbricidae, the reproductive apparatus has become even more complex due to rearward displacement of the gizzard and clitellum to allow the development of a strong anterior musculature (Bouché, 1972). This evolutionary advance improved the ability of the Lumbricidae to dig deep burrows and to drag leaves into them. Nonetheless, during mating, spermatozoa are transported from the male pore to the spermathecae of the partner via an external tegumental fold (Figure III.41).

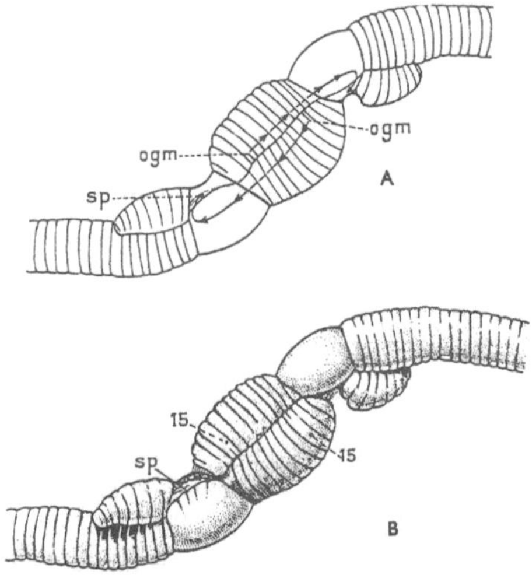

Figure III.41 Copulation in *Eisenia fetida* (Lumbricidae) with external transport of spermatozoa. A: arrows indicate the movement of spermatozoa from the genital pores of male (ogm) to the spermathecae (sp). B: segment 15 bears male genital pores (Avel, 1959).

Mating occurs when male apparatus is fully developed. Since the female organs generally mature some days to weeks later, spermatozoa are stored in spermathecae. Eggs are laid in a cocoon excreted by the clitellum, an annular or saddle-shaped conspicuous protuberance situated in the anterior part of the body. Cocoons vary in size from 1 to 25 mm, depending on the size of adults. In each cocoon, up to 10-12, but most frequently one or two embryos develop (Lavelle, 1981). Cocoons possess a thick external chitinous envelope with a gelatinous filling which provides food for the developing embryo(s) (Figure III.42). They have a remarkable ability to resist both drought and infection.

Figure III.42 Cocoon of the large African Megascolecidae (*Dichogaster terrae-nigrae* Omodeo and Vaillaud) (size: 1.5 cm).

Earthworms are semi-aquatic animals which extract water continuously from their environment and must maintain their cuticle in a moist state to facilitate gas exchange. They also add considerable amounts of water to the soil and litter they ingest to facilitate intestinal transit and digestive processes (de Puytorac, 1954; Barois and Lavelle, 1986; Daniel and Anderson, 1992; Trigo and Lavelle, 1993). Only part of this water is reabsorbed and the fresh casts and the cutaneous mucus that they excrete are moister than the surrounding soil. Consequently, soil moisture status is a major limitation to earthworm activities and distribution since they are unable to withstand prolonged periods of water potentials less than -0.10 MPa (pF 3) (Lavelle, 1971). To survive extended dry periods, certain species have developed such adaptations as quiescence or even diapause (Saussey, 1966; Bouché, 1984); worms empty their gut and coil up into a tight ball inside a chamber which has been previously coated with fine-textured casts. They may resist drought better, but the reduction in mortality resulting from this behaviour does not appear to be greater than 50 % in the case of the tropical endogeic species *Millsonia anomala* (Martin, 1991). Other species, such as the pantropical endogeic *Polypheretima elongata,* may have no resting stage and migrate down to depths of 1 to 2 metres to escape dry conditions in the upper soil horizons. Where active populations are periodically extinguished through water-logging (*e.g.*, in flooded tropical forests), or because of severe drought or frost, the more resistant cocoons may survive to form new populations when favourable conditions return (Bouché, 1977).

Fecundity is highly variable among species and populations; maximum values of 5.5 cocoons per adult per week have been reported for *Eisenia fetida*, a common species in the farm-yard manure and compost heaps of temperate regions (Hartenstein *et al.*, 1979; Venter and Reinecke, 1988). Minimum values of one or less cocoons per year are common in deep-dwelling tropical geophagous species (Lavelle, 1978). Tropical species generally have low fecundities (1 to 15 cocoons per year) with the exception of a few pantropical species with a parthenogenetic mode of reproduction, *e.g.*, *Pontoscolex corethrurus* and *Polypheretima elongata* which may produce up to 80-100 cocoons per adult per year (Lavelle, 1981; Senapati, 1980; Lavelle *et al.*, 1987). In temperate areas, most values calculated for Lumbricidae are in the range of 20 to 60, with maximum values for epigeic species and minimum values of *ca.* 10 for endogeic and large anecic species. Substantial differences also exist between generation times which vary from a minimum of 2-3 months in *Eisenia fetida* and the pantropical geophagous species *Pontoscolex corethrurus* to 20-24 months in the deep-dwelling geophagous tropical species. Such differences among species are a major consequence of the diverse adaptive strategies found among earthworm species.

Digestion in the earthworm gut is mediated by a mixture of enzymes produced by the gut wall and enzymes produced by the ingested soil microflora (Barois and Lavelle, 1986; Loquet and Vinceslas, 1987; Zhang *et al.*, 1993, see Chapter IV.4). Assimilation rates calculated as the difference in energy values between ingested food and excreta may be as high as 30 and 75 % respectively for the epigeic species *Lumbricus rubellus* fed alder (*Alnus* sp.) leaves and one year old larch (*Larix*) litter (Dickschen and Topp, 1987). Geophagous species have much lower assimilation rates of *e.g.*, 9 % or less in the tropical mesohumic species *Millsonia anomala* (Omodeo) (Lavelle, 1978) and 1 % in the temperate climate Lumbricidae *Allolobophora rosea* (Bolton and Phillipson, 1976).

Earthworms seem to move little; in conditions where new polder soils were opened to colonisation, dispersal distances were 4.5 m and 9 m $year^{-1}$, respectively, for *Apporectodea caliginosa* and *Lumbricus terrestris* (Hoogerkamp *et al.*, 1983). Similar figures of *ca.* 10 m $year^{-1}$ have been reported for *Dendrobaena rubida tenuis* (Eisen) invading the subantarctic island La Possession (Frenot, 1985) or marked earthworms introduced in a temperate pasture of France (Mazaud and Bouché, 1980). Nonetheless, passive dispersal in water currents, transportation by man and the active dispersal which occurs when earthworms come to the surface of water-logged soils eventually allow populations to colonise most suitable environments. This process may be slow and soils subjected to such heavy disturbance as cultivation and cropping, are rarely colonised by adapted species before their carrying capacity for these populations has dropped irremediably (Lavelle and Pashanasi, 1989; Lavelle *et al.*, 1994b).

4.3.1.2 *Ecological categories*

As large inhabitants of the litter-soil environment, earthworms face three major constraints including the need to: (i) live on relatively poor quality feeding resources; (ii) survive occasionally adverse microclimatic conditions; and (iii) to be able to move through a sometimes compact environment.

In the course of evolution, species have adapted to these constraints through the development of particular morphological, behavioural and physiological characteristics. A wide range of adaptive strategies have developed which are integrated as three main modal ecological types (Bouché, 1977; Lavelle, 1981; Lee, 1985) (Figure III.43):

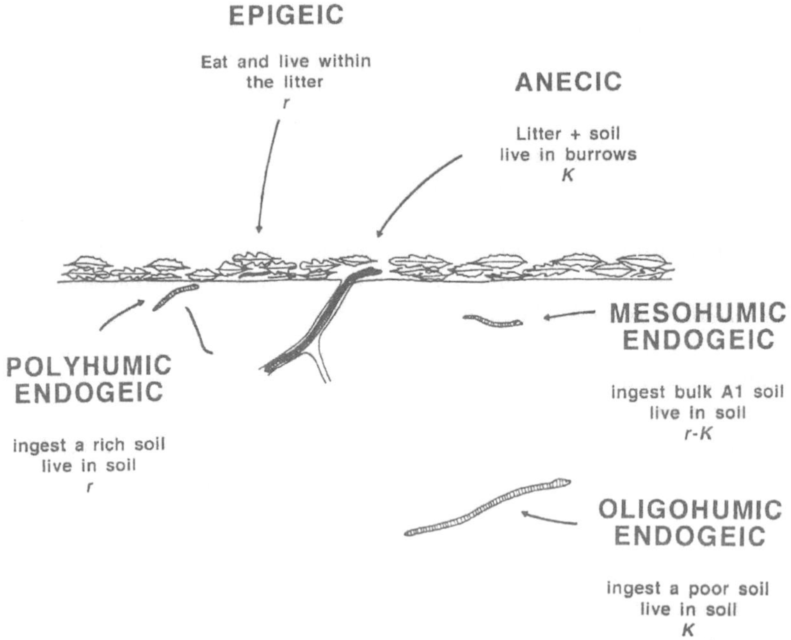

Figure III.43 Ecological categories of earthworms (after Bouché, 1977 and Lavelle, 1981); *r* and *K* refer to demographic strategies.

(i) epigeics live within the litter layers, a relatively harsh environment, where they are subject to occasional drought, extreme temperatures and high predator densities. They are small and homochromic worms, pigmented green or reddish depending whether they inhabit grassland or forest. Epigeics balance a high mortality by using a high-quality food (leaf litter) which permits rapid growth rates and a high fecundity. They are typical *r*-selected (prolific) species in the sense of Pianka (1970);
(ii) anecics feed on surface litter that they mix with soil but pass most of their time in subvertical subterranean galleries created within the soil. They are large worms with a dark antero-dorsal pigmentation and a strong anterior digging musculature. Their demographic profile is of the *K* type, with long lives, relatively slow growth rates but low mortalities;
(iii) endogeics are unpigmented geophagous worms that live and feed within the soil, a better buffered and more predictable environment than the litter. However, food resources, *e.g.*, decomposing roots and soil organic matter are generally of lower quality than leaf-litter and are concentrated within the upper ten centimetres of soil and around plant roots. Endogeics have developed different ways to use these resources (Lavelle, 1981); polyhumics exploit concentrations of organic matter by, *e.g.*, being small and selectively ingesting organic particles, especially in the rhizosphere, or feeding at the soil-litter interface; mesohumics are medium-sized worms that ingest the soil of the upper 10-15 cm without selecting organic particles. Finally, oligohumics are large, slow-moving worms that live deep in the solum in highly predictable and stable environments with very poor food resources. The demographic profiles of endogeic species change from *r* to *K* or *A* (*i.e.*, adaptation to adverse conditions, *sensu* Greenslade, 1983) the deeper they live in the soil or become larger in size (Lavelle, 1981).

As a result of this wide range of adaptations, earthworms have highly diverse functions in soils. Epigeics are efficient compost-makers but have no impact on the soil structure. Anecics build dense semipermanent gallery networks, deposit casts at the soil surface and translocate large amounts of leaf-litter into the soil. Finally, endogeics which do not usually form permanent galleries, are major agents of soil aggregation (at least in tropical soils where they are predominant) and soil organic matter stabilisation; they may deposit large amounts of casts at the soil surface, as a response to soil compaction.

4.3.1.3 *Density and biomass of communities*

Earthworms comprise the major animal biomass in the wide range of ecosystems that experience less than a few consecutive months of drought or frost in an average year. Density is often in the range of 100 to 500 individuals m^{-2} and may be as high as the 2000 ind. m^{-2} reported from such environments as temperate pastures in New Zealand or irrigated orchards in temperate Australia (Lee, 1985). Live biomass commonly ranges from 30 to 100 g m^{-2}. Bouché (pers. communication) using a combination of formalin extraction and wet sieving of soil evaluated at 110 g m^{-2} the average earthworm biomass in 55 sites representative of French soils. Maximum biomasses of 200 to 400 g have been recorded in fertilized pastures in temperate regions (see *e.g.*, Cotton and Curry, 1980) and tropical pastures (Barois *et al.*, 1988).

Markedly different patterns of abundance occur at decreasing spatial scales (see reviews in Edwards and Bohlen, 1996; Lavelle, 1983c; Lee, 1985):

(i) at a worldwide geographical scale there is a clear thermo-latitudinal gradient, as mean population density in natural environments tends to increase from a few tens on average in cold temperate areas to maximum values of several hundred in the tropics. In contrast, biomass increases from cold to mild temperate environments and then decreases again towards tropical latitudes (Figure III.44);

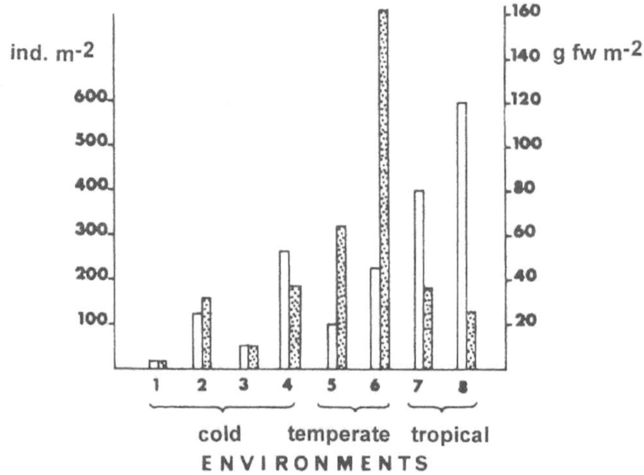

Figure III.44 Variation in mean abundance unfilled bars and biomass stippled bars of earthworm communities along a thermo-latitudinal gradient. Values are means for eight clusters isolated from a Principal Components Analysis of 52 communities from different climatic zones (after Lavelle, 1983c).

(ii) at a regional (landscape) scale, soil and vegetation regimes greatly influence earthworm abundance (Stöckli, 1928). Grasslands tend to have much larger populations than forests. This is particularly in the case of temperate or tropical pastures colonised either by exotic peregrine species, or by adapted local species. These may have biomasses of 1 t ha^{-1} or more whereas an adjacent forest may only have approximately half as much (Lavelle and Pashanasi, 1989). Large biomasses may also be associated with a high soil nutrient status as, for example, in tropical rainforest communities (Fragoso and Lavelle, 1992).

(iii) at a local scale, variation in soils, land-use and cultivation techniques greatly influence earthworm populations. Cultivation and the application of nematicides and fungicides generally depress earthworm abundances. In contrast, most herbicides have no significantly deleterious effects and cattle grazing in pastures has positive effects (Edwards and Lofty, 1982; Haines and Huren, 1990; House and Parmelee, 1985; Lavelle and Pashanasi, 1989; Lee, 1985; McKay and Kladivko, 1985; Parmelee *et al.*, 1990; Yule *et al.*, 1991).

4.3.1.4 *Community structure*

Species richness and diversity

Earthworm communities rarely comprise more than 8-10 species and there are no marked changes in diversity along latitudinal gradients. Such a low species richness is considered to result from the large functional plasticity that occurs within populations. The ecological roles of individuals within particular populations may be quite diverse, due to the important size differences between young and adult worms, and the adaptability of populations to variable environments (Table III.13). During the dry season in the savannas of Côte d'Ivoire, populations of the mesohumic endogeic species *Millsonia anomala* are dominated by young individuals that are better adapted than adults to the hot and dry conditions prevailing; during the rainy season, the environment becomes progressively more suitable for adults which are favoured by cooler temperatures and a better water supply (Lavelle, 1978). Adults live deeper in the soil than the young, ingest less soil per unit of fresh weight and have their optimum activity in moister soils.

Table III.13 Differences in the ecological characteristics of young and adult *Millsonia anomala* Omodeo from a humid African savanna environment (Lavelle, 1978).

		YOUNG	ADULT
WEIGHT	(g fresh wt)	0.10	4.0
AVERAGE DEPTH	(cm)	7.3	11.6
MAXIMUM INGESTION RATE	(g soil g^{-1} f wt d^{-1})	34	10
OPTIMUM TEMPERATURE	(°C)	29	26
OPTIMUM MATRIC POTENTIAL	(pF)MPa	2.5	1.5-2
		-0.035	*ca.* -0.005

At a geographical scale, ß diversity increases from the coldest to the warmest climates. This appears to result from two processes:

(i) the effect of past glaciation. This largely destroyed the fauna of the northern hemisphere above latitudes of 40-50°N (Michaelsen, 1903); endemic species were destroyed and only a few species with high capacities for colonization re-invaded the area following the retreat of the glaciers;

(ii) the exploitation of marginal niches by earthworms in the humid tropics. The considerable radiation that has taken place among the earthworms of the humid tropics has led to the development of highly diverse adaptations and the selection of such peculiar species as specialist arboreals, deep-dwelling oligohumic geophages and even carnivorous species, as in the African genus *Agastrodrilus* (Lavelle, 1983b; Fragoso and Lavelle, 1987). This may be due to: (a) the greater efficiency of the mutualistic system of digestion at high temperatures, which permits earthworms to live on particularly poor food resources, and (b) the longer time during which radiation, adaptation and selection have operated in the stable tropical environment.

Functional diversity (*i.e.*, the diversity of functions performed by the species of a community) increases notably with temperature, unlike species diversity as measured by standard mathematical indexes. This is clearly indicated by a lower niche overlap due to a broader range of sizes among the species, more diverse patterns of vertical distribution and the exploitation of a broader range of food-resources (Lavelle, 1983c; Fragoso and Lavelle, 1987). This increase in functional diversity with temperature is paralleled by a progressive change in the relative importance of individual ecological categories. Epigeics are dominant in cold environments, anecics in temperate areas and mesohumic endogeic species in humid tropical environments. Oligohumics are restricted to the warmest soils of subequatorial humid savannas (Figure III.45) (see also Chapter IV.4).

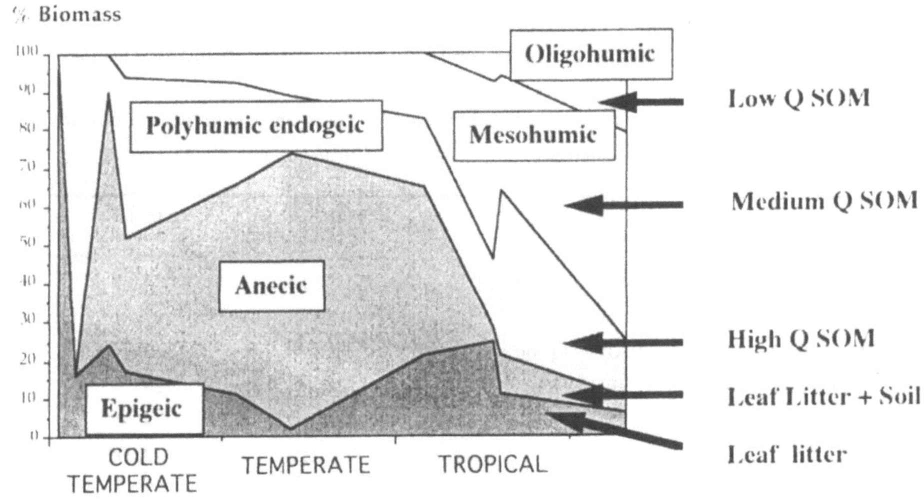

Figure III.45 Percentages of biomass in different ecological categories of earthworm communities along a thermo-latitudinal gradient (modified from Lavelle, 1983c). Arrows to the right indicate the main food source used.

Spatial distribution

At the scale of an hectare, earthworm populations may be concentrated to varying degrees within specific areas although intense aggregation is limited to epigeic populations exploiting resources with heterogenous distribution patterns such as cow pats. Differences in population densities have also been correlated with heterogeneity in such soil properties as texture and microtopography (Fragoso and Lavelle, 1987; Phillipson *et al.*, 1976). Nonetheless, the variability of demographic processes may also determine patchy distributions in metapopulations, irrespective of the heterogeneity of the soil. Rossi and Lavelle (1998) found that all earthworm species in the Lamto (Côte d'Ivoire) savannas had aggregated distributions, and that aggregation became increasingly intense as species became smaller and had higher annual fecundities. Simulations of the population dynamics of one of these species, the mesohumic species *Millsonia anomala*, have shown that natural populations are actually composed of subpopulations with distinct dynamics; these subpopulations form patches 20-30 m in diameter which appear and disappear in the course of a few years (Martin, 1991). The patchy distributions of earthworms and the variability of their demographic compositions observed within savanna lands tend to support the hypothesis of a relative independence of distribution from soil properties (Rossi, 1998). Similar patterns, though much more obvious were observed in populations of *Polypheretima elongata* in a pasture of Martinique (French West Indies): in a 25x65 m plot, populations were distributed in patches of 20-40 m in diameter; one patch mainly comprised young individuals and cocoons, while the other two were dominated by adults. In this plot, the presence of earthworm patches seemed to be independent of the distributions of such edaphic factors as soil depth and soil organic matter content (Rossi *et al.*, 1997) (Figure III.46).

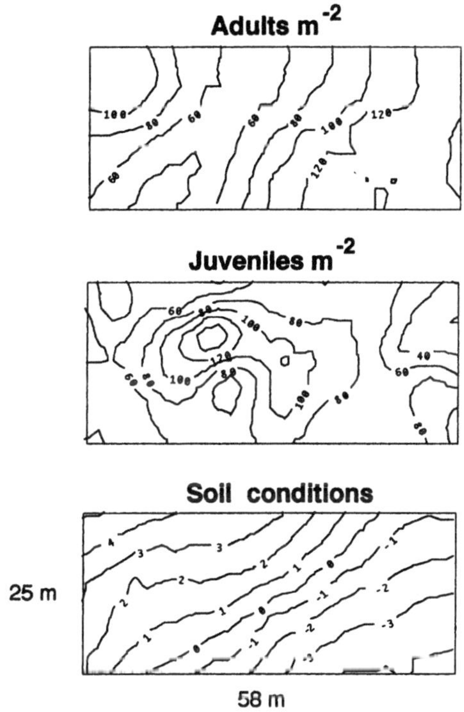

Figure III.46 Horizontal distribution of *Polypheritima elongata* and soil conditions in a tropical pasture in Martinique (French West Indies) (Rossi *et al.*, 1997). (A) Adult population (ind m^{-2}); (B) Juveniles (ind m^{-2}); (C) Soil conditions summarised in values of the first principal component of a multivariate analysis of soil parameters at every sampling point. Maps are derived from the geostatistical treatment of a random sampling.

Earthworm species stratify themselves vertically in the soil depending on their ecological strategies. Epigeics are concentrated in the litter layers, endogeics may have relatively superficial (polyhumics) or deep (in oligohumics) patterns of distribution (Gerard, 1963; Lavelle, 1978; Fragoso and Lavelle, 1987) (Figure III.47) whereas anecics may occur at diverse depths depending on thermal and hydrological gradients (Bouché, 1984).

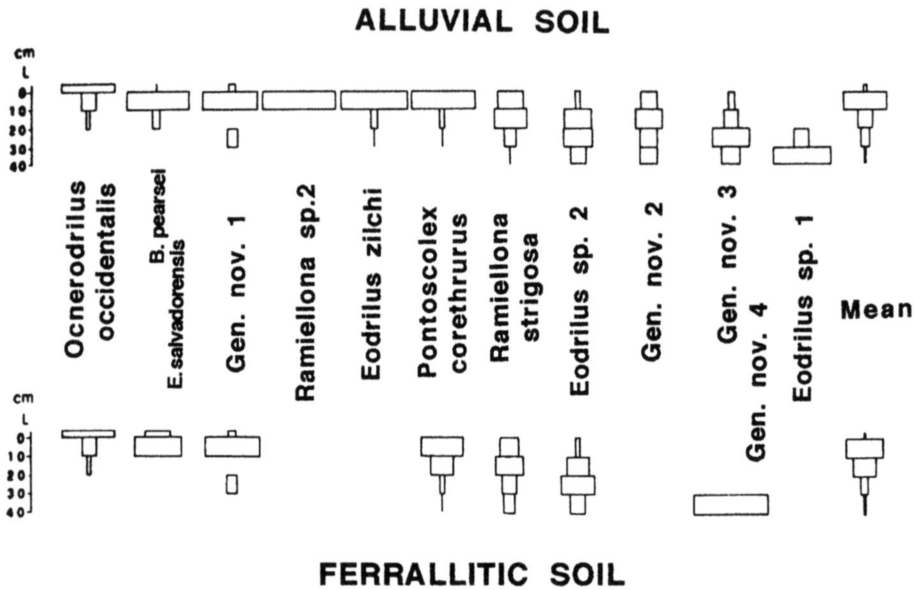

Figure III.47 Depth distributions of earthworm species in a tropical rainforest at Chajul (Mexico) (Fragoso and Lavelle, 1987).

Seasonal variation

Most earthworm populations have clear seasonal patterns of activity or breeding. Such periods generally reflect changes in any one of the major micro-environmental parameters that determine function, *i.e.,* moisture, temperature and food availability. As a result, activities among coexisting species from different ecological categories are not necessarily synchronous (Phillipson *et al.*, 1976). The type of drought resistance behaviour (*i.e.*, escape towards deeper horizons with high moisture content, quiescence or true diapause) also affects seasonal activity. In populations of *Martiodrilus carimaguensis,* a Glossoscolecid species from Colombian tropical savannas, juveniles enter into diapause early, in June, whereas adults remain active until October, two or three weeks before the onset of the dry season (Jimenez *et al.*, 1998b).

4.3.2 TERMITES

4.3.2.1 *Distribution and general biology*

The termites (Order Isoptera) are a group of hemimetabolous eusocial insects allied to the cockroaches (Order Blattodea). They occur between the approximate latitudes of 30-51°N and 40-45°S (Wood, 1979) and are represented in all biogeographic regions (Pearce and Waite, 1994) (Table III.14). The order comprises more than 2,400 currently-described species in 271 genera divided among seven families. Many species still remain to be described, particularly in the tropical and sub-tropical regions (Watson and Gay, 1991; Martius, 1994; Eggleton *et al.*, 1996). Peregrine termite species, with worldwide distributions do not exist.

Table III.14 Distributions of the families and subfamilies of the extant Isoptera with the number of known genera and the biogeographic regions in which they occur (Pearce and Waite, 1994).

Taxon	**Number of genera**	**Biogeographic Regions**							
		Palaearctic	Afro-tropical	Malagasy	Indo-Malayan	Papuan	Australian	Nearctic	Neotropical
1. **Mastotermitidae**									
(1 extant species, *Mastotermes darwiniensis*)	1						x		
2. **Kalotermitidae** (dry wood feeders)	21	x	x	x	x	x	x	x	x
3. **Hodotermitidae (harvesters)**	3	x	x		x				
4. **Termopsidae**	5		x		x		x	x	x
Termopsinae (damp wood feeders)	3				x			x	
Stolotermitinae	1		x				x		
Porotermitinae	1		x				x		x
5. **Rhinotermitidae**	15	x	x	x	x	x	x	x	x
Psammotermitinae	2	x	x	x	x				x
Heterotermitinae	3	x	x	x	x	x	x	x	x
Stylotermitinae	1				x				
Coptotermitinae	1	x	x	x	x	x	x	x	x
Termitogetoninae	1				x				
Rhinotermitinae	6		x		x	x	x		x
Prorhinotermitinae	1			x	x	x	x		x

Table III.14 (Cont.).

Taxon	Number of genera	Biogeographic Regions							
		Palaearctic	Afro-tropical	Malagasy	Indo-Malayan	Papuan	Australian	Nearctic	Neotropical
6. **Serritermitidae**									
(1 extant species, *Serritermes serrifer*)	1								x
7. **Termitidae**									
(Higher termites)	225	x	x	x	x	x	x	x	x
Apicotermitinae	45		x		x		x	x	x
Termitinae	86	x	x	x	x	x	x	x	x
Macrotermitinae	13		x	x	x				
Nasutitermitinae	81		x	x	x	x	x	x	x

In the evolutionary sense, the largest and most advanced family is the Termitidae. Members of this family are known as the 'higher' termites while those of the remaining six families are known as the 'lower' termites. A true worker caste does not occur in the lower termites and the digestive system also differs between the two groups. The hind gut of most higher termites is longer, more differentiated in structure and comprises a larger proportion of total gut length than it does in the relatively simple gut typical of the lower and the fungus-cultivating termites (subfamily Macrotermitinae) of the higher termites (Bignell, 1994). Lower termites possess protist gut symbionts that contribute to their nutrition whereas those of the higher termites are mainly bacteria. In an obligate symbiosis between the termite and the fungus, specific fungi are cultivated as a food source within the termite mounds of species in the subfamily Macrotermitinae.

Emerson (1955) estimated that *ca.* 73 % of living species belonged to the family Termitidae. However, over the last 20 years, most new genera described have been in the higher termites (Pearce and Waite, 1994) which are therefore likely to comprise rather more than the 73 % of existing species calculated by Emerson. Among the lower termites, the Kalotermitidae (*ca.* 16 % of species) and the Rhinotermitidae (*ca.* 8 % of species) are the largest families while the Hodotermitidae and the Termopsidae are small. The Mastotermitidae and the Serritermitidae have only one living species each (Table III.14).

Termites are most prevalent in the tropics and sub tropics and are scarce or absent at higher latitudes and at higher elevations within their normal latitudinal range of occurrence. Within this range, both species and generic richness decline regularly with distance from the equator in most continents. Australia appears to be an exception to this trend with no group showing a consistent decline in species richness from the tropics to the temperate latitudes (Abensperg-Traun and Steven, 1997b). Where it occurs, this trend may be due to the decline of insolation and rainfall. Figure III. 48a presents estimates of species richness for a number of sites in relation to latitude and across a range of different environments. While collecting methods and sampling areas differed between the studies, a clear reduction in species richness is apparent with increasing latitude. Figure. III 48b presents the mean number of genera within five-degree latitudinal bands and shows that generic richness also diminishes markedly with latitude but that the decline is steeper in the northern hemisphere (Eggleton, 1994). The lower termites are best represented at higher latitudes, most commonly by the families Termopsidae, Kalotermitidae and Rhinotermitidae with some Hodotermitidae in the Palaearctic region. Species richness also diminishes rapidly with increasing elevation. On a high mountain in Sarawak, the number of species declined from 58 in the lowland Dipterocarp forest to 10 in the montane forests above 1000 m; no termites were found above 1860 m (Collins, 1983).

The termites are an ancient group and, at the family and generic levels, are unequally represented in the various biogeographic regions of the world. Table III.14 lists the biogeographic distributions of termite taxa at the family and subfamily levels (Pearce and Waite, 1994). Members of subfamily Macrotermitinae (fungus cultivators) are distributed through Africa, the Malagasy and the Indo-Malayan regions but are absent from the Americas, Australia and the Palaearctic regions. Some genera are particularly widespread. Amongst the lower termites, *Coptotermes* and *Cryptotermes* occur in all biogeographic regions while *Heterotermes* is absent only from the Palaearctic region. In the higher termites, *Amitermes* occurs in all biogeographic regions except the Malagasy and

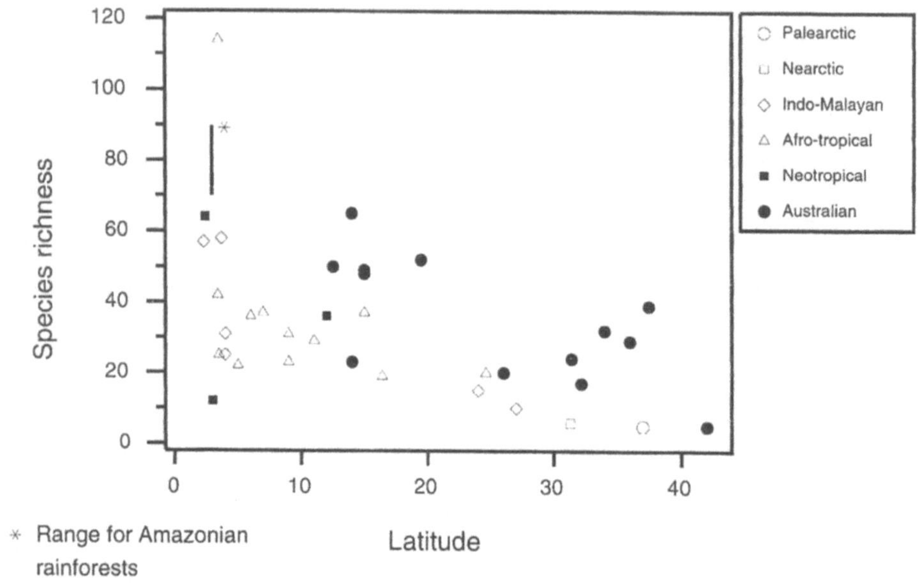

Figure III.48(a) Relationship between species richness in termite communities and latitude.

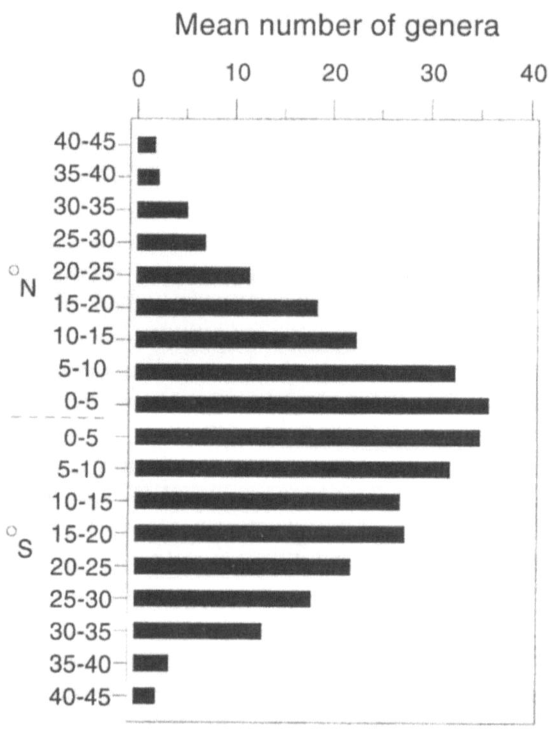

Figure III.48(b) The number of termite genera occurring in five-degree latitudinal bands (Eggleton, 1994).

Microcerotermes in all except the Nearctic.

Termites are generally small insects, most ranging in length from a few millimetres to *ca.* one centimetre and weighing from 1 to 15 mg fresh weight (Wood *et al.*, 1982). However, queens of the larger Macrotermitinae may occasionally attain nearly 12 cm in length (Noirot, 1990) and weigh up to 56 grams (live weight) (Darlington and Dransfield, 1987).

Termite colonies

As eusocial insects, termites are organised into colonies comprising a number of castes, each with a discrete set of morphological and physiological specialisations tailored to permit an efficient division of labour. Individuals pass through several larval and nymphal stages before they reach the adult stage. Polymorphism among larvae, the ability of larvae to participate in social life and neoteny (the ability of certain immatures to reproduce) are peculiarities of termites not found in other social insects (Noirot, 1989).

The workers are defined as 'a set of more or less distinctive individuals belonging to more than one developmental stage who contribute to such ordinary chores of the colony as nest construction, cleaning, nursing and foraging' (Wilson, 1971). Termite workers are also responsible for feeding the dependent castes in a process known as trophallaxis. The latter castes include the reproductives, the soldiers and the larvae which may be fed on salivary secretions, regurgitated food or specialised faeces. Individuals feeding on specialised faeces are inoculated with the symbiotic protists important in digestion in the lower termites (Waller and La Fage, 1987; Nalepa, 1994).

Termite workers change their roles as they mature (polyethism) and those in earlier developmental stages tend to restrict their activities to brood care and inner nest activities. In contrast, older workers are likely to be active in more 'hazardous' activities such as repairing nests and foraging (McMahan, 1979). In the African fungus-cultivating termite *Macrotermes michaelseni*, for example, the older workers forage for food materials outside the nest while younger workers process the forage, construct the fungus combs and feed the dependent castes (Veivers *et al.*, 1991).

The soldiers carry out defensive functions both attacking invaders of their nests and accompanying the workers on their foraging forays. Soldiers are completely dependent on the workers for their nutrition. The soldier head is typically elongated, heavily sclerotised and most possess large, forwardly-directed mandibles. In one subfamily of the higher termites (Nasutitermitinae), the head is elongate forming a snout from which irritant and entangling chemicals are sprayed onto their enemies. Defensive chemicals are produced from the frontal and salivary glands.

The queen and king are the primary reproductives. They normally develop in the parent nest as specialised reproductive nymphs (higher termites) and from the general worker nymphs (in the lower termites). At certain times of year, mature individuals leave the nest in one or several swarms and after pairing, establish new colonies. Following colony establishment and particularly in the higher termites, the queen's abdomen swells enormously and she becomes a veritable egg factory. Production rates as high as 18,000 to 43,000 eggs per day have been estimated for the queens of certain Macrotermitinae (Grassé, 1984); while such rates are probably variable over time, lifetime production must be many millions. Queens are less fecund in the lower termites (Noirot, 1990).

When the queen becomes moribund or dies, she may be replaced by secondary reproductives of various origins (worker line, reproductive line, very occasionally the soldiers) whose capacity to breed had hitherto been suppressed by the former queen.

In the lower termites, individuals must pass through up to nine different forms before reaching the primary reproductive stage (Figure III.49a). After completing the larval moults, the colony pheromone balance directs future development along either the soldier or replacement reproductive pathways, depending on the physiological state of the colony. Synchrony is achieved through such environmental cues as the onset of seasonal rains.

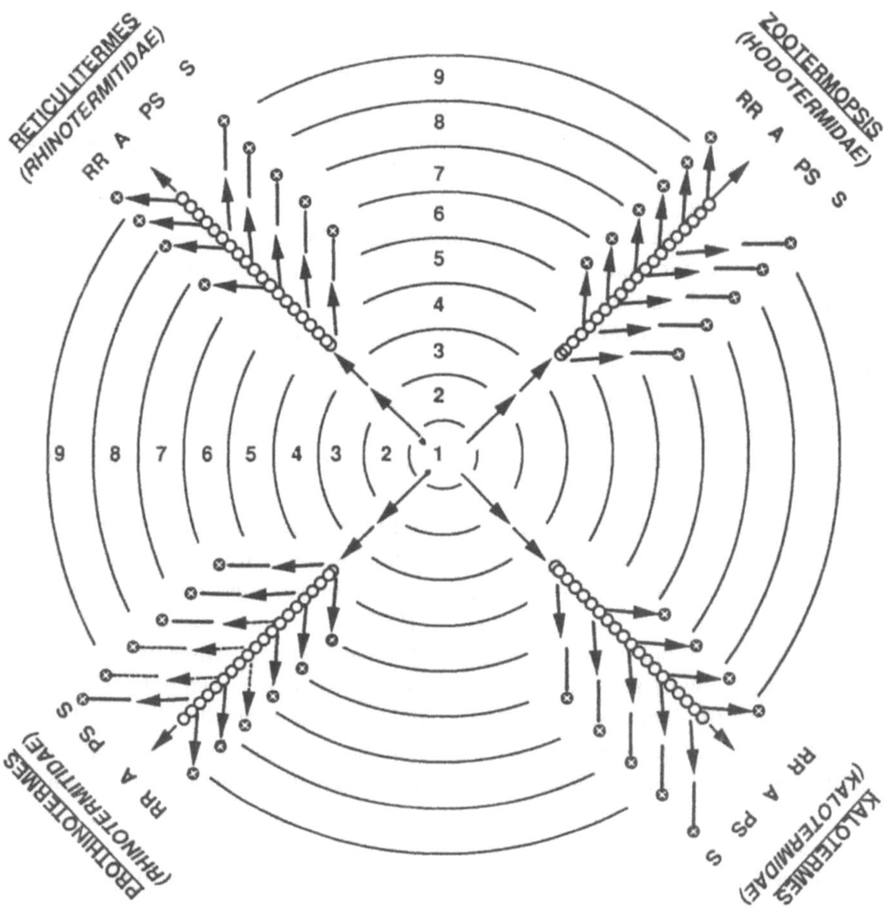

Figure III.49(a) The potential developmental pathways open to each newly-hatched larva in four lower termite species: (A) alate or primary reproductive; (PS) presoldier; (RR) replacement or supplementary reproductive; (S) mature soldier. After emerging from the egg, the termite must proceed through nine instars to reach the primary reproductive stage. At any instar after the third or fourth, it can be diverted to the soldier or replacement reproductive pathways. The central chain line represents the instars from which either forward or regressive moults are possible (Redrawn from Wilson, 1971).

Development in the higher termites is characterised by the direction of individuals into neuter or reproductive pathways during, or prior to the first larval stage (Figure III.49b). In the neuter line of the higher termites, in *Mastotermes darwiniensis*, the Hodotermitinae and some Rhinotermitidae, there is a greater separation between the castes and an increased developmental rigidity compared with the remaining lower termites, particularly the Termopsidae, Kalotermitidae and some Rhinotermitidae. The future reproductives grow to maturity through a series of developmental stages separated by moults.

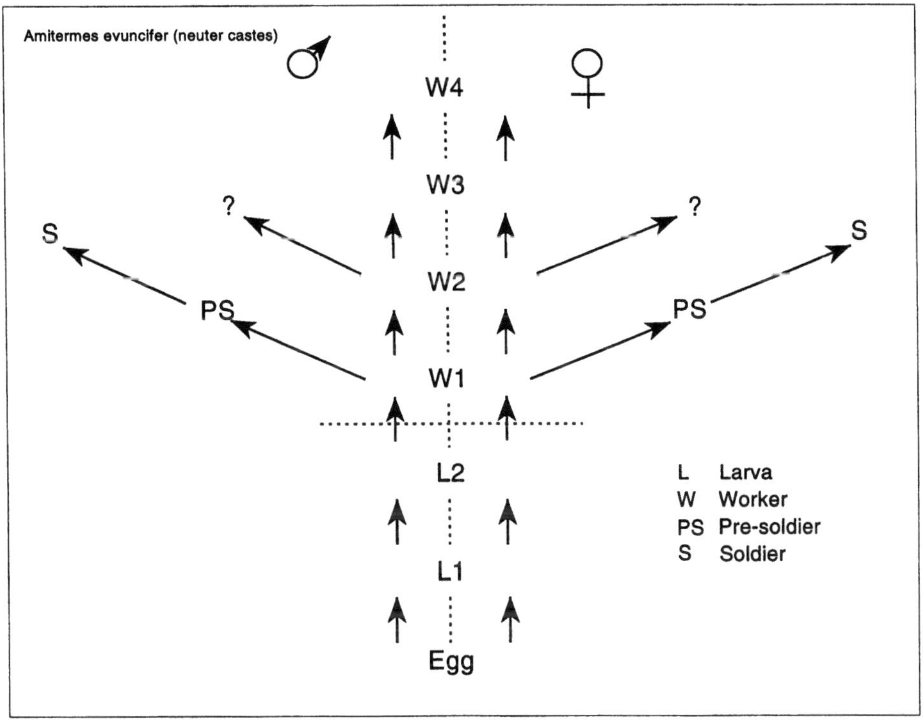

Figure III 49(b) Development of neuter castes of *Amitermes evuncifer* (Termitidae: Amitermitinae). In this relatively primitive "higher" termite, soldiers are formed from both sexes, and the development of males and females is closely parallel in both the soldier and worker castes: The larval and worker instars are numbered sequentially (Redrawn from Wilson, 1971).

Colony ontogeny and reproduction

Termite colonies reproduce in three ways (Grassé, 1984). The first and most common occurs through the periodic release of male and female reproductives from their parental nests (swarming). A small proportion of those released survive predation to form nuptial pairs and may establish new colonies separate from those within which they were reared. However, new colonies also establish from secondary reproductives where part of the original colony becomes isolated from its parent colony (budding). Occasionally, colonies may subdivide by a process know as sociotomy which involves the departure of a group of individuals from the parent colony to establish one or more new colonies elsewhere.

Colonies progress from juvenile to mature stages and eventually decline (Bodot, 1969). The juvenile stage is differentiated from the adult by the production of reproductives. Caste proportions alter in characteristic ways during the different stages of colony development. In Kenya, juvenile colonies of the litter-feeding and fungus-cultivating termite *Macrotermes michaelseni* grow rapidly and the population comprises 50 % or more of larvae. On reaching a population of *ca.* 1.2 million, the colonies become mature and start to release reproductives. The proportion of larvae then falls to *ca.* 41 % and the colony growth rate subsequently declines (Darlington and Dransfield, 1987). In the final or senile stage, the queen's fecundity declines and the colony vegetates and dies. In some species, supplementary reproductives may assume the former queen's role and these termite colonies are potentially longeval. Although the reason is unknown, all colonies eventually decline and die, even in the apparent absence of obvious causative factors, Colonies may exist for considerable periods; those of the Australian species *Drepanotermes perniger* may survive for substantially more than 50 years (Watson *et al.*, 1988).

Movements and dissemination

As stated above, the principal dispersal mechanism among the termites is the seasonal release of primary reproductives from established colonies that occurs during the wetter periods of the year. However, shorter distance dispersal may occur through sociotomy (Grassé, 1984).

Neither individual termites nor colonies normally travel long distances since they are constrained to live within their territory borders or, as in certain lower termites, within their food materials. Most termites only move in sheltered locations within galleries distributed vertically and horizontally in the soil or within their food materials. While a range of grass-harvesting and litter-feeding termites regularly forage unprotected on the surface, most species move and feed beneath thin covers or workings built of soil particles and plant materials which protect them from desiccation and from predators.

The scales of termite movement may be very small, as in the dry wood termites that feed and nest in small pieces of wood on a single tree. Litter-feeding and certain wood feeding termites may move up to radii of *ca.* 50 metres from the central nest, as recorded for *Macrotermes michaelseni* in Kenya (Darlington, 1982). The scale of movement of the processional termite (*Hospitalitermes hospitalis*) during foraging for lichens in Brunei may be more than 65 m (Jones and Gathorne-Hardy, 1995).

Feeding habits and digestive processes

A diversity of organic food materials is utilised by termites although most feed predominantly on lignocellulose-rich or partly-humified dead plant materials. Actual diet depends on the ecological strategies adopted by individual species and this is determined by the part of the decay continuum from which they select their food materials (Chapter IV.5.2.1). Certain termite species regularly attack a wide range of materials, including plastics, the softer metals and other non-assimilable materials (Gay and Calaby, 1970). The nutritional ecology of termites has been reviewed by Waller and La Fage (1987) and, more recently, by Lenz (1994) for wood-feeding species.

Termite food materials. A number of species feed on living plants, both above and below ground, and certain of these may become serious pests, particularly in agricultural systems where dead residues are scarce (see, for example, Wood, 1996), in forest plantations (Cowie *et al.*, 1989) and occasionally in native forest environments exploited for timber production. Most species feed on dead-plant materials above, at or below the soil surface. These may include the dead foliage of grasses and other types of vegetation, woody materials including roots, seeds, the faeces of higher animals and other materials.

Species of most termite families attack dead wood and perform a vital ecological role in breaking down this and other organic materials and recycling their mineral elements. A proportion of the wood-feeding species also attack man's wooden constructions (Edwards and Mill, 1986).

The species of two further groups, the soil-wood feeders and the soil feeders, ingest considerable mineral material. They derive their nutrition from, respectively, well-decayed wood and partly-humified soil organic matter, the latter sometimes in the mounds of other termites (Bignell, 1994). Evolution of the soil-feeding habit was an important advance for the termites since it allowed those so evolved access to a class of highly-degraded, resistant materials that would have otherwise been unavailable. The success of this advance is demonstrated by the estimate that 130 (58 %) of the 225 genera of the family Termitidae are soil-feeders (Noirot, 1992).

In addition to these, a few specialist feeders are known. Species of the higher termite genus *Hospitalitermes* feed on lichens, mosses and wood in South East Asian rainforests (Collins, 1979) and the Neotropical species *Anoplotermes pacificus* feeds on the tips of living roots that proliferate in its nest (Kaiser in Araujo, 1970) and on the root tips of a number of crop plants (Mill, 1992). Carrion feeding has been reported in *Nasutitermes nigriceps* from Panama (Thorne and Kimsey, 1983).

Recalcitrance of termite food materials. Termite food materials are rich in such highly-polymerised compounds as lignin, cellulose and hemicellulose and are thus relatively recalcitrant to breakdown. Despite this, assimilation rates in termites are high and published values range between 54 and 93 % of the food eaten (Wood, 1978). The recalcitrance of most termite food materials has lead to apparently-universal associations with micro-organisms. These latter, with their wide spectra of enzymatic capacities, contribute in different ways and to different degrees to the breakdown of the materials eventually assimilated, depending on the ecological strategies of the termite species.

Lignins are highly complex molecules and differ strongly from the other polymeric constituents of lignocellulose with which they are intimately linked (Breznak and Brune, 1994). Structural differences occur between lignins from various plant sources and, while all are difficult to degrade, those from grasses are slightly less recalcitrant. Lignins appear to be partially degraded by passage through the termite gut although the mechanisms are as yet unknown, since the currently-known methods of degradation are aerobic (Breznak and Brune, 1994). Evolution of labelled CO_2 from termites fed lignin labelled with ^{14}C has been demonstrated (Butler and Buckerfield, 1979). Both the higher and lower termites have some capacity to slightly degrade lignins although their ability to do so varies between lignins from different tree species (Cookson, 1987). However,

Breznak and Brune (1994) consider it unlikely that termites can substantially degrade the aromatic core of lignin molecules.

Almost all termite food materials are of very low protein content (Section IV.5.2.1) and the problem for termites in meeting their metabolic N requirements has been met at least partially through the acquisition of N-fixing symbionts which reside in the hind gut. N fixation has been reported from all termite families although not from the fungus-cultivating species of the sub-family Macrotermitinae (Slaytor and Chappell, 1994). Tayasu *et al.* (1994) estimated that from 30 to more than 60 % of the N in the tissues of the dry wood termite *Neotermes koshunensis* was derived from fixation. Bentley (1987) estimated that N fixation accounted for nearly 25 % of the dietary requirements of certain Venezuelan higher termites and considered that N-fixation rates were inversely proportional to food N concentrations. Further, N-conserving mechanisms have been described including the recycling of uric acid for possible reuse by the termites through the medium of the symbiotic gut biota, digestion of symbiotic intestinal organisms and such behavioural traits as oophagy, cannibalism and necrophagy (Breznak and Brune, 1994).

Digestive processes. Although it has been claimed that much of the breakdown of cellulose in termite food materials occurs through symbiotic associations with micro-organisms (see, for example, Cleveland, 1923; Breznack, 1984; Rouland *et al.*, 1988a, b and Figure III.50), work by Slaytor (1992) and his colleagues has cast considerable doubt on the generality of this view. All of the lower and higher termites studied by Slaytor (1992) could survive on crystalline cellulose and it was concluded that the necessary cellulases were secreted by the salivary glands, the foregut and midgut tissues. A range of other hydrolytic enzymes (amylase, maltase, invertase, hemicellulase) are also secreted in the same areas. Further, appreciable cellulolytic activity is largely absent from the hindgut except in the lower termites where it is associated with the presence of cellulolytic protists (Breznak and Brune, 1994).

Wood-feeding species occur in both the higher and lower termites, although digestive mechanisms differ between the two groups. In the lower termites, resistant materials appear to be digested partly through mutualistic associations with specialised protists; these obligate anaerobes inhabit a specialised region of the hind gut, although other prokaryotes also occur there. Higher termites lack the key glycolytic enzyme pyruvate dehydrogenase which is needed to convert glucose to acetyl CoA or acetate which is the major cellulose derivative that can be absorbed by the termite. The role of the bacteria may perhaps be to further break down materials from the initial stages of cellulose dissimulation and to mediate the conversion of enzymic degradation products to acetate (Breznak and Brune, 1994). In the many species of wood-feeding termites that feed on wood or stored grasses previously attacked by saprotrophic fungi, it is not known whether enzymes acquired from the ingested fungi play a functional role in the insect gut, if the fungi are used solely as a dietary item or whether there is an energetic advantage to the termites in attacking partly-degraded materials. A number of species appear to be nutritionally dependent on certain fungi (Sands, 1970).

The workers of soil-feeding termites have mandibles specialised for crushing and a substantial gut flora of prokaryotes. Adoption of the soil-feeding habit has lead to large changes in the structure of the intestinal tract and a system of digestion different from that of the wood-feeding termites. Digestive mechanisms are uncertain although the alkaline pH of the hind gut is considered to promote the alkaline hydrolysis of tannin-protein and tannin-cellulose complexes. Fermentation products found in the hind gut include acetate and methane (Bignell, 1994).

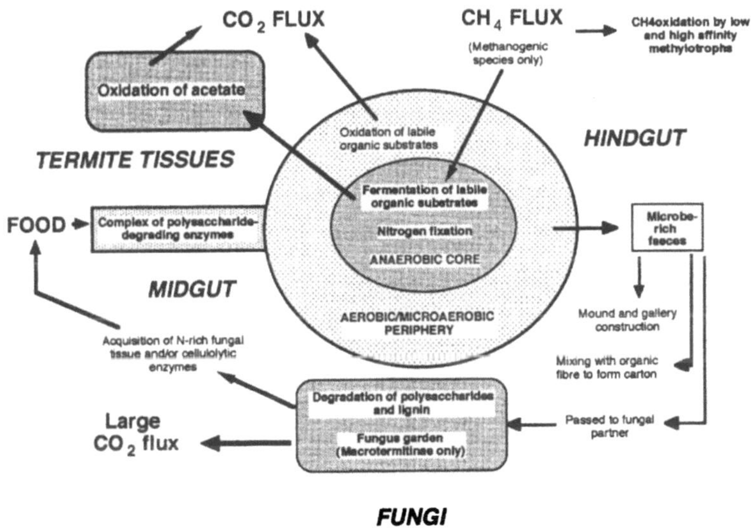

Figure III.50 Digestive systems in termites (Bignell, 1994).

Methane production. Methane may be produced in considerable quantities during the fermentative breakdown of food materials (Seiler *et al.*, 1984 and Figure III.50) and termite production of this gas has been considered to make a significant contribution to global production (Schlesinger, 1991). However, most estimates are based on very little information. Recent estimates by Martius *et al.* (1997) consider that methane production by termites may account for *ca.* five *per cent.* of the world's annual methane flux from all sources. Brauman *et al.*, (1992) and Rouland *et al.* (1993) have shown that differences in methane production between termites of separate ecological groupings are related to differences in their gut microbiota. Wood-feeding termites were found to have a dominance of acetogenic bacteria and to produce little methane. Soil-feeding and some fungus-cultivating termites were found to produce methane at much higher rates due to the presence of a well developed methanogenic bacterial flora in their hindguts.

Fungi and fungus cultivating termites. Fungi are particularly important in termite nutrition both directly as a source of energy and because of their ability to degrade resistant substrates such as wood (Figure III.50). It is clear that the many termites that feed on wood and other organic materials that are either hard or of high C:N ratio, find such materials more attractive following some degree of fungal attack. Figure III.51 relates the consumption of *Pinus radiata* wood blocks by the Australian termite *Coptotermes lacteus* to previous weight losses resulting from fungal decomposition (French *et al.*, 1987).

However, if decay has progressed beyond a certain point, the wood may become unpalatable to particular species (Lenz, 1994). The grass-harvesting termites that store the sectioned stems of dead grasses in their mounds may well be subjecting them to a similar period of fungal decomposition prior to their consumption (Holt, 1998).

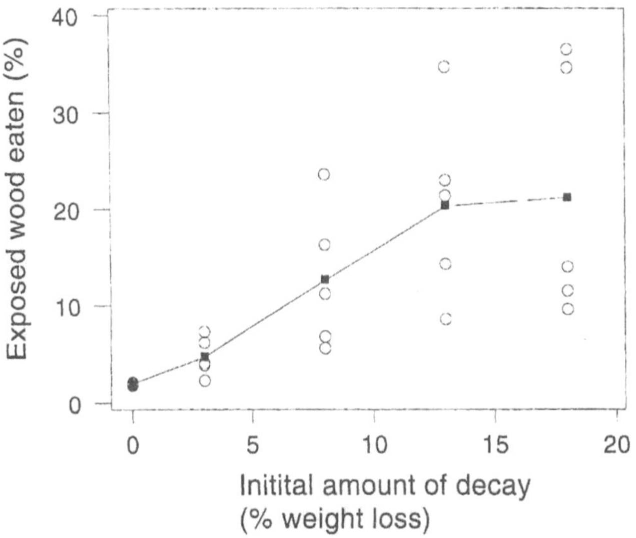

Figure III.51 Consumption of *Pinus radiata* blocks by the termite *Coptotermes lacteus* in relation to weight loss from prior fungal attack (French *et al.*, 1987)(circle: observations; square block: treatment means).

Termites of the subfamily Macrotermitinae feed on slightly decomposed dead plant materials including leaf litter, dead grasses, woody litter and standing dead wood (Darlington, 1994). With the exception of *Sphaerotermes sphaerothorax* considered below, they cultivate symbiotic fungi of the Basidiomycete genus *Termitomyces* in their nests which partly assist in the digestion of cellulose and other resistant materials (Thomas, 1987). The association between the termites and the fungus is obligate: the fungi of this genus grow nowhere else than in the termite nest and incipient colonies that fail to develop a comb do not survive (Darlington, 1994).

In colonies of one fungus-cultivator, *Macrotermes michaelseni,* the older workers return food materials to their mounds in the form of specialised faeces which have been only slightly degraded by residence in the gut. In the mound, these faecal pellets are deposited on special structures known as fungus combs where they are ingested by the younger workers. These latter workers deposit their own faeces onto the comb where they are then further decomposed by the cultivated fungus. Members of the reproductive line castes and the younger workers appear to feed on the glucose-rich spores of the fungus and on comb faecal materials. The older workers feed on senescent comb and other organic materials in which cellulose, hemicellulose, pectin and lignin are partly degraded and which have a higher N concentration than the original forage (Bignell *et al.*, 1994). In the gut of these termites, resistant materials are broken down by a combination of the termites' own enzymes (Rouland *et al.*, 1990), possibly enzymes acquired from their fungal symbiont (Martin, 1984; Slaytor, 1992) and perhaps from fermentation in the hind gut (Breznak and Brune, 1994). However, Veivers *et al.* (1991) found no evidence that micro-organisms play a significant role in digestion in these termites.

The African wood-feeding termite *Sphaerotermes sphaerothorax* is atypical of the Macrotermitinae in that it has lost or perhaps never developed an association with *Termitomyces* (Garnier-Sillam *et al.,* 1989). In its place, external associations with cellulolytic and N-fixing bacteria have evolved and these are considered further in Chapter IV.

In sum, the tripartite action of the workers' enzymes, those of the hind gut microflora and of the fungus *Termitomyces* combine to effect an almost complete breakdown of the plant materials harvested by the fungus-cultivating termites (see, for example, Garnier-Sillam *et al.*, 1988b). The high proportion of food material organic matter broken down has considerable ecological and pedological implications, as shown in Chapter IV.

Carbon and nitrogen stable isotopes in termite foods. The $\delta^{13}C$ values of termite tissues reflect the relative proportions of ^{12}C and ^{13}C stable isotopes in termite bodies and this provides useful information on their feeding habits. As occurs in other consumers, values for their tissues are normally close (within 2 $\delta^{13}C$ units) to those of the food that they assimilate (Fry and Sherr, 1984). This relationship has been used to discriminate between species that feed on wood, on tropical grasses and on a mixture of both (Figure III.52, Spain and Reddell, 1996). It has also been used to identify the proportions of grasses and broad leaf materials present in the diets of several African litter-feeding termites (Boutton *et al.*, 1983; Lepage *et al.*, 1993).

Unlike the situation in most animals, the $\delta^{15}N$ values of termite tissues vary over a small range in wood-feeding species because the N fixation referred to above leads to tissue values that are relatively close to that of the air (Tayasu *et al.*, 1994; Spain *et al.*, in prep., Figure III.52). However, termites that feed on partially-decomposed litter have slightly higher tissue $\delta^{15}N$ values (Tayasu *et al.*, 1997, 1998), possibly associated with a reduced requirement for N fixation consequent on the greater N concentrations of their food materials (Slaytor and Chappell, 1994).

4.3.2.2 *Ecological categories*

While it is difficult to devise non-overlapping ecological classifications, broad classifications have been proposed, based on the types and variety of food materials used and the locations of their nesting and feeding sites.

On the basis of their feeding habits, termites may be separated into five broad, somewhat-overlapping trophic groups (Josens, 1983; Eggleton *et al.*, 1996), namely:

(i) grass harvesters;
(ii) surface litter feeders;
(iii) wood-feeders;
(iv) soil-wood feeders; and
(v) soil-feeders (or humivores).

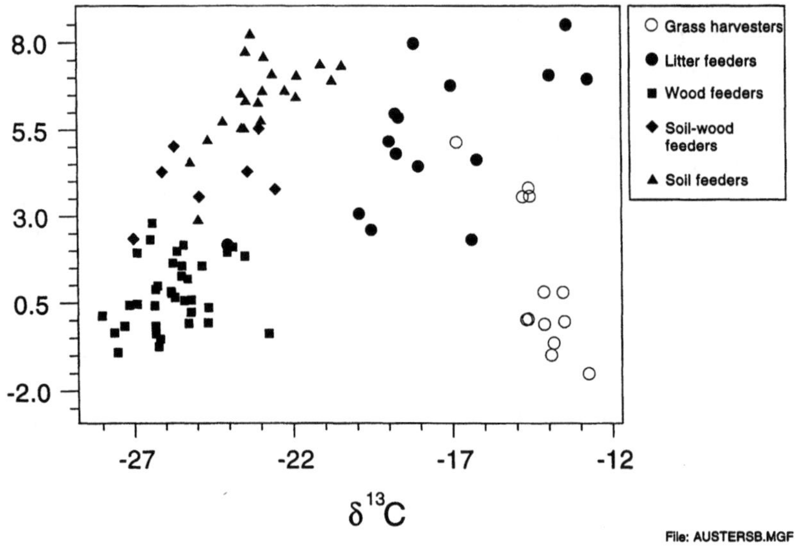

Figure III.52 Distributions of termite species of five feeding strategies on the basis of their C and N stable isotope ratios (Tayasu *et al.*, 1998; Spain *et al.*, unpublished).

Additional categories include the highly polyphagous species and the few specialist feeders referred to above. As considered in Chapter IV.5.2.1, substantial dietary overlaps occur between members of the first three groups and dietary shifts may result from food shortages.

Termite nests occur in the following six main locations:

(i) within the wood of living and dead trees, and in fallen timber;
(ii) subterranean locations;
(iii) epigeal nests;
(iv) intermediate nests on the lower parts of tree trunks, usually in contact with the soil;
(v) arboreal nests; and
(vi) within the nests of other species (Noirot, 1970; Martius, 1994) (Figure III.53).

Abe (1987) integrated the six 'life types' he defined into three broad categories, although transitional types also occur (Lenz, 1994):

(i) 'One-piece' types. These include the 'dry' and 'damp' or 'wet' wood feeding termites. In this type, the entire colony generally lives in one piece of wood for the whole of its existence and normally doesn't attempt to locate additional food resources. However, some dry wood species may extend their colonies into living wood. Examples include most Kalotermitidae, Termopsidae and species of the genus *Prorhinotermes* (Rhinotermitidae).

(ii) 'Intermediate' types. Species in this group nest predominantly in wood but construct galleries to exploit other food resources. Australian examples of this group include *Coptotermes acinaciformis* and *Mastotermes darwiniensis*.

(iii) 'Separate' types. This grouping comprises the many species that nest apart from

Figure III.53 The distribution of selected nest types in an hypothetical tropical Australian savanna termite community: a. *Coptotermes acinaciformis*; b. Species largely associated with the exteriors of trees (species: 1: *Neotermes insularis*, 2: *Schedorhinotermes actuosus*, 8: *Ephelotermes melachoma*, 11: *Nasutitermes graveolus*); c. Species attacking wood on the soil surface, dead shrubs and the bark of trees and soil-feeding species living in the mounds of other termites (species: 4: *Heterotermes vagus*, 6: *Microcerotermes serratus*, 9: *Amitermes perelegans*, 10: *Macrognathotermes errator*); d. Epigeal mound-building grass-harvesting and litter-feeding species (species: 5: *Amitermes laurensis*, 7: *Drepanotermes rubriceps*, 12: *Nasutitermes triodiae*).

their food resources; it includes many subterranean and epigeal nesting species and those that construct arboreal nests on the outsides of tree branches. Species included in this group represent a wide variety of feeding habits, including wood-feeding, grass harvesting and litter-feeding, soil-wood and soil-feeding species.

The initial primitive condition is considered to be the one-piece type since it resembles that of the presumedly ancestral wood-feeding cockroaches. Evolution of life types is considered to have proceeded to the intermediate and then the separate types.

The different ecological groupings and their environmental impacts are considered further in Chapter IV.5.2.1.

4.3.2.3 *Population densities and biomasses*

Termite colonies are difficult to sample accurately and most population estimates for termite colonies must be regarded as minima. Difficulties arise in enumerating whole colonies since a proportion of the individuals may be absent foraging or escape from the nest during sampling. Subterranean termites may have large diffuse colonies with patchy distributions and complete sampling would often require the separation of enormous amounts of soil. Finally, the small size of termites is a serious obstacle to their enumeration. Darlington (1984), for example, found that fumigation of termitaria prior to sampling increased colony population estimates substantially. Where fumigation was omitted, differences in the estimation of caste proportions also resulted; in mature mounds most of the eggs had disappeared and estimates of larvae were diminished by an order of magnitude. Separation of termites from mound materials using wet sieving methods gives biomass estimates up to 5-10 times higher than direct hand sorting (see, *e.g.*, Lavelle and Kohlmann, 1984).

Isotopic methods have been used to estimate populations, although with varied success. It appears that isotope dilution methods overestimate colony populations because of rapid elimination of the isotopic 'label' into the nest environment. However, the use of isotopes as labels in mark:recapture techniques is apparently more successful (Easey and Holt, 1989).

Colony population estimates

Despite the methodological problems considered above, colony sizes have now been estimated for a range of termite species revealing populations ranging from a few tens of individuals to several millions. Small colonies may simply be juvenile or they may be constrained by the size of their resources. However, the mature colony size is generally related to the size and quality of the food base. Large food supplies lead to bigger colonies, longer life cycle spans, later colony maturity and greater numbers of supplementary and replacement reproductives that may be present in addition to, or replace the original reproductive pair (Lenz, 1994).

The colonies of 'dry wood' termites are generally small, less than *ca.* 3000 individuals. Many Kalotermitidae have small nests, often formed in the dead branch stubs of trees and in similar locations. *Kalotermes flavicollis* is a Mediterranean dry wood feeding species that forms small colonies in the dead and diseased wood of living trees (Harris, 1970). Reported colony sizes for *K. flavicollis* range up to a maximum of 3250 workers (Noirot, 1990).

Damp wood-feeding termites have larger colonies, mostly in the range of 3000 to 8000 individuals. Medium-sized colonies include those of the Australian termopsid species *Porotermes adamsoni* which nests in living trees and is a notable pest of timber in the areas where it occurs. Nkunika (1990) reported small colonies (to *ca.* 5500) while elsewhere the larger trees present permitted colony expansion up to *ca.* 62,000 individuals (Lenz, 1985).

Colonies of the soil-feeding termite *Cubitermes fungifaber* had populations ranging from 7000 to 69,000 (Maldague, 1964) although Noirot (1990) states that worker populations rarely exceed 30,000 individuals. Easey and Holt (1989) reported estimates for four Australian grass- and litter-feeding species of the family Termitidae that ranged from 14,100 to 257,000 individuals. It should be noted that the colonies of some of these species were distributed throughout more than one mound (Holt and Easey, 1985).

Large termite colonies occur mainly in the species belonging to Abe's (1987) intermediate and separate categories and examples have been reported from the Afrotropical, Neotropical and Australian biogeographic regions. *Mastotermes darwiniensis* normally forms only small colonies but when substantial food resources become available, the species can rapidly expand its populations due to its ability to produce large numbers of neotenic (secondary) reproductives (Watson and Gay, 1991). In a plantation of the exotic conifer *Pinus caribaea* in northern Australia, Spragg and Paton (1980) estimated the population of a large diffuse colony of *Mastotermes darwiniensis* at seven million individuals including several nests and spread over one hectare. However, because of the methodological problems considered above, this may be an overestimate. Other examples of species with estimated colony populations of more than one million individuals include the Australian wood-feeding species *Coptotermes acinaciformis* (Greaves, 1967), *Coptotermes lacteus* (Gay and Greaves, 1940) and *Nasutitermes exitiosus* (Gay and Weatherly, 1970), some African fungus-cultivating termites of the genus *Macrotermes* (one of which, *M. michaelseni,* had a total population in excess of five million (Darlington and Dransfield, 1987) and the neotropical species *Nasutitermes macrocephalus* (Martius, 1994).

Despite the substantial estimates in the large colony category, most colonies are much smaller than this. Most termite colony population estimates are likely to fall within the low to medium end of the size range considered above but will also depend on the ecological strategies of the species under consideration.

Estimates on an areal basis

On the basis of surface area, estimated termite populations and biomasses range up to exceptional maxima of more than 10,400 individuals m^{-2} and 123 g m^{-2} live weight (Eggleton *et al.*, 1996). However, most population and biomass estimates are considerably lower than the above values and the medians presented by Wood and Sands (1978) and Collins (1983) for a total of 24 sites are, respectively, 1429 ind. m^{-2} (inter-quartile range 400 to 2939) and 3.0 g m^{-2} (inter-quartile range 1.5 to 8.7). However, as stated above, termite abundances are often grossly underestimated, due to their aggregated distributions, small sizes, high mobilities and the specificities of their behaviours (Eggleton and Bignell, 1995; Eggleton *et al.*, 1996).

As a group, termites tend to survive such disturbances as land clearing and cultivation better than, *e.g.*, litter arthropods or earthworms, although it is largely the species with deeper subterranean nests that remain (Wood and Pearce, 1991). Consequently, they often comprise a large proportion of animal biomass in annual crops (Dangerfield, 1990; Lavelle and Pashanasi, 1989). In five year old *Hevea* plantations in southern Côte d'Ivoire, termite communities (47 g f wt m^{-2}) comprised more than half of the total biomass of macroinvertebrate communities. This was due to the development of substantial populations of wood-feeding termites which utilised the dead woody material remaining following forest clearance. Ten years later, this woody material had been exhausted and termite biomass fell to 2.5 g m^{-2}, five times less than that of the original forest (Gilot *et al.*, 1995) (Figure III.54). However, substantial and damaging populations may remain in long-cultivated lands and Wood *et al.* (1977) reported economically-important populations (6825 ind. m^{-2}) in 24 year old fields cultivated for groundnuts (*Arachis hypogaea*).

Fire may depress both the abundance and diversity of termites and alter the balance of their communities. In Colombian savannas, Decaëns *et al.* (1994) observed a sharp decrease of termite populations following firing and cattle grazing while earthworm abundance was enhanced. Abensperg-Traun and Milewski (1995) compared burned and unburned areas of nine vegetation types in south-western Australia two years after an intense fire and noted a significantly-reduced incidence of wood-feeding species. They attributed this to the destruction of colonies nesting in surface wood, rather than food limitation. Species dwelling in earthen epigeal mounds were noted to be largely protected from the fire. However, in the northern Australian tropics, small carton-rich epigeal mounds of *Microcerotermes* sp. are often destroyed by low-intensity annual fires. Further evidence of shifts in community balances is available. In an African savanna environment, Benzie (1986) reported a rapid increase in the abundance of the mounds of grass-harvesting termites in the two years following protection from fire.

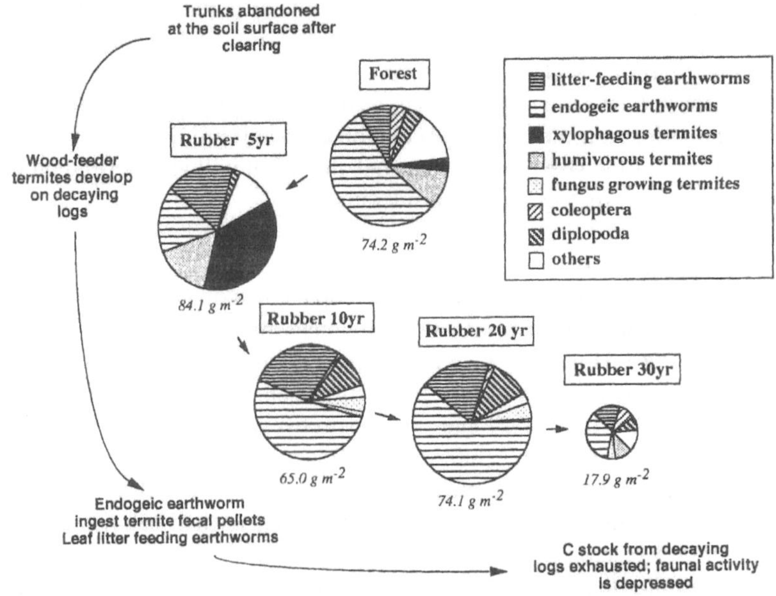

Figure III.54 Changes in soil macrofauna communities in a primary forest and in an age series of rubber plantations in the Côte d'Ivoire (Gilot *et al.*, 1995).

4.3.2.4 *Community structure*

Determinants of community structure

Termite communities comprise species of differing ecological strategies. Community complexity is determined by factors operating at different spatial scales. The large scale factors are those of biogeographic, climatic and vegetational variation, while topographic location and soil variability are influential at more local scales.

Differences in the presence or absence of certain taxa across biogeographic regions are clearly important since some termite groups are absent from regions where in terms of climate, they would be expected to readily survive. Examples include the absence of the fungus cultivating termites (sub-family Macrotermitinae) from the Australian and Neotropical biogeographic regions referred to above and the many incidences of the successful spread of exotic termites through commerce. The recent establishment of *Mastotermes darwiniensis* in Papua New Guinea is one example of this (Watson and Gay, 1991). However peregrine termite species equivalent to earthworms with wide geographical distribution do not exist (except as pests of buildings).

Climate varies systematically from the equator to the higher latitudes and termite trophic groups react to this differently. Wood-feeding species form a greater proportion of total species as latitude increases and may be the sole strategy represented at higher latitudes (Abe, 1987). Conversely, the grass-harvesting termites are better represented at lower latitudes.

The distribution of vegetation is clearly important to community structure at a range of scales. At the broadest scale, the various biomes of the world offer different opportunities to the various ecological groupings and this also pertains across vegetation types at a local scale. Comparison of the termites of five habitat types in the tropical Australian sites studied by Braithwaite *et al.* (1988) showed a different balance of trophic groups between environments that ranged from monsoon forest (a dry rainforest facies) to open forests, woodlands and two types of rockland. Commensurate with their vegetation structure, no grass and grass-litter feeding species occurred in the monsoon forest and wet rockland sites and less soil-feeding species were found in monsoon forests than in open forest and woodland environments.

A somewhat contrasting trend, however, has been observed in Africa since the abundance of humivorous species tends to decline from forest to moist savannas and drier areas; they are also most sensitive to disturbances, possibly due to their inability to fix nitrogen (Eggleton *et al.*, 1996; Lepage, pers. comm.).

At a regional and local scale, the physical and other properties of certain soils may also limit termite colonisation. In northern and central Queensland, few mound-building termite species can tolerate the expansive nature of certain vertisols which shrink on drying and swell when re-wet (Ratcliffe *et al.*, 1952). Mound building termites are virtually excluded from these soils although across abrupt lines of demarcation between the vertisols and the adjacent sandy alfisols, mounds built by grass and litter-feeding termites are populous. However, certain mound-building species are able to tolerate the expansive qualities of these soils and Leprun and Roy-Noël (1976) record the mounds of the fungus-cultivating species *Macrotermes subhyalinus* from vertisols in Senegal. At the coarser end of the texture spectrum, the only *Eucalyptus* community type from which the widespread

Australian wood feeding species *Coptotermes acinaciformis* is excluded are those formed on deep sands (Gay and Calaby, 1970). The influence of fertility on species richness is discussed below.

Species and generic richness

The most species-rich communities occur in certain humid tropical rainforests. Eggleton *et al.* (1996) found 114 species in the moist pre-montane Mbalmayo Forest Reserve in the Cameroon of which 48 species were recorded during the quantitative sampling of two 1 ha plots in little-disturbed rainforest. However, as noted below, the greatest species diversity occurred in a disturbed environment. In reviewing a number of studies from Amazonia, Martius (1994) considered that species densities of 70 to 90 species were likely to occur throughout the Amazonian rainforests reducing to 11-12 species in the seasonally-flooded forests. Most other rainforests studied have lower species richness and Collins (1983) found 43 species in South-East Asian rainforest in Sarawak. The mesophytic rain-forests of northeastern Australia are particularly species poor; for zoogeographic reasons, only four termite species occur regularly in these forests (Gay and Calaby, 1970). Species richness in savannas is generally somewhat lower than in rainforests and Josens (1983) recorded from 19 to 36 species at eight African locations studied, although slightly more were found in some Australian savanna sites (Figure III.48a), as discussed below.

Plant community structure and certain soil properties clearly influence species richness and the relative balance of ecological categories. Of the 50 species recorded from communities studied at Kakadu (tropical Australia), numbers varied from 16 to 36 over five environments which ranged from monsoon forest (a dry rainforest facies) which had 23 species to a wet rockland which had 16. The most diverse environments were the open forest and woodland savanna sites which had, respectively, 36 and 35 species followed by the dry rockland which had 26 (Braithwaite *et al.*, 1988).

Josens (1983) reported the existence of a positive relationship between species richness and rainfall in African savanna environments. However, this was not confirmed across the 30 tropical Australian sites studied by Braithwaite *et al.* (1988) and is unlikely to hold generally across northern Australian environments because of the reduced number of species in rainforest communities.

Termite species richness has been shown to be inversely related to soil fertility in Australian environments (Braithwaite *et al.*, 1988; Stafford-Smith and Morton, 1990) and Goodland (1965) reported a negative correlation between soil fertility and the number of termitaria in savanna environments in northern South America. Across 30 tropical Australian sites representing five habitat types, Braithwaite *et al.* (1988) found strong negative correlations between species richness and soil N and P concentrations. In contrast, species richness in the live-wood feeding termites of these sites was positively correlated with both these variables. Similarly, in the Malaysian rainforests studied by Salick and Tho (1984) species richness, guild richness and population densities were all positively correlated with a soil productivity index. Eggleton *et al.* (1996) found strong positive correlations between termite abundance and the concentration of 'available' soil N in a disturbance series of study plots in a rainforest environment in the Cameroon. Other soil-based limitations to termite community structure are considered in Chapter IV.

Changes in land use from indigenous to man-modified landscapes normally result in substantial reductions to termite species richness and large changes in community structure. In a Cameroon rainforest, Eggleton *et al.* (1996) reported a markedly-reduced species richness (also biomass and abundance) in plots recently cleared and either replanted with young trees or left as a weeded fallow, in comparison with nearby primary forest and older wooded sites. Over three samplings, species richness varied from 4 to 13 in the first category and 28 to 56 in the wooded plots. Soil-feeding termites were most severely reduced by clearing. Wood *et al.* (1982) found an influx of savanna species into the disturbed plots they studied in a rainforest environment in Nigeria. The study sites used by Eggleton *et al.* (1996) were located *ca.* 40 km from the forest:savanna border while that of Wood *et al.* was sited in a riparian forest within a savanna environment, where a large pool of savanna species was present nearby. Cultivation markedly reduced the diversity of termite communities in all locations studied (Wood, 1996).

4.3.2.5 *Interactions between termites*

Termites are territorial animals; colonies defend their nests and the surrounding areas within which they forage and have the ability to identify individuals from other colonies or species as 'not-self' (Shelton and Grace, 1996). Both inter- and intra-specific antagonisms have been demonstrated between termites of a number of species and there is good reason to believe that this is of general occurrence. Adams and Levings (1987) report inter- and intra-specific battles at territory borders and Darlington (1982) describes finding dead soldiers of *Macrotermes michaelseni* in underground galleries as evidence of battles near territory borders. However, not all inter- and intra-specific interactions result in aggressive behaviours (Shelton and Grace, 1996).

Territory area is clearly related to colony size. Areas reported range from 9.5 to 338 m^2 (average 92 m^2) for the grass-harvesting African termite *Hodotermes mossambicus* (Nel, 1968), from 500 to 800 m^2 depending on colony size for three species living in Panamanian mangrove forests (Levings and Adams, 1984; Adams and Levings, 1987) and was *ca.* 7800 m^2 for a colony of the fungus-cultivating termite *Macrotermes michaelseni* in Kenya (Darlington, 1982).

4.3.2.6 *The natural enemies of termites*

As concentrated parcels of energy and nutrient elements, termites are attractive prey items and are subject to considerable pressure from invertebrate and vertebrate predators, both within and outside their nests. The primary defence of the colony is that provided by the nest which few non-specialised predators can breach, either because of inaccessibility within arboreal or subterranean locations or through the protection afforded by strong epigeal nest structures. Nonetheless, a number of mammals are successful specialist termite predators (Table III.15) and common adaptations in the most highly specialised include the possession of strong claws, spatulate forelimbs, reduced teeth and a long, sticky and prehensile tongue which is extruded to extract termites from their galleries.

Table III.15 Selected specialised mammalian termite predators.

Order	Species	Common name	Distribution
Carnivora	*Proteles cristatus*	Aardwolf	Southern and eastern Africa
Edentata (Myrmecophagidae)	*Myrmecophaga tridactyla* *Tamandua tetradactyla* *Cyclopes didactylus*	Anteaters	Southern Mexico to Paraguay
Pholidota	Manidae *Manis* spp. (7 species)	Pangolins	Africa, Asia
Tubulidentata	*Orycteropus afer*	Aardvark	Africa, south of the Sahara and the Sudan
Monotremata	*Tachyglossus aculeatus*	Spiny anteater, Echidna	Australia
Polyprotodonta	*Myrmecobius fasciatus*	Numbat	South of West Australia
	Thylacomys lagotis	Bilby	South of West Australia, Central Australia

A range of other specialised vertebrate predators exist; examples include many lizards (Abensperg-Traun and Steven, 1997a) and small blind snakes of the family *Typhlopidae*.

Beyond the specialised *vertebrate* predators, termites are subject to predation by a wide range of general predators, particularly the swarming alates (Deligne *et al.*, 1981; Grassé, 1986). Ants are without doubt the most important invertebrate termite predators and Wheeler (1936) recognised the following five categories of ants in terms of their relationships with termites:

(i) ***termitharpactic*** ants raid termite colonies and are specialised termite predators;

(ii) ***cleptobiotic*** ants steal termite prey and other food materials from the termitharpactic and other ants;

(iii) ***termitolestic*** ants occur in termite nests where they prey on eggs and the termites, especially those that have been injured;

(iv) ***inquiline*** ants range from those occupying completely or partially abandoned termitaria to those that may attack and destroy the host termite population;

(v) ***termitoxenic*** ants live among the termites without any evidence of mutual hostility.

Mutualistic relationships between ants and termites are also known. In northeastern Australia, two ants of the genus *Camponotus* reportedly defend the epigeal mounds of the litter-feeding species *Amitermes laurensis* from attacks by the meat ant *Iridomyrmex sanguineus* (Higashi and Ito, 1989). The camponotine ants depend on the termitaria for their nesting sites and may receive food from the termites in exchange for their defensive activities.

4.3.3 ANTS

Ants may occur in great numbers in soils and on their surfaces. They are considered as ecosystem engineers since they build a large variety of structures in soils and at their surface. However, because of their feeding habits, they may perhaps be of less general importance in directly regulating soil function than the termites or earthworms. Tropical American *Atta* leaf-cutting ants (Tribe Attini) represent a spectacular exception in that they make huge subterranean nests and their harvesting may lead to the incorporation of large amounts of organic matter and nutrients into the soil. Many other ants nest in the soil although they may also form part of the aerial food-web. In some locations, ants may be important agents of bioturbation (Levieux, 1976; Cowan *et al.*, 1985; Lockaby and Adams, 1985). A number of species also concentrate plant nutrients in their nests and associated soils (*e.g.*, Wagner, 1997).

Ants have a wide latitudinal distribution extending from the arctic to the tropics; they occur in all but the most extreme of the world's ecosystems although native ants are absent from some isolated oceanic islands. Where significant populations occur, they often dominate the activities of the biota below, at and above the soil surface. Most species are effective predators although many also feed at least partially on materials of plant origin and a number of species are parasitic. In contrast to the termites, the ants have not specialised in directly utilising cellulose as a food resource.

Ants have achieved their ecological dominance through their wide taxonomic diversity and numerical abundance, their successful development of a stable sociality and the broad diversity and pliability of their ecological strategies (Wheeler, 1910). As effective predators, ants influence herbivore populations and therefore plant productivity. Hölldobler and Wilson (1990) attributed the success of the ants to a range of causes. They were the first predaceous, eusocial insects to live and forage primarily on the ground and within the litter layers. Their successful evolution of sociality during the Cretaceous and their later radiation within the Tertiary period thus pre-empted the development of other groups. Perhaps a major part of their successful adaptation to soil dwelling was the development of the metapleural gland, an organ that produces antibiotic secretions. These secretions are spread throughout the ants' nests and appear to protect them from pathogenic micro-organisms that might otherwise affect them in the humid soil environment.

A broad range of relationships exists between ants and plants. These vary in degree from mutualisms in which, for example, certain plants (myrmecophytes) provide shelter or food to a resident population of ants in exchange for possible protection from herbivores, to herbivory in which plants are damaged either directly by consuming their tissues or indirectly by eating their seeds. Ants of the tribe *Attini* culture fungi on collected food materials in a way an analogous to the macrotermitine Isoptera (Weber, 1982).

As with the termites, ants also modify soil chemical and physical properties by transporting food and soil materials during such activities as feeding and mound and gallery construction. These activities affect soil developmental processes and fertility and may modify the nature and distribution of the vegetation, at least on a local scale.

Through their manifold activities, ants frequently impinge on man and his interests. Many ants have highly-efficient defense mechanisms and use their effective stings and mandibles to repel man and other vertebrates from their areas of influence. In agricultural situations, ants may protect such crop herbivores as honeydew-secreting Homoptera or act directly as herbivores themselves. Nonetheless, their beneficial role as effective general predators should also be noted (Way and Khoo, 1992). In parts of the Neotropical region, underground nest excavation by attine ants may lead to the subsidence of such structures as roads and buildings (Nogueira and Martinho, 1983).

This section examines the diversity, distribution, development and community structure of the ants and their interactions with soils and vegetation. Ants are highly mobile animals and do not necessarily restrict themselves to particular strata in the landscape. Many soil-nesting species exploit the food resources available in trees and, conversely, at least some predominantly arboreal ants also forage on the ground surface thereby influencing the local ground ant community. Therefore, the roles played by the total ant community are important in understanding the impacts that ants have on soils and ecosystems.

4.3.3.1 *Taxonomy*

The ants belong to the sub-order Apocrita (Aculeata) of the insect order Hymenoptera. This sub-order also includes the wasps, the bees and a myriad of predatory and parasitic species important in many environments for their roles in controlling populations of herbivores and saprovores.

The ants are defined as the members of the super-family Formicoidea (which contains the Formicidae as its sole family) and are derived from the Vespoid line of evolution within the Apocrita. All are social species with winged male and mostly winged female reproductives (queens). The queens discard their wings after a nuptial flight in a manner analogous to that occurring in termite queens, while the workers remain apterous.

The family Formicidae is divided into sixteen subfamilies (with others known only as fossils) and is further sub-divided into a larger number of tribes (Bolton, 1994). The total number of described ant species is estimated at approximately 9-10,000 species distributed among 296 extant genera although Hölldobler and Wilson (1990) estimated that there may be as many as 20,000 species distributed among some 350 genera. An account of the morphology and systematics of the Formicidae is beyond the scope of this book but is available in Hölldobler and Wilson (1990) and Bolton (1994).

4.3.3.2 *Development*

As holometabolous insects, ants hatch from an egg stage which is followed by three to six (normally four) larval instars and a pupal stage during which the body is reorganised into the adult form. Over a range of species, development from the egg to eclosion from the pupa may take up to three months. Most species form a cocoon during the pupal stage.

As in other Hymenoptera, sex is determined by haplo-diploidy, a system in which the males are haploid and the females diploid. Parthenogenesis is uncommon but not unknown.

Castes and determination

As in the termites, ant castes may be broadly defined as groups of individuals that specialise in particular roles or groups of tasks within the colony. While castes are normally defined on a morphological basis, the roles performed may alter over time. To distinguish between these two forms of polyethism, the former are known as physical and the latter as temporal castes. Physiological castes are defined in relation to a major physiological state that may also be related to age or anatomy. With the exception of the males, castes appear to be determined on the basis of colony needs rather than a genetic basis.

The following castes and other adult forms are to be found among the ants.

Males. With few exceptions, the males fulfil no defined productive role within the ant colony and their role is purely a reproductive one.

Queen. The queen (or gyne) is the principal female reproductive of the colony and, in most species, is winged with a large gaster and a fully-developed hymenopteran thorax. Wings are shed early in the process of colony founding. In contrast, among the army ants, the queens are blind and wingless with a reduced thorax (dichthadiiform). The wings of certain social parasitic ants are also much reduced and non-functional.

Workers. The workers comprise the wingless female caste in the ants. With few exceptions, they are sterile and have reduced or no ovarioles. In contrast to the queen, the workers have a simple thorax consistent with their lack of wings.

The worker caste may be subdivided into major, media and minor sub-castes on the basis of size, each of which may have defined behavioural roles. Where the major worker is highly specialised for fighting, she is known as a soldier. Such specialisations may include modifications to the mandibles to form sickle- or hatchet-like organs more effective in fighting. A further specialisation is phragmosis, or the gross enlargement of the whole head leading to a massive structure which is used to block passages within the nests thereby denying access to enemies. In some species, the gaster may be similarly modified for blocking.

Ergatogynes. These are non-reproductive forms intermediate between the workers and the queen and occur in two forms namely, intercastes and ergatoid queens. The intercastes occur in a graded series and are present in conjunction with the workers and the queens; they lack spermathecae and are therefore unable to mate. In contrast, the ergatogynes or ergatoid queens can replace the queen and do not form part of a graded series. They are most common in primitive ants and possess functional spermathecae.

Gamergates. These comprise a true physiological caste that occurs in some of the ponerine ants where the queen is replaced by fertile, worker-like forms.

In addition to the gamergates, two further physiological castes are of interest. The first of these is the 'replete' or 'honey pot' ant in which some of the major workers have become specialised for the storage of liquid food (as sugar solutions) and water through enormous dilation of the gaster. These occur in the families Dolichoderinae and Formicinae and appears to be an adaptation to the unfavourable conditions pertaining in such harsh habitats as deserts. A physiological caste occurs in certain granivorous ant genera (*e.g.*, *Pheidole, Acanthomyrmex*) in which seeds constitute a substantial part of the diet. As a specialisation for milling seeds, the head is grossly enlarged relative to body size to accommodate the enormous mandibles and their associated musculature (Hölldobler and Wilson, 1990).

In addition to the physical and physiologically-defined castes and the other morphologically-defined forms listed above, temporal castes occur that are defined on the changing roles that they perform in the colony (temporal polyethism). For example, in the myrmicine ant *Pheidole dentata*, the youngest minor workers (0-2 days) care preferentially for the queen, the eggs and the youngest larvae. From 2 to 16 days, they care for the medium-sized larvae. After this, they continue to work in the nest but an increasing proportion of their time is spent in foraging outside the nest (Wilson, 1976).

4.3.3.3 *Colony ontogeny*

Ant colonies progress through a series of developmental stages analogous to those occurring in the termites. There is an initial founding stage, followed by a period of growth after which the colony achieves mature status, defined by the production of reproductive forms. As occurs in all social insect societies, the colony eventually declines and dies, often following the death of the founding queen and failure of other workers or ergatogynes to take over the deceased queen's role. The colony may also be destroyed at any stage through interactions with other ants, by predation, parasitism or pathogenesis or as a consequence of other disturbance.

While colony establishment normally follows the liberation of virgin queens from the parent nest, colonies may also establish through other swarming processes. This takes place where a group of workers, sometimes with a queen, leaves the parent nest together with one or more queens to establish a new colony.

The founding stage

Virgin emerging queens often leave the parent nest in response to a stimulus such as rain and, after attracting a male and being inseminated, proceed to shed their wings and establish their own nests. Following inital nest establishment, the queen produces her first batch of eggs. In order to feed the developing brood, the queen supplies them with trophic or non-fertile eggs, produced from the breakdown products of her fat bodies and wing muscles. In some species, the queen remains in the nest and the first workers are reared solely from the energy contained in her bodily secretions; this is known as claustral founding. The situation in which she emerges periodically from the nest to forage for food is known as partially-claustral founding.

The exponential, or ergonomic stage

This is the stage of colony growth. The workers produced from the initial brood are known as minims and are usually smaller than those found later in colony development. Subsequent to this, the queen devotes herself to egg production and the colony population proceeds to increase over time in an approximately sigmoidal progression.

The reproductive stage

The stage of the mature colony is achieved when the population has started to approach its asymptotic level. Brian (1957) has divided the reproductive stage into an adolescent period during which only males are produced and the mature stage during which queens are also produced. Winged queens and males are liberated from the maternal nest at characteristic times of year for each species, the proximal cue often being an environmental event such as a rainstorm. However, not all colonies produce queens in all years and individual colonies may not produce individuals of both sexes in any one year.

Colonies may live for considerable periods although this is only related to the longevity of the original queen in those species where queens are not replaced. Queens of many species (*Myrmica, Formica, Lasius*) have been shown to live for more than 20 years (Baroni-Urbani *et al.*, 1978) although life spans for many others are considerably less and a few species may live for longer periods (Hölldobler and Wilson, 1990).

The time scale of colony ontogeny is illustrated by that of the Nearctic seed-harvesting ant *Pogonomyrmex barbatus* (Gordon, 1995; Gordon and Kulig, 1996) (Figure III.55). Mating takes place in mid-summer with the release of the winged forms. Following mating, the males die and the females drop their wings and dig a nesting hole within which they remain sealed until the first workers emerge to forage on the surface, some 10 weeks later. Colonies grow steadily reaching *ca.* 3000 workers at 2 years, 4000-8000 at 3-4 years and reach maturity at about 5 years with the release of the first reproductives. Thereafter, mature colonies maintain a stable size of 10,000 to 12,000 workers. The colony dies after 15 to 20 years since dead queens are not replaced in this species.

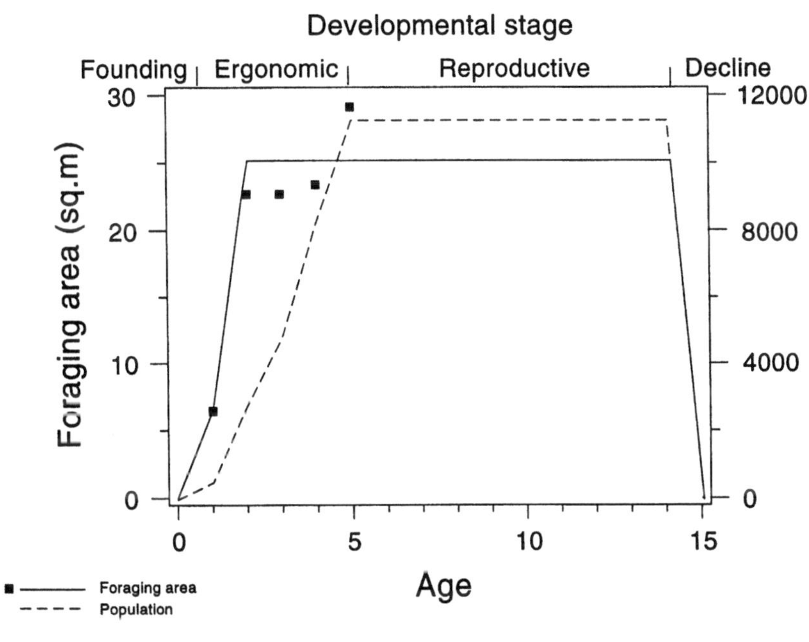

Figure III.55 Growth in colony size and foraging area with age in the Nearctic harvester ant *Pogonomyrmex barbatus* (redrawn from Gordon, 1995).

4.3.3.4 *Food materials and feeding habits*

In the presumed ancestral vespoid pattern of diet still retained by some ponerine ants, larvae are fed with protein-rich animal materials while the adults are largely nectivorous. Tobin (1994) considered that two distinct and opposing trophic shifts have occurred in the evolution of the ants. One has involved an increasing reliance on animal matter while the second has led to a greater degree of herbivory.

While most ant species are generalised predators and scavengers, a wide range of feeding habits occurs within the family. Tobin (1994) argued that the importance of herbivory has been greatly underestimated, in part due to the difficulties in quantitatively assessing the origins and amounts of liquid food carried by foraging ants. Most species accept a spectrum of food materials and only a minority have highly-specialised diets. Of 143 'New World' genera considered by Tobin (1994), only in 29 genera (20 %) were all species considered to derive most of their nutritional requirements from animal sources. The food resources utilised by a colony may alter qualitatively and quantitatively over time, depending on the stage of colony development and on changing seasonal requirements associated with such factors as reproduction (Stradling, 1987).

Many species active on the soil surface or underground depend largely on animal food materials. Of these, specialised predators are most common in the tropics and their incidence diminishes with increasing latitude. Lévieux (1983) reported that *ca.* 25 % of the 113 species found in the savanna of Lamto in the Côte d'Ivoire were thus specialised; some species, for example, specialise as consumers of Isopod crustacean prey, others of termites or earthworms. Although some dietary overlaps occur, most specialised feeders appear to rely on largely different staple food resources.

Some ants are extremely polyphagous. Cherrett (1968) reported that the leaves of 36 of 72 plant species present in his study area in Guyana were attacked by the leaf cutter ant *Atta cephalotes*. Fallen flowers were also attacked. Adams (1986) and Reagan (1986) summarised the diet of the highly-invasive introduced fire ants, mainly *Solenopsis invicta* (but also *S. richteri*), in the southern part of the United States. These ants predate upon a wide variety of invertebrates but also feed regularly on different tissues amongst a range of cultivated and other plant species (Table III.16). In addition, *S. invicta* feeds on honeydew excreted by sap-sucking insects (Homoptera) and enters into a mutualism with at least some species. Further, the species predates on the tissues of some higher animals and may effect a notable reduction of certain forms of wildlife. Adams (1986) cites attacks on a number of nestling birds and the eggs and newborn young of reptiles reported by a number of authors. Such invasive species may, over time, become more harmoniously integrated into local ant communities (Hölldobler and Wilson, 1990).

Workers may forage singly or in groups within the territory of their own colony and recruit other workers to assist in the exploitation of significant food resources. In the absence of group transport, such ants are limited in the size of the food items they may return to the nest by the carrying capacities of individual workers. Potential food items may range widely in size and, in species where workers differ in size, each size class of workers may transport separate size classes of food items. Other ants forage cooperatively in groups and this allows them to utilise food items that would normally be too large or too well defended to be available to individual workers.

Table III.16 A partial list of food items of the invasive fire ant, *Solenopsis invicta* in the southern part of the United States (Adams, 1986; Reagan, 1986).

Food Item	**Crop**
1. Plant food items	
Reproductive tissues	
Nectar	Citrus
Flowers	Citrus
Fruits	Citrus
Seeds	Corn, peanuts, beans, *Pinus palustris*
Vegetative tissues	
Sap	Citrus
Growing point	Egg plants
Leaves and stems	Corn, beans, cabbages, potatoes, sweet potatoes
Roots	Corn, okra, soyabeans, beans, potatoes, sweet potatoes, Cabbages
Tubers	Potatoes
Bark	Citrus
2. Animal food items	
Major taxon	**Minor taxa**
Coleoptera	Scarabaeid larvae and eggs, curculionid larvae
Lepidoptera	Eggs, larvae of several species
Diptera	Larvae of dung-breeding flies, Tabanidae
Dermaptera	
Hemiptera	Aphididae, Pentatomidae
Collembola	
Orthoptera	Gryllidae
Hymenoptera	Formicidae
Acari	Metastigmata, other mites
Araneae	
Annelida	Earthworms
Myriopoda	millipedes, centipedes

Army ants represent the greatest development of this foraging behaviour and are distinguished by a combination of frequent nest movement and the behavioural traits of group foraging and group prey retrieval (Gotwald, 1995). These ants send out very large numbers of workers and, through their co-ordinated activities, can subdue large prey items such as cockroaches, earthworms and even small vertebrates such as frogs. The food items are then subdivided where necessary and transported back to the nests, also through the concerted actions of multiple workers.

Foraging activity is dependent on current environmental conditions. Many ants forage within a temperature:humidity envelope and this is illustrated in Figure III.56

where the rate at which workers returned to a nest of the North American desert fungus feeding ant, *Trachymyrmex smithi* declined to zero at saturation deficits of greater than 35 g m^{-3} and at soil surface temperatures greater than 50 °C. In contrast, workers of *Formica perpilosa* at the same site continued to forage at temperatures of 55 °C and saturation deficits of 52 g m^{-3} because they were able to replace lost fluid from the honey dew that they feed on and because of their greater tolerance of water loss (Schumacher and Whitford, 1974).

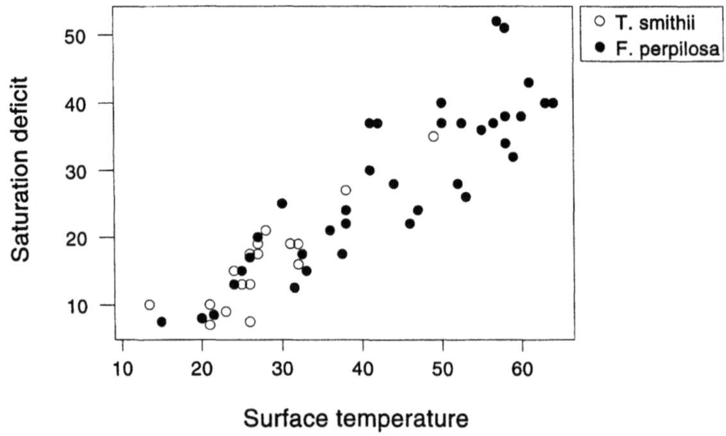

Figure III.56 Surface-soil temperature (°C) and saturation-deficit (g m^{-3}) envelopes within which foraging occurs in the North American desert ants *Trachymyrmex smithi* and *Formica perpilosa* (redrawn from Schumacher and Whitford, 1974).

Activities are also constrained by strong diurnal rhythms (Hunt, 1974; Roces and Nunez, 1989) and daily cycles of foraging activity may be modified to avoid predation and reduce competition (Hölldobler and Wilson, 1990). At a seasonal scale, foraging may be constrained by environmental conditions. In a Nearctic temperate environment, the three seed-harvesting species of *Pogonomyrmex* studied by McKay (1981) foraged most actively during the warmer summer months but were inactive during the winter.

Plant materials

Tobin (1994) argued that while the ants as a group have not specialised in the digestion of cellulose, their abundance and dominance in many environments is due to their capacity to exploit plant resources. Many species ingest a wide variety of plant materials, both vegetative and reproductive, in appreciable and perhaps hitherto unrecognised quantities (Tobin, 1994). In particular, many ants feed on seeds, seed-associated structures, plant secretions and other products or little-modified plant materials excreted by herbivores.

Few ants feed exclusively on plant materials although many use them on a seasonal or facultative basis. For example, the granivorous ant *Messor pergandei* may consume leaves, stems and flower parts when seed sources are in short supply (Rissing and Wheeler, 1976). Vegetative tissues are less commonly eaten although those taken tend to be the softer materials. Wheeler (1910) records that a number of hypogaeic ant species feed on underground plant parts including the cotyledons of germinating seeds. These materials are probably eaten more for their more readily-digestible components than for their cellulose.

Many of the relationships between ants and plants have co-evolved over time to produce an apparent mutual benefit to both groups. However, in reviewing a number of studies of this phenomenon, Becarra and Venable (1989) reported that only 43 % of the 28 associations they examined were beneficial to the plant and that many of these associations may be facultative.

Seed harvesting

Seed harvesting is a way of life for a number of species of the subfamilies *Ponerinae*, *Formicinae* and most notably the *Myrmicinae*, which feed wholly or largely on seeds which they collect and transfer to their nests. These are known as harvester ants and dominate in deserts and drier grasslands in temperate and tropical environments; seed harvesting may buffer them against temporal changes in the availability of food (Handel and Beattie, 1990). While most seeds gathered by harvester ants are destroyed, some are lost during transport and others may germinate in the nests or on refuse dumps. Seed-harvesting ants are therefore simultaneously agents of destruction and dispersal for the plant species whose seeds they feed on and they may substantially influence the distribution of vegetation in space, and over time.

Seed harvesting ants are also active in such humid environments as rainforests. Levey and Byrne (1993) reported that *Pheidole* spp. harvested small seeds from frugivore faeces on the forest floor of a Costa Rican lowland rainforest. Depending on the species, 25-32 % had seed cached in their nests and 24-32 % had discarded seed on their refuse piles. In glasshouse studies, seedlings of one genus grew substantially better on the refuse piles than on local topsoils, illustrating the advantage to the surviving plants.

Many other ant species occasionally use seeds as food materials but contrast with the more specialist seed feeders. Even such prominent carnivores as the army ants may collect and consume appreciable quantities of seeds (Gotwald, 1974).

Elaiosomes (arils)

Seed dispersal by ants is a widespread feature in all continents where they occur (Handel and Beattie, 1990) and plants whose seeds are spread largely by ants are termed myrmecochores. Berg (1975) lists more than 1500 species of these plants from Australia and similar numbers have been recorded from South Africa (Bond and Slingsby, 1983).

The evolution of structures on seeds that contain materials attractive to ants has occurred many times and the structures involved originate from a diversity of tissues. Elaiosomes are large bodies attached to seeds that contain such materials as lipids, proteins, starch and sugars that are attractive to ants (Beattie, 1985). On being shed from the plant, the seeds and their attached elaiosomes are transported by ants towards their nests, the elaiosomes eaten and the seeds discarded undamaged, often within or in the vicinity of the nests.

In addition to elaiosome-related transport by ants, compound seed dispersal mechanisms are known that also involve other animals. Clifford and Monteith (1989) report the three phase dispersal of seeds of *Petalostigma pubescens* (Euphorbiaceae) in northern Australia. The first or frugivorous phase is ingestion of the fruit by the large flightless emu, *Dromaius novaehollandiae*, which may retain the fruit within its intestinal tract for more than 72h. After the partially-degraded fruit is voided, it dries and the second phase takes place as an explosive dehiscence of the endocarp which scatters the seeds 1.5 to 2.5 m from the point of deposition. The final dispersal phase is transport of the seed by ants attracted to their elaiosomes. Up to 1200 seeds may occur in a single scat and the advantage to *P. pubescens* of the two final phases apparently lies in the reduced seedling competition consequent on wider dispersal of the seeds.

Nectar feeding

While certain ants may feed regularly on floral nectar (Haber *et al.*, 1981), few plants appear to be regularly pollinated by ants possibly because of the toxicity of their metapleural gland secretions (Peakall *et al.*, 1991). Perhaps of equal importance to the biology of both the plants and the ants is the utilisation by the latter of extra-floral nectaries. As their name suggests, these nectar-producing structures have evolved in a wide diversity of plants and vary from single to multiple celled structures; they may occur on many parts of the plants including the stems, leaves or petioles. Their secretions are largely carbohydrates although amino acids may also be present (Bentley, 1977; Beattie 1985; Baker *et al.*, 1978). Such secretions are attractive to the many ants that feed regularly on them and which may defend the extra-floral nectaries.

Food bodies

Food bodies are small epidermal structures of varied origin found on certain plants that provide nourishment to ants; they form one further component of the mutualisms that have developed between certain ants and plants. Beattie (1985) lists them as providing lipids, proteins and glycogen to the ants in return for their potential plant protection services.

Honeydew

Honeydew is an energy-rich product excreted by a range of homopterous insects. These insects feed by inserting their suctorial mouthparts (proboscis) into the phloem vessels of the host plant. The internal turgor pressure within these vessels obliges such insects to imbibe sugar-rich fluids substantially in excess of their own metabolic requirements (Chapman, 1982). This excess is excreted in the form of small droplets and it is this material that is imbibed by the ants.

The material excreted by the homopterans is only slightly modified by its passage through the insect gut. It contains a high proportion of sugars together with some nitrogenous components in the form of amino acids and amides. A range of insects (honey and other bees, wasps, flies, etc.) exploit this readily-available energy source although it is the ants that have specialised in its utilisation (Wheeler, 1910).

While ants feed on the honeydew, the homopterans are protected by the ants from predators and parasitoids; they may also be transported by the ants and sheltered in specially-built structures. The homopterans taking part belong to the families *Aphididae, Coccidae, Pseudococcidae, Psyllidae, Membracidae, Cicadellidae, Cercopidae, Fulgoridae,*

Eurymelidae and *Aleyrodidae* (Hölldobler and Wilson, 1990; Carver *et al.*, 1991). The ants included in this mutualism largely belong to the phylogenetically-advanced subfamilies *Myrmicinae*, *Dolichoderinae* and *Formicinae*. The variety of both ant and homopteran species involved indicates that this mutualism has developed many times and this is reflected in the degree of closeness of the relationship which may range from casual to occasional to obligatory. While this mutualism is most apparent on the aerial surface of plants, it also occurs widely in associations occurring on underground plant parts. In a calcareous grassland in England, Pontin (1963) recorded a total of 16 subterranean species of *Aphididae* and *Coccidae* tended by the formicine ants *Lasius flavus* and *Lasius niger*.

Fungivory

Members of the tribe *Attini* (subfamily *Myrmicinae*) are distributed from 44°S in South America to 40°N in the United States (Weber, 1982); the tribe includes approximately 190 species distributed among 12 genera (Hölldobler and Wilson, 1990). These ants partake in a unique obligate mutualism with a fungus that they culture on organic materials brought into their nests, in a similar association to that occurring in the macrotermitine Isoptera.

As Hölldobler and Wilson (1990) have stated, the *Attini* are a particularly successful group in the areas where they occur. For example, Cherrett (1986) has estimated that they may remove from 12 to 17 % of annual leaf production in the tropical rainforests where they occur and may influence the structure of the vegetation through selective herbivory (Vasconcelos and Cherrett, 1997). They are also destructive agricultural pests attacking a range of crops and grasses (including sugarcane) and may reduce the carrying capacities of the pastures in which they occur by up to 50 % (Fowler *et al.*, 1986).

Perhaps the best known species of this tribe belong to the genera *Atta* and *Acromyrmex* which are known as 'leaf-cutting' ants from their habit of cutting small sections of leaves from their food plants with their specialised mandibles. These are returned to the nest to be cultured with a symbiotic fungus. The remaining ants of this tribe culture their fungi on other plant parts, insect faeces or carcasses (Weber, 1982). The fungal symbiont associated with the attine ants appears to be *Leucocoprinus gongylophora*, a member of the Agaricales (Basidiomycota) (Hölldobler and Wilson, 1990). However, it is not entirely clear whether this is the only member of this genus routinely involved and one species *Cyphomyrmex rimosus*, cultivates a yeast-like fungus (Wetterer, 1994).

As materials harvested by leaf-cutting ants are brought into the nest, they are thoroughly chewed and moistened by the workers and, together with a drop of protease-containing anal fluid, are deposited onto the fungal body and inoculated. Worker ants eat the swollen tips of the growing hyphae although they derive most of their nutriment from plant sap exuded from the leaf surfaces cut while foraging and during preparation of the plant material for culturing (Quinlan and Cherrett, 1979).

The workers have a limited capacity to produce proteases and may acquire most of those needed from the tissues of their fungal symbiont (Martin, 1984). The larvae, in contrast, secrete their own proteases which they use to degrade proteins within the hyphal walls of the symbiotic fungus. Workers derive glycogen and starch from, respectively, the fungus and the food plant; degradation products from the pre-oral digestion of carbohydrates are passed on to the midgut where further enzymatic degradation occurs (Febvay and Kermarrec, 1986). Proteases acquired from the symbiont are passed through the worker gut largely unaltered (Martin, 1984) and are applied to the harvested leaf material facilitating its utilisation by the symbiotic fungus.

Apart from the above, the only non-pathogenic interaction between ants and microorganisms so far described is that between the harvester ant *Pogonomyrmex occidentalis* (and certain other *Pogonomyrmex* species) and the elevated abundances of the hyphae of vesicular-arbuscular mycorrhizal fungi that occur in the mound materials of this species (Friese and Allen, 1993).

P. occidentalis clears all vegetation from circular areas around its nest and such areas may cover up to 10 % of the ground surface. Through the return of food materials to their nests, these areas become enriched in nutrient elements and are heavily exploited by very dense mats of VA mycorrhizal-infected roots of the shrub *Artemisia tridentata*. Ants clip the VAM-infected root tips, presumably to prevent them from obstructing galleries although it is unknown whether the ants ingest any of this material. Following colony death, rapid invasion of the cleared areas by *A. tridentata* and a mycotrophic grass is facilitated by the presence of the highly infective VA inoculum of the mounds.

4.3.3.5 *Geographic distribution of ants*

Ants are spread widely throughout the world and the approximate numbers of species known from the major biogeographical regions of the world are indicated in Table III.17 (from Bolton, 1995). As stated by these authors, many species still remain to be described.

Table III.17 Distribution of the known ant species among the major biogeographical regions of the world (Bolton, 1995).

Region	Number of Species	Percentage of Total
Neotropical	358	24.7
Nearctic	555	5.8
Palaearctic	1153	12.1
Afrotropical	1686	17.7
Malagasy	251	2.6
Oriental	770	8.1
Australia	1057	11.1
Indo-Australian	1709	17.9
Total	**9539**	

Because of their marked thermophily, ants have radiated most widely in the tropics and in a parallel with the termites, the number of species declines steadily with increasing latitude. Ants are absent from Iceland, Greenland and Antarctica (Hölldobler and

Wilson, 1990). Kusnezov (1957) plotted the decline in species richness with latitude for the Neotropical and Nearctic regions, and for Europe (Figure III.57a). More recent work (see, for example, Cushman *et al.*, 1993) has confirmed this general latitudinal trend.

In addition to latitudinal variation, the number of species falls off drastically with increasing elevation and this is illustrated (Figure III.57b) with data from the Great Smoky Mountains National Park in the USA (Cole, 1940 in Kusnezov, 1957). Even in the tropics, ants are largely absent from forested mountains above 2500 m in elevation (Hölldobler and Wilson, 1990). However, above the tree line, the number of species may

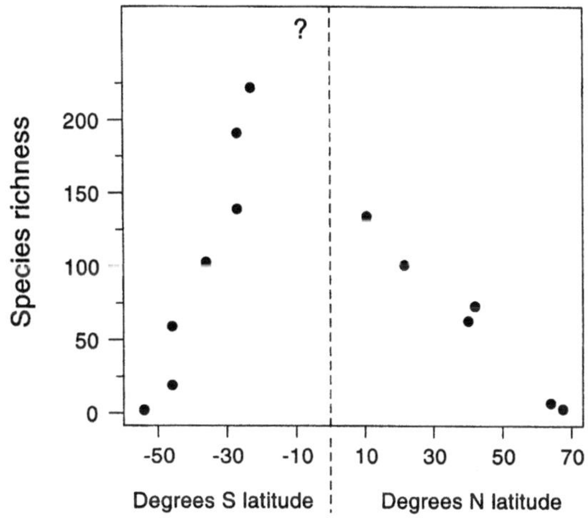

Figure III.57a Changes in ant species richness with latitude in the Neotropical and Nearctic regions and in Europe (redrawn from Kusnezov, 1957).

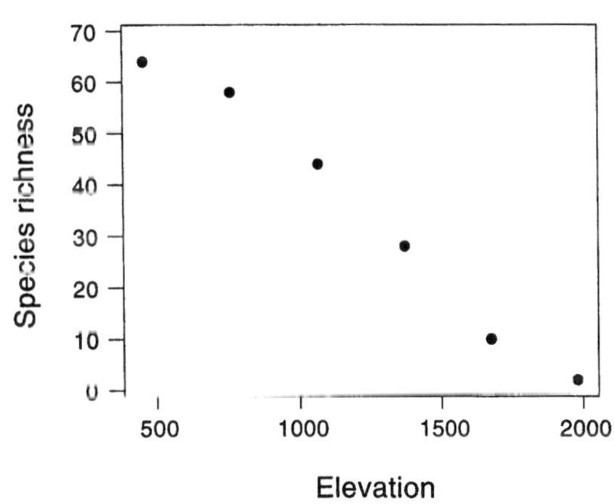

Figure III.57b Changes in ant species richness with elevation in the Great Smoky Mountains National Park, USA (from Cole, 1940 in Kusnezov, 1957).

increase again. The endemic formicine *Formica picea* is found at elevations from 2800 to 4800 m above sea level in the northwest Himalayas (Mani, 1962).

Figure III.57a includes few data from the inter-tropical areas. In a further parallel with the termites (Section III.4.3.2), substantially higher levels of species richness have been reported in tropical environments (Hölldobler and Wilson, 1990). At the local level, these authors cited local richnesses of 172, 219 and 272 species from areas of *ca.* 2.6 km^2 in, respectively, Papua New Guinea, Ghana and Peruvian Amazonia. In addition, Lévieux (1983) reported a total of approximately 230 species from the humid savanna and gallery forests of Lamto, in the Côte d'Ivoire. Within even smaller areas, a single tree in Peruvian Amazonia yielded 43 species (Wilson, 1988) while Room (1971) found 128 species in only 250 m^2 in a cocoa plantation in Ghana.

Ant species richness is limited by isolation and aridity. A few very isolated areas (Polynesia east of Tonga, some of the isolated islands of the Atlantic and Indian Oceans) have no native species of ants (Wilson and Taylor, 1967). Aridity also limits the species richness of ant communities and dry areas tend to have substantially fewer species than more mesophytic locations with similar temperature regimes. This was recorded for African savannas by Lévieux (1983) and for two South American locations by Kusnezov (1957).

In addition to the species indigenous to particular localities, a number of ant species have been spread by man to a broad range of countries throughout the world, sometimes with severe impacts on the local biota. Among these are included the 'tramp' species which live in close association with man.

4.3.3.6 *Spatial distribution of ants*

Perhaps the major driving force in ant communities is competition for resources, both between and within species (Hölldobler and Wilson, 1990). This competition is severe and unrelenting and results in continuing changes to the composition and community structure of ant populations and to their spatial dispositions. Competition operates over a wide spectrum of spatial and temporal scales ranging from the individual colony to whole faunas and over short and long time scales.

This severe competition has led to niche specialisation and saturation in such species-rich environments as tropical savannas. It has also led to the horizontal partitioning of the environment into territories and vertically into a series of partially-overlapping strata.

Territories

The defence of a foraging area containing resources (territoriality) appears to be a general phenomenon within the ants and may occur between ants of the same or different species (Hölldobler and Wilson, 1990). It is usually manifested by interference which may include predation and other competitive interactions leading to the exclusion of other ants from such resources as food and nesting sites.

In the ants, a territory comprises an area surrounding the nest within which members of the colony forage for food resources and which is defended against competing members of the same and other species. Territories may be **absolute** in which case they are permanently defended against almost all other ant species and very few others are permitted to remain with their borders. Figure III.58 presents an example of the distributions of the absolute territories of the dolichoderine ant *Iridomyrmex purpureus* at a four hectare study site in South Australia (Greenslade, 1975). Territories are classified as internal where they are bounded by those of other *I. purpureus* colonies and external territories which taper off into dense vegetation, an unsuitable habitat for this species which requires well insolated and well drained sites. Some colonies extend along tracks in the dense vegetation and may extend up to 200 m from nests. *I. purpureus* is a dominant species and excludes most other ants from within the area; the area of suitable habitat is almost fully occupied by territories of *I. purpureus*. The colony pattern is considered stable and the borders of the territories are maintained by frequent intraspecific confrontations.

Other territories have flexible borders and are known as **spatio-temporal.** In these, only that part of the territory containing resources currently being exploited is defended at any one time. However, the remaining area may occasionally be searched by scouting workers to permit the discovery of new resources or to guard it against the inroads of competing colonies. Ants that migrate regularly, such as the army ants, do not defend permanent territories (Gotwald, 1995).

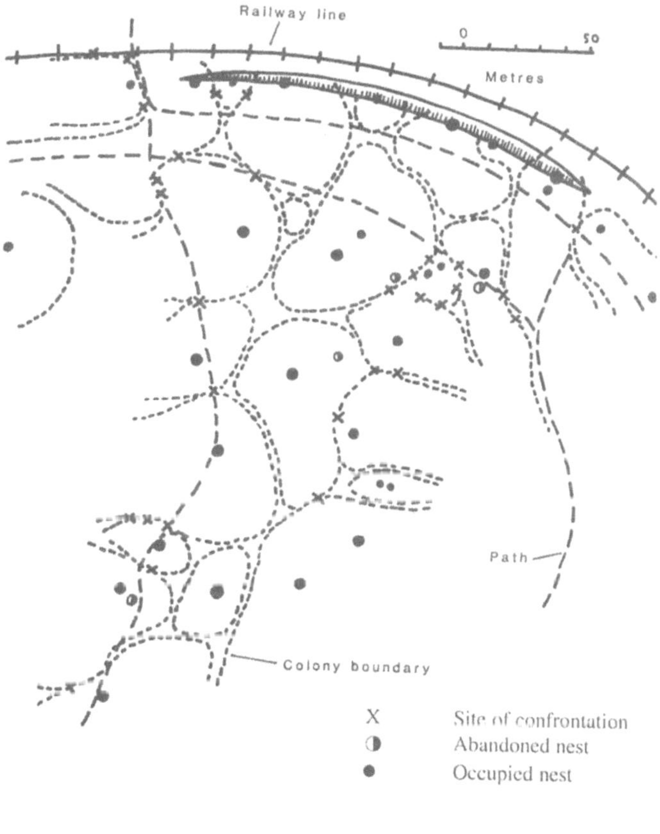

Figure III.58 Distributions of the colonies of *Iridomyrmex purpureus* (redrawn from Greenslade, 1975).

The territories of ground-dwelling ants normally comprise an area of the ground surface and the galleries in the soil beneath it. For any given species, territory size depends on the stage of colony development and its population size; territorial boundaries are continually being adjusted through interspecific or conspecific encounters. Gordon (1995) showed that in colonies of the Nearctic seed harvester *Pogonomyrmex barbatus* (Figure III.55), foraging area increases markedly from one to two years of age and differs little thereafter, despite the continuing increase in worker populations. Competition between adjacent colonies has been verified by the observation that unused areas are soon incorporated into the foraging spaces of neighbouring colonies (Gordon, 1992).

Where ants nest in trees, they may confine themselves entirely to the trees or may also utilise an area surrounding the base. The formicine weaver ant, *Oecophylla smaragdina*, has an arboreal nesting system and a colony territory that may extend over a number of trees (Hölldobler, 1983). However, this species also forages extensively on the ground surface between the trees (Wilson 1959). Even predominantly-terrestrial species may have territories that include the surrounding trees. Room (1975) reports that, in a number of forested and agricultural environments in Papua New Guinea, the territory of the dominant formicine ant *Anoplolepis longipes* has a three dimensional character in that it nests on the ground but forages over all strata, including the trees within its territories. However, this situation is apparently less common than that represented by *O. smaragdina* (Wilson, 1959) and most ants appear to forage downwards from arboreal nests rather than upwards into the canopy.

A clear vertical stratification has been found in the species-rich humid tropical savannas of Lamto in the Côte d'Ivoire. Here species are clearly stratified within the soil, the herbaceous layer and in the trees, with only a few species overlapping between adjacent layers (Lévieux, 1983). The ants of the Lamto savanna are divided into the following overlapping sets, based on the principal strata they exploit in the landscape:

(i) Soil dwelling species, largely specialised ponerines, that further stratify themselves according to the range of soil depths within which they forage;
(ii) The epigaeic species that forage on and near the soil surface;
(iii) Those that nest within the hollow stems of grasses;
(iv) The arboreal species that nest in hollow branches;
(v) The arboreal ants that construct leaf carton nests in the canopy.

Patterns of horizontal dispersion

The nests of ants that defend persistent nesting sites are frequently overdispersed in uniform habitats. That is, they are distributed horizontally in patterns that are nearer regularity than might be expected to occur by chance alone and this implies population limitation by competition for food and other resources. In reviewing the results of approximately 80 previous studies of ant nest dispersion, Levings and Traniello (1981) found that 83 % of those capable of being analysed statistically reported distributions significantly more regular than random. Traniello and Levings (1986) found that even in colonies with multiple nests, nest entrances were overdispersed within territories. Some evidence of an increased regularity in nest dispersion at higher population densities of *Lasius flavus* has also been found in Britain (Waloff and Blackith, 1962).

Not all colonies are overdispersed and Herbers (1989) found that the ant nests she studied in regrown North American forests were aggregated in 15 of 23 plots and were randomly distributed in the remainder. This was considered to result from the aggregated nature of suitable nesting sites in the forests that she studied and to environmental heterogeneity. On the basis of these results, Herbers concluded that the ants in her study area were limited more by a dearth of appropriate nesting sites than by food. Gordon (1991) also reported non-regular distributions of the nests of competing colonies of the Nearctic harvester *Pogonomyrmex barbatus*. She attributed this to the regular interdigitating shapes of the foraging areas and considered it likely that it is the latter rather than the nests that are evenly partitioned. In both of these studies, nest dispersion patterns altered over time.

Even established ant colonies may be considered relatively mobile entities. Some species apparently move their populations seasonally or at shorter intervals while others with large, climatically-buffered nests do so only when considerably provoked. Approximately 10 % of the colonies of the Nearctic seed-harvesting ant *Pogonomyrmex barbatus* moved annually in the area studied by Gordon (1992) and Gordon and Kulig (1996) while those of the other *Pogonomyrmex* spp. present moved more frequently. However, colonies of many mound-building species often remain at a single location for a number of years. Such influences as mechanical disruption of the nest, flooding, predation, competition and environmental changes may lead to the movement of whole populations of affected colonies (Hölldobler and Wilson, 1990). Army ants move their nest sites continuously during the nomadic phase of their activities.

4.3.3.7 *Ant populations and biomasses*

Ant colonies differ widely in their abundances depending on the subfamily they belong to, colony ontogeny and their environment. Where colonies are small, up to thirteen per square metre may occur (Baroni-Urbani and Pisarski, 1978). At the other extreme, Higashi and Yamaguchi (1979) reported an extremely large colony of *Formica yessensis* that contained an estimated 307 million individuals living in approximately 45,000 nests and covering an area of 2.7 km^2. Nonetheless, most colony population densities fall nearer to the lower end of this range and the median colony density for the more than 145 sites (mostly from areas of temperate climate) recorded in Baroni-Urbani and Pisarski (1978) was the equivalent of 1700 colonies per hectare (interquartile range 300-4800 colonies).

The population abundances and biomasses of ants are very poorly known on a unit area basis. The maximum population density estimates recorded in the summaries of Baroni-Urbani and Pisarski (1978) and Petersen and Luxton (1982) are *ca.* 7300 workers m^{-2} and this corresponds to a standing crop biomass of 2.4 g m^{-2} (dry weight) (most ants weigh between 0.1 and 1.0 mg dry weight although a few exceptional species reach 3 mg). Most populations are substantially less abundant than the maximum cited above. Ants attain their greatest importance in tropical savannas where they may comprise more than 16 % of the total invertebrate standing crop biomass; in other biomes, their biomasses attain maximum values of *ca.* 2 % of invertebrate biomass (Petersen and Luxton, 1982).

4.3.3.8 *Effects of ants on soils*

While arboreal ants have few direct effects on soils, many ants nest on and within soils where they may build epigeal mounds and construct systems of galleries, nesting chambers and other voids. Through their contributions to bioturbation (Levieux, 1976; Cowan *et al.* 1985; Lockaby and Adams, 1985), they add their effects to those of other soil-based ecosystem engineers and to the landscape level processes of soil and biomantle formation discussed in Chapter II.

Ground-dwelling ants, particularly the mound-building ants can be considered as ecosystem engineers (Folgarait, 1998) in that they modulate the availability of resources and alter the soil and surface environments in ways that affect other organisms (Jones *et al.*, 1994). Where abundant, ants modify the physical structure of the soil through the creation of systems of galleries and chambers and this influences soil porosity, aeration, infiltration and drainage and creates habitat for smaller soil organisms. Changes to texture and other soil properties occur through the vertical and horizontal transfer of materials of different particle sizes and the creation of new soil horizons (Alvarado *et al.,* 1981; Gotwald, 1986). They also produce characteristic structures in soils, different from those found in the surrounding soil matrix (Humphreys, 1994).

The activities of ants can also influence the chemistry of the soil, notably by increasing the amounts of organic matter, phosphorus, potassium and nitrogen in the mounds (Petal, 1978; Carlson and Whitford, 1991; McGinley *et al.*, 1994) although such increases are not universal (Lobry de Bruyn and Conacher, 1990). Wagner (1997) showed that soil materials associated with the workings of *Formica perpilosa* in a North American desert environment had a higher organic matter status and elevated phosphorus and nitrogen concentrations compared with the surrounding soils. Where these materials were associated with the bases of *Acacia constricta* shrubs, these plants set approximately twice as much seed as those without ant workings. In southeastern Australia, a nearly fifty *per cent.* increase in the growth of *Eucalyptus obliqua* seedlings was obtained in mound materials of *Aphaenogaster longiceps* compared with that in more distant surface soils (Andersen, 1988). The physical changes and elevated chemical status of many soil materials associated with ant mounds induce greater mineralisation activities by decomposers (McGinley *et al.*, 1994) and heightened root and mycorrhizal growth, as shown in mounds of the harvester ant *Pogonomyrmex occidentalis*. Plant communities differing from those of the general environment may occur on or surrounding ant mounds as a response to the altered physical and chemical conditions pertaining (King, 1977; Culver and Beattie, 1983; Carlson and Whitford, 1991).

4.3.4 MACROARTHROPOD COMMUNITIES

Soil macro-arthropod communities include a wide range of taxa. The most important groups are Myriapoda, Coleoptera, Isopoda, Diptera larvae and Araneae although some Homoptera (larval Cicadidae), Heteroptera (Gryllidae), Thysanura, Blattoidea, endogeic Hemiptera or lepidopteran larvae may occasionally be important. Most of these macro-invertebrates have a much lesser capacity to move through the mineral soil than earthworms, termites or ants. Thus, most live in the litter or in the upper few centimetres of soil; saprophages play a major role in comminution of surface litter and a diverse community of predators regulates their populations.

4.3.4.1 *General Biology*

Isopoda

Isopods are relatively consistent components of litter systems (Sutton, 1972; Warburg, 1987). Their size ranges from a few millimetres to 1-2 cm and their fresh weight is of the order of a few milligrams. Respiratory organs range from the gills of aquatic forms to closed, lung-like structures (Warburg, 1987). Most species are highly susceptible to water-loss and are thus restricted to moist, sheltered habitats although a few drought-resistant species colonise desert habitats. They also have little resistance to cold temperatures. As a result, they are absent from tundra soils and, in temperate environments, they have long periods of quiescence during winter.

Isopods may live for several years. They are saprophages that preferentially ingest decomposing material with a low phenolic content (Neuhauser and Hartenstein, 1978). Coprophagy is frequent in Isopoda and this external digestion seems to be obligatory for normal development (Hassal and Rushton, 1982; Gunnarsson and Tunlid, 1986). They need calcium for their exoskeleton and copper: a relationship between the concentration of this element in litter and isopod abundance has been established (Wieser, 1966). Many, but not all, species may be limited by the availability of these elements.

Daily ingestion rates (in mg dry matter ingested per mg fresh weight x100) are generally less than 5 % but may vary from 0.5 to 20 % of body weight, depending on the species and its diet (Bizé and Célérier, 1986). Assimilation rates may be relatively high with values in the range of 4 to 45 % with an average of 25-30 %, depending on the species, the stage of decomposition and the quality of the litter (Rushton and Hassal, 1983; Bizé and Célérier, 1986).

Myriapoda

The larger myriapods are divided into two separate classes with quite disparate ecological requirements and roles in the soil. The Chilopoda are mostly carnivorous whereas the Diplopoda are saprophages, with a few exceptions. Three main categories may be distinguished: (i) large sized macro-saprophages (Iulida, Glomerida and Polydesmida), (ii) small micro-phytophages (Craspedosomatida) or (iii) intermediate forms with a liquid alimentary regime (David, 1987). Some large tropical Diplopoda may ingest soil and egest compact faecal pellets (Aouti, 1978).

Diplopods and chilopods differ substantially in size with average individual weights varying from 0.5 mg f wt in a Swedish coniferous forest to 560 mg in the litter layer of a mixed deciduous forest in North Carolina. Some giant tropical Iulidae weigh up to 10 g fresh weight, although most values are in the range 10 to 25 mg (Petersen and Luxton, 1982).

Diplopods and chilopods are found on the soil surface, in the litter layers, in the sub-cortical habitats of tree stumps and decaying logs where their food occurs, and where the humid air protects them from desiccation. Most have a limited ability to penetrate the soil which they move through by displacing it in all directions ('bulldozing').

The geophilid chilopods and the ilulid diplopods are the only groups which possess adaptations for digging. Even so, they prefer soils which facilitate their movement such as those of sandy texture or with extended networks of fissures and earthworm galleries. Growth is often slow and myriapods may live for several years. In temperate climate French forests, two major groups may be distinguished: species with short developmental cycles (6 months to one year) (such as the Polydesmida and Craspedosomatida) and those with long cycles (2-4 years), for example, Iulida and Glomerida (Fairhurst, 1974; Geoffroy, 1981; David, 1982).

Diplopods have a thick cuticle encrusted with calcium carbonate which protects them from both water excess and desiccation. This adaptation to widely changing moisture conditions has made them a constant feature of forest leaf litter layers all over the world. Chilopods which have a thinner cuticle are more sensitive to water stresses.

Diplopods are often favoured by high soil calcium concentrations (Fairhurst, 1974). They feed on decomposing wood or leaf materials and they grow better on partially-weathered material and have a marked preference for litters with high Ca and low polyphenol contents. As a result, in temperate deciduous forests, birch (*Betula alba*) and ash (*Fraxinus excelsior*) litters are more favourable than beech (*Fagus sylvatica*) or oak (*Quercus robur*) litters (Raw, 1967; Marcuzzi, 1970; Edwards, 1974). Development is slow and may last for three years in tropical species (Lévieux and Aouti, 1978; Aouti, 1977) and 5-7 years in colder areas (Vachon, 1947; Blower and Miller, 1977).

Chilopoda include large Scolopendrida or Lithobiomorpha which hunt in the litter, and tiny Geophilidae which go deeper into the soil, following galleries and crevices.

Diptera

Dipteran larvae (Insecta) are an extremely diverse group both taxonomically and nutritionally. They may occasionally become a predominant component of the soil fauna and may effect a spectacular reduction of litter mass (see *e.g.*, Healey and Russell-Smith, 1971; Altmuller, 1979; Deleporte, 1987). In a French temperate-climate forest, the soil and litter community comprised 29 different families (Mollon, 1982). Dipteran larvae are predominantly saprophagous but also include predaceous, necrophagous, xylophagous, coprophagous, micro-and macro-phytophagous species (Brauns, 1954). The larvae live mostly in litter and some of them have developed remarkable resistance to drought (Delettre and Baillot, 1977) although soil texture and water regime may influence their distributions (Trehen, 1971; Blanchart *et al.*, 1987).

Coleoptera

The adults and larvae of the insect order Coleoptera are also a consistent feature of soil animal communities. They are diverse taxonomically and differ widely in size, and in the ecological roles they fulfil in soils and litters. Most of them live in the litter. They are saprophagous, phytophagous or predators. Two groups may have considerable local importance: Scarabaeinae larvae (dung-beetles) which are crucial in burying cow-dung in grasslands used for cattle grazing (see, *e.g.*, Bal, 1982; Brussaard and Hijdra, 1986; Lumaret and Kirk, 1987) and Melolonthinae whose larvae may occasionally be abundant in grasslands and affect crop production by feeding on living roots (*e.g.*, Villalobos and Lavelle 1990). Other important families in soils and litter are the super family Staphylinoidea, families

Curculionidae, Elateridae and Tenebrionidae.

The vast majority of Scarabaeidae have three larval instars and they complete their life cycle in 1 to 4 years. The European cockchafer (*Melolontha melolontha*), which used to be a common pest has a three-year life cycle: adults merge during spring (April) and lay eggs in the soil during the following 2-3 months before they die. The incubation period of eggs lasts about six weeks. The first larval instar lasts 2 months and the first moult takes place at the end of summer (September). The second instar larvae hibernate and resume devastating activity during the next spring. This is the time when the most severe damage occurs. The second moult takes place in June-July. Third instar larvae hibernate and further develop during spring up to metamorphosis which starts in June and finishes in October when adults hatch out and hibernate in soil before emerging in spring (Hurpin, 1962). An average 5 t ha^{-1} (*i.e.*, 20 %) decrease in sugar beet production occurred in France between 1925 and 1950 when one-year larvae were present in soils.

Beetles of the genus *Phyllophaga*, which may build up large populations in pastures of North America and tropical Mexico, usually complete their life cycle in 2-4 years (Tashiro, 1990; Villalobos, 1991) (Figure III.59).

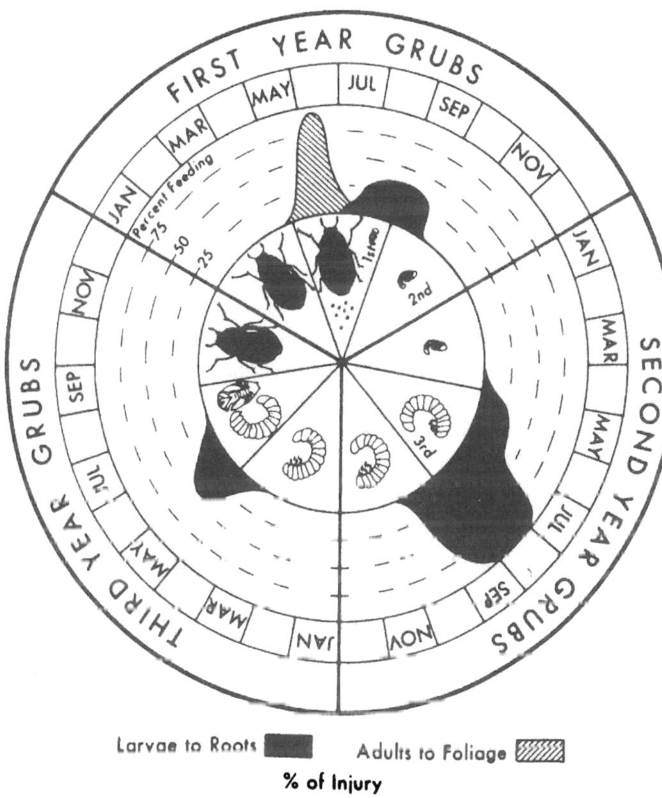

Figure III.59 Diagram of the 3-year life cycle of a June Beetle *Phyllophaga sp.* (Scarabaeidae) with periods of feeding damage (Dindal, 1990 after Hammond, 1940).

The superfamily Staphylinoidea comprises several families of importance to the litter and soil: Staphylinidae, Pselaphidae, Ptiliidae and Silphidae. Adults and larvae of Staphylinidae are mostly predators which digest their prey by extra-oral digestion, although populations from the two sub-families Osoriinae and Oxytelinae are fungal feeders. Staphylinidae may complete their life cycle in a relatively short period of a few weeks to months, and have several generations per year. Adults may live for several months to years (Newton, 1990).

Dung beetles (Scarabaeinae) may play a critical role in the burial of dung, especially in relatively dry environments. They dig subvertical galleries 10-15 mm wide, down to 50-70 cm depth, with a variable number of chambers which are further filled with large pellets of dung; one egg is laid in each pellet and the larva will complete its developmental cycle feeding on the pellet. Excavations are deposited at the soil surface and form small mounds a few centimetres high (Hurpin, 1962). The shape, depth and number of chambers of a nest are a species specific characteristic which allows evolutionary lineages to be discriminated (Halffter and Edmunds, 1981). Elateridae are another typical family of soil and litter dwelling Coleoptera which are constantly present in soils and may occasionally become serious pests of field crops (d'Aguilar, 1962).

Arachnida
Finally, a range of predators live largely within litters and, to a lesser extent, in soils. Arachnida (Opilionidae, Araneidae and some smaller groups) form an extremely diverse group, which mainly live in the litter with exception of a few burrowing species. They are predators of Collembola, Enchytraeidae, Acari and many other small litter and soil invertebrates.

4.3.4.2 *Population Densities and Biomasses*

Coleoptera generally represent the third or fourth rank of macro-invertebrate biomass in soils, after earthworms, termites or ants. In dry savannas of South Africa at Nilsvley, they are the most important group in terms of biomass with values of 7.15 to 8.15 g f wt m^{-2} depending on the type of savanna (Scholes and Walker, 1993); in humid tropical pastures of Mexico, mean annual biomass of Coleoptera was 19 g f wt m^{-2} with a peak value of 60 g m^{-2} at the end of the rainy season; Scarabaeidae Melolonthinae comprised 90 % of this biomass (Villalobos and Lavelle, 1990). Most biomass values are in the range 50-200 mg dry wt m^{-2} (Petersen and Luxton, 1982). Maximum densities have been recorded in an abandoned field in Sweden (1404 ind. m^{-2} ± 189 with a biomass of 2.86 ± 4.2 g dry wt equivalent to 10 g f wt, Persson and Lohm, 1977) and savannas at Lamto (Ivory Coast) (Athias *et al.*, 1975).

Isopoda. Where present, Isopoda generally have low densities of a few tens to a few hundreds of ind. m^{-2}. Maximum values of 1000 to 7900 ind. m^{-2} have been reported for some British grasslands and forests (Sutton, 1972; Petersen and Luxton, 1982; Molfetas, 1982; Mocquard *et al.*, 1987). Species diversity is generally limited to 4-5 species (Mocquard *et al.*, 1987). Densities are particularly low in acid soils with mor-type humus or where frost or drought occur.

Diplopoda frequently have high population densities of 200 to 500 ind. m^{-2} in temperate forests (Bornebusch, 1930; Dunger, 1964; Saulnier and Athias-Binche, 1986) with biomasses of 8-20 g f wt m^{-2}. They are consistent inhabitants of litter layers in tropical forests and woodlands where their populations may vary from mean values of a few tens (*e.g.*, 26 ind. m^{-2} weighing 5.7 g f wt in Miombo woodland, Dangerfield, 1990) to several hundred individuals weighing 0.3 to 4 g f wt (see, *e.g.*, Lavelle and Kohlmann, 1984; Lavelle *et al.*, 1981; Beck, 1971). In temperate and tropical grasslands, populations are generally limited to a few tens per square metre (see for example, Persson and Lohm, 1977; Athias *et al.*, 1974) with a low biomass of 0.1 g f wt m^{-2} to 0.30 in moist savannas at Lamto (Côte d'Ivoire) and 0.6 in tropical pastures of Mexico (Lavelle *et al.*, 1981).

Soil and litter-dwelling Diptera larvae vary widely in size with small individuals, *e.g.*, in the families Chironomidae or Ceratopogonidae, of a few millimetres long and large Tipulidae or Bibionidae which may attain three centimetres. They often have strongly aggregated distributions and are only present in litters for a few months until they become adults. They are seasonally important components of tundra soils with densities of 100-1000 ind. m^{-2} in most sites. The highest densities have been recorded in temperate forests (especially in moder and mor humus types) where densities of several thousand ind. m^{-2} are common (see, *e.g.*, Mollon, 1982; Petersen and Luxton, 1982). Maximum values of 12,000 (Healy and Russell-Smith, 1971) and even 370,000 individuals per m^2 have been found (Deleporte, 1981). Densities in temperate grasslands and tropical environments rarely exceed a few tens of individuals m^{-2}.

Biomasses may be as high as 16-40 g f wt m^{-2} in temperate forests and tundra sites; most data however range between 0.5 and 5 g m^{-2} in these environments (Petersen and Luxton, 1982).

Arachnida are the most abundant group of predators. Most values are in the range 20-200 ind. m^{-2} and maximum densities of 400-800 ind. m^{-2} have been recorded in temperate forests with a mor or moder type of humus. Biomasses are modest and rarely exceed 0.20 to 0.30 g f wt m^{-2}.

Chilopoda have densities of a few tens of individuals.m^{-2} and biomasses of less than 1 g f wt with maximum values in forests rather than grasslands and in temperate rather than tropical ecosystems.

4.4 Determination of Soil Faunal Communities

Soil faunal communities differ markedly in their toxonomic and funtional diversities at all scales of observation. These latter may vary from broad latitudinal gradients of temperature, to regional mosaics of different soils, vegetation and land-use and down to a single soil profile (see *e.g.*, Petersen and Luxton, 1982). Such variation results from population response to: (i) local environmental constraints represented by a suite of hierarchically-organised climatic and edaphic factors (see section IV.1.2.2), (ii) phylogenetic and biogeographical constraints, and (iii) their interactions with other soil organisms, especially micro-organisms.

Soil organisms, especially the invertebrates, differ from those of other strata within ecosystems in that they are exposed to unique sets of limiting factors. They tend to be more closely constrained by external environmental factors and the limitations of their own digestive systems than, *e.g.*, vertebrates or insects living in the aerial part of the ecosystem. They have therefore developed specific adaptive strategies whereby they rely more on mutualistic relationships and suffer less from antagonistic interactions with other species (Lavelle, 1985).

4.4.1 ADAPTIVE STRATEGIES OF SOIL INVERTEBRATES

4.4.1.1 *Principal constraints to soil invertebrate activities*

Soil invertebrates have adapted in different ways to the major constraints of their environment. Some major accommodations to their environment include: (i) the derivation of their nutrient energy requirements from a suite of largely poor-quality organic resources; (ii) adaptation of their locomotion to suit the dark, confined and sometimes compact soil environment; and (iii), modification of their metabolism to cope with the occasionally unfavourable temperature and moisture conditions.

Food resources
The nutritional value of most available soil resources is relatively poor. Most comprise decomposing materials with characteristically low nutrient contents and substantial concentrations of highly-polymerised and resistant molecules such as lignin, humic acids and tannin-protein complexes. Higher quality resources such as the microbial biomass or freshly-dead invertebrates are much less abundant, dispersed in the soil profile and rendered partly inaccessible by physical protection in microhabitats (see Chapter I.3.2).

Movement
The mineral soil environment is compact and movement for most species is only possible through a labyrinth of pores and channels. These comprise a relatively small proportion of the whole soil volume which decreases with increasing depth (see Chapter I.1.3.1).

Moisture regimes
The semi-aquatic soil environment is theoretically suitable for those faunal groups with high moisture requirements. However, water availability is highly variable in time and space due to climatic variation and capillary forces which, depending on the pore size distribution, may keep much of the water beyond the extraction capacity of many animals (see Chapter I.2.2.3.2).

Temperature
Through its overall extremes and short term changes, temperature is the ultimate regulator of metabolic activity in soil organisms which, with few exceptions, are poïkilothermic. Temperature and moisture regimes determine the intensities and temporal distributions of the seasonal periods of activity. In that respect, they largely determine the success of the various adaptive strategies. Strategies based on the use of low-quality resources will only be satisfactory where long and intense periods of activity are possible; if they are reduced, life-cycles must be completed in a shorter time, through the use of higher-quality resources.

Water availability and temperature regimes usually become more favourable and predictable with depth while food availability and quality, and exposure to predators all decrease. The existence of inverse gradients in environmental predictability and food availability down the soil profile is a constant feature of soil environments.

The way and the extent to which soil animals face these constraints largely determine their abundances and ultimately the roles that they play in the soil system. Adaptive strategies have been selected in response to the characteristic suites of environmental constraints present in the soil environment.

4.4.1.2 *Digestion*

Soil invertebrates generally have relatively restricted enzymatic spectra. Animal cellulases, for example, are rare and the cellulolytic activity detected in animal guts is generally due to micro-organisms (see, *e.g.*, Devigne and Jeuniaux, 1961; Parle, 1963; Loquet and Vinceslas 1987; Rouland *et al.*, 1988a; Urbasek, 1990; Deleporte and Rouland, 1991; Zhang *et al.*, 1993). An association between invertebrates and micro-organisms is thus necessary to allow digestion of resistant substrates. Depending on the intensity of this association, digestion may occur: (i) by direct digestion; (ii) through the 'external rumen'; (iii) through a mutualistic association with the soil microflora in the animal gut; and (iv) through a symbiotic association with a specialised gut microflora or microfauna.

(i) direct digestion occurs in predators but appears to be extremely limited among saprovores. Higher termites have cellulases (Potts and Hewitt, 1974; Martin and Martin, 1979; Rouland *et al.*, 1988b). Cellulases have also been reported in earthworms, but it is still uncertain whether they are actually produced by the worm itself, or by associated micro-organisms (Laverack, 1963; Loquet and Vinceslas, 1987) ;

(ii) in the external rumen type of digestion, invertebrates which do not produce endogenous cellulase, periodically re-ingest their faeces and thereby take advantage of the release of assimilable compounds due to external microbial activity (Swift *et al.*, 1979). Most litter-feeding epigeic arthropods use this digestive system. They play important roles in the comminution, fractionation and humidification of the ingested material, thereby enhancing its microbial activity. Macrotermitine termites and ants of the tribe Attini which cultivate fungi on specially prepared wood or leaf and other materials also belong to this group;

(iii) facultative mutualism with a non-specific microflora ingested with soil or litter material occurs where invertebrates provide suitable conditions for microbial activity in their gut. They then absorb the assimilable metabolites released by the enhanced microbial activity. A digestive system of this type has been described, *e.g.*, in endogeic geophagous earthworms (Barois and Lavelle, 1986). The ingested soil is mixed with an equivalent amount of water and 5 to 16 % of intestinal mucus, a readily-assimilable mixture of low molecular weight glycoproteins, amino acids and simple sugar molecules (Martin *et al.*, 1987). Intensive mixing in the gizzard disperses the soil and results in the formation of an homogenous suspension of bacteria, readily assimilable substrates (mucus) with soil mineral particles and soil organic matter whose physical protection within aggregates has been removed. The soil microflora first increases its activity by using mucus as an energy substrate. By the middle part of the gut, the mucus has been almost completely metabolised, microbial activity is high and bacteria, through a priming effect (see Chapter IV.1 and Jenkinson, 1966) are able to digest part of the soil organic matter. The digestive products are released within the gut and partly reabsorbed by the worm. This facultative symbiosis has been observed in the larvae of a number of wood and leaf-litter feeding Diptera and Coleoptera (Campbell, 1929; Hassall and Jennings, 1975), in termites (Breznak, 1984) and temperate climate earthworms (Trigo and Lavelle, 1993) (Figure III.60).

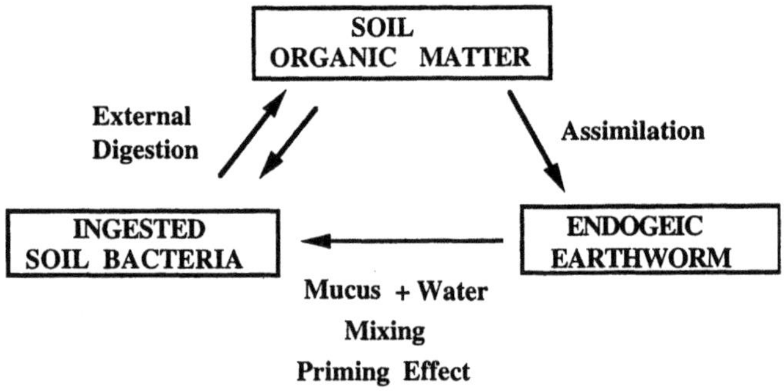

Figure III.60 Mutualistic digestive system in endogeic earthworms: an hypothesis (after Barois and Lavelle ,1986; Trigo and Lavelle, 1993).

(iv) symbiosis between soil animals and their specific gut microflora is the ultimate stage of an obligate association between invertebrates and micro-organisms which permits them to digest lignocellulose. Phylogenetically 'lower' termites, have flagellate protists and bacteria in a bulbous 'paunch' in the proctodeum. They are unique species, found only in the gut of these termites and are attached to the gut wall by a holdfast organelle which secures them to the epithelial tissue. These protists digest lignin by an aerobic process and release CO_2, H_2, acetate and sugars which are used by the termite (Kirk and Farrell, 1987) (see section III.4.3.2.1). Bacteria provide growth factors for protists, but termites can survive without these bacteria (Yamin, 1981). In higher termites, filamentous bacteria attached to cuticular spines partly fill the proctodeal 'paunch'.

4.4.1.3 *Locomotion*

Three main strategies are known and are related to the sizes of the animals involved:
(i) swimming through the water films which fill soil capillaries and pores and cover soil particles. The hydrobiont fauna moves in this way within the soil although such movements are limited to a few millimetres, due to the small size of the animals and the frequent discontinuity of the water phase;
(ii) movement through the air-filled pore space. This strategy is adopted by most meso-faunal species and some slender, worm-shaped macrofauna (*e.g.*, Chilopoda Geophilidae and some thin polyhumic earthworms). Epigeic macroinvertebrates do the same in the leaf-litter layers; a large number of litter invertebrates shelter in the upper centimetres

of the mineral soil using macropores or the galleries of burrowing animals;
(iii) tunnelling through the soil. Only the larger soil animals have sufficient strength to do this and this behaviour only occurs in anecic and endogeic earthworms, termites, ants and a few insects with fore-legs modified for digging, *e.g.*, nymphal Cicadidae, Gryllotalpidae (Orthoptera), Cydnidae (Hemiptera) or Scarabaeidae (Coleoptera). The ability to dig gives such invertebrates a great selective advantage. Like anecic earthworms or termites, these invertebrates are able to feed in leaf-litter (where the highest-quality resources are found) and avoid the frequently unfavourable environmental conditions and high predator densities that occur in this environment. Some of these invertebrates are also active root-feeders.

4.4.1.4 *Adaptations to temporarily unfavourable climatic conditions*

Smaller soil organisms are able to survive drought and sub-zero temperatures since they can shelter in favourable microsites when overall conditions become unfavourable. This is especially true of bacteria and protists which may remain in the water films or micropores 1 μm or less in diameter, that is in soils with matric potentials as low as -5 MPa (pF 5). Protists and nematodes have also developed such remarkable forms of drought resistance as anhydrobiosis and encystment.

Large invertebrates such as earthworms do not survive long in very dry soils since their water-conserving mechanisms are poorly developed in most cases. However, certain organisms modify their environments to obtain better protection. This is especially true of such social insects as ants and termites, which may build nests with highly-buffered temperature and moisture conditions.

4.4.1.5 *Response to soil constraints: adapt to, or modify the environment?*

The efficiency with which soil organisms exploit particularly low quality soil resources, their ability to move and their resistance to unfavourable environment and moisture conditions are largely determined by their sizes and their individual or social behavioural responses to the constraints considered above. Small animals cope better with temperature and moisture extremes although they are less able to use low-quality resources and modify their environment. Conversely, large soil animals make better use of low-quality food but are poorly adapted to extreme environmental conditions. However, some groups, notably the termites, have a great ability to transform their environment.

Ants are an exception to this in that they normally utilise high-quality resources through strategies of predation, nectar and seed feeding.

Individual size partly determines the ability to use microflora in mutualistic digestive systems. A minimum size is necessary to allow transport and temporary or semi-permanent storage of an associated microflora in the gut:
(i) the microfauna can only use micro-organisms as food;
(ii) the mesofauna also contains many microbivores. However they also transport micro-organisms and partly utilise an 'external-rumen' type of digestion;
(iii) the macrofauna has the ability to develop the external-rumen system on a large

scale. Alternatively, a large-scale utilization of soil resources by a facultative or obligate symbiotic microflora may occur in their guts.

As a result, the capacity of soil invertebrates to use low-quality resources normally increases with size, other environmental conditions being equal.

Size also determines the mobility of invertebrates. The smaller they are, the less effectively can they move and the more closely they depend on the distribution and continuity of water films and/or the pore space. Conversely, large animals may range much further provided that they can dig through the soil when their diameter is greater than the average pore size.

Finally, size partly determines the ability of invertebrates to survive temporarily unfavourable conditions and the larger invertebrates are generally less successful at this. These differing responses may partly explain why the distributions of micro- and, to a lesser extent, mesofaunal taxa are continuous whereas macrofauna, especially termites and earthworms have highly discontinuous distributions at both regional and geographical scales.

The behaviour of individuals and populations may modify this general pattern. Aggregation and social organisation are most important in this respect:

(i) aggregation is frequent in most groups of mesofauna; although sometimes interpreted as the result of purely biological features (concentration of egg-laying sites, response to the heterogeneity of resource distribution); its adaptive value has been pointed out by several authors (Joose, 1970; Usher, 1975; Verhoef and Nagelkenke, 1977). In contrast, the non-social large invertebrates which do not rely on external-rumen types of digestion (*i.e.*, predators and earthworms with an 'internal-rumen' type of digestion), generally do not have such highly-aggregated patterns of distribution.

(ii) social organisation is an efficient way to enlarge the scale at which individuals operate, and thus the range and abundance of resources they can utilise. Co-ordinated efforts among attine ant and macrotermitine termite workers allow the construction and maintenance of fungal gardens in which cellulose and lignin are transformed into secondary resources readily assimilable by individuals. Such fungal gardens represent the most highly evolved and efficient external rumens found among soil invertebrates. Social organisation also allows species to modify their environments to an extent that far exceeds the capacity of a single individual. Termites are individually extremely sensitive to low humidities but nevertheless survive well in semi-arid and arid environments. The nests, galleries and surface sheetings within which they live have a constantly humid atmosphere due to acquisition of water from the water-table which may sometimes be many metres deep.

4.4.1.6 *Functional guilds: microfoodwebs, litter transformers and ecosystem engineers*

Soil invertebrates have a continuum of strategies from the smallest microfauna that colonise the water-filled pore space in the same way as micro-organisms, to macrofauna that modify the soil environment to accommodate their own needs. The size of invertebrates and the aquatic or aerial nature of their respiration, reflect their way of adapting to spatial constraints. Three groups have been distinguished (Bachelier, 1978; Swift *et al.*, 1979):

1. Microfauna, which comprise aquatic invertebrates that live in the water-filled soil porosity. They are small, less than 0.2 mm on average, and include the protists and nematodes, together with other groups of lesser importance like tardigrades, rotifers;
2. Mesofauna, which comprise microarthropods (mainly Collembola and Acari) and the small Oligochaeta and Enchytraeidae; they have an average size range of 0.2-2 mm and live in the air-filled pore space of the soil and litter;
3. Macrofauna, which include invertebrates with an average length greater than 2 mm. Termites, earthworms and large arthropods are the main components of this group. They have the ability to dig or eat their way through the soil and create specific structures to accommodate their movements and living activities (*e.g,* burrows, galleries, nests and chambers) and produce casts and faecal pellets through their feeding activities. These organisms have also been called 'ecosystem engineers' (*sensu* Jones *et al.*, 1994) for their ability to profoundly affect the soil structure and hence influence major soil processes through the structures that they build (Stork and Eggleton, 1992).

Three major guilds of soil invertebrates may be distinguished on the basis of the relationship that they have with the soil micro-organisms and the kinds of excrement and other biogenic structures that they produce (Figure III.61).

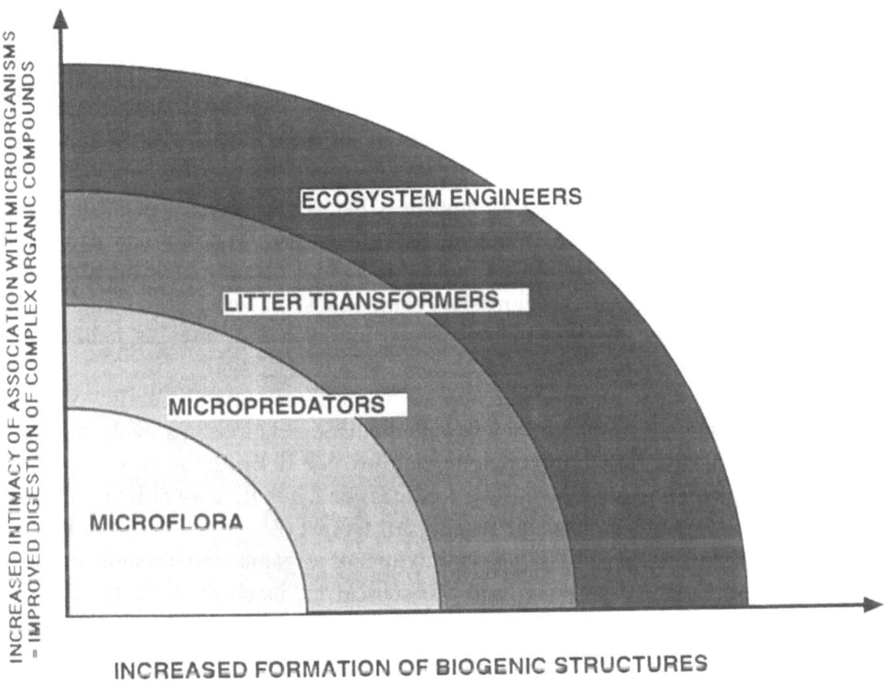

Figure III.61 Interactions among micro-organisms and macro-organisms in soils. As the size of organisms increases, their relationships with the microflora gradually shifts from predation to external digestion and mutualism, and they produce biogenic structures of increasing strength. When present, larger organisms tend to be more competitive than smaller ones, but their activity is more frequently limited by low temperature and moisture conditions (after Lavelle, 1997).

Micro-foodwebs mainly comprise microfauna that are predators of bacteria and fungi, and their predators. Microfaunal species do not produce recognisable solid excrements and the effects of these invertebrates on soil organic matter dynamics are not prolonged within structures that remain stable for significant periods after deposition. However, they have a significant impact on the population dynamics of micro-organisms and the release of nutrients immobilised within the microbial biomass (Trofymow and Coleman, 1982; Clarholm, 1985). This process is particularly well developed in the rhizosphere. Predatory Acari, Collembola, and even larger invertebrates (earthworms) may extend this foodweb over several trophic levels.

Litter transformers mainly comprise mesofauna and large arthropods which normally ingest purely organic material and develop an external ('exhabitational' *sensu* Lewis, 1985) mutualism with microflora characterised by an 'external rumen' type of digestion, dissemination of spores or comminution (Swift *et al.*, 1979). Litter arthropods may digest part of the microbial biomass or develop mutualistic interactions within their faecal pellets: in these structures, organic resources which have been fragmented and moistened during gut transit, are actively digested by the microflora. After some days of incubation, arthropods often re-ingest their pellets and absorb the assimilable organic compounds that have been released by microbial activity, and occasionally, part of the microbial biomass (see *e.g.*, Hassal and Rushton, 1982). This specific type of exhabitational mutualism is known as the 'external rumen' type of digestion (Swift *et al.*, 1979). Within this general adaptive strategy, a large diversity of behaviours may be identified (Vannier, 1985). Insects such as Diptera (Sciaridae) have a comparatively efficient enzymatic machinery and produce liquid faeces that are not re-ingested (Deleporte and Rouland, 1991). This digestion system is unable to digest tannin-protein compounds and highly polymerised polysaccharide-aromatic complexes (lignin) efficiently and these accumulate to hamper the progress of decomposition (Minderman, 1968; Gourbière, 1982; Toutain, 1987a) (see also IV.2.5.3). The acid organic compounds released in the course of decomposition are not flocculated in the presence of mineral particles; they behave as aggressive compounds that may leach and actively participate in mineral weathering, thereby favouring such processes as podzolisation (Berthelin *et al.*, 1979; Pedro, 1989).

Finally, the ecosystem engineers comprise macrofauna, mainly earthworms and termites that are large enough to develop mutualistic relationships with microflora within their own bodies. These interactions may involve obligate (such as the protists contained in the posterior pouch of lower termites) or facultative symbionts; the latter occur in the gut of higher termites and also in earthworms (Barois *et al.*, 1987; Breznak, 1984). These organisms usually ingest a mixture of organic and mineral elements. Organic acids produced by digestion and the subsequent incubation of organic matter in casts, are normally flocculated in the presence of clay minerals and have a high microbial activity. Digestion is efficient and complex organic compounds like cellulose, lignin and tannin-protein complexes are at least partly assimilated (*e.g.*, Butler and Buckerfield, 1979; Toutain, 1987b; Breznak, 1984; Rouland *et al.*, 1990). The larger faecal pellets (in the range of 0.1 to >2 cm) may form the component elements of macro-aggregate structures; their properties may contribute prominently to those of the stable structures through the regulation of porosity, aggregation, bulk density and surface features (Bal, 1982, Blanchart *et al.*, 1997). The large structures built by organisms, such as mounds and networks of galleries and chambers, have significant impacts on the evolution of soils at medium time scales. Ants may be considered as part of this group, although the vast majority of them only use soil as an habitat and have a limited direct impact on soil organic matter dynamics. They will not be considered in this section because of the small number of studies devoted to their effects on soil processes (review in Folgarait, 1998).

Whenever conditions are suitable for their activities, macrofauna, and especially earthworms and termites, become the major regulators of microbial activities within their spheres of influence (*i.e.*, the termitosphere of termites and the drilosphere of earthworms, Lavelle, 1984) in which they also determine the abundance and activities of the smaller groups of soil fauna (Dash *et al.*, 1980; Yeates, 1981). These 'Biological Systems of Regulation' include the rhizosphere in which roots are the major determinant (Lavelle *et al.*, 1993) (see also IV.1.4).

The apparent looseness of the soil trophic structure may be due to the juxtaposition of two fundamentally different types of relationships between invertebrates and micro-organisms *i.e.*, 1. a 'classical' foodweb in which organisms of a given size feed on the smaller organisms, at a lower level in the foodweb; and are eaten by larger organisms which comprise the higher levels and 2. Biological systems of regulation which are systems of interactions based on the mutualisms that occur between micro-organisms and invertebrates of different sizes.

The overall structure of the foodweb is further complicated by the effects of interactions between these three guilds. For example, certain local conditions may favour high levels of activity among termites or anecic earthworms and thereby significantly reduce the resources available for the development of litter transformers and micropredators.

4.4.2 DETERMINATION OF SPECIES RICHNESS

Within individual communities, species richness within particular taxa is determined by

(i) phylogenetic constraints,
(ii) biogeographical factors and
(iii) environmental conditions.

4.4.2.1 *Phylogeny and species richness*

Species richness differs greatly between taxa. Under favourable conditions, microarthropod communities may comprise hundreds of species with α diversity in Acari up to 400 to 500 species and that of Collembola 60 to 80 species. Under comparable conditions, nematodes may comprise up to 90 species; protists, up to 60; termites, 60; Enchytraeidae, 22; earthworms, 15-17; Diplopoda, 15.

Two general trends seem to exist:

(i) species richness decreases in parallel with size (May, 1986; Stork and Brendell, 1993);
(ii) species richness appears greater in arthropod than in non-arthropod groups.

Some hypotheses may be proposed to explain these features. The effect of size may be a consequence of the scale at which the organisms operate. A single soil macroaggregate may be the smallest environmental unit recognisable by an endogeic earthworm although it comprises a large variety of microenvironments at the scale of a protist or nematode. Thus, a greater diversity of potential niches for small organisms will sometimes result in an intense speciation. This effect of size on species richness has already been assessed for above ground terrestrial animals by May (1978).

The larger diversity of arthropods as compared to non-arthropods (*e.g.*, Oligochaeta) may possibly be interpreted as the effect of a greater niche specialisation due to their external chitinised skeleton. As a result, adult size is fixed and little variation occurs between individuals. The evolution of a chitinised exoskeleton has favoured the development of efficient locomotory and protective structures and allowed the evolution of a wide range of mouthpart types of fixed size and shape. Larval development permits some plasticity in the niches used by a particular species, although still much less than in non-arthropods. Conversely, during growth, Oligochaeta continually change their ecological roles. For example, young anecic earthworms, often behave like epigeics and progress deeper into the soil as they grow. Populations of a single earthworm species may thus cover a relatively wide range of niches. At the end of the dry season, populations of the tropical endogeic species *Millsonia anomala* comprise a majority of young individuals which are more resistant to drought and high temperatures than adults. In the midst of the rainy season, populations show a preponderance of adults more highly adapted to cool temperatures and a high soil moisture regime (Table III.13).

4.4.2.2 *Biogeographical effects*

For biogeographical reasons, certain taxa may be absent from environments in which they could readily exist. This is not uncommon in large invertebrates and two examples are the absence of the fungus-growing termites (Macrotermitinae) from Australia and the Americas. In the latter environment, ants of the tribe *Attini* have been considered as functional equivalents, even though the most prominent harvest is living leaves instead of using dead plant material as a primary resource (Wheeler, 1907; Weber, 1972). A further example is the striking differences between earthworm communities in tropical forests: in Western Africa and Central America, communities in oxisols occurring in regions with an average annual rainfall of 1600-2000 mm are dominated by endogeic populations (Lavelle, 1978; Fragoso and Lavelle, 1987). Comparable forests in South America have predominantly anecic populations (Nemeth, 1981; Lavelle and Pashanasi, 1989). In the former environment, most species belong to the family Megascolecidae which has a relatively primitive pattern of organisation and comprises a majority of endogeic species. In the Amazonian forest, most species belong to the family Glossoscolecidae, a group that includes numerous true anecic species and in this respect is similar to the Lumbricidae, the dominant earthworm family of most temperate regions of the world.

4.4.2.3 *Environmental determinants*

Species richness and diversity are highly dependent on variation in environmental factors. This may range from the small scale at which α-diversity is measured (a few square metres to 1 ha), to the much wider continuum of a thermo-latitudinal gradient.

Smal scale determination

At the small scale at which communities operate, species richness and diversity are determined by the diversity and abundance of resources. However, this determination is not simple and contradictory results may be obtained. In an English chestnut (*Castanea)* wood, a direct relationship was established between environmental diversity as measured by the occurrence of 23 variables describing microhabitats and resources and the diversity of species of Cryptostigmata (Acari) (Anderson, 1977) (Figure III.37). In contrast, similar studies of a mixed-species wood near Paris (France) showed that species richness may also be negatively correlated with the abundance (and hence diversity) of microhabitats occurring in leaf litter (Garay, 1981b). In that situation, the small standing crop biomass of litter resulted from rapid decomposition. The large flux of assimilable organic matter and nutrients thereby made available may account for the large observed diversity. Mechanisms of niche partitioning which operate in these communities are discussed elsewhere (Section III.4.2.1.2).

Large-scale determination: the first link hypothesis

Many explanatory - and sometimes controversial - hypotheses have sought to explain the outstanding species richness that occurs in many taxa in the humid tropics (see *e.g.*, Giller,1984; Solbrig, 1991). It seems likely that species richness is determined by several factors which can act at two levels: (i) determination of the corporate niche occupied by a community and (ii) separation into individual niches.

The main hypotheses that explain the possibly enlarged corporate niches in the humid tropics compared with those of cooler and less humid environments refer to environmental favourableness, spatial heterogeneity, temporal variability and enhanced plant production: these are considered to enlarge the resource base available to a given community. However, these hypotheses have never directly addressed soil processes or considered the effects of increased temperature. The latter is the one unquestioned difference that occurs between tropical and extra-tropical environments.

In the 'first link' hypothesis, it is proposed that increased soil temperatures give roots access to an enlarged nutrient resource base by increasing the efficiency of mutualistic associations in the rhizosphere (see Chapter IV.3). An enlarged corporate niche due to increased nutrient resources would result in a larger number of plant species once niche separation has taken place. This would be the first link of a cascade process in which species richness within the consumer and decomposer foodwebs would become larger than in colder areas, irrespective of such biogeographic and historical determinants as Pleistocene refuges, catastrophic events and the degree of ecosystem maturity (Odum, 1983) (Figure III.62).

This hypothesis originated from the changes noted in the structure of earthworm communities along a thermo-latitudinal gradient extending from Iceland to Southern Venezuela and Côte d'Ivoire (Lavelle, 1983c). The observed pattern was subsequently extrapolated to plants, based on the similarities noted in the general functioning of the drilosphere and rhizosphere systems (Lavelle, 1986).

Along this thermo-latitudinal gradient, earthworms become increasingly able to use food resources of lower quality: in the coniferous forests of northern Europe, earthworms only ingest litter (Figure III.44). In temperate-climate areas, part of the earthworm community additionally ingests soil with a high organic content (taken in the rhizosphere or from beneath the litter) or litter mixed with soil. In the latter case, ^{13}C labelling methods have demonstrated that little if any soil organic matter is assimilated (Martin *et al.*, 1992).

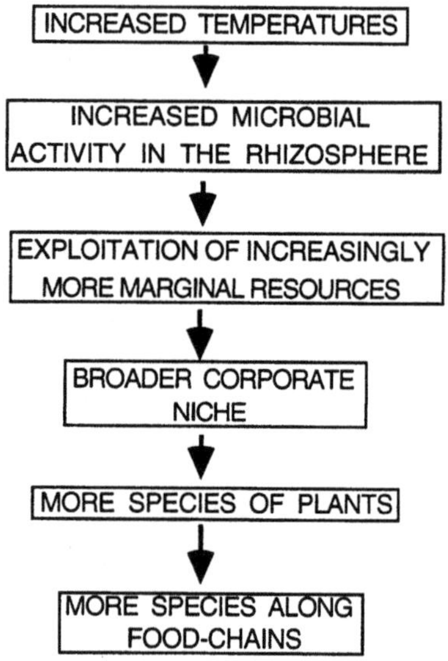

Figure III.62 Mechanism of increase of species diversity with increasing temperature: the first link hypothesis (Lavelle, 1986).

Within the tropical latitudes, certain earthworms have developed the capacity to live on soils with low organic matter contents. They comprise more than 70 % of earthworm biomass in the humid savannas of Lamto (Côte d'Ivoire) and can digest equally all particle-size fractions of SOM, including the finest and most decomposition-resistant fractions (Martin *et al.*, 1991). Thus, with increased temperatures, the resource base of earthworm communities is considerably enlarged even though alpha diversity is not significantly increased. However, functional diversity is greatly increased and niche overlap between species significantly decreased (Lavelle, 1983c).

There is a striking parallel between roots and earthworms straightforwardly expressed by Janzen (1985) who asserted that 'plants wear their guts on the outside'. Roots produce exudates, a mixture of assimilable carbohydrates and proteins, which trigger microbial activity and subsequently nutrient release in the rhizosphere (see Chapter IV.3). In the earthworm gut, intestinal mucus is the functional equivalent of root exudates (Lavelle *et al.*, 1983b; Martin *et al.*, 1987), and the movement of soil through the gut may be considered analogous to root elongation.

The digestive system described in these worms is analogous to the mechanisms of nutrient release in the rhizosphere postulated by Trofymow and Coleman (1982) and Clarholm (1983) and further demonstrated by Billes *et al.* (1986); earthworms disperse the ingested soil with large amounts of water (100 % of the weight of dry soil) and mucus (5-16 %). This leads to an intense microbial activity in the anterior part of the gut at the expense of the intestinal mucus, a mixture of highly assimilable compounds. By the time that the digesta has moved to the median part of the gut, the mucus has been entirely metabolised and the ingested microflora has become able to digest the more complex and physically unprotected organic matter of the ingested soil. This mechanism, typical of the priming effect defined by Jenkinson (1966, after Broadbent and Norman, 1946), is equivalent to that observed in the rhizosphere. In the rhizosphere, exudates are the functional equivalent of mucus and root elongation replaces the movement of soil through the gut.

It has been demonstrated that the mutualistic digestive system of earthworms becomes increasingly efficient at higher temperatures. Here it is hypothesised that the same phenomenon exists in the rhizosphere because of the broadly similar processes occurring in both systems.

The mutualism between termites and their intestinal microbiota (Section III.4.3.2.1) may also become more efficient at higher temperatures. These insects rely partially (although not entirely, as shown in this chapter) on micro-organisms for the digestion of the lignocellulose-rich plant materials that they feed on. Their species diversity and environmental significance are greatest in the tropics and this may result from the increased efficiency of their mutualistic digestive systems in these warmer climates (Lepage, 1983).

4.4.3 DETERMINATION OF SOIL INVERTEBRATE COMMUNITIES

4.4.3.1 *Local responses to the vertical gradient of food resources and environment predictability: trade-offs in adaptive strategies*

Within a single zoological group, possessing similar phylogeny (*i.e.*, with the same basic morphology, anatomical and physiological organisation), individual species may apply different solutions to the problem of exploiting the resources of a given soil. This is illustrated by the adaptations of three sympatric earthworm species to a sandy alfisol in the humid savannas of Lamto (Côte d'Ivoire) (Table III.18). This environment is characterised by the contrasting gradients of food of diminishing quality and increasing environmental predicability that exist from the surface to deep soil horizons. The species present have adapted to distinct sections of this gradient by having characteristically non-overlapping size ranges and utilising different nutritional resources. The three species discussed below show the most contrasting adaptive strategies.

Leaf litter, the highest quality nutritional resource, occurs largely on the soil surface and organic matter concentrations diminish markedly with depth. Further, mortality due to environmental unpredictability and exposure to predators tends to decrease with depth. Thus, *Dichogaster agilis*, a litter feeder, has a relatively rapid growth rate since it uses a high-quality resource although mortality is also high so close to the surface. Therefore, individuals must complete their life cycles in a short time to maximize their reproductive success; this is achieved by limiting their body size, which allows a higher basic metabolic rate.

In contrast, *Millsonia ghanensis* lives in the deeper horizons where the soil organic matter it feeds on is a very poor quality resource, although risks of mortality are reduced. It completes its life cycle over much longer periods and has a relatively-low metabolic rate. Its large size confers a selective advantage as these worms must possess considerable strength to penetrate the relatively dense soil. This and similar species ingest large quantities of soil since the organic matter at its preferred depth is both sparsely-distributed and of low energy density. *Millsonia anomala* is a mesohumic endogeic species which exploits an intermediate part of the gradient; size, preferred depth and demographic profiles are all intermediate between the two other species discussed above.

Table III.18 Size, depth distributions and demographic profiles of earthworm species in the savannas of Lamto (Côte d'Ivoire) (Lavelle, 1978).

Species	Maximum fresh w. adults (g)	Mean depth (cm)	Food (% OM in food)	Ecological category
Dichogaster agilis	0.50	6	soil OM + 20% litter	epi-endogeic
Millsonia anomala	6.00	8	soil OM (0-10 cm)	mesohumic endogeic
Millsonia ghanensis	16.00	32	deep soil OM (20-40 cm)	oligohumic endogeic

	Generation time (months)	Life expectancy at hatching (months)	Cocoon production (n y^{-1})
Dichogaster agilis	15	3.4	10.7
Millsonia anomala	20	6.2	6.2
Millsonia ghanensis	42	10.6	1.3

Further examples are found in many other groups such as Collembola whose life-forms are directly related to depth (*e.g.*, Gisin, 1943). Among the Collembola and Acari, body size tends to decrease with depth, in response to the smaller soil pores. However, some groups can colonise a wider part of the environmental spectrum than others and size and phylogenetic constraints are again the determining features. While the major groups of macrofauna *i.e.*, earthworms and termites are able to exploit most of the gradient, the mesofauna is generally restricted to the upper part of the gradient, with a distinction between epi- and hemiedaphic groups. The limited size and strength of these animals does not permit them to dig their way through the soil and they require relatively high-quality resources to sustain their high metabolic rates. Endogeic life is limited because movement is restricted to the larger pores and resources are highly dispersed and partly inaccessible. Only a few animals with specific adaptations (*e.g.*, Symphyla) living on tuberous roots or specialised root-feeding herbivores or saprophages are found. As a result, predators of endogeic macrofauna are also scarce.

4.4.3.2 *Regional scale: the effect of types of vegetation, soils and land use on communities*

Soil faunal communities clearly reflect differences in vegetation and management practices. In temperate and tropical areas, grasslands are always more favourable for earthworms than forest ecosystems. This is probably due to the better quality and greater abundance of organic resources available in grasslands, and also to a more favourable water regime since soil water losses by evapotranspiration are greater in woody than in herbaceous vegetation. Woody vegetation favours litter and wood-feeding invertebrates.

In tropical areas, an inverse relationship often exists between earthworm abundance and that of termites at the scale of a region (Senapati *et al.*, 1994; Decaëns *et al.*, 1994).

Soil invertebrate communities are reliable indicators of land management. The proportions of different groups of nematodes have been combined in an index which is highly sensitive to changes of the ecosystem (Bongers, 1990). Larger invertebrates react to changes brought about by soil management: cultivation, changes of vegetation, decrease of shade and inputs of fresh organic matter. For example, the soils of humid tropical agroforestry systems derived from the forest may conserve a large part of the diversity and abundance of invertebrate communities of the natural ecosystem (Figure III.63). At Yurimaguas (Peruvian Amazonia), palm-tree plantations with a legume cover had an overall biomass of soil macroinvertebrates of 93.9 g f wt m^{-2}, 74 % higher than the original forest (53.9). Conversion of the original forest to systems which comprise elements of both the original forest and an herbaceous strata often lead to larger biomasses than occur in the original system; this is often due to the proliferation of introduced exotic earthworm species (*e.g.*, *Pontoscolex corethrurus)* which develop large populations of 500 kg fresh weight ha^{-1} or more. Pastures also have high macroinvertebrate biomasses, because of their substantial earthworm populations.

In these systems, grazing transforms a significant part of the above-ground production into dung which is a high quality resource for soil fauna communities and favours the development of soil invertebrate communities, especially earthworms. When the original vegetation was a forest, native species normally do not adapt to the new environment and large populations develop through the immigration of opportunist colonists.

When savannas are transformed into pastures, local populations may develop and build up populations up to 5 times larger than in the original savanna; this is the case *e.g.*, in the Colombian llanos where earthworm biomass increased from 4 to 11 g f wt m^{-2} when the original savanna was grazed; when African grasses and an herbaceous legume were substituted for the original vegetation, earthworm biomass increased up to 42 g m^{-2}. Termite abundance showed an exactly inverse trend with a significant decrease of abundance along the aggrading sequence from the natural savanna to the improved pasture (Decaëns *et al.*, 1994).

Annual crops always have severely depleted macroinvertebrate communities; this has been noted for earthworms in a large variety of crops (rice, corn, beans, yam, yuca, etc.) and the original fauna may disappear within a few weeks of clearing and planting. This depletive effect of annual crops has been reported from a wide range of temperate and tropical environments (see, *e.g.*, Lee, 1985 for earthworms).

4.4.3.3 *Geographic scale: changes in overall community structure along a thermo-latitudinal gradient*

The relative abundance of the micro- and meso-fauna declines in tropical areas as a consequence of: (1) an increased abundance of macro-invertebrates because of favourable environmental conditions; and (2) a subsequent dominance of biological systems of regulation based on mutualistic relationships between roots, soil ecosystem engineers and the soil microflora. This contrasts with systems in which the microfauna (micropredators) and meso-fauna (litter transformers) have less-close associations with these micro-organisms (Figure III.61).

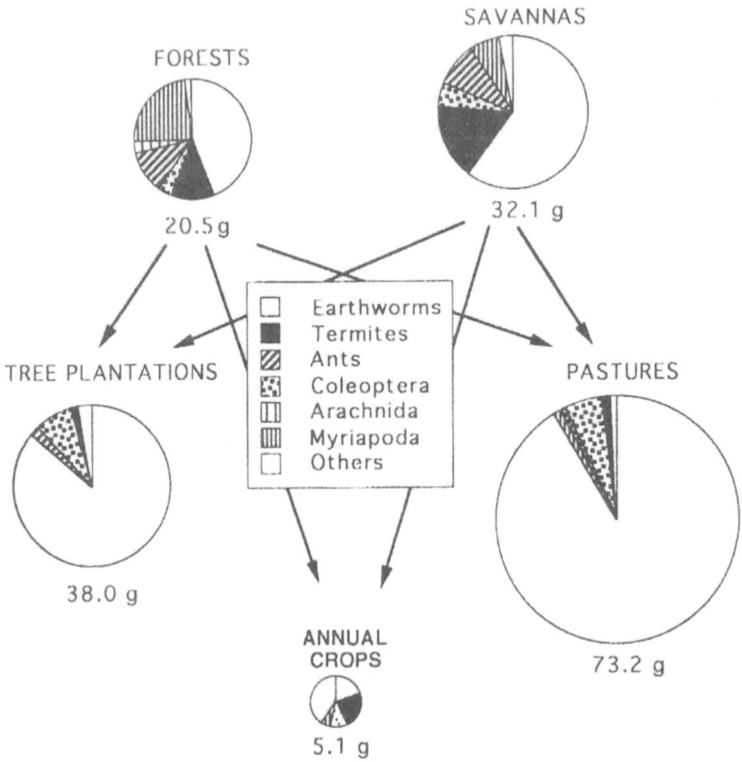

Figure III.63 Compositions of soil macro-invertebrate communities under different types of land use across 80 sites in tropical Africa and America (Lavelle *et al.*, 1994).

Micro- and meso-fauna dominate the soils of ecosystems in which climatic limitations strongly constrain biological activity. These invertebrates have highly-developed abilities to survive such harsh environmental conditions as low temperatures and drought. Under such conditions, the macrofauna is poorly represented (if at all) and does not compete for soil and litter resources with the micro- and meso-fauna. In the semi-arid grassland soils of North America (Hunt *et al.*, 1987) or in desert ecosystems (Freckman and Mankau, 1986), micro- and meso-faunal activity play key roles in determining microbial activity. Their selective removal may result in significant changes to decomposition and nutrient dynamics (*e.g.*, Ingham *et al.*, 1986a). In these ecosystems, micropredator food-webs (*i.e.*, food-webs based on predation of micro-organisms by microfauna) and mesofauna, or mutualistic relationships between mesofauna and micro-organisms (as defined, *e.g.*, by Moore, 1988) are the major interactive systems that control decomposition processes. They regulate decomposition and channel nutrients from decomposing resources to micro-organisms which are in turn consumed by protists, nematodes and microarthropods or Enchytraeidae. These animals release nutrients (*e.g.*, N as ammonium) as products of their metabolism or during the decomposition of their dead bodies (see also Chapter IV.1.5.3).

Mutualism between the litter transformers (*i.e.*, mesofauna and most arthropods living in the surface litter) and the microflora is mainly based on indirect relationships characterised by an 'external rumen' type of digestion, dissemination of spores or comminution. Introducing these animals into microcosms generally results in significant but rather limited increases in CO_2 evolution (*e.g.*, Coleman *et al.*, 1978; Setälä *et al.*, 1988; Persson, 1989). In systems where this type of interactive system predominates, nutrient cycling is generally slow. This is partly due to the limited capacity of these decomposers to digest resistant organic compounds, especially tannin-protein complexes and lignin. When abundant, litter arthropods may have an inhibitory effect on decomposition (Hanlon and Anderson, 1980).

Whenever climatic conditions favour **ecosystem engineers** that is, mainly earthworms (particularly anecic or endogeic species) and termites, their effects on microbial activity outweigh those of the smaller invertebrates (Seastedt *et al.*, 1987). Nematode populations decrease in the presence of earthworms (Dash *et al.*, 1980; Yeates, 1981; Boyer *et al.*, probably due to changes in the environment and in microbial populations. It is likely that these indirectly have a greater effect on nematode populations than direct predation by earthworms, which has not been conclusively demonstrated. Diminished nematode populations in the presence of earthworms further illustrates the competitive advantage of biological systems involving macro-invertebrates, when environmental conditions are not limiting.

In the presence of such macro-invertebrates as grass and litter-feeding termites, leaf litter is buried or transported to the nests at considerable rates and the biomasses and residence times of litter at the soil surface are significantly reduced. The abundance of the mesofauna and micro-arthropods decreases in response to the diminished resources resulting from competition with larger animals. The micro-predator food-chain is therefore reduced in importance in the litter system although it remains effective in the rhizosphere system where it plays an important role in plant nutrition (see, *e.g.*, Trofymow and Coleman 1982). The micro-predator food-chain may also exist in earthworm structures as some earthworms may partly feed on protists and stimulate their populations (Piearce and Phillips, 1980; Rouelle *et al.*, 1985) (see also Chapter IV.1.4.4).

As considered earlier in this chapter, soil animal communities differ structurally across sites with different temperatures regimes: along a gradient of increasing temperature, macrofaunal densities and diversities increase. Tundra soils have relatively few macro arthropods although their populations rapidly increase in temperate regions. This is because earthworm populations attain their highest biomasses in the temperate climate regions although the majority feed on litter rather than soil organic matter. It is largely in sub-tropical and tropical soils that termites become important and earthworm communities include all the existing ecological categories.

The mesofauna reaches its greatest abundance in non-extreme tundra soils and is fully developed, qualitatively and quantitatively, in cold temperate and temperate ecosystems with high organic matter accumulations. Where leaf-litter turnover is more rapid, mesofaunal abundance decreases perhaps as a result of increased temperature and/or earthworm and termite activity. The availability of decomposing litter is drastically reduced and the mesofauna may be limited to small euedaphic populations, except in some tropical forests where seasonal drought and/or acidity limit the development of sizeable populations of earthworms. This also applies to sites where termites do not consume all the available litter.

Thus, two major and contradictory effects of temperature become clear. Higher temperatures favour the increased activity of both micro-organisms and soil faunal populations. As a result, highly-efficient, mutualistic digestion systems become possible (Lavelle, 1986), permitting earthworms and termites access to a wide range of resources, including many that are recalcitrant and highly dispersed (*e.g.*, Martin *et al.*, 1992). The decreased litter and holorganic soil horizons and the diminished organic matter resources available limit the epigeic macrofauna and mesofauna associated with this part of the environmental gradient.

CHAPTER IV

FUNCTIONING OF THE SOIL SYSTEM

As discussed in the introduction to this work, three of the major integrated functions of soils are the provision of physical and nutritive support for plant production and the decomposition of dead organic materials. In addition to their ecosystem roles, these functions implicate soils in such major global processes as the hydrological cycle, climatic and stratospheric ozone regulation through the medium of greenhouse and trace gas emissions. Decomposition of organic residues and the maintenance of soil structure are largely complementary processes in most soils. This is because organic matter has important effects on soil structure both in colloidal form (cement) and as larger particles, (see Chapter I). Secondly, the energy released through decomposition processes is used by organisms for bioturbation, an important process in the creation and maintenance of soil structure, and in soil formation (see Chapter II) (Figure IV.1).

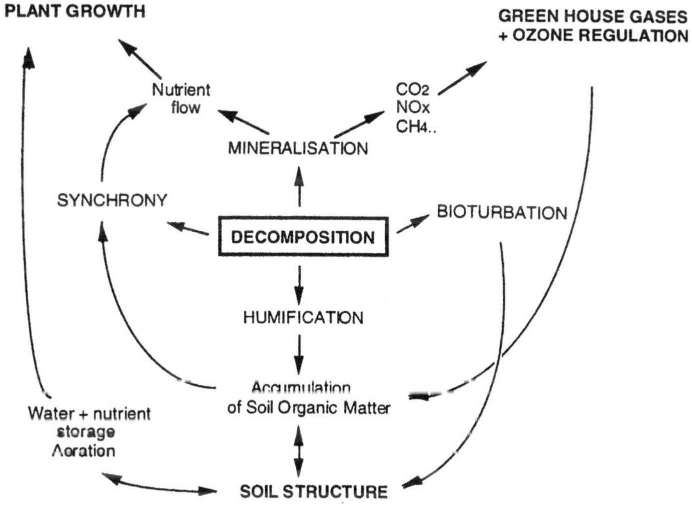

Figure IV.1 Relationships between decomposition and major ecosystem processes.

Soil organisms acting as functional interactive units - referred to as biological systems of regulation (BSR) - are the main mediators of soil functioning at micro- and meso-scales. Ecosystem engineers largely determine the activities of the other organisms, smaller invertebrates and micro-organisms; microbial activity is regularly stimulated in microsites where BSR's are active through the production of readily assimilable compounds such as root exudates or earthworm intestinal mucus (see the 'Sleeping Beauty paradox', Chapter III.2.4.3). However, their activities are mediated through a set of hierarchically-organised factors: climate, soil characteristics - especially the abundance and types of clays and nutrient status - and the quality of the organic materials input.

The functioning of the soil system is defined by:

(i) the decomposition rates of dead organic materials, and the balance between mineralisation (which releases nutrients available to plants and micro-organisms) and humification (which forms reserves of soil organic matter and colloidal organic compounds);

(ii) the degree of synchronisation and synlocalisation of nutrient release with plant demand (definitions in Swift, 1986; van Noordwijk and de Willigen, 1986);

(iii) the soil physical structure which determines the rates and patterns of gas exchange, soil water movement into and through the soil, and erosion rates.

This chapter first describes the general mechanisms and regulation of decomposition, the most important process of soil function. Subsequently, functioning of the major biological systems of regulation, the litter-systems, rhizospheres, drilospheres and termitospheres are then detailed, with particular emphasis on their compositions, structures and the effects of the major groupings of organisms on soil physical properties and organic matter dynamics.

IV.1 DECOMPOSITION

1.1 General processes

Decomposition is the general process whereby dead organic materials are transformed into simpler states with the concurrent release of energy and their contained biological nutrient - and other - elements in inorganic forms. Such forms are directly assimilable by micro-organisms and plants and the remaining soil organic matter may be stabilised through physical and chemical processes, or further decomposed. Two important features of decomposition processes are their overall rates and the transfers of organic materials to different macro- or microsites within the ecosystem.

1.1.1 THE ROLE OF ORGANIC MATTER IN SOIL FUNCTION

Decomposition may be defined as the sequence of organic matter transformations occurring after the death of organisms. These transformations involve two simultaneous but complementary processes: mineralisation and humification. Mineralisation is the catabolic process through which the elements contained in organic form within biological tissues are converted to inorganic forms such as nitrate, phosphate and sulphate ions. In the contrasting anabolic process, humification, organic molecules are condensed into degradation-resistant organic polymers which may persist little-altered for decades or centuries. Both processes occur simultaneously and are important aspects of soil fertility. Mineralisation determines the fluxes of plant- and micro-organism-available nutrients and their distributions in time and space; humification regulates the accumulation of stabilised organic matter within the soil.

Soil organic matter contributes to fertility in three ways:

(i) Cementation of soil within aggregates: part of the soil organic matter occurs in colloidal forms and cements soil particles together to create solid structural units known as aggregates (Section I.1.3.1). These are surrounded by inter-connected pore spaces that permit the movement of water, solutes and gases through the soil matrix and contribute substantially to erosion resistance.

(ii) Retention of cations: as a material possessing a pH-dependent, net negative electrostatic charge at soil pH values, soil organic matter aids the retention of positively-charged cations, especially in the acid pH range where soil minerals may retain lesser amounts.

Properties (i) and (ii) are common to soil organic matter and clay minerals (see Chapter I.1.1.1.2), although some differences exist. For example, in soils more acid than pH 4.2, soil minerals retain few cations whereas humic molecules still retain an appreciable cation exchange capacity down to pH 2.5 (Bonneau and Souchier, 1982).

(iii) Conservation of nutrients in organic forms: organic matter conserves both nutrients and energy in forms that are neither readily assimilable by micro-organisms nor susceptible to leaching. Low concentrations of nutrients frequently limit growth processes in soil and their conservation in organic forms is a key feature of decomposition. In addition, the synchronisation of nutrient release through decomposition with plant demand is of paramount importance in avoiding nutrient losses (Swift, 1984; van Noordwijk and de Willigen, 1986).

1.1.2 PHYSICAL ASPECTS: RATES OF DISAPPEARANCE AND TRANSFERS OF DECOMPOSING MATERIAL

Decomposition is the sum of three different processes (Figure IV. 2): i. Leaching, whereby water-soluble materials are transported into and through the soil by percolating water; ii. comminution, which is the physical reduction in the size of the plant material with little chemical change but a large increase in surface area; iii. microbial and enzymatic catabolism (mineralisation), which results in the release of mineral components, accumulation of resistant organic matter (HU) and the formation of decomposer tissue (DO). At each step, a resource is transformed into a set of secondary resources which will then undergo further transformation. Decomposition is therefore a series of sequential processes during which a resource is progressively transformed through ingestion and egestion by several successive organisms. As a result, the total ingestion of decomposing materials by all organisms may substantially exceed the overall primary productivity of an ecosystem. Decomposition may be physically characterised by the rate of disappearance of the resource and the transfer of the material to different locations within the soil system. Such transfers may increase the apparent rate of resource disappearance.

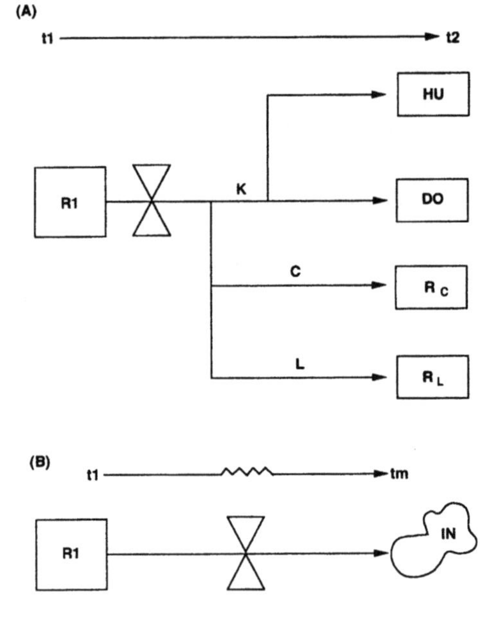

Figure IV.2 Decomposition of a primary resource R1 results in the production of four secondary resources *i.e.*, HU (humified material), DO (Decomposer biomass), R_c (comminuted resource), R_l (leached resource) and IN (inorganic resources), through three different processes namely, catabolism (K), comminution (C) and leaching (L) (Swift *et al.*, 1979).

Decomposition rates

The rates of disappearance of decomposing resources have been widely measured by exposing known amounts of leaf and other litter in 'litter bags' made from a mesh of an inert fibre (see Anderson and Swift, 1983; Anderson and Ingram, 1993; Bockheim *et al.*, 1991; Schowalter *et al.*, 1991; Montagnini *et al.*, 1993; Thomas and Asakawa, 1993; Basaguren and Pozo, 1994 in terrestrial ecosystems, Irons *et al.*, 1994 in fresh water ecosystems). Weighing the material remaining at several time intervals permits the calculation of decomposition rates. To avoid the problems resulting from litter confinement (lack of faunal access, altered water status) some authors prefer to study the decomposition of individual leaves tethered by nylon line (Swift *et al.*, 1979; Vitousek *et al.*, 1994). Another common and simpler method compares annual inputs with the accumulated decomposing leaf litter. In a system at steady state, the annual input of dead organic matter (I) is assumed equal to the amount decomposed annually (Olson, 1963). Under such conditions, and assuming a simple exponential pattern of decomposition, the decomposition constant (*k*) may be calculated:

$$k = I/Xss$$

where *k* is the % of resource decomposed annually, I is the annual input and Xss the mean annual weight of accumulated, decomposing material. Time for decomposition of a given proportion of the residue is calculated from the equation:

$$X/X0 = e^{-kt}$$

where X/X0 is the proportion of the resource decomposed in a given period of time (t) with a decomposition rate *k* (Bernhard-Reversat, 1982). However, this equation is unsuitable for litters with complex decomposition patterns. The X/X0 value varies greatly with a maximum of approximately four (half the input mass decomposed in 3 months) in tropical rainforests and a minimum of 0.01 (half the mass decomposed in 100 years) in tundra ecosystems (Figure I.43). The data compiled by Heal *et al.* (1981) show a clear climatic effect, with *k* values decreasing from tropical to tundra ecosystems.

Other more complicated models have been used to characterise litter decomposition rates. The more realistic of these estimate different decomposition constants (*k* values) for each of the major litter components (Jones, 1990) although few models cope satisfactorily with the important interactions that occur among them in many plant materials.

Transfers

Decomposition is often associated with the movement of part or all of the decomposing resource although roots and most under-ground resources, are little affected by this process. In contrast, above-ground litter is progressively translocated deeper into the soil as it decomposes. Organic matter leaching through the profile is usually metabolised (sugars, amino-acids) or immobilised in the soil (phenolic compounds) by iron cations (Toutain, 1974). The latter are precipitated as organic and organo-mineral concentrations or become associated with clays and other mineral particles. The older, partly-decomposed litter is buried under successive layers of more recently-deposited material (Figure I.42)

(Garay *et al.*, 1986a). Finally, rapid burial of slightly-decomposed material may be effected by physical processes such as the opening of cracks in drying vertisols and biological activities, particularly those of anecic invertebrates. Anecic earthworms, leaf and seed harvesting ants and litter-feeding termites all accumulate dead plant materials within their burrows, nests and galleries; substantial proportions of annual primary productivity may be translocated and sequestered within these biogenic structures (Jones, 1990; Martin, 1991).

1.1.3 CHEMICAL ASPECTS: RESOURCE QUALITY AND CHEMICAL TRANSFORMATION DURING DECOMPOSITION

As shown in Chapter I, decomposing resources are made up of a wide range of chemical substances. They differ in the relative proportions of their major constituents *i.e.*, lipids and waxes, water-soluble carbohydrates, hemicellulose, cellulose, lignin, proteins, phenols and other secondary plant compounds. Each of these compounds is considered to have its own specific decomposition rate, and decomposition of any resource will depend - as a first approximation - on their relative abundances. Minderman (1968) calculated annual loss rates ranging from 99 % for sugars to 10 % for phenols (Figure IV.3).

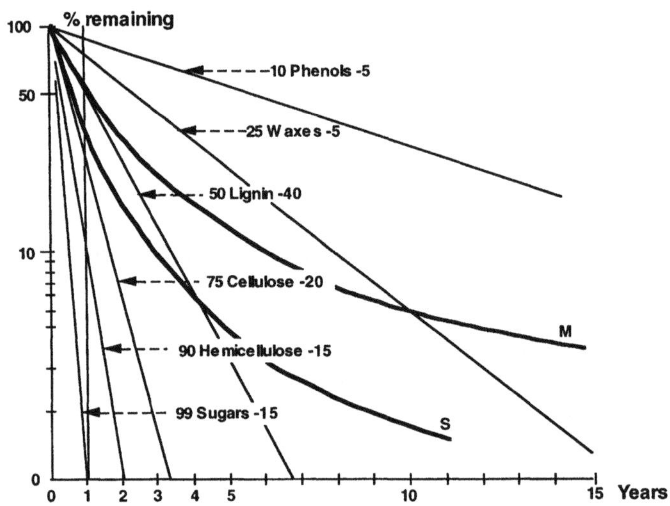

Figure IV.3 Decomposition curves of different constituents of litter material (logarithmic units). The numbers preceding and following the name of each component indicate, respectively, the percentage lost over one year and its weight as a percentage of that of the original litter. The curve S shows the theoretical overall decomposition curve obtained by summing the curves for individual components; curve M is that actually observed (Minderman, 1968).

Decomposition rates therefore decrease over time as the more-readily decomposed materials disappear. However, the observed decomposition rate (M) was slower than that predicted (S) by simply summing the decomposition rates of the individual components. Decomposition rates may be reduced by such interactions as the tanning of cytoplasmic proteins by phenols released from vacuoles at leaf death and by physical constraints to microbial colonisation posed by lignin-rich structures and waxy cuticles. The effects of these materials on decomposition dynamics are detailed in Section IV.1.3.3.

Decomposition rates measured, for example, as the respiration rates of microbial decomposers, or C losses from decomposing residues, may be described as decreasing functions of time. Jenkinson and Ayanaba (1977) calculated the following double exponential equation to describe mass loss in uniformly ^{14}C labelled Rye Grass (*Lolium perenne)* mixed with soil in England and Nigeria:

$$y = 0.9\ e^{-2.83t} + 29.1\ e^{-0.087t}$$

where y is % C remaining and t is time in years.

This double exponential model describes a two-compartment model, in which about 70 % of the plant material decomposes with a half-life of 0.25 years and the remainder with a half-life of 8 years. Interestingly, the authors found that decomposition curves for each location only differed in their time scales, rates being four times higher in the tropical site (Figure IV.4). In these experiments, the Rye Grass samples were decomposing only under the influence of micro-organisms; roots and macro-invertebrates were not allowed to enter the tubes within which the grass samples were decomposing.

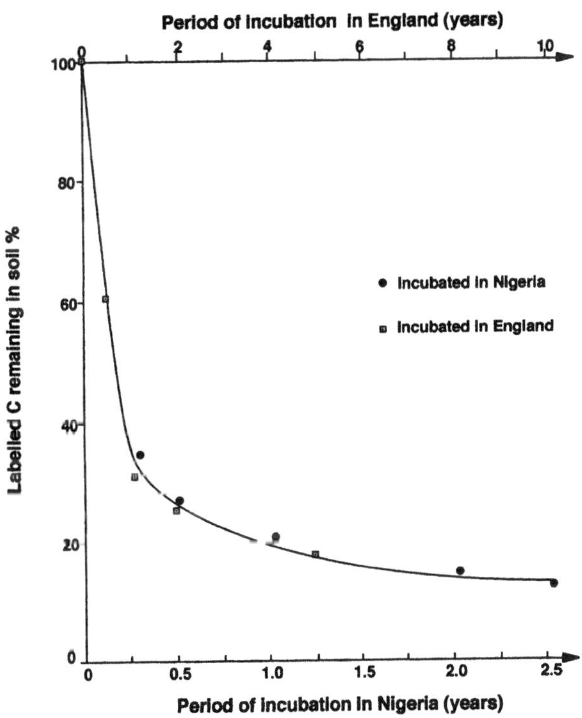

Figure IV.4 Decomposition of uniformly labelled rye grass *(Lolium perenne)* in England (Rothamsted) and Nigeria (Ibadan) (Jenkinson and Ayanaba, 1977).

1.2 An hierarchical model of the abiotic and biotic factors determining decomposition

Decomposition rates are determined by a set of physical (climate), chemical (decomposing resource quality) and biological (micro- and macro-organism communities) factors that operate at different spatial and temporal scales. A general hierarchical model of the factors controlling decomposition is presented that classifies these factors by the scales at which they operate.

1.2.1 DYNAMICS OF DECOMPOSITION

Decomposition is described as a 'cascade' process during which a given resource (R) is progressively transformed into a set of secondary resources which, in turn, are transformed into resources of tertiary or higher order (Bunnell and Scoullar, 1975). These resources include a variety of materials derived from above- and below-ground ecosystem components. They include plant litter, invertebrate faecal pellets, living and dead micro-organisms, invertebrates and other materials ranging downwards in size to amorphous soil organic matter and comprising fractions of different age and chemical composition.

Decomposition does not usually occur at even rates. Its time course may be represented as a succession of active phases separated by periods of inhibition during which physical, chemical or biological factors limit or completely inhibit decomposition processes (see, for example, Swift, 1976; Heal and Dighton, 1985; Toutain, 1987b). The time course of decomposition processes undergone by Spruce *(Picea abies)* needles in temperate ecosystems clearly illustrates this point (Gourbière, 1982) (Figure IV.5) (see also Chapter IV.2.5.4).

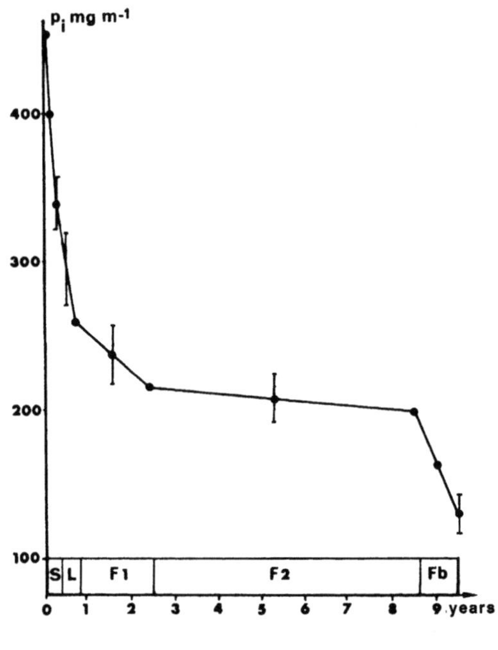

Figure IV.5 Decomposition dynamics of the needles of *Abies alba* showing an initial phase of rapid decomposition (in surface litter (S) and L layer), a second phase of inhibition (in layers F1 and F2) due to the relative accumulation of tannin-protein complexes, and a third phase corresponding to its release through the activities of white-rot fungi in the Fb layer (Gourbière, 1982). p_i = dry weight of one metre of needles; Fb: F layer invaded by white rot fungi.

Decomposition is essentially a biological process. Abiotic chemical oxidation is unlikely to account for more than 20 % of total CO_2 evolution from soils (Scharpenseel *et al.*, 1984; Seastedt, 1984; Moorhead and Reynolds, 1989; Warneke *et al.*, 1999). At least 80 %, and normally more than 95 % of CO_2, may therefore be derived from organismal respiration. Nutrients taken up by plants in natural environments are derived largely from decomposition processes (Table IV.1 and I.2) although the proportions derived from mineral weathering and other sources differ among soil and ecosystem types, and with the nutrient considered.

Micro-organisms are by far the major contributors to soil respiration and are responsible for 80 to 95 % of the total CO_2 respired and consequently of the organic C mineralised (Satchell, 1971; Ryszkowski, 1975; Reichle *et al.*, 1975; Persson and Lohm, 1977; Lamotte, 1975, 1989). Consideration of their ecological traits is of paramount importance in understanding decomposition dynamics, since low levels of micro-organism activity may result directly in complete or partial inhibition of decomposition (see *e.g.*, Trofymow and Coleman, 1982). This is mainly due to their relative lack of mobility, depression of their activities under extreme temperature and moisture conditions, metabolic specificity and susceptibility to chemical inhibition and partial inactivation when covered with crystalline clay particles (see the 'Sleeping Beauty paradox', Chapter III.2.4.3).

Decomposition processes are therefore determined by interactions among three factors, namely, organisms, environmental conditions (particularly the climate and the suite of minerals present in the soil) and the quality of the decomposing resources (Swift *et al.*, 1979; Anderson and Flanagan, 1989). These three factors are not equally important since they operate at different spatial and temporal scales and may have opposing influences on decomposition processes. It therefore seems more appropriate to use an hierarchical model to describe the factors determining decomposition than the triangular scheme proposed by Swift *et al.* (1979).

Table IV.1 Sources of nutrients taken up by plants (Chapin, 1991).

	Source of plant nutrient (% of total)		
Nutrient	**Atmosphere**	**Weathering**	**Recycling**
Tundra (Barrow, Alaska)			
N	4	0	96
P	4	<1	>95
Temperate Forest (Hubbard Brook)			
N	7	0	93
P	1	<10?	>89
K	2	10	88
Ca	4	31	65

Table IV.2 Participation of soil organisms in decomposition of litter and soil organic matter (%).

Nature of the ecosystem	Authors	C mineralised (%)			
		Micro-organisms	Earthworms Invertebrates	Other Invertebrates	Total Invertebrates
Deciduous forest (USA)	Reichle *et al.*, 1975	99.7	0.02	0.28	0.30
Arable land (Poland)	Ryszkowski,1975	96.0	0.30	3.70	4.0
Deciduous forest (England)	Satchell, 1971	83.0	2.50	14.50	17.0
Abandoned field (Sweden)	Persson & Lohm, 1977	93.5	2.60	3.90	6.5
Savanna (Côte d'Ivoire)	Athias *et al.*, 1974	80.0	9.50	10.50	20.0

1.2.2 HIERARCHY OF THE FACTORS GOVERNING DECOMPOSITION

The factors determining microbial activity, and therefore decomposition rates, operate at varying scales of space and time and differ in relative importance. They act in a hierarchical fashion with the higher-level factors dominating those acting below them (Allen and Starr, 1982; Di Castri *et al.*, 1988). The following four hierarchical levels may be identified (Figure IV.6): (i) climatic factors (moisture and temperature regimes); (ii) edaphic properties such as the clay minerals present; (iii) the physical and chemical characteristics of the decomposing resources; and (iv) biological regulation through interactions between macro- and micro-organisms. In this model, all hierarchical levels interact and the position of a factor only indicates the probability that it may have a dominating effect over a factor operating at a lower level.

The appropriateness of this hierarchy is confirmed by such general models of decomposition as that of Heal *et al.* (1981) which demonstrate the dominant effect of climate in controlling decomposition rates at broad geographical scales. Meentemeyer (1978) and Aerts (1997) also provided evidence for the dominance of climate over chemical factors as major determinants of decomposition rates in ecosystems ranging from the arctic to tropical forests. Similarly, at a regional scale, Vitousek *et al.* (1994) found a strong relationship between decomposition rates and elevation (and hence mean annual temperature) on the slopes of Mauna Loa volcano (Hawaii).

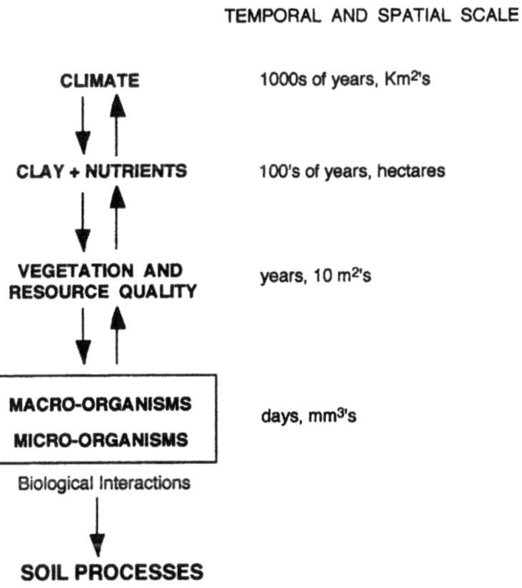

Figure IV.6 An hierarchical model of the factors determining soil processes in terrestrial ecosystems (modified from Lavelle *et al.*, 1993).

Despite the above, considerable differences still occur at a regional scale, demonstrating that other factors may also be important, or even predominate. Anderson and Swift (1983) have shown that considerable overlap in leaf litter decomposition rates occurs between temperate and tropical forests. Further, intra-site variation may be larger than inter-site variation and this emphasises the importance of local edaphic and biological factors (Figure IV.7). At this scale, the additional effects of clay mineralogy, resource quality and invertebrate activity have also been confirmed (see, for example, Meentemeyer, 1978; Anderson *et al.*, 1985; Darici *et al.*, 1986; Laishram and Yadava, 1988; Nicolai, 1988; Seastedt *et al.*, 1988). However, the effects of the higher-level factors are so powerful that they may override the effects of those, such as the soil fauna, that act at lower levels making them difficult to detect (Andrén *et al.*, 1988, Persson, 1989).

In other situations, the hierarchy of factors may be less clear. In aquatic environments, for example, Irons *et al.*, (1994) found no relationship between latitude and leaf litter decomposition. Factors specific to this environment (*i.e.*, oxygen tension or water nutrient status) may have confounded the stimulatory effect of higher temperatures on decomposition. Similarly, Coleman *et al.* (1990) showed that reducing nematode and micro-arthropod populations explained most of the differences in decomposition rates in soils where climate and resource quality were expected to be the main determinants.

Interactions between levels II (soil nutrient status) and III (litter quality) of the proposed hierarchy have been demonstrated several times. Plants growing on unproductive soils often produce poor-quality litter with lower N and greater concentrations of secondary compounds (polyphenols, terpenoids) than those growing on more fertile soils (Janzen, 1974; Waterman, 1983; Baas, 1989). The same situation has been observed with plants growing in atmospheres enriched in CO_2 (Coûteaux *et al.*, 1991; Woodin *et al.*, 1992;

Cotrufo and Ineson, 1995). Conversely, plants also modify soils in ways that improve conditions for subsequent plant growth (Boettcher and Kalisz, 1992; van Breemen, 1993; Berendse, 1994).

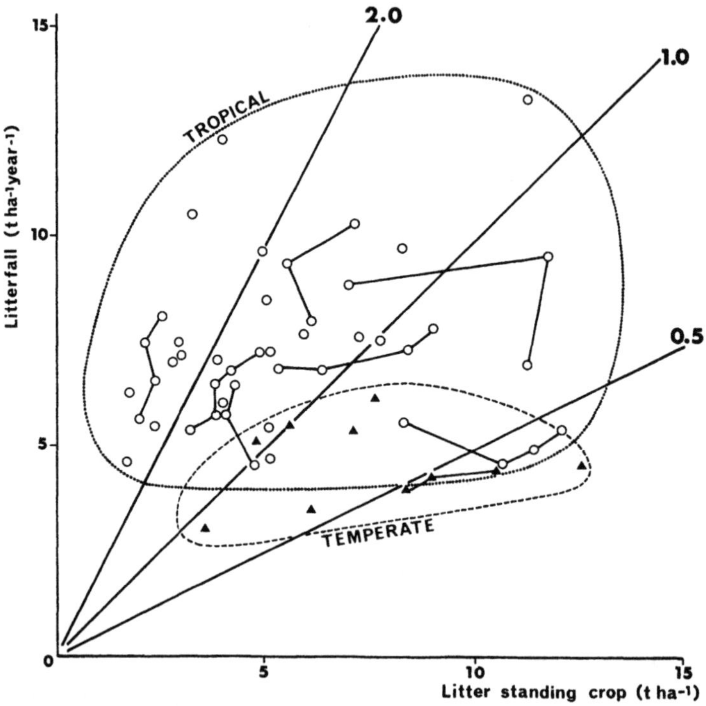

Figure IV.7 Relationships between litterfall and litter standing crop in tropical rain forests and temperate -climate deciduous forests (after Anderson and Swift, 1983). Bold continuous lines delimit portions of the figure with litterfall/standing crop (k) quotients ranging from less than 0.5 to greater than 2.0. Studies carried out in the same locality are linked by lines.

There are two important corollaries to the hypothesis that factors governing decomposition rates are organised hierarchically:

(i) Where higher-level factors depart little from optimal levels, the effects of factors operating at the next lowest level of organisation will become more apparent. This implies that the postulated hierarchy presented above has some plasticity and may be altered locally or regionally when constraints corresponding to a particular level are unimportant, or non-existent. The large variation in decomposition rates observed in the humid tropics at local and regional scales considered above, clearly illustrates this point (Anderson and Swift, 1983). In those systems where climatic limitations are largely absent and clay minerals are present in small quantities or are of low activity (such as kaolinite), the quality of resources and composition of macro-invertebrate communities may become the determining factors.

(ii) Disturbance may alter the relative importance of these factors, creating new environments with completely modified constraints to decomposition. One such example is the eradication of the original soil invertebrate communities and perennial roots through the clear felling of forests and the cultivation of annual crops and grasses (Coleman *et al.*, 1990).

1.3 Processes involved at each level of the hierarchy

Despite their dependence on other abiotic and biotic factors, both the natural histories and life strategies of micro-organisms impact on the nature and dynamics of decomposition processes. Such traits directly determine decomposition rates and the release of soluble metabolites; they also control the succession of active and inactive phases and affect their interactions with macro-organisms.

1.3.1 MICROBIAL PROCESSES (SEE ALSO CHAPTER III.2)

Micro-organisms may have huge population densities and biomasses, and wide functional diversities. Densities of up to 10^{10} bacteria and several kilometres of fungal hyphae per gram of dry soil have been measured over a wide range of tropical and temperate climate soils. This represents a microbial biomass of several hundred kg ha^{-1}, *e.g.*, 370 to 760 kg dry weight ha^{-1} in early- and late-stage secondary Nigerian forests (Ayanaba *et al.*, 1976) and 150-400 kg dry weight ha^{-1} in pasture soils from tropical India (Dash *et al.*, 1985). Despite this, microbial biomass comprises only a small percentage of total soil carbon (1.7 to 4 % in the examples cited) with a considerable turnover time, of the order of a year (see for example, Jenkinson and Ladd, 1981; Voroney, 1983; Chaussod *et al.*, 1988). Microbial communities are thus both numerous and diverse, but largely dormant. The slight ability of most non-filamentous prokaryotes and protoctists to move within the compact soil environment largely limits their individual spheres of activity to their immediate microsites.

The huge discrepancy between high potential metabolic rates and the slow turnover of biomass has been defined in Chapter III.2.4.3 as the 'Sleeping Beauty' paradox. Soil macro-organisms act as 'Prince Charming' in that they provide assimilable substrates (root exudates, earthworm mucus and other materials) which initiate their metabolic capabilities. They may also transport propagules to locations where organic substrates are favourable for colonisation and development.

Finally, the microbial biomass has a C:N ratio in the range of 5:1-7:1 (bacteria) to 7:1-25:1 (fungi) (Chapter I.3.1.4.3). Depending on the C:N ratio of the resource that they exploit, mineral-N may be released during decomposition (net mineralisation) - and possibly supplied to plants - or retained within microbial tissues (immobilisation or reorganisation) and thereby made unavailable to plants (see Section IV.1.3.3).

1.3.2 ABIOTIC LIMITATIONS TO DECOMPOSITION

Climatic factors

In many parts of the world, some combination of low temperature and drought severely limits the seasonal extent and rates of decomposition processes (see, for example,

Moore, 1986; Swift *et al.*, 1981; Garay *et al.*, 1986a, Andren *et al.*, 1993). Meentemeyer (1978) and Berg *et al.* (1993) found a significant relationship between the actual evapotranspiration (AET) and decomposition constants (k) applicable across a climatic range extending from arctic to tropical areas. Within individual climatic areas, however, this relationship was only significant in the Mediterranean region (Aerts, 1997). In a Scots pine (*Pinus sylvestris*) forest in central Sweden, the combined effects of different temperature and moisture regimes explain 95-99 % of the variation in decomposition rates (Jansson and Berg, 1985). This is mainly due to the inactivation of micro-organisms, although they may be active under much more extreme conditions than are plants and invertebrates (Dommergues and Mangenot, 1970; 1980; Billès *et al.*, 1975; Legay and Schaefer, Singh and Shekhar, 1986).

In deserts, decomposition processes may be severely drought limited. In an arid coastal desert in Chile, Cepeda-Pizarro (1993) estimated annual biomass losses of *Atriplex* (Chenopodiaceae) litter at 11 to 18 %, the lowest values ever recorded in litter bag experiments.

Although temperature and moisture conditions are suitable much of the time for near-optimal biological activity in the humid tropics, locations differ in the length and intensity of their dry seasons. Furthermore, under comparable conditions of temperature and rainfall, the type of vegetation present influences the soil moisture regime; savanna soils are less prone to desiccation than those of forests, because grasslands transpire at lower rates than forests (Anderson and Swift., 1983; Lavelle, 1983b; Vitousek and Sanford, 1986).

The effects of clay minerals and the ionic environment

Decomposition processes may be inhibited simply by the physical inaccessibility of resources to decomposers within soil structural aggregates. This occurs (1) when micro-organisms are included within micro-aggregates and have no contact with decomposing substrates and (2) when micro-organisms or organic substrates are coated with clay minerals thereby limiting access to adjacent substrates and organisms, respectively.

Vitousek and Sanford (1986) showed a clear relationship between patterns of nutrient cycling in tropical forests and soil type, and this ultimately depends on the amounts and types of the clay and other minerals present. Clay minerals may directly limit decomposition by forming coatings on organic substrates and micro-organisms. This may also occur through the adsorption of organic molecules (*e.g.*, on allophanes) or perhaps by the sequestering of organic matter between the layers of phyllosilicate clays, particularly the smectites and vermiculites, or within tactoids. This results in the interposition of further barriers between micro-organisms and potential substrates (Figures I.11 and IV.8a.). Consequently, clear inverse relationships have been established between the abundances of certain clay minerals (of an homogeneous nature) and mineralisation rates in soils (see, for example, Kobus and Pacewiczowa, 1966; Stotsky and Rem, 1966; Chaussod *et al.*, 1986; Darici *et al.*, 1986; Schäfer *et al.*, 1993).

In some environments, clay minerals may inhibit microbial activity, and therefore decomposition, by adsorbing enzymes (McLaren, 1975; Haider and Martin, 1980; Sarkar *et al.*, 1989; Burns, 1990). Since this effect depends on a number of other factors (Burns, 1990) its expression is variable and, in some cases, enzymes immobilised on clay minerals both retain their activity and even have increased thermostability (Rowell *et al.*, 1973). Clay minerals may also adsorb inhibitors and thus prevent the inactivation of particular microbial functions (Gilmour, 1984; Bernhard-Reversat, 1996).

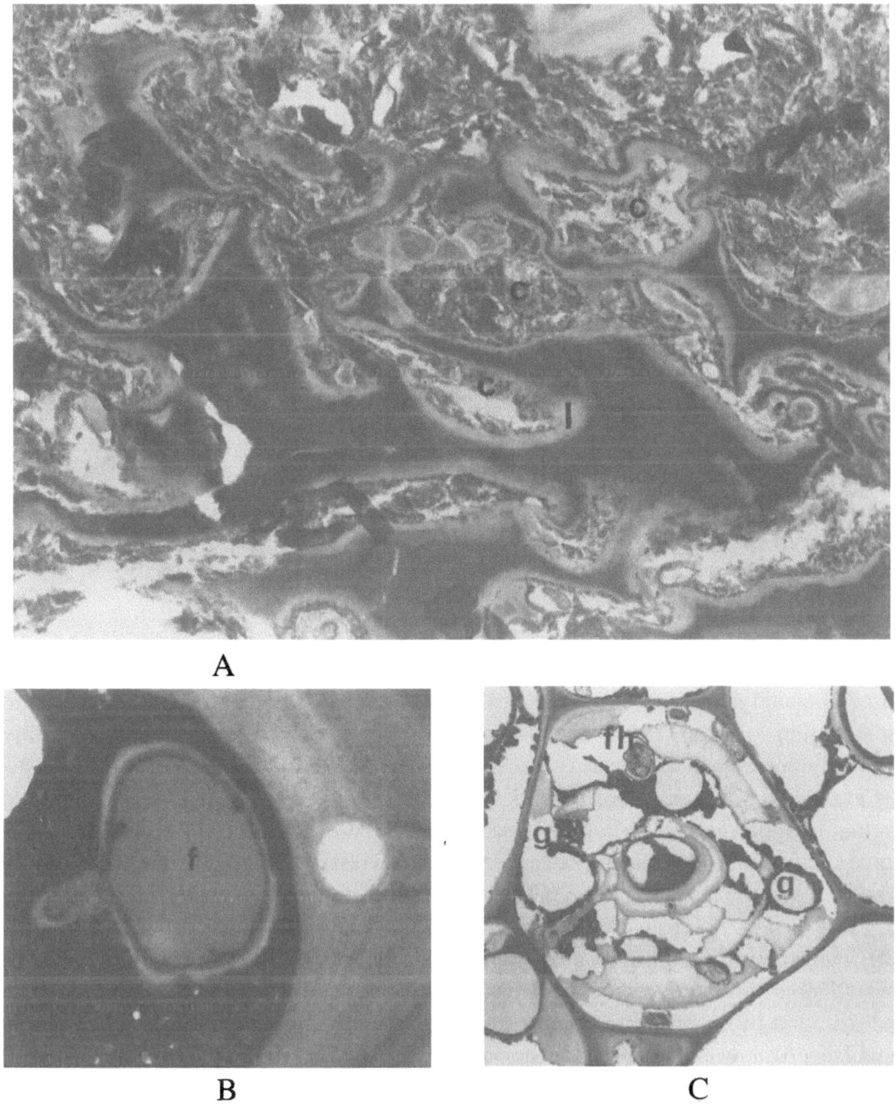

Figure IV.8 Electron micrograph of decomposing material showing **A**. penetration of clay minerals (c) into decomposing leaves. Lysis has occurred at the clay/leaf interface (l) (F. Toutain, x 15,000); **B**. Tannin-protein material lysed by a white-rot fungal mycelium (f) in a parenchyma cell of *Treculia obovoides* N. E. Br (Moraceae) (F. Toutain, x 9,000); **C**. Sclerenchymatous fibre of a *Treculia obovoides* N. E. Br (Moraceae) leaf colonised by lignolytic fungal hyphae (fh). (g) indicates remnants of tannin-protein granules (F. Toutain, x 4,500).

The types and amounts of clay minerals present may greatly influence the effects mentioned above. Minerals with low specific surface areas and low charge densities have relatively low overall electrostatic charge levels and are therefore less effective in forming structural aggregates (see Chapter I). This is the case for many soils in which kaolinite is the dominant clay mineral (Uehara, 1982), as in the widespread ultisols and oxisols (Table I.5) which comprise about 60 % of tropical soils (Sánchez and Salinas, 1983). Despite this, in oxisols where the iron supply is high (as in soils derived from basaltic parent materials) and organic matter concentrations are also high, the considerable surface areas and reactivities of iron oxides may lead to high levels of complexation with soil organic matter (Schwertmann *et al.*, 1986).

Physical protection in biological structures
Many common soil invertebrates produce compact organic and organo-mineral faecal pellets which have a low internal porosity. There is some evidence that mineralisation of plant material is depressed in the faecal pellets of some arthropods and enchytraeid Oligochaeta, especially when they are small (see, for example, Toutain *et al.*, 1982; Hanlon and Anderson, 1980; Bernhard-Reversat, 1993).

This process is particularly well developed in rounded earthworm casts. In the Lamto savannas (Côte d'Ivoire), endogeic geophagous earthworms may produce stable casts from soils with high (>80 %) sand contents (Blanchart, 1992); decomposition of soil organic matter is substantially reduced within these structures (Martin, 1991) (see also development in Chapter IV.4.3).

1.3.3 RESOURCE QUALITY AND DECOMPOSITION PROCESSES

Decomposition rates are highly dependent on the chemical quality of the decomposing resource. In analysing data from 192 different sites distributed across temperate to tropical regions, Aerts (1997) found that the litter decomposition rate was highly significantly correlated with quality, as assessed by the lignin to nitrogen ratio (r^2 estimated at 0.24). However, litter decomposition rates were much more closely related to the actual evapotranspiration rate (r^2 estimated at 0.46). As decomposition proceeds, resource quality rapidly changes because the readily-assimilable substrates are rapidly metabolised and resistant compounds tend to accumulate (see, for example, Minderman, 1968; Lobo *et al.*, 1974; Rapaire and Turenne, 1977; Martin and Haider, 1986).

There are three principal types of chemical limitations to decomposition processes: an initial inhibition in recently-dead tissues due to the formation of phenol-protein complexes, and those resulting from the accumulation of lignin and of resistant humic compounds. At any stage, imbalances in the ratios of carbon to the other nutrient elements present may also inhibit decomposition processes.

Phenol-protein complexes
When plant tissues die, polyphenolic compounds accumulated in vacuoles are released into the cell contents. They react with the cytoplasmic proteins to form phenol-protein complexes which impart the brown pigmentation to dead roots and leaves (Handley, 1954, 1961; Toutain, 1987b). (Figure IV.8b). These complexes usually constitute 10 to

22 % of total carbon and 22 to 72 % of the nitrogen of freshly-dead leaves (Toutain, 1987a; François *et al.*, 1986) and they may comprise 25 % of the weight of Beech (*Fagus sylvatica*) leaves (François *et al.*, 1986).

Phenol-protein complexes are also an important component of decomposing roots; in Beech *(Fagus sylvatica)* roots, they make up 15 % of the overall weight, 17 % of the total carbon and 85 % of the total nitrogen (Baraer and Toutain, 1987). However, due to the large variety of tannins and cytoplasmic proteins, it is probable that a wide range of phenol-protein associations exist, all with potentially-different degrees of resistance to decomposition (Waterman and McKey, 1989).

Concentrations of condensed tannins and phenols varied respectively, in the ranges 1.08-9.60 % and 3.64-10.10 % of the dry weight of mature leaves over 10 forest sites distributed throughout the tropics (Waterman and McKey, 1989). There is some evidence to suggest that where these products occur at high concentrations in leaves, they may act as an antiherbivore defence. In soils of poor nutrient status, plants grow slowly and may allocate more carbon to the production of secondary compounds (Baas, 1989) since this is less limiting than nutrients,

The degradation of phenol-protein complexes is usually extremely slow, unless particular organisms capable of decomposing them are present, namely, certain earthworms, termites and 'white-rot' basidiomycote fungi (Figure IV.8b and c). In the absence of such organisms, degradation of these brown pigments occurs through progressive hydrolysis caused by percolating water (the so-called 'infusion' effect) (Toutain, 1987a). In a litter-bag experiment at Yurimaguas, in Peruvian Amazonia, Palm and Sánchez (1990) found a negative relationship between the ratio of polyphenolic to nitrogen contents and the release of mineral nitrogen from decomposing legume leaf litter (Figure IV.9).

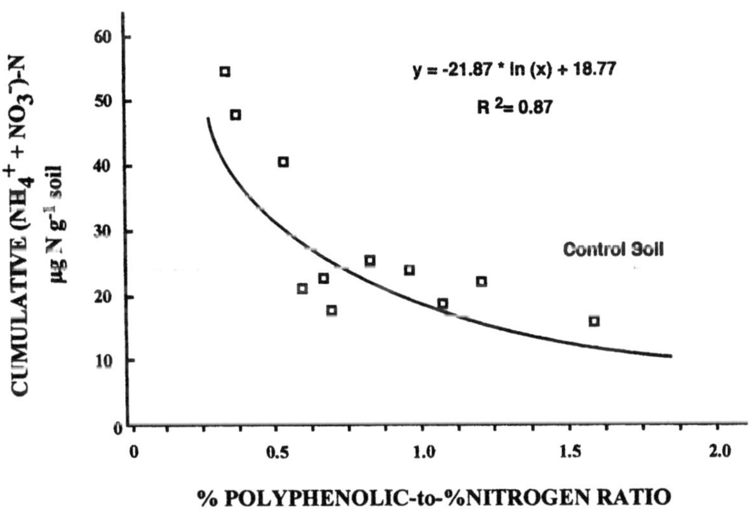

Figure IV.9 Relationship between the ratio of phenolic compounds to nitrogen in the leaves of selected tropical legumes and their decomposition rates under standardised conditions (Palm and Sánchez, 1991).

High lignin contents
Lignin is one of the plant components most resistant to decomposition and its relative concentration, usually expressed as lignin to nitrogen ratio, is often inversely and closely linked to decomposition rates (Meentemeyer, 1978; Melillo *et al.*, 1982; Laishram and Yadava, 1988). Lignin degradation of is also highly dependent on the presence of a specific microflora, mainly the 'white-rot' fungi (discussed in sections 2.2 and 2.5.4), and may accumulate in ecosystems where these fungi are lacking (Bernhard-Reversat and Schwartz, 1997). The lignin to nitrogen ratio (L/N) of decomposing material has been used to describe decomposition rates in the Century model of soil organic matter dynamics (Parton *et al.*, 1983). Lignin and phenolic compounds may have similar inhibiting effects on decomposition rates, although different mechanisms are involved.

Ratios of carbon to nutrient elements
At any stage of decomposition, nutrient-element deficiencies may limit microbial activity and thereby block the release of nutrient elements available to plants and other micro-organisms. This occurs when the C:nutrient ratio of the decomposing resource is high compared with that of the living micro-organisms and nutrients remain immobilised within the microbial biomass. In the case of the C:N ratio, this is 5-7:1 for bacteria and 7-25:1 for fungi (Swift *et al.*, 1979).

In these situations, decomposition is slowed or halted if sufficient quantities of the particular nutrient are not available in the immediate environment (Figure IV.10). This often occurs in nutrient-limited soils and the C:nutrient ratio is adjusted through the progressive elimination of carbon as CO_2 respired by micro-organisms leading to a mobilisation of the nutrients contained within their biomass. In the case of nitrogen, a C:N ratio of *ca.* 20-25 represents this limit and other nutrient elements may be immobilised or mineralised following the same rule. Therefore, depending on the initial values of the C:nutrient ratios, some nutrients may be released while others are immobilised.

A relationship between decomposition rates and soil nutrient contents has often been postulated for the humid tropics. However, apparent exceptions exist, possibly due to the dominant effect of a determinant from a higher level in the hierarchy (see, for example, Spain and Lefeuvre (1987) and reviews by Anderson and Swift (1983). The unexpected significant positive correlation found between the C:N ratio and the rate of N mineralisation from heavy organic fractions is another example of such effects (Sollins *et al.*, 1984). In this case, the organic matter fractions with the lowest C:N ratios were the most prone to physical protection through adsorption onto the surfaces of clay minerals and their mineralisation rates were subsequently decreased. They may also have had a greater chemical complexity than SOM from the light fraction since this has a higher C:N ratio.

Resistant humic compounds
At all stages of decomposition, humification results in the synthesis or relative accumulation of highly-condensed humic molecules. They may be derived from microbial syntheses or the insolubilisation of water-soluble organic matter when associated with clay and sesquioxides (permanent or variable charge clay minerals). They may also contain plant components that have resisted decomposition (see Chapter II). Some humic

polymers (fulvic acids) have molecular weights from 1 to 10 kDa and humic acids range from 5 to >1,000 kDa (Swift *et al.*, 1979).

One current three dimensional model of soil organic matter molecules suggests a complex structure with an elemental composition of $C_{349}\ H_{401}\ N_{26}\ O_{173}\ S_1$ and an approximate elemental composition of 54 % C, 5.2 % H, 4.7 % N, 35.7 % O and 0.4 % S (Schulten and Schnitzer, 1997). A feature of these macromolecules is their association with a range of simpler materials (proteins, polypeptides, amino acids, etc.) trapped and occluded within the voids in the macro-molecular structures, or held physically or chemically at their surfaces (Schulten and Schnitzer, 1998).

The high molecular weights of these molecules, their abundant aromatic structures, frequent bonds with clay minerals and complex formation with metal cations make them only slowly susceptible to microbial degradation (see review by Theng *et al.*, 1989). Some results suggest that the constitution of organic matter in many tropical soils would be dominated by alkyl-C, a particularly recalcitrant form of soil C with a polymethylene structure. In acid soils, these compounds would acquire a random coil shape which may facilitate their protection within the micropores of clay structures (Schnitzer, 1986; Theng *et al.*, 1989). The aromaticities and molecular weights of humic acids do not seem to differ significantly between temperate and tropical soils (Arshad and Schnitzer, 1989).

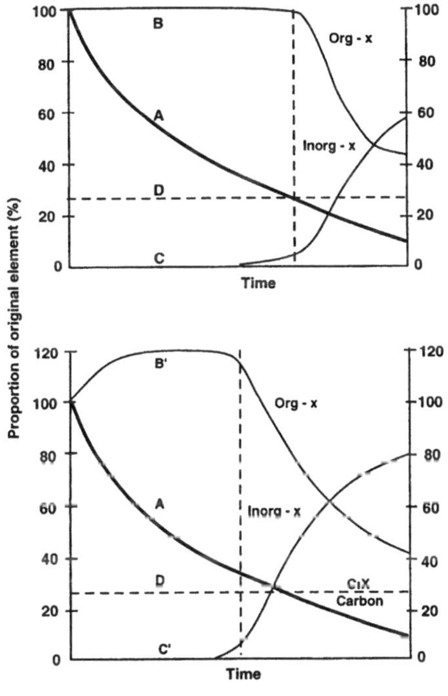

Figure IV.10 Immobilisation and mineralisation of nutrients as a function of the C:nutrient ratio in the decomposing resource (Swift *et al.*, 1979). This simple time-course model illustrates the importance of C:nutrient ratios as indices of the equilibrium between immobilisation and mineralisation of an element X. Curve A = C:X ratio of an organic resource having an initial ratio of 100:1. Curve B = proportion of X in organic form (immobilised), (Curve B' = case where X is imported from sources external to the resources being decomposed). Curve C = proportion of X in inorganic form (mineralised), (Curve C' = mineralisation in response to alteration of C:X ratio by import of X as in case B'). D = C:X ratio of decomposer organism. After this point net mineralisation occurs and curve A flattens off although carbon loss may continue on the same gradient.

1.4 Biological systems of regulation

1.4.1 DEFINITION AND GENERAL PROPERTIES

Soil organisms directly control decomposition processes and rates (Lavelle, 1987). Microbial respiration makes by far the most important contribution to overall respiration contributing between 80 and 95 % of the total (see, *e.g.*, Satchell, 1971; Ryszkowski, 1975; Reichle *et al.*, 1975; Persson and Lohm, 1977; Lamotte, 1975 (Table IV.2)). However, microbial activity is limited by their relative immobilities and their high sensitivities to environment constraints (see the 'Sleeping Beauty paradox' in Chapter III.2.4.3). Their activities are therefore largely determined by processes operating at larger spatial scales, particularly those associated with the larger soil organisms, or macro-organisms. Biological systems of regulation include three components: (i) a decomposing resource, *e.g.*, leaf or root litter, or soil organic matter; (ii) micro-organisms, which mediate most chemical transformations; and (iii) macro-organisms, which create suitable conditions for microbial activity, at defined scales of space and time (Figure IV.11).

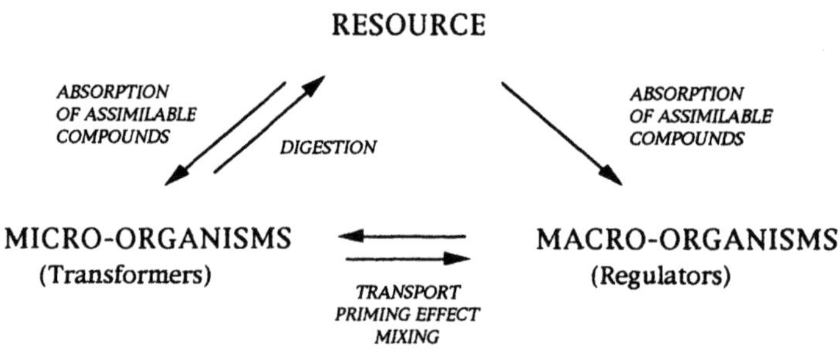

Figure IV.11 The composition of biological systems of regulation in the litter and soil systems (Lavelle, 1984).

With the exception of termites and leaf-cutting ants, the direct effects of macro-organisms on decomposing resources are generally limited (see Chapter IV.5). The transport of micro-organisms to new substrates and the priming of their activities by the production of limited amounts of readily-assimilable organic matter (*e.g.*, root exudates and earthworm intestinal mucus) are key processes in the functioning of these systems. They generally enhance microbial activity and hence accelerate decomposition through a priming effect (see details in Section IV.1.5.2) although inhibitory feed-back mechanisms may sometimes develop, either in parallel or at different scales of time. For example, increased humification may chemically limit mineralisation rates or the compact structure of earthworm casts and the walls of termite nests may physically protect organic matter from digestion (see, for example, Okello-Oloya *et al.*, 1985; Garnier-Sillam *et al.*, 1987; Martin, 1989 and Sections IV.4 and 5).

The functioning of biological systems of regulation depends directly on the physical, chemical and biological characteristics of all three major components. It is therefore essential to separately consider (Figure IV.12):

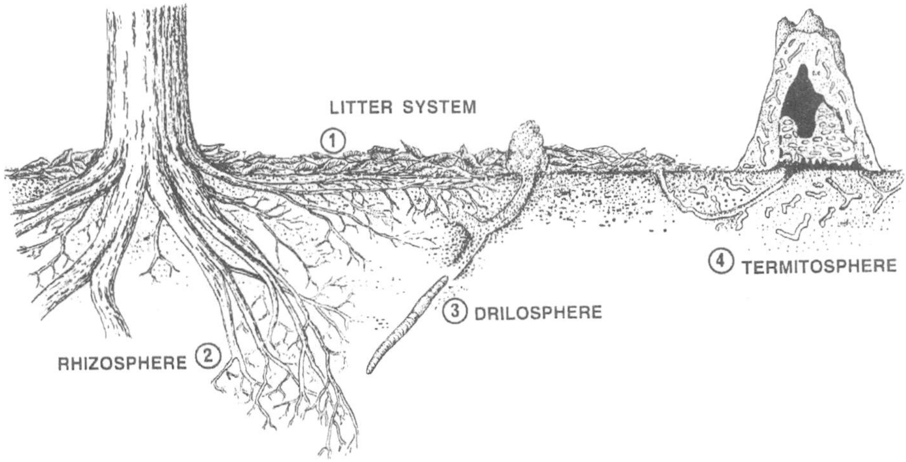

Figure IV.12 Major biological systems of regulation in soils (Lavelle, 1984).

(i) the **litter system**, which comprises leaf litter as a food supply, surface lateral roots, epigeic invertebrates (mainly arthropods) and microbial communities dominated by fungi;
(ii) the **rhizosphere**, *i.e.*, living subterranean roots and the soil and microflora that they influence;
(iii) the **drilosphere**, which include earthworms, soil organic matter as a food source, and free living soil microflora dominated by bacteria; and
(iv) the **termitosphere** *i.e.*, the whole volume of the soil and organic resources influenced by termites, both directly and through their associations with obligate and facultatively mutualist micro-organisms. Processes in each of these systems are detailed in Sections 3 to 6 of this chapter, respectively.

In the future, when their roles are more precisely known, it may be possible to define a 'myrmecosphere'. In a parallel with the termites, this could include those ecosystem components, micro-environments and processes affected or mediated directly or indirectly by ants in those environments where these animals exert a major regulatory influence on ecosystem processes (see, for example, Folgarait, 1998). Except possibly for the leaf-cutting ants, ants appear to have little direct effect on organic matter decomposition. However, they have strong effects on soil development through void formation and other processes (Paton *et al.*, 1995) and exert a strong indirect control on the rest of the biota, notably on many of the major species involved in soil formation processes.

The *biological systems of regulation* (BSR) have a number of basic properties in common:
(i) their functioning is largely determined by mutualistic relationships mediated by highly assimilable compounds (*e.g.*, root exudates, earthworm intestinal mucus) which may initiate a priming effect on the microflora (see Section I.3.2.5);
(ii) they juxtapose organisms which operate at widely-different scales of time and space *e.g.*, weeks to months and centimetres for fine roots, and hours to days and micrometres for the associated bacteria;

(iii) not all BSR's are equally important in all ecosystems; they may develop mutualistic and antagonistic (competitive) relationships. For example, litter systems dominate in forest ecosystems, especially where anecic earthworms and termites are absent since these animals often export litter to their own functional domain. In contrast, in savannas, important drilosphere and termitosphere components are normally present provided that climatic constraints, biogeographic barriers or anthropogenic effects do not exclude or eliminate them;

(iv) organisms of very different sizes are likely to react differently to temperature changes. The adjustment of microbial activity to improved environmental conditions (*e.g.*, addition of a high quality organic substrate and/or water) is likely to be much faster at high temperatures than at low. The mutualistic digestion system of the geophagous endogeic earthworm *Pontoscolex corethrurus* is highly efficient at 25-27 °C but no longer functions when the temperature is reduced to 15 °C. At this temperature, microbial activity in the hindgut is only slightly greater (1.5 times) than that of control soil values and is five times lower than that occurring at 27 °C (Barois, 1987); in consequence, *P. corethurus* ceases to grow at 15 °C (Lavelle *et al.*, 1987).

Consequently, under the temperature conditions pertaining in soils of the humid tropics, mutualistic relationships between soil micro- and macro-organisms are likely to be much more frequent and better developed than in colder environments. The increased importance of mutualism on the structure and functioning of food-webs is considered to be a key issue in understanding the functioning of humid tropical soils (see Chapter III.4.4.2).

1.4.2 LITTER-SUPERFICIAL ROOTS SYSTEM

The principal energy source for this system is the above-ground litter (see Chapter IV.2). It also comprises a dominant community of arthropods as macro-organisms, a microflora dominated by fungi and an occasionally-dense mat of fine roots and fungal hyphae (mycorrhizal and saprotrophic) which serve as a sink for the nutrients released from the decomposing litter. In certain situations, the internal cycling of nutrients may be almost closed, as in some Amazonian forests occurring on spodosols (Herrera *et al.*, 1978), although water-soluble organic materials and mineral elements are normally leached downwards into the underlying soil horizons. Litter may be exported to other systems of decomposition by the anecic fauna (*i.e.*, species that feed on litter but live in burrows or galleries built within the soil: earthworms, some termites and species living at its surface or in trees (some termites) (see Sections IV.4 and IV.5).

Direct mineralisation of litter by the root-fungal (mycorrhizal) association has apparently been observed in certain forests on sandy, highly leached and acidic soils such as the tropical spodosols of Amazonia (Herrera *et al.*, 1978; Leroy *et al.*, 1992). Smith and Read (1997) among others, however, consider that AM fungi do not have the enzymatic capacity to directly mineralise litter. It also occurs in the litter of plant species involved in symbioses with ericoid mycorrhizae (Dighton, 1991) and in temperate climate litters colonised by certain ectomycorrhizal fungi. In more fertile soils, this system of direct cycling may be less important and the thickness of surface root mats is correspondingly smaller. Other processes operating include direct decomposition by free-living micro-organisms, principally fungi, together with fragmentation and digestion by invertebrates. In general, an important part (10 to 20 %) of the minerals released and water-soluble elements are eluviated downwards into the mineral soil. They may then be absorbed by deeper soil roots, their mycorrhizal fungi or the free-living microflora; they may also be insolubilised or adsorbed onto the surfaces of soil colloids.

All the non-predatory 'epigeic' fauna (*i.e.*, invertebrates that feed on and live in the surface litter, Bouché, 1977) participate in fragmentation: large myriapods (iulids and polydesmids), isopods and epigeic earthworms are the main macrofaunal groups while micro-arthropods and Enchytraeidae represent the mesofauna. The microfauna is also represented in the litter system, but is relatively unimportant, except in extremely arid or cold environments.

1.4.3 THE RHIZOSPHERE

The rhizosphere is defined as that part of the soil which is influenced by roots (see Chapter IV.3). Soil organic matter resources are exploited through active mutualistic relationships with that section of the soil microflora that is associated with living roots. These relationships are mediated by root exudates and regulated by the micropredator food-web (protists and nematodes with bacteria as prey) (see Section IV.1.5.3). In this system, roots regulate microbial activity by providing readily-assimilable carbon sources (rhizodeposited materials) which stimulate the activities of the free-living soil microflora and the community of obligate (or occasionally facultative) microbial symbionts.

Rhizodeposition may represent from 5 to 30 % of total photosynthates *i.e.*, several Mg ha^{-1} (Hale *et al.*, 1981; Trofymow *et al.*, 1987). Micro-organisms initially develop utilising exudates although they also digest the more complex substrates contained within the soil organic matter through a priming effect (*sensu* Jenkinson, 1966) leading to a significant mineralisation of soil organic matter (Sallih *et al.*, 1987) (see Section IV.1.5.2 on priming effects). These micro-organisms may directly release metabolites that are used by the plant. However, much of this release may result from the activity of micropredator food-webs which operate in the rhizosphere (Trofymow *et al.*, 1987). These food-webs include bacteria as a food source, with protists and free-living nematodes as predators. Micropredators release nutrients from bacteria through their own metabolism or following their death which may be associated with the diminution of food resources that occurs as the roots mature (Clarholm, 1985).

The rhizosphere is also characterised by the intense activity of obligate or semi-obligate root symbionts, mycorrhizal and nitrogen fixing micro-organisms (including Rhizobia and *Frankia*) which play substantial roles in ecosystem processes through the provision of nutrients to their host plants.

1.4.4 THE DRILOSPHERE

The drilosphere (from the Greek 'drilos'= earthworm) is defined as that part of the soil that is influenced by earthworm activities (casts and galleries); the energy source may be surface litter or soilorganic matter, depending on the ecological category of the earthworms involved (Bouché, 1977) (see Chapter IV.4). The microflora is activated by the mixing of litter and soil (anecic litter-feeding species) or by breaking the physical protection of soilorganic matter within soil aggregates during gut transit.

The production of large amounts of intestinal mucus in the anterior part of the gut initiates an intense microbial activity which develops using this mucus as an energy source; in the midgut, the microbiota breaks down the soilorganic matter which was previously inaccessible to microbial degradation in the soil. This intense activity results in an increased microbial biomass in the incubated casts of anecic earthworms (*e.g.*, Binet, 1993; Brown, 1995); after re-ingestion of these casts, part of this biomass may be digested through the microfaunal food-chain (*i.e.*, the earthworm digests protists which have fed on bacteria or fungal hyphae growing in the casts) (Piearce and Phillips, 1980; Rouelle *et al.*, 1985).

In the strictly geophagous endogeic earthworms, an enhanced microbial activity in the gut results in the release of assimilable organic matter in the posterior part of the gut and its further assimilation by both the worm and soil bacteria (Barois and Lavelle, 1986). True endogeic species do not appear to re-ingest their own casts thereby avoiding the possibility of feeding on the contained microbial biomass (Howe, 1984).

1.4.5 THE TERMITOSPHERE

The termitosphere is that portion of the soil and ecosystem influenced by termite activities. This system may have a wide variety of configurations due to the large number of species that exist (more than 2,400 described species, Grassé, 1984) and their diverse ecological functions (see, *e.g.*, Lee and Wood, 1971b, Lavelle *et al.*, 1992a).

Five major trophic groups may be distinguished (Chapter III.4.3.2): (i) grass and leaf litter harvesting species, (ii) wood feeders, (iii) fungus cultivators, (iv) soil-wood feeders that consume highly decayed wood but also ingest some inorganic soil materials, (v) humivorous species that feed on soilorganic matter. These groupings are not mutually exclusive since some species feed on more than one of the above categories and a few species are highly polyphagous. These differences in feeding regime are accompanied by equally diverse ecological habits; species in a range of groups may build very large epigeic termitaria (grass harvesters, litter feeders, fungus cultivators, soil organic matter feeders, some xylophagous species) whereas others make carton or soil nests on tree trunks or within sound or decomposing wood (some xylophagous species). A last group makes purely subterranean nests with a complex array of chambers and galleries (some xylophagous and humivorous species and some fungus cultivators) (Grassé, 1984). Digestive systems also differ greatly between species. All termites have communities of symbiotic micro-organisms in their intestinal tracts: protists in the most primitive wood-feeding lower termites, largely bacteria in the other groups. Systems of digestion may involve complex combinations of the termites' own enzymes, enzymes acquired from cultivated fungi, fermentation and bacterial digestion (Breznak, 1984; Rouland *et al.*, 1990; Bignell *et al.*, 1994) (Chapter III.4.3.2.1).

1.5 Basic processes within the biological systems of regulation

The structure and function of biological systems of regulation is illustrated in Figure IV.11. Characteristic physical and biological structures form part of each of these systems and are centres for the particular processes that characterise, respectively, the litter, rhizosphere, drilosphere and termitosphere systems (Sections IV.3 to IV.6). Three processes common to all these systems are the synchronisation of nutrient release with plant root uptake, priming effects on organic matter and the functioning of micropredator foodwebs.

1.5.1 SYNCHRONY AND SYNLOCALISATION OF NUTRIENT ELEMENT RELEASE AND UPTAKE

The activities and the intensities of the processes operating within the biological systems of regulation determine the spatial and temporal patterns of nutrient release from decomposing resources. The restriction of nutrient-element release to locations and times where they may be taken up by plants and other organisms is of great importance for the conservation of nutrients in ecosystems where they are limited (Swift, 1986) (Figure IV.13). It is hypothesised that, in natural ecosystems, this would occur in the following ways: (i) by a synchronous stimulation of the plant and decomposer activities by such high level determinants as climate; (ii) by plant responses to existing temporal and spatial variation in decomposer activities through adaptive rooting strategies; (iii) through the presence of a wide diversity of plants and decomposers with differing patterns of spatial and temporal activity; and (iv) a broad diversity in the quality of the decomposing resources, leading to extended temporal patterns of nutrient release (see Section IV.2.6.3).

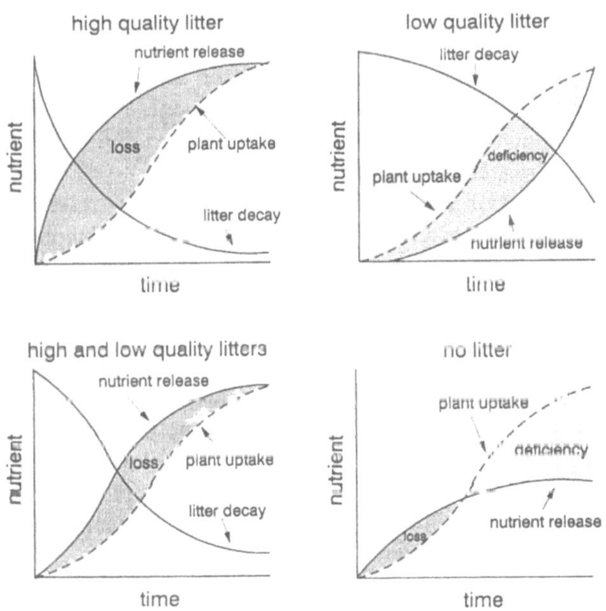

Figure IV.13 Comparison of the expected timing of nutrient release from decomposing resources of different qualities and their uptake by plants: the synchrony model (Swift, 1986).

1.5.2 THE ROLE OF PRIMING EFFECTS: AN HYPOTHESIS

The priming effect was first described by Bingeman *et al.* (1953, in Jenkinson *et al.*, 1985) as the stimulation of soil organic matter decomposition by the addition of fresh organic material. In more general terms, the priming effect was defined by Jenkinson (1966) as a positive or negative change in the decomposition rate of soil organic matter caused by the addition of fresh organic matter. A simple experiment illustrated this effect: a 125 mg sample of Rye Grass (*Lolium perenne*) foliage uniformly labelled with ^{14}C was mixed with 100 g of soil and subsequent carbon dioxide evolution was monitored during an incubation period of 78 days. Production of unlabelled carbon dioxide resulting from mineralisation of the soil organic carbon was greater in the mixture than in the control soil. The difference observed was attributed to a positive priming action.

Most examples of priming effects described in the literature are positive: the addition of fresh organic matter to soils as green manure, or the addition of mineral nitrogen as fertiliser, usually stimulates the mineralisation of soil organic matter (reviewed in Jenkinson *et al.*, 1985). Nonetheless, negative effects have been observed in some circumstances. For example, during the first few days following the addition of glucose to an organic soil (Bingeman *et al.*, 1953, in Jenkinson *et al.*, 1985). The observed priming effects may be apparent or real. Apart from experimental artifacts (*e.g.*, the exchange of labelled for unlabelled carbon in calcareous soils, or errors due to an heterogeneous labelling of the introduced material), some observed priming effects may be due to environmental changes associated with the introduction of the organic material. In this last circumstance, changes in soil organic matter mineralisation may result from an altered pH or oxygen supply, irrespective of the effect of the fresh organic materials (see *e.g.*, Barrow, 1960 and Parr and Reuszer, 1959 in Jenkinson, 1966). Jenkinson (1966) attributes real priming effects to three different causes:

(i) An increased rate of spore germination leading to greater overall microbial activity.
(ii) Interactions between compounds derived from the added and 'native' organic matter, which render the latter more labile (Mandl and Neuberg, 1956, in Jenkinson, 1966).
(iii) An increased production of extra-cellular enzymes by micro-organisms, leading to the accelerated organic matter decomposition.

In soils, several classes of water-soluble organic materials (referred to as 'soluble resources' in Chapter I.3.2.5) are normally available to prime the decomposition of resistant organic matter. The former materials have been named 'ecological mediators' for their unique roles in mediating soil function (Lavelle *et al.*, 1994b). They include:

(i) High-energy, water-soluble organic leachates percolating from the above-ground vegetation and leaf litter.
(ii) Root exudates are water-soluble organic substances, readily assimilated by micro-organisms, that become intimately mixed with the soil and penetrate the very fine pores within micro-aggregates. Real priming effects on CO_2 evolution due to maize root mucilages have been measured by Mary *et al.* (1992).
(iii) Earthworm cutaneous and intestinal mucus. Intestinal mucus produced by earthworms is considered to be responsible for much of the priming of soil micro-organism activity observed during transit through the gut (Lavelle and Gilot, 1994). The remaining part being an apparent priming due to the addition of water and an intense mixing.

At least some soil macro-organisms appear to stimulate microbial activity by producing water-soluble organic molecules and these may be considered as the mediators which activate mineralisation processes wherever and whenever they are produced. This activation may be intense and microbial activity eight times greater than that of a wet sieved control soil has been measured in the gut of tropical geophagous earthworms (Barois and Lavelle, 1986). From this, it appears that roots and invertebrates are important regulators of soil organic matter decomposition in the rhizo- and drilospheres, through their capacities for initiating priming effects. Priming is a key process by which these macro-organisms may enhance microbial activity within their BSR's.

The addition of ecological mediators to soils induces highly specific effects on the structure and activity of microbial communities. Earthworm intestinal mucus initiates a massive microbial activation within 45 minutes whereas similar effects are only obtained 15-30 days after root mucilage is added to the soil (Lavelle and Gilot, 1994; Mary *et al.*, 1992). These processes therefore appear to operate at time scales compatible with other activities. For example, the digestion that occurs during a one-hour transit of soil material through the earthworm gut in the first case, and activation of the rhizosphere microflora along the tip of a relatively slowly growing root in the second case (see details of these processes in Sections 3 and 4 of this chapter).

1.5.3 MICRO-FOODWEBS

The vast majority of soil invertebrates rely on microbial activities to digest the organic resources available in the soil system (see Chapter III.4.4.1, Figure III.61). The nature of the relationships that particular invertebrate groups develop with the microflora depends largely on their size. Large invertebrates interact with elements of the microflora within structures that they create including faecal pellets, gallery walls and their own gut contents. Within such structures, the relationships are predominantly mutualistic although evidence exists for some degree of invertebrate predation on the microflora. The smallest invertebrates, *i.e.* protists and nematodes are mainly predators of micro-organisms since their small sizes prevent them developing internal mutualistic relationships or utilising 'external rumen' structures. Micro-organisms interact within clearly identified foodwebs which may be further extended to include some larger mesofaunal elements. With increasing size, the relationship between the microflora and fauna gradually shifts from predation to mutualisms of increasing efficiency. The food web concept becomes progressively more difficult to apply as the trophic structure becomes increasingly 'fluid and interactive with individual species operating on several levels which might be distinguished as trophically different' (Swift *et al.*, 1979; Wardle and Lavelle, 1997).

It seems therefore both logical and practical to distinguish micro-foodwebs, *i.e.*, the foodwebs that links micro-organisms to their predators, from systems based on mutualistic relationships between larger organisms and the microflora. Furthermore, micro-foodwebs are often included within the complex, globally mutualistic, interactions that occur between macro- and micro-organisms and which operate at larger scales. At the root surface, for example, micro-organisms and their micropredators appear to play critical roles in releasing mineral nutrients available to the plant. In this case, predation in the micro-foodweb is just one process in the overall functioning of the rhizosphere which is broadly based on mutualism (see Section IV.3.2). As explained in Chapter III.4, it is possible to identify three successive levels of interactions between micro-organisms and invertebrates, based on the size of the latter organisms.

Among the larger invertebrates, interactions with the microflora shift from predation to external and internal mutualisms, while the invertebrates increasingly affect the structure of the litter and soil system. Micro-foodwebs adapt to the soil structure and their development and activity is largely dependent on the nature and size of the system of soil pores and the proportion of these that are filled with water. Litter-transformers are unable to dig the soil or build nests and rely on the relatively-inefficient external rumen type of digestion. Consequently, their activities are largely restricted to the litter system. At the next larger size level, earthworms, termites and many ants have well developed digging and tunnelling abilities and the capacity to create such diverse structures as nests, galleries, burrows, surface sheathings and casts that greatly modify soil processes. For this reason, they have been called 'ecosystem engineers' (Stork and Eggleton, 1992; Jones *et al.*, 1997). The first two groups of these organisms have developed efficient internal mutualistic digestion systems with the microflora and, where populous, they play dominant roles in soil function (see Sections IV.4 and IV.5).

The composition of microfoodwebs
The soil micropredator foodwebs include micro-organisms, mainly bacteria and fungi, protists, nematodes - and some predaceous Acari (Figure IV.14). Several levels of complexity exist since the system includes microbial grazers, predators and superpredators.

MICROFOODWEBS

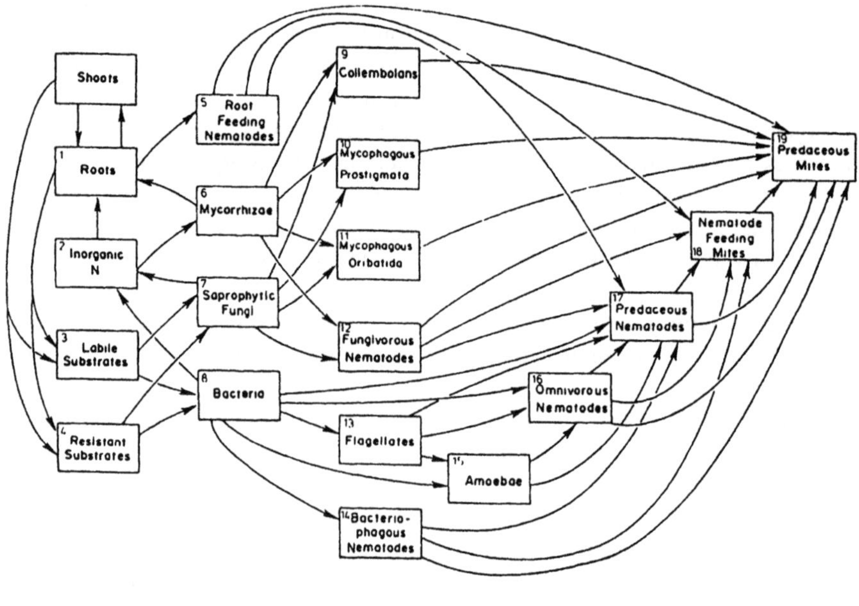

Figure IV.14 Representation of micro-foodwebs in a shortgrass prairie (Hunt *et al.*, 1987).

The microflora is generally divided into bacteria and fungi, which perform distinct functions and often colonise quite separate microsites (see Chapter III.2). Mycorrhizal fungi have roles that are usually critical to the success of the plants with which they are associated and are sometimes considered separately. Small populations of predaceous bacteria and fungi may be found. These are the bacterial predators of other bacteria, *e.g.*, of the genus *Bdellovibrio* (Casida, 1988) or nematode-trapping fungi (Cooke and Godfrey, 1964; Mitsui, 1985; Kendrick, 1992). The latter are considered beneficial due to their potential to control populations of phytoparasitic nematodes. Other fungi of the genus *Dactyella* have been reported to capture and consume amoebae.

Predatory protists mainly feed on bacteria; cases of mycophagy exist, but they are rare (Petz *et al.*, 1986). They are active consumers: a single heterotrophic nanoflagellate may ingest three to nine bacteria per hour (Bloem *et al.*, 1989) and Bryant *et al.* (1982) estimated that 8300 bacteria were consumed for each new amoeba produced in soil microcosms. Stout and Heal (1967) estimated that protists consume 150-900 g bacteria m^{-2} $year^{-1}$, equivalent to 15-85 times the standing crop (estimated by the plate counting technique). Bamforth (1988) distinguished two ecological types of protists: a group which may serve as food for a large variety of organisms and the others which are endosymbionts (commensal or parasitic). The latter are regularly found in the gut of some termites and earthworms (Astomata) and may have endosymbiotic bacteria within their cytoplasm. Protists of the former group are often considered as transformers of bacterial protoplasm into higher trophic levels.

Mycophagous amoebae and colpodid ciliates can also ingest fungal spores and may thereby play a role in the control of populations of phytopathogenic, mycorrhizal and other fungi (Chakraborty, 1985; Coûteaux, 1985).

Nematodes are another constant component of micro-foodwebs, generally one trophic level above the protists. A few adult nematodes may actually ingest flagellates and some amoebae (Elliott *et al.*, 1980) although the vast majority are either bacterial- or fungal-feeders (Chapter III.4.1.2). Nematodes are attracted to their food sources by a number of stimuli including pheromones, CO_2 and temperature gradients (Freckman, 1988). Bacterial feeders are active predators that can ingest up to 5000 cells per minute or 6.5 times their own weight per day when fully active. Overall consumption may be as high as 800 kg bacteria ha^{-1} $year^{-1}$ and the amount of N turned over is in the range of 20-130 kg ha^{-1} $year^{-1}$ (Coleman *et al.*,1984b). This estimate suggests a dramatic role of microfaunal predation in microbial biomass turnover. Finally, mycophagous nematodes have been reported to suppress plant diseases by ingesting pathogenic fungi (Curl, 1988).

The major components of the mesofauna *i.e.*, Acari, Collembola and Enchytraeidae, also form part of micro-foodwebs (Moore, 1988), although most probably rely more on mutualistic relationships than predation. Predatory mites can feed on nematodes (*e.g.*, Martikainen and Huhta, 1990) and have been reported to regulate nematode density in North American desert ecosystems (Elkins and Whitford, 1982). Fungivorous mites are divided into two groups depending on their growth rates (Moore, 1988).

Almost all macrofaunal invertebrates rely on mutualistic relationships with the microflora for their digestion although some degree of predation exists, especially on protists. For example, earthworms can digest ciliates (Piearce and Phillips, 1980) and amoebae (Rouelle, 1983). Protists may also participate in the external rumen type of digestion in association with earthworms: incubation in faeces results in the excystment and development of protist populations which are further digested by invertebrates. In a microcosm experiment, five days after earthworms were introduced into sterilised soil, soil protist population density was 3-70x10^6 g^{-1} soil, dominantly amoebae (50-60 %) (Shaw and Pawluk, 1986).

The structure of micro-foodwebs has been elucidated for a number of sites (Hendrix *et al.*, 1986; Elliott *et al.*, 1988; Ingham *et al.*, 1989) and models have been built to simulate the effects of such assemblages on soil processes (Hunt *et al.*, 1987; de Ruiter *et al.*, 1993). The characteristics of such foodwebs are largely influenced by the relative dominance of fungi or bacteria and this, in turn, is determined by abiotic factors and management practices. In dry environments, most bacteria, protists and nematodes are inactive and soil foodwebs appear to be fungal-based, with abundant fungivorous, moderately drought-tolerant mites (Whitford, 1989). As found in conventionally-tilled, temperate-climate agroecosystems in the USA, plant residues are distributed throughout the plowed layer thereby promoting bacterial activities and bacterial based foodwebs. In 'no-tillage' systems, plant residues accumulate on the soil surface, promoting fungal growth and immobilisation of nutrients in the decomposing plant material and the fungal biomass. Populations of fungivorous faunal groups increase in the surface layers and decomposition rates are slower (Hendrix *et al.*, 1986).

Distribution and dispersal of microsites

In soil micro-foodwebs, interactions occur mainly within the water-filled soil pore space and the water films that cover solid particles. These microsites are discrete in time and space, and depend largely on soil moisture content and porosity which, in turn, depend on texture and overall biological activity. Finally, the presence of adequate carbon resources determines the size of the normally large microbial biomass which forms the nutritive base for the micropredators. The pore size distribution determines population size at the microsite level. Flagellates and small amoebae colonise pores with a minimum diameter of 8 μm; the bigger protists and nematodes live in the larger pores and most nematodes inhabit the external medium. Soil texture and porosity are therefore critical determinants of micro-foodweb structure (Hassink *et al.*, 1993). The importance of porosity is demonstrated, for example, by an experiment in which the bulk density of humus was increased from 0.25 to 0.41 Mg m^{-3}; pores larger than 10μm decreased and the growth of ciliates and thecamaoebae was inhibited which would reduce predation pressure on the bacteria (Coûteaux, 1985). Soil ecosystem engineers that may largely determine soil structural porosity are thus likely to determine micro-foodweb communities.

The presence of organic resources and the capacity of the microbiota to move between soil microsites are critical to the definition of microsites favourable to micropredator foodweb activities. Such microsites are regularly found in the rhizosphere where root growth provides energy for the bacteria and their micropredators, and disseminates their communities throughout the volume of soil explored by the root system. Micropredator foodwebs also occur in other larger systems of regulation organised around the activities of such soil macro-organisms as earthworms and termites. These are just some of the many structures which contribute to the overall soil function.

Evidence for the functional importance of micro-foodwebs

The functional importance of micro-foodwebs has been demonstrated several times in small-scale laboratory chambers called 'microcosms'. Their effects on carbon flows and ecosystem-level nutrient cycles have been observed and quantified.

C flows: Although microfaunal respiration only represents a small percentage of overall soil respiration (0.6 to 2 % for nematodes according to Freckman, 1988), their effects are disproportionate to their sizes and respiration rates. The microfauna significantly stimulates the growth and turnover of microbial populations thereby promoting more rapid rates of decomposition, mineralisation and thus nutrient turnover (Wasilewska, 1979; Hendrix *et al.*, 1986; Hunt *et al.*, 1987; Sohlenius *et al.*, 1987, 1988; Moore *et al.*, 1993).

The N cycle is particularly affected by micro-foodweb interactions since bacteria have average C:N ratios of 6:1, close to that of the protists (5:1 according to Reich, 1948, in Stout and Heal, 1967) and slightly lower than that of nematodes (10:1). Consequently, predation on micro-organisms results in the release of mineral-N that may be further used by plants. In the presence of nematodes, ammonification by bacteria is greatly increased (Clarholm, 1981; Ingham *et al.*, 1986b). N-fixation may be greatly enhanced through the maintenance of predominantly young populations, provision of stimulatory compounds or decreased oxygen concentrations (Darbyshire and Greaves, 1973).

The P cycle is also affected although contrasting results have been found. Coleman *et al.* (1984b) observed that, in the presence of amoebae, a significant part of the P accumulated within the microbial biomass is transformed into bicarbonate-extractable P. However, the flux of P affected by this transformation can be small and this may explain why few effects on the P cycle have been noted, even in microcosms (Bååth *et al.*, 1980).

Positive effects of micro-foodweb activities on plant growth have been measured in microcosms. The productivity of wheat plants grown in sterilised soil into which bacteria and protists had been introduced, increased by 80 % and N mineralisation from soil organic matter was increased by 59 % in relation to a control soil without plants. This emphasises the role of root-derived carbon in providing energy to the system (Clarholm, 1984). In another 35-day microcosm experiment, bacterial biomass was reduced eightfold and N uptake by plants was increased by 20 % through the activity of protists. The shoot:root ratio of plants was increased and more than 65 % of bacterial ^{15}N was taken up by the plants (Kuikman and van Veen, 1989).

IV.2 THE LITTER SYSTEM

The litter system is that part of the ecosystem within which above-ground litter accumulates and decomposes. Most studies of decomposition and nutrient cycling have been conducted in this specific environment and the basic decomposition processes described in Section IV.1 are directly relevant to the functioning of this system. Litter systems differ widely in their morphologies and in their community compositions. This is reflected in the modal types recognised in past classifications of humus types and by the distinct microbial and invertebrate communities that characterise litters of different quality. In this section, particular emphasis is placed on: (i) the relationships between biotic communities and the morphology and structure of the litter system; and (ii) on the spatial and temporal patterns of the processes taking place within them.

2.1 Composition

The litter system includes the above-ground litter which serves as the energy source, a rich microflora dominated by fungi and the epigeic invertebrates and surface roots that act as regulatory macro-organisms. As seen in Chapter I (Section I.3.2.2) above-ground litter is an heterogeneous resource. It comprises a mixture of relatively high quality resources such as fresh leaf-litter, flowers, fruits, seeds, dead micro-organisms and animals and structures of lower quality, mainly woody materials. As decomposition proceeds, fresh litter falls to cover the previously deposited materials; these older materials have greater resistance to further decomposition and the litters of certain plant species accumulate in deep layers forming a gradient of substrates of diminishing quality with increasing depth. However, transfers of organic materials and bioturbation processes may alter this distribution.

Litter inputs are also highly variable in time (seasonal variations) and they change with climate resulting in clear patterns in variation of inputs along a thermo-latitudinal gradient. At individual sites, trees of different ages (Bernier and Ponge, 1994) and different species (Boettcher and Kalisz, 1992; Grandval, 1993) may produce litters of contrasting qualities thereby inducing heterogeneity in the litter system at scales that may range from metres to decameters.

Readily-assimilable carbohydrates may be released from decomposing litter and percolate through the underlying litter layers (see, *e.g.*, Carlisle *et al.*, 1966) to prime local microbial activity. The microflora is dominated by the fungi which possess a pseudo-mobility resulting from their capacity for rapid growth and an ability to translocate cytoplasm. As shown in Chapter III, their ability to perforate cell-walls and feed preferentially on carbohydrates makes them efficient early litter colonisers.

Litter-dwelling, or epigeic invertebrate communities may be rich and genetically diverse, with large populations of micro-arthropods (Collembola and Acari), enchytraeids, epigeic earthworms, macro-arthropods including ants, termites, beetles and other animals. This fauna includes a large proportion of saprophages which feed on decomposing litter and sometimes microflora and the products of their external digestion. A significant part of this fauna is predatory and regulates the dynamics of saprovore populations. Roots are an important, though often neglected component of the litter system. They grow principally in the lower parts of the system (*i.e.*, the F and H layers) and absorb the nutrients released in these layers either directly or through the hyphae of their mycorrhizal fungal associates (*e.g.*, Herrera *et al.*, 1978; Soma and Saito, 1979; Ponge, 1990).

2.2 Classification

The structure of litter-systems differs substantially among plant communities and depends on:
(i) The quality of inputs (*e.g.*, woody, leafy or grassy materials of differing chemical quality and physical structure).
(ii) The nature of the microbial communities present (particularly the presence or absence of the 'white-rot' fungi (Basidiomycota) which have the capacity to decompose lignins and phenol-protein complexes) (Section IV.2.5.4.)
(iii) The composition and abundance of the macro-invertebrate communities present, with particular emphasis on the presence or absence of anecic invertebrates that can rapidly transfer litter to other systems of decomposition.

In (ii) above, the 'white-rot' fungi are a functionally-defined group of basidiomycote fungi belonging largely to the families Agaricaceae, Hydnaceae, Corticiaceae, Polyporaceae and Thelophoraceae although a few species also occur in the ascomycote family Xylariaceae. These fungi have the capacity to extensively degrade all the important structural components of wood and other polyphenolic materials, including lignins, through the production of extra-cellular phenoloxidases (Crawford, 1981; Kendrick, 1992).

Litter systems differ markedly in physical structure *i.e.*, the number of recognisable layers, their relative thickness and composition, and the abundance and composition of the microbial and faunal communities present. A number of major types have been recognised in past studies and the abiotic and biotic factors (*i.e.*, the quality of decomposing resources) determining their formation have been identified.

2.2.1 LITTER SYSTEMS AND HUMUS FORMS

Considerable research has been devoted to the description and classification of humus forms defined in the upper part of the soil where organic litter is decomposed by the soil biota (Müller, 1887; Kubiëna, 1953; Zachariae, 1962; Bal, 1982). Müller defined three groups of humus forms: Muld (now called mull), Muldagtig Mor (moder) and Mor, based on the increasing thickness of the holorganic layers and the properties and morphology of the upper mineral horizon. Based on a morphological description of organic layers of soil, these concepts introduced a dynamic view of decomposition, special attention being paid to the transfer of organic matter into the mineral soil profile, and the relative importance of climatic, edaphic, resource-quality, microbial and zoological determinants.

These ideas have been developed and refined to include a wide range of environments and litter types and to define the processes involved in organic matter transformations. Bal (1982), for example, defined 'humon' as the continuum of organic materials typical of the different stages of decomposition of organic residues. This concept emphasises the point that the decomposition of leaves is a continuous process that starts before abscission, continues at the soil surface where litter is deposited and may be completed in different microsites or soil layers when transfers occur (for example, the passive transfer of decomposing leaves when seasonal litter fall adds new strata, Figure I.42). Since these pioneering studies, several classification systems have been proposed (see *e.g.*, Delecour, 1980; Green *et al.*, 1993; Brethes *et al.*, 1995).

As stated by Satchell (1974), 'The concept of the humus type as a biological system is implicit in Müller's descriptions of mull and mor' and the concept of humus forms may be used to describe litter systems. However, unlike humus forms, the litter system as defined in this chapter comprises only the holorganic layers - the O horizon (Chapter II.1.1) - that overlie the predominantly mineral soil horizons. Whenever litter is exported to different sites and controlled by other biological systems of regulation, it is no longer considered to form part of the litter system. This is especially true of the litter and faecal pellets that initially accumulate in the holorganic layers but are later transferred to other systems within the mineral horizons by anecic and endogeic earthworms (drilospheres) or by termites (termitospheres).

2.2.2 MULLS

In mull litter systems, decomposing leaves and other litter materials do not accumulate at the soil surface, either because they are completely decomposed in less than one year, or because they are exported to different systems of decomposition, such as the drilospheres or the termitosphere. In mull systems, phenol-protein complexes are efficiently broken down and organic acids are neutralised through the intense mixing of organic compounds and clay minerals. In consequence, the holorganic layer usually only consists of an L layer with an abrupt transition to the A1 mineral horizon. In most cases, decomposition is initiated in the litter system and is completed following well defined pathways within the drilo- or termitospheres (see Chapter IV, Sections 4 and 5).

Mulls generally develop in sites where climatic conditions favour biological activity for at least a significant part of the year and edaphic conditions must also be suitable for anecic earthworms or termites. Typical soil properties include a pH close to neutrality and a frequently loamy texture which facilitates the ingestion of small soil particles by decomposers. Under such conditions, the litter is generally of high quality, easily decomposed and palatable to large decomposers.

In temperate-climate areas, mull litter systems are largely created through the activities of anecic earthworms. Litter is transferred into drilosphere structures including the large casts of the anecic earthworms comprising the macro-aggregates of the A1 layer and the walls of burrows and chambers. In some situations, edaphic conditions allow a rapid decomposition of litter by white-rot fungi in the absence of anecic earthworms and mulls may therefore develop on acid soils (Toutain, 1981). In humid tropical areas, decomposition is generally rapid due to the warm temperatures and favourable moisture regimes and mulls occur frequently, despite the large litter biomasses input to the soil surface.

In both areas, any limitations that exist among the climatic, edaphic or resource-quality determinants of biological activity may impede the formation of mull litter systems. In the Amazonian rainforest, decomposition rates may be depressed where drainage is impeded, where much of the incident rainfall is rapidly lost through runoff (Lucas and Chauvel, 1992), where soils are acid and nutrient-poor (Herrera *et al.*, 1978) and in places where low-quality litter is deposited at the soil surface (Leroy *et al.*, 1992).

2.2.3 MODER

In moder systems, decomposition is slow due to climatic, edaphic or trophic (resource quality) conditions that limit or preclude the activities of anecic decomposers (in neutral mull) and active white-rot fungi (in acid mulls) (Toutain, 1987a). Epigeic saprophages may be abundant and exploit the decomposition products of fungal activity. A large number of predators also exploit the wide range of microhabitats and food resources present within the substantial litter layers. The litter system is thick since decomposing material from previous years accumulates at the soil surface forming discrete layers of decomposing material. These may comprise an Ol layer formed of poorly decomposed leaves and an Of that consists of plant fragments of recognisable structure and the large faecal pellets of millipedes and epigeic earthworms. The underlying Oh layer consists of the accumulated faecal pellets of smaller invertebrates (micro-arthropods and Enchytraeidae) that feed on larger pellets and other debris accumulated within the Oh layer (Figure IV.15).

In this system, phenol-protein complexes are largely decomposed by white-rot basidiomycote fungi and the litter mass is progressively reduced through the activities of arthropods and enchytraeid worms. Since these invertebrates are largely confined to the litter layers and rely mainly on the external-rumen type of digestion, they progressively transform the decomposing material into an Of and subsequently an Oh horizon. This occurs through continuing processes of ingestion, egestion and re ingestion of litter materials until the nutritive value of the decomposing materials for invertebrates has been largely exhausted, deep within the Oh layer. In a two-year experiment studying leaf decomposition in litter bags, Garay *et al.* (1986a) observed that leaves were incorporated into the Of1 layer after 6-9 months still retaining 78 % of their initial mass; one year later (18-21months) they were found in the Of2 layer (*ca.* 40 % of initial mass), and after 2-3 years, they were incorporated into the Oh layer and colonised by fine roots (20 % of initial mass) (Figure I.42).

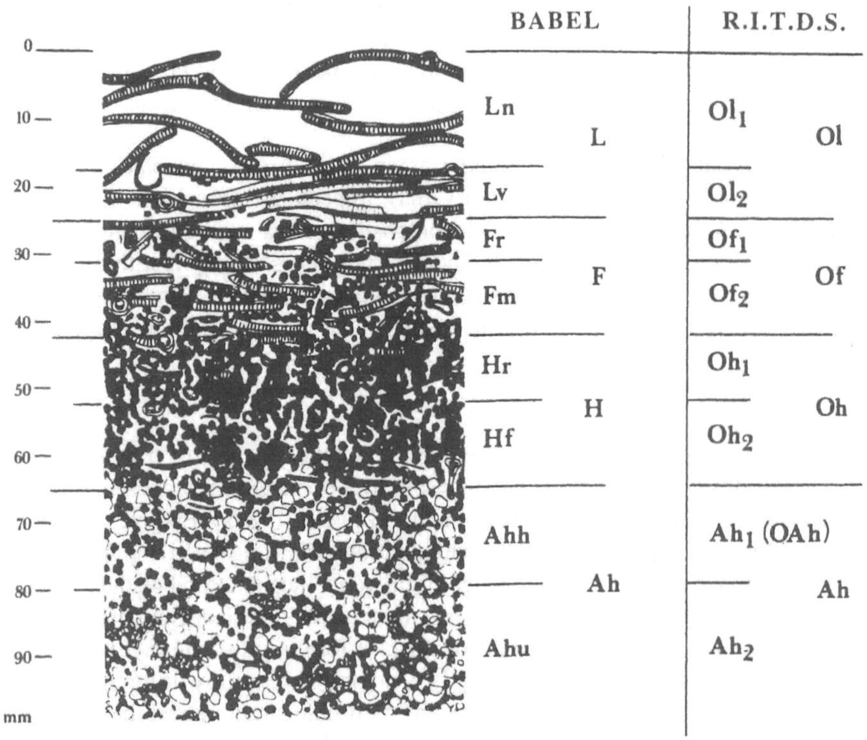

Figure IV.15 Arrangement of successive organic layers in a moder type of humus (Delecour, 1980). Note the fragmented leaf debris in the F layer and the accumulation of macro-arthropod faecal pellets in the F and H layers. Classified according to Babel (1971) and international nomenclature (R.I.T.D.S.).

2.2.4 MOR

Mor litter systems occur at sites with unfavourable climatic conditions (low temperatures and impeded drainage), often with nutrient-deficient soils and low-quality litter. Under such conditions, the activities of arthropod decomposers and white-rot fungi are extremely limited. In the absence of efficient decomposers of the phenol-protein complexes, matted layers of decomposing litter accumulate on the surface of the mineral soil.

In some mor forms, litter may still be recognisable after 50 years or more (Wittich, 1952) and the identification of litter residues down through a vertical profile section of the O horizon may provide an history of site vegetation (Bernier and Ponge, 1994). Hyphae, mainly those of mycorrhizal fungi, form an extremely dense network and the Oh layer is composed of a structureless mass of decomposing hyphae and the shapeless excrements of Enchytraeid worms, a ubiquitous component of the invertebrate communities of these systems (Bal, 1982).

Chemical limitations to decomposition slowly disappear and the end products of this process (*i.e.*, humic colloids, nutrient elements and organic acids) are eventually released. Nutrients released into the superficial holorganic layers may be absorbed by the large biomass of fine roots and the hyphae of mycorrhizal fungi. Such roots are relatively more abundant in infertile soils with a limited ability to retain and accumulate nutrients.

In such systems of decomposition, soil minerals are not mixed with the decomposing material, and thus aggressive organic acids are not neutralised in the early stages of their formation. As a result, they infiltrate into the mineral horizons where they have important roles in weathering minerals and promoting the translocation of iron and aluminium into the underlying horizons (see Chapter II, Section 3.1).

2.3 Structure of litter systems

Litter systems have a clearly layered vertical structure because of the burial of old litter by that more recently fallen. Lateral variation in litter systems may result from the distributions of structures such as bark and fruits that fall close to the plants that produce them, the presence of particular tree species that produce litters of contrasting qualities, and from topographic heterogeneity. A number of important litter-system properties are directly derived from the vertical and horizontal heterogeneity present.

Seasonal and successional çycles in climate and vegetation also affect litter systems and lead to changes in the intensities and types of processes operating at different temporal scales. The major of these processes are the seasonal changes that occur in litter inputs and those influencing the long term evolution of litter morphological, physical and chemical characteristics across the different humus types present at a given site.

2.3.1 VERTICAL STRUCTURE

Macro-scale features

As discussed above, and in Chapter II.1 in the context of horizon designation, litter systems may comprise up to three distinct layers of variable thickness. These three layers were first designated L, F and H by Hesselmann (1926); all were subsequently further subdivided into two sublayers by Babel (1971) (Figure IV.15).

The L layer comprises unfragmented leaves that have fallen within the last year; it may be further divided into an Ln layer with only a slight level of microbial attack and an Lv layer thoroughly colonised by fungi. The thickness of this layer depends on the amounts and seasonal patterns of litter fall.

The F layer includes fragmented leaves intensively colonised by fungi together with the accumulated faecal pellets of invertebrate decomposers. F layers are often sub-divided into two layers. The Fr (or F1) layer is composed mainly of large fragments of leaves and a few faecal pellets of the invertebrates (*e.g.*, large millipedes, Diptera larvae and epigeic earthworms) that feed on freshly fallen leaves. The second layer, the Fm (or F2), consists mainly of the smaller faecal pellets of Diptera and micro-arthropods mixed with fine leaf fragments. In these litter systems, white-rot fungi generally develop most conspicuously in the Fm layer (Gourbière, 1982).

The H layer is principally an accumulation of the smaller faecal pellets of micro-arthropods and Enchytraeidae, and amorphous decomposing fungal material. This layer may in turn be subdivided into an Hr in which plant remains may still be identified, and an Hf layer in which plant remains are no longer identifiable and are admixed with mineral materials.

Layer thickness may differ considerably depending on the type of litter system (Figure IV.16). In mulls, the litter system is restricted to an L layer comprising leaves aged less than one year. A notable exception to this is acid mull litters that develop in the absence of anecic earthworms through the intense activity of white-rot basidiomycote fungi; where this occurs, the L layer may be substantially thicker and consist of slowly decomposing leaves accumulated over the previous 2-3 years (Toutain, 1974).

In moders, the L layer is often subdivided into Ln and Lv sublayers to distinguish new (n) and old (v) leaves (Brun, 1978). The F layer is the most developed and may be several centimetres deep with well defined Fr and Fm layers. In tropical moders, the H layer can be colonised by a dense root mat that may be 10-30 cm thick in extreme situations (Cuevas and Medina, 1988; Jordan, 1989; Leroy *et al.*, 1992). In mors, the litter system is thick and the H layer is often fibrous and composed of humified fungal and other materials.

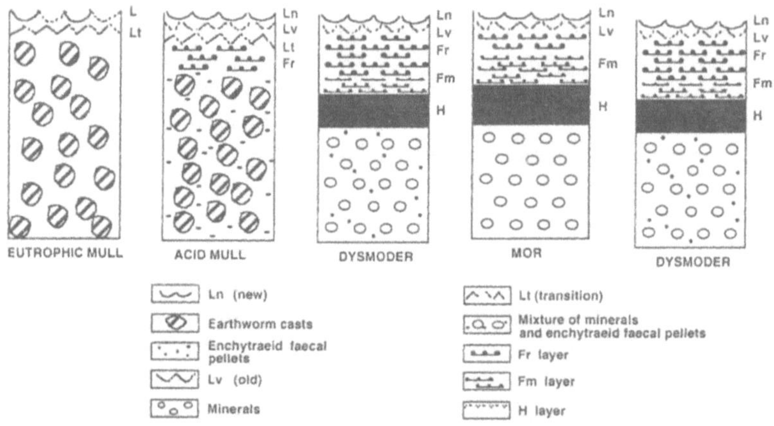

Figure IV.16 Schematic representation of litter systems in a sequence of humus forms on soils developed from acid rock parent materials in Eastern France (Brun, 1978).

Meso- and microscale features

At a finer scale of resolution, microscopic observations of litter layers may indicate their composition and structure. Anderson (1977) made a detailed description of microstructures within the L, F and H layers of a moder litter in an English Chestnut (*Castanea sativa)* forest. Twenty-three different structural units were recognised and quantified on sections of the litter system embedded in gelatine (Table IV.3). A positive correlation was found between the diversity of the environment (as measured by the Shannon diversity index) and that of the oribatid communities. The F layers were found to be the most diverse followed by the L and H layers (Figure III.3).

Table IV.3 A list of the different structural units used in the quantitative analysis of litter sections (Anderson and Hall, 1977).

1.	Intact leaves	13.	Hyaline fungal hyphae
2.	Leaf fragments > 5 mm	14.	Melanised fungal hyphae
3.	Leaf fragments > 1 mm <5 mm	15.	Faeces: free living macrofauna
4.	Leaf fragments < 1 mm	16.	Faeces: endophagous macrofauna
5.	Twigs	17.	Faeces: free living mesofauna
6.	Wood fragments >5 mm	18.	Faeces: endophagous macrofauna
7.	Wood fragments >1 mm <5 mm	19.	Cavities > 5 mm
8.	Wood fragments <1 mm	20.	Cavities >1 mm <5 mm
9.	Roots without mycorrhizae	21.	Cavities < 1 mm
10.	Roots with living mycorrhizae	22.	Mineral material
11.	Roots with dead mycorrhizae	23.	Animal remains
12.	Other macrophyte material		

Using a similar approach but employing a much larger number of characteristics (see *e.g.*, Figure IV. 17), Ponge (1985, 1988) made a detailed description of the successive layers of a moder litter developed in a *Pinus sylvestris* forest. The proportions of different items were measured and the casts and excrements of different invertebrates were accurately located and identified under a light microscope. A combination of these methods was later used to describe the vertical structure of humus layers of Spruce (*Picea abies*) forests in the northern French Alps (Bernier and Ponge, 1994) and permitted reconstruction of the vegetation successions that had occurred on the site.

2.3.2 HORIZONTAL AND TEMPORAL PATTERNS

Litter systems are highly heterogeneous in the horizontal dimension. At scales between those of a watershed catchment and a hectare, forests consist of 'eco-units' *i.e.*, component elements of a mosaic at different stages of regeneration after gap formation. At a smaller scale, wind and slope may contribute to the creation of specific areas with accumulations or deficits of litter. Finally, in multispecies communities, each tree creates a gradient of heterogeneity referred to as a 'single-tree effect' (Zinke, 1962; Spain, 1973). In consequence, the three types of litter systems may coexist, and form a mosaic in the forest floor (Figure IV.18).

Litter systems vary also in time, at scales differing from a few months (seasons) through decades to centuries (successions). Significant changes in the seasonal thickness of litter systems have been reported in both temperate (Mollon, 1982) and tropical systems (Healey and Swift unpublished in Swift *et al.*, 1979, Spain, 1984)(Section IV.2.6.1). At the broader scale of vegetation successions, changes in litter systems parallel those of soil organic matter status, communities of decomposers and vegetation. Changes are not necessarily synchronous: in the alpine forest considered below, differences in litter palatability determine the presence or absence of anecic and endogeic earthworms, and the formation of a mull

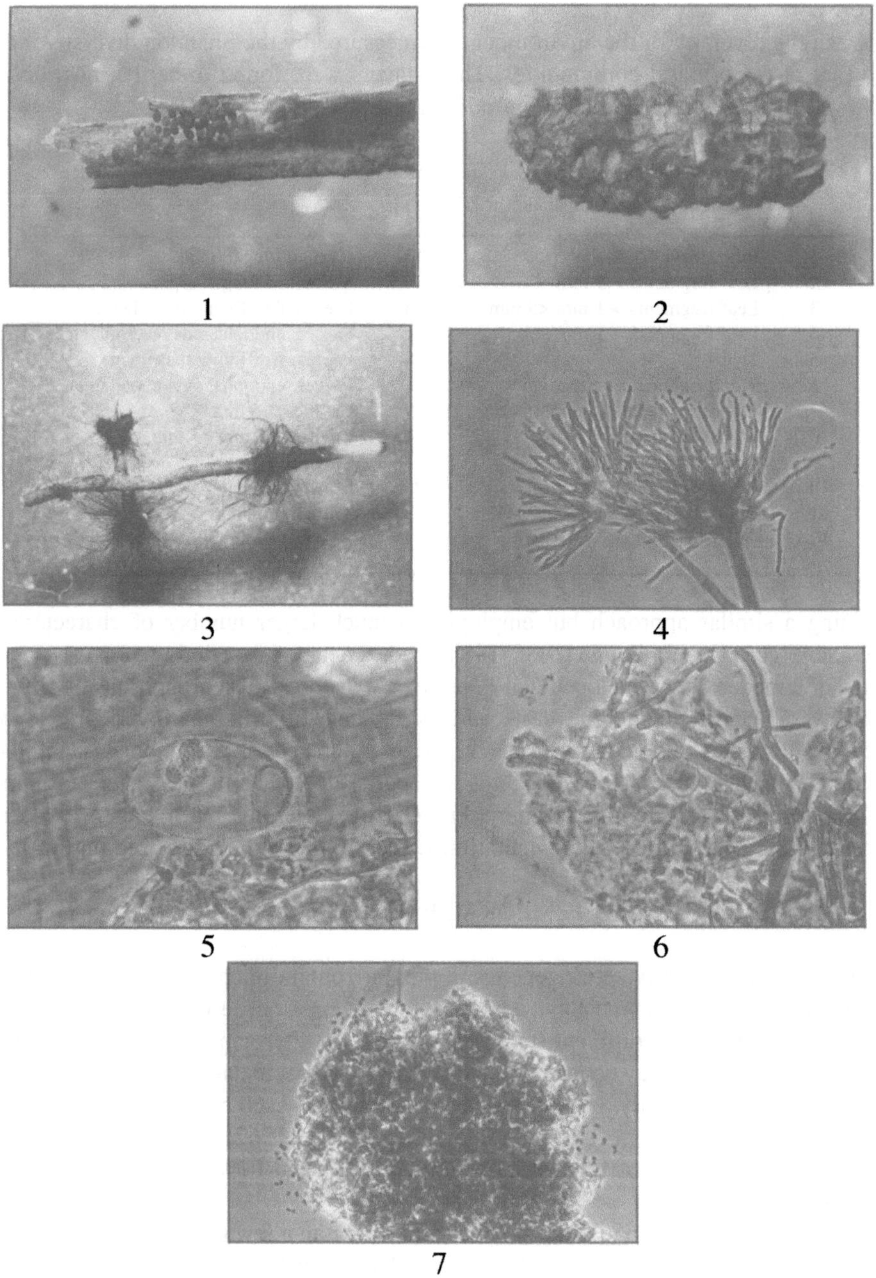

Figure IV.17 Selected components isolated from microscopic observation of small volumes of a *Pinus sylvestris* forest humus (Ponge, 1984, 1985, 1988): 1. Cryptostigmatid faecal pellets in a *P. sylvestris* needle; 2. Isopoda faecal pellet; 3. long root of *P. sylvestris* with hyphae of the ectomycorrhizal fungus *Cenococcum geophilum* behind the elongation zone, producing two short roots infected by the same fungal species; 4. Coniodophores of *Verticicladium trifidum*, micromycete growing in pine needles responsible for their blackening; 5. Empty test of the protist *Corythion dubium* colonised by bacteria; 6. Fungi and bacteria growing on a Collembolan faecal pellet; 7. Development of Cyanobacteria at the surface of a faecal pellet of an oribatid mite.

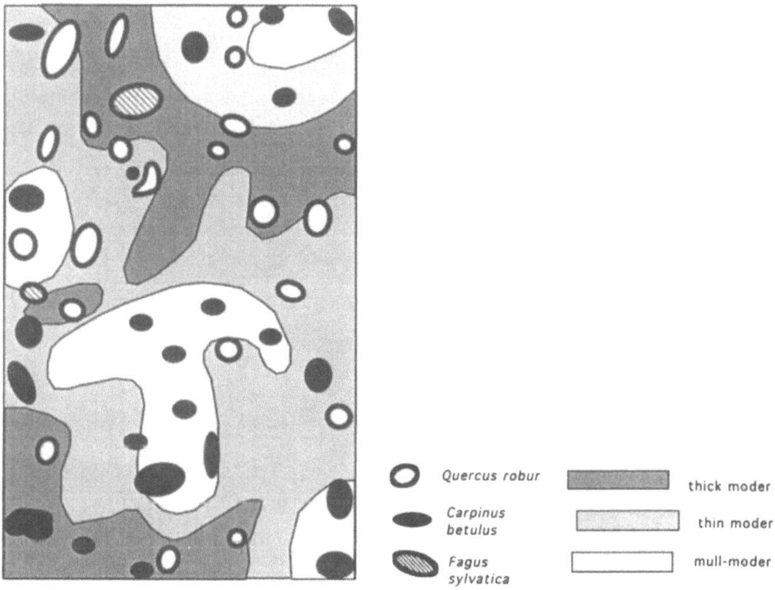

Figure IV.18 Distribution of humus forms and vegetation in a 25 x 40 m plot in a mixed-species French forest (modified from Mollon, 1982). *Carpinus betulus* produces a rapidly decomposing litter, in contrast to those of *Quercus robur* and *Fagus sylvatica*.

rather than moder system. Large time lags (10-30 years) may occur between vegetation change and the responses of the humus type and decomposer communities; the lengths of such lags also vary with climatic conditions (Bernier and Ponge, 1994).

Quantification of the different structural elements found in thin sections of the litter layers can give important information on the dynamics of decomposition. In moder and mor litters, successions of past vegetation types may be detected. Fern litter is deposited at specific seasons and may serve to mark the ages of the annual layers (Ponge, 1985). In Alaska, for example, Wagener and Anderson (unpublished data) noted that two litter layers are deposited each year: one of Spruce (*Picea sp.*) needles that fall during the growing season and another that forms when a thick cover of snow weighs the aerial parts of dead ferns to the ground. Thus, by counting the overlying layers of ferns, the age of materials found at different depths can be estimated.

Eco-units

Eco-units (Oldeman, 1990) are the component units of a forest mosaic defined by their ages (that is, the time passed since the last major disturbance such as a forest fire or the formation of a gap from which regeneration and growth started) and architecture. In a natural alpine forest, at 1550 m above sea level, Bernier and Ponge (1994) distinguished 9 different eco-units: grass vegetation, Spruce (*Picea abies*) stands of 30, 55, 60, 160 and 190 years, a Larch (*Larix decidua*) stand (215 years), moss and *Oxalis acetosella* vegetation and Bilberry (*Vaccinium myrtillus*) heath.

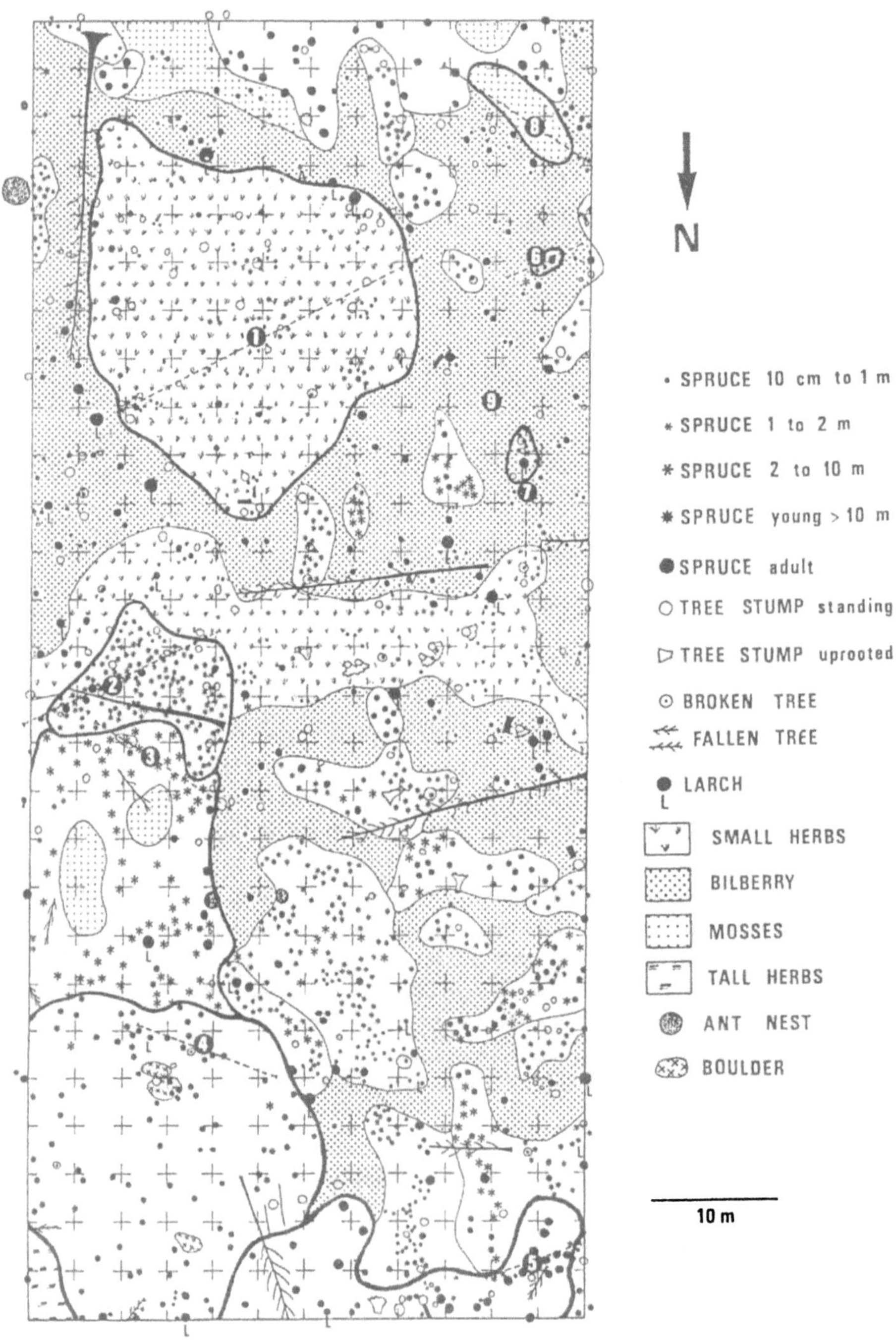

Figure IV.19 A mosaic of successional stages in an alpine *Picea abies* forest (after Bernier and Ponge, 1994) 1. Regeneration phase (or innovation phase); 2. Early aggradation or young growth phase of spruce at 30 years; 3. Intense growth or aggradation phase of spruce at 50 yr; 4. Late growth or aggradation phase of spruce at 60 Years; 5. Mature or biostatic phase of spruce at 160 yr; 6. Mature or biostatic phase of spruce at 190 years; 7. Collapse phase with herbaceous development; 8. Collapse phase with the development of a mossy and ericaceous vegetation; 9. Ericaceous heathland.

These units occur as a mosaic and represent different stages in the succession of vegetation types which proceeds in the order: herbaceous units within *Picea* forest of 30-190 years, formation of gaps with a return to herbaceous vegetation, or the development of a moss cover followed by *Vaccinium* heath. The latter occur especially in places that have experienced human disturbance (such as canopy openings caused by selective timber exploitation) (Figures IV.19 and 20). At this site, Bernier and Ponge (1994) were able to reconstruct the dynamics of the vegetation over the last three to five decades by analysing the changing composition of the moder litter layers, down to a depth of 12-15 cm in the eco-units that represented the different seral stages.

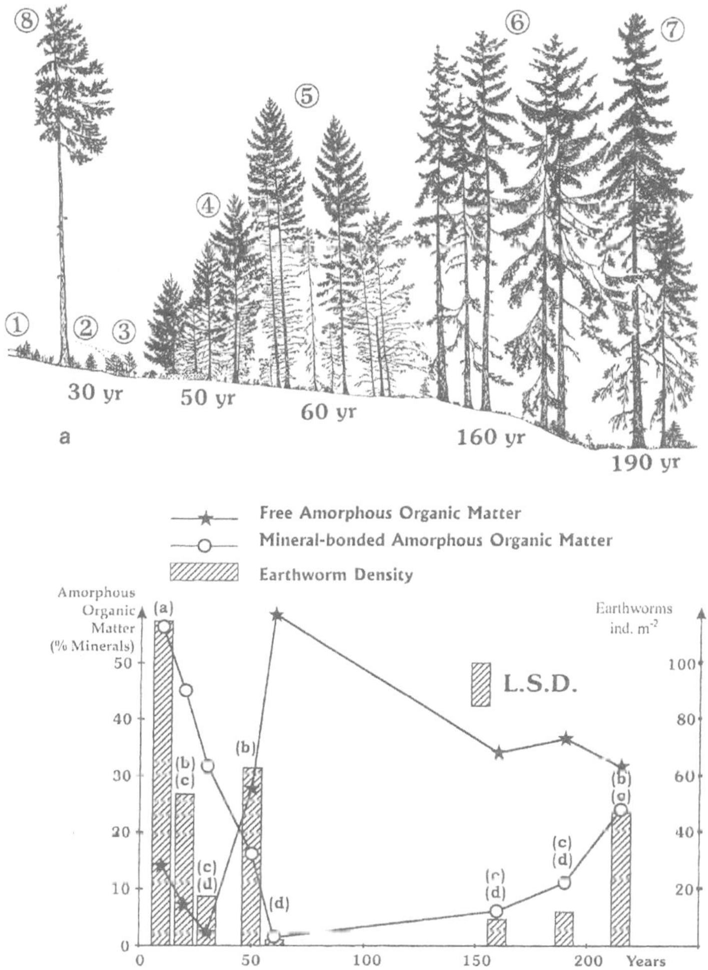

Figure IV.20 a.: Transect through different eco-units in an alpine spruce forest (after Bernier, 1996): 1: Regeneration phase; 2: Early aggradation phase before canopy closure; 3: Aggradation phase after canopy closure; 4: Aggradation phase; 5: Late aggradation phase; 6: Adult phase; 7: Late adult phase; 8: Senescent phase.
b.: Changes in earthworm densities (m^{-2}) and amorphous organic matter (% minerals).

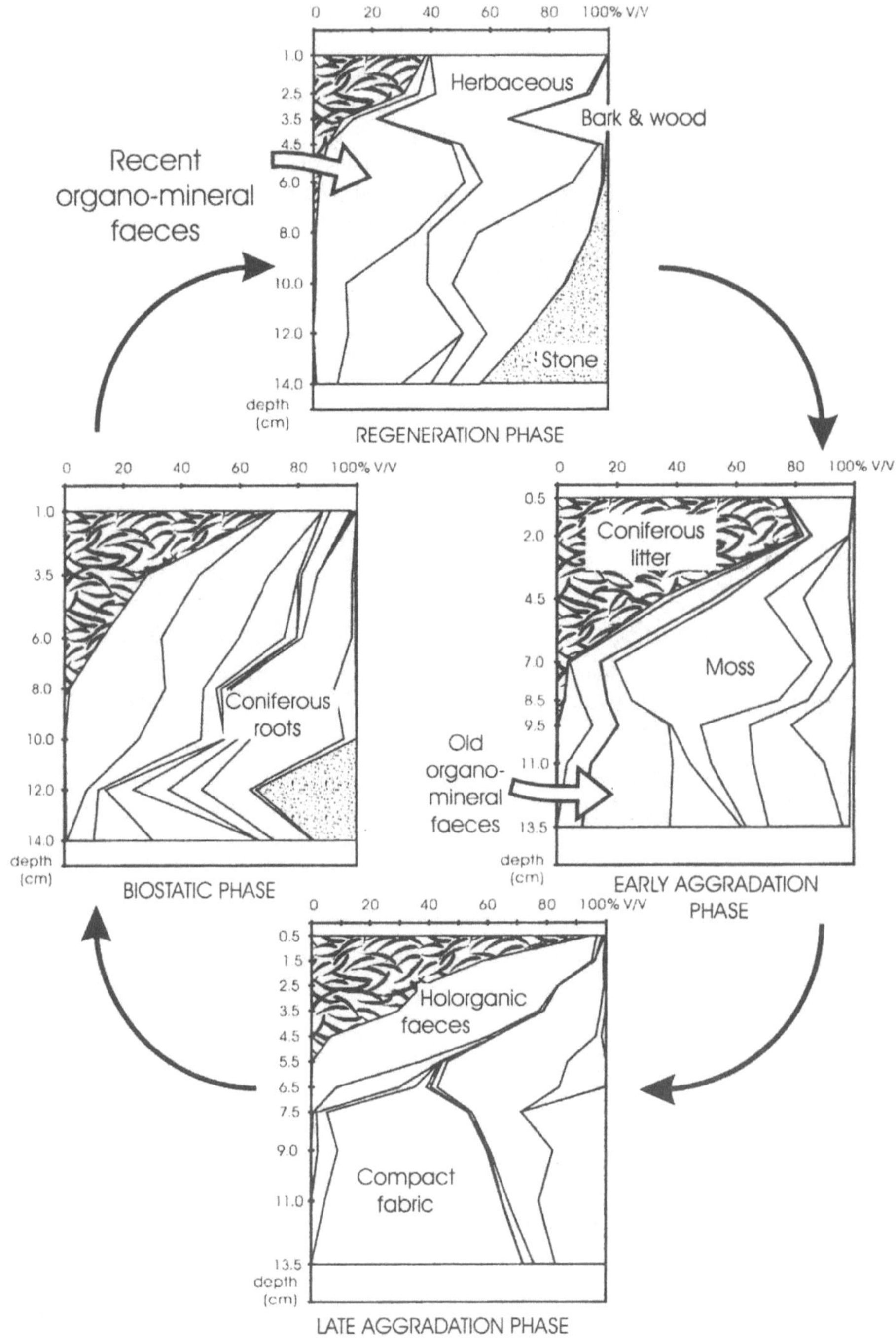

Figure IV.21 Variation of humus composition in four eco-units in a French alpine forest: a: regeneration phase; b: 30-y *Picea* eco-unit (note accumulation of moss debris at 6-9 cm and bark at 9-12 cm that give evidence of changes in vegetation during successional processes); c: 60-y *Picea* eco-unit. Bulk humus material observed at x40 magnification, with depth on the ordinate and cumulated percentages on the abscissa (Bernier and Ponge, 1994).

The profile sketched in Figure IV.21 (top), was observed in a grass unit (regeneration phase). These units have a mull litter system in which litter is rapidly exported to the drilosphere within the casts of the endogeic and anecic earthworms. These casts form most of the macro-aggregate structure of the underlying A1 horizon (5 to 12 cm). Bark found at 3-4 cm depth and dead coniferous roots at 6-12 cm indicated that this gap had only recently been formed. In a 30-year Spruce *(Picea abies)* stand (early aggradation phase, Figure IV.21 right), only mixed Spruce and Larch *(Larix decidua)* litters and mosses were present in the upper part of the litter system. At 10-15 cm depth, fragments of dead Spruce roots comprised two thirds of the organic fragments and the remaining third consisted of dead grass components (*Deschampsia flexuosa* and *Luzula sylvatica*) and mosses. At 8-12 cm depth, bark residues were abundant. Above this deposit, large amounts of dead mosses occured. From 3 to 6 cm depth, decomposing Spruce needles were derived from the present plant community.

This sequence suggests that before the present canopy of young Spruce trees closed, the soil was first covered by grasses and then by a thick moss cover. At the same time, large inputs of bark and woody material indicated gap formation within a mature Spruce stand. The presence of stumps of old (200-year) Spruce trees and a few living individuals confirm these observations.

As the Spruce stands grow, the canopy progressively closes, the density of endogeic and anecic earthworms decreases sharply and the litter system shifts to a dysmoder (*i.e.*, intermediate between moder and mor) in the 60-year old Spruce stand (Fig. IV.21, bottom). The H layer between 4.5 and 7.5 cm depth mainly comprises holorganic invertebrate faecal pellets. At 6-7 cm depth, large amounts of woody materials and bark fragments in the soil evidence the collapse of a previous Spruce stand. Below these organic deposits a 1 cm thick bleached eluvial horizon is present (Bernier and Ponge, 1994) indicating podzolisation. In 160-190 year old Spruce stands (biostatic phase Figure IV.21 left), small populations of anecic earthworms (*Lumbricus terrestris)* are restored and re-incorporate organic matter from the micro-podzol created in the previous phase (Figure IV.21 bottom). The litter system is a dysmull, an intermediate stage between the moder and the mull that will develop later in the grass eco-unit.

At different elevations, the correspondence between phases of the vegetation succession and the appearance and disappearance of anecic earthworms or the time of their peak density is slightly different. At 900 m elevation, the peak density of anecic and endogeic earthworms occurred in the old Spruce stand, whereas it was delayed up to the phase when grass covers the soil at 1550 m, and to an even later phase at 1700 m, *i.e.*, that of young (30-50 year) Spruce stands (Bernier, 1996). Differences in the quantity and quality of organic matter accumulated in the litter system may explain these differences.

Actively growing Spruce stands produce large quantities of low quality litter that discourage drilosphere activities and result in the formation of moder litter systems. In mature stands, the return of anecic and endogeic earthworms may be attributed to an overall change in the quantity and quality of the organic matter available in the environment. Old trees may release large amounts of twig material and organic matter may become more palatable to the earthworms after approximately a century of maturation. Anecic earthworms may thus return and their communities become fully mature at the next stage, in the grassy vegetation where Spruce also re-establishes itself. At 1700 m elevation, temperatures are lower and maturation of the organic matter may be delayed up to the stage when the Spruce forest first matures.

This example illustrates the importance of carbon cycling and energy fluxes as determinants of succession in the above-ground vegetation and litter system. It also demonstrates the importance of climatic factors in the synchronisation of plant successions with changes in the organic matter status of the system. Considerable energy flows occur through the litter system. Ovington (1961) estimated that the energy contained in the needle litter of a 55-year-old *Pinus sylvestris* stand was *ca.* 251 GJ ha^{-1}, and that which had already been released was 3.8 TJ. Part of this energy may be used by anecic earthworms and other 'ecosystem engineers' to fuel bioturbation processes (Villenave *et al.*, 1999).

Accumulation of litter and single-tree effects

Macro- and microtopographic features and the disposition of trees may lead to considerable variation in the thickness of litter cover, especially of the L and F layers. In a French temperate-climate forest with a moder litter system, the thickness of the litter layer (L + H) was significantly correlated with microtopographic features. Litter was unexpectedly less abundant in small depressions than between them, perhaps due to the effects of the white-rot fungi that develop preferentially in thick F layers (Garay, 1989) (see Section IV.2.5.4).

The distributions of trees, and differences in the quality of the litters that they produce locally determine the litter system. Zinke (1962) proposed the concept of the 'single-tree circle of influence' to substantiate the idea that forests are a mosaic of distinct soil and litter systems that are associated with individual trees. In a Bornean tropical forest, Burghouts (1993) reported large variation in the thickness of the L and F layers. Surprisingly, positive correlations were found among nutrient contents, litter inputs, and the thickness of the litter system. In that situation, litter accumulation was not due to slow decomposition rates, but to the rapid cycling of nutrients which favours the production of large quantities of relatively high quality materials.

In a tropical forest in French Guyana, Leroy *et al.* (1992) and Grandval (1993) described the patterns of distribution of leaf-fall around the trunk of a single tree. Due to the high diversity of tree species in the rainforest, the proportion of leaves deposited close to and derived from this particular tree only accounted for 18-22 % of the overall litter collected; the remaining 80 % comprised leaves from the surrounding trees but belonging to different species. The computation of semivariograms based on an intensive sampling showed that the mass of leaves deposited at the foot of the tree decreased regularly with increasing distance away from the trunk and became negligible at approximately 30 m from the trunk. In systems where litter quality differs among species, a mosaic of litter patches of different quality may form (Figure IV.22).

Such 'single-tree influence circles' have been found to affect soil properties and decomposer communities in temperate-climate forests where trees produce litters of contrasting quality (Boettcher and Kalisz, 1991, 1992). Similarly, the introduction of conifers or *Eucalyptus* spp into broadleaf deciduous or conifers into *Eucalyptus* spp. forests may alter the litter system so that it shifts from a mull to a moder humus type (*e.g.*, Turner and Franz, 1985; Garay, 1989).

Finally, differences in litter production by individual trees, and especially, asynchrony in leaf fall at the level of populations or individual trees may reinforce the patchiness of litter layers. Such processes are especially relevant in tropical rainforests (Sabatier and Puig, 1986; Burghouts, 1993).

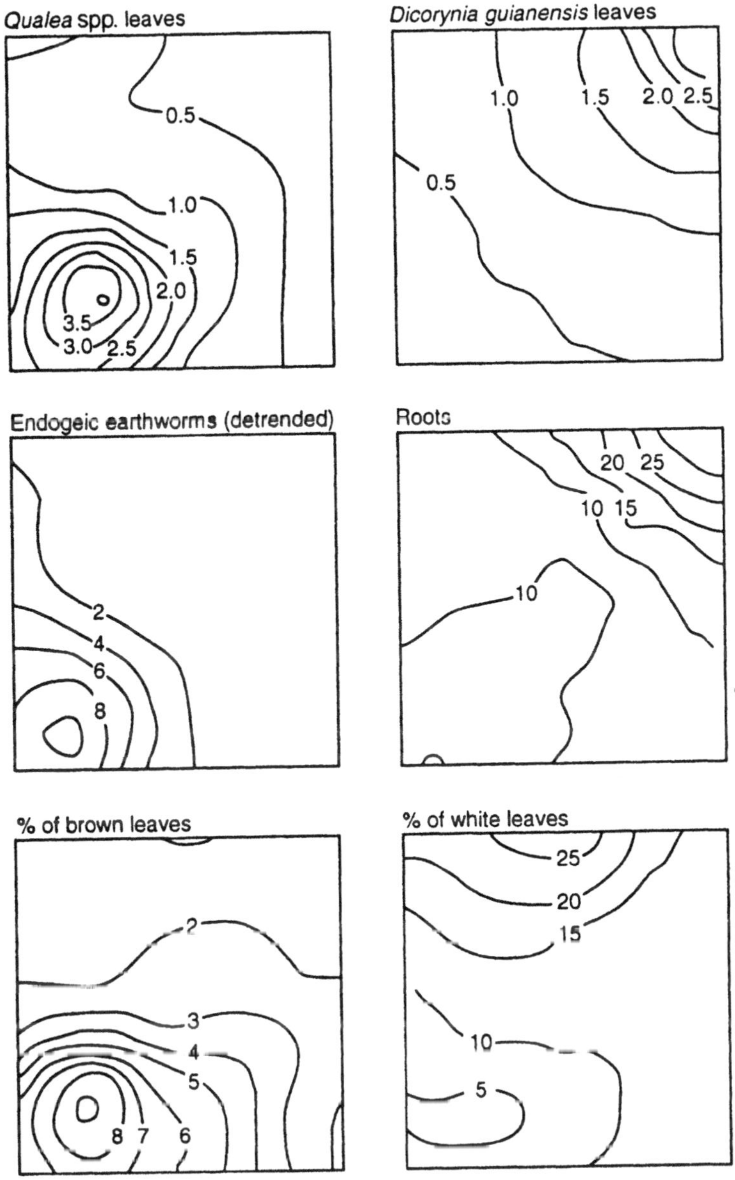

Figure IV.22 Krigged maps of distribution of leaf abundance of two individual trees, endogeic earthworms, leaves attacked by white-rot fungi and root density in a tropical rainforest in French Guiana (Charpentier *et al.*, 1995) (relative units). Note the respective concentration of leaves of *D. guianensis*, a thick root-mat and bleached leaves at the foot of *D. guianensis*; leaves of *Qualea sp.*, brown leaves and endogeic earthworms at the foot of *Qualea sp.* Tree trunks are located respectively at the upper right and lower left corners.

2.4 Decomposer communities

2.4.1 MICRO-ORGANISMS

In the early stages of decomposition, well before leaf fall, colonies of bacteria, yeasts and other fungi invade leaf surfaces. Amycelium then develops and, after some weeks, penetrates the leaf tissues (Minderman and Daniels, 1967).

Fungi dominate the microflora of the litter system. As discussed earlier, these micro-organisms have a pseudo mobility resulting from their capacity for rapid growth and ability to translocate cytoplasm. They are also adapted to the rapid changes in temperature and moisture conditions that may occur in the litter system. Their ability to perforate cell-walls and feed preferentially on carbohydrates makes them efficient early litter colonisers. Details of community structure and temporal changes are presented in Chapter III (Section III.2.4.3) and the activities of white-rot fungi are considered in Section IV.2.5.4 below.

After a succession of fungi belonging to different functional groups has taken place (Table IV.4), the remaining secondary resources comprise a large proportion of dead fungal material. By this stage, the remaining organic materials are located in the H layer or A1 horizons and further decomposition of this material will continue mainly through bacterial activities (see Chapter III.2).

2.4.2 ROOTS

Roots are an important but often neglected component of the litter system in humid climates. They grow principally in the lower parts of the system (*i.e.*, the F and H layers) and absorb the nutrients released in these layers, either directly or through the medium of the hyphae of their associated mycorrhizal fungi (Herrera *et al.*, 1978). In a Beech (*Fagus sylvatica)* forest growing on an inceptisol (acidic brown soil), fine roots actively colonised the F2 and H layers of a moder humus system: 53.7 and 21.7 % of root-tips were located in the F2 and H layers, respectively; 69.7 % of fine root biomass but only 34.1 % of small roots were concentrated in the same layers (Table IV.5) (Meyer and Göttsche, 1971). In spodosols, all root tips were located in the F and H layers, whereas in an inceptisol (eutrophic brown soil), root tips were less numerous but more evenly distributed throughout the soil profile.

In soils with a particularly low nutrient status, roots may concentrate at the surface to form a thick mat which may absorb virtually all the nutrients released from decomposing leaves and dead roots (Cuevas and Medina, 1988). In an Amazonian spodosol, less than 0.1 % of the labelled calcium and phosphorus sprayed onto the surface of a litter system with a thick root mat was leached through it (Stark and Jordan, 1978).

In most natural environments, certain zygomycote, ascomycote and basidiomycote fungi (Section IV.3.1.2.2) enter into close relationships with plant roots to form largely-mutualistic arbuscular mycorrhizal (AM) or ectomycorrhizal associations (ECM) (Section III.3.2.1.2); these act as extensions of the root system and may create direct links between decomposing leaves and roots. In a *tierra firme* or upland forest in Amazonia, for example, St. John and Uhl (1983) estimated the proportion of infected roots at 71.6 % with significant mean differences among species ranging from <25 % to 100 %. In a *caatinga* (low and relatively dry) forest on an extremely nutrient-deficient spodosol, all roots were infected. The mycorrhizas formed were largely AM.

There is some evidence that certain tree species specifically promote the formation of root mats near their bases, as in the Guyanese tree species *Dicorynia guianensis* (Figure IV.22). In this situation, a significant proportion of roots that accumulated in the mat appeared to belong to another tree species (*Eperua falcata*) that was growing a few metres distant (Grandval unp. data, in Wardle and Lavelle, 1997). Similar effects have been observed by Griffiths *et al.* (1996) in coniferous forests in Oregon (USA); the distribution of ectomycorrhizal mats (*i.e.*, dense accumulations of ectomycorrhizal roots) had a clear spatial structure that could be related to the presence of a few species of understory trees.

Table IV.4 Succession of fungal species on decomposing leaves in a Canadian Aspen (*Populus tremuloides*) forest (Visser and Parkinson, 1975).

	Living leaves	**Net-caught leaves**	**Litter**	**F-layers 1-4 cm**	**Humus 4-6 cm**
Pleurophomella spermatiospora	+				
Aureobasidium pullulans	+	+			
Sterile dark form	+		+		
Cladosporium sp.	+	+	+		
Penicillium janthinellum		+			
Sterile hyaline forms		+			
Paecilomyces ochraceus		+			
Penicillium spp.		+		+	+
Sterile hyaline form			+		
Beauveria bassiana			+		
Phialophora sp.			+		
Discula sp.			+	+	
Mortierella sp.			+		+
Mucor sp.				+	
Penicillium syriacum				+	+
Trichoderma spp.				+	+
Phoma sp.				+	+
Absidia sp.				+	+
Acremonium sp.				+	+
Penicillium farinosus				+	+
Cylindrocarpon spp.					+
Vollutella ciliata					+

Table IV.5 Distributions of root tips, fine and small roots down a soil profile in a German Beech (*Fagus sylvatica*) forest with a Brown Acidic soil (Inceptisol) (Meyer and Göttsche, 1971). Note that depth intervals for the 2-5 mm root diameter fraction differ slightly from those of the fine roots.

Layer	Depth (cm)	Number of root tips 100 ml^{-1} of soil (±1SE)	Fine roots (<2 mm) mg 100 ml^{-1}	Depth (cm)	Small roots (2-5 mm) mg 100 ml^{-1}
F2	2-3.5	6 259 (±1 063)	342.4	2-3.5	99.9
H	3.5-5	2 523 (±665)	226.7	3.5-5	39.3
A1	5-7	961 (±431)	70.0	5-15	150.4
B	13-17	439 (±95)	46.9		
B	19-23	309 (±50)	44.8	15-30	50.1
B	29-33	328 (±61)	28.7		
B	35-39	346 (±66)	27.7		
B	43-47	194 (±42)	13.5	30-50	60.3
B	52-56	123 (±95)	8.8		
B	61-65	67 (±82)	2.9		
B	72-76	75 (±57)	3.1	50-80	7.3
B	83-87	13 (±15)	0.7		

2.4.3 INVERTEBRATES

The large variety of microhabitats and food resources present in slowly-decomposing litter favours the establishment of highly-diverse communities (see *e.g.*, Anderson, 1977). This fauna includes a large proportion of saprophages which feed on decomposing litter and sometimes the colonising microflora and the products of its external digestion. A significant part of this fauna is predatory and regulates the population dynamics of the saprovores.

These invertebrates are 'litter transformers' (see Chapter III.4.4.1.6) and are generally unable to egrade phenol-protein complexes. Although certain groups may ingest some mineral components from the soil and mix them with organic matter (*e.g.*, Collembola, Touchot *et al.*, 1983; Isopoda, Mocquard *et al.*,1987; Enchytraeidae, Albrecht, 1984, Wolters, 1991), most litter transformers produce purely organic faecal structures and do not participate in the transfer of decomposing litter into the sub-soil or to other systems of decomposition, such as the drilosphere or termitosphere. Consequently, their direct contribution to decomposition through respiration is limited to a small percentage of overall mineralisation. For example, in a Swedish coniferous forest, Persson *et al.* (1980) calculated that a faunal community with a biomass of 1.7 g m^{-2} dry weight contributed only 4 % of heterotroph respiration; however, through the grazing of 30-60 % of microbial production in the litter and humus layers, they directly contributed between 10 and 49 %

of total N-mineralisation.

Despite their limited digestive efficiencies and inability to incorporate litter into the inorganic soil matrix, communities of epigeic invertebrates contribute to decomposition through comminution, grazing on microbial populations and the dissemination of fungal propagules.

Invertebrate communities of the soil:litter environment may be populous and taxonomically diverse. Communities may typically comprise several hundred species and 50,000-200,000 individual Enchytraeidae and micro-arthropods per square metre (Collembola and Acari) and several tens of species and hundreds to thousands of epigeic macroarthropods such as Myriapoda, Isopoda, Coleoptera and Diptera and epigeic earthworms per square metre (see, for example, the pioneering studies by Bornebush, 1930; Petersen and Luxton, 1982; Garay, 1980; Flogaitis, 1984; David *et al.*, 1993). In a German Beech (*Fagus sylvatica)* forest with a moder humus type, Schaefer and Schauerman (1990) found more than 360 species of saprophagous invertebrates in the meso and macrofauna with annual average population densities of 275,000 m^{-2} and 7800 m^{-2} for the meso- and macrofaunal components, respectively. Thirteen predatory species were present in the mesofauna (10,800 individuals m^{-2}) and 249 in the macrofauna (960 individuals m^{-2}). This fauna largely occurred within the litter system.

In an adjacent mull humus type, the saprophagous fauna consisted of a similar number of species whereas predators comprised 232 species in the macrofauna and 67 species in the mesofauna. Populations of litter-feeding saprophages were lower than in the moder with densities of 92,000 m^{-2} in the mesofauna and 3500 m^{-2} in the macrofauna. Similarly, smaller predator populations were noted and they occurred deeper within the soil, at an average depth of 3-5 cm in the A1 horizon. The overall biomass of litter saprophages averaged 4.9 g ash-free dry mass m^{-2} in the moder and 4.7 g m^{-2} in the mull. In the latter system, a large earthworm biomass (10.7 g ash-free dry mass m^{-2}) was found and much of the litter was probably transferred to the drilosphere system by these invertebrates (Schaefer and Schauerman, 1990).

Comparable results have been obtained along a marked soil pH gradient in a Beech (*Fagus sylvatica*) forest near Paris (P. Lavelle and A. Faille, unpublished data, Figure I.44 and IV.23). At this site, soil parent materials change from limestone to a sandy outcrop over 500 m and the soil pH correspondingly declines from 7.8 to 3.8. In the acid upper areas of the gradient, the humus is a thick moder; as estimated by the TSBF method (Lavelle, 1988), overall population density of the macrofauna is 300 individuals m^{-2} and 80 % of the macrofauna occurs in the litter (L+F) layer. As pH increases, the number of macro-invertebrates also rises to 818 individuals m^{-2} at pH 4.6 and then progressively decreases down to 526 individuals m^{-2} at pH 7.6; the proportion of individuals in the L + F layers decreases continuously to 78 %, 34.4 %, and 5 % as the humus progressively alters to a mull. Further, the macrofauna becomes increasingly dominated by earthworms and the litter biomass in the L + F layers diminishes .

Similar results were obtained by Garay (1980) for micro-arthropod communities in a mixed species forest near Fontainebleau (France). In that environment, patches consisting of litter derived from each species had significantly-different micro-arthropod population densities - with distinct vertical distributions - and their own patterns of seasonal change (Table IV.6).

In tropical environments, litter communities also include a wide diversity and abundance of invertebrates (*e.g.*, Lavelle *et al.*, 1981; Lavelle and Kohlmann, 1984: M'Boukou-Kimbatsa *et al.*, 1998). In these environments, litter systems are often limited in magnitude because of high decomposition rates and massive exportation of litter to the termito- and drilospheres. Consequently, the population densities and biomasses of other invertebrates are frequently lower than in temperate or boreal litters. In these forests, however, epiphytic litter systems frequently occur in, for example, the leaf crowns of Bromeliaceae, and these systems may have highly abundant and active invertebrate communities. In a tropical rain forest in southeastern Mexico, the population density of macro-invertebrates per gram of litter in Bromeliaceae was 12 times as high, and the biomass 3-fold higher than the equivalent values in ground litter. Where water accumulates within the leaf bases of the Bromeliaceae, the decomposer community may include such normally-aquatic components as dragon fly (Odonata: Zygoptera) larvae or those of the family Eristalidae (Diptera) (Lavelle and Kohlmann, 1984).

Relationships between litter biomass and invertebrate abundance are not straightforward. In some circumstances, a clear relationship may exist between the weight of litter accumulated at the soil surface and the abundance of the invertebrates that exploit the feeding resources and habitats of this system (*e.g.*, litter feeding arthropods in Figures I.44 and IV.23). However, in other circumstances, an inverse relationship may result, as when rapid litter consumption leads to substantial population increases (Garay, 1980).

Table IV.6 Distributions of Prostigmatid mites within the different layers of the O and A1 horizons of a temperate-climate deciduous forest, in two types of litter with contrasting qualities (Garay, 1980) (densities in 10^3 individuals m^{-2}).

	Carpinus betulus		*Pinus sylvestris*	
	WINTER	**SUMMER**	**WINTER**	**SUMMER**
L	5.2 ± 1.3		4.4 ± 2.2	
F	13.8 ± 4.7		34.4 ± 9.7	
L + F	18.9 ± 5.1	14.4 ± 3.7	38.7 ± 11.0	64.4 ± 14.8
H + A1	163.9 ± 31.0	270.0 ± 50.0	3.7 ± 1.7	139.0 ± 32.0
TOTAL	**182.9 ± 32.0**	**288.6 ± 4.70**	**42.4 ± 33.7**	**196.2 ± 32.0**

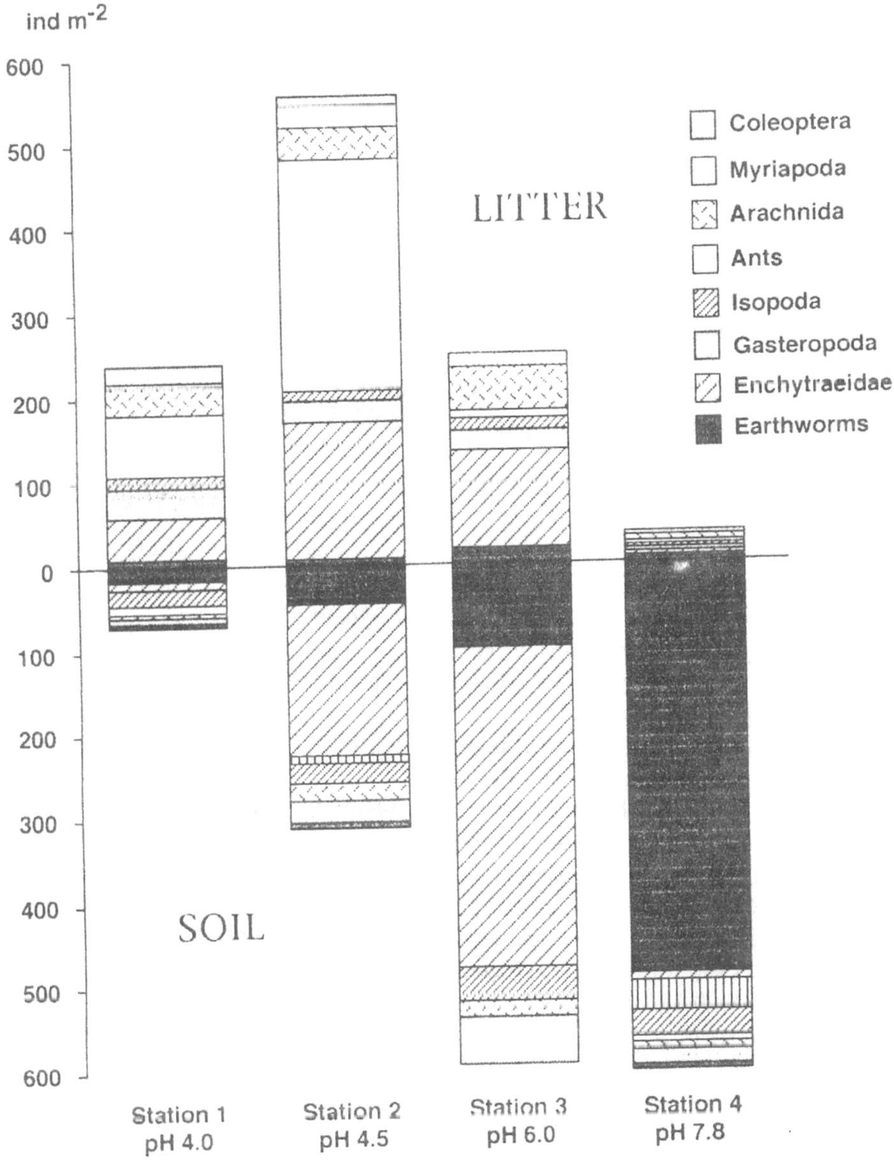

Figure IV.23. Distribution of major components of the litter- and soil- fauna at four stations along a pH gradient in the *Fagus sylvatica* forest at Butte Montceau, France, described in section 1.3.2.2.4 (Lavelle and Faille, unpublished data).

2.5 Processes in the litter system

2.5.1 COMMINUTION

Some epigeic invertebrates ingest large amounts of litter. In an Oak *(Quercus robur)* wood in western France, Isopoda with a mean annual biomass of 20 kg live weight ha^{-1} annually ingest 210 kg ha^{-1} or 7 % of the dry mass of leaf litter and egest 160 kg ha^{-1} of faecal pellets (Mocquard *et al.,* 1987). Larvae of the sciarid fly *Bradysia confinis* were estimated to have ingested 2.5 % of the leaf litter accumulated in an Oak-Beech *(Quercus robur, Fagus sylvatica)* wood in Brittany over a period of two months, *i.e.,* 121 to 219 kg dry mass ha^{-1} (Deleporte, 1987). The activities of these larvae are concentrated in small areas where all the 'old' litter of the F and H strata may be transformed into faecal pellets in a few weeks. Slugs also ingest significant amounts of litter, commonly in the range 1.3-8.4 % of total accumulated litter mass (Jennings and Barkham, 1979; Phillipson, 1983) although overall litter ingestion may be as high as 23 % (Kurcheva, 1967).

In a Mediterranean forest, the millipede *Glomeris marginata* (Diplopoda) may ingest 2.6 g of dry litter per gram of live body mass. As a result, a population of 50-100 individuals m^{-2} weighing 7.2 to 10.8 g may annually ingest 250-280 kg dry litter ha^{-1}, *i.e.* 8-11 % of annual litter production (Marcuzzi, 1970). In a temperate climate forest near Orleans (France) consumption of Oak (*Quercus robur*) litter by the millipede *Cylindroiulus nitidus* was estimated at 10-14 g m^{-2} $year^{-1}$, *i.e.*, 3-4 % of annual litterfall (David, 1987). In relatively dry environments, these invertebrates may ingest a much larger proportion of litter: in miombo woodland in Zimbabwe, for example, Dangerfield and Telford (1989) estimated annual consumption by millipedes at 88.1 g dry matter m^{-2}, equivalent to 30.6 % of annual litter fall.

From these few examples, it is clear that a variable but significant proportion of annual litter production is ingested by epigeic animals and re-ingested by the same or increasingly-smaller organisms. Diversity of species implies some diversity in feeding preferences; the palatability of litter to saprophagous invertebrates differs greatly among species, with different responses to the type of litter and its state of decomposition. In temperate forests, for example, Diptera larvae are among the earliest colonisers of freshly fallen litter (Mollon, 1982) whereas anecic earthworms usually prefer litter that has already undergone some degree of fungal decomposition (Cooke and Luxton, 1980; Cortez and Hameed, 1988).

As a result of fragmentation, the overall surface area of the decomposing litter is increased by a factor of 50 to 200 after having been egested and re-ingested by a sequence of increasingly smaller animals (Bachelier, 1978; Bertrand *et al.*, 1987). These animals may range from large saprophages such as Diplopoda to smaller micro-arthropods and even nematodes.

2.5.2 TRANSFERS

Transfers within the litter system involve: (i) leaching of water soluble elements and other materials towards deeper horizons or laterally in runoff; (ii) passive burial of decomposing litter under freshly fallen leaves (Figure I.42); and (iii) active exportation by anecic earthworms or termites to, respectively, the drilosphere or termitosphere (see Sections IV.4 and 5). Decomposing material is leached into the soil where it is flocculated and precipitated in contact with permanent-charge clay minerals and sesquioxides (Toutain, 1981; Duchaufour, 1982) (Figure IV.24).

In an English pasture, 20 years of pesticide (phorate) application eliminated earthworm populations. This resulted in a dramatic increase in bulk density, shear strength, penetrability and depth of leaf litter, but greatly reduced soil organic matter content, pH and soil moisture content (Clement *et. al.*, 1991).

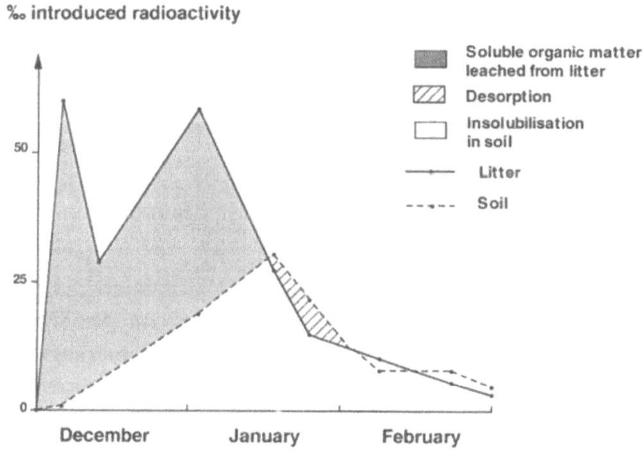

Figure IV.24 Evidence for the insolubilisation of water-soluble organic matter in soil (Toutain, 1981).

2.5.3 REGULATION OF MINERALISATION BY INVERTEBRATES

Invertebrates usually have limited direct impacts on mineralisation. In contrast, their indirect effects on microbial activities through, *e.g.*, grazing and mechanical litter transformation litter may be far more important. Verhoef and Brussaard (1990), for example, estimated the approximate contribution of fauna to nitrogen mineralisation to be 30 %, over a range of natural ecosystems and agroecosystems. However, microbial activity may be inhibited for long periods in the faecal pellets of certain invertebrates.

Changes in mineralisation rates

Digestion generally results in the direct assimilation by invertebrates of 5 to 50 % of the ingested litter (Bocock, 1963; Petersen and Luxton, 1982; Dickschen and Topp, 1987; Deleporte, 1987). However, faecal pellets egested by the saprophagous macro- and mesofauna may be resistant to degradation and accumulate to form a progressively increasing proportion of the H layers in moder litter systems (Kubiëna, 1953; Rusek, 1985; Toutain, 1987b). Invertebrate feeding thus significantly accelerates decomposition in comparison with situations where these animals have been eliminated by such techniques as naphthalene exclusion (Crossley, 1977). Removal of the fauna from decomposing maple litter through the application of the insecticide carbofuran reduced its decomposition rate by 40 % (Weary and Meriam, 1978).

In microcosm studies, decomposition is always accelerated by the addition of litter invertebrates in comparison with situations where micro-organisms are the only decomposers present. Increases of 14.6 to 32 % in carbon dioxide evolution from decomposing birch (*Betula* sp.) litter have been recorded following the addition of epigeic animals (Setälä *et al.*, 1988). Over eight months, Seastedt (1984) observed an increase of *ca.* 75 % in the decomposition rate and Anderson *et al.* (1983a) found increases of 1.6, 2-3 and 10-15 times in the ammonium released from Oak leaf-litter in the L, F and H layers, respectively, in the presence of the millipede *Glomeris marginata*. The biomass of the epigeic fauna is therefore a critical factor in the modeling of decomposition processes. However, this effect is generally delayed in time (by *ca.* 2 months in the case of a British deciduous woodland), and only significant when the biomass comprises at least 10 % of the weight of the decomposing litter. In the latter case, cumulative losses of mineral nitrogen are given by the equation:

$$NA = a + bT + cB + dBT$$

where NA is nitrogen mineralised (g N g^{-1} resource), T is temperature (°C) and B is the invertebrate biomass (g fresh weight) to resource quotient (Anderson *et al.*, 1985). In similar field experiments, Anderson *et al.* (1983a) found increases in ammonium leaching of 23 % during the first 32 weeks of decomposition but followed by a decrease of 38 % over the subsequent 32 weeks of the experiment (Figure IV.25). Where the same experiment included living roots, the amount of ammonium leached diminished by 12 to 48 %. Such experiments have shown that, in addition to increased mineralisation through faunal activities, interactions with other biological components and/or the decrease in mineralisation rates within faecal pellets may lead to a net immobilisation (Hanlon and Anderson, 1980).

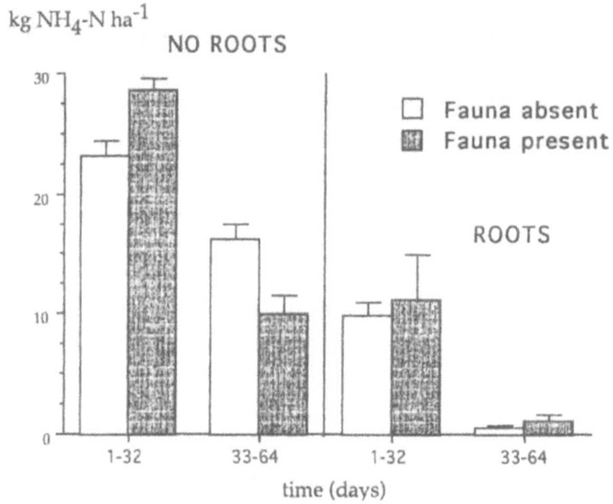

Figure IV.25 Release of ammonium and nitrate from decomposing litter in the presence and absence of roots and invertebrates in a deciduous temperate-climate forest (Anderson *et al.*, 1985).

Activation mainly results from fractionation and comminution of litter during transit through the invertebrate gut, and by the activation and dissemination of micro-organisms.

Regulation of microbial activity
Two further basic processes, both mediated by invertebrates and leading to the enhancement of decomposition rates are the grazing of microbial populations and dissemination of their propagules. In litter-systems, most invertebrates appear to feed at least partly on fungal material. Some Diptera and most Collembola are almost entirely fungivorous (see *e.g.*, Anderson and Healey, 1972; Kilbertus and Vannier, 1979; Saur and Ponge, 1988; Bardgett *et al.*, 1993). Furthermore, some specificity in feeding has been noted: Collembola, for example, have demonstrated preferences for mycorrhizal fungi (Schultz; 1991; Thimm and Larink, 1995). Grazing has two complementary effects:
(i) The release of nutrients immobilised in the microbial biomass, through excretion by the grazing organisms (see *e.g.*, Fenchel and Harrison,1976; Ingham *et al.*, 1986b). As an example, Persson *et al.* (1980) calculated that invertebrates consumed 30 to 60 % of the annual production of micro-organisms in a coniferous forest. This resulted in the release of 10 to 49 % of total mineral nitrogen, 70 % of which was directly excreted by micro-predators;
(ii) The maintenance of micro-organismal diversity by keeping population densities at sufficiently low levels that competition is avoided (Paine, 1966; Wauthy, 1982). Grazing is known to be selective (Proth, 1978) and may partially explain the high diversity of fungal species observed in moder in comparison with mull litter systems (see *e.g.*, Martinez *et al.*, 1980).

Propagule dissemination may be:
(i) Passive, when spores and other propagules are temporarily attached to the body surfaces of soil animals and are later released (Touchot *et al.*, 1983);
(ii) Active, when spores which have survived intestinal transit are released in faecal pellets and subsequently germinate. This may involve a range of micro-organisms including heterotrophs, autotrophs and organisms that form deleterious and beneficial associations with plants and animals. For example, Reddell and Spain (1991a, b) reported the successful transmission of viable spores of ecto and endomycorrhizal fungi and of the actinobacterium *Frankia* sp. through the intestinal tracts of a number of earthworm species. However, this process is selective with differential survival between the spores of different species and many of them may loose their ability to germinate (see, *e.g.*, Cervek, 1971; Kozloskaja, 1971, Wynn-Williams, 1983). This selectivity has been used in the control of some fungal plant pathogens (Ulber, 1983).

The micro-environments of the faecal pellets of epigeic species are generally conducive to microbial digestion of their components (although some exceptions to this rule exist, *e.g.*, in Enchytraeidae). Faeces therefore represent privileged incubation sites and act in part as external rumens in that decomposition proceeds progressively as they undergo ingestion and egestion by a succession of invertebrates. The invertebrate gut may itself be a favourable, though temporary micro-environment for micro-organisms. In one example, Hanlon (1981) observed a nine-fold increase of microbial biomass during the passage of food through the gut of isopods and glomerid millipedes.

Spores are not the only propagules to retain viability after passage through the earthworm gut. Reddell and Spain (1991a, b) also showed that root fragments infected with arbuscular mycorrhizal fungi still retained the capacity to infect the roots of *Sorghum bicolor* after passage through the gut of *Pontoscolex corethrurus*. This form of transmission may be important in the spread of infection of mycorrhizae, plant pathogenic fungi and animal parasites where active earthworm populations are present.

In some circumstances, microbial activity may proceed through two distinct phases. These include an initial phase of intense activation in fresh pellets followed by a longer term inhibition, which may result from either the relative accumulation of resistant compounds or unfavourable moisture and aeration conditions within the compact structure of the pellets (Figure IV.26) (Hanlon and Anderson, 1980). The accumulation of faecal pellets in the H horizon of moder and mor humus types shows that in certain circumstances, they are resistant to decomposition. Fractionation of organic matter from the 0-2 cm stratum of soil under a specific tree species in a Congolese forest showed that the dominant 200-500 μm fraction is rich in faecal pellets and has a lower mineralisation rate than the same fraction extracted from litters beneath another tree species where faecal pellets are rare (Bernhard-Reversat, 1993). This suggests that faecal pellets may have significant roles in the long-term conservation of organic matter.

As decomposition proceeds, mineral nutrients are released and redistributed into the microbial and invertebrate biomass. The relative abundances of epigeic invertebrates and differences in climatic regimes may guide the differentiation of litters into either moder or mor systems. The faecal pellets of micro-arthropods, Enchytraeidae and Diptera larvae often dominate the mass of the H layer in moder litters (Kubiëna, 1953; Rusek, 1985; Toutain, 1987b).

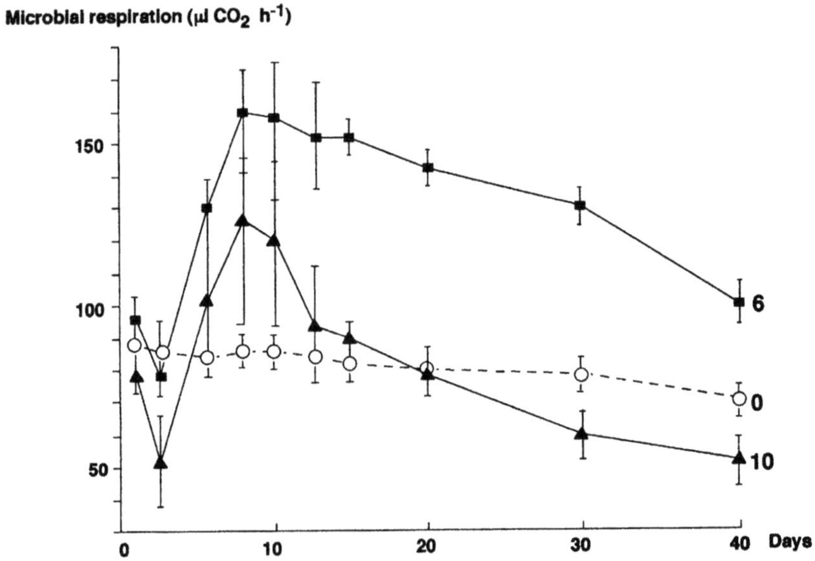

Figure IV.26 Effects of feeding by the isopod *Oniscus* sp. on the microbial respiration of 1g samples of oak leaf litter in microcosms with 0, 6, or 10 individuals introduced (Hanlon and Anderson, 1980).

2.5.4 EFFECTS OF WHITE-ROT FUNGI

Soil and litter acidity and the climatic regime are the major factors determining whether moder or mor humus types with deep and layered litter systems will form. However, even in these humus types, rapid decomposition may ensue when conditions become suitable for the development of basidiomycote white-rot fungi (Section IV.2.5.4). In mixed stands of Oak *(Quercus robur)* and Beech *(Fagus sylvatica)* in eastern France, Toutain (1987b) found both moder and acid mull humus types. Soils with an acid mull had twice as much iron as those with moder litter regimes. The higher iron levels permitted the development of dense populations of white-rot Basidiomycota, the release of large amounts of soluble organic matter from the decomposing litter and their insolubilisation in soil as iron: organic matter complexes. White-rot fungi readily decompose phenol-protein complexes thereby obviating this chemical limitation to decomposition characteristic of acid environments (Figure IV.10c) (see *e.g.*, Kirk and Moore, 1972; Eriksson *et al.*, 1980; Blanchette, 1984; Adaskaveg and Gilbertson, 1986).

Electron microscope studies have shown that the whitening of decomposing resources results from the digestion of brown polyphenol-protein compounds (Reisinger *et al.*, 1978). White-rot Basidiomycota can penetrate and progressively digest cellulosic structures and condensed protoplasmic residues, releasing large amounts of ammonium and soluble carbohydrates. White-rot fungi may be present in all litter systems and in decomposing wood. They initially grow as patches but later form expanding annuli with characteristic 'fairy ring' growth patterns (Figure IV.27). They generally grow in the F layer, and notably in locations where the F2-layers have accumulated to a significant depth (Gourbière, 1982; Garay, 1989). Litter decomposes much more rapidly as a result of their activity (Figure IV.5). In temperate forests, they are restricted to places that are within the reach of growing 'fairy-ring' shaped colonies (Gourbière, 1982).

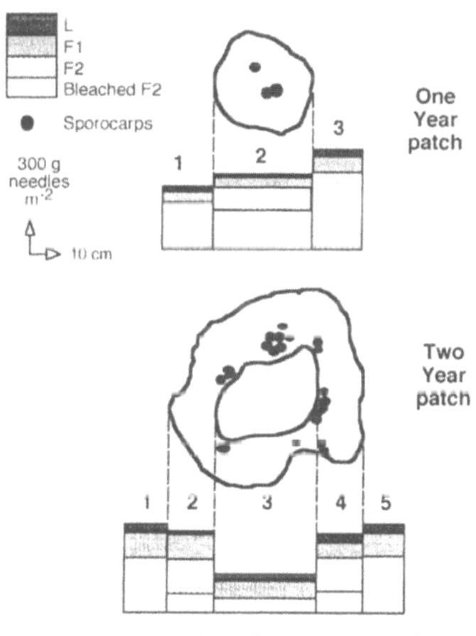

Figure IV.27 Horizontal and vertical representation of two patches of litter colonised by white-rot fungi of *Collybia maculata* for one and two years, respectively (Gourbière, 1982).

In a French temperate-climate forest with a mixed vegetation cover of *Quercus sessiliflora* and *Carpinus betulus*, white-rot fungi developed preferentially in the thick F layers and the structure of the acarine communities present appeared to depend on the occurence of these fungi (Garay, 1986b). Overall population density of mites in the F horizon of sites where litter had accumulated was 100 500 m^{-2} in patches colonised by white-rot fungi compared with 60 800 m^{-2} in non-infected areas; such population density increases were unevenly distributed among taxonomic groups. In a French Guyanese tropical rain forest, the occurrence of white-rot fungi may be closely related to litter quality; again, different invertebrate communities have been found in litters colonised and not colonised by white-rot fungi: earthworms colonised patches with no white rot fungi whereas Phalangidae (Opiliones) and Coleoptera tended to have an opposite distribution (Leroy *et al.*, 1992; Grandval, 1993). In tropical forests and tree plantations from Congo and French Guyana, the dependence of white-rot fungi on litter quality has been confirmed by Bernhard-Reversat and Schwartz (1997) and Leroy *et al.* (1992).

2.5.5 ROOT ACTIVITIES

Roots act as sinks for nutrients and water. As a result, the quantities of nutrients leached from the litter system are significantly diminished in microcosms with roots in comparison with those where roots are absent (Figure IV.26) (Anderson *et al.*, 1983a). In litterbag experiments conducted in Central Amazonia, roots colonised the decomposing litter and had no significant effect on the rate of decomposition (Luizao, 1994). Roots may be colonised by a wide range of different mycorrhizal fungi that interact with the non-symbiotic microflora. Through their effects on soil physical conditions (reductions in moisture and oxygen contents), and biological interactions with both the free-living and symbiotic microflora, roots may have a positive (priming) or negative (conserving) effect on decomposition.

Studies in which roots have been isolated from the litter system by trenching, have demonstrated the substantial capacity for tree roots to absorb soil moisture (*e.g.*, Slavikova, 1958; Staff, 1988) and to take up nutrients from the litter system (Aber *et al.*, 1983; Harmer and Alexander, 1985). Gadgil and Gadgil (1971) hypothesised that mycorrhizal tree roots suppress the decomposition of forest litter. In a New Zealand *Pinus radiata* plantation, in which root activity - and thus mycorrhizal activity - were reduced by trenching the soil to a depth of 30 cm to isolate blocks of soil and litter from root activity, litter decomposition rates were significantly increased. This increase was attributed to the suppression of competition for nutrients between mycorrhizal fungi and saprotrophic micro-organisms. However, subsequent similar experiments have not confirmed this effect (Staff, 1988). This may be due to the complex synergistic or competitive interactions that can take place between tree roots, mycorrhizal and saprotrophic fungi (Dighton *et al.*, 1987). The factors determining the direction and intensity of these interactions are far from being clearly understood.

2.6 Organic matter fluxes in the litter system

Nutrient and carbon fluxes are major consequences of litter acquisition and breakdown. Major issues influencing ecosystem function and the global environment are the storage and conservation of C and nutrients and the release and recycling of litter components. The total and relative amounts of nutrients that are recycled annually through the litter system are highly variable.

Nutrient release occurs at characteristic times and places in different litter systems. A major challenge in the management of degraded ecosystems, is to restore or enhance litter systems that provide nutrients to plants and energy to litter transformers and ecosystem engineers, while conserving nutrients and storing carbon (Woomer and Swift, 1994).

General patterns of carbon and nutrient fluxes in litter systems are clearly influenced by the hierarchical suite of factors that govern decomposition processes (Figure IV.8) and the specific properties of individual nutrient elements. Consequently, three general trends occur in litter-system dynamics: (i) a marked dependence of accumulation and decomposition rates on climatic conditions, soil properties and resource-quality attributes; (ii) the accumulation of nutrients in plant biomass and litter rather than the soil in nutrient-deficient systems; and (iii) markedly different turnover times for individual elements.

2.6.1 CARBON MINERALISATION IN THE LITTER SYSTEM

The amounts of carbon stored in litter systems differ greatly between humus types. In mull humus environments with shallow litter systems, little carbon accumulation occurs and this contrasts with the increasingly larger amounts found in moder and mor systems, respectively (Heal *et al.*, 1981; Figure IV.3).

CO_2 release from the litter system may be slow and even, or undergo marked fluctuations over different scales of time. The seasonal changes of litter mass recorded in temperate and tropical systems suggest that associated changes in temperature and moisture availability are responsible for differences in their rates of C release (Figure IV.28) (see Section IV.2.3.2). At the broader temporal scale of ecosystem development, changes in litter mass may also occur. In an age series of floristically-similar Australian *Eucalyptus* forests, Polglase and Attiwill (1992) reported a progressive increase of litter mass at the soil surface, mainly due to an increase in the amount of woody litter present (Figure IV.29). Significant changes were also noted in the successional process described by Bernier and Ponge (1994) in an alpine Spruce *(Picea abies)* forest (Figures IV.20 and 21). Organic matter accumulated in the litter layers during the mature phase of the forest; when the old trees died and new trees started to grow, the altered litter quality allowed earthworms to invade the unit. Through this, a significant proportion of the accumulated organic matter was decomposed and incorporated into the soil as mineral-bonded organic matter while the large flush of nutrients boosted the growth of the young trees.

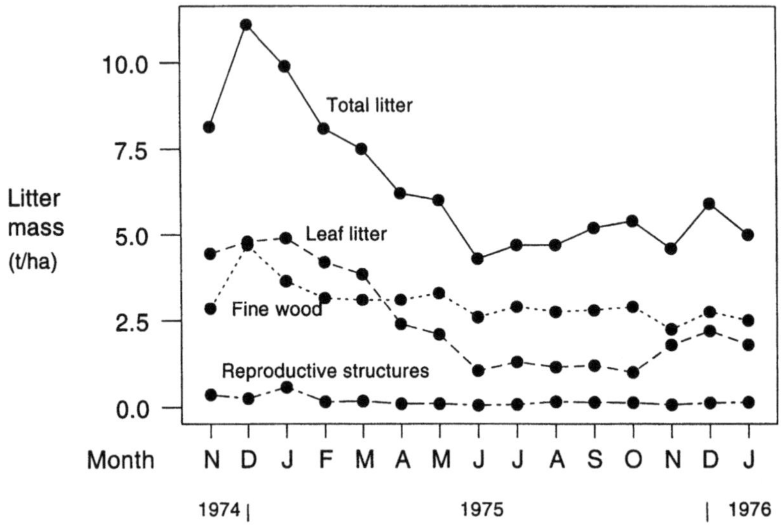

Figure IV.28 Seasonal changes in the composition of litter layers of a tropical Australian rainforest (Spain, 1984).

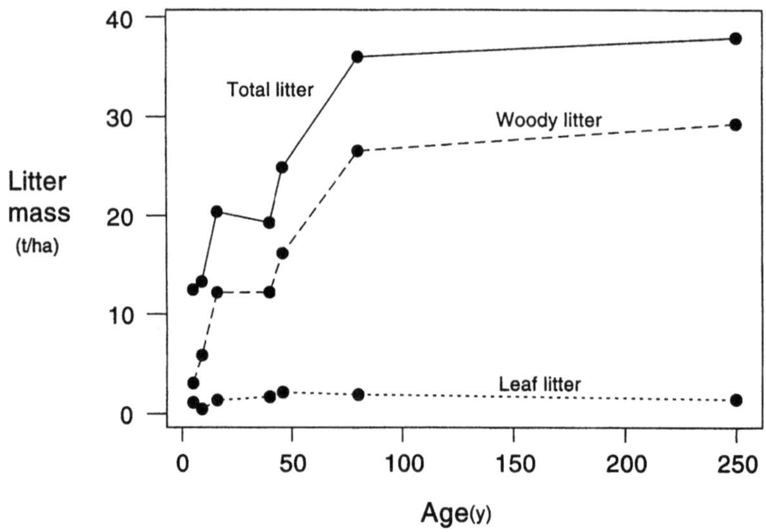

Figure IV.29 Changes in the composition of the litter layer in an age series of similar Australian *Eucalyptus* forests (Polglase and Attiwill, 1992).

The chemical composition of organic matter clearly differs between humus types. In a sequence of humus types from mull to moder to mor, Brun (1978) found great differences in the quality of organic matter accumulated in the litter and A1 horizon, as reflected in the distribution of C in different particle size fractions, and chemical quality assessed by extraction techniques and C:N ratios. In the A1 horizon, for example, the proportion of total humin decreased whereas inherited humin increased from the eutrophic mull to the dysmoder. Inherited humin corresponds to the resistant organic matter accumulated as lignin and as invertebrate faecal pellets; the difference between total humin and inherited humin comprises insolubilised humin *i.e.*, soluble organic compounds leached out of the litter system, flocculated and associated with clay minerals in organo-mineral complexes. This trend was confirmed in a review by Zech *et al.* (1996) who stressed that organic matter accumulates in mor litters as strongly-humified, decomposition-resistant plant residues, whereas in mull systems, organic matter is bound to minerals in less humified forms.

2.6.2 NUTRIENT CYCLING

Decomposition and mineralisation processes release nutrient elements largely in soluble forms that may be taken up by plants and other soil organisms, or leached. The regulation of nutrient release rates is of paramount importance in nutrient conservation since most nutrient elements occur in limited stocks. Only small external inputs occur in rainfall, dry deposition and through N-fixation, which generally accounts for a few kilograms ha^{-1} y^{-1}. In tree plantations in the Congo, for example, N-fixation rates have been estimated at respectively 14-21 kg ha^{-1} y^{-1} in stands of *Eucalyptus* sp. (Bernhard-Reversat, 1993) and 120-141 kg ha^{-1} y^{-1} in stands of N-fixing *Acacia* sp. (Bernhard-Reversat *et al.,* 1993).

Inputs in rainfall and as 'dry' deposition may represent significant amounts on an annual basis. For example, 2-21 kg of mineral and organic N, 0.2-0.5 kg P, 4-9 kg K and 3-16 kg recorded annually per hectare from tropical forests in Sénégal, the Côte d'Ivoire and North Cameroon (Bernhard-Reversat *et al.*, 1978; 1987; Harmand, 1997).

In highly-industrialised areas, air pollution from industrial and agricultural sources may lead to the deposition of much larger amounts of nutrients from aerial sources, particularly N and S. High N deposition rates (*e.g.*, >18 kg ha^{-1} y^{-1} wet deposition over a ten year period in England, Tamm, 1991) results in N saturation, acidification and cation leaching in poorly-buffered soils and this, in turn, leads to nutrient inbalances and changes in plant community composition. N saturation is defined as the situation in which the physiological N requirements of the primary producers are satisfied and considerable nitrate leaching takes place (Tamm, 1991).

At a given site, rates of release differ between nutrients: K and Mg are generally lost rapidly, whereas the release of P, N and Ca has variable kinetics (Figure IV. 30). Depending on the local availability of nutrients, N, P or Ca may be immobilised for longer periods than other nutrients. In a French temperate-climate forest (Foljuif), N is retained in fungal biomass for a longer period than other nutrients (Fig. IV.30a). During the summer period (months 6 to 9 and 18 to 21), intense fungal growth leads to the accumulation of N beyond the stocks initially present in litter: this N may have been translocated from deeper horizons, or input with rainfall and throughfall (Garay *et al.*, 1986a) or through N fixation which has been reported to range from not detectable to relatively small amounts (less than1 to more than 5 kg ha^{-1} y^{-1}) in different litters (see, for example, Heath *et al.*, 1988). N fixation also occurs in the harvest residues of crops such as sugarcane (Patriquin, 1982).

In a rainforest in Sarawak, P and N were immobilised over a 10 month experimental period, emphasising the effects of P and N limitations on decomposition processes (Figure IV.30b) (Anderson *et al.*, 1983b). In a Nigerian bush-fallow Ca appeared to be the most limiting nutrient (Swift *et al.*, 1981); during the first few months Ca was strongly immobilised whereas P and N were regularly released. Decomposition rates in litter bags varied with mesh size. In fine mesh bags, relative confinement and the exclusion of large invertebrates resulted in reduced rates of release of all nutrients, except for K which is highly mobile and readily leached (Figure IV.30c). In an Amazonian rainforest, Cornu *et al.* (1997) showed that mineral elements that are not actively taken up by plants (Fe, Si, Al, Ti) tend to accumulate in decomposing litter.

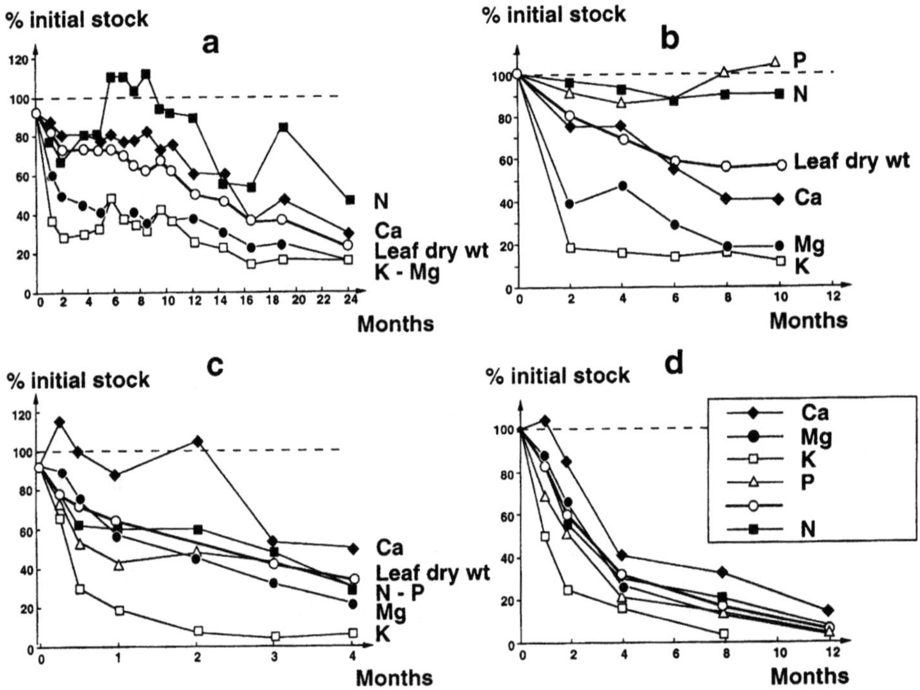

Figure IV.30 Changes in dry mass and nutrient contents of leaf litter decomposing in litter bags at different sites: a. temperate-climate deciduous forest at Foljuif (France)(Garay *et al.*, 1986a); b. rain forest at Sarawak (fine-mesh bags) (Anderson *et al.*, 1983b); c. bush-fallow in Nigeria (fine-mesh bags that exclude macrofauna); d. same with coarse-mesh bags.

There is some evidence that, if all other constraints (climatic, edaphic, litter quality) are satisfied, nutrient availability in soil is the main regulator of decomposition rates in the early phases. In the later stages of decay, lignin, lignocellulose and probably phenol:protein complexes emerge as limiting factors (Berg and Staaf, 1980; Gourbière, 1982; McClaugherty and Berg, 1987).

The relative importance of the various determinants of decomposition rates may change with time. In a latitudinal series of 14 Swedish Scots Pine (*Pinus sylvestris*) forests, losses in litter mass occurring during the first year of decomposition were correlated with climatic factors and site productivity indices; during the later stages of decomposition, such losses were correlated with the litter holocellulose to lignocellulose ratio (Johansson, 1994). During the early phase of decomposition, high concentrations of cellulose, hemicellulose and nutrients in litter are other factors that accelerate decomposition. After two years, 60 % of the litter had decomposed and decomposition rate became increasingly independent of climatic variables. However, a significant relationship was found between decomposition rate and the combined (holocellulose + lignin) contents of the remaining material.

Different patterns and processes of nutrient release result in variations in the turnover of elements and changes in their relative concentrations in plant biomass and soil. In a tropical lower montane rainforest at El Verde (Puerto Rico), the turnover rates of different elements varied greatly: the ratio of total stock (S in g m^{-2}) to annual cycling rate (C in g m^{-2} $year^{-1}$ released through decomposition) differed considerably between elements with values of 0.08, 0.14 and 0.16 years for K, Cu and P respectively contrasting with 13.3, 47.5 and 81 years respectively for N, Ca and Mg (Odum, 1970; Table IV.7). A ranking of turnover times follows the order: K <Cu <S <P <Na <Fe <Mn <C <N <Ca <Mg.

Table IV.7 Distributions of nutrients in plant parts, litter and soil of different forest systems (Exchangeable bases in $cmol.he^{-1}$).

Element	**Total stock (S)** g m^{-2}	**Cycling rate (C)** g m^{-2} yr^{-1}	**S/C** (years)
Carbon	21,225	3,000	7.0
Nitrogen	136	10.2	13.3
Magnesium	114	1.4	81
Calcium	242	5.1	47.5
Potassium	107	1,340	0.08
Phosphorus	67	474	0.15
Manganese	2.5	1.6	1.6
Iron	10.3	11.4	0.9
Copper	0.49	3.6	0.14

In a range of temperate and tropical forests, the proportions of nutrients stored within the plant biomass differ substantially between sites and nutrient elements (Edwards and Grubb, 1982; Lemée; 1978; Bernhard-Reversat *et al.*, 1978; Klinge, 1975; Greenland and Kowal, 1960) (Table IV.8). K is the most labile element and tends to accumulate in plant biomass, with a minimum value of 40 % of the total stock of the ecosystem in a temperate-climate Belgian forest, and a maximum of 88 % in a tropical

Amazonian forest. P, Ca and Mg may also be significantly concentrated in plant biomass. In the Banco forest (Côte d'Ivoire) effectively all of the Ca is found in biomass and litter, and virtually none remains in the highly-weathered soil. Stocks of N in plants and litter expressed as a proportion of that in the soil are much lower than for other elements.

Temperate forests have larger soil nutrient reserves than tropical forests. In the tropics, forests tend to accumulate much larger proportions of nutrients in biomass and litter when they grow on nutrient-poor oxisols than on richer soils. In the former soils, nutrient losses from litter systems are extremely limited because of the conservation mechanisms that ensure synchrony between nutrient release and their uptake by plants.

Table IV.8 General parameters of nutrient cycling in litter systems of temperate and tropical forests (exchangeable bases in cmol kg^{-1})

Temperate forests						
Belgium (Duvigneaud, 1971)	Dry weights t ha^{-1}	N kg ha^{-1}	P tot kg ha^{-1}	K exch	Ca exch	Mg exch
Above-ground plant parts	316 (total)	712	52	403	1639	5
Below-ground plant parts		143	9.7	75.7	337.2	12.0
Litter	n d	66.0	6.0	27.0	87.0	9.0
Soil (0-30 cm)		6,400	4,500	771	5,900	1003
Plants/Total (%)		12.6	1.5	39.6	25.9	2.5
France (Fontainebleau, forest); Lemée, 1978	Dry weights t ha^{-1}	N kg ha^{-1}	P tot kg ha^{-1}	K exch	Ca exch	Mg exch
Above-ground plant parts	230	1,361	34.3	546.6	1639.2	227.3
Below-ground plant parts	40	220	6	75	284	42
Litter	7.2	147	5	46.4	110	15
Soil (0-30 cm)		7,480	292	982	3,730	340
Plants/Total (%)		17.2	11.9	37.7	33.3	43.2
Lower montane tropical rain forest						
New Guinea Grubb and Edwards, 1982	Dry weights t ha^{-1}	N kg ha^{-1}	P (fluoride sol uble extract)	K exch	Ca exch	Mg exch
Above-ground plant parts	310	683	37	668	1270	187
Below-ground plant parts	40	137	6.4	186	333	61
Litter	20.7	170	10	33	158	28
Soil (0-30 cm)		19,200	16	403	3750	682
Plants/Total (%)		4	63	66	29	26

Table IV.8 (cont.)

Lowland tropical rain forests						
Ivory Coast Bernhard-Reversat, 1978 Banco I	Dry weights t ha^{-1}	N kg ha^{-1}	P exch	K exch	Ca exch	Mg exch
Above-ground plant parts	513	1548	118	711	1427	619
Below-ground plant parts	49					
Litter	11.9	161	9.8	28	57	44
Soil (0-50 cm)		6500	350	80	100	80
Plants/Total (%)		20.8	26.9	90.3	93.7	89.2
Ghana, Greenland and Kowal, 1960	Dry weights t ha^{-1}	N kg ha^{-1}	P tot	K exch	Ca exch	Mg exch
Above-ground plant parts	233	1690	112	753	2,370	320
Below-ground plant parts	54	326	24	143	268	65
Litter	74	264	20	46	569	57
Soil (0-30 cm)		4950	13	649	2,580	295
Plants/Total (%)		28	80	56	46	52
Tropical tree plantation						
Acacia mangium plantation (Congo), Bernhard-Reversat *et al.*, 1993	Dry weight t ha^{-1}	N kg ha^{-1}	P tot	K exch	Ca exch	Mg exch
Above-ground plant parts	117	663	52	58	169	52
Below-ground plant parts	22	169	18	13	21	13
Litter (annual input)	8.6	170	25	18	38	26
Soil (0-10 cm)		600	nd	58	33	e
Plants/Total (%)		52	nd	48	73	71

2.6.3 SYNCHRONY IN THE LITTER SYSTEM

It is generally believed that 'natural grasslands and forests are nutritionally conservative in the sense that nutrients losses are minimal (see also Section IV.1.5.1). The small losses that do occur are balanced by equivalent inputs from the weathering of soil parent materials, from atmospheric deposition or, in the case of nitrogen, from fixation' (Myers *et al.*, 1994). This was illustrated above (Section IV.2.4.2) by the retention of almost all of the Ca and P applied to the surface of an Amazonian spodosol within the thick litter:

root mat layer. In a two-year study conducted in the same region, Grimaldi and Chauvel (unpublished) used lysimeters to assess nutrient losses through leaching. They confirmed the efficient recycling of most ions since only small amounts of Ca, K, Mg, and Si were found at 70 cm depth, and only traces reached the river. Nonetheless, some ions were occasionally more concentrated in leachates sampled during the dry (K^+, NH_4^+) or wet seasons (SO_4^{2-}, SiO_4^{4-}).

In deforested and cultivated areas, the newly-established juvenile ecosystems typically lose substantial amounts of nutrients (Odum, 1983). A great challenge is to restore synchrony through manipulation of the major variables that control decomposition and mineralisation, notably microclimate, litter quality and decomposer communities. In an experimental plot at San Carlos de Rio Negro, Jordan (1989) established a balance of the main nutrients in a traditional slash-and-burn agriculture system (Figure IV.31). He observed: (i) an increase in all soil nutrient stocks after burning that was maintained throughout cultivation; (ii) low nutrient stocks in the crop and successional vegetation in comparison with those contained in the biomass of the original forest; (iii) greater losses from the entire ecosystem than from the soil alone resulting from the replacement of nutrients leached from the soil by nutrients leached down from decomposing remains of the primary forest; (iv) depletion of nutrient stocks from the original forest biomass, including the root mat.

Developing synchrony in agricultural systems

Agricultural systems lack several attributes that are believed to promote synchrony *i.e.*, a large diversity of decomposing resources with different chemical and physical structures; a diverse and abundant community of decomposers, especially invertebrates of the different functional groups (micropredators, litter transformers and ecosystem engineers); an extended and perennial root system able to capture nutrients released within the soil volume. Manipulation of any of these ecosystem components may potentially contribute to the restoration of synchrony.

Mixing organic resources of different qualities. Decomposition rates are determined by the chemical quality of the litter material (especially their concentrations of polyphenols, lignin and nitrogen) and such physical properties as size of the individual litter units and physical strength (see Section IV.1.3.3). It is hypothesised that mixing litters of 'high' and 'low' quality may lead to release patterns that are intermediate between the rapidly and slowly decomposing components. Adjusting the proportions of each material may allow accurate calibration of nutrient release patterns to plant needs, under given conditions of climate, soil and macro-organism activities. Some but not all experiments support this view (*e.g.*, Bandara and Anderson unpublished data in Myers *et al.*, 1994; Becker and Ladha, 1997). In an experiment in which the decomposition of 21 mixtures of litters of different types was monitored, interactions between litter types occurred in 17 cases (McTiernan *et al.*, 1997). In 12 cases, decomposition of the mixture was slower than predicted from a simple additive model. Much still remains to be understood before reliable predictions are possible.

Utilisation efficiency is the proportion of nutrients released that is actually taken up by the crop and is lower for organic than inorganic fertilisers. In three cropping systems in the humid tropics, Sisworo *et al.* (1990) calculated efficiencies of respectively 9 to 40 % for inorganic, and 2 to 27 % for organic fertilisers.

Litter placement. Nutrient release may differ considerably between litter deposited as a surface mulch, or buried within the soil. The superposition of a stratum of high quality material over one of lower quality material may prevent losses by temporarily immobilising nutrients leached from the high-quality material within the underlying lower-quality material.

Manipulation of soil macrofauna and root systems. Experiments have shown that a significantly higher proportion of nutrients mineralised from a decomposing surface mulch may be found in crops in the presence of earthworms (Spain *et al.*, 1992). Mechanisms for this process may involve accelerated burial of litter under surface casts, concentration of earthworm activity in the rhizosphere and modified pathways for water flow within soil.

The spatial distribution and seasonal occurrence of root growth are key components of synchrony. In some cases, nutrients that have been leached into the soil during the early stages of plant growth may be taken up when the rooting system reaches the horizons within which these nutrients are located.

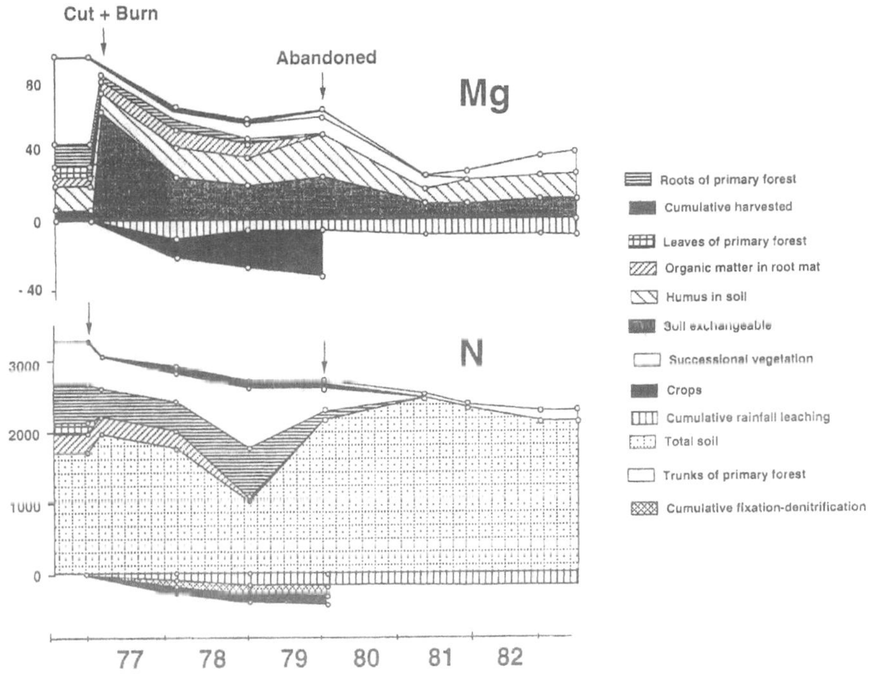

Figure IV.31 Stocks and cumulative losses of magnesium and nitrogen in relation to time after clearing an Amazonian rainforest on an oxisol subjected to 'slash-and-burn' agriculture (San Carlos de Rio Negro, Venezuela) (Jordan, 1989).

IV.3 RHIZOSPHERE

The concept of the rhizosphere was initially propounded by Hiltner (1904) to describe the specific environment comprising the soil-root interface. The rhizosphere may be defined as 'that narrow zone of soil subject to the influence of living roots, as manifested by the leakage or exudation of substances that affect microbial activity' (Curl and Truelove, 1986) or 'the sum of the microsites where interactions, often mutualistic, develop between living roots and micro-organisms' (Bottner and Billes, 1987). The concept has been recently extended to the whole volume of soil that is directly influenced by roots, and includes all the biological, chemical and physical processes associated with soil functioning. This definition is retained in this chapter.

Such interactions may occur within the root (endorhizosphere), outside it (exorhizosphere) or at the surface of the root (rhizoplane). The term 'mycorrhizosphere' has also been used to describe the extended rhizosphere of a root colonised by mycorrhizal fungi (Garbaye, 1991). The rhizosphere may comprise a significant proportion of the soil volume and be an important source of functional diversity while the mycorrhizosphere may form a substantial extension to the rhizosphere, especially in those species with reduced root systems.

Rhizosphere effects are most clearly expressed in the fine-root fraction where exudation is at a maximum. Nonetheless, dead roots are an essential energy source for subterranean organisms and provide another focus for so-called secondary rhizosphere effects.

The rhizosphere is the site of a range of complex interactions between the living and dead soil components which may be grouped into the following three main processes: (i) the extraction of nutrients from the soil solution, organic reserves and from the general and soil atmospheres (N), to be later absorbed by the roots and their associated microflora and fauna; (ii) the production of modifier substances which promote plant growth (hormones and growth factors), mediate the level of competition with other plants or prevent attacks by pathogenic micro-organisms (allelopathy); and (iii) the formation and maintenance of soil macro- and microstructure. In addition, the development of such heterotrophic microbial activities as ammonification, denitrification, methanogenesis and sulphate reduction has great relevance to biogeochemical cycles and environmental issues.

3.1 General characterisation

The rhizosphere system includes the following components: (i) the soil organic matter and root litter, which provide energy and nutrient elements; (ii) living roots as macro-organisms, which promote microbial activity; (iii) free-living micro-organisms dominated by heterotrophic bacteria and such predators as nematodes and protists, (iv) symbiotic micro-organisms including nitrogen-fixing bacteria, mycorrhizal fungi and actinobacteria such as *Frankia* spp.

In terms of function, living roots 'fuel' microbial activity through the production of readily-assimilable substances (exudates) and exfoliated cortical cells. These are released into the immediate environment of the roots in a process known as rhizodeposition; such materials promote microbial activity at the root surface through priming effects.

3.1.1 SPATIAL EXTENSION

3 1.1.1 *Longitudinal characterisation of the rhizosphere*

Interactions between roots and micro-organisms are confined to microsites which differ in the nature of the organic materials produced and in their microfloral and microfaunal communities. The following five main zones are located distally to proximally along the growing root (Balandreau and Knowles, 1978; Rovira *et al.*, 1979; Coleman *et al.*, 1984a; Guckert, 1985): (i) root cap and meristem zone; (ii) elongation zone; (iii) root hair zone; (iv) mature region; and (v) root litter (Figure III.25).

The root tip may be considered as the distal point of the rhizosphere. As described by Rovira *et al.* (1979), it is a 1 to 2 mm zone of intense cell division which comprises a meristematic zone protected by a root cap. Energy is mainly provided as root exudates, *i.e.*, polysaccharides of high molecular weight often comprising chains of galactose and galacturonic acid, bearing blocks of neutral sugars (glucose, galactose, arabinose and fucose) (Balandreau and Knowles, 1978) secreted by the root cap, together with low-molecular weight exudates and sloughed cells from the short-lived tissues of the root cap.

Mucilage was formerly considered as a lubricant that would facilitate penetration and the establishment of a close contact between the soil and the growing root. This region is poorly colonised by micro-organisms. Root-cap cells have short life spans, estimated at one day for maize (Clowes, 1976). These cells actively secrete mucilage from their Golgi apparatus and, each day, approximately 10,000 cells from the root cap of a single maize root tip produce mucilage before being sloughed off. Estimated secretion rates are as high as 300 to 1250 $m^3\ ha^{-1}$ and 3 to 12.5 Mg $ha^{-1}y^{-1}$ (Samtsevich, 1971).

The root elongation zone occurs just behind the root cap and is a microsite several millimetres long where mucilage is secreted together with low molecular-weight exudates. Intense bacterial growth occurs in this zone resulting in catabolism of the exudates and the sloughed cells of the root cap and epidermis. Some bacteria may produce polysaccharide slimes which mix with the mucilage to form mucigel. The microfauna is also abundant with large populations of protists and nematodes; the mesofauna is also concentrated in that zone (Section IV.3.1.2.3).

Root hairs occur in a zone commencing some 10 mm behind the root cap. Low molecular-weight exudates are secreted in this zone and it is also the site of maximum nutrient absorption by the root. The zone of maturation found just proximal to the zone of root hairs is suberised and no longer produces exudates. Arbuscular mycorrhizal fungi colonise roots in this zone.

Secondary roots grow through ruptures in the primary root; the increased exudation associated with such microsites substantially stimulates microbial activity. Further, secondary roots may grow through areas that have already been activated by primary roots thereby greatly enhancing subsequent activity.

3.1.1.2 *Radial extension of the rhizosphere*

Biological effects in the rhizospheres of most plants extend outwards from the root for only a few tens of micrometres in most plants, diminishing with increasing distance from the root surface through the inner and outer zones (Foster and Rovira, 1978).

The rhizoplane is that part of the rhizosphere, approximately 1 µm thick, in contact with the root surface. Microfloral communities are usually dominated by bacteria and in the rhizoplane of *Trifolium subterraneum*, small compact colonies about 5 to 10 µm in diameter sometimes form; they are surrounded by a capsule of polysaccharides different from the mucilage produced by the root. Microbial communities have their greatest diversity (*e.g.*, 11 taxa recognised under the experimental conditions of Foster and Rovira, 1978) and density in the rhizoplane although Kiffer *et al.* (1977) observed that ectomycorrhizal infection of roots significantly depletes bacterial populations in the rhizoplane.

The inner rhizosphere includes the rhizoplane and extends up to 10 µm from the root surface. Community diversity and density are still high in the 5 µm closest to the root surfaces but start to decline steadily in the 5-10 µm layer, with values of one third to one half of those in the rhizoplane.

The outer rhizosphere extends from 10 to 20 µm from the root surface. In the example presented by Foster and Rovira (1978), species richness fell to two taxa and micro-organism density to $13x10^9$ cells cm^{-3} soil, or *ca.* a ninth less than the rhizoplane value ($120x10^9$). Nonetheless, this is still an order of magnitude greater than the highest values for soils external to the rhizosphere ($13x10^9$ cells cm^3 at 15-20µm from the root). Micro-organisms may colonise the root cortex by penetrating the epidermal tissue or entering the superficial root tissues through occasional wounds. Such colonised cells and decaying cortical cells constitute the **endorhizosphere**. Microbial density may be greater in this part of the rhizosphere than in the mucilage, although turnover is much slower (Knowles, 1978).

The zone of rhizosphere soil extends beyond the rhizosphere *sensu stricto* and may be described as the zone in which microbial activity is modified by the diffusion of root products and metabolites. In this zone, extending a few millimetres to centimetres from the root surface, soil structure is modified and such decomposition-resistant materials as cell debris and mucilage may preferentially accumulate, often associated with clays (Breisch *et al.*, 1975; Goss *et al.*, 1984; Callot *et al.*, 1984; Morel *et. al.*, 1991; Misra *et. al*, 1986; Miller and Jastrow, 1990; Dexter, 1991; Pojasok and Kay, 1990). The hyphae of the mycorrhizal fungi associated with the roots may extend for distances of several centimetres from the root surface substantially enlarging the effective volume of the rhizosphere (see Section IV.3.1.2.2).

3.1.1.3 *Micro-environment of the rhizosphere*

Micro-environmental conditions in the rhizosphere may differ from those of the external medium. The atmosphere is enriched in CO_2 by root respiration, clay minerals may accumulate along the root and the movement of water, exudates and ions creates specific conditions. The soil of the rhizosphere is relatively dry and nutrient-poor compared with the soil external to this zone.

Roots can induce pH changes at the root-soil interface and gradients of H^+ concentration in the rhizosphere can be very steep (*e.g.*, Nye, 1981; Marschner and Römheld, 1983; Kotyk *et al.*, 1991; Gahoonia and Nielsen, 1992). Plants maintain neutral electric conditions in the rhizoplane and thus induce unequal uptakes of anions or cations depending on pH of the bulk soil (Hedley *et al.*, 1982). Release of CO_2 through root respiration, enhanced efflux of H^+ resulting from iron deficiency and the secretion of organic acids are other mechanisms whereby roots change the pH of their rhizosphere. The pH is often, but not always, higher in the root tip region than close to root hairs, it becomes more acid when N is supplied as ammonium rather than nitrates and differs greatly between plant species and with soil type (Gijsman, 1990; Marschner and Römheld, 1983; Marschner *et al.*, 1986; Youssef *et al.*, 1989; Dormaar, 1988). Haynes (1990) and Jaillard *et al.* (1996) particularly emphasise the effects of proton (H^+) excretion into the soil solution during cation absorption (the 'proton pump'): where more cations than anions are absorbed, the soil is acidified; where more anions are absorbed, the opposite pattern occurs (Figure IV.32).

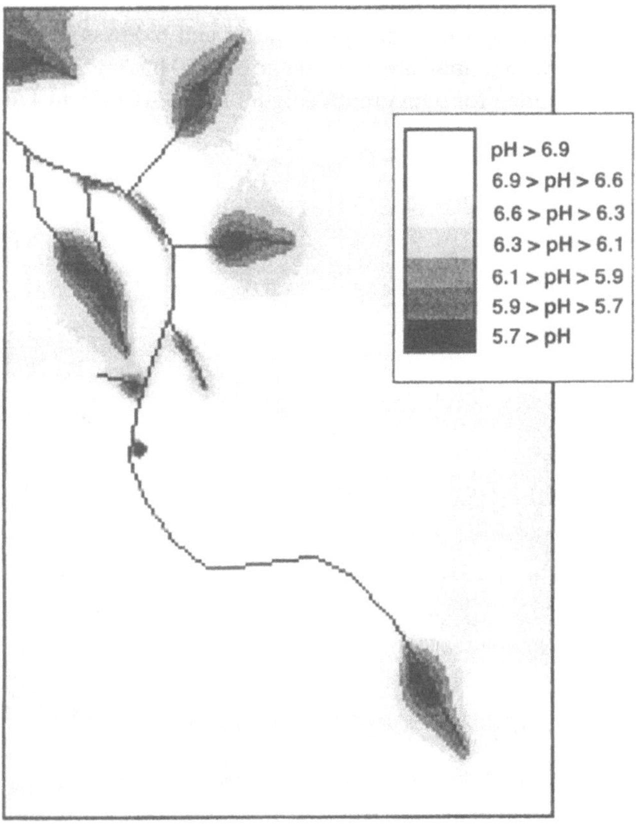

Figure IV.32 Map of pH in the rhizosphere of a Colza (*Brassica napus*) plant grown in a P-deficient soil. Under these growing conditions, root tips acidify the environment whereas the pH approaches neutrality in other parts (Jaillard and Hinsinger, 1993).

3.1.2 RHIZOSPHERE COMMUNITIES

A number of microsites in the rhizosphere are relatively dry and nutrient-poor even though they are well supplied with energy-rich resources that are exploited by large populations of micro-organisms and invertebrates. Furthermore, pH is significantly influenced and roots create channels along which movement is facilitated. As a result, microbial communities are active and numerous, and large populations of saprovores and micropredator invertebrates exploit the root litter and microbial biomass. The ratio R/S (abundance in the rhizosphere to non-rhizosphere soil) is often much greater than one for a wide range of organisms but differs widely with plant species and organisms.

Microbial and faunal activities in the rhizosphere may have positive and negative effects on plants. Beneficial processes include enhanced provision of nitrogen, phosphorus and iron, biocontrol of diseases and deleterious organisms and stabilisation of soil structure (Lynch, 1990).

3.1.2.1 *Non-symbiotic micro-organisms and the micropredator foodweb*

As roots grow, exudates and exfoliates are produced in the vicinity of the root tips leading to substantial increases in microbial numbers and biomass in these microsites. Values of R/S for bacteria and fungi are in the range of 2-20 but maximum values greater than 100 have been recorded for both groups (Figure IV. 33) (Curl and Truelove, 1986).

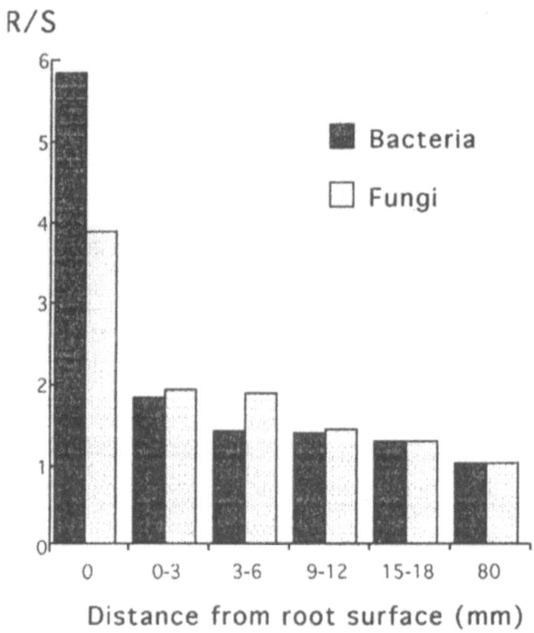

Figure IV.33 Relative abundances of micro-organisms at various distances from the root surface of lupin seedlings grown in unamended Elsinboro sandy loam (after Papavizas and Davey, 1961 in Curl and Truelove, 1986) (R/S: density in rhizoplane/density in soil external to the rhizosphere).

On a given plant, bacterial populations generally respond more rapidly than fungi to root products and have higher R/S ratios. Bacteria with *r* demographic strategies develop initially, followed by *K* selected bacteria; the latter have the capacity to digest the most recalcitrant organic compounds, including cellulose and lignin (de Leij *et al.*, 1994). Maximum populations generally occur in the rhizoplane although, beyond a distance of 3 mm, population densities are little different from those of the extra-rhizosphere soil. Nonetheless, individual populations of micro-organisms may have different distributions, depending on their ecological requirements (Bazin *et al.*, 1990) (Figure IV.34).

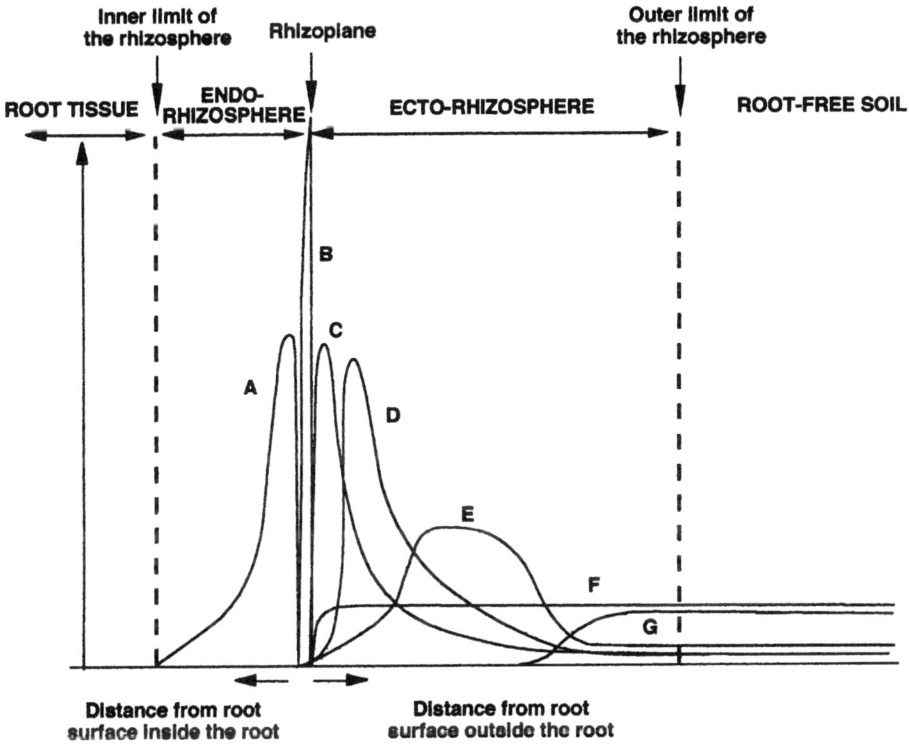

Figure IV.34 Distribution of different types of micro-organisms in the rhizosphere. A: root-penetrating species growing in the intracellular space. B: species growing exclusively in the rhizoplane but not root penetrating. C: physically excluded from the rhizoplane by type B organisms but which can utilise organic materials from both the root and type B micro-organisms. D: organisms physically or biochemically excluded from the root surface by type B and C organisms but living on their metabolites or utilising organic material that type B and C organisms cannot use. E: organisms which utilise secondary products originating from the rhizosphere. F: organisms excluded from niches on or close to the rhizoplane, but not sensitive to the inhibitory products of organism types A to E. G: organisms excluded from the rhizosphere by the activities and inhibitory products of rhizosphere organisms.

Microbial communities of the rhizosphere also differ qualitatively from the non-rhizosphere soil. They appear to possess higher proportions of bacteria and free living fungi requiring amino acids for maximal growth and lower proportions of those requiring nutrients supplied in soil extracts (Curl and Truelove, 1986). Bacteria capable of synthesising growth factors for plants and other micro-organisms, and also toxic metabolites, have been reported to be more abundant in the rhizosphere than in the non-rhizosphere soil. Nonetheless, Curl and Truelove noted that 'presently available methodology precludes elucidation of the true balance between these groups in natural environments'. Rhizosphere communities seem to be less diverse than those external to it and two genera of bacteria and fungi respectively, *Pseudomonas* and *Fusarium* have been reported several times to comprise up to 60-90 % of total rhizosphere populations (Kreutzer, 1972; Vancura, 1980). Mavingui *et al.* (1992) noted a diminished genetic diversity in *Bacillus polymixa* populations in the rhizoplane of wheat as compared to rhizosphere and non-rhizosphere soil.

3.1.2.2 *Symbiotic micro-organisms*

Symbiotic associations between roots and micro-organisms involve mycorrhizal fungi, the Rhizobial bacteria that form nodules on legume roots and the actinobacteria (*Frankia* spp.) associated with the roots of the Casuarinales and a number of other plants.

Mycorrhizal fungi
Frank (1885) was the first to recognise mycorrhizae as associations of mixed plant and fungal origin. These composite organs have three components: the fine roots of their host plants, the fungal structures that interface with the host plant and an often extensive network of external nutrient- and water-absorbing hyphae that explore the surrounding soil environment. In the internal or arbuscular mycorrhizae, the first two components are associated with the root hair regions of plant roots but do not occur on very young or lignified root parts. In the external or ectomycorrhizae, the final order lateral roots may be covered by a dense sheath of fungal hyphae. Mycorrhizal associations are normally mutualistic: they enhance the acquisition of nutrients by the host plant and may provide other benefits (as discussed in Section IV.3.2.1.2) while the fungi involved receive energy in the form of organic carbon compounds.

Mycorrhizae are of two basic types: endomycorrhizae in which the fungus penetrates the cell wall (but not the plasmalemma) and ectomycorrhizae where they do not. Harley (1989) recognised seven different mycorrhizal types and these are listed in Table IV.9 together with information on the host and fungal taxa involved in each and other selected properties of these associations. However, the internal arbuscular (or vesicular-arbuscular) and ectomycorrhizal types are the most widespread and clearly the most important in terms of both agricultural and forestry production, and in their influences on the natural communities within which they occur. It is these two types that is given most prominence here.

The majority of plant species enter into mycorrhizal associations and this is considered to be the primitive condition. Fungi similar to the arbuscular types are thought to have originated 353-362 million years ago (Simon *et al.*, 1993) and this is consistent with the hypothesis that symbiotic fungi were instrumental in the colonisation of land by ancient plants (Pirozynski and Dalpé, 1989).

Table IV.9 The major mycorrhizal types, their fungal symbionts and selected of their properties (Harley, 1989; Smith and Read, 1997)

	Mycorrhizal type						
	AM	**ECM**	**Ectend omycor rhizae**	**Arbutoid**	**Mono-tropoid**	**Ericoid**	**Orchid**
Fungal hyphae septate	-	+	+	+	+	+	+
Fungal hyphae aseptate	+	-	-	-	-	-	-
Hyphae enter cells	+	-	+	+	+	+	+
Fungal sheath	-	+	+ or -	+ or -	+	-	-
Fungal taxa	Z	B, A, Z	B, A	B	B	A	B
Plant taxa	Bryo. Pterido. Gymno. Angio.	Gymno. Angio.	Gymno. Angio.	Ericales	Monotro-paceae	Ericales	Orchidaceae

Key:
i. Fungal divisions: Z, Zygomycota; A, Ascomycota; B, Basidiomycota
ii. Plant taxa: Bryo., Bryophytes; Pterido., Pteridophytes; Gymno., Gymnosperms; Angio., Angiosperms

Plant families may be predominantly ectomycorrhizal or predominantly endomycorrhizal although both forms occur in families such as the Salicaceae (Newman and Reddell, 1987). Even in families considered largely mycorrhizal, species occur that may not form these organs and, conversely, mycorrhizal associations may also occur within families that are generally considered non-mycorrhizal (Tester *et al.*, 1987). Of 200 families of vascular plants from all over the world, 23 have been identified within which mycorrhizal colonisation is generally absent or poorly developed. This last group includes the important families Chenopodiaceae, Brassicaceae and the Proteaceae, the last widely represented in Australia and South Africa (Tester *et al.*, 1987). Trappe (1987) stated that approximately 12 % of the angiosperm species examined so far are facultatively mycorrhizal, 18 % are typically non-mycorrhizal leaving 70 % as probably obligatorily mycorrhizal. Effective mycorrhizal associations are often absent from or much reduced in highly disturbed, aquatic and poorly drained environments.

A large number of fungi are known to enter into mycorrhizal associations with higher plants although by far the greatest diversity occurs in those forming the external or ectomycorrhizal type. Molina *et al.* (1992) estimated the number of fungi forming ectomycorrhizal and ectendomycorrhizal associations at more than 5400 species, of which *ca.* 25 % fruit underground. Most of the fungi entering into these associations belong to the Basidiomycota (*ca.* 92 %) but a number of Ascomycota (*ca.* 8 %) and a few Zygomycota of the genus *Endogone* (Haselwandter and Bowen, 1996) are also involved. These fungi are rarely obligate symbionts since they can grow in the absence of roots; however, most only fruit when symbiotically associated with plant roots (Molina *et al.*, 1992). In contrast to the arbuscular mycorrhizal fungi, ectomycorrhizal fungi exhibit a considerable degree of variation in their host specificities. Some are relatively narrow in their host range and are restricted to single genera of plants while most have much broader host ranges (Molina *et al.*, 1992). Multiple fungi may colonise the roots of a single host plant.

The fungi that form arbuscular endomycorrhizal associations are obligate and probably asexual symbionts belonging to the orders Glomales and Endogonales, tentatively placed in the Zygomycota (Morton and Benny, 1990). More than 160 species are distributed between eight genera in both orders (Walker and Trappe, 1993). Some species are very widespread, none are host specific and most species fruit underground or in the litter layers. As with the fungi forming ectomycorrhizal associations, multiple colonisation of single host plants may occur.

The principal types of mycorrhizae. As indicated above, the two principal types of mycorrhizae are the external or sheathing mycorrhizae and the internal arbuscular types although the ericoid and orchid mycorrhizae are important in particular habitats and in certain plant taxa. The remaining minor types are considered in Smith and Read (1997) (Figure IV.35).

i. Ectomycorrhizae are associations in which hyphae form a dense external hyphal sheath or mantle over the fine lateral roots of their host plants. The hyphae also grow inwards between the cells of the epidermis and outer cortex of the root to form the Hartig network; while the hyphae do not penetrate the host cells, the latter become modified on contact with the network (Smith and Read, 1997). Hyphae also grow outwards from the sheath into the soil to form an external mycelial network. The individual hyphae of this network are small and may penetrate micropores and micro-aggregates inaccessible to root hairs. Such hyphae may ramify very extensively: values ranging from 300 to 8000 $m\ m^{-1}$ of root have been reported across a number of studies (Smith and Read, 1997).

Many species form diffuse or anastomosing hyphal networks although more specific structures are also known. Such external hyphal structures include hyphal mats and the circular 'fairy ring' distributions of hyphae and fruiting bodies also found in saprophytic species. The former comprise dominant, single-species aggregations of hyphae that locally dominate the microbial biomass, often at the interface of the mineral soil and the litter layer. They are known to secrete organic acids such as oxalate which act as chelating agents and are active in mineral weathering and in solublising nutrient and other elements. These mats are therefore of nutritional and pedological significance (Griffiths *et al.*, 1994).

Rhizomorphs are long-distance compound conducting structures composed of many hyphae that may extend for considerable distances from the root cylinder. Their overall diameter may exceed 100 µm. Molina *et al.* (1992) report that *Pisolithus tinctorius* will sometimes fruit on a road shoulder on the opposite side of the road from its host tree and be joined to it only by rhizomorphs.

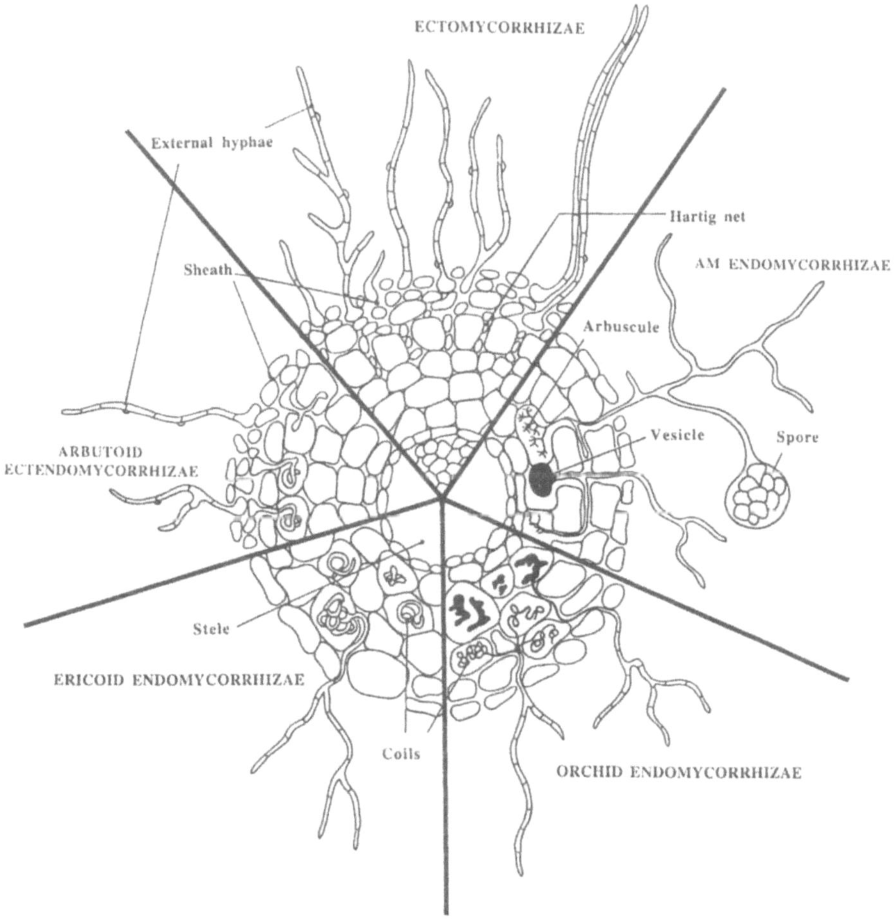

Figure IV.35 Schematic representation of mycorrhizal types (Chalot *et al.*, 1988 after Le Tacon, 1985). Upper part, ectomycorrhizae characterised by absence of penetration of root cells by fungal hyphae and the formation of a sheath around the root; reproductive organs of these fungi may be found at the soil surface. The lower part of the figure shows three types of endomycorrhizae and an intermediate type, ectendomycorrhizae (to the left). From the right, 1. endomycorrhizae with vesicles and arbuscules (AM) are the most widespread. Arbuscules are always intracellular; vesicles may be inside or outside cells; 2. Orchidaceous endomycorrhizae form coils inside the cells; 3. other endomycorrhizae.

The ectomycorrhizal mantle is a dense mass of fungal hyphae that may partly or wholly cover the final order terminal lateral roots. It may comprise 20-40 % of the mass of the colonised length of the root, not including the mass of the Hartig network within the root (Smith and Read, 1997).

The fungi that form ectomycorrhizal associations include a number of well-known edible species belonging to such common genera as *Boletus*, *Cantharellus*, *Tricholoma*, *Russula* and the important truffle fungi *Tuber melanosporum* and *Tuber magnatum*

(Ascomycota) which are able to grow in the absence of a root host, but do not produce sporocarps (Le Tacon, 1985). Sporocarps are an important dietary item for small mammals in different parts of the world; the contained spores largely survive passage through their intestinal tracts and the mammals thereby act as spore vectors for these fungi (see, for example, Maser *et al.*, 1978; Bennett and Baxter, 1989). Annual production may be substantial (*e.g.*, up to 523 kg ha^{-1} y^{-1} in an Australian *Eucalyptus* forest, Johnson, 1994) and ecologically important to wildlife.

Ectomycorrhizae mainly occur in association with trees, largely of the families Pinaceae, Betulaceae, Fagaceae, Myrtaceae, Dipterocarpaceae and some Leguminosae, (Newman and Reddell, 1987) together with a number of shrubs and a few herbs. Perhaps only 3 % of plant species enter into ectomycorrhizal associations but they include many of the species used in the commercial production of wood (Meyer, 1973) and may dominate the sites within which they occur (Read, 1991).

ii. Unlike ectomycorrhizae, arbuscular mycorrhizae are internal structures and are not visible to the naked eye. This form of association is particularly widespread and involves *ca.* 80 % of plant genera, including the agriculturally-important families Gramineae, Leguminosae, Solanaceae and Compositae. It occurs across a very wide range of plant taxa including ferns, lycopods, bryophytes as well as the angiosperms and gymnosperms. Arbuscular mycorrhizae are particularly common in herbaceous plants but are also associated with a range of trees and shrubs, including many plant species of economic interest, such as *Hevea*, *Prunus*, *Citrus*, *Fraxinus*, *Acer*, *Sequoia* and *Podocarpus*.

The fungi that form arbuscular mycorrhizal associations are obligate, probably asexual symbionts with very broad host ranges. The three components of this type of mycorrhiza are the plant root, the structures internal to the plant root (internal inter- and intracellular hyphae, arbuscules or organs of interchange and, in some groups, storage organs, the vesicles) and the network of hyphae (2-27 μm in diameter, Smith and Read, 1997) that extend beyond the plant root.

The fungal hyphae grow inside the root and penetrate the cell walls (although not the plasmalemma) to form the microscopic arbuscules which increase the area in contact with the host. The plant reacts by developing specific fibrillar structures at the point of contact with the fungus; the cytoplasmic volume is increased, nuclei are enlarged and starch reserves are reduced (Dexheimer *et al.*, 1985). Arbuscules are short-lived structures and are digested by the host after periods ranging from a few days to a few weeks. Where vesicles occur, they become filled with lipid droplets during maturation. Degraded fungal structures appear to be progressively replaced by new ones and the host may absorb the mineral nutrients released through degradation of the fungal cytoplasm (Kinden and Brown, 1976).

The fungi associated with these mycorrhizae develop an important network of both thicker runner hyphae and smaller absorbing hyphae in the soil, although they are less extensive than those of the ectomycorrhizal association. Estimated field values range from 0.03 to 13.9 m per gram of soil and up to 46 metres per metre of colonised root, with higher values obtained from glasshouse studies (Smith and Read, 1997). Such external hyphae markedly increase the soil volume explored over that which would be explored by the uncolonised root system (Harley, 1989).

iii. Less widespread types. Ericoid mycorrhizae are endomycorrhizal associations formed largely by ascomycote fungi with plants from a number of families within the Ericales, particularly the Ericaceae and the Epacridaceae and common in heathlands throughout the world. These fungi form associations with the fine roots of species in the above families and have significant saprophytic ability consistent with the often infertile soils within which they and their host plants occur.

Ectendomycorrhizae mainly occur in conifers and have many of the properties of ectomycorrhizae. They are characterised by the presence of an often-reduced external fungal sheath and a well-developed Hartig net of internal hyphae. In contrast to the ectomycorrhizal associations, the internal hyphae of these fungi penetrate the cells of the host plant. The term ectendomycorrhiza has no functional significance and certain single fungal strains may form ectencomycorrhizae on one host and ectomycorrhizae on another (Smith and Read, 1997).

Orchid mycorrhizae. All orchids are achlorophyllous when young and some remain so throughout life. Orchid seeds are very small - some 50-400 μm in length (Arditti and Ghani, 2000) (See also Table I.10) - and the young orchid is completely dependent on colonisation by a compatible mycorrhizal fungus for development. Adult orchids, both chlorophyllous and achlorophyllous, usually have mycorrhizal roots or tubers and the terrestrial forms may be very heavily colonised (Smith and Read, 1997).

While a number of fungi are associated with orchid roots, it is the *Rhizoctonia* species (Basidiomycota) that are best known. Most fungi are fast-growing saprophytes capable of using complex polymers as well as more soluble carbohydrates as energy sources. One species, *Rhizoctonia solani*, is a well-known parasite of commercial crops. The relationship between the fungus and the orchid may be unstable since in some cases the fungus may be rejected or become parasitic on its host.

During the heterotrophic stage of orchids that later become chlorophyllous and for the life of the achlorophyllous species, carbon and other nutrient elements are supplied by the fungus to the host orchid either from decaying plant materials or through parasitism in the epiphytic species. It is therefore difficult to understand what benefit is received by the fungus and these relationships may not be mutualistic.

The geographic distributions of mycorrhizae. The different types of mycorrhizae are not distributed at random over the terrestrial surface of the earth. At the biome scale, different vegetation types are dominated by particular mycorrhizal associations. Thus, the arctic tundra is dominated by ericoid associations and the northern conifer forests by ectomycorrhizal associations. Arbuscular associations are particularly widespread in many tropical rainforests, savannas and other environments although the tropical areas dominated by the largely ectomycorrhizal Dipterocarpaceae and *Eucalyptus* species (Myrtaceae) are a clear exception to this. Ericoid mycorrhizae dominate the arctic tundra regions, alpine and other heathlands (Read, 1991).

Symbiotic and free-living N-fixing bacteria

Micro-organisms capable of N fixation are distributed widely between the Eubacteria and the Archaea and the capacity to fix N has been considered as an ancestral attribute, although many species have subsequently lost this capacity (Young, 1992).

Some N-fixing species are free-living in the soil and other terrestrial environments, including certain bacteria within the genera *Clostridium, Azotobacter, Beijerinckia* and several tens of bacterial species from other genera. Members of a range of groups within the Eubacteria form ecologically - and agronomically - important symbiotic associations with fungi and other organisms within the plant kingdom (Table IV.10). They derive their energy from carbohydrates formed through photosynthesis, and provide their associates with significant amounts of nitrogen.

Table IV.10 Symbiotic N-fixing prokaryotes and selected examples of their hosts (Sprent, 1984; de Faria *et al.*, 1989).

N-fixing prokaryotes		**Plant host**	
		Grouping	**Genera and comments**
Cyanobacteria	*Anabaena*	Pteridophyta	*Azolla*
	largely *Nostoc*, possibly *Anabaena, Calothrix*	Liverworts, Hornworts, Cycadophyta, Angiosperms	all cycad genera, *Gunnera* spp.
Actinobacteria	*Frankia*	Betulaceae	*Alnus, Arctostaphylos*
		Casuarinales	*Casuarina, Allocasuarina, Gymnostoma*
		Coriariaceae	*Colletia, Coriaria*
		Dasticaceae	*Dastica*
		Eleagnaceae	*Eleagnus, Hippophae*
			Sheperdia
			Comptonia
		Myricaceae	*Ceanothus,*
		Rhamnaceae	*Discaria, Trevoa*
			Cercocarpus, Chaemaebatia, Cowania, Dryas
		Rosaceae	*Purshia, Rubus*
Other Eubacteria	*Bradyrhizobium*	Ulmaceae	*Parasponia*
	Rhizobia	Leguminosae	
		sub-family Papilionoidea	97% of species*
		sub-family Mimosoidea	90% of species*
		sub-family Caesalpinoidea	23% of species*
	Non-nodulating forms	Mostly Graminaceous species	rhizosphere organisms and endophytes

* Estimates from de Faria *et al*, 1989.

Within the Eubacteria, the most important groups involved in forming symbiotic associations with plants are species of the genus *Rhizobium* and related genera (known collectively as 'Rhizobia'). These bacteria form close symbiotic associations almost entirely with legumes in which the bacteria inhabit nodules associated with the roots, although not all legumes form such associations. As discussed below, N-fixing associations occur with other plant groups, notably grasses, although they do not involve the formation of nodules. Members of the actinobacterial genus *Frankia* fix N in large nodule-like structures ('actinorhizae') that form on the roots of plants from a number of families and genera. Certain cyanobacteria such as *Anabaena* and *Nostoc* also form N-fixing associations with plants (Table IV.10). Rhizobia and *Frankia* only form nodulating N-fixing associations with the relatively primitive dicotyledons. Such associations do not appear to occur in more recent dicotyledons and monocotyledons (Dommergues *et al.*, 1985).

The 'Rhizobia' are symbiotic bacteria that initiate and occupy nodules on the roots of many legumes. Dinitrogen from the atmosphere is reduced by nitrogenase enzymes (N-fixation) into forms available to the host. Rhizobia belong to four genera: *Rhizobium, Bradyrhizobium, Sinorhizobium* and *Azorhizobium* (Martinez-Romero, 1994) including approximately 13 species. Although they are widespread in soils, indigenous Rhizobia may not ensure effective nodulation and N-fixation, because of the specificity of many host legumes (Lie *et al.*, 1987). *Bradyrhizobium* forms root nodules in an efficient N-fixing symbiosis with pioneer species of the genus *Parasponia* (Ulmaceae) which is distributed from the Malay Archipelago to Polynesia (Soepadmo, 1977); this is the only known case of nodulation with a non-legume. The substantial N fixation abilities of this symbiosis provides these pioneer plants with a capacity to invade disturbed sites effectively (Trinick and Hadobas, 1988).

Nodulation can be impeded by low pH, Al toxicity, nutrient deficiencies, salinity or genetic incompatibility with the host (Dart, 1974; Munns, 1978; Singleton and Bohlool, 1983). Rhizobia are facultative symbionts with some saprophytic ability and can survive in soils and proliferate within the rhizospheres of both host and non-host plants (Giller and Wilson, 1991). Rhizobia are stimulated by the presence of the host plant and this leads to infection, nodulation and nodule senescence, a series of events which serves to maintain or increase rhizobial populations (Woomer and Bohlool, 1989). Rhizobia enter the roots of the host plant through the root hairs and through cracks in the epidermis (Giller and Wilson, 1991).

In commercial environments, the legume-rhizobia symbiosis is ensured before planting by applying inoculum to the exterior of seeds as liquid formulations or pellets (FAO, 1984). Nitrogen fixation rates as high as 584 kg ha^{-1} $year^{-1}$ have been reported for Leucaena (*Leucaena leucocephala*) and 354 kg ha^{-1} $year^{-1}$ for Cowpea (*Vigna unguiculata*) (FAO, 1984); however, the more usual range of rates reported lies between 50 and 200 kg ha^{-1} $year^{-1}$ (see Giller and Wilson, 1991).

N fixing Actinobacteria belonging to the genus *Frankia* form so-called actinorhizal nodules on roots of about 200 plant species belonging to 29 genera and eight families (Table IV.10) (Huss-Danell, 1997). *Casuarina* and *Alnus* are well-known examples of plant genera with species infected by *Frankia*. Nodules may remain active for several years and grow to more than 20 cm in diameter. N-fixation in these associations may be substantial: for example, Dommergues (1997) reports that *Casuarina equisetifolia* may fix from 15 to 94 kg ha y^{-1} although this high potential may be limited by unfavourable soil, climatic and other factors. N-fixation rates are also highly variable between species and even between clones of such host plants as *Casuarina equisetifolia* grown in West Africa (Sougoufara *et al.*, 1987).

A number of symbiotic N-fixing associations occur between non-leguminous plants and bacteria of the genera *Azospirillum*, *Azotobacter*, *Acetobacter*, *Azoarcus* and others (Baldani *et al.*, 1997) and more are likely to be found. These associations have been mainly described in graminaceous species although they also occur in palms and other woody taxa; some are highly specific to particular hosts. They do not involve the formation of nodules.

Three basic groups of non-nodulating N-fixing associations are known, depending where the association occurs in the plant and on the properties of the organisms involved. The rhizosphere organisms survive well in soil and colonise the root surface. An example is the highly specific association between *Azotobacter paspali* and *Paspalum notatum* cv batatais which, on an areal basis, has been estimated to fix *ca.* 20 kg ha^{-1} y^{-1} (Boddey *et al.*, 1983). Endophytes are organisms that may colonise the internal tissues of their host plants. The facultative endophytes survive well in soils and colonise the rhizosphere and root interior; they are mainly *Azospirillum* species. Yield increases from these associations are undoubted although the modes of action have been questioned. Apart from N-fixation, yield increases may also be due to the stimulation of plant growth by phytohormone production or an improved capacity of the plant to metabolise assimilated N. Obligate endophytes do not survive well in soils and colonise the internal plant tissues, including roots, stems and leaves. Their hosts include many plants of economic importance, including sugarcane and other graminaceous species, coffee and oil palms.

Fixation in non-leguminous plants may make an important contribution to total N inputs in tropical grasslands, as shown above (*e.g.*, Robertson and Rosswall, 1986). In Guinean savannas at Lamto (Côte d'Ivoire), Balandreau and Villemin (1973) estimated non-symbiotic N fixation at 12 kg ha^{-1} $year^{-1}$. However, this estimate may be an underestimate since it was based on the acetylene reduction technique (Boddey, 1987; Abbadie *et al.*, 1992).

Heterocystous cyanobacteria form N-fixing associations with a diverse range of fungi and members of the plant kingdom. The cyanobacteria involved appear to belong largely to the genera *Nostoc*, *Anabaena* and possibly to *Calothrix* and some others. These organisms enter into N-fixing associations with fungi to form certain types of lichens and with liverworts (Hepatophyta) and hornworts (Antherophyta) (Rodgers and Stewart, 1977; Stewart and Rodgers, 1977). They also associate with the *ca.* 135 species of primitive Gymnosperms of the Cycadophyta (Grobbelaar, 1993). *Anabaena* enters into N-fixing associations with the floating fern *Azolla* and this is of importance in the N-nutrition of wetland rice crops where it supplies an average of 15 Mg ha^{-1} biomass or 30 kg ha^{-1} N to the rice crop as a green manure (Roger, 1995). *Nostoc* also enters into an unusual intra-cellular association with the *ca.* 50 known species of the largely tropical and southern-hemisphere herbaceous angiosperm genus *Gunnera* (family Haloragaceae) (Bergman *et al.*, 1992a). In the cycads, *Nostoc* is the most common symbiont and, whereas mixtures of several strains can associate with a given plant host, single strains appear to predominate in individual hosts.

N-fixation may be significant on an areal basis in these associations: amounts up to 18 kg ha^{-1} y^{-1} have been reported for cycad-cyanobacterial associations in Australian *Eucalyptus*-dominated forests (Grobbelaar, 1993).

There is broad scope to manipulate N-fixing micro-organisms since natural selection has often resulted in strains with an ability to survive under specific environmental conditions, rather than fix N efficiently (Bowen, 1980). Considerable research efforts have and are still being made to develop appropriate technologies that can be used at a field scale.

3.1.2.3 *Faunal communities*

Roots attract a large number of soil invertebrates. While species belonging to the macrofauna are too large to live in the rhizosphere and specifically adapt to this environment, certain species can exploit roots as a food resource and become serious pests. Interestingly, there seems to be a rather low proportion of macro-invertebrates that are adapted to feed on either living or dead roots (*e.g.*, Athias *et al.*, 1974) and only six of the 26 insect orders are represented by active below-ground herbivores (Brown and Gange, 1990). Certain taxa including Coleoptera larvae, Gryllotalpidae (Orthoptera) or earthworms may ingest roots and future research using adequate methods may demonstrate that root consumption is more widespread than currently known (Baylis *et al.*, 1986; Villalobos and Lavelle, 1990). There is also some evidence that endogeic earthworms might, at least in some circumstances, preferably ingest rhizosphere soil and feed on exudates and root litter (*e.g.*, Spain *et al.*, 1990).

Microfauna and, to a lesser extent mesofauna, are specific inhabitants of the rhizosphere. Amoebae are common components of the rhizosphere where they feed on bacteria and fungi (*e.g.*, Darbyshire and Greaves, 1967; Stout and Heal, 1967; Ingham, 1981; Chakraborty *et al.*, 1983). They live close to aggregates and can produce long pseudopodia that penetrate micropores to catch their bacterial prey (Darbyshire *et al.*, 1989; Foster and Dormaar, 1991). Flagellates and ciliates are also found in the rhizosphere, although in lesser quantities. Free-living nematodes and micro-arthropods predate on protists and may comprise a further level in the foodweb (Elliott *et al.*, 1979; Moore, 1988; Elliott *et al.*, 1988).

In general terms, the relative concentration of fauna in the rhizosphere varies inversely with the soil organic matter content.

Root parasitic fauna

Nematodes. A variable percentage (up to 83 %) of soil nematodes are root feeders (see Chapter III.4.1.2). Environmental disturbance and the presence of host plants generally stimulate the activity of this functional group (Ferris, 1993).

Complex biochemical interactions occur between phytoparasitic nematodes that are attracted to potential food supplies by substances released into the rhizosphere. These include volatile and gaseous compounds (CO_2, ethylene), soluble and diffusible substances such as amino acids, aliphatic and organic acids, salts, ions and mucigel (Cohn and Spiegel, 1991). Egg hatching is further stimulated by specific substances produced in the rhizosphere (Jones, 1959; reviewed in Curl and Truelove, 1986; Norton,

1978; Yeates, 1981; Bilgrami *et al.*, 1985). Although these substances have not been clearly identified, they are known to be water-soluble and dialysable (Blake, 1962). The close relationship between the abundance of some nematode species and plant distributions clearly demonstrates their attraction to the rhizosphere (Nusbaum and Ferris, 1973; Rossi *et al.*, 1996) (Figure III.31).

Parasitic nematodes preferentially attack roots in the elongation zone, young lateral apices and through tissue injuries inflicted by other agencies (Wyss, 1981). Some nematodes are ectoparasites and feed by inserting their stylets into the root while others are endoparasitic and most or all of their body may be located within the root. Plants react in three different ways to nematode attacks: (i) by production of a local swelling and 'stubby root' symptom and the immediate suppression of root growth; (ii) by cell necrosis at sites where wounds and lesions have occurred; and (iii) through the formation of root galls resulting from cell hypertrophy.

Evolution in nematodes has tended to result in a sedentary life-style, progressing from ecto- to endoparasitism, with an increasing emphasis on female hypertrophy, enhanced reproductive capacity and improved mechanisms of egg protection (Figure IV.36).

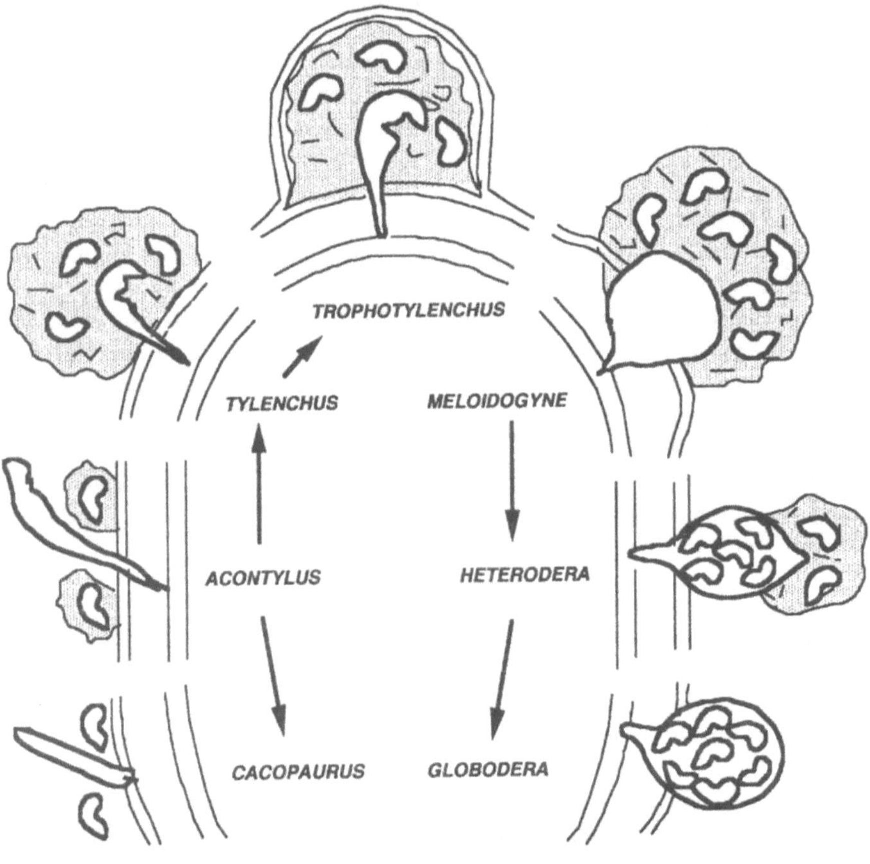

Figure IV.36 Sedentary nematode parasites of roots: progressive trends toward feeding on internal root tissues, increased female obesity, enhanced egg production, and improved mechanisms for egg protection (Cohn and Speigel, 1991).

In response to this evolutionary trend, plants have developed progressively more active cell responses, including the formation of specialised *transfer* cells induced in the host tissue by the nematode to sustain its feeding for long periods of time (Wyss, 1981).

Once nematode populations have developed at a site where a particular host plant has been cropped, they may survive as resistant cysts for periods up to 10 years, as in the genus *Heterodera*. Population density generally increases steadily for five years and then stabilises. If the abundance attained at that stage is detrimental to crop yields, the usual recommendation is to cease growing the particular crop and leave the soil to fallow, or to cultivate non-susceptible crops until the cysts are eliminated. Certain gall-forming nematodes such as *Meloidogyne sp.* may significantly reduce crop production after only two or three years.

Free-living micro- and mesofauna in the rhizosphere

Microfoodwebs comprising the microflora, a number of microflora predators and their own next-level predators are well developed in the rhizosphere where they play specific roles in mediating plant nutrient uptake (see Section IV.3.2.1.1). They comprise protists (ciliates, amoebae and flagellates), a range of root-parasitic nematodes and Acari, or mites.

Protists and free living nematodes are significantly concentrated within the rhizosphere. Rouatt *et al.* (1960) reported a two-fold increase of protozoan population density in the wheat rhizosphere, and more evidence of such concentrations have been provided by Darbyshire and Greaves (1967), Clarholm (1989), Foster and Dormaar (1991) and others. Darbyshire and Greaves (1967) estimated populations of 3000 to 14,000 amoebae g^{-1} soil in the rhizosphere of *Lolium perenne*, with no evidence for the selection of particular species in the rhizosphere as compared with root-free soil. In a microcosm study, Clarholm (1985) noted a three-fold increase in Ciliates, a 2.4-fold increase in flagellates and a 25-fold decrease in amoebae in the wheat rhizosphere as compared with a soil with no plants. In a North American short-grass prairie, Ingham (1981) reported that 20 to 40 % of populations of microbial feeding nematodes were concentrated in the rhizosphere although this represents only 3 % of the soil in the upper horizons.

In a laboratory study, Parmelee *et al.* (1993) demonstrated a significant concentration of nematodes (1.6 to 8.2-fold increase) and micro-arthropods (2.2-2.5-fold) in the rhizosphere of coniferous seedlings grown in a mineral soil. Increased micro-arthropod populations were only found where root density was high and no significant effect was observed in the case of seedlings grown in an organic soil. In field studies in a humid African savanna (Lamto, Côte d'Ivoire), nematodes and micro-arthropods were found to be more populous (3.7 and 2.2-fold, respectively) in the soil below grass tussocks than elsewhere (Malcevski, 1978; Athias, 1974).

It is not yet clear whether the concentrations of organisms noted in the rhizosphere occur within specific microsites as proposed by Clarholm (1985) or are distributed throughout the rhizosphere. Microscopic observations by Foster (1981), and Foster and Dormaar (1991) offer support to the hypothesis that bacterial populations initially develop in the root cap zone where mucilage is produced, and then as protists start to feed on these bacteria, they move into the space freed by the bacterial consumption of exudates.

Macro-invertebrates: root and exudate grazers and litter feeders

Rhizophagy is not a common feeding habit in soil macro-invertebrates and only a limited number of studies have been conducted on the abundance and impact of rhizophagous organisms on root production in the absence of severe attack on roots; such studies are particularly rare in natural ecosystems. In most cases, estimates are based on assumed feeding regimes. In the humid savannas of Lamto (Côte d'Ivoire), rhizophagous invertebrates only comprise 0.3 % of the total biomass of communities (Lavelle, 1983a). Higher estimates have been made for a range of diverse ecosystems: 3.8 % of biomass in forest patches in tropical Mexico and 40.3 % in adjacent pastures where larval Coleoptera may become serious pests (Lavelle *et al.*, 1981); 2.55 g dry weight m^{-2} (23.6 % of biomass) in a Swedish grassland soil (Lohm and Persson, 1977). Scott *et al.* (1979) estimated the total consumption of underground parts by soil organisms at 7-26 % of root biomass. In agro-ecosystems where organic resources are scarce, relative consumption is generally enhanced.

Rhizophagous groups include a large number of larval Coleoptera and Diptera, Myriapoda, Homoptera, some Acari, and a number of other micro-arthropods. Major groups of soil invertebrates like termites, ants or earthworms do not seem to feed on living roots to a significant extent. Baylis *et al.* (1986), however provide some isotopic evidence for root feeding in temperate earthworms. Spain *et al.* (1990) using natural abundance of ^{13}C in biomass of the earthworm *Pontoscolex corethrurus* in a sugarcane plantation concluded that these worms feed partly on fresh sugarcane root material which may comprise exudates and dead or living root biomass. However, as indicated in Chapter I, root litter is not a good quality resource for geophagous earthworms. The endogeic tropical earthworm *Millsonia anomala* was fed soil mixed with 1 % organic matter of different origins. When fed fresh and two-week-decomposed root litter of savanna Gramineae, *M. anomala* had a lower growth rate than in the control 0-10 cm soil (Lavelle *et al.*, 1989). In this case, root material had high concentrations of secondary compounds that were detrimental to growth.

In an experiment where clover (*Trifolium repens*) had been labelled with ^{32}P *in situ* for six hours on six different occasions, a number of soil invertebrates became radioactive, thus demonstrating feeding on living roots. An unexpected 22.9 % of earthworms appeared to feed on live roots. Other root feeders included larval Coleoptera (23.1 %), 'heat-extracted Annelids' *i.e.*, Enchytraeidae plus earthworms (13.8 %), Collembola (0.9 %) and nematodes (0.3 %), whereas Acari, adult Coleoptera and Araneae did not show any labelling (Baylis *et al.*, 1986). Observations by Piearce (1978) and Ferrière (1980) on the gut contents of lumbricid earthworms occasionally report the occurrence of root debris in guts of some temperate Lumbricidae. In tropical soils, specialised root feeding species have not yet been found. However, root litter is the major source of young organic matter in a number of ecosystems and sapro-rhizophages comprise a large proportion of the endogeic invertebrates considered to be geophagous or humivorous.

A wide taxonomic spectrum of arthropods may achieve crop pest status through their severe effects on crop yields. They include Myriapoda (Chilopoda, Diplopoda and Symphyla), Acari, Collembola, Hemiptera, Orthoptera (Gryllotalpidae), Coleoptera, Diptera, Lepidoptera and some Isoptera (termites). Calculated yield losses from these pests may range from less than 10 % up to 40-50 % (Schwartz and Classen, 1981).

3.1.3 CLASSIFICATION OF RHIZOSPHERES

Although relatively little is yet known about rhizosphere processes, it is likely that different types of rhizosphere will be identified when more plant species are considered. For example, in the rhizosphere of six different herbaceous plants in temperate-climate soils grown in two different German soils, enzymatic activities, sugar and amino-acid contents and CO_2-evolution differed significantly, indicating clear differences in the rhizosphere environment and in function (Bachman and Kinzel, 1992). Rhizosphere characteristics also differed between soil types (Figure IV.37).

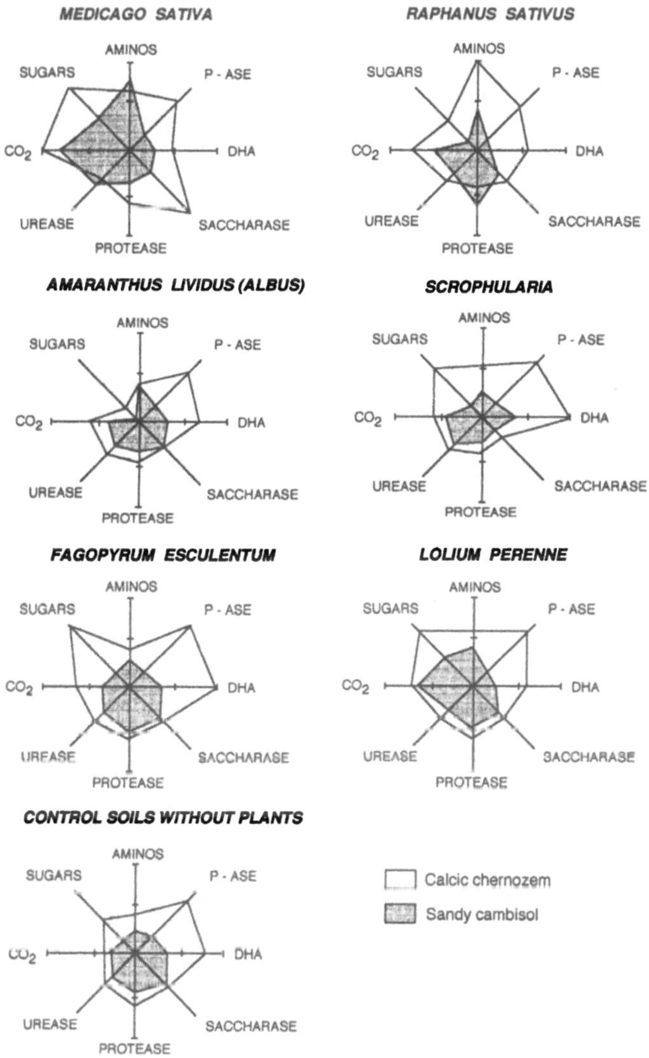

Figure IV.37 Star plots of the measured parameters-influences of the plants on their associated soils. Sugars and amino acids: µmol^{-1} soil volume; enzyme activities and CO_2-evolution: mg product h^{-1} 200 ml^{-1} soil volume. P-ASE: phosphatase; DHA: dehydrogenase (Bachman and Kinzel, 1992). AMINOS: amino-acids.

Perennial plants in natural ecosystems are likely to differ markedly from the cultivated annual species which have been mainly studied to date. There is good evidence that the patterns of root distribution and nutrient acquisition strategies differ between species. Such differences extend to the chemical properties of their exudates and mucilages and to their microbial and microfaunal communities. It is likely that the rhizosphere processes operating in species that are unlike in such fundamental ways, will also differ and provide a basis for their functional classification.

Within a single plant species, phenological and seasonal climatic change may also lead to a different balance of processes as root activity varies. Billes and Bottner (1981) distinguished two separate effects in the rhizosphere of wheat comprising:
(i) living rhizosphere effects from germination to flowering and characterised by an active production of exudates and mineralisation of nitrogen from soil organic matter; and (ii) dead rhizosphere effects that occur during root senescence and decomposition, with nitrogen immobilisation occurring during decomposition of dead roots with high C:N ratios.

3.2 Rhizosphere processes

3.2.1 NUTRIENT UPTAKE FROM THE SOIL

Nutrient uptake by roots is not normally constrained by their capacity to absorb nutrients, but rather by their limited capacity to explore sufficient volumes of the dilute soil solution, and to mobilise mineral nutrients associated with the soil solid phase. Two general processes allow roots to use nutrient sources other than those contained in the soil solution they make contact with. These are: (i) the rhizosphere interactions that permit nutrient mobilisation from mineral pools and soil organic matter reserves and (ii) the exploitation of plant nutrients in soil solutions distant from the roots through associations with mycorrhizal fungi.

3.2.1.1 *Mobilisation of nutrients from soil organic matter: the role of microfoodwebs*

The priming of microfloral activity in the rhizosphere by root exudates results in the mineralisation and assimilation of nutrients from the soil organic matter pool by free-living micro-organisms (Trofymow and Coleman, 1982; Clarholm, 1983). Death of these micro-organisms may result from energy shortages, predation by protists and nematodes or attack by bacteriophage viruses leading to the transfer of some portion of their contained nutrients to plants. These processes operate at single microsites, in which microfloral and microfaunal communities progressively alter during root growth (Figure IV.38). The mucilage produced in the root-cap region alters the soil physical structure and the energy available at the microsite:

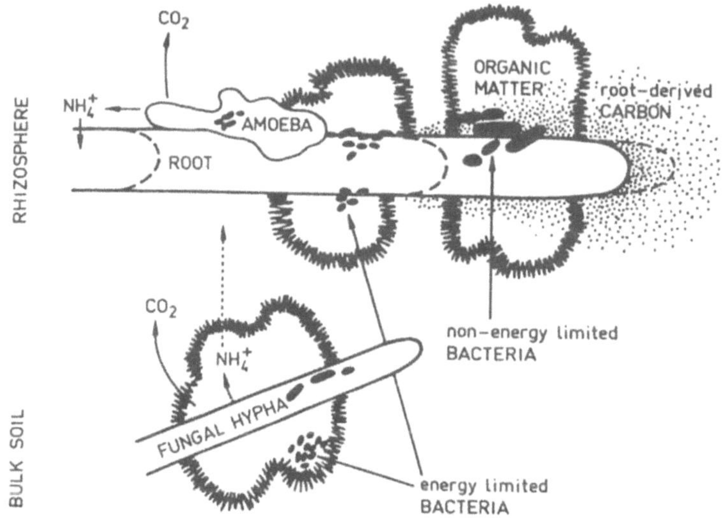

Figure IV.38 Model of interactions in the rhizosphere and in the bulk soil. A root is growing through the soil from left to right. Under the influence of root-derived carbon (dots), bacteria on organic matter are temporarily not energy limited and start to mineralise nitrogen from the organic matter and grow; they are consumed by naked amoebae that are attracted to the site. In digesting the bacteria, the protists release part of the bacterial nitrogen as ammonium at the root surface, where it can be taken up by the root. Below the root, in the bulk soil, a fungal hypha is decomposing organic material. Ammonium will be released as a waste product and can diffuse towards the root as ammonium or, after nitrification, as NO^-_3 (Clarholm, 1983).

(i) Aggregates are disrupted by the penetration of mucilage and organo-mineral linkages are established between clay platelets and the gel matrix and also with organic particles detached from the mucilage sheath (Breisch *et al.*, 1975; Morel *et al.*, 1991). Such linkages occur in the absence of microbial processes and are important features in the development of rhizosphere soil and soil-particle aggregation (Lespinat *et al.* 1975; Guckert, 1985; Gouzou *et al.*, 1993).

(ii) Micro-organisms encountered by the growing root tip become enclosed in the mucilage and those unable to lyse this material remain inactive. At the margins of the mucilaginous slime, bacteria are attracted by chemotactic stimuli (Mandimba *et al.*, 1986). However, this response varies with the plant species, bacterial species and strain, thus demonstrating a degree of specificity in the establishment of plant-bacteria associations (Guckert, 1985). Fungal spores have been shown to germinate at distances up to 10 mm from the root (Stanghellini and Hancock, 1971; Short and Lacy, 1974). Mucilage therefore acts as a selective filter for microbial communities, and may protect the root from massive colonisation by micro-organisms, particularly those of pathogenic habits (Samtsevich, 1971; Harris and Hartley, 1976).

(iii) The micro-organisms responding to the chemotactic stimuli start to lyse the mucilage and soon establish microcolonies within it. The generation times of these bacteria may be shortened to as little as 7 % of those measured in non-rhizosphere soil, although large differences occur between microbial species (Bowen and Rovira, 1976). The presence of these bacteria stimulates root exudation which may be twice as intense as that in a sterile environment (Barber and Martin, 1976; Prikryl and Vancura, 1980; Heulin *et al.*, 1987). Microbial activity is therefore not limited by available carbon since the concentration of available C is several times higher in the rhizoplane than in the soil beyond the rhizosphere (Cheng *et al.*, 1994). Again, different strains and species of bacteria have different capacities to stimulate exudation.

(iv) As root growth continues, full colony development now occurs in the elongation zone of the root. At this stage, many microcolonies, 5 to 10 μm in diameter have been observed, closely packed in the rhizoplane and inner rhizosphere on the roots of *Trifolium subterraneum*. As the average carbon:nitrogen ratio of exudates is approximately 30 (Lynch, 1982), nitrogen is deficient in terms of optimal microbial activity and must be supplemented. It may be supplied through N-fixation, and from the external soil solution through the activity of enzymes diffusing into the mucigel. Comparable activity, although less intense, may also occur in the rhizosphere soil.

Micro-organisms appear to be able to partly digest the organic matter of the rhizosphere soil by priming microfloral activity with mucilage. In laboratory experiments using ^{13}C labelled mucilage, Mary *et al.* (1992) measured the priming effect as the increased soil organic matter mineralised due to the addition of the mucilage (Figure IV.40).

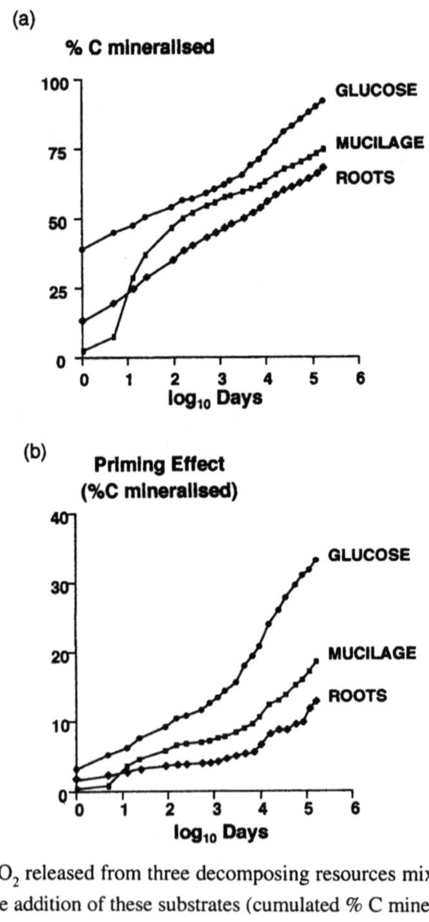

Figure IV.39 Kinetics of CO_2 released from three decomposing resources mixed with a mineral soil (a) and priming effect induced by the addition of these substrates (cumulated % C mineralised (b) (Mary *et al.*, 1992).

The priming effect started after an incubation period of one day and after seven days, 40.4 % of the C in the mucilage had been mineralised together with C from soil organic matter equivalent to 5.8 % of the C added as mucilage. In a comparative experiment with glucose, 44.7 % of C from glucose was mineralised, plus an extra mineralisation due to priming on the soil carbon, equivalent to 9.2 % of the C in the added glucose. Dead root material decomposed at a much slower rate (30.8 %) and priming was limited to 3.5 % of C introduced as root material.

Beyond these general results, an interesting feature is the form of the microbial response to the introduction of mucilage: mineralisation is very slow during the first two days and suddenly accelerates to attain maximum values between the third and the sixth days. This contrasts with the addition of glucose and roots that triggers a large and immediate response (Figure IV.39). Such a response may be interpreted as an adaptation to the specific conditions in the rhizosphere: during the few days following deposition of mucilage in the root cap region, the root will have grown and maximum microbial activity will occur in the elongation zone behind the root tip. This microbial activity is therefore likely to lead to nutrient release in close proximity to the root hair zone where absorption will occur.

This priming effect probably occurs throughout the rhizosphere soil, associated with the mucilage. Although the sites of the priming effect are as yet unclear, bacteria accumulated in the rhizosphere mucilage may release enzymes which diffuse to the zone of contact with the rhizosphere soil. They digest the soil organic matter at this location thereby releasing mineral nutrients that will accumulate initially in the microbial biomass. Priming has been clearly demonstrated by Sallih *et al.* (1987) in a pot experiment lasting 700 days. They observed a high microbial biomass in the presence of living roots while there was a 25 % increase in mobilised nitrogen (*i.e.*, mineral + root nitrogen + microbial-biomass nitrogen) in treatments where living roots were continuously present, compared with a control treatment without plants. A large microbial biomass was maintained by the living plant providing clear evidence that the mineralisation of soil organic matter and the subsequent release of mineral nitrogen was enhanced through the priming effect of root exudates (Figure IV.40).

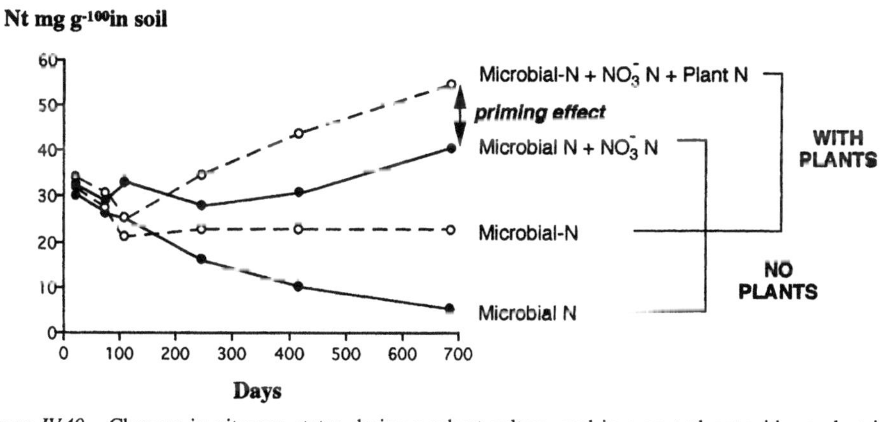

Figure IV.40 Changes in nitrogen status during a wheat culture, and in a control pot with no plant in a 'fersiallitic' soil. Nitrogen accumulated in microbial biomass, mineral-N (NO^-_3 N) and plant biomass. Rhizosphere activity increased the amount of N immobilised in labile (mineral + biomass) forms by *ca.* 50% through a priming effect induced by exudates (Sallih *et al.*, 1987).

(v) At a later stage of root growth, the microsite considered occurs in the region of the root hairs. Since exudation has decreased sharply, bacteria are deprived of nutritional substrates and either die or return to protected, semi-dormant states. At this stage, the slowly-increasing populations of micropredators of bacteria are most active. As a result, part of the nutrients taken up from soil organic matter through priming effects (*e.g.*, ammonium nitrogen) will be released as metabolites and assimilated by plants. Micropredation in food chains has been demonstrated to increase nutrient release, and hence plant production, in microcosm experiments (Ingham, 1981; Anderson *et al.*, 1983c; Setälä, 1990).

3.2.1.2 *Roles of mycorrhizae*

In nearly all environments, the arbuscular and ectomycorrhizal associations are of most significance and much of what is discussed below refers to these two types. As indicated above, dependency on mycorrhizal associations is widespread and the roots of dependent species often have low surface areas, longer life spans and other properties that differ from those not so dependent. Mycorrhizal associations generally lead to improved plant success in both natural and man-influenced environments (Brundrett, 1991; Francis and Read, 1994) although ineffective and semi-parasitic mycorrhizal associations are also known.

Perhaps the major effect of the mycorrhizal associations is to increase the volume of soil explored over that capable of being accessed by the root system alone. This increase may alter the water relationships of the plant and lead to a greater uptake of water. The greater soil volume explored also enhances the uptake of such immobile and slowly diffusing elements as P, NH_4^+, K, Cu and Zn. Functional differences occur between mycorrhizal types and ectomycorrhizal and ericoid mycorrhizal fungi may access organic forms of N and P that are unavailable to arbuscular mycorrhizal fungi, and to uncolonised roots (Marschner and Dell, 1994).

The formation of mycorrhizae may also protect the host plant against root diseases and parasites including bacteria, nematodes and fungi although virus diseases may be more severe in mycorrhizal than non-mycorrhizal plants (Perrin, 1990). Protection may occur in both arbuscular and ectomycorrhizal associations although it is not universal and may be expected to alter with different combinations of pathogens, mycorrhizal fungi and hosts. In ectomycorrhizal associations, protection may result through the provision of a physical barrier between the host plant and the soil environment although it is more likely that antibiosis and host-plant mediated effects are also involved (Smith and Read, 1997). Interactions between pathogens and potential host plants are strongly influenced by plant nutrition.

Mycorrhizal associations influence the absorption and uptake of metals that may occur in potentially toxic concentrations and both increases and decreases in tissue concentrations have been noted in both arbuscular and ectomycorrhizal associations (Haselwandter and Bowen, 1996). The fungal associates of ericoid mycorrhizae are considered to protect their hosts against the potentially toxic concentrations of Al, Zn and Cu common in their often acid environments (Smith and Read, 1997).

The benefits received by the host plants in mycorrhizal associations are counterbalanced by the energy costs involved in supporting the fungal partner. Up to 20 % of current photosynthate may be used to support mycorrhizal structures in both arbuscular and ectomycorrhizal plants (Marschner and Dell, 1994; Smith and Read, 1997).

Mycorrhizal fungi interact both positively and negatively with a wide range of other soil organisms and thereby influence the success of their host plants (Fitter and Garbaye, 1994). Such interactions may occur within the root, in the rhizosphere or in the external soil matrix and at all stages of mycorrhizal development. A range of herbivorous nematodes, arthropods and protists browse the hyphae and spores of mycorrhizal fungi although Klironomous and Kendrick (1996) found that the mites and collembolans they studied preferred the hyphae of non-mycorrhizal fungi which may represent some selection for browse resistance in the mycorrhizal species.

Other organisms may stimulate mycorrhizal formation and function. Garbaye (1994) discusses the positive effects of the 'mycorrhization helper bacteria' which are defined as 'bacteria associated with mycorrhizal roots and mycorrhizal fungi which selectively promote the establishment of mycorrhizal symbioses'. They are best known in ectomycorrhizae, poorly known in arbuscular associations and not at all in the minor types. These bacteria selectively associate with particular fungal species and may be antagonistic to non-compatible fungi. However, where they are compatible, they may markedly increase mycorrhizal development, although the mechanisms by which this is achieved remain uncertain.

Ecology of the association

Initiation of colonisation. Colonisation may derive from resting spores, previously colonised roots or hyphae (Smith and Read, 1997). Once hyphae from these propagules come into contact with the root, the probability of successful colonisation initially depends on several factors (Hayman, 1983):

(i) The inoculum potential *i.e.*, the number of propagules present and the number of germination tubes formed;

(ii) The host response, which is mediated through exudate production and which may stimulate or inhibit colonisation;

(iii) The compatibility of the fungus with the tissues of the host plant;

(iv) Environmental factors, particularly light and temperature which influence host plant vigour and hence the flux of exudates towards the roots. Due to their low oxygen status, water-logged soils have few mycorrhizal associations. Extreme pH, salt or carbon dioxide concentrations, fungicides and high inputs of phosphorus or nitrogenous fertilisers all adversely affect mycorrhizal development;

(v) The hormonal equilibrium at the root surface. The establishment and regulation of mycorrhizae may be dependent on the production of fungal hormones, notably auxins, cytokinins and giberellins although other factors have also been implicated, including nutritional effects and the presence of other rhizosphere micro-organisms (Gogala, 1991). These hormones have been considered to be responsible for mediating morphogenetic changes in roots, including the suppression of root hairs and induction of short roots (Slankis, 1973). Metabolic changes also occur within infected tissues: ultrastructural and metabolic observations (Gianinazzi-Pearson and Gianinazzi, 1986) indicate a greater metabolic activity in host tissues. These reactions differ substantially between species and strains and may be reversible; they may depend on complex equilibria between auxin oxidases produced by roots and inhibitors of auxin oxidases produced by fungi which allow fungal hormones to act on the roots. In consequence, the hormonal equilibrium and its effect on the association, may vary along the root system and change over time (Moser and Haselwandter, 1983). Roots also produce non-specific materials in their exudates which stimulate spore germination on contact (Smith and Read, 1997). Finally, soil fertility and particularly phosphate levels, may greatly influence the relationship.

Spread of propagules. While the majority of ectomycorrhizal fungi fruit above ground and their spores are spread by air currents, arbuscular mycorrhizal fungi and many ectomycorrhizal fungi fruit underground or in the litter layers. As indicated above, spores may largely survive passage through the intestinal tract of the animals that ingest them, either purposely or unintentionally with their food (Section IV.3.1.2.2). Smaller organisms may be important for more localised spore spreading and these and other propagules are commonly spread by earthworms of different ecological strategies (Allen, 1991; Reddell and Spain, 1991a). The propagules of both arbuscular and ectomycorrhizal fungi are also spread by wind and water, notably through the activities of organisms or other agencies that expose the surface soil (Allen, 1991, Allen *et al.*, 1989).

Colonisation. Within the soil, the hyphae of arbuscular mycorrhizal fungi grow along the root and enter it at several sites from where the fungal network ramifies internally throughout the root. The extent to which colonisation extends and the time over which it occurs vary greatly although colonisation never affects more than 50-80 % of the root (Sutton, 1973). Observations suggest that mycorrhizal roots have longer life spans than non-mycorrhizal roots, and that infectivity in excised mycorrhizal roots may persist for several months after connection with the parent plant has been severed (Persson, 1982; Ferrier and Alexander, 1985). Mycorrhizae will only provide nutrients to the root as long as the root is able to provide carbohydrates.

The hyphae of both arbuscular and ectomycorrhizal fungi can also ramify from one root to another and establish networks between plants of the same and different species (Chiariello *et al.*, 1982; Read *et al.*, 1985). While the ecological significance of these networks is not yet clear, inter-plant transfer of C (Graves *et al.*, 1997; Simard *et al.*, 1997), N, water and other nutrients has been noted. Successions have been observed in the colonisation of roots by mycorrhizal fungi and 'early-' and 'late-stage' species have been reported from mycorrhizal sequences, notably in plantations and in rehabilitated minesites (see, for example, Gardner and Malajczuk, 1988). However, Molina *et al.* (1992) consider such models inadequate for predicting the species distribution of fungi in natural forests, especially those subject to periodic disturbance.

Water uptake. Mycorrhizal colonisation alters the water relationships of both arbuscular and ectomycorrhizal host plants although the mechanisms of this are not completely clear. Interactions with nutrient status are also involved (Smith and Read, 1997).

The greater soil volume explored following mycorrhizal colonisation may also increase water uptake, since hyphae and ectomycorrhizal rhizomorphs are able to absorb water and transport it to the roots (Read *et al.*, 1979; Boyd *et al.*, 1985; Mousain, 1983). Duddridge *et al.* (1980) have shown that mycelial rhizomorphs can absorb water up to 20 cm away from the root surface and transport it to the root at a rate equivalent to that of xylem vessels (*i.e.*, 20-30 cm h^{-1}). Mycorrhizal plants have a greater resistance to water stress (Bowen and Theodorou, 1973; Dixon *et al.*, 1980) due to the higher tolerance of low water potentials shown by certain fungi and through a possibly diminished resistance during water flow to the roots. For example, the resistance to water flow into soybean (*Glycine soja*) plants infected with arbuscular mycorrhizal fungi is 50 % less than that of uninfected plants (Safir *et al.*, 1972). Finally, changes in stem histology following mycorrhizal colonisation may increase the resistance of trees to drought or water-logging (Mason *et al.*, 1977).

Phosphorus uptake. Mycorrhizae have a remarkable ability to take up phosphorus from soils within which this element is poorly available. In experimental studies, the zone of phosphate uptake around the roots of *Trifolium repens* (White Clover) colonised by the arbuscular mycorrhizal fungus *Glomus mossae* was estimated to extend to 11.5 cm from the rhizoplane compared with 1 cm in uncolonised plants (Li *et al.*, 1991). The increased soil volume explored was reflected in the greater mass of the colonised plants at harvest and their lower ratios of root- to shoot mass. These authors calculated that in colonised plants, phosphorus uptake was increased by more than four times and that 76 to 79 % of this uptake occurred through the fungal hyphae. Similarly, in ectomycorrhizal *Pinus pinaster* plants, Mousain and Lamond (1978) found that the phosphorus content was 2 to 3.5 times greater than in non-colonised plants. Plant phosphorus uptake rates differ greatly, depending on the type of mycorrhiza, the species and strain of the fungal partner and soil and environmental factors (Marschner and Dell, 1994).

In plants colonised by arbuscular mycorrhizal fungi, this improved phosphorus uptake appears to result from a more complete exploitation of soluble phosphorus sources and, to a much lesser extent, from the exploitation of less soluble phosphorus sources. The increased volume of soil explored and longer life spans of mycorrhizal as compared with non-infected roots, result in a more complete exploitation of slowly diffusing phosphorus. The phosphatases of mycorrhizal fungi may play an important role in the release of phosphorus fixed in organic forms allowing its transport and absorption by the plant root. However, the quantitative significance of this requires more attention (Haselwandter and Bowen, 1996).

Nitrogen uptake. Mycorrhizal formation generally improves nitrogen uptake in plants although high soil nitrogen concentrations can depress their formation. Differences exist between the major mycorrhizal types in the forms of nitrogen that they may access. Both arbuscular and ectomycorrhizal fungi appear to take up inorganic N as NH_4^+ although at least some ectomycorrhizal species may also utilise both NO_3^- and organic (amino acids) forms of N and transfer them to the plant (Marschner and Dell, 1994; Smith and Read, 1997). Ericoid mycorrhizal fungi have well developed capacities to utilise organic nitrogen sources (Smith and Read, 1997).

As eukaryotes, mycorrhizal fungi do not fix nitrogen although free-living, nitrogen-fixing micro-organisms (*e.g.*, *Beijerinckia* or *Clostridium*) may be active in the mycorrhizosphere. In phosphorus deficient soils, nitrogen fixation by legumes is indirectly enhanced by the formation of arbuscular mycorrhizae largely through an improved uptake of phosphorus and other deficient elements (Marschner and Dell, 1994).

Uptake of other elements. A large number of macro- and micronutrient elements (and some non-nutrient elements) may be taken up by external hyphae and transferred to the plant: nitrogen, calcium, potassium, zinc, sulphate, copper, amongst others (Marschner and Dell, 1994). As indicated above, mycorrhizae may increase the uptake of heavy and other potentially toxic metals or exclude them from their tissues.

3.2.1.3 *N-fixation by prokaryotes*

Symbiotic nitrogen fixation is a unique process whereby plants acquire atmospheric nitrogen to meet their nutritional requirements through associations with prokaryotes that possess the *nif* (nitrogen fixing) genes (see Section IV.3.1.2.2). It is therefore a highly challenging area for research and technical development since, in favourable circumstances, certain prokaryote-plant associations can routinely fix up to 200 kg N ha^{-1} per cropping cycle and sometimes more (see *e.g.*, Rinaudo *et al.*, 1983; Toomsan *et al.*, 1995). On a world-wide basis, an overall estimate for biological N-fixation is 10^5 Mt per year (Sprent, 1984).

Almost all of the higher plants forming nodulating N-fixing associations with Rhizobial bacteria (Eubacteria) (Table IV.10) occur in the family Leguminosae or in the taxonomically-diverse range of trees and shrubs associated with N-fixing Actinobacteria of the genus *Frankia. Alnus* (Betulaceae) and *Casuarina* (Casuarinales) are perhaps the best-known genera associated with *Frankia.*

While most legumes fix N in root-associated nodules, species in a number of genera, including *Aeschynomene* (20 species), *Sesbania* (4 species) and several others may have active N-fixing nodules on their stems (Duhoux *et al.*, 1993). The number of nodules may thus be 5 to 10 times higher than in other plants and populations of this shrub are able to fix up to 200 kg ha^{-1} over a cropping period (Rinaudo *et al.*, 1983). Similarly, two species of *Casuarina* also form active actinorhizae on their stems which make a notable contribution to their N nutrition. In both these groups, the structures are associated with the formation of adventitious roots and are most important in humid environments (Duhoux *et al.*, 1993).

Establishment of symbiotic associations

The mutualistic association between legumes and rhizobial bacteria starts as an infection of the root tissues which later becomes a mutualistic relationship. Not all legumes can be infected by Rhizobia, and both tropical and temperate plants form associations with different species of *Rhizobium* and related genera (Sprent, 1984). Bacteria enter the root through the tips of absorbing hairs, and multiply to form infection cords which grow through the hair down to the root cortex (Figure IV.41). These cortex cells then actively multiply to form meristem cells which are colonised by the Rhizobia. At this stage, the cells have changed into so-called bacteroids which are aggregated into colonies of up to 20 cells covered by an envelope called the peribacteroid membrane. Bacteroids have the ability to transform N_2 into NH_3. The colonised cells eventually form aggregations called cortex nodules, usually a few millimetres in diameter. The formation of shoot nodules in *Sesbania rostrata* follows a similar pattern although the initial phase differs (Dommergues *et al.*, 1985.).

Frankia organisms penetrate the root via the root hairs or by inter-cellular penetration and form infectious hyphae which extend to the cortex. Once in the cortex, these organisms form subspherical vesicles within which N-fixation occurs (Huss-Danell, 1997).

The biochemistry of N-fixation

Biological nitrogen fixation (N-fixation) is the reduction of atmospheric nitrogen (N_2) to ammonia in a reaction controlled by the nitrogenase enzyme. Nitrogenase consists of two components. Component I is dinitrogenase, the site of N_2 reduction, and is a molybdo-ferro-protein of about 230 kDa. Component II is dinitrogenase reductase, an iron protein (molecular weight about 60 kDa) which provides electrons to Component I for N_2 reduction. Dinitrogenase is associated with the iron-molybdenum cofactor which is believed to contribute a surface for N_2 reduction (Sprent, 1984; Giller and Wilson, 1991). The nitrogenase enzyme is similar among all N-fixing organisms although alternative systems have been identified for *Azotobacter* spp. that do not contain Mo in the cofactor (Pau, 1989).

Nitrogenase is extremely sensitive to oxygen, with both enzyme components becoming irreversibly deactivated upon exposure to atmospheric concentrations. Consequently, free-living, nitrogen-fixing bacteria only fix nitrogen in anaerobic or low oxygen environments (Eady, 1992). Some cyanobacteria develop thick walled heterocysts, within which nitrogenase occurs and where it remains isolated from aerobic processes (Elkin, 1992). Nitrogenase is also restricted to the heterocysts in the cyanobacteria-plant symbioses (Bergman *et al.*, 1992b). Oxygen is excluded from nitrogenase in the legume-rhizobia symbiosis by two mechanisms. The nodule cortex acts as a physical barrier to oxygen diffusion (Witty and Minchen, 1990) and leghaemoglobin binds oxygen and transports it to respiratory sites while excluding it from nitrogenase (Gallon and Chaplin, 1987). Leghaemoglobin is the red pigment which colours the interiors of legume root nodules.

Nitrogen fixation is a highly energy-demanding process, whether carried out biochemically within a prokaryote, or within a pressure vessel in fertiliser factory. Sprent (1984) estimated that about 15 tonnes of photosynthate are required to assimilate 1 tonne of nitrogen by biological fixation. The energy requirements for N fixation are met by the host plant and are estimated to be 2.5 mg C mg N^{-1} (Warembourg and Roumet, 1989), about 16 % less efficient than nitrate assimilation (see Neves, 1982). Saari and Ludden (1986) estimated that 28 moles of ATP are required for each mole of N fixed.

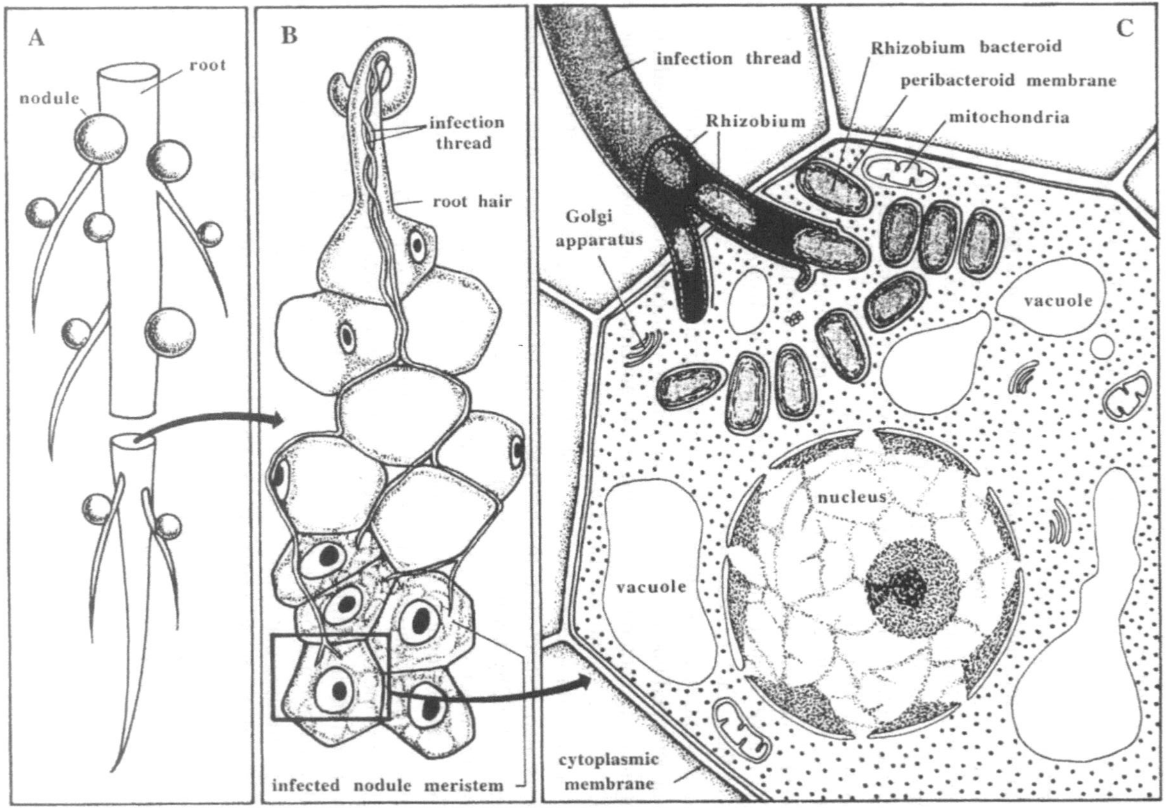

Figure IV.41 Initiation of root nodules after infection by Rhizobia (Dommergues *et al.*, 1985): infecting micro-organisms penetrate the bent tip of a root hair. One or two infection threads filled with bacteria form in each hair, extend to the cortical cells and release bacteria inside the cells. Rhizobia surround themselves with a membrane and as they become bacteroids with different properties from those of free cells, they become able to fix atmospheric nitrogen and transform it into ammonium.

Environmental influences on N-fixation
N-fixation rates are highly dependent on environment conditions. For example, N-fixation in soybean crops varies from 10 to 250 kg ha^{-1} per crop, and some legumes may fix no N at all (Dommergues *et al.*, 1985). Factors that adversely influence N fixation rates are: (i) edaphic, particularly drought and water logging, and (ii) the presence of assimilable mineral-N as NO_3^- or NH_4^+; (iii) environmental effects, notably insufficient light, (iv) climatic, especially very high temperatures, (v) biological, including the presence of root parasites such as nematodes, and (vi) the energetic constraints mentioned in the section above. In this last respect, N fixation is clearly dependent upon an efficient photosynthesis.

3 2.2 STIMULATION OF GROWTH AND DEFENCE AGAINST COMPETITORS AND PATHOGENS

3.2.2.1 *Plant growth promoting substances*

Rhizosphere organisms produce a diversity of substances that stimulate plant growth. Bacteria and fungi produce vitamins of the B group, and plant hormones including auxins, gibberellins, kinetins and indole-3-acetic acid (IAA) (Brown, 1972; Rempe, 1972; Rademacher, 1994). The stimulatory effects of earthworm activities on plant growth have also been attributed to a 'hormone-like' effect (Tomati *et al.*, 1988).

Other substances may indirectly influence plant growth through allelopathic effects on competing plants and the suppression of pathogens.

3.2.2.2 *Allelopathy*

Allelopathy includes any directly or indirectly harmful effect by one plant upon another organism through the production of chemicals released into the environment. These chemicals are secondary compounds and may be produced through exudation or the decomposition of plant litter.

Chemical compounds implicated in allelopathy
Rice (1974) has identified 14 different groups of allelochemical compounds based on their composition and metabolic pathways. The most important groups are phenolic acids and their derivatives: terpenoids, steroids, coumarins, flavonoids, alkaloids, cyanohydrins and tannins (Table IV.11). They are generally water soluble, except for a few volatile compounds that diffuse from roots and leaves (Jordan *et al.*, 1993).

Effects of allelopathic compounds
Secondary plant compounds are chemicals produced by plants and micro-organisms that are not essential to the growth of the organisms involved although they may play important roles in mediating interactions with other organisms. They may directly inhibit seed germination or the growth of other nearby plants; they may affect water uptake (Einhellig *et al.*, 1985) or nutrient supply through, for example, the inhibition of nitrification or nitrogen fixation (Balke, 1985; Hoagland and Williams, 1985).

Table IV.11 Sources and types of allelochemicals reported from higher plants (Hoagland and Williams, 1985).

Method of Isolation	Species and Tissue	Chemical Class
Extraction - ethanol	Cyperus esculentus, tubers and leaves	Ferulic acid and other phenolics
	Cyperus rotundus, tubers and leaves	Ferulic acid and other phenolics
- water	Abutilon theophrasti, leaves	Phenolics
	Agropyron smithii, litter	Phenolics
	Ambrosia artemisifolia, shoots	Chlorogenic and caffeic acids
	Aster pilosus, shoots	Chlorogenic and caffeic acids
	Avena fatua, dead litter	Ferulic and coumaric acids
	Bidens pilosa, leaves	Phenylheptatriyne
	Brachiaria mutica, dried leaves	Vanillic, p-hydroxybenzoic, o-hydroxyphenylacetic acids
	Chloris gayana	Phenolics - ferulic acid
	Cynodon dactylon, dried leaves	Phenolics - p-coumaric acid
	Digitaria decumbens, dried leaves	Ferulic acid and other phenolics
	Digitaria sanguinalis, whole plant	Chlorogenic, isochlorogenic and sulfosalicylic acids
	Erica arborea, leaves	Salicylic acid, scopoletin, p-hydroxybenzaldehyde
	Helianthus annuus, root	Chlorogenic, isochlorogenic acids; scopoletin
	Kochia scoparia, leaves	Ferulic acid, myricetin, quercitin
	Panicum maximum, dried leaves	o-Hydroxyphenylacetic acid
	Paspalum plicatulum, dried leaves	Ferulic acid, other phenolics
	Polygonum orientale, roots, stems, leaves	Flavone glycosides
	Rumex crispus, leaves	Phenolics
	Setaria sphacelata, dried leaves	Ferulic acid, other phenolics
	Sorghum halepense, leaves, rhizomes	Chlorogenic, p-coumaric acids, p-hydroxybenzaldehyde
	Tripsacum laxum, dried leaves	Ferulic acid, other phenolics
Exudates - root	Ambrosia artemisifolia	Chlorogenic and caffeic acids
	Chenopodium album	Oxalic acid
	Helianthus annuus	Chlorogenic, isochlorogenic acids; scopoletin
	Hemarthria altissima	Cinnamic and benzoic acids and derivatives
	Polygonum aviculare	Phenolic glucosides, fatty acids
- root and rhizome	Sorghum halepense,	Chlorogenic, p-coumaric acids; p-hydroxybenzaldehyde
Leachates - leaves	Cyperus esculentus	Ferulic acid, other phenolics
	Datura stramonium	Scopolamine, hyoscyamine
	Polygonum aviculare	Phenolic glucosides, fatty acids
- fronds	Pteredium aquilinum	Phenolics
	Salsola kali	Quercitin, ferulic acid, others
- fruits	Ammi majus	Xanthotoxin, (furanocoumarin)
- seeds	Abutilon theophrasti	Phenolics
	Datura stramonium	Scopolamine, hyoscyamine
- roots and rhizomes	Agropyron repens	Acetic, butyric acids
	Cyperus esculentus	Ferulic acid, other phenolics
Volatiles - leaves	Artemisia tridentata	Terpenes, camphor, pinene,
	Salvia leucophylla	Various volatiles

Plant species have specific responses to allelochemicals; some may even be significantly suppressed by products from their own species and tolerate those produced by others (Newman and Rovira, 1975). Allelochemicals are considered to be an important driving force in successional processes since they influence the growth and competitive ability of the plants present at the different seral stages of this process. It has been shown, for example, that regrowth of *Picea abies* in montane forest of French Alps may be inhibited by secondary compounds produced by the heath shrub *Vaccinium myrtillum* (André *et al.*, 1987). Allelopathy may also be responsible for the decline of legumes in pastures (Halsall *et al.*, 1995) and for difficulties in the establishment of agroforestry systems that attempt to combine plantings of trees and herbaceous plants of economic interest (Leigh *et al.*, 1995).

Inhibition of nitrification

In chemically-infertile soils, nutrients are mainly conserved within the resistant soil organic matter fractions and a lack of assimilable C may severely limit microbial activities (Darici *et al.*, 1986). N mineralisation rates may be extremely low in such diverse ecosystems as natural savannas, artificial pastures or coniferous forests (Scholes and Sánchez, 1990; Lensi *et al.*, 1992; Stienstra *et al.*, 1994). Inhibition of nitrification by water-soluble substances produced by roots has been demonstrated in nutrient-poor savannas (*e.g.*, Meiklejohn, 1962; Munro, 1966; Lensi *et al.*, 1992). Conservation or the immobilisation of NH_4-N rather than nitrification is an efficient way of conserving N and preventing C loss from infertile soils (Marrs *et al.*, 1991).

3.2.2.3 *Inhibition and spread of pathogens by micro-organisms and invertebrates*

Soil invertebrates may have a significant role in the suppression of root diseases, through their direct and indirect effects on the microbial and invertebrate pathogens that affect roots (Curl, 1988). Dormant fungal propagules freed from fungistasis by root exudates become able to initiate infection although, when communities of fungal feeding protozoans and nematodes are abundant, these germinands may be consumed before this can occur (Griffin and Roth, 1979). In soils considered as 'suppressive', infection does not develop, despite the presence of the root-infecting fungi.

In the case of 'take-all' disease of wheat caused by *Gaeumannomyces graminis tritici*, suppressiveness or reduced severity of disease has been related to the presence of mycophagous amoebae (Chakraborty *et al.*, 1983; Chakraborty and Warcup, 1984); or lumbricid earthworms (Stephens *et al.*, 1994). Other cases of reduced fungal attack have been related to the activities of mycophagous nematodes, although very high densities may be necessary to achieve a significant effect (Cayrol *et al.*, 1978; Barnes *et al.*, 1981).

In certain circumstances, Collembola may also reduce the infection rate of plants by grazing on pathogenic fungi. Curl *et al.* (1983) showed that the ratio of mycofloral density in the rhizosphere to non rhizosphere (R/S) diminished from 2.41 to less than unity (0.41) after the introduction of 500 Collembola into glass tubes containing 140 g soil and cotton seedlings. Laboratory experiments have demonstrated that the 'damping-off' disease of cotton seedlings due to fungal infection may be partly suppressed through the addition of Collembola (Curl, 1988) (Table IV.12).

Phytoparasitic nematodes may also to be controlled by a number of biological factors in the rhizosphere. The bacterium *Bacillus penetrans* (Stirling, 1984), the predaceous fungi (certain zygomycote and other fungi, also protists) including species that trap nematodes in constricting hyphal loops and against adhesive knobs (Kendrick, 1992) and a few other nematotrophic fungi (Kerry, 1980) are spectacular examples of nematode parasites or predators. Earthworms may be effective predators on parasitic nematodes since their populations are often reduced in the presence of earthworms (Dash *et al.*, 1980; Yeates, 1981) (see also Chapter III.4).

Table IV.12 Effect of Collembola on the emergence and disease index of cotton seedlings inoculated with the pathogenic fungus *Rhizoctonia solani* (Curl, 1988). Values within a column followed by a common letter are not significantly different ($P > 0.05$);
[1] NSS = Non sterile soil; CO: Collembola; RS = *Rhizoctonia solani* (0.02 g infected oats g^{-1} soil).
[2] Disease index: 0 = healthy to 5 = dead plant; index calculated as mean for all plants.

	1000 insects kg^{-1} soil		2000 insects kg^{-1} soil	
Treatments[1]	**Emergence %**	**Disease index (0-5)**[2]	**Emergence %**	**Disease index (0-5)**[2]
NSS (No treatment)	66.2 b	2.34 c	75.0 ab	2.25 b
NSS + CO	91.2 a	0.94 d	95.0 a	0.90 c
NSS + RS	47.5 c	4.65 a	48.3 c	4.10 a
NSS + RS + CO	75.0 b	3.51 b	88.3 ab	2.20 b

However, the same mechanisms that suppress disease may also have adverse effects on plants, either by reducing the effectiveness of root symbionts, or promoting the active dissemination of plant pathogens. The nematodes (Shafer *et al.*, 1981) or Collembola (Warnock *et al.*, 1982) that graze on pathogenic fungi may also affect mycorrhizal fungi reducing the extent of infection and consequently plant growth. In a number of experiments, mainly conducted under laboratory conditions, mycorrhizal infection was either impeded or rendered ineffective in the presence of mycophagous amoebae (Chakraborty, 1985; Sutherland and Fortin, 1968) and Collembola (Warnock *et al.*, 1982).

3.2.3 EFFECTS ON SOIL STRUCTURE

Root distribution and abundance are strongly influenced by soil physical structure (see Chapter III.3). In turn, growing roots significantly affect the soil physical structure by three different mechanisms: (i) the formation of macropores and channels; (ii) physical enmeshment of soil particles and aggregates (Tisdall and Oades, 1982; Miller and Jastrow, 1990); and (iii) aggregation through the production of exudates mixed with clay and other mineral particles (Table IV.13). The importance of these effects differs with the type of root system. For example, root systems with lignified tap roots heavily influence soil porosity whereas aggregation is effectively mediated by the dense, fasciculated root systems of many grasses.

Table IV.13 Effects of plant biological features on soil physical structure (after Goss, 1991).

Plant biology	Effects on soil
Elongation	Compression/shear
Expansion	Enmeshment
	Macroporosity
	Aggregation
Death and decay	
Exudation	Change Organic matter
Sloughing of cells	

Porosity: root death leaves holes

In the early phases of pedogenesis, roots colonise the weathered material of the C horizon where they create channels that remain open after their death. In calcareous substrates, Jaillard (1983) noted that limestone may be transformed into calcareous sands as a result of solubilisation and the further crystallisation of $CaCO_3$ in root cells.

In fully-developed soils, root tips may exert maximum pressures of 0.61 to 1.3 MPa, with a few values greater than 2 MPa leading to compression of the soil at the point of root contact. The penetration of small aggregates requires lower axial root growth pressures than large aggregates and the strength of these larger aggregates may impede root penetration. Such common plants as pea (*Pisum sativum*), cotton (*Gossypium hirsutum*) or sunflower (*Helianthus annuus*) may only be able to exert pressures of 0.1-0.3 MPa (Misra *et al.*, 1986). Reorientation of particles along the root results in slight (2-7 %) increases of bulk density (Goss, 1991).

Roots may thus break down large aggregates into small, and create channels that can have important roles in facilitating water infiltration and gas exchange (Figure IV.42).

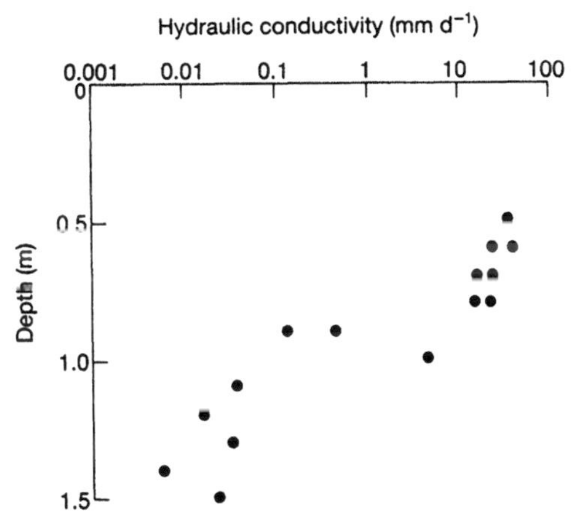

Figure IV.42 Variation in hydraulic conductivity with depth in a clay soil associated with root distribution. The decrease in conductivity below 0.9 m depth is associated with a considerable reduction in the amount of water taken up by roots (Goss, 1991).

This process is known as 'biological drilling' and its effectiveness varies greatly between plant species. The tap root systems of woody perennial species ameliorate soil porosity more effectively than the fibrous root systems of grasses (Hulugalle and Lal, 1986). However, certain perennial grasses (*e.g.*, deep rooted African species) and herbaceous legumes (*Medicago sativa)* may penetrate 2-3 m into the soil and produce extensive networks of soil macropores (Meek *et al.*, 1991; Fisher *et al.*, 1994). This process is of utmost importance in the restoration of degraded soils.

Direct observation in the field shows that once root channels are freed following root death and decomposition, they may be used preferentially by new roots (Dexter, 1986) leading to an improvement in subsequent crop yields (Elkins, 1985). This enhanced capacity for deep root penetration will only benefit the plant if the nutrients and water in these pores are readily available (Cresswell and Kirkegaard, 1995).

Aggregation: effects of root enmeshment and SOM addition

Bartoli *et al.* (1993) showed that the surface fractal dimension Ds (as determined by mercury porosimetry) is better than porosity at discriminating between the root zone and non-root zone. This result indicates that the soil heterogeneity created by roots influences both porosity and aggregation.

Roots may enhance soil aggregation by two complementary mechanisms. Firstly, the release of mucilages by the root and the rhizosphere microflora 'glue' small soil particles into micro-aggregates <250 μm (Gouzou *et al.*, 1993), and secondly, these small aggregates become entangled by fine roots and the hyphae of mycorrhizal fungi to combine as compound or macro-aggregates, greater than 250 μm (Tisdall and Oades, 1982) (Section I.1.3.3). This model is supported by the results of Thomas *et al.* (1986) who demonstrated positive relationships between root abundance and aggregate size and abundance. Gouzou *et al.* (1993) also found that inoculation of *Bacillus polymyxa* into the rhizosphere of wheat increased the amount of soil adhering to roots by 59 % and the amount of aggregates in the size class 0.2 to 2 mm by 100 %.

Miller and Jastrow (1990) observed a correlation between the overall mean diameter of water-stable aggregates and such biological parameters as root length and the development of the hyphae of mycorrhizal fungi. However, the magnitude of this effect is likely to differ between soils, with plant species and with the associated mycorrhizal fungi. This is due to differences in the architecture and diameter of roots, in the type of mycorrhizal association formed, the hyphal density of the mycorrhizal and other fungi present and the amount and quality of exudates that they produce.

Aggregation in the rhizosphere may also result indirectly through the accumulation of the faecal pellets of earthworms and other invertebrates that feed in the rhizosphere. In pots containing the less than 2 mm fraction of a sandy A horizon soil (alfisol) from the Côte d'Ivoire, macro-aggregate formation (>2 mm) after three months was limited to *ca.* 10 % of the soil mass in the presence of the tropical grass *Panicum maximum*, whereas 80 % of the soil mass was aggregated in treatments to which endogeic earthworms had also been added (Blanchart, 1992).

IV.4 THE DRILOSPHERE

Earthworms are ecosystem engineers *sensu* Jones *et al.* (1994) in that they effectively modify their environments. Their large size and their capacity to move substantial quantities of soil mean that their activities dramatically change the distribution patterns of organic resources and the fluxes of water and gases in soil. Both the distributions and activities of micro-organisms are modified by earthworms and through this, the dynamics of soil organic matter and nutrient release.

Bouché (1975) introduced the concept of the drilosphere (from the Greek 'drilos' = earthworm) to describe the 2 mm thick zone surrounding the walls of earthworm galleries. This term will be used here in the broader sense of describing those components of the soil (including microbial and invertebrate populations) that are affected by earthworm activities. The drilosphere therefore includes galleries and burrows, but also extends to surface and subterranean casts, 'middens' (accumulations of litter surrounding the openings of burrows) and the peculiar micro-environment represented by the gut and its contents. Earthworms accumulate organic matter within their structures as a consequence of their burrowing and feeding behaviours. Active communities can therefore incorporate a large part of litter fall and root litter into the drilosphere and transform a substantial proportion of the upper soil horizons into casts of different ages and origins (Section II.3.3.2).

The drilosphere may be considered the functional equivalent of the rhizosphere and litter systems both of which are micro-environments with a localised and intense microbiological activity. This activity is stimulated by the fine subdivision and thorough mixing of litter debris with soil inorganic materials, and by the production of cutaneous and intestinal mucus. The latter materials have analogous effects to the root exudates of the rhizosphere and the leaf leachates of the litter system.

4.1 Classification of drilospheres

The effects of earthworms on soil properties differ markedly, depending on the ecological categories of the species involved (see Chapter III.4.3.1.2):

(i) Epigeic species are part of the litter-system and their roles are similar to those of other epigeic invertebrates. They have little or no impact on the physical properties of the mineral soil horizons and their effects are thus marginally 'drilospheric';

(ii) Anecic species strongly influence soil properties by translocating leaf litter and mixing it with the soil, and through the construction of their subvertical burrow systems;

(iii) Endogeic species affect soil organic matter dynamics and physical structure by building and maintaining a macro-aggregate structure, and a dense network of macropores.

4.1.1 DETERMINANTS OF EARTHWORM COMMUNITY STRUCTURE

The regional abundances of earthworms and the relative importances of the different ecological categories are determined by large-scale climatic factors (mainly temperature and rainfall), their phylogenetic and biogeographical histories together with such regional parameters as vegetation type and soil characteristics. As a result, drilospheres differ greatly at the scale of broad climatic areas and regions, and between patches with different soil and vegetation.

Comparative analysis of a large number of earthworm communities of wet areas has revealed that temperature is the main large-scale factor determining the nature of earthworm communities (Lavelle, 1983c). Along a thermo-latitudinal gradient, population density increases from a few tens to several hundreds m^{-2} while biomass achieves a maximum in temperate latitudes. The most striking feature is the gradual shift in dominant community feeding habits: these change from largely litter-feeding (epigeics) at high latitudes, to a co-dominance of litter and soil feeding (anecics) at temperate latitudes, to a dominance by soil feeding (meso- and oligohumic endogeics) in the inter-tropical areas (Figure III.45). Thus, with rising temperature, the activities of earthworm communities become increasingly concentrated at greater depths as they become capable of exploiting resources of lower quality; this starts with the high-quality resources found at the soil surface (litter) and is lowest with the extremely poor-quality soil organic matter used by the oligohumic species characteristic of deeper horizons (see Chapter III.4.3.1.2).

Phylogenetic and biogeographical factors may influence the functional structure (*i.e.*, the proportions of the different ecological categories) of earthworm communities. Oligohumic endogeic species do not occur in such families as the Lumbricidae and Glossoscolecidae although the anecic earthworms - species with deep, subvertical burrows and a resistance to seasonal drought derived from a true diapause - are well represented. In contrast, anecic species are not found within the more primitive families such as the Megascolecidae although endogeic earthworms are more diverse, including those of oligohumic and mesohumic feeding habits. Overall, it is difficult to determine whether the phylogenetic differences do not simply reflect selection for the different soil environmental conditions pertaining over the geographical ranges of these families.

At a regional scale, forests generally have smaller populations than grasslands. In Europe, for instance, forests have live earthworm biomasses in the range of 400 to 680 kg fresh weight ha^{-1} whereas in grasslands, values of 500 to 2000 kg ha^{-1} or more are common (Lee, 1985; Edwards and Bohlen, 1996). In the introduced pastures of humid tropical areas, communities dominated by the peregrine endogeic species *Pontoscolex corethrurus* and *Polypheretima elongata* may have live biomasses of 1000-1500 kg ha^{-1} in contrast to the species-rich communities of much lower biomass (100-200 kg ha^{-1} fresh weight) characteristic of natural forest soils (Fragoso and Lavelle, 1992).

Soil characteristics and rainfall are also important determinants of earthworm communities within a given biome. In western Africa, earthworms do not occur where average yearly rainfall is less than 800 mm (Lavelle, 1983a). In tropical rainforests in America and Africa, earthworm biomass increases with rainfall up to values of *ca.* 3,000 mm and then decreases with higher values, probably due to the low nutrient status of soils typical of these high rainfall conditions (Figure IV.43). Nutrient status and pH determine the dominance of both litter-feeders (epigeics and anecics) in nutrient poor and/or acid soils, and of soil organic matter feeders (endogeics) in richer soils with higher pH. In humid (5000 mm annual average rainfall) Indian monsoon forests, large earthworm biomasses (up to 1200 kg fresh weight ha^{-1}) were found, although the soil was young with high pH and nutrient status (Ferry, 1994). Nutrient contents are also correlated with soil pH and local rainfall (Fragoso and Lavelle, 1992). In general, tropical earthworms seem to be more acid-tolerant than those of temperate climates (Lavelle *et al.*, 1995). Comparison of species distributions in France (Bouché, 1972) with those at a number of tropical sites (Lavelle *et al.*, 1995) - including soils with a wide pH range - showed that modal species diversity occurred at pH values of 6 to 7 in France and at 5 to 6 in the humid tropics, respectively.

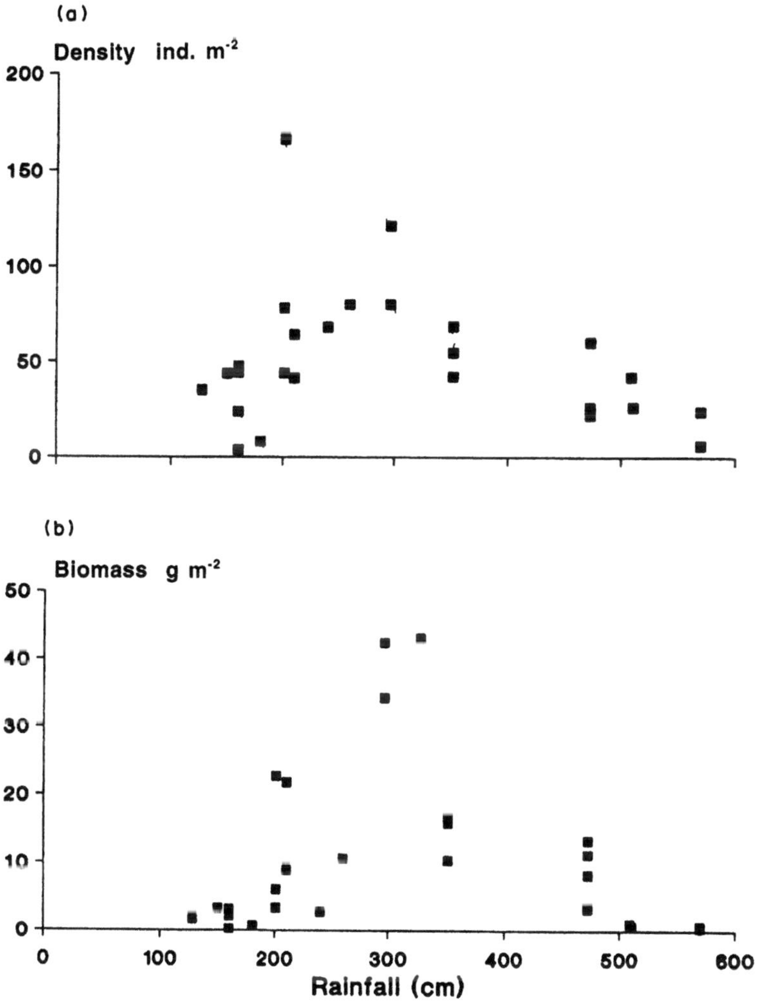

Figure IV.43 Relationship between annual rainfall and abundance of earthworm communities in tropical rainforests in America and Africa (Fragoso and Lavelle, 1992).

The type of land-use also greatly affects earthworm communities. The main factors that severely deplete populations are intensive cultivation, pesticides and soil denudation (see, *e.g.*, Edwards, 1983; Lofs-Holmin, 1983; Lavelle and Pashanasi, 1989; Lavelle *et al.*, 1994b; Edwards *et al.*, 1995; Edwards and Bohlen, 1996). In the humid tropics, cropping of soils recently cleared from their native vegetation depletes the stocks of assimilable soil organic matter and reduces the carrying capacity for earthworm communities. In terms of assimilable C, the cost of earthworm communities has been estimated at 1-1.2 t ha^{-1} in the humid savannas of Lamto (Côte d'Ivoire) (Lavelle, 1978); this C is mostly, but not exclusively, derived from fresh residues largely concentrated in the coarse organic fractions of soil organic matter (Martin *et al.*, 1991; 1992). In tropical agroecosystems, the cost of maintaining active earthworm communities of *ca.* 400 kg fresh weight ha^{-1} has been estimated at 1.4 t ha^{-1} (Charpentier, 1996; Gilot *et al.*, in press).

Chemicals also affect earthworms in diverse ways: fertilisers and a wide range of pesticides do not affect them, provided concentrations are not unduly high. However, when N fertilisation leads to soil acidification, communities may be significantly depressed (Potter *et al.*, 1985) and a few pesticides, notably the organochlorine compounds, are extremely toxic (see, *e.g.*, Lee, 1985; Edwards and Bohlen, 1996).

Neutralisation of acid pH following liming always enhances earthworm communities. In a Finnish spruce forest, lime and the litter of deciduous species were applied and earthworms (*Allolobophora caliginosa*) were introduced (Huhta, 1979). After two years, earthworm biomass had significantly increased in all treatments. The application of litter alone had a similar effect although it only resulted in a slight increase in pH. After three years, earthworm biomass significantly decreased in the litter treatment, probably because the resource was exhausted (Figure IV.44).

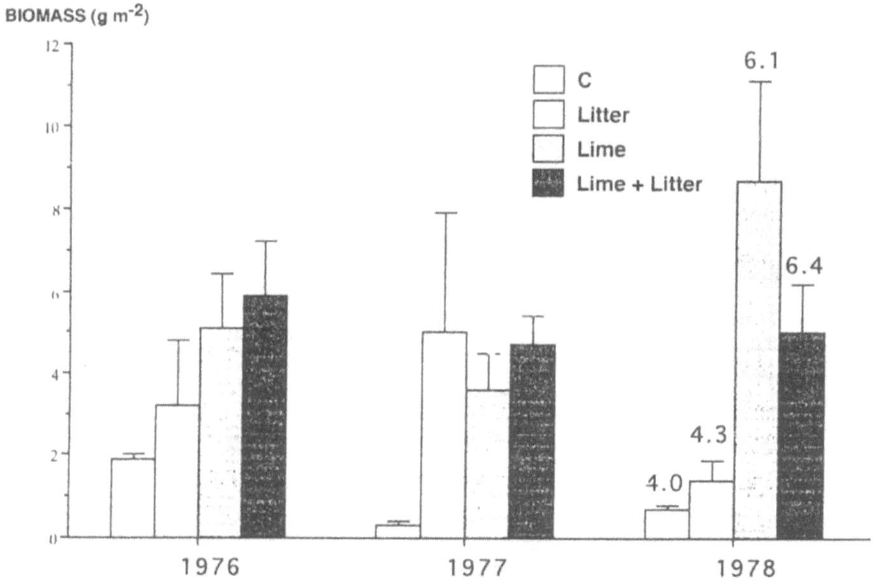

Figure IV.44 Effects of litter and lime applications on the biomass of earthworm populations in a spruce *(Picea abies)* forest in Finland (Huhta, 1979) (numbers attached to the 1978 data indicate soil pH).

These results have been confirmed by Persson (1988) who observed a strong response of the soil fauna to the liming of a 40-year-old spruce stand in southern Sweden. Three years after the application of lime (1.55 and 8.75 t ha^{-1}), earthworm numbers had increased tenfold with no significant effect attributable to the amount of lime applied.

Finally, the overall effect of plant community change on earthworm communities depends closely on the type of replacement vegetation and the earthworm community that was present in the original ecosystem. Three examples are used to illustrate this point: the establishment of pastures in Europe following the clearing of deciduous forests, pastures established in Peruvian Amazonia following slashing and burning of the primary forest, and pastures of African grasses and herbaceous legumes established in a natural savanna environment in the eastern plains of Colombia (Table IV.14).

Table IV.14 Effects of the conversion of natural ecosystems into pasture on earthworm communities in three contrasting environments.

		ORIGINAL ECOSYSTEM				PASTURE		
Location of site and original ecosystem	**References**	**N° species (mean)**	**Density (m^{-2})**	**Biomass g f w m^{-2}**	**Total N° species**	**N° sp from original system**	**Density m^{-2}**	**Biomass g m^{-2}**
Southern Sweden (deciduous forest)	Nordström and Rundgren, 1974	6-11 (m=9)	58-227	12-89	6	6	109	59
Peruvian Amazonia (tropical rainforest)	Lavelle and Pashanasi, 1989	4	120	23	2	0	740	140
Eastern Plains of Colombia (savannas)	Decaëns *et al.*, 1994; Jimenez *et al.*, 1998 ab	6	157	4.7	6	6	139	51.1

Species richness is low in the largely forest-derived earthworm communities of European pastures although it still remains in the upper range of values normally found in forests. In Amazonia, all the species indigenous to the forest disappeared after clearing and a new community of purely exotic species was established. Earthworm populations had twice the biomasses of those of the primary forest, due to colonisation by the highly vagile pantropical endogeic species *Pontoscolex corethrurus*. In contrast, in the Colombia savannas, conversion of native savanna into pastures dominated by African grasses and herbaceous legumes lead to substantial increases in the populations of the existing earthworm communities.

Biogeographical factors explain these observations. In Northern Europe ancient communities were destroyed by the Quaternary glaciations. Following glacial retreat, recolonisation occurred through the invasion of a suite of ubiquitous, vagile, southern-European species. Pastures provide conditions favourable for the development of quite populous communities and, in middle European sites, such communities generally have much greater biomasses than adjacent forests of similar species composition (Lee, 1985). Nonetheless, detailed studies show that the relative dominance of each species depends on the type of ecosystem (Scheu, 1992) (Figure IV.45).

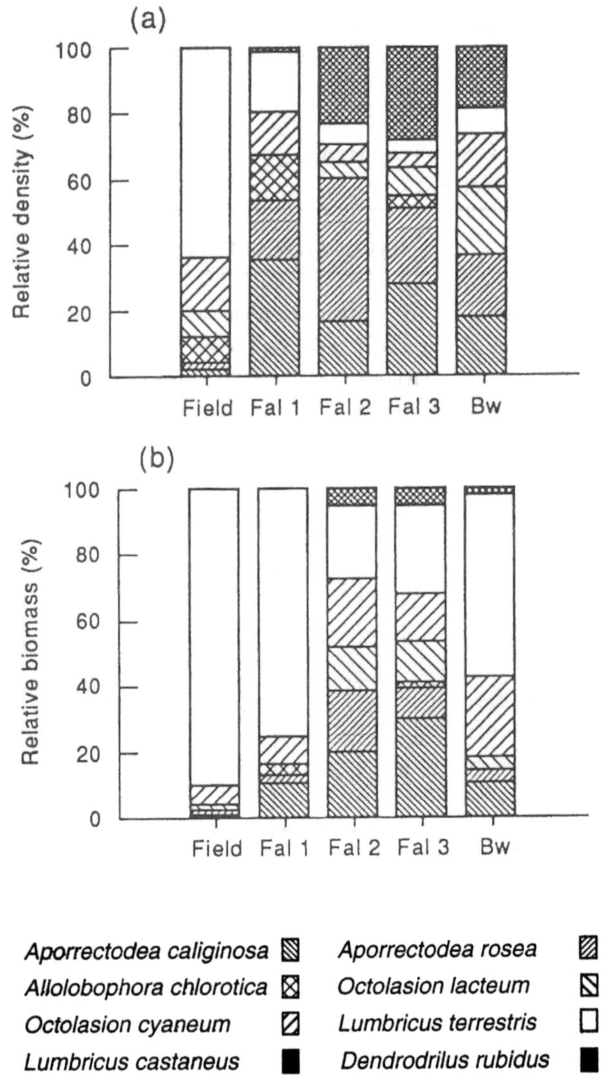

Figure IV.45 Dominance structure in the population density (a) and biomass (b) of lumbricid earthworms at five different sites representing separate stages in a secondary succession in Northern Germany (Scheu, 1992).

A similar situation occurs in the Colombian llanos when natural savanna is transformed into pasture. In that environment, local species are adapted to grassland conditions and benefit from the substantially-increased plant production, and the return of a significant proportion of this production as cow dung (Decaëns *et al.*, 1994; Jimenez Jaen, unpublished data). Again, not all species respond equally to the new environmental conditions.

In Peruvian Amazonia, deforestation eliminates the native species and communities establish comprising one or two highly-adaptable, aggressive, colonist species. These species build up large populations with biomasses greater than 1 Mg fresh weight ha^{-1}: maxima of 3-4 Mg ha^{-1} have been found in a rich vertisol in Martinique (West Indies) colonised by the pantropical species *Polypheretima elongata* (Barois *et al.*, 1988).

4.1.2 EARTHWORM DIGESTIVE SYSTEMS

A large part of the drilosphere effect on soil organic matter dynamics is due to the mutualistic digestion of organic matter by earthworms acting in association with micro-organisms. Although detailed information is still lacking, digestion in the earthworm gut appears to result from a combination of 'direct' processes including enzyme production by the worm and indirect processes involving some degree of mutualism with the microflora (see Chapter III.4.4.1.2).

Epigeic lumbricids possess a rather diverse enzymatic capacity, including cellulases (see *e.g.*, van Gansen, 1962; Parle, 1963; Neuhauser and Hartenstein, 1978), and this has led to the hypothesis of direct digestion of litter by these worms. *Lumbricus rubellus* is a common representative of this group in temperate regions and appears to digest living bacteria, even when they are protected by polysaccharides and clay particles (Kristufek *et al.*, 1994). However, it is still unclear which of these enzymes are produced by the worm, and which by the ingested microflora (Loquet and Vinceslas, 1987).

Anecic earthworms appear to combine a form of direct digestion based on the production of a few basic enzymes, with a mutualist digestion that occurs in combination with the soil and litter microflora. There is good evidence that the 'external rumen' type of mutualist digestion is important, and perhaps essential, for these worms. Anecic earthworms drag dead leaves from the litter into their burrows and ingest them some weeks later, once microbial incubation has rendered them more palatable (see *e.g.*, Wright, 1972; Cooke, 1983, Cortez *et al.*, 1989). Other species concentrate litter around the openings of their galleries thus creating 'hot spots' of increased activity where meso-fauna and microbial populations are more numerous and active; such accumulations are called 'middens' (*e.g.*, Nielsen and Hole, 1964; Dozsa-Farkas, 1978; Hamilton and Sillman, 1989).

Tropical endogeic earthworms have a mutualistic digestive system based on interactions that occur during gut transit between the worm and the microflora ingested with the soil. Bacteria stimulated in the gut produce some enzymes that are not secreted by the earthworm itself. Experiments on several tropical earthworm species have shown that the enzymes cellulase and mananase (that digest important components of root material) found in the gut content are produced by ingested bacteria and not by cells of the gut wall (Zhang *et al.*, 1993; Lattaud *et al.*, 1997a, b) (Table IV.15).

Table IV.15 Enzymatic activities in the foregut of three tropical earthworm species and sterile tissue cultures (Lattaud *et al.*, 1997ab; Zhang *et al.*, 1993) (values expressed as µg glucose mg^{-1} protein $minute^{-1}$).

Substrate	***Pontoscolex corethrurus***		***Millsonia anomala***		***Polypheretima elongata***	
	Gut content	Culture	Gut content	Culture	Gut content	Culture
Maltose	16.3 ± 2.5	++	48.6 ± 4.6	39.0 ± 3.3	79.4 ± 4.5	82.6 ± 0.3
N-acetyl glucosamine	26.3 ± 4.9	+	87.4 ± 9.0	16.9 ± 1.1	26.3 ± 4.9	167.5 ± 2.5
Starch	26.5 ± 3.5	+	0.99 ± 0.05	0.36 ± 0.03	197.7 ± 7.7	88 ± 1.1
Cellulose	1.9 ± 0.02	**0**	0.49 ± 0.01	**0**	6.3 ± 1.1	0.4 ± 0.01
Mannan	7.9 ± 1.7	**0**	5.60 ± 0.88	**0**	2.96 ± 0.09	0.99 ± 0.01

Favourable conditions for micro-organism activity occur in the anterior part of the gut: high water contents (100 to 150 % of the dry weight of soil), neutral pH and, above all, high concentrations of readily assimilable mucus (5-16 % depending on the species) (Barois and Lavelle, 1986). This mucus is a mixture of low molecular weight (*ca.* 200 Da) amino-acids and oxides and a glycoprotein of 40 to 60 kDa (Martin *et al.*, 1987). In the mid-gut, micro-organisms whose metabolism has been sharply increased by feeding on mucus, become able to digest soil organic matter. Their activity at 28 °C is then 6 to 10 times higher than in the control, and perhaps up to 30 times that in undisturbed field conditions. The products of their external digestion are partly reabsorbed with water in the posterior part of the gut.

Through this process, an estimated 3 to 19 % of soil organic matter may be assimilated during a gut transit which may last from 30 minutes to 2-4 hours (Lavelle, 1978; Martin, 1989; Barois, 1987). Evidence also exists for a mutualistic digestive system in the temperate-climate earthworms *Allolobophora molleri* and *Octolasium tyrtaeum* (Trigo and Lavelle, 1993; Trigo and Lavelle, 1995).

The efficiency of this digestive system is highly dependent on temperature. When the temperature falls from 27 °C to 15 °C, increased microbial respiration in the posterior gut of *Pontoscolex corethrurus* is limited to twice the control value, with maximum rates diminished respectively by 70 and 85 % at 20 and 15 °C in comparison with respiration rates measured at 27 °C (Figure IV.46). In temperate-climate earthworms, the microflora is more active at 20 °C than at 27 °C, although mutualistic digestion seems to be less efficient. The lower efficiency may be compensated for by a higher-quality food resource, and an increased production of intestinal mucus in the gut (Figure IV.47). This may explain why earthworms tend to feed more on litter in cooler climates. Drought, by reducing the periods of activity, is likely to have a similar effect on communities. Finally, nutrient deficiencies, especially of nitrogen, may also constrain earthworms to use resources with higher nutrient contents *i.e.*, the organic matter of the uppermost centimetres of the soil and leaf litter. This is supported by the dominance of litter-feeding earthworms in tropical forests with acid, nutrient-poor soils (Fragoso and Lavelle, 1992).

The relative contributions of the three digestive systems are thus expected to vary with microclimate and the quality of resources available.

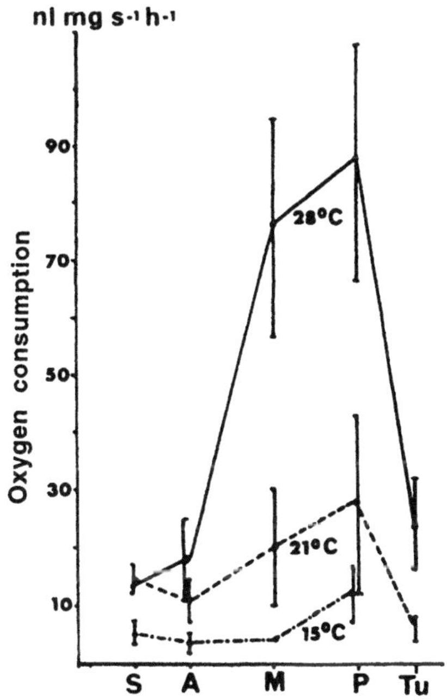

Figure IV.46 Effect of temperature on average oxygen consumption of soil from different parts of the gut of *Millsonia anomala* (A: anterior; M: midgut; P: posterior), fresh casts (Tu) and a control soil (S) sieved (<2 mm) and maintained at field capacity (Barois, 1987).

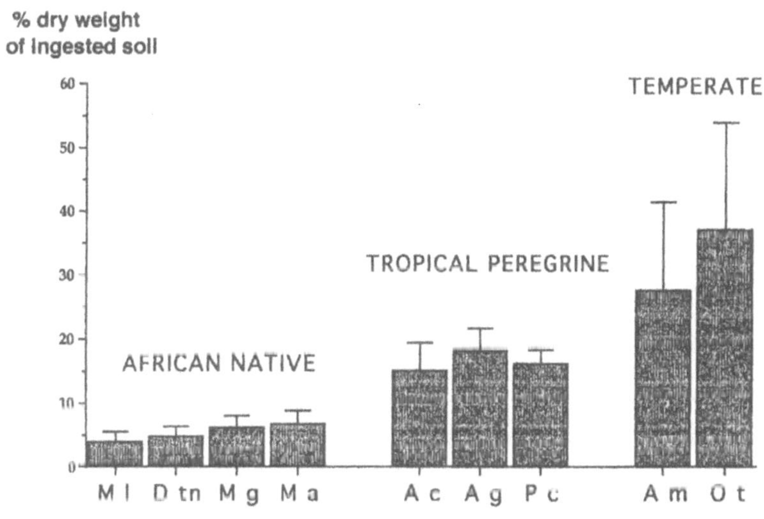

Figure IV.47 Proportion of intestinal mucus mixed with the soil ingested by different earthworm species. *M.l.: Millsonia lamtoiana; D.tn.: Dichogaster terrae-nigrae; M.g.: M. ghanensis; M.a.: M. anomala; A.c.: Amynthas corticis; A.g.: A. gracilis; P.c.: Pontoscolex corethrurus; A.m.: Allolobophora molleri; O.t.: Octolasium tyrtaeum.*

4.2 The anecic drilosphere

4.2.1 DISTRIBUTION OF ANECIC EARTHWORMS

Anecic earthworms are unevenly distributed throughout the world for both climatic and biogeographical reasons. They are absent from the acid soils of cold temperate regions and coniferous forests in warmer temperate soils. They are dominant in most temperate grasslands and deciduous forest soils where they may comprise up to 80 % of earthworm biomass. In arable land, the proportions of anecics are much lower, although highly variable, with average values of 10-30 % (Boström, 1988; Cuendet, 1983; Binet, 1993). In tropical soils, they may be completely absent or have small populations. Savannas of the Columbian llanos and, to a lesser extent, Amazonian rainforests at Yurimaguas, are exceptions to this and anecic Glossoscolecidae may comprise from a few *per cent.* in the forest to 15 % of the biomass in natural savannas, and 80 % in pastures derived from these savannas (Lavelle, 1983a; Fragoso and Lavelle, 1987; Lavelle and Pashanasi, 1989; Jimenez Jaen *et al.*, 1998).

In cold regions, anecics are clearly limited by climatic and edaphic factors and they may rapidly colonise soils when pH is increased to neutrality by an amendment (Toutain, 1987b).

Their absence from African and Central American forests seems to have a phylogenetic reason *i.e.*, the lack of an anecic adaptive strategy in the families Acanthodrilidae and Eudrilidae that predominate in these regions. In contrast, in South American tropical forests and savannas, the dominant family Glossoscolecidae, contains a large number of true anecic species, as does the Lumbricidae in temperate areas. However, as indicated previously, these distributions may also be explained in terms of soil nutrient concentrations and it is not yet possible to discriminate between the two effects.

4.2.2 INGESTION OF ABOVE-GROUND LITTER BY FIELD COMMUNITIES

Anecic earthworms ingest a mixture of litter and soil although the relative proportions of both components vary between species, showing a clear gradient from epigeics to anecics to polyhumic endogeics (Figure IV.48; Table IV.16). Anecics generally ingest large debris whereas epigeics tend to feed on smaller debris, probably due to their smaller size. The proportion of plant debris in the casts of anecics may be highly variable since they ingest large proportions of litter when they are actively feeding at the soil surface, and ingest a preponderance of soil when they are active deeper in the soil, creating new galleries or avoiding unsuitable surface conditions. Tropical species of a given ecological category generally ingest a much lower proportion of organic debris. Earthworms also ingest less soil when litters are more palatable (Cortez and Hameed, 1988) and contain more nitrogen (Abbott and Parker, 1981; Lavelle *et al.*, 1989). Ferrière (1980) observed that co-existing populations of different species may separate their niches by selectively ingesting litters of different plant species. Finally, tropical anecics seem to ingest a smaller proportion of litter than their temperate equivalents (Kanyonyo ka Kajondo, 1984). The large anecic Acanthodriline *Millsonia lamtoiana*, a common species in the savannas of Côte d'Ivoire, ingests 10 to 15 % organic debris which consists of 69.2 % large grass particles, 19.3 % unidentified debris, 3.9 % seeds,

3.4 % tree leaves, and smaller proportions of roots (1.7 %), stems and other debris. Comparison of the sizes and relative proportions of these dietary items indicates that grasses and leaves are more completely digested than seeds, roots or stems.

Re-ingestion of casts is frequent in temperate climate anecic earthworms. It accounts for 50 % of the diet of *Nicodrilus velox*, a large anecic of eastern France; the remaining 50 % comprises 20 % of leaf material and 30 % of mineral soil (Bouché *et al.*, 1983). This behaviour has also been noted in laboratory studies (Binet, 1993; Cortez and Hameed, 1988) and clearly separates anecic from endogeic species since the latter do not re-ingest their casts until their compact structure has been broken down.

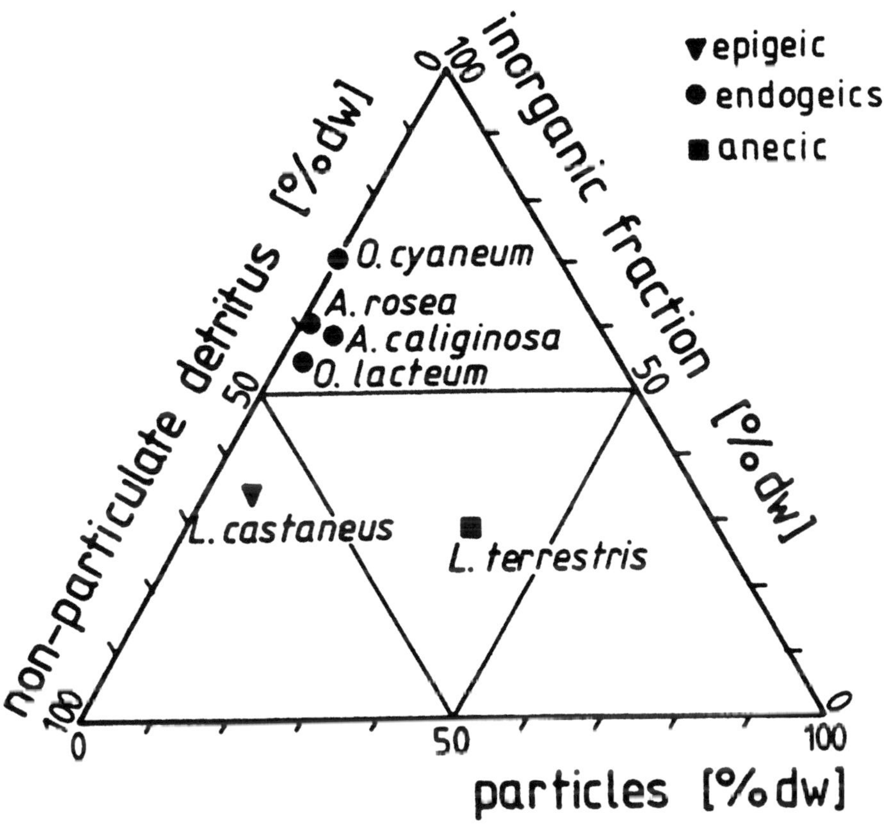

Figure IV.48 Compositions of the midgut contents of six earthworm species from a temperate deciduous forest (Judas, 1992). O: *Octolasion*, A: *Aporrectodea*, L: *Lumbricus*.

Table IV.16 Organic matter concentrations (%) in earthworm gut contents (after Judas, 1992). Data in parentheses indicate the proportion of gut contents present as plant particles >0.5 mm.

Biotope	Grassland	Woodland	Beech wood	Pasture	Savanna
Reference	Bouché and Kretz-schmar (1974)		Judas (1992)	Piearce (1978)	Kanyonyo (1984)
Method	Ultrasonic/ centrifuge	Ultrasonic/ centrifuge	Ultrasonic/ centrifuge	Loss on ignition	Ultrasonic/ centrifuge
Origin	Crop/Gizzard	Crop/Gizzard	Crop/Gizzard	Faeces	Crop/Gizzard
Anecics					
Lumbricus terrestris		39	76 (38)		
Allolobophora longa	38			20	
Millsonia lamtoiana					9-14
Epigeics					
Lumbricus castaneus			61 (+)	26	
Lumbricus rubellus				17	
Dendrodrilus rubidus		26			
Endogeics					
Aporrectodea chlorotica	62	15		16	
Allolobophora caliginosa		8	39 (+)	15	
Allolobophora rosea	60		40 (0)		
Octolasium lacteum			44 (0)		
Octolasium cyaneum			29 (0)		

4.2.3 ANECIC STRUCTURES

Anecic earthworms mainly produce surface and subterranean casts, and excavate burrow systems.

4.2.3.1 *Casts*

Anecic earthworms cast around the openings of their galleries, at the soil surface. Cast size depends on that of the worm and may range from some millimetres to several centimetres in diameter. Their structure is also variable. They generally comprise an accumulation of somewhat isolated round or oval-shaped pellets, one to several millimetres in diameter, which may coalesce into 'paste-like slurries' that form large structures of rounded shape (Lee, 1985). These casts have relatively high proportions of organic matter, especially large particles of plant material (Graff, 1971; Aldag and Graff, 1975) and a larger proportion of small mineral components than in the surrounding soil (see *e.g.*, Bolton and Phillipson, 1976).

Anecics also produce subterranean casts which may partly fill their burrows, or are deposited into other sorts of soil macropores. The abundance, structure and composition of such casts has not yet been documented. Some results suggest that they are different from casts produced by endogeics: in a microcosm study, Shaw and Pawluk (1986) showed that the endogeics *Octolasium tyrtaeum* and *Allolobophora turgida* created granular structures in sandy loam and clay loam soils whereas the anecic *Lumbricus terrestris* created a fused structure. According to these authors, the combination of both functional groups allows the formation of the most favourable structure.

The annual production of surface casts has been frequently estimated with average values of *ca.* 5 Mg ha^{-1} in temperate areas where anecic earthworms are dominant (reviewed in Lee, 1985). However, these represent only a fraction of the overall material actually passed through the earthworm gut, and the casts produced by anecics and endogeics are rarely distinguished.

4.2.3.2 *Galleries*

Abundance and structure

Anecic earthworms may create relatively sophisticated burrow systems (Lamparski *et al.*, 1987). In French temperate climate pastures, the burrow system comprises a large number of galleries with total lengths of 142 to 890 m m^{-2} and overall surface areas of 1.6 to 12 m^2 m^{-2}, depending on the season (Kretzschmar, 1982). Galleries are concentrated in the upper 60 cm of soil while an average of 16.3 % occur below this; they may also be found at depths of 2-3 m. The majority (56.9 %) of galleries are subvertical whereas subhorizontal galleries only account for 18.9 % of the total. Most galleries are rather narrow and their average diameter is in the range 2-4 mm (Table IV.17). Similar observations were made in a Dutch polder recently colonised by anecic earthworms (especially *Allolobophora*): density of galleries rapidly increased up to a maximum of *ca.* 800 m^{-2} at 5 cm depth after four years and then diminished slightly following a decrease in earthworm biomass. Most burrows were found in the upper 40 cm of soil (Ligthart, 1996). Estimates of 'active galleries', opening at the soil surface, at seven French sites gave much lower figures, ranging from 6.2 to 66.6 m m^{-2}, with average distances between active burrows ranging from 10 to 30 cm in the upper 40 cm of soil (Lopes-Assad, 1987). This indicates that a large part of the functional galleries may not open at the soil surface, either because they have never done so or because they have been partly refilled, especially when earthworms have become inactive or have died.

Burrow systems are dynamic entities which are periodically renewed. In temperate-climate regions of the northern hemisphere, burrows are actively dug and maintained during winter, from October to February whereas in late Spring and Summer, surface activity diminishes and burrows are partly filled in as anecic worms enter diapause (Kretzschmar, 1982). Ligthart (1996) found that cattle trampling and earthworm casting in burrows were the main reasons for burrow destruction. Estimates of annual rate of destruction were respectively 1172 ± 401 burrows m^{-2} at 2.5 cm depth and 193 ± 55 m^{-2}

at 22.5 cm. This implies that burrows may exist for a few months close to the soil surface but for several years deeper in the soil.

Table IV.17 Mean characteristics of an earthworm gallery system in a French pasture (after Kretzschmar, 1982).

	Abundance m^{-2} (range)	Distribution in depth: % per stratum	Diameter mm: % in size class	Inclination ° to horizontal: % per class
Length (m)	435 (142-890)	0-20:20.6 20-40:31.2 40-60:31.9 60-80:12.6 80-100:3.2 >100:0.5	0-2:3.1 2-3:37.3 3-4:46.8 >4:12.8	<30:19.9 30-60:24.2 60-90:56.9
Volume (l)	4.9 (1.4-9.2)			
Number	14,280 (4000-30,000)			

Overall developed surface: 1.6-12m^2 m^{-2}

Micromorphology of burrows

Galleries are cylindrical and their walls are regularly recoated with cutaneous mucus each time the worm passes through. Sometimes, worms deposit casts on the walls of their galleries; this seems to occur more frequently in deep soil strata rather than close to the surface. As a result, there is a continuity between the soil porosity and earthworm galleries in the upper centimetres of soil and a discontinuity in the deeper soil. Such specific features as iron-stained linings are not uncommon; they evidence the important physico-chemical changes which can influence soils close to earthworm galleries (Jeanson, 1979; Kretzschmar, 1987; Lamparsky *et al.*, 1987).

4.2.4 ASSOCIATED MICROFLORA AND FAUNA

The soil microflora is remarkably concentrated at the surface of gallery walls and within the adjacent 2 mm of the surrounding soil. This micro-environment comprises less than 3 % of the total soil volume but contains 5 to 25 % of the whole soil microflora with a predominance of certain functional groups. Typical figures are 13 % of anaerobic N-fixers, 16-40 % of denitrifiers, over 40 % of proteolytic and aerobic N-fixers and more than 60 % of hemicellulolytic and pectinolytic bacteria. However, the microbial communities of galleries and the surrounding drilosphere vary markedly with depth and between seasons (Bhatnagar, 1975; Rouelle, 1977). In a microcosm experiment with an earthworm biomass equivalent to *ca.* 800 kg fresh weight ha^{-1}, the densities of bacteria and protists were 100 times greater on burrow walls than in the soil, whereas fungal biomass decreased in litter deposited at the soil surface (Binet, 1993).

The anecic drilosphere community is also characterised by an increased proportion of actinobacteria and protists with lesser (or adverse) effects on fungal populations. Hendrix *et al.* (1986) observed that minimum tillage stimulates earthworm activities and the microbial communities that were dominated by fungi in conventional systems, become dominated by bacteria in 'no-till' plots, as a consequence of earthworm activities. Nonetheless, a detailed survey of the available literature shows that the results differ between soils and with earthworm species; the hierarchy of determinants that influence the composition of microbial communities is not yet fully understood (Brown, 1995; Edwards and Bohlen, 1996).

The fate of the ingested micro-organisms varies broadly; some groups may be digested by the worms, especially protists, algae and some fungi or bacteria whereas a large proportion remains unharmed (see *e.g.*, Cooke and Luxton, 1980; Piearce and Phillips, 1980; Rouelle, 1983; Brusewitz, 1959). Marked differences seem to exist between species since bacterial densities increase during gut transit in *Lumbricus terrestris* and *Lumbricus rubellus*, but decrease in the gut of *Aporectodea caliginosa* (Kristufek *et al.*, 1992; Parle, 1963). Finally, some astomatous ciliates are permanent residents of the gut of anecic earthworms (de Puytorac, 1954; Piearce and Phillips, 1980; Rouelle, 1983).

The effects of earthworms on other soil faunal groups are variable. Earthworm activities tend to depress nematode populations, especially those of phytoparasitic species (Roessner, 1986; Boyer). This effect is probably due to changes in the soil environment (Yeates, 1981) and positive effects on nematophagous fungi (Edwards and Fletcher, 1988), rather than direct feeding on nematodes (Dash *et al.*, 1980). Litter arthropod populations are generally depleted when anecic earthworms ingest most of the litter present and mix it with the soil. Nonetheless, their populations may increase locally in litter accumulated as 'middens' around the gallery openings into the soil (see, for example, Szlavecz, 1985).

4.2.5 FLUX OF LITTER AND SOIL THROUGH THE ANECIC DRILOSPHERE

Where abundant, anecic earthworms may incorporate the total annual litter mass produced into the soil within as little as two to three months (Lee, 1985). Their activity seems to be limited more by food availability than by the environmental determinants of their activity. In a microcosm experiment conducted at 12 °C with a population of *Lumbricus terrestris* equivalent to *ca.* 800 kg fresh weight ha^{-1}, 85 % of litter deposited at the soil surface had disappeared in three months (Binet, 1993). In southern France, Bouché (1983) estimated that earthworms had assimilated 30 % of the carbon available in litter in 17 weeks. Such a figure is only acceptable if earthworms re-ingest the litter contained in their casts several times.

In the humid savannas of the Côte d'Ivoire, anecic earthworms annually incorporate the equivalent of 180 to 510 kg of dry litter into the soil, in regularly burned savannas. This represents *ca.* 30 % of the weight of litter annually decomposed (*i.e.*, mineralised or incorporated into the soil instead of being destroyed by fire), but less than 10 % of the annual production. In unburned savannas, ingestion increases up to 980 kg ha^{-1}, which represents 9 % of the annual litter biomass decomposed (Lavelle, 1978).

4.2.6 FUNCTIONING OF THE ANECIC DRILOSPHERE

Earthworm activities influence two major soil processes: (i) the dynamics of litter and soil organic matter decomposition and (ii), the physical structure of the soil. Anecic and endogeic drilospheres are distinguished, although they may sometimes have comparable characteristics.

4.2.6.1 *Effects of anecic earthworms on C and nutrient cycling*

C and nutrient cycles in the anecic drilosphere are especially characterised by the:
1. relatively low 'direct impact' of digestion on litter decomposition;
2. rapid turnover of C and N in the earthworm biomass;
3. important translocation of litter into the subsoil and mineral soil close to the surface;
4. accelerated turnover of SOM;
5. further transformation of organic matter mediated by microbial activity within casts.

Direct effects of digestion

Assimilation of litter by anecic earthworms seems to be highly variable. Maximum assimilation rates between 30 and 75 % have been measured in *Lumbricus rubellus,* depending on litter quality and temperature (Dickschen and Topp, 1987; Daniel, 1991) although much lower values (11.6-28.5 %) are reported by Crossley *et al.* (1971) for a number of other species. The degree of previous litter decomposition, ambient temperature and moisture conditions are other important determinants of litter assimilation in earthworms, irrespective of species (Cooke and Luxton, 1980; Cooke, 1983; Cortez *et al.*, 1989). Specific rates of assimilation of N by anecic earthworms have not been measured.

The flux of energy through anecic earthworms has rarely been estimated. Populations of *Lumbricus terrestris* may assimilate *ca.* 10 % of litter fall in English woodlands (Lakhani and Satchell, 1970). Based on only a few more estimates, the total assimilation of litter by anecic earthworms is not considered likely to exceed 5 % of overall plant production (Lee, 1985). The dramatic effects of anecic earthworms on litter decomposition and soil functioning are therefore mainly due to the reduction they effect in litter particle size, its translocation into the soil and the large and selective stimulatory influences they have on microbial communities.

The mineral soil that is mixed with litter is mostly derived from the upper 15 cm and its organic matter is poorly assimilated; 95 % of the C assimilated by the anecic worm *Nicodrilus velox* originates directly from ingested leaf litter, or after a period of incubation in old casts (Bouché *et al.*, 1983). This finding is corroborated by observations based on $\delta^{13}C$ measurements (Scharpenseel *et al.*, 1989; Martin *et al.*, 1991) and microcosm experiments (Cortez *et al.*, 1989) which show that anecic earthworms only assimilate young organic matter from the litter.

As a result of digestion, a significant proportion of the organic matter present may be directly cycled through earthworms and the contained nutrients released. Estimates range from a few kilograms to 100 kg N ha^{-1} yr^{-1} released in temperate soils (equivalent to 20 % of the total amount recycled annually). Differences may be due to variation in earthworm abundance, but can also result from inaccuracies in the assumptions made in estimating the sizes of those compartments that are difficult to measure. These include such factors as the amounts of N released in urine and cutaneous mucus (Syers *et al.*, 1979; Keogh, 1979; Scheu, 1987; James, 1991). Most of the nutrients thus released have originated from casts which have high concentrations of assimilable nutrients.

Turnover of C and N in earthworm biomass

The rapid turnover of C and N in the earthworm biomass is a striking feature of the biology of these animals. Under laboratory conditions, the C and N of anecic earthworm bodies may be turned over in a few months (Bouché and Ferrière, 1986; Cortez *et al.*, 1989; Binet, 1993). Part of this N is released in casts as ammonium, but more may directly flow into the soil, through nephridiopores as urine (Needham, 1957), readily assimilable cutaneous mucus or decaying bodies. Laverack (1963) assumed that the amount of N contained in urine as ammonium and urea represents about half of that excreted through the intestinal wall and the tegument; the amount of N secreted in urine may therefore be greater than that in cutaneous mucus. Few estimates of the influence of earthworm populations on soil nitrogen cycling include such effects which may lead to the release of 10-20 kg mineral N ha^{-1} $year^{-1}$ together with a further similar amount released from decomposing dead bodies. In particular, the chemical nature and flow of cutaneous mucus are still poorly known (see Cortez and Bouché, 1987; Scheu, 1991) although they certainly play a major role in the enhanced microbial activity observed, for example, on burrow walls.

Transfers of C and nutrients in the soil system

Anecic activity results in a rapid and complete mixing of litter with the mineral soil and an accelerated organic matter turnover. Litter does not accumulate at the soil surface and the humus formed is a neutral mull. In the presence of anecic earthworms, ^{14}C released into the atmosphere during nuclear tests and incorporated in litter, has been thoroughly mixed with the soil down to a depth of 18 cm. In the absence of earthworms no ^{14}C was found below 10 cm (Stout and Goh, 1980). Field and laboratory experiments using artificially-labelled grass litter deposited at the soil surface have confirmed the ability of earthworms to rapidly incorporate litter into the soil (Binet and Curmi, 1992; Hameed *et al.*, 1993). The half-time of apparent decomposition of ^{14}C-labelled litter of perennial ryegrass (*Lolium perenne)* deposited at the surface of a temperate soil was 5.7 weeks in the presence of earthworms and 20 weeks in their absence (Dietz and Bottner, 1981).

Organic matter removed from the litter system through feeding by anecic earthworms accumulates in drilosphere structures. Microcosm experiments have shown that, in the presence of *Lumbricus terrestris,* three times more N from litter was incorporated into the soil in comparison with a treatment without earthworms. Seventy-two per cent of this earthworm-incorporated nitrogen was located within drilosphere structures (Figure IV.49) (Binet, 1993). Another important effect was the favouring of bacterial over fungal proliferation.

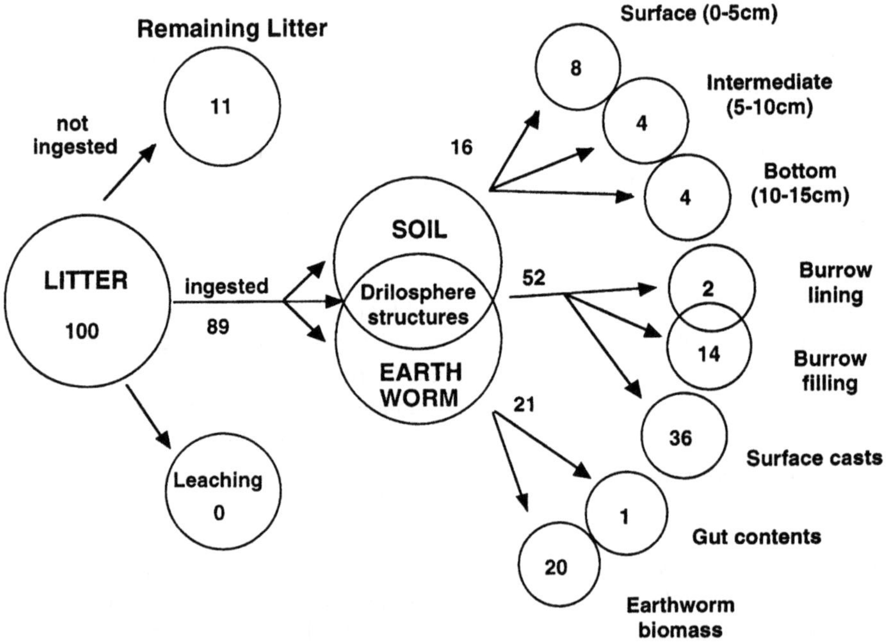

Figure IV.49 Distribution of ^{15}N from litter deposited at the soil surface after 85 days in a microcosm inoculated with *Lumbricus terrestris* (Binet, 1993).

Mixing of surface-litter with the mineral soil, and its incubation in casts

The intimate admixture of organic and mineral components in the casts of anecic earthworms results in the formation of stable organo-mineral structures. As a result, soil organic matter in casts tends to be more highly humified than the originally ingested plant material (Rafidison, 1982). The importance of this process in soil formation has been emphasised elsewhere (see Chapter II). Microbial incubation in casts during the few hours to days following their deposition may result in a significant release of assimilable P (*e.g.*, Sharpley and Syers, 1977; Satchell *et al.*, 1984), nitrogen fixation (Shaw and Pawluk, 1986; M'ba, 1987) or denitrification (Elliott *et al.*, 1990), depending on micro-environmental conditions.

4.2.6.2 *Soil physical properties*

Porosity

Both anecic and endogeic earthworms contribute generally to the maintenance of elevated soil porosity through the creation of galleries. Deposition of casts at the soil surface regulates soil porosity since the volume of voids created in the soil is at least equivalent to that of the casts and this may represent the annual formation of volumes of tens of cubic metres of voids per hectare. Such voids may be created preferentially at times and at sites where compaction is occurring. Seasonal variation in soil properties has sometimes been related to changes in the abundance of earthworm populations (Hopp, 1973; Marinissen, 1995). Soil porosity increases of the order of 75 to 100 % may follow the colonisation of new soils by earthworms (Hoeksema and Jongerius, 1959) with a predictable increase in the proportion of large pores (Edwards and Lofty, 1982; Knight *et al.*, 1992).

Infiltration of water and gas exchange
Burrows have a significant effect on water infiltration despite their small contribution (generally less than 5 %) to the total soil volume (Edwards and Bohlen, 1996). Experiments in arable land have demonstrated that when direct-drilling is substituted for ploughing, anecic earthworms are favoured. Hydraulic conductivities at the interfaces of the A and B horizons may be doubled through their enhanced activities (Douglas *et al.*, 1980; Urbanek and Dolezal, 1992; Joschko *et al.*, 1992; Ligthart 1996) although many other authors have emphasised the fact that burrow systems may only affect the flux of water through the soil if ponding occurs at the soil surface. When this occurs, infiltration rates will depend on the connectivity of the pore space and the properties of the burrow walls. These may have linings of fine particles, a compressed zone around the periphery, or no modifications compared with the surrounding soils. Gas diffusion is affected in the same way although diffusion into the soil surrounding the galleries is greater than direct diffusion along the burrows (Kretzschmar, 1989).

In microcosms inoculated with the epi-anecic species *Lumbricus rubellus,* Kladivko *et al.* (1986) concluded that these earthworms could diminish surface crusting through the formation of stable macro-aggregates.

4.3 The endogeic drilosphere

4.3.1 DISTRIBUTION OF ENDOGEIC EARTHWORMS

Endogeic earthworms form part of most earthworm communities in environments distributed from the cold temperate areas to the humid tropics (Figure III.45). Polyhumic endogeic species are the most populous worms in temperate and semi-dry tropical areas. They represent a functional group intermediate between true endogeics and anecics since some species actually feed on surface-litter that they drag from underneath and mix with the soil. The large proportion of plant remains sometimes found in their guts (up to 50-70 %) attests this behaviour. These earthworms also create galleries which are mainly subhorizontal (Kretzschmar, 1982).

Mesohumic and oligohumic endogeics are most typical as they feed solely on soil organic matter and do not build burrow systems. Gut content analysis has shown no selective ingestion of dead plant tissues.

At a broad geographical scale, the distribution of true endogeics is mainly determined by temperature. Since they feed on low quality soil organic matter, they can only thrive where high temperatures allow an efficient digestion through a mutualistic association with the soil microflora (see Section IV.4.1.2).

At a regional scale, for example in tropical rainforests, endogeics mostly occur in humid soils with a near-neutral pH and relatively high concentrations of nutrient elements and organic matter (Fragoso and Lavelle, 1992). They usually comprise 75 to 99 % of earthworm biomass in most humid tropical pastures and some tropical forests of tropical America and western Africa (Lavelle, 1983c).

Finally, there is some evidence that endogeics and anecics may occur at different times during successional processes. Endogeics may feed on the faecal pellets of litter invertebrates and litter-feeding earthworms, and thus predominate in successions after litter feeders have converted large amounts of plant material into pellets that the endogeic species may ingest. This situation has been observed *e.g.*, by Cluzeau (1992) in successional processes triggered by the input of domestic wastes into natural heathland in Brittany (France). Similar successional processes seem also to occur in rubber plantations where the carbon from trunks left at the soil surface after clearing the forest, is first used by xylophagous termites, and then transferred to humivorous termites and endogeic earthworms that feed on the faecal pellets deposited in the upper few centimetres of the soil (Gilot *et al.*, 1995).

4.3.2 ENDOGEIC STRUCTURES

Endogeics produce casts within the soil and at the soil surface. Annual production of surface casts may be as high as 14-15 Mg ha^{-1} as shown in Costa Rican tropical pastures colonised by endogeic populations (Fraile, 1989). Even higher figures of 25-30 Mg ha^{-1} have been recorded from the moist savannas of the Côte d'Ivoire and the Cameroons (Kollmannsperger, 1956; Lavelle, 1978). However, at Lamto (Côte d'Ivoire), surface casts only represent 1.7 to 3.5 % of the 1,000 t ha^{-1} of dry soil that endogeic earthworms annually ingest. Interestingly, this proportion varies seasonally, probably as a response to soil compaction and is at a maximum when the overall soil ingestion is at its lowest (Figure IV.50).

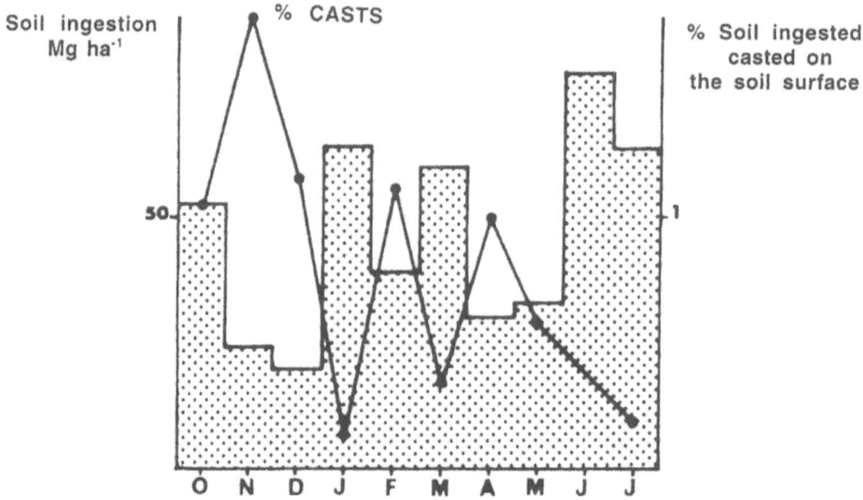

Figure IV.50 Monthly variation in the production of surface casts (solid line) as a percentage of the whole soil ingested by earthworms (blocks) in a Guinean savanna (Lamto, Côte d'Ivoire) (after Lavelle, 1978).

Two major kinds of casts may be distinguished:

(i) globular casts comprising coalescent round or flattened units (Figure I.17);

(ii) granular casts formed from an accumulation of small, fragile and fine-textured pellets.

Production of either type of casts appears to be related to soil texture and the size of the worm. Specific anatomic features have been identified that allow the worm either to release large units at regular intervals that will form the globular cast, or a thin constant flux of cast material that will form the poorly structured granular casts (Lapied and Rossi, unpublished data).

4.3.3 ASSOCIATED MICROFLORA

The microbial communities of earthworm guts and casts differ little from those of uningested soils, at least in terms of their species composition. Under favourable conditions, the endogeic drilosphere may comprise a large proportion or even the entire volume of the soil (Section II.3.3.2). This may result from the deposition of casts over long periods, by different earthworm populations.

Soil micro-organism populations tend to decline slightly in the anterior part of the gut of the tropical earthworm *Pontoscolex corethrurus*. In this preliminary phase of the reorganisation of microbial communities, some micro-organisms may be killed and digested by the worm. In the median and posterior parts of the gut, and in the fresh casts, microbial activity is highly enhanced, although no significant changes are observed in the density and composition of microbial communities (Barois, 1987). However, the overall microbial biomass is significantly greater in fresh casts (Figure IV.51). These results indicate that, broadly, the microflora of the earthworm gut differs little from that of the non-ingested soil. Detailed microscopic observation of gut walls has demonstrated the existence of specific micro-organisms although it is possible that the environmental conditions that prevail in the gut differentially stimulate microbial populations of distinct species or strains. The gut contents of endogeic earthworms and their casts must be considered as specific microhabitats, fully continuous with the soil. In the first two days following deposition microbial biomass decreases steadily and stabilises at values close to those of the control soil (Figure IV.51).

4.3.4 FUNCTIONING OF THE ENDOGEIC DRILOSPHERE

Effect on soil physical properties

Endogeic earthworms in particular, may favourably affect soil structure by promoting the formation of large stable macro-aggregates greater than 2 mm in diameter (see Chapter I). Their effects on water infiltration and bulk density may also be important although both positive and negative effects have been reported. In Western Africa, Casenave and Valentin (1989) found a clear relationship between the presence of earthworms, mainly endogeics, and rates of water infiltration. However, endogeic subsystems may not be as important as those of the anecic drilosphere due to the even distribution of macropores and the lack of real galleries (Figure IV.52).

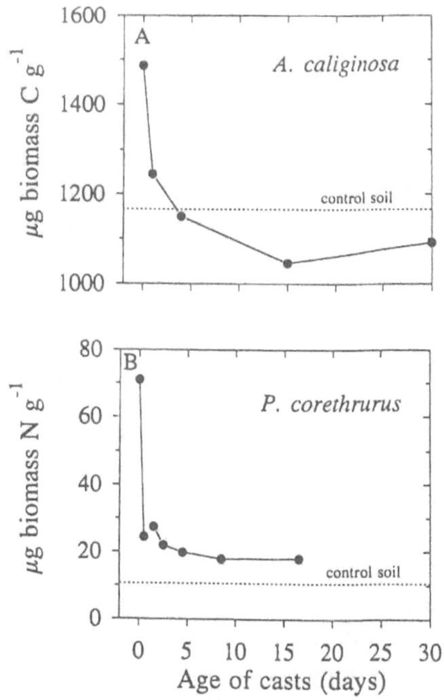

Figure IV.51 A. Changes in microbial biomass C over time in the casts of *Aporrectodea caliginosa* (Scheu, 1987); and B. Changes in microbial biomass N over time in casts of *Pontoscolex corethrurus* (Blair *et al.*, 1995, after Lavelle *et al.*, 1992b).

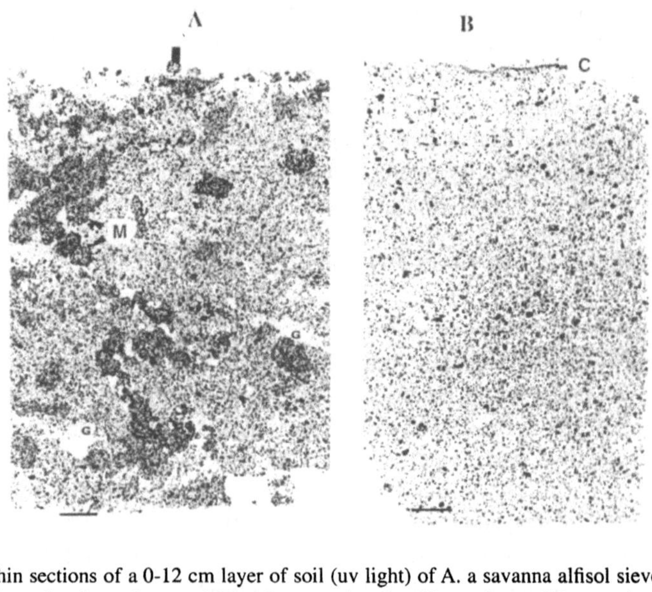

Figure IV.52 Thin sections of a 0-12 cm layer of soil (uv light) of A. a savanna alfisol sieved at 2-mm and inoculated with the endogeic earthworm *Millsonia anomala*, after 30 months, and B. an uninoculated control, after 30 months. Note surface crust (C) in A, compact casts of *M. anomala* (M) and subhorizontal galleries (G) in B (Blanchart, 1992) (bars = 1 cm).

Selection of particles

Most endogeic earthworms exhibit some degree of selection in their choice of both organic and mineral soil particles. In temperate regions, Bouché and Kretzschmar (1974) found concentrations of plant debris of up to 15 % thus demonstrating that these endogeic earthworms actively select for this type of material. Most earthworm casts actually have greater concentrations of organic matter than the surrounding soil in field conditions and in cultures with homogeneous soil (Lee, 1985). The rate of relative concentration, however, varies with the species, and with the soil organic matter concentration (Barois *et al.*, 1999) (Figure IV.53).

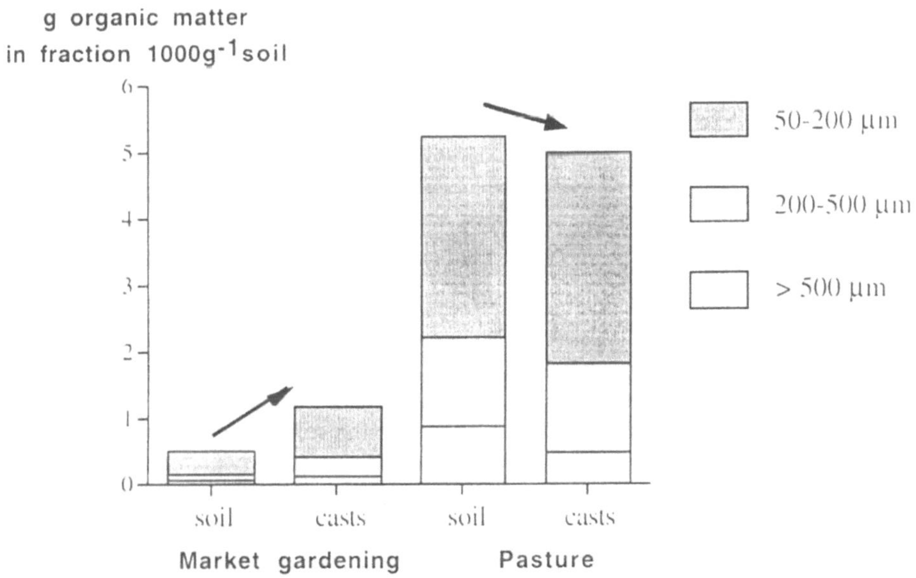

Figure IV.53 Proportion of plant debris > 50 µm in casts of the tropical endogeic earthworm *Polypheretima elongata* fed the same soil but with contrasting organic contents (Duboisset in Barois *et al.*, 1999).

All earthworms select mineral materials of particular size classes, depending on species and soils. Standard measurements made for nine species of anecic and endogeic tropical species show a clear preference for clay-sized materials in soils with clay contents less than 30 %; a few cases of significant selection for sand particles have been indicated in highly clay-rich soils (Figure IV.54).

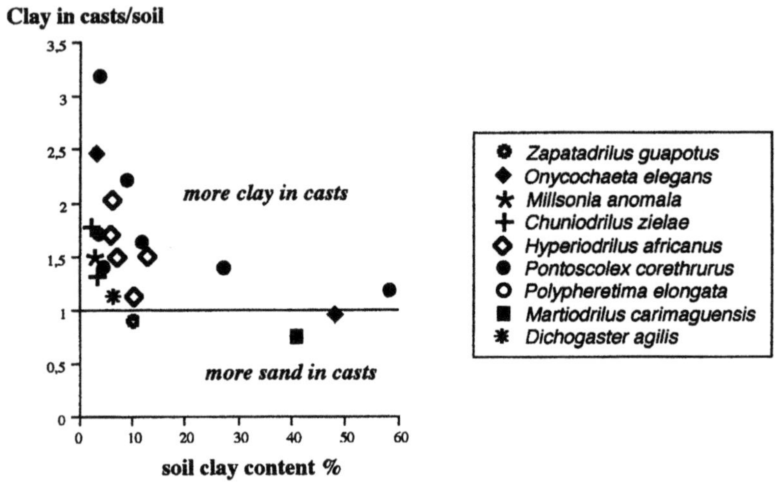

Figure IV.54 Effect of soil clay content on selective ingestion of clay particles by nine different tropical earthworm species (Barois *et al.*, 1999).

Soil aggregation

Endogeic earthworms may annually ingest the moist equivalent of several hundred to more than one thousand Mg dry soil per hectare (Lavelle, 1978; Boström, 1988). In four arable soils with, respectively, unfertilised and fertilised barley, grass and lucerne crops, annual cast production by *Aporrectodea caliginosa* was estimated to be between 36 and 108 Mg ha^{-1}, of which 20 to 50 % was deposited at the soil surface. In humid tropical grasslands, endogeic earthworms may ingest daily 5-6 (*Pontoscolex corethrurus*) and up to 25-30 (*Millsonia anomala)* times their own weight of soil (Lavelle, 1975; Lavelle *et al.*, 1987). Overall annual cast production may therefore commonly range from several hundreds to a maximum of 1250 Mg per hectare. As indicated above, only a small proportion of these casts (5 to 20 % depending on species) are deposited at the soil surface and subterranean casts form the component units of stable macro-aggregate structures (Figure IV.52).

Experiments have demonstrated that such structures may be built in surprisingly short periods of time suggesting that factors other than the simple production of globular casts contribute to aggregation. Blanchart *et al.*, (1989) found that in 33 days, the activity of 5 g of *Millsonia anomala* resulted in the formation of 2883 g of aggregates (*i.e.*, 42.4 % of the soil in the experimental container), as compared to 906 g (13.3 %) in a control treatment and 1075 g (15.8 %) in a treatment with the grass *Panicum maximum.* Cast production by the earthworms was estimated at 1815 g, that is, 26.7 % of the aggregates formed; the remaining 15.7 % may have resulted from the proliferation of fungal hyphae, perhaps those of mycorrhizal fungi associated with the roots. Similar levels of aggregation have been obtained with the endogeic earthworm *Aporrectodea caliginosa* (Hopp and Hopkins, 1946). After four days, two worms kept in 75g of soil had formed 11.6 g of stable aggregates (15.5 %) *i.e.,* more than three times the control without worms.

When earthworms were excluded from a soil in which they had built a macro-aggregate structure, this structure remained intact for a long period (at least several years) due to the stabilisation of aggregates that takes place over time. However, the introduction of earthworms producing granular casts destroys the structure as large aggregates are split into much smaller and more fragile units. Soil aggregation is regulated through this process with small earthworms breaking down large aggregates thereby preventing an excessive accumulation of large compact aggregates which negatively affects plant growth (Rose and Wood, 1980; Chauvel *et al.*, 1999).

Blanchart *et al.* (1997) have proposed the classification of endogeic earthworms into, respectively, 'compacting' and 'decompacting' species, depending on their effects on soil macro-aggregation and bulk density (Figure IV.55). In a series of experiments conducted at Lamto (Côte d'Ivoire), undisturbed monoliths of soil dug out from the savanna were de-faunated by temporary flooding and replaced in the field; some were isolated from the surrounding soil with a fine stainless steel mesh and reinoculated either with large-sized *Millsonia anomala*, or filiform *Chuniodrilus ziclae;* in another treatment, monoliths were reintroduced with no metallic mesh allowing recolonisation by the soil fauna from the surrounding soil. After 20 months, aggregation and bulk density were lowest in the treatment with filiform Eudrilid worms thereby confirming their effect as soil 'decompactors'. In the absence of fauna, soil was significantly less aggregated, whereas inoculation with *Millsonia anomala* significantly increased the bulk density and the proportion of macro-aggregates >2 mm in diameter.

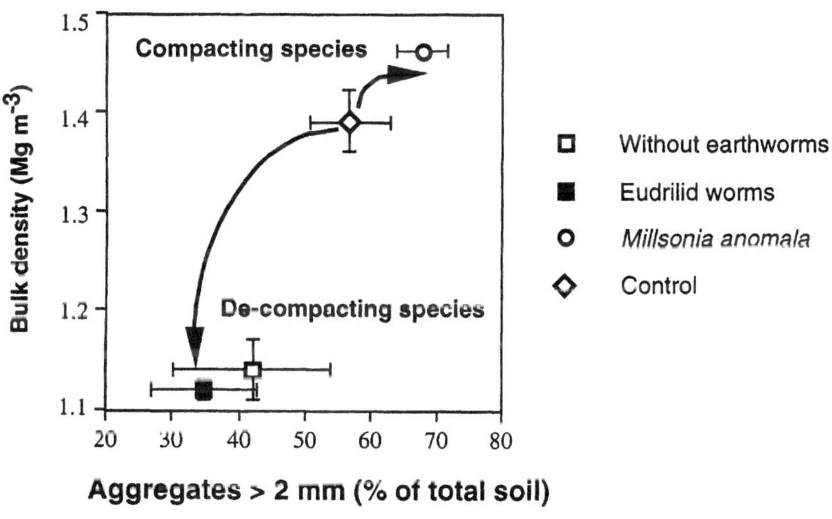

Figure IV.55 Effects of compacting and decompacting earthworm species on aggregate size distribution and bulk density in a savanna soil (Lamto, Côte d'Ivoire) Bars indicate S.E. of means (adapted from Blanchart *et al.*, 1999).

The effects of 'compacting species' may also depend on factors operating at larger scales of time and space, notably organic matter supply and the types and abundances of the different clay materials present. In a three-year field experiment conducted in Peruvian Amazonia, the earthworm *Pontoscolex corethrurus* was introduced into the soil of an annual cropping system with significant positive effects on plant production, and evidence of soil compaction in some treatments (Alegre *et al.*, 1996). Compaction only occurred in treatments with no organic inputs; in these, earthworms produced casts of low structural stability which dispersed to form a compact, continuous surface crust. In contrast, in soil treatments where stubble mulch and legume green manure were applied, casts retained a solid structure and the soil of the upper 4 cm had a large proportion of macropores and macro-aggregates, with much better conditions for infiltration (Duboisset, 1995).

Regulation of SOM dynamics
Endogeic earthworm populations in soils of the humid tropics have been shown to exert a significant role on SOM dynamics at different scales of time and space (Martin, 1991; Lavelle *et al.*, 1998)(Figure IV.56).

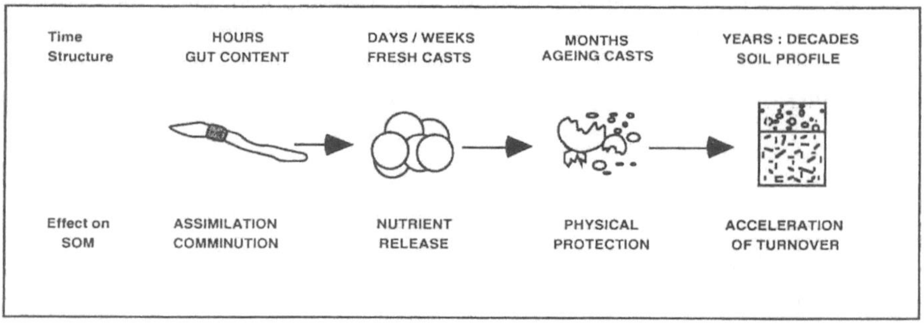

Figure IV.56 Effects of earthworms on soil organic matter at different scales of time and space (Lavelle, 1997).

In the short term. Earthworm activity results in the assimilation of a relatively small proportion of SOM, ranging from 2-6 % to a maximum of 18 % (Martin *et al.*, 1990; Barois *et al.*, 1987). Clear changes in the distribution of SOM between particle size fractions have also been found following digestion by the endogeic earthworm *Millsonia anomala*: that in the coarse fractions decreased sharply (*e.g.*, 25 % in the 250-2000 μm fraction) while SOM in the finer fractions tended to increase. Similar changes were noted with *Polypheretima elongata* in a Martiniquan vertisol (Figure IV.53). By selectively feeding worms with particle size fractions naturally labelled with ^{13}C, Martin *et al.* (1991) demonstrated that this effect was largely due to comminution as all particle-size fractions were equally digested under the conditions of her experiments.

Medium-term effects. Mineralisation of SOM decreases sharply in casts of the compacting species *Millsonia anomala* and stabilises at a minimum level after a few days. As a result, the SOM contents of casts remained higher than values measured in a 2 mm-sieved, non-ingested control soil incubated under similar conditions (Figure IV.57). After one year, the total amount of SOM mineralised in casts of *Millsonia anomala* in a sandy alfisol was 30 % less than that of the control and mineralisation rate was highly limited. Similar results have been obtained with European Lumbricidae, namely the endogeic species *Aporrectodea caliginosa* and the anecic *Lumbricus terrestris* (McInerney and Bolger, 1996). Physical protection probably occurs in the massive structure of compact casts and it is likely that SOM dynamics in the thin casts of 'de-compacting' species would be different. No data currently exist to test this hypothesis.

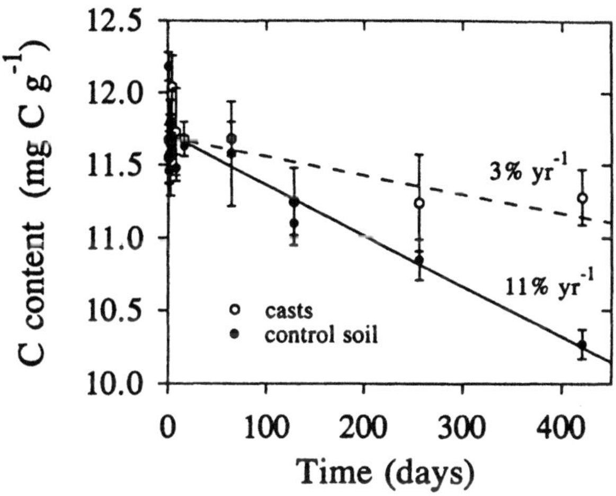

Figure IV.57 Changes in the C content of *Millsonia anomala* casts and of non-aggregated control soil (2 mm sieved) incubated for 420 days at 15 % moisture and 28°C (from Martin, 1991).

Longer-term effects. At scales of several years to decades, the effects of earthworms on SOM dynamics are determined by the relative importances of short-term acceleration of mineralisation during gut transit, and further effects in ageing casts. There is some evidence that earthworms accelerate the overall turnover of SOM while the total soil stock remains smaller but is, on average, more accessible. In a 7-year experiment in a continuous maize crop, inoculation of the endogeic earthworm *P. corethrurus* increased the loss of C due to cropping by 3.2 Mg C ha^{-1} and decreased the rate of incorporation of new C from the crop. Earthworms significantly decreased the proportion of C in coarse organic fractions (>50 µm) and increased C in the smallest (<2 µm) fraction. In a similar experiment conducted on a sandy alfisol at Lamto (Côte d'Ivoire), the rate of incorporation of new C from the crop residues was increased from 2.9 to 6.2 % (Villenave *et al.*,1999).

Experiments comparing the functioning of soil in the presence and absence of earthworms have never been conducted for periods of more than a few months to years. Consequences of the long-term accumulation in soils of earthworm biogenic structures can only be assessed through modeling. Two attempts have been made using the CENTURY model of soil organic matter dynamics of Parton *et al.* (1983). In the first exercise, the predicted effect of SOM protection in casts was for positive long term effects on SOM contents, especially in the slow pool (Martin and Parton, unpublished data in Lavelle *et al.*, 1998). In the second exercise, modeling was used to examine the consequences of surface litter incorporation below surface casts of *Pontoscolex corethrurus*, with the effects of bioturbation simulated as a gentle cultivation. The model adequately described the decline in organic matter concentration observed in a three year-experiment of a low-input agricultural system at Yurimaguas (Peruvian Amazonia), with higher losses in soils inoculated with earthworms than uninoculated soil. However, after some 12 years, the effect was reversed and predicted C stocks were 28 % greater after 30 years in systems where earthworms were present (Lierman and Woomer, unpublished data) (Figure IV.61).

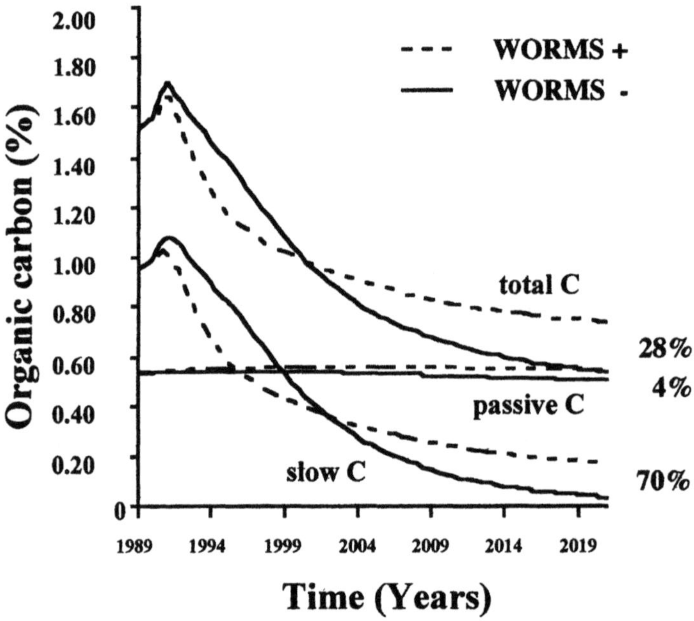

Figure IV.58 Simulation of changes in the C content of an Amazonian ultisol cropped to maize in the presence and absence of earthworms, using the CENTURY model (Lierman, Woomer and Lavelle, unpublished data).

Release of nutrients

Increase of concentrations in casts. The passage of SOM through the earthworm gut leads to the release of significant amounts of assimilable nitrogen and phosphorus. Nitrogen is secreted by the nephridia as ammonium into the gut (when nephridia open in the gut, endonephridia), or at the surface of the body (ectonephridia). Part of the ammonium found in casts may also result from microbial mineralisation before ingestion, or inside the gut. Under standardised laboratory conditions (T = 27 °C, soil previously sieved to less than 2 mm and moistened to field capacity), the amounts of ammonium released differ between soils which suggests that part may result from the stimulation of microbial activity within the gut (Lavelle *et al.*, 1992b) (Table IV.18).

Table IV.18 Concentration of mineral nitrogen in the fresh casts of the endogeic tropical earthworm *Pontoscolex corethrurus* after Lavelle, Chotte, Laurent and Bouet unpublished data (1), Lavelle *et al.*, 1992b (2) and Barois *et al.*, 1987 (3).

Depth (cm)	N content % *(s.e.)*	NH_4^+ µg g^{-1} *(s.e.)*	NO_3^- µg g^{-1} *(s.e.)*	TOTAL min. N µg g^{-1} *(s.e.)*	% mineral-isation *(s.e.)*	NH_4^+ in casts/ control
ANDISOL						
Martinique[1]						
0-5	1.52 *(0.15)*	1052 *(69.6)*	43 *(11.6)*	1095	6.7	2.26
5-10	1.26 *(0.15)*	906.1 *(71.6)*	17.1 *(6.0)*	923	6.8	2.37
15-20	0.60 *(0.06)*	531 *(29.1)*	5.7 *(2.1)*	537	8.9	2.73
VERTISOL						
Martinique[1]						
0-10	0.181 *(0.056)*	712 *(167.7)*	46.6 *(18.1)*	759	29.5	3.77
ULTISOL						
Peruvian Amazonia[2]						
0-10	0.156	133	41	174	10.0	8.9
INCEPTISOL						
Tropical Mexico[3]						
0-10	0.32 *(0.05)*	67 *(28.3)*	82 *(9.3)*	149	4.4	3.7

In fresh casts of the pantropical species *Pontoscolex corethrurus* kept in standardised laboratory cultures, concentrations of ammonium-N varied from 67 to 1052 µg g^{-1} soil. Maximum absolute values were found in an andisol with a high N concentration and minimum values in a much poorer inceptisol from Mexico. In the Martiniquan andisol, ammonium-N contents of casts decreased when the worms were fed soil from increasingly deeper soil strata with lower organic-N contents. The maximum absolute mineralisation rate between ingested soil and casts was found in a vertisol (29.5 %) while most other values were in the range 4-10 %. Finally, comparison of NH_4^+ N concentrations in control soils (in which N mineralisation was artificially enhanced by sieving the soil to less than 2 mm and keeping it at field capacity for the duration of the cultures) and casts indicates that the effect of earthworms becomes increasingly important in soils with a low organic nitrogen content.

In laboratory cultures of the earthworms *Pontoscolex corethrurus* and *Polypheretima elongata* in Martiniquan andosols, 45 to 65 % more ammonium was found in casts of the former species and suggests that N mineralisation rates differ between species (Lavelle *et al.* unpublished data). Other results suggest that tropical native species have greater mineralisation capabilities than exotics (Barois *et al.*, 1999).

Extrapolation of these results to the scales of field populations and a year shows that mineral-N released by earthworm populations may represent a significant proportion of plant requirements in grasslands. In the moist savanna of Lamto (Côte d'Ivoire), the overall release of mineral-N in casts of the endogeic earthworm *Millsonia anomala* was estimated at 13-18 kg ha^{-1} $year^{-1}$ by using the DRILOTROP model which simulates the dynamics and effects on soil of populations of this species (Martin, 1990). With the addition of N from decomposed dead earthworms and cutaneous mucus, a total amount of 21 to 39 kg assimilable N ha^{-1} is annually released in this savanna. This represents a significant proportion of the 70 kg of mineral-N ha^{-1} annually incorporated into the plant biomass. Nonetheless, the proportion of earthworm-derived N actually incorporated into plant biomass is not known.

In the absence of plants, ammonium-N released in casts rapidly disappears, either through nitrification or reorganisation into the microbial biomass (Syers *et al.*, 1979; Lavelle *et al.*, 1992b)(Figure IV.51 and IV.59). Eight days after deposition, casts of *M. anomala* had a mineral-N content equal to uningested soil control values. In soils with higher clay contents, reorganisation is slower and a few weeks may be required to achieve values equivalent to the control soils (Lavelle *et al.*, 1992 op. cit.). In those soils where nitrification does not occur, *e.g.*, the savanna soils at Lamto, ammonium may be taken up by plants or directly transformed to microbial tissues.

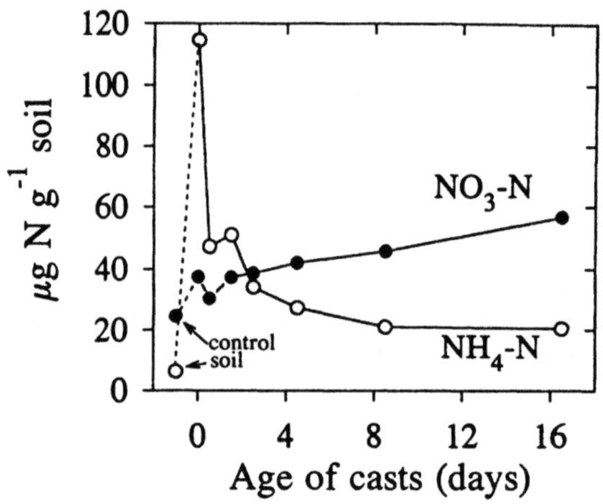

Figure IV.59 Temporal changes in the concentrations of ammonium and nitrate nitrogen in aging casts of *Pontoscolex corethrurus* fed an Amazonian ultisol (from Lavelle *et al.*, 1992b).

Phosphorus availability is also influenced by endogeic earthworm activities. In casts of *P. corethrurus*, ^{32}P exchangeable and water-extractable inorganic P increase over time with maximum values, respectively, 2-8 times and 3-8 times those of a control after one to four days (Table IV.19) (López-Hernández *et al.*, 1993). In a vertisol from Martinique (West Indies), fresh casts of *Polypheretima elongata* contained 43 % more exchangeable P than non-ingested soil, mainly due to digestion processes. Extrapolation of these results to field populations suggested that 14 to 28 kg ha^{-1} year^{-1} of assimilable P are released in these exotic pastures (Brossard *et al.*, 1996).

Table IV.19 Forms of 'available' P in the casts of *Pontoscolex corethrurus* and in control soils (López-Hernández *et al.*, 1993).

Soil	Water extractable inorganic P (μg ml^{-1})	Exchangeable P (μg g^{-1})
Ferric acrisol (Lamto, Côte d'Ivoire)	0.06 (0.05)	2.00 (0.11)
Casts (12h)	0.158 (0.05)	6.00 (1.78)
Casts (96h)	0.499 (0.10)	15.89 (1.78)
Vertisol Laguna Verde (Mexico)	0.026	15.00 (0.78)
Casts (12h)	0.021	10.11 (1.78)
Casts (24h)	0.053	43.33 (4.22)
Casts (96h)	0.032	11.44 (1.11)

Synchrony
Little information is yet available to determine whether the release of nutrients contained within the casts of endogeic worms is synchronous with plant demand. In a microcosm experiment, ^{15}N-labelled, endogeic worms placed in a non-labelled soil supporting the grass *Panicum maximum* rapidly eliminated their ^{15}N as ammonium in their casts; 45 % of this ^{15}N was transferred to the plant (Spain *et al.*, 1992). In short-term field experiments in which ^{15}N-labelled plant residues were applied at the soil surface, the incorporation of ^{15}N into plants was always greater in the presence of endogeic earthworms (Gilot, 1994; Pashanasi and Lavelle, unpublished data).

At a larger scale, evidence for synchrony is still lacking. In the African savanna at Lamto (Côte d'Ivoire), seasonal changes in the release of mineral-N in the drilosphere have been simulated using the DRILOTROP model (Martin, 1990). Results indicate that a significant proportion of ammonium-N is released over the last few weeks of the rainy season, when earthworm activity is at a maximum. By this time, plants have completed their growth and do not appear to take up further N from the soil. Furthermore, Abbadie *et al.* (1992) analysed the natural abundance of ^{15}N in grasses and a control rhizosphere soil and concluded that the grasses appear to make little use of N from soil organic matter since they recycle their own nitrogen within plant tussocks. The apparent contradiction between these results emphasises the need to consider processes at different scales in order to understand field observations. It is also likely that a better knowledge of the feeding behaviour of endogeic earthworms would facilitate interpretation of the results and allow an increased predictability of the effects of earthworms on nutrient cycling. ^{13}C measurements of soil organic matter, plant material and earthworms in a sugarcane plantation in northeastern Australia suggest that, in this sandy infertile soil, earthworms may feed preferentially in the rhizosphere, possibly on root exudates and root litter (Spain and Hodgen, 1994). This is consistent with the attraction of *P. corethrurus* to roots where it may feed and release nutrients that are readily used by plants.

4.4 Influence of the drilosphere on plant growth

Positive effects of earthworms on plant growth have been observed with many species and in many environments. Most of these studies have been 'pot-experiments' in which earthworms and plants are maintained together in a relatively small volume of soil. Suitable conditions are provided for both the plant and earthworms, and earthworms are forced to conduct their activities in the rhizosphere itself. Under these conditions, spectacular increases in plant production may occur over short-time periods. For example, plant biomass increases of up to 300 % over a three-month period followed the introduction of the mesohumic endogeic earthworm *Millsonia anomala* into pots containing the tropical grass *Panicum maximum* (Spain *et al.*, 1992). The N and P contents of plant material (especially roots) were significantly increased in treatments with earthworms. The increase in production was proportional to the earthworm biomass added up to a maximum biomass equivalent to *ca.* 150 g m^{-2}; above this, the soil became compacted and water infiltration was impeded.

The responses of tropical tree seedlings to inoculation with *P. corethrurus* were highly variable: *Bixa orellana* (Bixaceae) which has a fine, dense root system responded vigorously leading to biomass increases of 14-24 times the control value after 120 days. *Eugenia stipitata* (Myrtaceae) responded less (+1.6 to 2.5 times) and the palm tree *Bactris gasipaes* was negatively affected by the inoculation of earthworms (-1.8 to 2.7 times). In a similar experiment conducted with the epi-anecic species *Lumbricus rubellus* in a soil taken from a Finnish coniferous forest, above-ground production of seedlings of *Betula pendula* was enhanced by *ca.* 30 % whereas root production was slightly decreased, probably as a result of a better soil nutrient availability (Haimi *et al.*, 1992). Further experimentation has confirmed the existence of species-specific responses by plants to earthworm presence.

In a series of 240 experiments conducted in pots or intermediate scale field designs in tropical and subtropical areas, with 30 different earthworm species or associations of species and 15 plant species, Blakemore (1995) and Brown *et al.* (1999) found positive effects of earthworm inoculation in 72 % cases and measured an average increase of 63 % in production (Figure IV.60). Increases were most marked in soils that had low organic contents, high sand concentrations and were of intermediate pH. Not all plants responded in the same way, and there was some evidence that not all earthworms had the same impact. Response depended on the amount of earthworm biomass: average effects on plant growth were significant at biomass levels greater than 30-40 g m^{-2}. Graminaceous species such as rice (*Oryza sativa*) and maize (*Zea mays*), and trees such as *Bixa orellana,* tea (*Camellia sinensis*) or *Eugenia stipitata* are highly responsive whereas legumes and the palm tree *Bactris gasipaes* do not seem to benefit from earthworm activities.

The carrying capacity of soils for earthworms is highly variable. The availability of assimilable organic residues in the soil appears to be one of the major factors that support the maintenance of earthworm biomasses at levels that significantly stimulate plant production (Gilot, 1994).

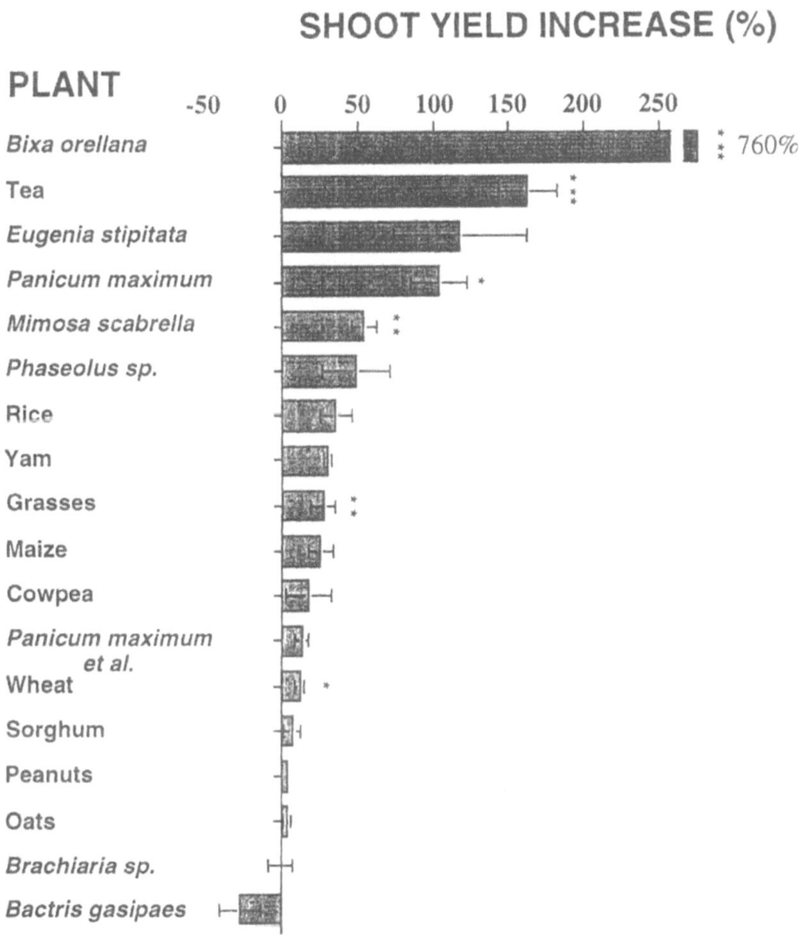

Figure IV.60 Average % increase in plant biomass compared with controls following earthworm inoculation of crops tested in small scale field experiments (from a total of 240 data points) (Brown *et al.*, 1999)

The reasons why plants grow better in the presence of earthworms are still poorly understood. The enhanced growth probably results from a combination of increased nutrient release from above ground litter and soil organic matter in the rhizosphere, a reduction in the adverse effects on plant parasitic nematodes (Dash *et al.*, 1980, Yeates, 1981), enhanced mycorrhizal infection of roots (Reddell and Spain, 1991ab; Ydrogo, 1994) and 'hormone-like' effects (Tomati *et al.*, 1988).

At larger scales, the introduction of endogeic and anecic earthworms into soils where they are absent, or represented by less active populations, often results in significant increases in plant production. In New Zealand, the native fauna is dominated by endogeic Megascolecidae which are adapted to a natural environment where grasses produce low quality organic matter. The introduction of European anecic and endogeic species following the introduction of European grasses, had dramatic effects on grass production (Stockdill, 1982). Similar effects have been described in Dutch polders where patches of soil colonised by earthworms could be readily identified on aerial photographs from their brighter green colouring (Hoogerkamp *et al.*, 1983).

IV.5 THE TERMITOSPHERE

5.1 Introduction

The termitosphere is defined as the zone of influence of termites within the biosphere. It comprises that part of the soil and above-ground ecosystem components influenced by the termite communities present and a rather specific microbial community that, in combination with the termites' own enzymes, allows them to use almost all the plant resources available in the ecosystem (see Chapter III.4.3.2.1). The social organisation of termite populations also allows them to carry out considerable engineering activities. They produce four categories of organo-mineral biogenic structures that profoundly affect soil function: epigeic and subterranean nests, surface sheathings and runways, subterranean galleries and chambers, and aggregates. Some termites also produce holorganic structures in the form of near-surface and arboreal carton nests, and fungal combs.

Where termite communities are diverse, their functional domain may extend throughout and below the soil horizons to the tree tops. It may include the external and internal parts of standing trees and shrubs (both living and dead), standing dead grasses and other herbaceous plants, the layer of litter on the soil surface (including such materials as dead leaves and other small plant parts), fallen timber and higher animal faeces. Below ground, it particularly encompasses the upper soil horizons and, diminishing in influence with increasing depth, the weathering zone beneath and intermittently down to the phreatic zone.

Termites strongly modify soil physical and chemical properties within their zones of influence. Over long periods, they substantially influence soil profile development and some termite groups effect changes to soil organic matter concentration and quality. The parts of the weathering zone affected will depend on the depth distributions of roots, whether the termites gather fine particles for their mounds from the weathering zone and the depth of the water table. The termitosphere also includes plant roots, both living and in all stages of decomposition.

Termites influence vegetation distribution and productivity directly through their effects on individual plants and indirectly through the creation of mounds and other constructs and through their influences on soil properties. Epigeal mounds, subterranean nests and their associated systems of galleries modify drainage patterns in the immediate environs of their mounds while certain groups influence soil fertility and thereby affect plant distributions. While a few species are spectacularly destructive of mans' constructs, food and other stored materials, few termite species are able to effectively attack healthy, undamaged plants in natural ecosystems. Nonetheless, a range of termite species causes severe economic losses to crop and other cultivated plants.

Relationships with micro-organisms are important for termites in several ways. As considered in Chapter III, their nutrition is partly dependent on interactions with internal prokaryotes and protists and, in the Macrotermitinae, with an external cultivated fungus. The nest environment creates a habitat for a range of saprophytic and other micro-organisms and for mycorrhizal fungi and the roots of surrounding plants.

5.2 Subdivisions of the termitosphere

The termitosphere particularly includes systems of nests and galleries, the food resources exploited and some unusual and highly specific microbial communities. It may be subdivided into a number of different but spatially, and even functionally-overlapping termitospheres, depending on the various ecological strategies represented among the species present in particular environments. The functional structure of termite communities is determined by the suite of abiotic and biotic factors described in Chapter III.4.3.2.4.

5.2.1 ECOLOGICAL CATEGORISATION

As discussed in Chapter III.4.3.2.2, termites may be categorised by the materials on which they feed, where they feed and by their nest types and locations (Abe, 1987). Many termites are mobile animals and individual colonies may utilise organic resources distributed across several strata of the biosphere. Species living within trees may either exploit the resources found there or in the surrounding area, depending on their ecological strategies. Conversely, species with subterranean nesting habits may also exploit underground, surface and arboreal resources. Their considerable mobilities, the broad dietary ranges of certain species and the capacity of many species to alter their dietary regimes, together with the lack of correspondence between particular feeding habits and nest characteristics makes it difficult to erect a general and mutually-exclusive classification of termitospheres (Table IV.20).

Table IV.20 Feeding habits and nesting behaviours in the different ecological categories of termites

Food / Nest	Live grass	Surface litter	Dry Wood	Soil/Wood	Soil	Fungi
arboreal						
wood			**Wood**	**Soil-wood feeders**		
epigeal	**Grass harvesting**	**Litter feeding**	**feeding**	**Soil-wood feeders**		**Fungus**
soil				**Soil-wood feeders**	**Soil feeders**	**Growers**
inquiline commensal						

Despite the above and the considerable variation in dietary items that occurs within each group (Sleaford *et al.*, 1996), the following broad trophic groupings have emerged as useful at the ecosystem level (Josens, 1983; Eggleton *et al.*, 1995):

i. *Grass-harvesting species.* These species largely harvest the dead leaves of grasses which may be stored in their nests; they belong to the families Hodotermitidae and Termitidae. Unusually, the northern Australian rhinotermitid species *Schedorhinotermes derosus* harvests standing grasses, mostly dead but some living, in addition to the diet of wood characteristic of this family (Watson, 1969). The Neotropical termitid harvester *Syntermes praecellens* also harvests green leaves (Roonwal, 1970). A number of species in this category also collect litter: the tropical Australian species *Drepanotermes rubriceps* stores a wide range of materials in its low epigeal mounds including the leaves of grasses, broad leaved species and a diversity of seeds and small woody litter (Gay and Calaby, 1970).

ii. *Litter-feeding species.* Species in this category feed on a broad variety of litters on the soil surface in diverse habitats ranging from savannas to tropical rainforests. Food materials in this class include the partly-decomposed and finely subdivided litter of grasses and broad-leaved species, leaves, seeds, twigs and animal faeces. While a few species of this group belong to the lower termites, most species of this category belong to the Termitidae, including the fungus cultivators of the subfamily Macrotermitinae.

Amitermes laurensis is a widespread and common species in northeastern Australia where it builds low approximately-conical mounds in well-drained areas but elongate, wedge-shaped mounds where drainage is impeded (Gay and Calaby, 1970). Populations of mounds may exceed 500 ha^{-1} in favourable locations. Its food comprises small seeds, pieces of dead grass, the dung of domestic animals and other finely-divided vegetable matter (Lee and Wood, 1971a).

In the humid West African rainforest, the fungus-cultivating termite *Macrotermes mülleri* collects leaf litter which it shreds into small pieces. On returning the subdivided leaf materials to its nest, they are moistened with saliva and stored for a period prior to adding them to the fungus comb, possibly in an initial fermentation stage (Garnier-Sillam *et al.*, 1988a).

iii. *Wood-feeding species.* These may be separated into three smaller groups comprising species feeding on a decay series:

- wood in living trees;
- sound, largely-undecomposed wood;
- partly-decomposed wood.

The wood-feeding habit occurs in species of all termite families except the Serritermitidae and includes certain of the fungus cultivators (subfamily Macrotermitinae). Many wood-feeding termites include materials from more than one of the above categories in their dietary range, sometimes even within a single colony (Lee and Wood, 1971a). The Australian rhinotermitid species *Coptotermes acinaciformis*, for example, attacks a wide range of living and dead trees, predominantly *Eucalyptus* species (Calaby and Gay, 1956), and dead wood lying on the ground; it is also a severe pest of timber in buildings (Gay and Calaby, 1970). Tree species also differ widely in their attractiveness to termites because of differences in the chemical composition of the wood and perhaps hardness (Waller and LaFage, 1987). Differences in wood quality may influence the growth and development of termites, their caste differentiation and reproductive success (Lenz, 1994).

Certain species attack the wood of living trees and these are considered below (Section IV.5.4.1). With the exception of the Hodotermitidae, the lower termites are dominantly wood-feeders although this habit is also widespread within the higher termites. The dead wood feeders include the 'dry-wood' feeders of the family Kalotermitidae that colonise the dead branch stubs of trees and develop independently of the ground. Species in this family also feed on largely-undecomposed wood, both in dead (and sometimes living) standing trees and in logs on the ground. Wood-feeders include species that nest in subterranean, epigeal and arboreal locations both within and on the outsides of trees. 'Damp-wood' feeders of the family Termopsidae occur in temperate areas and the *Prorhinotermes* spp. (Rhinotermitinae) that feed on decaying wood and logs in wet situations are largely limited to tropical and subtropical areas, often islands (Abe, 1987).

iv. *Soil-wood feeders.* Members of this group (Eggleton *et al.*, 1995) feed at the interfaces of decomposing logs and the soil, or within highly-decayed and friable wood. Species in this category are intermediate between the wood and soil feeders although their gut contents include substantial quantities of mineral materials. Species in this group belong entirely to the higher termites.

v. *Soil-feeders.* The soil-feeders or humivores subsist on humified organic materials and are characterised by the presence of substantial masses of mineral soil particles within their gut (Sleaford *et al.*, 1996). These termites are distributed particularly in the upper soil profile where they ingest organic-rich, humified materials. However, Eggleton *et al.* (1995) have postulated that some species in this group may be fine root feeders. Some soil feeders, exemplified by species of the Afro-tropical genus *Cubitermes,* build epigeal or partly epigeal mounds. However, many species have subterranean nests or live in the abandoned or occupied mounds of other termites where they may subsist on the carton of the host termites nest (Miller, 1991). As with the soil-wood feeders, soil feeders belong entirely to the higher termites and representatives are found within all subfamilies except the Macrotermitinae.

vi. *Polyphagous species.* A number of species are highly polyphagous and their feeding activities may cross several of the group boundaries given above; many cause considerable economic damage. For example, the primitive, largely-tropical Australian species, *Mastotermes darwiniensis* attacks a variety of living plants ranging from herbaceous agricultural crops such as maize (*Zea mays*) to trees such as *Pinus caribaea, Eucalyptus* and other indigenous species, sound and partly-decayed wood, animal faeces and a variety of stored organic materials, including plastics (Gay and Calaby, 1970). In Hawaii, the wood-feeding termite *Coptotermes formosanus* is recorded as being the most serious structural pest. It is reported to attack 47 species of plants distributed between 27 plant families in addition to such materials as paper, cardboard and fabric (Lai *et al.*, 1983). Dietary flexibility is also present in other groups. In the Chihuahuan Desert, the subterranean higher termite *Gnathamitermes tubiformans* prefers to eat fragments of dead grasses and forbs but will also consume leaves of the shrub *Larrea tridentata* and the faeces of large herbivores when its preferred food is in short supply (Whitford, 1991). Similarly, the African grass harvester *Trinervitermes trinervius* may change to feeding on dead wood during drought conditions (Lepage, 1974).

Over all environments, the wood and litter feeders are probably the most important ecologically because of their widespread distributions and numerical preponderance in terms of both species and individuals. In addition, with the grass harvesters, they are the major initial processors of the dead organic matter produced by plants. Substantial overlaps occur between the first three groups in the food materials they gather from their environments. However, it seems desirable to separate them since their biological and pedological effects may differ substantially. Because of their dietary flexibility, many economic pests are represented among the polyphagous species. No overlap in feeding habits occurs between the soil feeders and those of the first three groups (Noirot, 1992) although some members of the soil-wood feeder group are reported to feed partly on relatively sound wood (Miller, 1991).

The fungus cultivators (Macrotermitinae) include both litter and wood feeders (Darlington, 1994). As considered below, they break down their food materials in a much more complete way than other groups and have different effects on soils. This group may be worthy of separate categorisation because of the distinctive ecological and pedological effects of its members.

5.2.2 BIOGENIC STRUCTURES IN THE TERMITOSPHERE

5.2.2.1 *Nests and nest materials*

Nest types

Termites nest in a range of locations (Chapter III.4.3.2.2 and Table IV.20). Many species nest and feed within their foodstuffs while others build nests in locations often, but not necessarily, closely associated with their food materials and exploit the resources of their surrounding territories. As shown below, the nests of certain species may not be confined to one location and a single colony may have more than one nest within its territory.

Termite nests and other constructions are built by a combination of excavation and construction (Noirot, 1970). Excavation in wood is considered the most primitive since it most closely resembles the behaviour of the presumedly-ancestral, wood-feeding cockroaches. Some form of construction is always present in termite nests. This may range from a simple lining of galleries with faeces to the construction of highly-elaborate nest structures. The materials used in the construction of termite mounds include earthen and little-digested plant materials and these may be diluted, softened or cemented by saliva or faecal materials. In certain types of construction, faecal materials may comprise much of the mass.

The materials comprising nests and other termite constructions may be of four major types (Wood, 1996):

(i) Those built largely of repacked, orally-transported mineral soil materials with

little or no admixed faeces. This group is represented by the Macrotermitinae and some others which use no or minimal faeces in their constructions but build their constructions with earthen materials softened with saliva (Noirot, 1970; Lee and Wood, 1971a).
(ii) As for (i) above but the earthen materials are admixed with faeces, mainly as gallery linings or special structures;
(iii) Those dominated by faecal material, as in the soil and soil-wood feeding species;
(iv) Those consisting almost entirely of faeces derived from ingested plant materials (carton), as in the nests of the two Australian arboreal *Nasutitermes* species discussed below. Carton is a common component of many nest structures.

Compound structures also occur. Mounds of the widespread Australian rhinotermitid (lower) termite *Coptotermes acinaciformis* have a soft carton centre with some masticated wood and little mineral material, as in (iv) above. Their mounds also have a hard, protective, dominantly-earthen outer covering some 10 cm thick. Mounds of the Australian species *Microcerotermes nervosus* and certain others contain composite earth:carton structures with intermediate mineral contents (Lee and Wood, 1971b).

As illustrated in Chapter III, nests occur in the following five major locations (Noirot, 1970; Martius, 1994):

i. *In the wood of living and dead trees*

Examples of species nesting in these locations include the 'dry-wood' termites (family Kalotermitidae) that nest and explore for food in dead branch stumps on living trees. Other members of this family, such as the *Neotermes* species, may live and nest in logs and stumps, and within the wood of living trees. In all these circumstances, colonies are established in the aerial parts of the trees and have no contact with the ground. Economic damage to the affected trees often results where stems are attacked and this may happen through devaluation of the wood or by weakening the vigour of the tree. *Neotermes insularis* is a large termite (soldier length to 15 mm) common in near-coastal northern and eastern Australia where it causes economic damage to a wide variety of trees ranging from native *Eucalyptus* species to ornamental and exotic fruit trees such as Mango (*Mangifera indica*) and Tamarind (*Tamarindus indica*) (Hill, 1942).

Other termites establish nests within living trees. In temperate Australia, colony founding pairs of the widespread termite *Coptotermes acinaciformis* may enter the tree near the base or through shallow exposed roots. Colony development occurs within the tree and the termites eventually create a central pipe that can occupy much of the length of the tree stem, often extending into the branches. Large colonies extend through underground galleries to the roots of neighbouring trees which are also exploited. In the tropics, this species builds large earth-capped mounds, often associated with the base of the tree and within which much of the colony occurs.

This group also includes species that nest and feed in fallen wood lying on the ground surface. The 'damp-wood' termites of the genus *Stolotermes* (Termopsidae) live and feed in decaying wood in damp situations, often logs lying on the soil surface or in the bases of trees.

ii. *Subterranean locations*

Many termites nest underground with no surface expression of their presence while others such as the harvester *Hodotermes mossambicus* betray their presence by surface soil dumps and other structures (Coaton, 1958). Nests may be diffuse with little differentiation between the galleries with transitional forms culminating in a clearly-recognisable nest area and differentiated structures (Noirot, 1970). *Heterotermes* spp. provide examples of species with diffuse, little-differentiated nest systems which occur in the soil, in or under logs and stumps and in the mounds of other termites (Watson and Gay, 1991).

In contrast, some groups, such as the fungus-cultivators (Macrotermitinae), build highly complex subterranean nests that may comprise multiple nesting chambers, food stores and fungus chambers with interlinking galleries (Darlington, 1994).

iii. *Epigeal nests*

Epigeal termite mounds of diverse types are a common feature of many tropical landscapes. The mounds are normally built by higher termites belonging to a range of ecological strategies, including grass harvesters, litter feeders, wood-feeding species and, in some locations, soil-wood and soil feeders. A number of fungus-cultivating termites also build epigeal mounds. The epigeal and semi-epigeal mounds of some soil-wood and soil feeders are occasionally numerous and the species of this group are therefore of ecological and pedological importance, although their biological role is currently ill-defined. Certain Australian *Coptotermes* species (family Rhinotermitidae) are among the few lower termites to build epigeal mounds, although they may not do so at all locations within their ranges.

The distinction between subterranean and epigeal nesting locations may be somewhat arbitrary. Epigeal mounds normally originate from subterranean beginnings or from a colony established in wood and most of these mounds have at least some part of their structure below the ground surface. A number of intermediate forms occur between epigeal and the wholly subterranean forms. Miller (1991) reports that the nest structure of the Australian soil-wood feeder *Paracapritermes secundus* (family Termitidae) is largely underground but that the tip protrudes above the litter within its rainforest habitat.

Epigeal mounds occur in a range of sizes from a few centimetres to almost 9 m tall in the Australian grass harvester, *Nasutitermes triodiae* (Lee and Wood, 1971a). Noirot (1970) distinguishes two major types: those comprising a series of essentially similar chambers and those with more differentiated internal structures. The mounds of the tropical Australian litter-feeder *Amitermes laurensis* and *Noditermes aburiensis* (both Termitidae) from the Côte d'Ivoire are of the first type while the fungus-cultivating termites of the sub-family Macrotermitinae may build complex nests that include their fungus chambers (Noirot, 1970).

iv. *Intermediate nests*

These nests are usually associated with the bases of trees and have clear connections to the ground. The builders of such structures are often soil or soil-wood feeders in contrast to the predominantly wood-feeding termites that construct arboreal nests (Martius, 1994).

v. *Arboreal nests*

Arboreal nests are built on the outsides of trees, frequently in the angles between branches. They are built only by the Termitidae and by species in all subfamilies except the Macrotermitinae. They are most frequent in the Nasutitermitinae, common in the widespread genus *Microcerotermes,* but also occur in *Speculitermes* (Indomalayan) and *Anoplotermes* (Neotropical) (Noirot, 1970). The latter has small fragile nests that may hang from foliage although little is known of their function (Araujo, 1970). Australian examples of arboreal-nesting termites include *Nasutitermes graveolus* and *Nasutitermes walkeri*, two species that build nests constructed of carton on the exterior of trees, often many metres above the ground (Chapter III.4.3.2.2).

vi. *In the nests of other termites*

Inquiline termites occur in both occupied and abandoned mounds built by many termites although no communication normally takes place between the portions occupied by different species. Noirot (1970) reported that as many as 10 species may occur in a few cubic decimetres of the mounds of the soil feeder *Cubitermes* sp.. Many soil-feeding species occur as inquilines within the mounds of other termites (see, for example, Miller, 1991).

Another common situation is the acquisition of space within an existing mound by a termite different from the original builder which then constructs its own mound on top of the original. In northern Australia, a common compound mound type is that formed by the construction of a conical mound of the litter-feeder *Amitermes vitiosus* on the surface of a low mound of the grass harvester *Drepanotermes rubriceps.*

Abundance and composition

Epigeal mounds may be abundant: Aloni and Soyer (1987) report populations of more than 5000 mounds per hectare from Zaïre, principally belonging to the soil-feeding *Cubitermes* species. The above-ground sections of termite mounds may weigh many tonnes per hectare and occupy considerable proportions of the soil surface, depending on their type. In tropical Australia, mound populations may number up to 1100 ha^{-1} representing approximately 62 Mg ha^{-1} of soil and covering 1.7 % of the sampled area (Wood and Lee, 1971). Median mound mass for the 58 Australian sites recorded by Spain *et al.* (1983b) was, however, only 20 Mg ha^{-1} (inter-quartile range 15-25) and the median area covered by their bases just 0.8 % of the sampled sites (inter-quartile range 0.5-1.1). Similar estimates from African sites range up to 10 % of the area sampled although Meyer (1960) estimated that the bases of mounds at one site covered an exceptional 33 % of the soil surface, representing an estimated dry mass of 2400 Mg ha^{-1}.

The spatial distributions of termite mounds may be random, aggregated or over-dispersed. Where mound populations are high, they frequently tend to over-dispersion (Spain *et al.*, 1986) suggesting that such termite populations can utilise a substantial proportion of the organic resources available to them in the landscape. This is reinforced by the denudation and dietary shifts that occur during drought years (Lepage, 1981a; Watson and Gay, 1970) or as a consequence of altered or intensified land use (Wood and Pearce, 1991).

Different species build their epigeal mounds with characteristic proportions of inorganic materials. They may be composed predominantly of soil materials, as are the mounds of many grass-harvesting and litter-feeding termites. Some have a hard earthen capping with a softer, more-organic interior while others possess a largely-organic structure with little or no included soil material. The largely-earthen parts of the mounds are normally of finer texture than the surrounding soils since termites exclude the larger particles from their nests (Lee and Wood, 1971a). As discussed below, the mound materials of many grass-harvesting and litter-feeding termites are substantially enriched in nutrient elements compared with the local surface soils; this is ascribed to the transport of harvested materials to their nests where they may be partly digested and their breakdown products incorporated in mounds or galleries as linings or infillings.

5.2.2.2 *Galleries*

Termite galleries vary in shape from circular to narrowly-elliptical and, in size, from less than 1 mm to more than 20 mm in their greatest diameters (Grassé, 1984) (see also Table I.10); they are therefore all in the large macropore size range. Gallery lengths of up to 7.5 km ha^{-1} (Darlington, 1982; McKay and Kladivko, 1985; McKay and Malcolm, 1988) have been estimated for soils associated with the mounds of *Macrotermes michaelseni* although Wood (1988) felt that this figure could perhaps have been doubled to account for the galleries of the subterranean termite species also present. The equivalent of 90 000 storage chambers per hectare also occurred in the soils of Darlington's study area. Assuming that these structures were largely confined to the top 20 cm of the soil and applying Wood's multiplier, it may be calculated that termite galleries and related structures formed voids occupying approximately 0.4 % of the soil volume to this depth. Some Macrotermitinae may open access holes at the surface which they use for foraging at night; the overall surface of these openings has been estimated at 2-4 m^2 ha^{-1} in a dry savanna of Kenya (Lepage, 1981a).

It is clear that the sizes, types of wall construction, depth distributions and extents of termite gallery systems differ markedly, depending on the ecological strategies of the species involved and the characteristics of the soils present. The galleries of soil-feeding termites, for example, are usually more superficial than those of the Macrotermitinae and their walls are dark and plastered with fluid excreta. In contrast, those of the Macrotermitinae are smooth and massive due to compression of the soil and the clay-saliva mixture used to line them (Kooyman and Onck, 1987).

In certain Australian soils (alfisols) with coarse-textured A and massive, clay-rich B horizons, termite galleries occur principally in the A horizon and descend into the B horizon largely in old root channels and adjacent to rocks where local changes in the soil matrix have occurred (A.V. Spain, unpublished). Where termite populations are substantial, their gallery systems are considered to strongly influence soil physical properties (Lee and Wood, 1971a). While few quantitative data are available, as shown below (Section IV.5.3.1.2), their effective penetration of surface crusts is of agronomic relevance.

5.2.2.3 *Aggregates, voids and other structures*

In addition to the biofabrics discussed in Chapter II, a wide range of micromorphological structures have been described from the mounds and galleries of a range of termites (see Sleeman and Brewer (1972) for those of a number of Australian species and Mermut *et al.* (1984) for four species of Macrotermitinae). The structures of the fungus-cultivating species (Macrotermitinae) were shown to be distinct from those of most other termites in that they do not incorporate faeces into their mound structures. Wielemaker (1984) and Kooyman and Onck (1987) used similar methods to categorise termite-derived structures in gallery systems distant from mounds. Garnier-Sillam *et al.* (1985) described organo-mineral micro-aggregate structures from the faeces of four species of termites of different ecological strategies and Eschenbrenner (1986) noted the similarity of aggregates found in certain termite-inhabited soils and those from termite mounds.

In addition to the gallery systems described above, termites create a range of voids both in epigeal mounds and within the soil. In both locations, these voids differ widely in size and in shape (Table I.10) and may serve as nests, food storage areas, burial chambers and, in some species of the subfamily Macrotermitinae, as chambers within which the colony's fungus combs are located. Such voids are not static: they are created by the termites to serve their designated purposes and, following this may be actively backfilled with organic or inorganic materials or subject to infilling by soil particles moving under the influence of gravity or in flows of water. Such voids may also be occupied by termites distinct from the original creators or other animals, and re-modelled according to their specific designs.

In stable environments, the continued cycles of void creation and infilling to which termites contribute have a considerable influence on landscape dynamics and soil formation. Humphreys (1994) described the systems of voids and related biologically-formed aggregates of termites, ants and other animals from sandy ultisols and entisols in southeastern Australia. He concluded that approximately 8 % of the surface A and E horizons consisted of recognisable faunally-derived structures and that the topsoil at the site formed a biomantle (Chapter II.3.3.2).

Other important structures built by termites include surface covers and runways. The former are temporary structures built over food materials to provide shelter for the termites from predators and from desiccation; most of these structures are constructed predominantly of inorganic soil materials. Runways are hollow linear structures formed by termites to cover their movements over exposed areas and may be constructed predominantly of either organic or inorganic materials.

Two examples of predominantly organic runways are those built by the higher Australian wood-feeding termites *Nasutitermes graveolus* and *Nasutitermes walkeri*. Their nests are built on the exteriors of trees, often many metres above the ground, although workers forage for woody materials around the bases of their host trees. The nests are joined to the ground by runways which are normally built over the bark of most tree species, but also within grooves cut through the soft flaky bark of *Melaleuca* and *Lophostemon* (Myrtaceae). Runways constructed by these species may be 1-2 cm across and are constructed of carton; they appear to contain little inorganic material.

Runways constructed largely of earthen materials are built by a wide variety of species and are used to protect foraging worker and soldier termites while they explore for or exploit arboreal or soil surface resources. Such runways, and sometimes more substantial structures, are associated with termite attacks on wooden materials within man's built environment. These often originate from external nests located in the soil.

5.2.3 ASSOCIATED ORGANISMS

As ecosystem engineers, termites greatly influence communities of micro-organisms, plants and invertebrates within their functional domains. The structures created offer a diverse range of chemical and physical environments that differ strongly from those present in the general soil mass.

As seen in Chapter III (Section III.4.3.2.1), specific microbial communities are associated with termite digestion, inside the gut and outside, in fungus gardens and other biogenic structures (Section IV.5.2.2, Table IV.21). The termitosphere as a whole has large microbial communities. For example, Holt (1995) calculated that microbial biomass in a mound of the Australian litter feeder *Amitermes laurensis* comprised 16 g C compared with 31 g for termite biomass carbon. As shown below (Section IV.5.4.2.3), mycorrhizal fungi may contribute to this biomass.

Table IV.21 Microbial RNA and biomass, microbial C and C and N contents in different parts of the termitosphere of the soil feeder *Cubitermes fungifaber* in a forest soil from Cameroon (Boulaud and Brauman, 1998) (values for S.E. in parenthesis).

	Faeces	Fresh constructions	Internal wall	External wall	Soil below nest	Control 0-5 cm	Control 5-10 cm
A R N ng mg^{-1}	177.7 (9.7)	94.1 (14.2)	20.3 (3.0)	6.2 (1.5)	3.4 (0.5)	1.3 (0.0)	not detectable
Organic C mg g^{-1}	nd	40.6	34.0	29.2	26.8	22.3	12.3
Organic N mg g^{-1}	nd	3.42	3.47	2.74	2.10	1.83	1.18
C: N	nd	11.9	10.0	10.9	12.9	12.2	10.4
C (%)	nd	nd	16	0.4	2.2	0.3	not detectable
CO_2 evolution μl g^{-1} h^{-1}	nd	21.0 (2.46)	14.0 (0.55)	10.5 (1.61)	20.8 (1.46)	11.8 (1.40)	5.1 (0.54)
Qr	nd	1.41 (0.06)	1.56 (0.24)	1.46 (0.16)	1.13 (0.09)	1.19 (0.19)	1.12 (0.02)

Termites also interact with a large variety of invertebrate and vertebrate animals through positive (mutualism, symbiosis), neutral (commensalism) or negative (predation, parasitism) associations (Sections III.4.3.2.7 and IV.5.4). Finally, the structure of plant communities and their growth are profoundly modified in the termitosphere (Section IV.5.4).

5.3 Interactions between termites and the non-living environment

The major ecological role of termites is the breakdown of dead plant materials (Figure IV.61). In ecosystems where the termitosphere is well developed, termites strongly influence energy flows and nutrient cycling characteristics by consuming much of the dead secondary production of the higher plant communities. Further, part of this energy is expended in modifying the soil and above-ground environments through the creation of biogenic structures located in the soil, at the surface within dead or living plant materials or at various heights in the arboreal stratum (Chapter III.4.3.2.2).

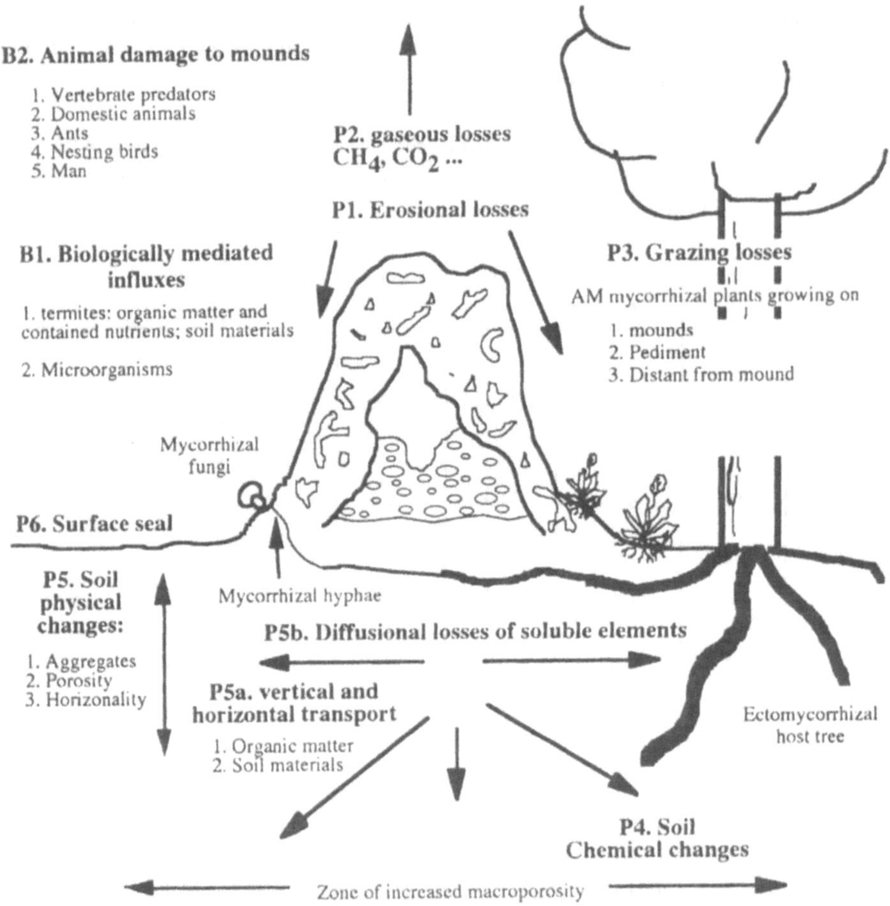

Figure IV.61 Main abiotic (P) and biotic (B) influences and processes associated with epigeal termitaria.

Termites strongly influence soil organic matter dynamics. They sequester organic matter and biological nutrient elements for considerable periods in their nests, mounds and in the walls of their galleries and other structures. The breakdown and erosion of above-ground structures through the agencies of organisms, rainfall and perhaps wind constantly return their contained nutrients to the soil surface. Also, roots of surrounding plants frequently abstract nutrients from the often-concentrated stores held in the bases of termite mounds and within decomposing organic materials in back-filled subterranean voids (Figure IV.62).

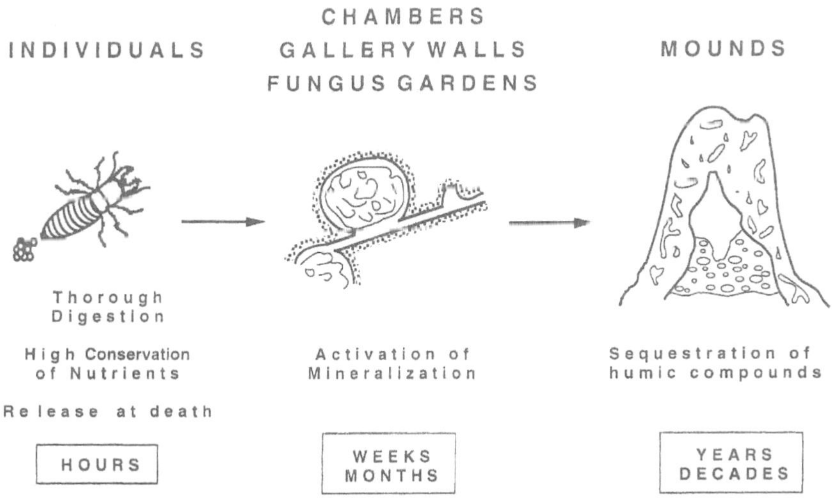

Figure IV.62 Effects of termites on soil organic matter dynamics at different scales of time and space.

Where nutrient concentrations in the termite-derived organic matter are high, the sequestration of organic matter within termite mounds leads to a nutrient patchiness in the landscape which helps to maintain ecological diversity. In many infertile savanna environments, the only places where plants requiring a high nutrient status may grow satisfactorily is on and adjacent to termitaria. The underground gallery systems created by termites improve gas exchange with the atmosphere and facilitate water infiltration and movement through the soil. Termites may also help to restart nutrient cycling processes in degraded ecosystems through mechanisms of nutrient concentration and organic matter breakdown, and their considerable effects on soil physical processes.

Because their diversities and populations are highest in the inter- and sub-tropics, it is there that termites have their greatest impact. As discussed below, the effects of termites on soil processes are significant at spatial scales ranging from the landscape to the catena to the micro-aggregate and even to the clay mineral level. In addition, their effects are expressed in soils at time scales ranging from the geological, to years, to a few months. Termites have also played significant roles in the formation and development of certain palaeosols (see, for example, Tardy and Roquin, 1992) whose soils may therefore be considered as forming biomantles.

5.3.1 SOIL PHYSICAL CHARACTERISTICS

The presence of substantial termite populations in a landscape affects most aspects of soil organisation and function, both at coarse and fine spatial scales (Figure IV.64). Over long periods, much of the soil mass may be modified directly by termite activity through the faunal pedoturbation (Hole, 1961) processes discussed in Chapter II. These activities include the continuing cycle of void formation (nests, galleries, storage structures) in the soil. It also includes the construction of above-ground structures together with their subsequent destruction by natural processes of erosion and the infilling of soil voids through biological activity, gravity- and water-mediated sediment transport.

In soils where termites are populous, their effects on profile morphology are clearly apparent in the form of the lower parts of epigeal mounds, subterranean nests, systems of small and large galleries radiating from nests, and food stores associated with the galleries. Their activities may lead to the formation of new horizons (Nye, 1954) or, in some instances, interfere with profile differentiation. At a landscape scale, the surface expression of faunal pedoturbation contributes to toposequence formation (Nye, 1954, 1955abc) and at even larger scales to the formation of biomantles (Johnson, 1990) (Chapter II.3.3.2).

At shorter time scales, termite activity influences most aspects of soil function including water movement, gas exchange, organic matter processing and biological nutrient flows. Even the spatial dispositions of the biota (micro-organisms to higher plants) and the stocks of organic matter, nutrient and other elements are affected. Fourteen years after experimental eradication of termites from a North American desert rangeland through pesticide application, soil N concentrations had increased although water infiltration and storage had diminished in comparison with untreated plots (Whitford, 1991).

5.3.1.1 *Termites and soil textural relationships*

Termite distributions and activity both influence and are affected by soil texture. As considered in Chapter III, mound-building termites are largely excluded from certain vertisols in northern Australia through the instability induced by the expansive smectitic clays. Shortages of clay in deep sands may equally limit the colonisation of certain species.

Termites substantially affect surface processes by transporting soil from various depths in the profile to the surface, where it is built into epigeal mounds and other structures. Surface structures erode continuously particularly under the influence of intense rainfall leading to the redistribution of soil materials in the landscape. In locations with dispersive B horizon clays, the fine-textured materials eroded from termite mounds may contain sufficient clay and silt to form a thin surface seal.

Because termites generally select the smaller particles from within the profile, materials brought to the surface are commonly finer in texture and may have different elemental and clay mineral compositions than those predominating at the original surface (Boyer, 1973b, 1982). Thus, as discussed in Chapter II.3.3.2, abstraction of finer particles from various depths in the profile and their deposition on the surface often lead to the gradual build up of a fine-textured surface soil which may sometimes be underlayed by stone lines, or by concentrations of particles too large to be transported by the termites (*e.g.*, Williams, 1968; Wielemaker, 1984; Johnson, 1990).

Alternatively, where surface erosion removes the fine materials brought to the surface, a coarse-textured surface soil may result (Wielemaker, 1984; Humphreys, 1994). On sloping surfaces, downhill transport and sorting of materials is hastened by the actions of termites and other burrowing animals through the exposure of soil materials to rainsplash and overland water flows. Where prevalent, termites therefore contribute materially to the formation of catenas (Nye, 1954, 1955abc; Wielemaker, 1984).

Net annual increments in the thickness of fine-textured, termite-derived surface soils have been estimated to range from 0.01 to 0.10 mm y^{-1} (Lee and Wood, 1971a). This is an important pedological and ecological process in the large areas of the tropics and sub-tropics where the termite fauna is dominated by species that build soil materials into epigeal mounds and other above-ground structures. Beneath the surface, changes in texture - and therefore water holding capacity - and the distributions of organic and mineral matter also reflect the composition and functioning of local termite communities.

5.3.1.2 *Effects on soil structure*

Termites have profound effects on soil structure through their mound-building activities, the formation of aggregates and other structures and through the creation of extensive systems of subterranean nests, galleries and storage chambers. Epigeal mounds are highly-visible structures of substantial importance in terms of pedogenesis and plant growth, both directly and in terms of their erosion products. However, the many species that nest underground also have substantial effects on soil structure. Humphreys (1994) considered that the four termite species present in the sandy ultisols and entisols at his study site in southeastern Australia were more important as sub-surface bioturbators than mound builders.

While a range of termite-derived structures are known from epigeal mounds and gallery walls (see, for example, Sleeman and Brewer, 1972; Garnier-Sillam *et al*, 1985), less is known of their contribution to soil aggregation away from the immediate environs of the mounds, or of the role of subterranean-nesting termites in this regard. Termite galleries act as networks of horizontal and vertical macropores influencing bulk density, aeration and water infiltration and forming conduits for the movement of materials upwards, downwards and laterally within the soil (see, *e.g.*, Wielemaker, 1984). Movement of soil materials through the galleries is mediated both by the activities of the termites and by physical processes.

Termites influence soil structural stability in the vicinity of their mounds although the effect is dependent on the ecological strategies and nesting locations of the termites involved. Garnier-Sillam *et al.* (1988b) recorded the contrasting effects of two species in a humid African forest environment. The soil-feeding species *Thoracotermes macrothorax* increased the structural stability of the soil in its area of influence while the fungus-cultivating species *Macrotermes muelleri* reduced it. The effects of termite activity on drainage properties are considered below.

Effects of subterranean nests and epigeal mounds

Epigeal termite mounds are usually externally hard, massive structures and shed almost all of the water that impinges on them. Through the energy of rainfall impact and runoff, a steady erosion of mound surfaces occurs, even while they still contain active populations (Bonell *et al.*, 1986). There is a considerable sorting of the eroded material: finer materials are transported furthest leaving coarser-textured materials close to the mounds (Janeau and Valentin, 1987) (Figure IV.64). Sometimes, sufficient fine-textured dispersive material is deposited close to the mounds to form surface seals which may continue to modify local infiltration and drainage patterns until disrupted through the activity of the soil fauna (Janeau and Valentin, 1987), or other agencies.

The bioturbing effects of termites have been used as part of systems for rehabilitating certain degraded soils in Burkina Faso. Roose *et al.* (1996) found that surface mulches attracted termites, locally disrupting surface crusts. In experimental studies of these crusted soils, Mando (1997) found that these burrowing activities increased infiltration rates and saturated hydraulic conductivity, reduced bulk density and cone penetration resistance. These effects were not evident in mulched soils from which the termites had been removed by pesticide application.

Mound erosion may exclude some epigeal mound-building termites from high rainfall areas, particularly those species whose mounds lack specific adaptations for water shedding. The repair of these losses must entail substantial energy costs for the termite colony, both during mound building and repair. The soil materials added to epigeal termitaria during both processes are initally soft and particularly susceptible to erosion, at least until hardened by curing or internal repacking processes.

After the colony has died, or the mound has been broken up, the resistant mound materials erode slowly and may continue to modify local drainage patterns for some time. Where mounds are common, an area equivalent to the entire soil surface will have been covered by their bases over quite short geological times. As stated above, the median area occupied by the bases of the mounds recorded from the northern Australian sites discussed above is 0.8 % of the land surface area. If this percentage were to remain constant over time, an area equivalent to the whole surface will have been covered by the bases of termite mounds in approximately 125 generations of mounds. If the standing mass of the mounds is assumed to be a constant 20 Mg ha^{-1} and turnover time of the mound materials is assumed to be of the order of 30-50 years, this is equivalent to the distribution of 2500 Mg ha^{-1} (or a layer of soil *ca.* 20 cm thick with bulk density 1.3 Mg m^{-3}) over the surface during a period of 3730-6250 years, a short time in terms of soil formation.

Simple calculations of this type patently underestimate the effects of termite mounds on the soil. Many incipient colonies build small mounds but soon die out through the effects of predation, competition or other factors. Further, the masses of soil materials brought to the surface to form runways and to cover food supplies have only occasionally been quantified. The few data available suggest that this may be in the order of 1 Mg ha^{-1} y^{-1} (see, for example, Lepage, 1981b, Bagine, 1984).

The turnover times of mound soil materials are poorly known but are substantially longer than the life spans of the colonies that construct the mounds. An individual mound of the Australian grass-harvesting termite *Nasutitermes triodiae* in northern Australia is known to have existed for approximately 100 years (Watson, 1972) and medium-sized mounds from a range of species and localities are recorded as having existed for 15-115 years (Grassé, 1984); however, the estimated ages of most mounds are likely to be towards the lower end of this range.

Bonell *et al.* (1986) measured the rates of erosion of abandoned mounds of the grass and litter-feeding termite *Amitermes vitiosus* in the semi-arid tropics of northeastern Australia. Erosion rates were most strongly correlated with high-intensity (>90 mm h^{-1}) rainfall although lesser intensities also resulted in lower mound erosion rates. These authors considered it likely that erosion rates would accelerate due to the exposure of internal galleries and the complete breakdown of these mounds through rainfall was estimated to require *ca.* 30 years. A possible 30 years of colony existence may be added to this to give turnover times of *ca.* 60 years for populations of these mounds.

For the soil materials of a population of small mounds (principally soil feeders of the genus *Cubitermes*) in Zaïre, Aloni and Soyer (1987) estimated a turnover time of approximately a decade. In addition to erosion by raindrop impact and runoff water, mounds are also subject to damage by cattle, other large herbivores and by specialised termitophilous vertebrates, including the birds that create nesting holes in their mounds (Hindwood, 1959). Over the period of existence of a mound, the total amount of sediment eroded may substantially exceed the mass of the mature mound, due to the continuing cycles of erosion and repair of the mound that take place during the period that the colony occupied it.

The large mounds built by African Macrotermitinae ('high termitaria') exist for very long periods, possibly many thousands of years. They undergo long periods of transformation, erosion and increase and develop their own distinctive fauna and flora (see below), long after the builders of the original mound have died out. Their soils are often distinctively different from those surrounding them; many have a different texture, a higher pH and base status (Aloni, 1975; Pullan, 1979) and their organic matter may differ in its stable C isotopic composition, as shown below.

Galleries and water infiltration

In an arid (350 mm annual rainfall) Australian woodland environment, sub-circular surface pavements up to 2 m diameter and occupying *ca.* 1.6 % of the soil surface area are the expression of subterranean nests of harvester termites of the genus *Drepanotermes* (Eldridge, 1994). Ponded infiltration rates on the pavement surfaces are very low and much of the incident water is shed to a surrounding annular area which exhibits enhanced plant growth with respect to the inter-pavement area more distant from the mounds. At 113.2 mm h^{-1}, ponded infiltration rates in the annular areas are more than 16 times higher than those of the pavement (6.7 mm h^{-1}) and are slightly higher again than those of the inter pavement area (103.6 mm h^{-1}). No differences were apparent in infiltration under tension and the higher rates of ponded infiltration in the annular zone were ascribed to the greater densities of termite galleries. In an experimental study of a desert rangeland ecosystem, Elkins *et al.* (1986) reported higher bulk densities, lower infiltration rates and higher bedloads in runoff waters from areas in which subterranean termites had been eliminated by insecticide application.

Soil aggregation

The overall effects of termites on soil aggregate formation and stability depend on the functional community structure, or the spectrum of ecological strategies represented among the species of the local termite community. While higher percentages of water-stable aggregates have been reported by several authors from termite-affected soils this is not always the case. Garnier-Sillam and Harry (1995) demonstrated differences in the stabilities of mound materials formed by four sympatric soil-feeding termite species from Congo rainforests. Judged on the basis of a stability index, materials from the mounds of two species were more stable, one was similar and one less stable than those of the adjacent surface soils. These authors noted a much greater structural stability in materials - in both the mounds and surrounding topsoils - reworked by the three species that incorporate faecal organic matter into their constructs.

In the same study, mounds of the fungus-cultivating species *Macrotermes mülleri* and their surrounding surface soils, were highly clay-enriched and had the low organic matter concentrations typical of structures built by these termites. Both the mound wall materials of *M. mülleri* and associated A1 horizon-soils were markedly less permeable to water and less stable than the control surface soils (Garnier-Sillam *et al.*, 1988b; Garnier-Sillam, 1989). The surface-sealing effect considered above was also noted in association with the dispersion of clay-enriched materials from the mounds of *M. mülleri*.

Effects on pedogenetic processes

The clay minerals present in termitaria are largely derived from the soil and sometimes the alteration zone below. In addition to these, 2:1 clay minerals have been found regularly in the mounds of certain African termites in environments where these are otherwise uncommon or absent (Boyer, 1982). Boyer reported the widespread presence of an artificial illite in the mounds of a number of wood-feeding termites although such neoformation is apparently less common in the soil-feeding termites. The neoformation of these minerals was considered to result from the direct effects of grinding and working of mica and other particles with the mandibles in the saliva-rich environment of the buccal cavity. An indirect effect was also considered to result from placement within the altered chemical and physical environment of the mounds. Other clay minerals, calcite, dolomite and certain oxide minerals have also been reported to be associated with certain African termitaria although Lee and Wood (1971b) could find no differences between the clay minerals present in the mounds of Australian termites and in their adjacent soils.

5.3.2 TERMITES AND SOIL ORGANIC MATTER

Termites may profoundly affect SOM dynamics at different scales of time and space. Short-term effects are associated with digestive processes, whereas storage within biogenic structures may lead to longer term effects that may extend over scales of years to decades (Figure IV.65). Such stores may however be partially depleted through exploitation by the roots of surrounding plants and their associated mycorrhizal fungi.

5.3.2.1 *Organic matter fluxes associated with the termitosphere*

The ecological impact of termite feeding may be large. Lepage (1972) reported that termites in low-lying parts of Soudano-Sahelian savannas may locally consume up to 49 % of annual herbage production, although the overall figure for the area was estimated to be 5.4 %. In further studies of pastures in semi-arid areas of Kenya, Lepage (1981b) estimated that termites and grazing mammals had similar impacts, both consuming approximately 1 Mg ha^{-1} y^{-1}.

In an Australian tropical savanna, termites and their associated microflora are considered responsible for *ca.* 20 % of annual C mineralisation (Holt, 1995). Lee and Wood (1971a) reported that *Nasutitermes exitiosus* consumed 17 % of the total estimated fall of leaves and small woody materials, or 4.9 % of the total annual litter fall in a dry temperate-climate Australian forest dominated by *Eucalyptus* spp. Similar estimates for Malaysian rainforests range from 1 % for a site prone to flooding, to 16 % for a relatively dry forest whose termite fauna was dominated by species of the subfamily Macrotermitinae (Collins, 1983). In other Malaysian rainforests, Matsumoto and Abe (1979) estimated that the termite *Macrotermes carbonarius* consumed *ca.* 22-32 % of daily litterfall.

Termite overgrazing may have substantial effects on soil stability where populations are high or where drought limits the productivity of the vegetation. With increasing severity of drought, *Macrotermes* spp. may progressively shift their diet from litter to standing herbaceous vegetation to roots, thereby disturbing the soil surface and rendering it more susceptible to erosion and degradation (Lepage, 1981a). However, in many of the situations where denudation has been reported, over-grazing by domestic animals may have been primarily responsible (Watson and Gay, 1970; Wood and Pearce, 1991); the effects of the termites may become evident only when most of the standing vegetation and litter has been previously removed by vertebrate grazing. Despite this, it is still unclear whether there is a direct competition with termites, even where stocking levels exceed carrying capacity.

In areas where Macrotermitine termites are dominant, a higher than normal proportion of annual litter fall may be recycled by termites. This has been ascribed to the greater efficiency of decomposition that results from their symbiotic relationship with fungi of the genus *Termitomyces*, described in Chapter III (Garnier-Sillam, 1988a, 1989).

5.3.2.2 *Short-term effects: the importance of adjustment of the C:N ratio*

Digestion by termites in mutualistic associations with specialised micro-organisms is highly efficient. Assimilation rates are by far the highest recorded for soil invertebrates, despite the elevated C:N ratios of the food materials in the grass, litter and wood feeding species. According to Higashi *et al.* (1992), these micro-organisms allow the termites to compensate for the unfavourable C:N ratios of their ingested food materials by increasing their N intake or by selectively eliminating C. N intake may be increased in a number of possible ways: increasing N intake through fixation, synthesis of organic N from N excreted as uric acid, application of excreted N as 'fertiliser' to fungus gardens prior to ingestion of the fungus; selective feeding on N-enriched litter humus, and the extraction of N from the most recalcitrant materials of the dead plant (lignin) or ingested soil organic matter. Selective elimination of C may occur through methane (CH_4) emission in species possessing methanogenic bacteria, or as CO_2 through respiration of the fungus gardens of the Macrotermitinae (Figure IV.63).

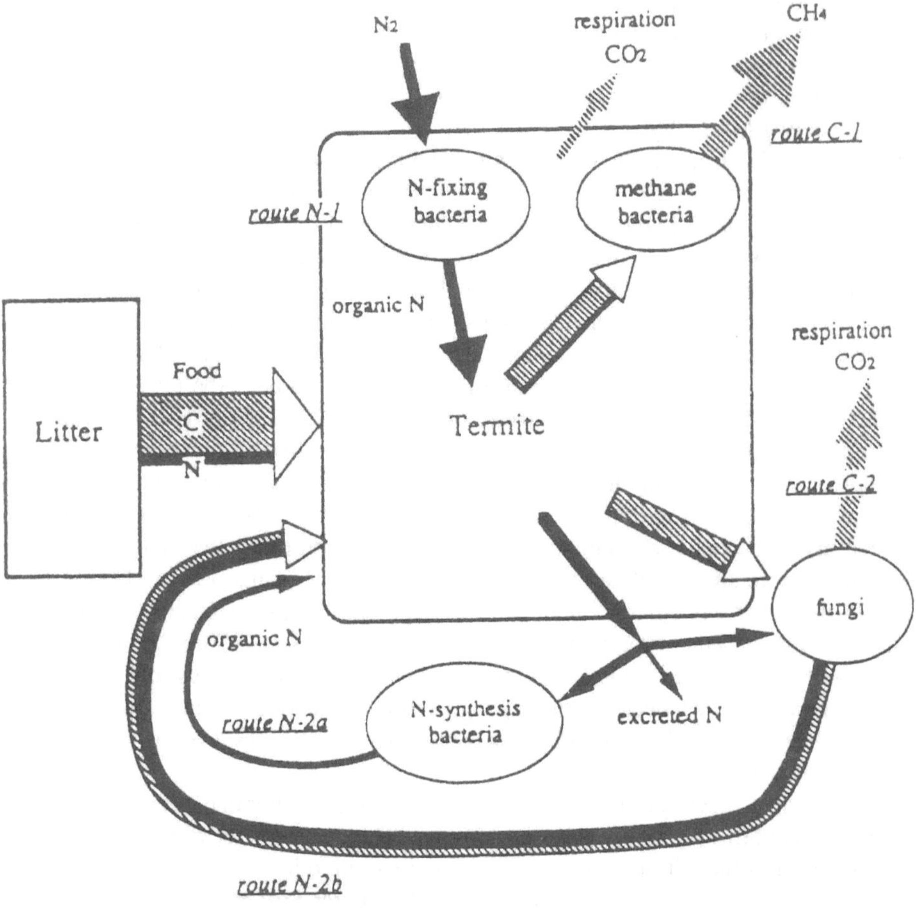

Figure IV.63 The carbon:nitrogen balance problem in termite digestion. Termites adjust the C:N ratio of ingested food through five different routes: Nitrogen fixation in the gut (N1), N2a retention of excreted N by cultivated fungi, N2b production of fungal biomass with a low C:N ratio or elimination of C through methanogenesis (C1), respiration of fungi (C2).

The end products of termite digestion may have high C and N contents due to their selection for C-rich food materials and N fixation in the gut . For example, the N content of the excreta of the Australian termite *Amitermes laurensis* is 1.7 % in spite of a diet low in nitrogen (<0.10 %). Some termite mounds that contain large amounts of faecal material may have high inorganic N concentrations. Nitrate concentrations in the mounds of grass and litter-feeding termites are generally less than 100 mg g^{-1} (Okello-Oloya *et al.*, 1985; Barnes, 1992) although Barnes *et al.* (1992) found maximum values up to 2000 μg g^{-1} at the surface of the mounds where it had been concentrated by evaporation. Soil-feeding termites usually feed on organic accumulations in soils and the faecal pellets used to build their nests have higher nutrient and c concentrations than the surrounding soil (Wood *et al.*, 1983; Anderson and Wood, 1984; López-Hernández *et al.*, 1989) and Table IV.22. Faecal pellets of some termites may also have high microbial biomasses: Boulaud (1996) measured very high concentrations of microbial ribonucleic acid (in the faeces of the African soil-feeding termite *Cubitermes fungifaber*.

Table IV.22 Carbon concentrations and C:N ratios in selected termitaria and their adjacent surface soils.

Species	Major food	mound sample location	Mound		Soil		Author
			C	C:N	C	C:N	
Nasutitermes triodiae	grass	outer galleries	3.0	28.1	0.3	9.6	Lee and Wood, 1971a
Amitermes laurensis	litter	upper middle base	6.9 3.9 1.7	14.2 19.7 14.0	0.5	12.8	Okello-Oloya *et al.*, 1985
Coptotermes acinaciformis	wood	carton centre earthen capping	50.0 2.4	64.2 24.0	0.5	13.9	Lee and Wood, 1971b
Trinervitermes geminatus	grass	mound	10.8	5.2	0.6	8.4	Lopez-Hernandez and Febres, 1984
Macrotermes bellicosus	litter	wall royal cell	0.7 1.4	7.2 4.8	2.0	5.3	*ibid.*
Nodotermes lumanlanus	soil feeder	nest wall	4.4	10.8	3.9	11.5	Garnier-Sillam and Harry, 1995
Cubitermes fungifaber	soil feeder	nest wall	5.6	12.4	5.3	12.9	*ibid.*
Crenetermes albotarsalis	soil feeder	nest wall	3.2	14.8	3.3	14.0	*ibid.*

5.3.2.3 *Intermediate scale effects*

In addition to the specialised fungi of the genus *Termitomyces,* a wide range of micro-organisms is found in termite mounds including nitrogen fixers, denitrifiers and nitrifiers (Lee and Wood, 1971a). Active populations of heterotrophs have been found in the mounds of *Amitermes laurensis* from Australian savannas (Holt, 1987) and these are probably widespread in many other termite structures. Microbial biomass was *ca.* four times, and respiration as high as six times greater in mound materials than in an external control soil (Holt, 1995). In a mound of the soil-feeding species *Cubitermes fungifaber,* microbial biomass represented 16 % of organic carbon compared with 2.2 % in soil below the nest and 0.3 % in the surrounding soil (Boulaud and Braumann, 1998) (Table IV.21). However, respiratory activity per unit of biomass is much lower in the walls of termitaria than in recently-completed constructs or in soils sampled below the nests that have a higher soil moisture status. According to Boulaud (1996), termite mounds of this species may function as 'external rumens' in which fermentative activities favoured by micro-aerophilic conditions may release organic compounds that could be assimilated by termites when re-ingesting soil from the internal mound walls.

Such microbial heterotrophs may even play a role in the nutrition of certain grass-harvesting and litter-feeding termites by partly degrading some of the resistant materials in stored food, and by contributing the nutrients and energy from their own tissues. In the Macrotermitinae, and prior to degradation by *Termitomyces*, a pre-fermentation stage of generalised heterotroph activity may partially degrade food stored in the mounds of certain species (Darlington, 1994). Subsequent breakdown occurs on the fungus combs and is due solely to *Termitomyces*. In mounds with active colonies, it is the only species capable of growing on the combs of *Macrotermes* spp., possibly due to an inhibitor produced by the termites (Thomas, 1987). In contrast, Lee and Wood (1971a) reported the inhibition of microbial activity in mounds of the wood-feeding species *Nasutitermes exitiosus.*

A significant activation of C and N mineralisation has also been observed in the walls of the fungus comb chambers of the West African termite *Ancistrotermes cavithorax* (Abbadie and Lepage, 1989). In a 31-day laboratory incubation, CO_2 evolution was 3.9 times greater in soil from the first 5 mm of the chamber walls on the first day than in a control soil. This activation also occurred at greater depths (a three-fold increase between 5 and 15 mm) and was maintained for several weeks (+50 % at day 31). This may be due to improved conditions of moisture and aeration and a slightly higher N concentration (from 0.48 ‰ in control soil to 0.54 ‰ and 0.58 ‰ respectively in soil from 0-5 mm and 5-15 mm).

Significant nitrogen mineralisation also occurred in the same experiment: after two days, NH_4^+ N produced in the 0-5 mm layer was more than twice that of the control soil (69 compared with 33 $\mu g\ g^{-1}$). This effect still remained significant after 31 days of incubation.

5.3.2.4 *Large scale effects*

In some circumstances, a relationship between termite presence and soil organic matter concentrations has been demonstrated. Parker *et al.* (1982) found increased concentrations of organic matter in a desert soil following experimental eradication of the subterranean termites present. Similarly, the absence of fire and an effective termite fauna from the vertisol studied by Moore *et al.* (1967) in sub-tropical Australia, were considered to be the causes of the large standing crop biomass of above-ground dead organic materials. These authors estimated that there were 75 Mg ha^{-1} of litter together with 58 Mg ha^{-1} of standing dead organic materials on the site, in addition to the relatively-elevated levels of soil organic matter and nitrogen characteristic of these soils, when undisturbed.

Sequestration of organic matter in termite constructs

Carbon concentrations in most epigeal termite structures are higher than those of the soils from which they are formed. This is due to the incorporation during mound construction of salivary or - more importantly - faecal materials that are frequently used as adhesives, or as linings to chamber and gallery walls. Waste materials may also be used as void infillings in mounds, subterranean galleries and other structures. A salivary adhesive from the mound wall of the wood-feeding species *Coptotermes acinaciformis* has been identified as a glycoprotein (Gillman *et al.*, 1972). In structures formed mainly from soil materials, the degree of organic matter enrichment depends primarily on whether faeces are included. The Macrotermitinae do not incorporate their faeces into the mound fabric which consequently has a lower C concentration than that of the soil-feeding species.

The C:N ratios of the constructs built by termites are often higher than those of the surrounding surface soils (Tables IV.22, IV.23), although the differences change between trophic groups. As discussed above, the macrotermitine fungus-feeding termites do not build faeces into their mound structures and their C:N ratios may be similar to or lower than those of the surrounding soils (Garnier-Sillam, 1989).

Table IV.23 Selected properties of the mound materials of *Amitermes laurensis* and their adjacent surface (0-10 cm) soils from northern Australia together with the masses of the above-ground parts of an annual grass and a pasture legume grown for six weeks on these materials in a glasshouse (Okello-Oloya *et al.*, 1985; Okello-Oloya and Spain, 1986).

Material	**Chemical properties**					**Oven-dry (60°) masses (g)**	
	Organic carbon (%)	**Nitrogen (%)**	**Acid extr. P (0.005M H_2SO_4)**	**CEC (cmol kg^{-1})**	**Sum of exch. Ca, Na, Mg, K**	***Digitaria ciliaris***	***Stylosanthes hamata***
Soil	0.5	0.04	2	2.49	1.74	0.02	0.03
Mound materials							
Upper	6.9	0.37	27	6.32	17.83	2.36	0.67
Centre	3.9	0.28	22	6.32	18.15	2.06	0.42
Lower	1.7	0.11	8	6.32	8.04	0.09	0.06

Figure IV.64 presents the C:N ratios of the mound materials of 44 Australian termitaria built by termites of five trophic groups (Lee and Wood, 1971a, Okello-Oloya *et al.*, 1985; Tayasu *et al.*, 1998). In the Australian data set, the C:N ratios of the soil:wood interface feeders and the wood feeders have markedly higher values than those of the litter, grass and soil-feeding species, the last of which differ only slightly from those of their surrounding surface soils. It seems likely that the higher values in mounds of the wood and soil-wood interface feeders may have been due to the incorporation of undigested wood in their faeces (Lee and Wood, 1971a).

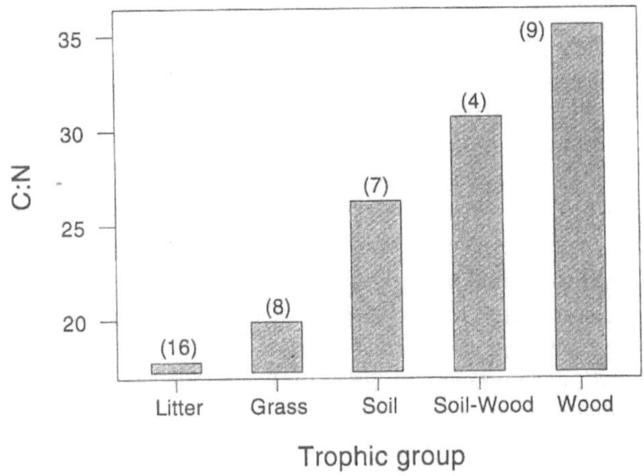

Figure IV.64 Mean C:N ratios in the mound materials of Australian termites of differing ecological strategies (sample size indicated in brackets for each group) (Lee and Wood, 1971a, Okello-Oloya *at al.*, 1985, Tayasu *et al.*, 1998).

As Lee and Wood (1971a,b) noted, Australian termitaria have higher C:N ratios than those of equivalent ecological strategies recorded from African environments. For example, a sample of eleven mound materials of African soil-feeding species had a mean C:N ratio of 11, less than half that of the seven mounds of the equivalent Australian species presented in Figure IV.64. The reasons for such differences are unknown.

The entire nests of a number of species and the interiors of the mounds of certain wood- and grass-feeding termites consist of carton, a relatively soft mixture of excreta, some inorganic material and fragments of undigested, comminuted plant material. Organic matter concentrations in carton are high and that in the mounds of *Coptotermes acinaciformis*, a wood-feeding termite from northern Australia ranged from 83 to 93 % loss-on-ignition (Lee and Wood, 1971b). Intermediate materials also occur; a mixture of carton-like material and inorganic matter (5-15 % C) occurs in the mounds of *Microcerotermes nervosus* from northern Australia (Tayasu *et al.*, 1998).

Soils closely associated with termitaria are also modified, often by such physical processes as the erosion of materials from the mounds and diffusion of materials and elements from the mound into the soils below and adjacent to the mounds. In tropical Australia, the mounds of litter and grass-feeding termites (*Amitermes* spp.) are surrounded by erosional pediments which have carbon concentrations similar to those of the mound materials, or intermediate between these and soils distant from the mounds (Okello-Oloya *et al.*, 1985).

There have been few detailed studies of the chemical characteristics of the organic matter of termite mounds. Lee and Wood (1971b) presented information on the concentrations of organic matter and selected organic and inorganic components of mound materials and the carton of a number of Australian wood-, grass- and litter-feeding termites. Carton from the Australian mounds studied was shown to be high in lignin and humic acids, with higher lignin, total carbohydrate and C:N ratios but lower nitrogen concentrations in the wood-feeding than in the grass-feeding species. Other studies have concentrated on the mounds of the Macrotermitinae. Arshad *et al.* (1988) studied organic matter from the mounds of two African Macrotermitinae (*Macrotermes michaelseni*, *M. herus*). Generally, humic acids from soils were shown to have higher molecular weights and less aromatic material than those from the mounds. In the mounds of *Macrotermes* spp., the concentrations of CO_2H groups increased while those of carbohydrates and other O-substituted compounds decreased in the order: outer wall, nursery, royal chamber. Some inter-site variation in the composition of the organic matter of the *Macrotermes* mounds was also noted.

The fungus comb of *M. michaelseni* was studied by Arshad and Schnitzer (1987) and Arshad *et al.* (1988) who found it to contain approximately 14 % of inorganic material. The organic matter contained 40 % carbohydrate (all of which was present as polysaccharides which, on hydrolysis, were shown to be dominated by glucose), 10 % proteinaceous material (all as peptides or longer-chain structures). Appreciable concentrations of aliphatic and aromatic plus phenolic materials were also noted. Using fractionation techniques commonly used for soil materials, the base-insoluble humin (40 % by mass of organic matter) was found to be mostly carbohydrate while the humic acid (40 %) and the fulvic acid (20 %) components contained most of the aromatic material.

As discussed in Chapter III.4.3.2.1, termite whole-body $\delta^{13}C$ values are closely related to those of their diet (Figure III.52). The $\delta^{13}C$ values of termite-mound organic materials are also related to those of the termite diet (Figure IV.65) (Spain and Reddell, 1996) because of the incorporation of faeces and secretions of limited $\delta^{13}C$ ranges into the mound materials during construction, as gallery linings and as infilling materials in voids of all types. Mound organic matter may therefore lie in the C3 range, the C4 range or somewhere in between, depending on the diet of the termite building the mound and the proportion and $\delta^{13}C$ value of the incorporated soil (Spain and Reddell, 1996).

In savanna environments, the organic materials included in the mounds of the predominantly-earthen, grass-harvesting termites are largely of C4 origin and raise the $\delta^{13}C$ values of the mound materials above those of the mound interspace soils. In contrast, the inclusion of materials of C3 origin depresses the $\delta^{13}C$ values of the mounds of the wood-feeding termites below those of the surrounding soils. The diet of the litter-feeding species may include a mixture of C3 and C4 materials; the $\delta^{13}C$ values of organic matter in their mounds depends on that of the predominant materials incorporated and this may alter with time and location. However, mounds in C3-dominated rainforest situations have similar values to those of the adjacent surface soil. In the high termitaria built by fungus-cultivating termites in African savannas, the C3 vegetation growing on the mounds may contribute to the substantial differences (up to 5.5 ‰) that may occur between the mounds and the adjacent C4-dominated surface soils (Spain and Reddell, 1996).

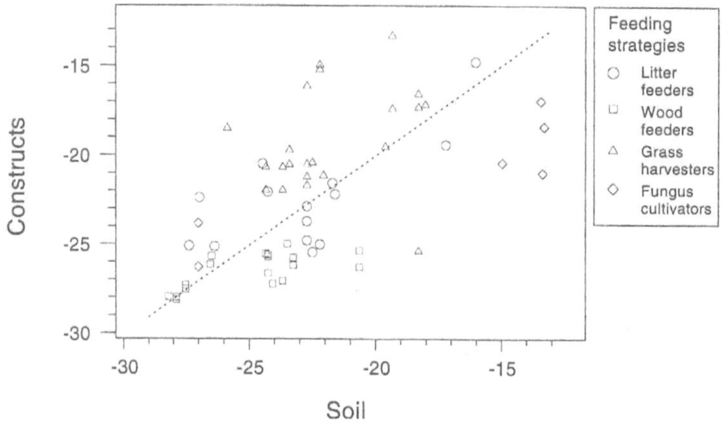

Figure IV.65 Relationships between the $\delta^{13}C$ values of termite constructs and those of the surface soils on which such constructs occur (Spain and Redell, 1996).

5.4 Interactions between termites, higher vegetation and micro-organisms

Over long periods, termites may considerably influence the distribution and community structure of higher plants. This influence is often indirect and may be mediated through alteration of soil properties or through interactions with other environmental factors. In the short time scales associated with agriculture and forestry, termites may directly attack plants causing damage that may kill them or reduce their growth and productivity. This is clearly of importance in agricultural and forestry enterprises but may also be of relevance in disturbed but near-natural environments. Other, less direct biological interactions also occur.

5.4.1 DIRECT EFFECTS

Termites are rarely sources of extensive plant mortality in undisturbed natural ecosystems although in disturbed and man-made environments, a wide range of crop and other plants may be severely affected.

Termites may be severe pests in cropping systems in contrast to the earthworms that are almost always considered beneficial, sometimes unjustifiably. Wood (1996) has noted that termites rarely fit into the common pest category of 'introduced species'. While exotic termite species may cause problems in a few places, species that damage crops (herbaceous or woody species) are nearly always present in the environment at sub-economic population levels and their elevation to pest status normally results from changes in the availability of food. However, to become pests they must also be able to survive the physical disruption and other stresses (*e.g.*, greater diurnal temperature fluctuations and periodic drought) normally associated with the conversion of land to arable use, or be able to colonise the disturbed areas rapidly and effectively.

Termite damage is often greatest to exotic plant species (commonly species of such genera as *Pinus* and *Eucalyptus*), to stressed plants, particularly young plants and those subject to drought, as at the end of the intertropical dry season. Such stress may also derive from the alteration of drainage regimes. Termites frequently gain entry to trees through areas that have been physically damaged: such injury may result from the effects of large mammals, through pruning scars and incisions (Wood, 1996) or through fire scars, all of which allow fungi to attack the damaged tissues and make them vulnerable to termite attack (Perry *et al.*, 1985).

As indicated in Section IV.5.3.2.1, a few examples exist of grass-harvesting termites apparently denuding areas subject to drought or overgrazing although they usually have only small effects on long term plant cover. Few species attack the leaves of living plants: species of the Neotropical genus *Syntermes* may defoliate nursery trees, sugarcane and other plants (Mill, 1992) and the northwestern Australian harvester species *Schedorhinotermes derosus* (family Rhinotermitidae) unusually harvests living grasses (Watson, 1969). Most grass-harvesting and litter-feeding species feed on hayed-off, nutrient-depleted, above-ground materials. Similarly, litter feeders may not cause severe damage directly to plants unless their feeding habits are broad or flexible enough to extend into other food materials, as illustrated in Section IV.5.3.2.1.

Most of the termites that directly damage plants are wood-feeding or polyphagous species. Damage may be minor to severe and can occur in a variety of ways, often being expressed through feeding on the roots, hollowing, severing or ring barking the stems. Colonies of many species may establish in larger trees and gradually eat out the centres of the stems; in species of the intermediate life type, colony members also extend their attacks to adjacent trees. The polyphagous Australian termite *Mastotermes darwiniensis* may kill mature trees by ring barking, often forming a characteristic cincture around the stem (Hill, 1942). In crop species, damage may occur directly to the plant or the product may be attacked, as illustrated below.

Termite attack may entrain damage by other organisms. In groundnut (*Arachis hypogea*) crops, damage to subterranean pods may lead to infection and spoilage by various fungi to such an extent that affected crops become unsaleable (Wood, 1996).

The biology and population dynamics of termites attacking crops is now well documented, especially in Africa and parts of South America (Black and Wood, 1989; Cowie and Wood, 1989; Cowie *et al.*, 1989; Mill, 1992; Sen-Sarma, 1986; Wood and Pearce, 1991; Wood, 1996). A very wide range of crops are at risk, including cereals (especially maize), groundnuts, dryland rice, cotton, yam, cassava, tea, cocoa, coconut, coffee, rubber and a range of fruit producing species. Seedlings of fast-growing exotic tree species, including softwoods, are also highly vulnerable and termites may therefore adversely impact timber plantations and agroforestry schemes.

The soil-wood feeders and the soil feeders cause little damage to living plants or their products. However, some soil-wood feeding species may occasionally damage timber and a few soil feeders may browse the root tips of certain crop species (Mill, 1992).

5.4.2 INDIRECT EFFECTS

5.4.2.1 *Termites and the local distribution of vegetation*

Termites interact with vegetation to influence its composition, distribution and productivity, although this interaction is also modified by other factors including herbivores, saprophytic organisms and environmental factors such as rainfall and fire. However, an important local factor is the presence of epigeal termitaria, particularly those of grass-harvesting and litter-feeding termites. The mound materials of these species may be substantially enriched in plant nutrient elements in comparison to their surrounding soils. Where these mounds are populous, it is partly through the dynamics of their construction and breakdown that termites influence the nature and distribution of the vegetation present. A major effect of the presence and erosional reduction of these termitaria is an increased nutrient patchiness in the landscape (Salick *et al.*, 1983); in some savanna environments this provides sites for species requiring a higher soil nutrient status than that available in the general environment (Spain and McIvor, 1988). In degraded areas, the same process may provide locally fertile sites for the initial regeneration of vegetation and the re-establishment of nutrient cycles.

Such erosional breakdown of mounds is not always beneficial. In parts of northern Australia, erosion of mounds of the litter-feeding termite *Amitermes laurensis* may lead to the deposition of an annulus of fine-textured soil materials around the mound bases. These deposits dry to form an impermeable crust, inhibiting the growth of the surrounding vegetation.

Subterranean termites may also influence vegetation growth. Whitford (1991) reported increased soil N concentrations 14 years after experimental eradication of the termite *Gnathamitermes tubiformans* from plots in a Nearctic desert environment. Further, the reduced water infiltration and storage reported apparently led to the eradication of the perennial grass *Erioneuron pulchellum* and higher litter production in the deep-rooted shrub *Larrea tridentata*. The increased litter mass present on these plots has also provided a habitat and food source for selected micro-arthropods, notably cryptostigmatid mites and psocopterans.

In sandy soils (alfisols) overlying laterite in Niger, areas of both poor and enhanced growth of the woody species *Faidherbia albida* have been found (Miedema *et al.*, 1994). The greatest growth occurred on soils close to abandoned *Macrotermes* mounds, but not on the mound materials themselves. The structure and texture of the materials supporting the greatest growth had higher clay and organic matter contents and therefore greater water-holding capacity and an enhanced soil nutrient status.

As shown in semi-arid environments of Burkina Faso, termitaria may influence both long and short term landscape dynamics in sites with banded vegetation distributions (Ouedraogo, 1997). In this environment, termite mounds cover, respectively, 2.7 % and 2.3 % of the soil surface in unvegetated and vegetated strips. Termite activities and structures:

- affect soil porosity by creating galleries that open at the soil surface (375 holes m^{-2} in the vegetated zones where termites are highly active; almost none in the unvegetated zones) and considerably loosen the soil;

- concentrate nutrients that are later released with the erosion of termitaria that may be deserted or contain active colonies;
- increase water infiltration (up to eight-fold in comparison with surfaces not affected by termites) especially in the vegetated strips where termitaria with active colonies are most common;
- increase water runoff (up to a maximum of 98 %) on unvegetated surfaces between strips where fine-textured soil eroded from termitaria tends to form an impermeable surface crust;
- enhance plant species richness and density of woody species such as *Combretum micranthum* (increased three fold) and *Boscia senegalensis* (increased ten fold).

The combination of the above effects may enhance or even determine, the dynamics of this vegetation system by favouring germination and growth of trees on the uphill part while mortality of trees on the downhill part is followed by death of termitaria and accelerated erosion of their structures.

5.4.2.2 *The vegetation associated with epigeal termite mounds*

While they may strongly influence the nature and productivity of the vegetation surrounding them, few termitaria support an extensive growth of vegetation on their surfaces. A notable exception to this generalisation is the distinctive and specialised vegetation of African 'high termitaria' which are considered separately.

Despite the general lack of actively-growing vegetation on the surfaces of most epigeal termitaria, the materials comprising certain of these termitaria are well provided with plant nutrient elements. Table IV.23 illustrates selected properties of the mound materials of the tropical Australian litter-feeding species *Amitermes laurensis* (Okello-Oloya *et al.*, 1985). As shown in the Table, organic C, total N and an index of 'plant-available' P were all substantially elevated compared with equivalent values for the surface soils of the inter-mound spaces. A feature of these mound materials is the substantial excess of the sum of the 'exchangeable' Ca, Mg, Na and K ions present over the cation exchange capacity indicating the large quantity of soluble bases present.

If the mound materials of many grass-harvesting and litter-feeding termites are broken up and supplied with water, commensurate with their enhanced nutrient status they may support substantially greater plant growth than their adjacent, unaffected surface soils (Okello-Oloya and Spain, 1986) (Table IV.23, p. 517). This may not be true of the wood-feeding species whose mound materials often differ little from the surrounding soils in their concentrations of plant nutrient elements (Lee and Wood, 1971b).

It appears that the densely-packed arrangement of soil particles in most mounds creates a compact surface structure whose physical condition is unfavourable to plant establishment and growth. Most mound materials, except during the brief seasonal periods when they are highly hydrated, may be too strong and their water held too firmly to constitute a satisfactory physical growth medium for most plants. However, despite their normally compact surfaces, the mounds of some grass-harvesting and litter-feeding termites may support a sparse growth of plants for periods during the wet season. In such humid environments as rainforests, termite mounds often support the growth of algae, mosses and lichens on their surfaces.

Mound surfaces are subject to erosional attrition through the action of water and wind and are therefore unstable surfaces, at least during the seasons when rainfall is high. The products of this erosion may form a chemically-fertile pediment (Okello-Oloya *et al.*, 1985), slightly elevated above the surrounding soils and without the physical conditions inimical to plant growth possessed by the termitaria. If, as shown above, the materials built into the mounds are easily dispersed, surface crusts may develop around the bases of the mounds.

The moisture contents of the soils surrounding termite mounds and other pavement-like structures may be elevated by runoff (Eldridge, 1994) leading to a higher moisture status in the materials forming the bases of termite mounds, and their surrounding erosional pediments. Some combination of the more favourable moisture status and the higher nutrient contents of the materials available to the root systems of the surrounding plants often leads to enhanced plant growth on and close to the mound base. In sites subject to seasonal flooding, the slightly-increased elevation of the pediment may permit a better aeration of these soils during the wetter part of the year. Figure IV.66 illustrates the substantial decline in vegetation biomass that occurs with increasing distance from the bases of the mounds of *Amitermes laurensis* in savanna environments in northeastern Australia (Spain and McIvor, 1988). Similar situations pertain in the soils surrounding certain African termite mounds, although the spatial scales may be different (see, for example, Arshad, 1982).

The growth strategies of the plant species that occur on and close to the mounds of grass and litter-feeding termites differ from those unaffected by the dynamics of mound erosion and water runoff. The annual grasses that grow on the pediment and close to the mounds compete well under the higher nutrient conditions present and grow and set seed rapidly. Similarly, many of the herbaceous dicotyledonous plants growing in this situation are biennial species but have the capacity to complete their life cycles within a single season and consequently behave functionally as annuals. In contrast, herbaceous vegetation not influenced by the mounds is normally dominated by slow-growing perennial grasses and sedges and the few annual grass species present are small. These effects are illustrated by the changes that occur, as a percentage of plant biomass, in four categories of plants with increasing distance from the mounds at a site in northeastern Australia (Figure IV.67) (Spain and McIvor, 1988).

5.4.2.3 *Termitaria, roots and mycorrhizal fungi*

In both savanna (Okello-Oloya *et al.*, 1985) and rainforest (Salick *et al.*, 1983) environments, the termitaria of grass-harvesting and litter-feeding termites may form 'islands' of enhanced fertility within the generally nutrient-poor soils of the surrounding landscape. While the mound exteriors are normally hard and impenetrable, the interiors are not. A network of galleries radiates from the bases of most termitaria and a proportion of these and the voids within the mounds may be backfilled with organic materials. Such openings into the mounds provide access for the roots of surrounding plants which frequently exploit the often elevated nutrient resources contained within and beneath the mounds.

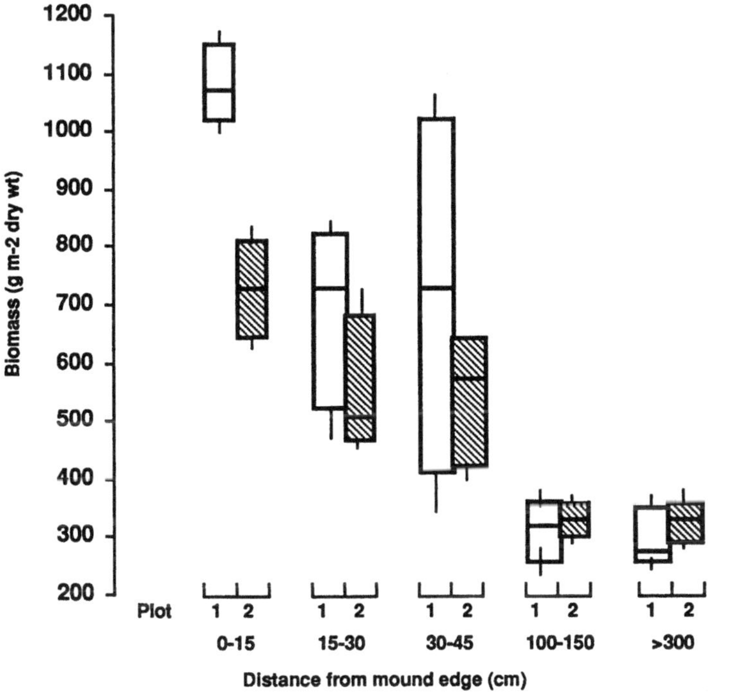

Figure IV.66 Changes in plant biomass at different distances from termite mounds of *Amitermes laurensis*, medians, inter-quartile and overall ranges) (Spain and McIvor, 1988).

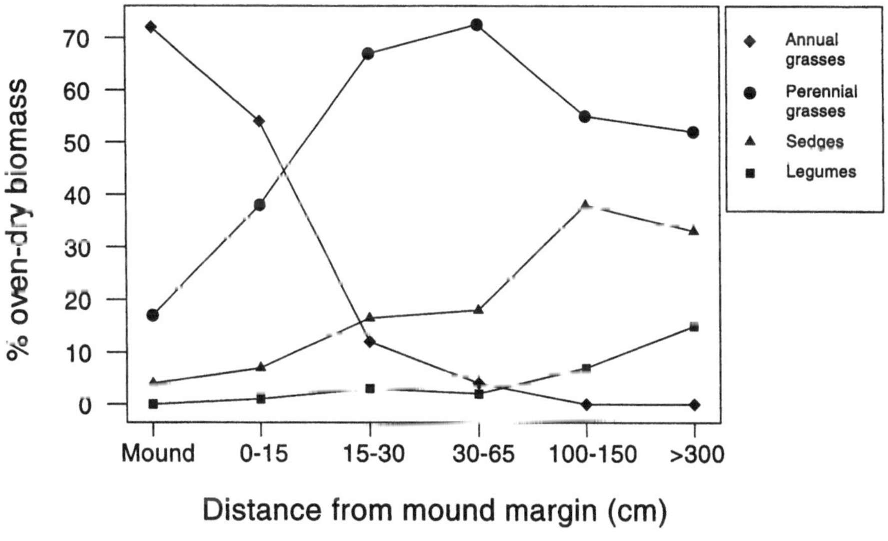

Figure IV.67 Changes in the relative biomass of four categories of plants with increasing distance from mounds of the litter feeding termite *Amitermes laurensis* at a site in northeastern Australia (Spain and McIvor, 1988).

Roots from woody plants some distance from the mounds have been consistently noted to enter the soil beneath the epigeal mounds and frequently to penetrate the mounds themselves. In a humid West African savanna environment, Mordelet *et al.* (1996) found that the roots of the palm *Borassus aethiopum* could extend as far as 20 m to exploit such nutrient-enriched sites. Roots have also been reported to follow old infilled termite galleries in subsoils beneath Kenyan coffee crops (Robinson, 1958), presumably because of their lower material strengths and greater nutrient contents.

Kaiser (1953, in Araujo, 1970) described a specific root-feeding relationship with *Anoplotermes pacificus* (Apicotermitinae) and Mill (1992) also referred to this species as attacking the root tips of certain crop plants (Section III.4.3.2.1). Eggleton *et al.* (1995) have suggested that more generalised feeding on fine roots by soil-feeding termites may take place in certain environments. In some Amazonian forests, rapid seedling growth in mounds follows their abandonment because the termites are no longer present to control their growth (Salick *et al.*, 1983). Further, the often extensive proliferation of roots within unoccupied, degrading termitaria suggests that the termites may play an active role in preventing the ingress of roots, while the colony remains active.

Recent Australian work has demonstrated the presence of infective propagules of both ectomycorrhizal and arbuscular mycorrhizal fungi in the mounds of grass-harvesting, litter-feeding and wood-feeding termites at concentrations similar to those in the surrounding soils (Spain *et al.*, in preparation). Also, ectomycorrhizal fungi (notably *Pisolithus* sp.) have been noted to fruit on the surfaces of termitaria at some locations.

The presence of roots, fungal propagules and by implication hyphae, suggests that termitaria are being explored either directly by the roots of ectomycorrhizal host species (largely *Eucalyptus* and other sclerophyllous species that may be located some metres distant from the mounds), or indirectly through their associated mycorrhizal fungi (Spain *et al.*, in preparation). Host plant species of vesicular-arbuscular mycorrhizal fungi (grasses, forbs) occasionally grow on the surfaces of the epigeal termitaria of grass harvesting and litter-feeding termites but occur more regularly and in greater profusion on the erosional pediment surrounding the mounds. Thus, the inward movement of nutrient elements to the mounds in transported termite food materials may be at least partly balanced by a countervailing outward movement back to the adjacent herbaceous vegetation and, at a larger radial scale, the surrounding trees.

A possibly quadripartite association between the widespread African wood-feeding termite *Sphaerotermes sphaerothorax* (Macrotermitinae), cellulolytic and N-fixing bacteria and the roots of the surrounding trees was described by Garnier-Sillam *et al.* (1989). This termite has either lost or never developed a relationship with *Termitomyces* and instead forms comb-like structures containing an active cellulolytic and N-fixing microflora. Roots from the surrounding trees grow into the nest and the combs rest on a mass of small roots and root hairs. Food materials in the form of small spheres of masticated wood are broken down by some combination of the termites' own enzymes, bacteria located in the hind gut or externally in the combs. Faeces in the lower part of the nest are explored by root hairs and it seems likely that interchanges of materials may occur between the roots, the termites and the two types of bacteria forming the association.

5.4.2.4 *High termitaria*

High termitaria are a feature of many parts of the African continent. They may be up to 10 m in height above the surrounding plain and as much as 30 m in diameter (Pullan, 1979). These termitaria may have been originally built by species of the genus *Macrotermes* but have undergone complex cycles of development and regression. At various stages of their development they may be partly occupied by the original builders, other species, or by none at all.

Since they are formed by different termite species from a variety of soil materials and exist under a range of environmental conditions, they may develop in various ways. Pullan (1979) describes how their development is influenced by the presence of fire, erosional influences and the activities of man and large mammals (notably elephants).

These high termitaria may support a diverse (more than 336 species of higher plants) and regionally-variable flora which may include cyanobacteria, grasses, forbs, shrubs, vines and trees, including palms. This flora is characterised by a higher incidence of thorniness, xeromorphy, succulence and eutrophy than occurs in that of the immediate surrounds. The diaspores of many of the woody species of the high termitaria are dispersed by animals (Malaisse, 1978).

The vegetative cover of these termitaria protects the surface from erosion and while it remains intact, termitaria may exist for long periods. Based on the vegetation, complex food webs may be associated with these termitaria. In addition, secondary termite species and a range of other soil animals may also be present (Malaisse, 1978). When regularly fired, these termitaria can degrade rapidly under the influence of raindrop erosion and surface water flows.

5.5 Interactions between termites and higher animals

The interactions between termites and higher animals may be direct - as in predation - or indirect, as in the provision of habitat or other effects. However, apart from predation on termites by the vertebrates mentioned in Chapter III, few direct relationships are known.

A number of indirect relationships have been recorded, most of which benefit the vertebrates. Provision of habitat is an important way in which vertebrates interact indirectly with termites and a widespread example of this is the use of both epigeal and arboreal termite nests as nesting sites by a range of bird species (Hindwood, 1959) and other vertebrates, particularly reptiles (see, for example, Araujo, 1970). The high termitaria of Africa provide habitat for a wide range of vertebrate and invertebrate animals, including secondary termites (Malaisse, 1978).

In a more general relationship, Braithwaite *et al.* (1988) demonstrated a positive relationship between termite species richness and the species richness and abundance of grasshoppers, but negative relationships with macropodid species richness and abundance. These authors also reported a positive correlation between the species richness of termites feeding on living trees and the abundance of the introduced water buffalo (*Bubalus bubalis*) in parts of northern Australia. It was considered that soil compaction caused by the buffaloes led to water stress in the trees and made them more susceptible to termite attack.

REFERENCES

Abbadie, L. (1983) *Contribution à l'étude de la production primaire et du cycle de l'azote dans les savanes de Lamto (Côte d'Ivoire)*, thèse de 3e cycle, Université Paris VI/ENS.

Abbadie, L. and Lepage, M. (1989) The role of subterranean fungus comb chambers (Isoptera, Macrotermitinæ) in soil nitrogen cycling in a preforest savanna (Côte d'Ivoire). *Soil Biology & Biochemistry*, **21** (8), 1067-71.

Abbadie, L., Mariotti, A. and Menaut, J.C.C. 1992) Independence of savanna grasses from soil organic matter for their nitrogen supply. *ecology*, **73** (2), 608-13.

Abbott, I. and Parker, C.A. (1981) Interactions between earthworms and their soil environment. *Soil Biology & Biochemistry*, **13**, 191-7.

Abe, T. (1987) Evolution of life types in termites, in *Evolution and Coadaptation in Biotic Communities* (eds S. Kawano, J.H. Connell and T. Hidaka), University of Tokyo Press, Tokyo, pp. 125-48.

Abensperg-Traun, M. and Milewski, A.V. (1995) Abundance and diversity of termites (Isoptera) in unburnt versus burnt vegetation at the Barrens in Mediterranean Western Australia. *Australian Journal of Ecology*, **20** (3), 413-7.

Abensperg-Traun, M. and Steven, D. (1997a) Ant- and termite-eating in Australian mammals and lizards: a comparison. *Australian Journal of Ecology*, **22** (1), 9-17.

Abensperg-Traun, M. and Steven, D. (1997b) Latitudinal gradients in the species richness of Australian termites (Isoptera). *Australian Journal of Ecology*, **22**, 471-6.

Aber, J., Melillo, J.M. and McClaugherty, C.A. (1983) Potential sinks for mineralized nitrogen following disturbances in forest ecosystems, in *Environmental Biogeochemistry* (ed. R. Hallberg), Ecological Bulletins/NFR, Stockholm, pp. 179-92.

Aber, J.D., Melillo, J.M., Nadelhoffer, K.J. *et al.* (1985) Fine root turnover in forest ecosystems in relation to quantity and form of nitrogen availability: a comparison of two methods. *Oecologia*, **66**, 317-21.

Abrahamsen, G. (1971) The influence of temperature and soil moisture on the population density of *Cognettia sphagnetorum* (Oligochaeta, Enchytraeidae) in cultures with homogenized raw humus. *Pedobiologia*, **12**, 26-82.

Adamo, P. and Violante, P. (1991) Weathering of volcanic rocks from Mt Vesuvius associated with the lichen *Stereocaulon vesuvianum*. *Pedobiologia*, **35**, 209-17.

Adams, C.T. (1986) Agricultural and medical impact of the imported fire ants, in *Fire Ants and Leaf-cutting Ants* (eds C.S. Lofgren and R.K. Van der Meer), Westview Press, Boulder, Colorado, pp. 48-57.

Adams, E.S. and Levings, S.C. (1987) Territory size and populations limits in mangrove termites. *Journal of Animal Ecology*, **56**, 1069-81.

Adaskaveg, J.E. and Gilbertson, R.L. (1986) *In vitro* decay studies of selective delignification and simultaneous decay by the white rot fungi *Ganoderma lucidum* and *G. tsugae*. *Canadian Journal of Botany*, **64**, 1611-9.

Aerts, R. (1997) Climate, leaf litter chemistry and leaf litter decomposition in terrestrial ecosystems: a triangular relationship. *Oikos*, **79**, 439-49.

Association Française pour l'Etude du Sol (1998) *Référentiel Pédologique, Principaux Sols d'Europe*, INRA, Paris.

Agren, G.I., Axelsson, B., Flower-Ellis, J.G. *et al.* (1980) Annual carbon budget for a young Scots pine.

Ecological Bulletins, **32**, 307-13.

Alabouvette, C., Lemaître, I. and Pussard, M. (1981) Densité de population de l'amibe mycophage *Thecamoeba ganifera ssp minor* (Amoebida, Protozoa). Mesure et variations expérimentales dans le sol. *Revue d'Ecologie et de Biologie du Sol*, **18**, 179-92.

Albrecht, A. (1984) *Role des Enchytraeidae dans l'agrégation et l'humification d'humus de type moder de l'Est de la France*, thèse de 3e cycle, Université Nancy I.

Albrecht, A., Brossard, M. and Feller, C. (1986) Etude de la matière organique des sols par fractionnement granulométrique. 2-Augmentation par une prairie à *Digitatria decumbens* du stock organique de vertisols cultivés en Martinique, in *XII Congress of the International Society of Soil Science*, Hambourg, pp. 214-5.

Aldag, R. and Graff, O. (1975) N-Fraktionen in Regenwurmlosung und deren Ursprungsboden. *Pedobiologia*, **15**, 151-3.

Alef, K., Beck, T., Zelles, L. *et al.* (1988) A comparison of methods to estimate microbial biomass and N-mineralization in agricultural and grassland soils. *Soil Biology & Biochemistry*, **20** (4), 561-5.

Alegre, J., Pashanasi, B. and Lavelle, P. (1996) Dynamics of soil physical properties in a low input agricultural system inoculated with the earthworm *Pontoscolex corethrurus* in Peruvian Amazonia. *Soil Science Society of America Journal*, **60** (5), 1522-9.

Alexandre, A., Meunier, J.D., Colin, F. *et al.* (1997) Plan impact on the biogeochemical cycle of silicon and related weathering processes. *Geochimica et Cosmochimica Acta*, **61**, 677-82.

Allen, T.F.H. and Starr, T.B. (1982) *Hierarchy. Perspectives for ecological complexity*, The University of Chicago Press, Chicago.

Allen, J.C. (1985) Soil response to forest clearing in the United States and the tropics: geological and biological factors. *Biotropica*, **17**, 15-27.

Allen, B.L. and Hajek, B.F. (1989) Mineral occurrence in soil environments, in *Minerals in Soil Environments* (eds J.B. Dixon and S.B. Weed), Soil Science Society of America, Madison, Wisconsin, pp. 199-278.

Allen, M.F., Hipps, L.E. and Woodridge, G.L. (1989) Wind dispersal and subsequent establishement of VA mycorrhizal fungi across a successional arid landscape. *Landscape Ecology*, **2**, 165-7.

Allen, M.F. (1991) *The Ecology of Mycorrhizae*, Cambridge University Press, New York.

Almendros, G., Martin, F. and Gonzales-Villa, F.J. (1988) Effects of fire on humic and lipid fractions in a Dystric Xerochrept in Spain. *Geoderma*, **42**, 115-27.

Aloni, K. (1975) Le sol et l'évolution des termitières géantes du Haut-Shaba Rép. Zaïre). *Pédologie*, **25**, 25-39.

Aloni, K. and Soyer, J. (1987) Cycles des matériaux de construction des termitières d'humivores en savane au Shaba méridional (Zaïre). *Revue Africaine de Zoologie*, **101**, 329-357.

Altmuller, R. (1979) Untersuchungen über den Energieumsatz von Dipterenpopulationem im Buchenwald (Luzulo-fagetum). *Pedobiologia*, **19**, 245-78.

Alva, A.K., Edwards, D.G., Asher, C.J. *et al.* (1986) Relationships between root length of soybean and calculated activities of aluminium monomers in nutrient solution. *Soil Science Society of America Journal*, **50**, 959-62.

Alvarado, A., Berish, C.W. and Peralta, F. (1981) Leaf-cutter ant *(Atta Cephalotes*) influence on the morphology of Andepts in Costa Rica. *Soil Science Society of America, Proceedings*, **45**, 790-4.

Amato, M. and Ladd, J.N. (1988) Assay for microbial biomass based on ninhydrin-reactive nitrogen in extracts of fumigated soils. *Soil Biology & Biochemistry*, **20** (1), 107-14.

Amato, M. and Ladd, J.N. (1994) Application of the ninhydrin-reactive N assay for microbial biomass in acid soils. *Soil Biology & Biochemistry*, **26**, 1109-15.

Amundson, R.G. and Davidson, E.A. (1990) Carbon dioxide and nitrogenous gases in the soil atmosphere. *Journal of Geochemical Exploration*, **38**, 13-41.

Andersen, A. (1988) Soil of the nest-mound of the seed-dispersing ant *Aphaenogaster longiceps*, enhances seeding growth. *Australian Journal of Ecology*, **13**, 469-71.

Anderson, J.M. (1971) Observations of the vertical distribution of *Oribatei* (Acarina) in two woodland soils, in *Organismes du Sol et Production Primaire* (eds J. d'Aguilar, C. Athias Henriot, A. Bessard *et al.*), INRA, Paris, pp. 237-72.

Anderson, J.R. and Westmoreland, D. (1971) Direct counts of soil organisms using a fluorescent brightener and a europium chelate. *Soil Biology and Biochemistry*, **3**, 85-7.

Anderson, J.M. and Healey, I.N. (1972) Seasonal and interspecific variation in major components of the gut contents of some woodland Collembola. *Journal of Animal Ecology*, **41**, 359-68.

Anderson, J.M. (1975) Succession, diversity and trophic relationships of some soil animals in decomposing leaf litter. *Journal of Animal Ecology*, **44** (2), 475-95.

Anderson, J.M. and Hall, H. (1977) Cryptostigmata species diversity and soil habitat structure, in *Soil Organisms as Components of Ecosystems* (eds U. Lohm and T. Persson), Ecological Bulletins/NFR, Stockholm, pp. 473-4.

Anderson, J.M. (1977) The organization of soil animal communities, in *Soil Organisms as Components of Ecosystems* (eds U. Lohm and T. Persson), Ecological Bulletins/NFR, Stocholm, pp. 15-23.

Anderson, J.P.E. and Domsch, K.H. (1978) A physiological method for the quantitative measurement of microbial biomass in soils. *Soil Biology & Biochemistry*, **10**, 215-21.

Anderson, J.P.E. and Domsch, K.H. (1980) Quantities of plant nutrients in the microbial biomass of selected soils. *Soil Science*, **130** (4), 211-6.

Anderson, D.W., Saggar, S., Bettany, J.R. *et al.* (1981) Particle size fractions and their use in studies of soil organic matter: I. The nature and distribution of forms of carbon, nitrogen and sulphur. *Soil Science Society of America, Journal*, **45**, 767-72.

Anderson, J.M. and Swift, M.J. (1983) Decomposition in tropical forests, in *Tropical Rain Forests: Ecology and Management* (eds S.L. Sutton, T.C. Whitmore and L.C. Chadwick), Blackwell Scientific Publications, Oxford, pp. 287-309.

Anderson, J.M., Ineson, P. and Huish, S.A. (1983a) Nitrogen and cation mobilization by soil fauna feeding on leaf litter and soil organic matter from deciduous woodlands. *Soil Biology & Biochemistry*, **15** (4), 463-7.

Anderson, J.M., Proctor, J. and Vallack, H.W. (1983b) Ecological studies in four contrasting lowland rain forests in Gunung Mulu National Park, Sarawak. III. Decomposition processes and nutrient losses from leaf litter. *Journal of Ecology*, **71**, 503-27.

Anderson, R.V., Gould, W.D., Cambardella, C. *et al.* (1983c) Organic and inorganic nitrogenous losses by microbivorous nematodes in soil. *Oikos*, **40**, 75-80.

Anderson, J.M. and Wood, T.G. (1984) Mound composition and soil modification by two soil-feeding termites (Termitinae, Termitidae) in a riparian Nigerian forest. *Pedobiologia*, **26**, 72-82.

Anderson, J.M., Leonard, M.A., Ineson, P. *et al.* (1985) Faunal biomass: a key component of a general model of nitrogen mineralization. *Soil Biology & Biochemistry*, **17** (5), 735-7.

Anderson, T.H. and Domsch, K.H. (1985a) Determination of ecophysiological maintenance carbon requirements of soil micro-organisms in a dormant state. *Biology and Fertility of Soils*, **1**, 81-9.

Anderson, T.H. and Domsch, K.H. (1985b) Maintenance carbon requirements of actively-metabolizing microbial populations under *in situ* conditions. *Soil Biology & Biochemistry*, **17** (2), 197-203.

Anderson, J.M. and Flanagan, P. (1989) Biological processes regulating organic matter dynamics in tropical soils, in *Dynamics of Soil Organic Matter in Tropical Ecosystems* (eds D.C. Coleman, J.M. Oades and G. Uheara), NifTAL project, University of Hawaii., Honolulu, pp. 97-125.

Anderson, J.M. and Ingram, J.S.I. (eds) (1993) *Tropical Soil Biology and Fertility: A Handbook of Methods*, 2nd edition, CAB International, Wallingford.

Andrew, P.A. and Nicholas, W.L. (1976) Effect of bacteria on dispersal of *Caenorhabditis elegans* (Rhabditida). *Nematologica*, **22**, 451-61.

André, J., Gensac, P., Pellissier, F. *et al.* (1987) Régénération des peuplements d'épicea en altitude: recherches préliminaires sur le rôle de l'allélopathie et de la mycorhization dans les premiers stades du développement. *Revue d'Ecologie et de Biologie du Sol*, **24** (3), 301-10.

Andrén, O., Lindberg, T., Boström, U. *et al.* (1990) Organic carbon and nitrogen flows, in *Ecology of Arable Land - Organisms, Carbon and Nitrogen Cycling* (eds O. Andren, T. Lindberg, K. Paustian *et al.*), Munksgaard, Copenhagen, pp. 84-125.

Andrén, O., Rajkai, K. and Kätterer, T. (1993) Water and temperature dynamics in a clay soil under winter wheat: influence on straw decomposition and N immobilization. *Biology and Fertility of Soils*, **15**, 1-8.

Andrén, O., Paustian, K. and Rosswall, T. (1988) Soil biotic interactions in the functioning of agroecosystems. *Agriculture, Ecosystems and Environment*, **24**, 57-67.

Andriese, J.P. (1988) *Nature and Management of Tropical Peat Soils*, FAO, Rome.

Anglade, P. (1967) Etude des populations de Symphyles en sols cultivés et de l'influence des traitements du sol, in *Progress in Soil Zoology* (eds O. Graff and J.E. Satchell), North Holland Publishing Co., Amsterdam, pp. 372-81.

Aouti, S.A. (1977) *Données écologiques, biologiques et histologiques sur le Diplopode Iulide* Pachybolus laminatus *Cook*, thèse d'Etat, Université d'Abidjan.

Aouti, A. (1978) Etude comparée des peuplements de Myriapodes Diplopodes d'une forêt hygrophile et d'une plantation *d'Hévéa* en Basse Côte d'Ivoire. *Annales de l'Université d'Abidjan, Série Ecologie*, **11**, 8-32.

Araujo, R.L. (1970) Termites of the neotropical region, in *Biology of Termites*, 2. (eds K. Krishna and F.M. Weesner), Academic Press, New York, pp. 527-76.

Arditti, J. and Ghani, A.K.A. (2000) Numerical and physical properties of orchid seeds and their biological implications. *New Phytologist*, **145**, 367-421.

Arnold, P.W. (1958) Potassium uptake by cation exchange resins from soils and minerals. *Nature*, **182**, 1594-5.

Arpin, P., Kilbertus, G., Ponge, J.F. *et al.* (1980) Importance de la microflore et de la microfaune en milieu forestier, in *Actualités d'Ecologie Forestière* (ed. P. Pesson), Gauthier-Villars, Paris, pp. 87-150.

Arpin, P. (1983) Quelques aspects de l'écologie des nématodes à vie libre du sol et des relations trophiques avec la microflore en milieu forestier. *Feuillets de l'UER Physique-Chimie-Biologie Nancy*, **1** (7), 1-35.

Arpin, P., Kilbertus, G., Ponge, J.F. *et al.* (1985) Réactions des populations animales et microbiennes du sol à la privation des apports annuels de litière: exemple d'une rendzine forestière. *Bulletin Ecologique*, **16** (1), 95-115.

Arpin, P., David, J.F., Guittonneau, G.G. *et al.* (1986) Influence du peuplement forestier sur la faune et la microflore du sol et des humus. 2. Microbiologie et expériences au laboratoire. *Revue d'Ecologie et de Biologie du Sol*, **23** (2), 119-53.

Arshad, M.A. (1982) Influence of the termite *Macrotermes michaelseni* (Sjöst) on soil fertility and vegetation in a semi-arid savannah ecosystem. *Agro-Ecosystems*, **8**, 47-58.

Arshad, M.A. and Schnitzer, M. (1987) The chemistry of a termite fungus comb. *Plant and Soil*, **98**, 247-56.

Arshad, M.A. and Schnitzer, M. (1989) Chemical characteristics of humic acids from five soils in Kenya. *Zeitschrift für Pflantzenernährung, Düngung, Bodenkunde*, **152**, 11-6.

Artaxo, P., Storms, H., Bruynseels, F. *et al.* (1988) Composition and sources of aerosols from the Amazon Basin. *Journal of Geophysiological Research*, **93**, 1605-15.

Asikidis, M.D. and Stamou, G.P. (1991) Spatial and temporal patterns of an oribatid mite community in an evergreen-sclerophyllous formation (Hortiatis, Greece). *Pedobiologia*, **35**, 53-63.

Asner, G.P., Seastedt, T.R. and Townsend, A.R. (1997) The decoupling of terrestrial carbon and nitrogen cycles. *Bioscience*, **47** (4), 226-34.

Athias, F. (1974) Les conditions microclimatiques dans le sol de la savane de Lamto, in *Analyse d'un Ecosystème Tropical Humide: la Savane de Lamto* (ed. M. Lamotte), Publication du Laboratoire de Zoologie de l'ENS, Paris, pp. 5-25.

Athias, F., Josens, G. and Lavelle, P. (1974) Le peuplement animal des sols de la savane de Lamto, in *Analyse d'un Ecosystème Tropical Humide: la Savane de Lamto* (ed. M. Lamotte), Publication du Laboratoire de Zoologie de l'ENS, Paris, pp. 45-53.

Athias, F., Josens, G. and Lavelle, P. (1975) Traits généraux du peuplement animal endogé de la savane de Lamto (Côte d'Ivoire), in *Progress in Soil Zoology. Proceedings of the 5th International Colloquium on Soil Zoology* (ed. S.J. Vanek), Czechoslovak Academy of Sciences, Prague, pp. 389-97.

Athias, F. (1975) Données complémentaires sur l'abondance et la distribution verticale des microarthropodes de la savane de Lamto (Côte d'Ivoire). *Bulletin du Muséum d'Histoire Naturelle*, **308**, 1-28.

Athias, F. (1976) Recherche sur les microarthropodes du sol de la savane de Lamto (Côte d'Ivoire). *Annales de l'Université d'Abidjan, série Ecologie*, **IX**, 193-270.

Athias-Binche, F. (1985) Analyse démographique des populations d'Uropodides (Arachnides: Anactinotriches) de la hêtraie de la Massane, France. *Pedobiologia*, **28**, 225-53.

Atkinson, D. (1985) Spatial and temporal aspects of root distribution as indicated by the use of a root observation laboratory, in *Ecological Interactions in Soil. Plants, Microbes and Animals* (eds A.H. Fitter, D. Atkinson, D.J. Read *et al.*), Blackwell Scientific Publications, Oxford, pp.43-65.

Attiwill, P.M. and Adams, M.A. (1993) Nutrient cycling in forests. *New Phytologist*, **124** (4), 561-82.

Ausmus, B.S. (1973) The use of the ATP assay in terrestrial decomposition studies. *Modern Methods in the Study of Microbial Ecology. Bulletins from the Ecological Research Committee (Stockholm)*, **17**, 223-34.

Avel, M. (1959) Classe des Annélides Oligochètes, in *Traité de Zoologie. V. Embranchement des Annélides* (ed. P.P. Grassé), Masson, Paris, pp. 225-470.

Avery, B.W. (1980) *Systems of Soil Classification for England and Wales (Higher Categories)*, Rothamsted Experimental Station, Harpenden.

Avery, B.W. (1990) *Soils of the British Isles*, CAB International, Wallingford.

Axelsson, B., Lohm, U. and Persson, T. (1984) Enchytraeids, lumbricids and soil arthropods in a northern deciduous woodland - a quantitative study. *Holarctic Ecology*, **7**, 91-103.

Ayanaba, A., Tuckwell, S.B. and Jenkinson, D.S. (1976) The effects of clearing and cropping on the organic reserves and biomass of tropical forest soils. *Soil Biology & Biochemistry*, **8**, 519-25.

Azcon, R. and Ocampo, J.A. (1984) Effect of root exudation on VA mycorrhizal infection at early stages of plant growth. *Plant and Soil*, **82**, 133-8.

Baas, W.J. (1989) Secondary plant compounds, their ecological significance and consequences for the carbon budget. Introduction of the carbon/nutrient cycle theory, in *Causes and Consequences of Variation in Growth Rate and Productivity of Higher Plants* (ed. H. Lambers), SPB Academic Publishing bv, The Hague, pp. 313-40.

Bååth, E., Berg, B., Lohm, U. *et al.* (1980) Effects of experimental acidification and liming on soil organisms and decomposition in a Scots Pine forest. *Pedobiologia*, **20**, 85-100.

Bååth, E. and Söderström, B. (1982) Seasonal and spatial variation in fungal biomass in a forest soil. *Soil Biology & Biochemistry*, **14**, 353-8.

Babel, U. (1971) Gliederung und Beschreibung des Humusprofils in mitteleuropïschen Wäldern. *Geoderma*, **5**, 297-324.

Babiuk, L.A. and Paul, E.A. (1970) The use of fluorescein isothiocyanate in the determination of the bacterial biomass of a grassland soil. *Canadian Journal of Microbiology*, **16**, 57-62.

Bachelier, G. (1978) *La Faune des Sols. Son Ecologie, son Action*, ORSTOM, Paris.

Bachman, G. and Kinzel, H. (1992) Biochemical screening of the soil-root system of six different plants. *Soil Biology & Biochemistry*, **24**(6), 543-52.

Bae, H.C., Costa-Robles, E.H. and Casida, L.E. (1972) Microflora of soils as viewed by transmission electron microscopy. *Applied Microbiology*, **23**, 637-48.

Bagine, R.K.N. (1984) Soil translocation by termites of the genus *Odontotermes* (Holmgren) (Isoptera: Macrotermitinae) in an arid area of northern Kenya. *Oecologia (Berlin)*, **64**, 263-66.

Baker, H.G., Opler, P.A. and Baker, I. (1978) A comparison of the amino-acid complements of floral and extrafloral nectars. *Botanical Gazette*, **139**, 322-32.

Baker, A.J.M. and Brooks, R.R. (1989) Terrestrial higher plants which hyperaccumulate metallic elements - a review of their distribution, ecology and phytochemistry. *Biorecovery*, **1**, 81-126.

Baker, A.J.M. and Walker, P.L. (1990) Ecophysiology of metal uptake by tolerant plants, in *Heavy Metal Tolerance in Plants: Evolutionary Aspects* (ed. A.J. Shaw), CRC Press, Boca Raton, Florida, pp. 155-77.

Bakonyi, G. (1989) Effects of *Folsomia candida* (Collembola) on the microbial biomass in a grassland soil. *Biology and Fertility of Soils*, **7**, 138-41.

Bal, L. (1973) *Micromorphological Analysis of Soils: Lower Levels in the Organization of Organic Soil Materials*, Netherlands Soil Survey Institute, Wageningen.

Bal, L. (1982) *Zoological Ripening of Soils. 1. The concept and impact in pedology, forestry and agriculture. 2. The process in two Entisols under developing forest in a recently reclaimed Dutch polder*, Centre for agricultural Publishing and Documentation, Wageningen.

Balandreau, J. and Villemin, G. (1973) Fixation biologique de l'azote moléculaire en savane de Lamto (Côte d'Ivoire). Résultats préliminaires. *Revue d'Ecologie et de Biologie du Sol*, **10**, 25-33.

Balandreau, J. and Knowles, R. (1978) The rhizosphere, in *Interactions Between Non-Pathogenic Soil Micro-organisms and Plants* (eds Y.R. Dommergues and S.V. Krupa), Elsevier, Amsterdam, pp. 243-68.

Baldani, J.I., Caruso, L., Baldani, V.L.D. *et al.* (1997) Recent advances in BNF with non-legume plants. *Soil Biology & Biochemistry*, **29**, 911-22.

Baldock, J.A., Oades, J.M., Waters, A.G. *et al.* (1992) Aspects of the chemical structure of soil organic matter as revealed by solid-state ^{13}C NMR spectroscopy. *Biogeochemistry*, **16**, 1-42.

Balesdent, J., Mariotti, A. and Guillet, B. (1987) Natural ^{13}C abundance as a tracer for soil organic matter dynamics studies. *Soil Biology & Biochemistry*, **19** (1), 25-30.

Balesdent, J., Wagner, G.H. and Mariotti, A. (1988) Soil organic matter turnover in long-term field experiments as revealed by Carbon-13 natural abundance. *Soil Science Society of America, Journal*, **52** (1), 118-24.

Balke, N.E. (1985) Effects of allelochemicals on plant-water relationships, in *The Chemistry of Allelopathy. Biochemical Interactions among Plants* (ed. A.C. Thompson), American Chemical Society, Washington D.C., pp. 161-78.

Ball, D.F. (1964) Loss-on-ignition as an estimate of organic matter and organic carbon in non-calcareous soils. *Journal of Soil Science*, **15**, 84-92.

Ballkwill, D.L., Rucinsky, T.E. and Casida, L.E. (1977) Release of micro-organisms from soil with respect to transmission electron microscopy viewing and plate counts. *Anatomie van Leeuwenbroek*, **43**, 73-87.

Bamforth, S.S. (1988) Interactions between protozoa and other organisms. *Agriculture, Ecosystems and Environment*, **24**, 229-34.

Banage, W.B. (1966) Nematode distribution in some British upland moor soils with a note on nematode parasitizing fungi. *Journal of Animal Ecology*, **35**, 3149-61.

Banerjee, M.R., Chapman, S.J. and Killham, K. (1993) Factors influencing the determination of microbial biomass sulphur in soil. *Communications in Soil Science and Plant Analysis*, **24**, 939-50.

Baraer, M. and Toutain, F. (1987) Importance des pigments bruns dans les racines superficielles du hêtre (*Fagus sylvatica* L.) et biodégradation dans le sol. *Revue d'Ecologie et de Biologie du Sol*, **24** (3), 431-42.

Barber, D.A. and Martin, J.K. (1976) The release of organic substances by cereal roots in soil. *New Phytologist*, **76**, 69-80.

Bardgett, R.D., Whittaker, J.B. and Frankland, J.C. (1993) The effect of collembolan grazing on fungal activity in differently managed upland pastures: a microcosm study. *Biology and Fertility of Soils*, **16**, 255-62.

Barnes, G.L., Russell, C.C. and McNew, R.W. (1981) *Aphelenchus avenae*, a potential biological control agent for root rot fungi. *Plant Disease*, **65**, 423-32.

Barnes, C.J., Jacobson, G. and Smith, G.D. (1992) The origin of high-nitrate ground waters in the Australian arid zone. *Journal of Hydrology*, **137**, 181-97.

Barois, I. and Lavelle, P. (1986) Changes in respiration rate and some physicochemical properties of a tropical soil during transit through *Pontoscolex corethrurus* (Glossoscolecidæ, Oligochæta). *Soil Biology & Biochemistry*, **18** (5), 539-41.

Barois, I., Verdier, B., Kaiser, P. *et al.* (1987) Influence of the tropical earthworm *Pontoscolex corethrurus* (Glossoscolecidæ) on the fixation and mineralization of nitrogen, in *On Earthworms* (eds A.M. Bonvicini and P. Omodeo), Mucchi, Bologna, pp. 151-8.

Barois, I. (1987) *Interactions entre les Vers de Terre (Oligochaeta) tropicaux géophages et la microflore pour l'exploitation de la matière organique du sol*, thèse, Université Paris VI / ENS.

Barois, I., Cadet, P., Albrecht, A. *et al.* (1988) Systèmes de culture et faune des sols. Quelques données, in *Fertilité des Sols dans les Agricultures Paysannes Caribéennes. Effets des Restitutions Organiques* (ed. C. Feller), ORSTOM, Fort-de-France, pp. 85-96.

Barois, I., Villemin, G., Lavelle, P. *et al.* (1993) Transformation of the soil structure through *Pontoscolex corethrurus* (Oligochaeta) intestinal tract. *Geoderma*, **56**, 57-66.

Barois, I., Lavelle, P., Brossard, M. *et al.* (1999) Ecology of species with large environmental tolerance and/or extended distributions, in *Earthworm management in tropical agroecosystems* (eds P. Lavelle, L. Brussaard and P. Hendrix), CAB International, Wallingford, pp. 57-86

Baroni-Urbani, C. and Pisarski, B. (1978) Density of ant colonies and individuals, in *Production Ecology of Ants and Termites* (ed. M.V. Brian), Cambridge University Press, Cambridge, pp. 336-9.

Baroni-Urbani, C., Josens, G. and Peakin, G.J. (1978) Empirical data and demographic parameters, in *Production Ecology of Ants and Termites* (ed. M.V. Brian), Cambridge University Press, Cambridge, pp. 5-44.

Barros, M.E., Blanchart, E., Neves, A. *et al.* (1996) Relaçao entre a Macrofauna e a agregaçao do solo em tres sistemas na Amazonia central, in *Solo Suelo, Resumos XII Congresso Latino Americano de Ciencia do Solo*, Aguas de Lindoia, S P, Brazil

Barrow, N.J. (1989) The reaction of plant nutrients and pollutants with soil. *Australian Journal of Soil Research*, **27**, 475-92.

Bartoli, F., Burtin, G., Philippy, R. *et al.* (1993) Influence of fir root zone on soil structure in a 23 m forest transect: the fractal approach. *Geoderma*, **56**, 67-85.

Basaguren, A. and Pozo, J. (1994) Leaf litter processing of Alder and *Eucalyptus* in the Aguera stream system (Northern Spain). 2. Macroinvertebrates associated. *Archiv Fur Hydrobiologie*, **132** (1), 57-68.

Batjes, N.H. (1996) Total carbon and nitrogen in the soils of the world. *European Journal of Soil Science*, **47** (2), 151-63.

Baylis, J.P., Cherrett, J.M. and Ford, J.B. (1986) A survey of the invertebrates feeding on living clover roots

(*Trifolium repens* L.) using ^{32}P as a radiotracer. *Pedobiologia*, **29**, 201-8.

Bazin, M.J., Markham, P., Scott, E.M. *et al.* (1990) Population dynamics and rhizosphere interactions, in *The Rhizosphere* (ed. J.M. Lynch), John Wiley & Sons, Chichester, pp. 99-127.

Beattie, A.J. (1985) *The Evolutionary Ecology of Ant-Plant Mutualisms*, Cambridge University Press, New York.

Beccara, J.X.I. and Venable, D.C. (1989) Extrafloral nectaries: a defense against ant-Homoptera mutualisms. *Oikos*, **55**, 276-80.

Beck, L. (1971) Bodenzoologische gliederung und Characterisierung des Amazonischen Regenwaldes. *Amazoniana*, **3** (1), 69-132.

Becker, M. and Ladha, J.K. (1997) Synchronizing residue N mineralization with rice N demand in flooded condition., in *Driven by Nature: Plant Litter Quality and Decomposition* (eds G. Cadisch and K.E. Giller), CAB International, Wallingford, pp. 231-8.

Beever, R.E. and Burns, D.J.W. (1980) Phosphorus uptake, storage and utilisation by fungi. *Advances in Botanical Research*, **8**, 127-219.

Begon, M., Harper, J.L. and Townsend, C.R. (1990) *Ecology. Individuals, Populations and Communities*, Blackwell Scientific Publications, Boston.

Belfield, W. (1956) The Arthropoda of the soil in a West African pasture. *Journal of Animal Ecology*, **25**, 275-87.

Bell, A.W. (1959) *Enchytraeus fragmentosus*, a new species of of naturally fragmenting oligochaete worm. *Science*, **129**, 1278.

Bennett, A.F. and Baxter, B.J. (1989) Diet of the long-nosed Potoroo *Potorous tridactylous* (Marsupalia: Potoridae) in south-western Victoria. *Australian Wildlife Research*, **16**, 263-71.

Bentley, B.L. (1977) Extrafloral nectaries and protection by pugnacious bodyguards. *Annual Review of Ecology and Systematics*, **8**, 407-27.

Bentley, B.L. (1987) N fixation by *Nasutitermes* and *Velocitermes* in Venezuela, in *Chemistry and Biology of Social Insects* (eds J. Eder and H. Rembold), Verlag J Peperny, Munich, pp. 635.

Benzie, J.A.H. (1986) The distribution, abundance, and the effects of fire on mound building termites (*Trinervitermes* and *Cubitermes* spp., Isoptera: Termitidae) in northern Guinea savanna, West Africa. *Oecologia*, **70**, 559-67.

Berendse, F. (1994) Litter decomposability - a neglected component of plant fitness. *Journal of Ecology*, **82** (1), 187-90.

Berg, R.Y. (1975) Myrmecochorous plants in Australia and their dispersal by ants. *Australian Journal of Botany*, **23**, 475-508.

Berg, B. and Staaf, H. (1980) Decomposition rate and chemical changes of Scots pine needle litter, in *Structure and Function of Northern Coniferous Forests* (ed. T. Persson), Ecological Bulletins/NFR, Stockholm, pp. 163-78.

Berg, B. (1986) The influence of experimental acidification on nutrient release and decomposition rates of needle and root litter in the forest floor. *Forest Ecology and Management*, **15**, 195-213.

Berg, B., Bottner, P., Box, E. *et al.* (1993) Litter mass loss rates in pine forests of Europe and Eastern United States: some relationships with climate and litter quality. *Biogeochemistry*, **20**, 127-59.

Bergman, B., Johansson, C. and Söderbäck, E. (1992a) The *Nostoc-Gunnera* symbiosis. *New Phytologist*, **122**, 279-400.

Bergman, B., Rai, A.N., Johansson, C. and Söderbäck, E. (1992b) Cyanobacterial-plant symbiosis. *Symbiosis*, **14**, 61-81.

Bernard, E.C. (1992) Soil nematode biodiversity. *Biology and Fertility of Soils*, **14**, 99-103.

Bernhard-Reversat, F., Huttel, C. and Lemée, G. (1978) La forêt sempervirente de Basse Côte d'Ivoire, in *Problèmes d'Ecologie: Ecosystèmes Terrestres* (eds M. Lamotte and F. Bourlière), Masson, Paris, pp. 313-45.

Bernhard-Reversat, F. (1982) Measuring litter decomposition in a tropical forest ecosystem: Comparison of some methods. *International Journal of Ecology and Environmental Sciences*, **8**, 63-71.

Bernhard-Reversat, F. (1987) Les cycles d'éléments minéraux dans un peuplement à *Acacia seyal* et leur modification dans un peuplement *d'Eucalyptus* au Sénégal. *Acta Œcologica. Œcologia Generalis*, **8**, 3-16.

Bernhard-Reversat, F. (1993) Dynamics of litter and organic matter at the soil-litter interface in fast-growing tree plantations on ferralitic soils (Congo). *Acta Œcologia*, **14**, 179-95.

Bernhard-Reversat, F., Diangana, D. and Tsatsa, M. (1993) Biomasse, minéralomasse et productivité en plantation d'*Acacia mangium* et *A. auriculiformis* au Congo. *Bois et Forêts des Tropiques*, **238**, 35-44.

Bernhard-Reversat, F. (1996) Nitrogen cycling in tree plantations grown on a poor sandy savanna soil in Congo. *Applied Soil Ecology*, **4**, 161-72.

Bernhard-Reversat, F. and Schwartz, D. (1997) Change in lignin content during litter decomposition in tropical forest soils (Congo): comparison of exotic plantations and native stands. *Comptes Rendus de l'Académie des Sciences, Paris*, **325**, 427-32.

Bernier, N. and Ponge, J.F. (1994) Humus form dynamics during the sylvogenetic cycle in a mountain spruce forest. *Soil Biology & Biochemistry*, **26**(2), 183-220.

Bernier, N. (1996) Altitudinal changes in humus form dynamics in a Spruce forest at the montane level. *Plant and Soil*, **178**, 1-28.

Berthelin, J. and Kogblevi, A. (1974) Influence de l'engorgement sur l'altération microbienne des minéraux d'une arène granitique. *Revue d'Ecologie et de Biologie du Sol*, **11**, 499-509.

Berthelin, J. (1976) *Etudes expérimentales de mécanismes d'altération des minéraux par des microorganismes hétérotrophes*, thèse d'Etat, Université Nancy I.

Berthelin, J., Souchier, B. and Toutain, F. (1979) Intervention des phénomènes biologiques dans l'altération. *Sciences du Sol, Bulletin de l'AFES*, **2** (3), 175-87.

Bertrand, M., Janati-Idrissi, A. and Lumaret, J.P. (1987) Etude expérimentale des facteurs de variation de la consommation de la litière de *Quercus ilex* L. et *Q. pubescens* Willd. par *Glomeris marginata* (V.) (Diplopoda. Glomeridae). *Revue d'Ecologie et de Biologie du sol*, **24** (3), 359-68.

Betsch, J.M. and Vannier, G. (1977) Caractérisation des deux phases juvéniles d'*Allacma fusca* (Collembola, Symphypleona) par leur morphologie et leur écophysiologie. *Zeitschrist für Zoologische Systematik und Evolutionforschung*, **15**, 124-41.

Betsch, J.M. and Betsch-Pinot, M.C. (1983) Recolonisation d'une coupe papetière par les microarthropodes du sol, in *New Trends in Soil Biology* (eds P. Lebrun, H.M. André, A. De Medts *et al.*), Dieu-Brichart, Ottignies, pp. 519-33.

Betsch-Pinot, M.C. (1977) Les parades sexuelles primitives chez les Collemboles Symphypléones. *Revue d'Ecologie et de Biologie du Sol*, **14** (1), 15-9.

Bettucci, L., Rodriguez, C. and Roquebert, M.F. (1990) Fungal communities of volcanic ash soils along an altitudinal gradient in Mexico. III. Seasonal variation. *Pedobiologia*, **34**, 61-7.

Bevege, D.I. (1978) Biomass and nutrient distribution in indigenous forest ecosystems. *Queensland Department of Forestry, Technical Paper*, **6**.

Bhatnagar, T. (1975) Lumbricids and humification: A new aspect of the microbial incorporation of nitrogen induced by earthworm, in *Biodégradation et Humification* (eds G. Kilbertus, O. Reisinger, A. Mourey *et al.*), Pierron, Nancy, pp. 168-82.

Biczok, F. (1965) Protozoa in the rhizosphere, in *Progress in Protozoology, London International Congress No.91*, Excerpta Medica Foundation, p. 120.

Bignell, D.E. (1994) Soil-feeding and gut morphology in higher termites, in *Nourishment and Evolution in Insect Societies* (eds J.H. Hunt and C.A. Nalepa), Westview Press, Boulder, pp. 131-58.

Bignell, D.E., Slaytor, M., Veivers, M. *et al.* (1994) Functions of symbiotic fungus gardens in higher termites of the genus *Macrotermes*, evidence against the acquired enzyme hypothesis. *Acta Microbiologica et Immunologica Hungarica*, **41**, 391-401.

Bilgrami, A.L., Ahmad, I. and Jairajpuri, M.S. (1985) Factors influencing attraction of adult *Hirschmaniella oryzae* towards cabbage seedlings. *Revue de Nématologie*, **8**, 67-75.

Billès, G., Cortez, J. and Lossaint, P. (1975) Etude comparative de l'activité biologique des sols sous peuplements arbustifs et herbacés de la garrigue méditerranéenne I. Minéralisation du carbone et de l'azote. *Revue d'Ecologie et de Biologie du Sol*, **12** (1), 115-39.

Billès, G. and Bottner, P. (1981) Effet des racines vivantes sur la décomposition d'une litière racinaire marquée au ^{14}C. *Plant and Soil*, **62**, 193-208.

Billès, G., Gandais-Riollet, N. and Bottner, P. (1986) Effets d'une culture de graminées sur la décomposition d'une litière végétale marquée au ^{14}C et ^{15}N dans le sol en conditions contrôlées. *Acta Œcologia, Œcologia Plantarum*, **7**, 273-86.

Billès, G., Bottner, P. and Texier, M. (1990) *Régulation du cycle de l'azote au niveau de la rhizosphère de plantes non fixatrices d'azote. Cas du blé*, CNRS - CEFE Centre Louis Emberger, Paris.

Binet, F. and Curmi, P. (1992) Structural effects of *Lumbricus terrestris* (Oligochaeta: Lumbricidae) on the soil-organic matter system: micromorphological observations and autoradiographs. *Soil Biology & Biochemistry*, **24** (12), 1519-23.

Binet, F. (1993) *Dynamique des peuplements et fonctions des lombriciens en sols cultivés tempérés*, Doctorat 3ᵉ cycle, Université de Rennes I.

Bingle, W.H. and Paul, E.A. (1986) A method for separating fungal hyphae from soil. *Canadian Journal of Microbiology*, **32** (11), 62-6.

Birkeland, P.W. (1984) *Soils and Geomorphology*, Oxford University Press, New York.

Birkeland, P.W. (1992) Quaternary soil chronosequences in various environments - extremely arid to humid tropical, in *Weathering, Soils and Paleosols* (eds I.M. Martini and W. Chesworth), Elsevier, Amsterdam, pp. 261-81.

Bissett, J. and Parkinson, D.S. (1979) Fungal community structure in some alpine soils. *Canadian Journal of Botany*, **57**, 1630-41.

Bizé, V. and Célérier, M.L. (1986) Etude d'un écosystème forestier mixte. XI.-Contribution à la connaissance du bilan énergétique de *Philoscia muscorum* (Seopoli, 1763) (Crustacé, Isopode). *Revue d'Ecologie et de Biologie du sol*, **23** (3), 259-69.

Black, C.A. (1957) *Soil-Plant Relationships*, John Wiley, New York.

Black, H.I.J. and Wood, T.G. (1989) The effects of cultivation on the vertical distribution of *Microtermes* spp. (Isoptera; Termitidae; Macrotermitinae) in soil at Mokwa, Nigeria, *Sociobiology*, **15**, 133-8.

Blackman, J.D. (1992) Seasonal variation in the aggregate stability of downland soils. *Soil Use and Management*, **8**, 142-50.

Blair, J.M., Parmelee, R.W. and Lavelle, P. (1995) Influences of earthworms on biogeochemistry, in *Earthworm Ecology and Biogeography in North America* (ed. P.F. Hendrix), Lewis Publishers, Boca Raton, Florida, pp. 127-58.

Blake, C.D. (1962) Some observations on the orientation of *Ditylenchus dipsaci* and invasion of oat seedlings. *Nematologica*, **8**, 177-92.

Blakemore, R.J. (1995) *Improving productivity of Brigalow soils through the introduction of earthworms*, Ph.D. thesis, University of Brisbane, Queensland.

Blanchart, E., Frenot, Y. and Tréhen, P. (1987) Signification biologique du potentiel hydrique dans la distribution des diptères à larves hydrophiles. *Pedobiologia*, **30**, 333-44.

Blanchart, E., Lavelle, P. and Spain, A.V. (1989) Effects of two species of tropical earthworms (Oligochaeta: Eudrilidae) on the size distribution of aggregates in an African Soil. *Revue d'Ecologie et de Biologie du Sol*, **26** (4), 417-25.

Blanchart, E. (1992) Restoration by earthworms (Megascolecidae) of the macroaggregate structure of a destructured savanna soil under field conditions. *Soil Biology & Biochemistry*, **24** (12), 1587-94.

Blanchart, E., Bruand, A. and Lavelle, P. (1993) The physical structure of casts of *Millsonia anomala* (Oligochaeta: Megascolecidae) in shrub savanna soils (Côte d'Ivoire). *Geoderma*, **56**, 119-32.

Blanchart, E., Lavelle, P., Braudeau, E. *et al.* (1997) Regulation of soil structure by geophagous earthworm activities in humid savannas of Cote d'Ivoire. *Soil Biology & Biochemistry*, **29** (3/4), 431-9.

Blanchart, E., Brussaard, L., Alegre, J. *et al.* (1999) Effects of earthworms on soil structure and physical properties, in *Earthworm Management in Tropical Agroecosystems* (eds P. Lavelle, L. Brussaard and P. Hendrix), CAB International, Wallingford, pp. 149-72.

Blanchette, R.A. (1984) Screening wood decayed by white rot fungi for preferential lignin degradation. *Applied and Environmental Microbiology*, **48** (3), 647-53.

Blanquaert, J.P. and Mertens, J. (1977) Mating behaviour in *Sphaeridia pumilis* (Collembola). *Pedobiologia*, **17** (5), 343-9.

Blaszak, C., Niedbala, W., Rajski, A. *et al.* (1977) The differentiation of Acari and Apterygota in Pineto-*Vaccinium myrtilli typicum* in Puszcza Kampinoska national Park. Preliminary report. *Ecological Bulletins*, **25**, 508-11.

Bleeker, P. and Sageman, R. (1990) Surface charge characteristics and clay mineralogy of some variable charge soils in Papua New Guinea. *Australian Journal of Soil Research*, **28**, 901-7.

Block, W., Young, S.R., Conradi-Larsen, E.M. *et al.* (1978) Cold tolerance of two Antarctic terrestrial arthropods. *Experientia*, **34**, 1166-7.

Block, W. (1983) Low temperature tolerance of soil arthropods - some recent advances, in *New Trends in Soil Biology* (eds Ph. Lebrun, H.M. André, A. de Medts *et al.*), Dieu-Brichart, Ottignies, pp. 427-31.

Bloem, J., Ellenbroek, F.M., Bär-Gilissen, M.J. *et al.* (1989) Protozoan grazing and bacterial production in stratified Lake Vechten estimated with fluorescently labeled bacteria and by thymidine incorporation. *Applied and Environmental Microbiology*, **55** (7), 1787-95.

Bloemers, G.F., Hodda, M., Lambshead, P.J.D. *et al.* (1997) The effects of forest disturbance on diversity of tropical soil nematodes. *Oecologia*, **111** (4), 575-82.

Blower, J.G. and Miller, P.F. (1977) The life-history of the iulid millipede *Cylindroiulus nitidus* in a Derbyshire wood. *Journal of Zoology, London*, **183**, 339-51.

Bockheim, J.G. (1979) Relative age and origin of soils in eastern Wright Valley, Antarctica. *Soil Science*, **138**, 145-52.

Bockheim, J.G. (1980a) Properties and classification of some desert soils in coarse-textured glacial drift in the Arctic and Antarctic. *Geoderma*, **24**, 45-69.

Bockheim, J.G. (1980b) Solution and use of chronosequences in studying soil development. *Geoderma*, **24**, 71-85.

Bockheim, J.G., Jepsen, E.A. and Heisey, D.M. (1991) Nutrient dynamics in decomposing leaf litter of four tree species on a sandy soil in northwestern Wisconsin. *Canadian Journal of Forest Research*, **21**, 803-12.

Bocock, K.L. (1963) The digestion and assimilation of food by *Glomeris*, in *Soil Organisms* (eds J. Doeksen and J. Van der Drift), North Holland, Amsterdam, pp. 86-91.

Boddey, R.M. (1987) Methods for quantification of nitrogen fixation associated with Gramineae. *CRC Critical Reviews in Plant Sciences*, **6** (3), 209-66.

Boddey, R.M., Chalk, P.M., Victoria, R.L. *et al.* (1983) The use of the ^{15}N dilution technique to estimate the contribution of associated biological nitrogen fixation to the nitrogen nutrition of *Paspalum notatum* cv batatais. *Canadian Journal of Microbiology*, **29**, 1036-45.

Boddy, L., Walting, R. and Lyon, A.J.E. (eds) (1988) *Fungi and Ecological Disturbance*, The Royal Society of Edinburgh, Edinburgh.

Bodot, P. (1969) Composition des colonies de termites: ses fluctuations au cours du temps. *Insectes Sociaux*, **16**, 39-54.

Boettcher, S.E. and Kalisz, P.J. (1990) Single-tree influence on soil properties in the mountains of eastern Kentucky. *Ecology*, **71** (4), 1365-72.

Boettcher, S.E. and Kalisz, P.J. (1991) Single-tree influence on earthworms in forest soils in eastern Kentucky. *Soil Science Society of America, Journal*, **55** (3), 862-5.

Bohn, H.L. (1976) Estimate of organic carbon in world soils. *Soil Science Society of America Journal*, **40**, 468-70.

Bolton, P.J. and Phillipson, J. (1976) Burrowing, feeding, egestion and energy budget of *Allolobophora rosea* (Savigny) (Lumbricidæ). *Œcologia (Berlin)*, **23**, 225-45.

Bolton, B. (1994) *Identification Guide to the Ant Genera of the World*, Harvard University Press, Cambridge.

Bolton, B. (1995) A taxonomic and zoogeographical census of the exact ant taxa (Hymenoptra: Formicidae). *Journal of Natural History*, **29**, 1037-1056.

Bond, W.J. and Slingsby, P. (1983) Seed dispersal by ants in shrublands of the Cape Province and its evolutionary significance. *South African Journal of Science*, **79** (6), 231-3.

Bonell, M., Gilmour, D.A. and Cassells, D.S. (1983) A preliminary survey of the hydraulic properties of rainforest soils in tropical north-east Queensland and their implications for the runoff process, in *Rainfall Simulation, Runoff and Soil Erosion* (ed. J. de Ploey), Catena (supplement 4), Braunschweig, pp. 55-78.

Bonell, M., Coventry, R.J. and Holt, J.A. (1986) Erosion of termite mounds under natural rainfall in semiarid tropical northeastern Australia. *Catena*, **13**, 11-28.

Bongers, T. (1990) The maturity index: an ecological measure of environmental disturbance based on nematode species compostion. *Oecologia*, **83**, 14-9.

Bonne, M., Andreux, F., Vedy, J.C. *et al.* (1982) Etude hydro-chimique saisonnière dans trois sols acides: composition des eaux gravitaires et des solutions extraites à pF 4,4. *Science du Sol, Bulletins de l'AFES*, **4**, 275-92.

Bonneau, M. (1979) La température du sol, in *Constituants et Propriétés du Sol* (eds M. Bonneau and B. Souchier), Masson, Paris, pp. 324-8.

Bonneau, M. and Levy, G. (1982) Assembly and physical organization of particles, in *Constituents and Properties of Soils* (eds M. Bonneau and B. Souchier), Academic Press, London, pp. 268-87.

Bonneau, M. and Souchier, B. (eds) (1982) *Constituents and Properties of Soil*, Academic Press, London.

Bonnet, L. (1975) Types morphologiques, écologie et évolution de la thèque chez les Thécamoebiens. *Protistologica*, **11** (3), 363-78.

Bonnet, L., Cassagnau, P. and Deharveng, L. (1976) Un exemple de rupture de l'équilibre biocénotique par déboisement: les peuplements de Collemboles édaphiques du Piau d'Engaly (Hautes Pyrénées). *Revue d'Ecologie et de Biologie du Sol*, **13**, 337-51.

Bornebusch, C.H. (1930) The fauna of forest soil. *Det Forstlige Forsogsvæsen i Danmark*, **11**, 1-244.

Boström, U. (1988) *Ecology of earthworms in arable land. Population dynamics and activity in four cropping systems*, Doctorate, University of Uppsala.

Boto, K.G. and Wellington, J.T. (1984) Soil characteristics and nutrient status in a northern Australian mangrove forest. *Estuaries*, **7**, 61-9.

Bottner, P. (1985) Response of microbial biomass to alternate moist and dry conditions in a soil incubated with ^{14}C and ^{15}N labelled plant material. *Soil Biology & Biochemistry*, **17** (3), 329-37.

Bottner, P. and Billes, G. (1987) La rhizosphère: site d'interactions biologiques. *Revue d'Ecologie et de Biologie du Sol*, **24** (3), 369-88.

Bouché, M.B. (1972) *Lombriciens de France. Ecologie et Systématique*, INRA, Paris.

Bouché, M.B. and Kretzschmar, A. (1974) Fonction des lombriciens. II. Recherches méthodologiques pour l'analyse du sol ingéré. *Revue d'Ecologie et de Biologie du Sol*, **11** (1), 127-39.

Bouché, M.B. (1975) Action de la faune sur les états de la matière organique dans les écosystèmes, in *Biodégradation et Humification* (eds G. Kilbertus, O. Reisinger, A. Mourey *et al.*), Pierron, Sarreguemines, pp. 157-68.

Bouché, M.B. (1977) Stratégies lombriciennes, in *Soil Organism as Components of Ecosystems* (eds U. Lohm and T. Persson), Ecology Bulletin/NFR, Stockolm, pp. 122-32.

Bouché, M.B., Rafidison, Z. and Toutain, F. (1983) Etude de l'alimentation et du brassage pédo-intestinal du lombricien *Nicodrilus velox* (Annelida, Lombricidae) par l'analyse élémentaire. *Revue d'Ecologie et de Biologie du Sol*, **20** (1), 49-75.

Bouché, M.B. (1983) Ecophysiologie des lombriciens: acquis récents et perspectives, in *New Trends in Soil Biology* (eds Ph. Lebrun, H.M. André, A. de Medts *et al.*), Dieu-Brichart, Ottignies, pp. 321-33.

Bouché, M.B. (1984) The subterranean behaviour of the earthworm, in *On Earthworms* (eds A.M. Bonvicini Pagliai and P. Omodeo), Mucchi, Modena, pp. 159-69.

Bouché, M.B. and Ferrière, G. (1986) Cinétique de l'assimilation de l'azote d'origine lombricienne par une végétation prairiale non perturbée. *Comptes Rendus de l'Académie des Sciences, Paris*, **301** (III-17), 789-94.

Boulaud, D. and Braumann, A. (1998) Influence des termites humivores et de leur microflore digestive associée sur le compartiment organique et la microflore des termitières et du sol environnant. *Actes des Colloques Insectes Sociaux*, **11**, 37-41.

Boulaud, J. (1996) *Influence d'un Termite Humivore Supérieur de Forêt Tropicale,* Cubitermes fongifaber, *sur la Microflore de la Termitosphère*, thèse de 3e cycle, Université de Rennes.

Boulet, R., Chauvel, A. and Lucas, Y. (1984) Les systèmes de transformation en pédologie, in *Livre Jubilaire du Cinquantenaire* (ed. A.F.E.S.), Maugein-Lachaise, Malermont, pp. 167-79.

Boulet, R., Pessenda, L.C.R., Telles, E.C.C. *et al.* (1995) Une évaluation de la vitesse de l'accumulation superficielle de matière par la faune du sol à partir de la datation des charbons et de l'humine du sol. Exemple des latosols des versants du lac Campestre, Salitre, Minas Gerais, Brésil. *Comptes Rendus de l'Académie des Sciences, Paris*, **320** (II a), 287-94.

Boutton, T.W., Arshad, M.A. and Tieszen, L.L. (1983) Stable isotope analysis of termite food habitats in East African grasslands. *Oecologia*, **59**, 1-6.

Bowen, G. and Theodorou, C. (1973) Growth of ectomycorrhizal fungi around seeds and roots, in *Ectomycorrhizae: their Ecology and Physiology* (eds G.C. Marks and T.T. Kozlowski), Academic Press, New York, pp. 119-26.

Bowen, G.D. and Rovira, A.D. (1976) Microbial colonization of plant roots. *Annual Review of Phytopathology*, **14**, 121-44.

Bowen, H.J.M. (1979) *Environmental Chemistry of the Elements*, Academic Press, London.

Bowen, G.D. (1980) Misconceptions, concepts and approaches in rhizosphere biology, in *Contemporary Microbial Ecology* (eds D.C. Ellwood, J.N. Hedger, M.J. Latham *et al.*), Academic Press, London, pp. 283-304.

Bowen, G.D. (1984) Tree roots and the use of soil nutrients, in *Nutrition of Plantation Forests* (eds G.D. Bowen and E.K.S. Nambiar), Academic Press, London, pp. 147-79.

Bowles, D.J. and Northcote, D.H. (1974) The amounts and rates of export of polysaccharides found within the membrane system of maize root cells. *Biochemical Journal*, **142**, 139-44.

Bowman, G.M. (1987) Pedochronology: potential for age determination of archaeological sites, in *Archaeometry:*

Further Australasian Studies (eds W.R. Ambrose and J.M.J. Mummery), Australian National University/Australian National Gallery, Canberra, pp. 97-109.

Boyd, R., Furbank, R.T. and Read, D.J. (1985) Ectomycorrhiza and the water relations of trees, in *Proceedings of the 1st European Symposium on Mycorrhizae*, INRA, Paris, pp. 689-93.

Boyer, P. (1973a) Quelques aspects de l'action des termites sur les argiles. *Clay Mineralogy*, **17**, 453-62.

Boyer, P. (1973b) Action de certains termites constructeurs sur l'évolution des sols tropicaux. *Annales des Sciences Naturelles, Zoologie*, **15**, 329-498.

Boyer, J. (1982) *Les Sols Ferralitiques. Tome X. Facteurs de Fertilité et Utilisation des Sols*, ORSTOM, Paris.

Brabant, P. (1991) *Le Sol des Forêts Claires du Cameroun*, ORSTOM-MESIRES, Paris.

Brady, N.C. and Weil, R.R. (1996) *The Nature and Properties of Soils*, Prentice-Hall, New Jersey.

Braithwaite, R.W., Miller, L. and Wood, J.T. (1988) The structure of termite communities in the Australian tropics. *Australian Journal of Ecology*, **13**, 375-91.

Brauman, A., Kane, M.D., Labat, M. *et al.* (1992) Genesis of acetate and methane by gut bacteria of nutritionally diverse termites. *Science*, **257**, 1384-7.

Brauns, A. (1954) *Terricole Dipterenlarven. "Musterschmidt"*, Wissenschaftlicher Verlag, Gottingen.

Bray, J.R. (1963) Root production and the estimation of net productivity. *Canadian Journal of Botany*, **41**, 65-72.

Breisch, H. (1974) *Contribution à l'étude du rôle des exsudats racinaires dans les processus d'agrégation du sol*, thèse d'Etat, Université Nancy I.

Breisch, H., Guckert, A. and Reisinger, O. (1975) Etude au microscope électronique de la zone apicale des racines de maïs. *Bulletin de la Société Botanique de France, collection Rhizosphère*, 55-60.

Brethes, A., Brun, J.J., Jabiol, B. *et al.* (1995) Classification of forest humus: a French proposal. *Annales des Sciences Forestières*, **52**, 535-46.

Brewer, R. (1954) Soil parent material. *The Australian Journal of Science*, 134-8.

Brewer, R. and Sleeman, J.R. (1988) *Soil Structure and Fabric*, CSIRO Australia, Division of Soils, Adelaide.

Breznak, J.A. (1984) Biochemical aspects of symbiosis between termites and their intestinal microbiota, in *Invertebrate-Microbial Interactions* (eds J.M. Anderson, A.D.M. Rayner and D.W.H. Walton), Cambridge University Press, Cambridge, pp. 173-204.

Breznak, J.A. and Brune, A. (1994) Role of microorganisms in the digestion of lignocellulose by termites. *Annual Review of Entomology*, **39**, 453-87.

Brian, M.V. (1957) Caste determination in social insects. *Annual Review of Entomology*, **2**, 107-20.

Brierley, W.B., Jewson, S.T. and Brierley, M. (1928) The quantitative study of soil fungi, in *Transactions of the1st International Congress on Soil Science*, ISSS, Washington D.C., pp. 48-71.

Bristow, K.L. (1988) The role of mulch and its architecture in modifying soil temperature. *Australian Journal of Soil Research*, **26**, 269-80.

Broadbent, F.E. and Norman, A.G. (1946) Some factors affecting the availability of the organic nitrogen in soil: A preliminary report. *Soil Science Society of America, Proceedings*, **11**, 264-7.

Brockmeyer, V., Schmid, R. and Westheide, W. (1990) Quantitative investigations of the food of two terrestrial enchytraeid species (Oligochaeta). *Pedobiologia*, **34**, 151-6.

Brook, G.A. and Box, E.O. (1983) A world model of carbon dioxide. *Earth Surface Processes and Landforms*, **8**, 79-88.

Brook, G.A. and Box, E.O. (1984) A world model of carbon dioxide: a reply. *Earth Surface Processes and Landforms*, **9**, 85-7.

Brookes, P.C., Powlson, D.S. and Jenkinson, D.S. (1984) Phosphorus in the soil microbial biomass. *Soil Biology & Biochemistry*, **16**, 169-75.

Brookes, P.C., Powlson, D.S. and Jenkinson, D.S. (1985) The microbial biomass in soil, in *Ecological*

Interactions in Soil ; Plants, Microbes and Animals (eds A.H. Fitter, D. Atkinson, D.J. Read *et al.*), Blackwell Scientific Publications, Oxford, pp. 123-5.

Brooks, R.R. (1983) *Biological Methods of Prospecting for Minerals*, John Wiley & Sons, New York.

Brossard, M., Lavelle, P. and Laurent, J.Y. (1996) Digestion of a vertisol by the endogeic earthworm *Polypheretima elongata*, Megascolecidae, increases soil phosphate extractibility. *European Journal of Soil Biology*, **32** (2), 107-11.

Brown, M.E. (1972) Plant growth substances produced by microorganisms of soil and rhizosphere. *Journal of Applied Bacteriology*, **35**, 443-51.

Brown, G. (1990) Structure, crystal chemistry, and origin of the phyllosilicate minerals common in soil clays, in *Soil Colloids and Their Association in Aggregates* (eds M.F. de Boodt, M.H.B. Hayes and A. Herbillon), Plenum Press, New York, pp. 7-38.

Brown, V.K. and Gange, A.C. (1990) Insect herbivory below ground. *Advances in Ecological Research*, **20**, 1-58.

Brown, G.G. (1995) How do earthworms affect microfloral and faunal community diversity? *Plant and Soil*, **170**, 209 - 31.

Brown, G.G., Pashanasi, B., Gilot-Villenave, C. *et al.* (1999) Effects of earthworms on plant growth, in *Earthworm Management in Tropical Agroecosystems* (eds P. Lavelle, L. Brussaard and P. Hendrix), CAB International, Wallingford, 87-148.

Brownell, P.F. (1979) Sodium as an essential micronutrient for plants and its possible role in metabolism. *Advances in Botanical Research*, **7**, 117-224.

Bruckert, S. and Rouiller, J. (1979) Méchanismes de régulation du pH de sols, in *Constituants et Propriétés du Sol* (eds M. Bonneau and B. Souchier), Masson, Paris, pp. 354-65.

Bruckert, S. and Kilbertus, G. (1980) Fractionnement et analyse des complexes organo-minéraux de sols bruns et de chernozems. *Plant and Soil*, **57**, 271-95.

Brun, J.J. (1978) *Etude de quelques humus forestiers aérés acides de l'est de la france ; critères analytiques, classification morphogénétique*, thèse de 3e cycle, Université Nancy I.

Brundrett, M. (1991) Mycorrhizas in natural ecosystems. *Advances in Ecological Research*, **21**, 171-313.

Brundrett, M., Bougher, N., Dell, B. *et al.* (1996) *Working with Mycorrhizas in Forestry and Agriculture*, ACIAR, Australian Capital Territory, Canberra.

Brusewitz, G. (1959) Untersuchungen über den Einfluss des Regenwurms auf Zahl, Art und Leistungen von Mikroorganismen in Boden. *Archiv für Mikrobiologie*, **33**, 52-82.

Brussaard, L. and Hijdra, R.D.W. (1986) Some effects of scarab beetles in sandy soils of The Netherlands. *Geoderma*, **37**, 325-30.

Bryant, R.J., Woods, L.E., Coleman, D.C. *et al.* (1982) Interactions of bacterial and amoebal populations in soil microcosms with fluctuating moisture contents. *Applied Environmental Microbiology*, **43**, 747-52.

Bunnell, F.L. and Scoullar, K.A. (1975) ABISKO II. A computer simulation model of carbon flux in tundra ecosystems, in *Structure and Function of Tundra Ecosystems (Ecological Bulletins No 20)* (eds T. Rosswall and O.W. Heal), Swedish Natural Science Research Council, Stockholm, pp. 425-448.

Buol, S. (1995) Sustainability of soil use. *Annual Review of Ecology and Systematics*, **26**, 25-44.

Burges, N.A. (1960) Dynamic equilibra in the soil, in *The Ecology of Soil Fungi* (eds D. Parkinson and J. Wald), Liverpool University Press, Liverpool, pp. 185-91.

Burghouts, T. (1993) *Spatial heterogeneity of nutrient cycling in Bornean rain forest.*, Ph.D. thesis, Vrije Universiteit.

Burke, D., Kaufman, P., McNeil, M. *et al.* (1974) The structure of plant cell walls. VI: a survey of the walls of suspension-culture monocots. *Plant Physiology*, **54**, 109-15.

Burns, R.G. (1983) Extracellular enzyme-substrate interactions in soil, in *Microbes in Their Natural Environment* (eds J. H. Slater, R. Whittenbury and J.W.T. Wimpenny), Cambridge University Press, New York, pp. 249-98.

Burns, R.G. (1986) Interactions of enzymes with soil mineral and organic colloids, in *Interactions of Soil Minerals with Natural Organics and Microbes.* (eds P.M. Huang and M. Schnitzer), Soil Science Society of America, Madison, Wisconsin, pp. 429-451.

Burns, R.G. (1990) Microorganisms, enzymes and soil colloid surfaces, in *Soil Colloids and their Associations in Aggregates* (eds M.F. de Boodt, M.H.B. Hayes and A. Herbillon), Plenum Press, New York, pp. 337-61.

Butler, J.H.A. and Buckerfield, J.C. (1979) Digestion of lignin by termites. *Soil Biology & Biochemistry*, **11**, 507-13.

Butler, B.E. and Churchwood, H.M. (1983) Aeolian processes, in *Soils, An Australian Viewpoint*, CSIRO/Academic Press, Melbourne/London, pp. 91-9.

Butler, D.R. (1995) *Zoogeomorphology*, Cambridge University Press, New York.

Buyanovsky, G.A. and Wagner, G.H. (1983) Annual cycles of carbon dioxide level in soil air. *Soil Science Society of America, Journal*, **47**, 1139-45.

Cadet, P. and Debouzie, D. (1990) Evolution spatio-temporelle d'un peuplement de nématodes parasites de la canne à sucre. *Revue de Nématologie*, **13** (1), 79-88.

Cadet, P. and Van den Berg, E. (1992) Les nématodes phytoparasites de la montagne Pelée à la Martinique. *Comptes-Rendus de la Société de Biogéographie*, **68** (3), 89-104.

Calaby, J.H. and Gay, F.J. (1956) The distribution and biology of the genus *Coptotermes* (Isoptera) in western Australia. *Australian Journal of Zoology*, **4**, 19-39.

Caldwell, M.M. and Camp, L.B. (1974) Belowground productivity of two cool desert communities. *Oecologia*, **17**, 123-30.

Callot, G., Chamayou, H., Maertens, C. *et al.* (1982) *Mieux Comprendre les Interactions Sol-Racine*, INRA, Paris.

Callot, G., Chamayou, H., Bullock, P. *et al.* (1984) Observations de concentrations argileuses dans les structures racinaires au cours des transformations pédologiques. *Comptes Rendus de l'Académie des Sciences, Paris*, **299** (II, 12), 825-30.

Callot, G., Chauvel, A., Arvieu, J.C. *et al.* (1992) Mise en évidence de kaolinite et de silice dans les structures cellulaires de l'épiderme et du cortex de racines de palmier, en forêt amazonienne. *Bulletins de la Société de Botanique, France*, **139**, 7-14.

Camargo, M.N., Klamt, E. and Kauffman, J.H. (1986) *Soil Classification as Used in Brazilian Soil Surveys*, International Soil Reference and Information Centre, Wageningen.

Campbell, W.G. (1929) The chemical aspect of the destruction of oakwood by powder-post and death-watch beetles, *Lyctus spp* and *Xestobium* sp. *Biochemistry Journal*, **23**, 1290-93.

Campbell, G.S. and Harris, G.A. (1977) Water relations and water-use patterns for *Artemisia tridentata* Nutt. in wet and dry years. *Ecology*, **58**, 652-9.

Campbell, I.B. and Claridge, G.G.C. (1987) *Antarctica: Soils, Weathering Processes and Environment*, Elsevier, Amsterdam.

Campbell, D.J., Kinninburgh, D.G. and Beckett, P.H.T. (1989) The soil solution chemistry of some Oxfordshire soils: temporal and spatial variability. *Journal of Soil Science*, **40**, 321-9.

Campbell, I.B. and Claridge, G.G.C. (1992) Soils of cold climate regions, in *Weathering, Soils and Paleosols* (eds I.P. Martini and W. Chesworth), Elsevier, Amsterdam, pp. 183-201.

Canada Soil Survey Committee (ed.) (1978) *The Canadian System of Soil Classification*, Canada Department of Agriculture, Ottawa.

Cancela da Fonseca, J.P., Poinsot, N. and Vannier, G. (1967) Essai comparatif sur les microarthropodes et la taille des échantillons de sol. *Revue d'Ecologie et de Biologie du Sol*, **IV** (3), 331-55.

Cancela da Fonseca, J.P. (1975) Observations préliminairess sur la colonisation par les microorganismes et par les microarthropodes de la litière fraîche de deux sols d'une Hêtraie (Forêt de Retz). *Pedobiologia*, **15**, 375-81.

Cancela da Fonseca, J.P. (1987) About the ecological niches of three Beech litter panphytophage oribatid mites, in *Soil Fauna and Soil fertility. 9th International Colloquium on Soil Zoology* (ed. B.R. Striganova), Moscow "Nauka", Moscow, pp. 289-95.

Cannon, M.G. and Coventry, R.J. (1989) Soils of the CSIRO Experimental Area, 'Redlands', Balfes Creek, North Queensland. *CSIRO Australia, Division of Soils, Divisional report*, **99**.

Cannon, M.G., Smith, C.D. and Murtha, G.G. (1992) Soils of the Cardwell-Tully Area, North Queensland. *CSIRO Australia, Division of Soils, Divisional report*, **115**.

Capelin, M.A. and Prove, B.G. (1983) Soil conservation problems of the humid coastal tropics of north Queensland, in *Proceedings of the Australian Society of Sugar Cane Technologists* (ed. B.T. Egan), Watson, Ferguson and Co., Brisbane, pp. 87-93.

Carlisle, A., Brown, A.H.F. and White, E.J. (1966) The organic matter and nutrient elements in the precipitation beneath a Sessile Oak *Quercus petraea* canopy. *Journal of Ecology*, **54** (1), 87-98.

Carlson, S.R. and Whitford, W.G. (1991) Ant mound influence on vegetation and soils in a semiarid mountain ecosystem. *American Midland Naturalist*, **126**, 125-39.

Carver, M., Gross, G.F. and Woodward, T.E. (1991) Hemiptera, in *The Insects of Australia* (eds I.D.Naumann, P.B. Carne, J.F. Lawrence *et al.*), CSIRO/Melbourne University Press, Carlton, pp. 429-509.

Casenave, A. and Valentin, C. (1989) *Les États de Surface de la Zone Sahélienne, Influence sur l'Infiltration*, ORSTOM, Paris.

Casida, L.E.J. (1988) Minireview: non-obligate bacterial predation of bacteria in soil. *Microbial Ecology*, **15**, 1-8.

Catroux, G. and Schnitzer, M. (1987) Chemical, spectroscopic and biological characteristic of the organic matter in particle size fractions separated from an Aquoll. *Soil Science Society of America, Journal*, **51** (5), 1200-7.

Cayrol, J.C., Dubos, B. and Guillaumin, J.J. (1978) Etude preliminaire *in vitro* de l'agressivité de quelques nématodes mycophages vis à vis de *Trichoderma viride*, Pers., *T. polysporum* (Link ex Pers.) Rifai et *Armillarriella mellea* (Vahl.) Karst. *Annales de Phytopathologie*, **10**, 177-85.

Cepeda-Pizarro, J.G. (1993) Litter decomposition in deserts - an overview with an example from coastal arid chile. *Revista Chilena de Historia Natural*, **66** (3), 323-36.

Cerri, C., Feller, C., Balesdent, J. *et al.* (1985) Application du traçage isotopique naturel au ^{13}C à l'étude de la dynamique de la matière organique dans les sols. *Comptes-Rendus de l'Académie des Sciences, Paris*, **300** (11-3), 423-8.

César, J. (1971) *Etude quantitative de la strate herbacée de la savane de Lamto (Côte d'Ivoire)*, thèse de 3ᵉ cycle, Université Paris VI.

Cervek, S. (1971) Fertility of conidiospores of *Penicillium sp.* in the excrements of *Entomobrya purpurascens* (PACKARD) Collembola, Entomobryidae, in *Organismes du Sol et Production Primaire . IV. Colloquium Pedobiologiae* (eds J. d'Aguilar, C. Athias-Henriot, A. Bessard *et al.*), INRA, Paris, pp. 69-72.

Chaboud, A. and Rougier, M. (1981) Sécrétions racinaires mucilagineuses et rôle dans la rhizosphère. *L'Année Biologique*, **XX** (4), 21-34.

Chakraborty, S., Theodorou, C. and Bowen, G. (1983) Amoebae from a take-all suppressive soil which feed on *Gaeumannomyces graminis* tritici and other soil fungi. *Soil Biology & Biochemistry*, **15**, 17-24.

Chakraborty, S. and Warcup, J.H. (1984) Populations of mycophagous and other amoebae in take-all suppressive and non-suppressive pasture soils. *Soil Biology & Biochemistry*, **16**, 197-99.

Chakraborty, S. (1985) Survival of wheat take-all fungus in suppressive and non-suppressive soils. *Pedobiologia*, **28**, 13-8.

Chalupsky, J.J. and Leps, J. (1985) The spatial pattern of Enchytraeidae (Oligochaeta). *Oecologia*, **68**, 153-7.

Chapin, F.S.III (1988) Ecological aspects of plant mineral nutrition. *Advances in Mineral Nutrition*, **3**, 161-91.

Chapin, F.S. III (1991) Effects of multiple environmental stresses on nutrient availability and use, in *Response of Plants to Multiple Stresses* (eds H.A. Mooney, W.E. Winner and E.J. Pell), Academic Press, San Diego, pp. 67-88.

Chapin, F.S. III, Moilanen, L. and Kielland, K. (1993) Preferential use of organic nitrogen for growth by a non-mycorrhizal arctic sedge. *Nature*, **361**, 150-3.

Chapman, R.F. (1982) *The Insects, Structure and Function*, Hodder and Stoughton, London.

Charpentier, F. (1996) *Effets de l'inoculation d'un ver de terre endogé sur la dynamique de la matière organique d'un sol ferrallitique cultivé de l'Amazonie Péruvienne*, thèse, Université Paris VI.

Chartres, C.J. and Walker, P.H. (1988) The effect of aeolian accessions on soil development on granitic rocks in south-eastern Australia. III. Micromorphological and geochemical evidence of weathering and soil development. *Australian Journal of Soil Research*, **26**, 33-53.

Chaussod, R., Nicolardot, B., Catroux, G. *et al.* (1986) Relations entre les caractéristiques physico-chimiques et microbiologiques de quelques sols cultivés. *Science du Sol, Bulletins de l'AFES*, **2**, 213-26.

Chaussod, R., Houot, S., Guiraud, G. *et al.* (1988) Size and turnover of the microbial biomass in agricultural soils: laboratory and field measurements, in *Nitrogen Efficiency in Agricultural Soils* (eds D.S. Jenkinson and K.A. Smith), Elsevier Applied Sciences, London, pp. 312-26.

Chauvel, A. (1976) *Recherches sur la Transformation des Sols Ferrallitiques dans la Zone Tropicale à Saisons Contrastées. Travaux et Documents 62*, ORSTOM, Paris.

Cheng, A.C., Lu, N.C., Briggs, G.M. *et al.* (1979) Effects of particulate materials on population growth of the free-living nematode *Caenorhabditis briggsae*. *Proceedings of the Society for Experimental Biology and Medicine*, **160**, 203-7.

Cheng, W.X., Coleman, D.C., Carroll, C.R. *et al.* (1994) Investigating short-term carbon flows in the rhizospheres of different plant species, using isotopic trapping. *Agronomy Journal*, **86** (5), 782-8.

Chenu, C. (1993) Clay- or sand-polysaccharide association as models for the interface between microorganisms and soil: water related properties and microstructure. *Geoderma*, **56**, 143-56.

Cherrett, J.M. (1968) The foraging behaviour of *Atta cephalotes* L. (Hymenoptera, Formicidae) I. Foraging pattern and plant species attacked in tropical rain forest. *Journal of Animal Ecology*, **37**, 387-405.

Cherrett, J.M. (1986) History of the leaf-cutting ant problem, in *Fire Ants and Leaf-Cutting Ants* (eds C.S. Lofgren and R.K. Van der Meer), Westview Press, Boulder, pp. 10-17.

Chernov, Y.J. (1975) Geozoological characteristics of the region of Taimyr biogeocenological station, in *Soil Invertebrates in the Tundra of Western Taimyr, International Tundra Biome Translation 12*, Tundra Biome Center, University of Alaska, Fairbanks, pp. 39-53.

Chesworth, W. (1973a) The parent rock effect in the genesis of soil. *Geoderma*, **10**, 215-25.

Chesworth, W. (1973b) The residual system of chemical weathering. A model for the chemical breakdown of silicate rocks at the surface of the earth. *Journal of Soil Science*, **24**, 69-81.

Chesworth, W. (1992) Weathering systems, in *Weathering, Soils and Paleosols* (eds I.P. Martini and W. Chesworth), Elsevier, Amsterdam, pp. 19-40.

Chiariello, N., Hickman, J.C. and Mooney, H.A. (1982) Endomycorrhizal role for interspecific transfer of phosphorus in a community of annual plants. *Science*, **217**, 941-3.

Chiba, S., Abe, T., Aoki, J.I. *et al.* (1975) Studies on the productivity of soil animals in Pasoh forest reserve, west Malaysia. 1. Seasonal change in the density of soil mesofauna: acari , collembola and others. *Scientific Report from the Hirosaki University*, **22**, 87-124.

Christensen, H. and Funck-Jensen, D. (1989) Growth rate of rhizosphere bacteria measured directly by the tritiated thymidine incorporation technique. *Soil Biology & Biochemistry*, **21** (1), 113-7.

Christiansen, K. (1964) Bionomics of Collembola. *Annual Review of Entomology*, **9**, 147-78.

Clarholm, M. and Rosswall, T. (1980) Biomass and turnover of bacteria in a forest soil and a peat. *Soil Biology & Biochemistry*, **12**, 49-57.

Clarholm, M. (1981) Protozoan grazing of Bacteria in soil. Impact and importance. *Microbial Ecology*, **7**, 343-50.

Clarholm, M. (1983) Dynamics of soil bacteria in relation to plants, protozoa and inorganic nitrogen. *Sveriges Lantbrukuniversitet*, **17**, 1-19.

Clarholm, M. (1984) Microbes as predator or prey. Heterotrophic, free-living protozoa: neglected micro-organisms with an important task in regulating bacteria populations, in *Current Perspectives on Microbial Ecology* (eds M.J. Klug and C.A. Reddy), American Society for Microbiology, Washington D.C., pp. 321-6.

Clarholm, M. (1985) Interactions of bacteria, protozoa and plant leading to mineralization of soil nitrogen. *Soil Biology & Biochemistry*, **17** (2), 181-7.

Clarholm, M. (1989) Effects of plant-bacterial-amoebal interactions on plant uptake of nitrogen under field conditions. *Biology and Fertility of Soils*, **8**, 373-8.

Clark, F.E. and Paul, E.A. (1970) The microflora of grassland. *Advances in Agronomy*, **22**, 375-435.

Clarke, A.L. and Barley, K.P. (1968) The uptake of nitrogen from soils in relation to solute diffusion. *Australian Journal of Soil Research*, **6**, 75-92.

Clarkson, D.T. and Hanson, J.B. (1980) The mineral nutrition of higher plants. *Annual Review of Plant Physiology*, **31**, 239-98.

Claudius, G.R. and Mehrotra, R.S. (1973) Root exudates from lentil (*Lens culinaris* Medic.) seedlings in relation to wilt disease. *Plant and Soil*, **38**, 315-20.

Clements,.R.O., Murray, J.P., Sturdy, R.G. (1991) The impact of 20 years absence of earthworms and three levels of N fertilizer on a grassland soil environment. Agriculture, Ecosystems and Environment, **36**, 75-85.

Cleveland, L.R. (1923) Symbiosis between termites and their intestinal protozoa. *Proceedings of the National Academy of Sciences of the USA*, **9**, 424-8.

Clifford, H.T. and Monteith, G.B. (1989) A three phase seed dispersal mechanism in the Australian quinine bush (*Petalostigma pubescens* Domin.). *Biotropica*, **21**, 284-6.

Clowes, F.A.L. (1976) Cell production by root caps. *New Phytologist*, **77**, 339-407.

Cluzeau, D. (1992) *Structure et dynamique des peuplments lombriciens dans des sytèmes tempérés anthropisés*, thèse de l' Université de Rennes.

Coaton, W.G.H. (1958) The Holotermitid harvester termites of South Africa. *Union of South Africa, Department of Agriculture, Scientific Bulletin*, **375**.

Cohn, E. and Spiegel, Y. (1991) Root-Nematode interaction, in *Plant Roots - The hidden Half* (eds Y. Waisel, A. Eshel and U. Kafkafi), Marcel dekker, New York, pp. 789-805.

Coineau, Y. and Seely, M.K. (1983) Mise en évidence d'un peuplement de microarthropodes dans les sables fins des dunes du Namib central, in *New trends in Soil Biology* (eds Ph. Lebrun, H.M. André, A. de Medts *et al.*), Dieu-Brichart, Ottignies, pp. 652-4.

Coleman, D.C. (1970) Nematodes in the litter and soil of El Verde rain forest, in *A Tropical Rain Forest* (eds H.T. Odum and R.F. Pigeon), Division of Technical Information, US Atomic Energy Commission, Oak Ridge, Tennessee, pp. 103-4.

Coleman, D.C., Anderson, R.V., Cole, C.V. *et al.* (1978) Trophic interactions in soils as they affect energy and nutrient dynamics. IV. Flows of metabolic and biomass carbon. *Microbial Ecology*, **4**, 373-80.

Coleman, D.C., Ingham, R.E., McClellan, J.F. *et al.* (1984a) Soil nutrient transformations in the rhizosphere via animal-microbial interactions, in *Invertebrates-microbial Interactions* (eds J.M. Anderson, A.D.M. Rayner and D.W.H. Walton), Cambridge University Press, Cambridge, pp. 35-58.

Coleman, D.C., Anderson, R.V., Cole, C.V. *et al.* (1984b) Roles of protozoa and nematodes in nutrient cycling, in *Microbial-plant Interactions* (eds R.L. Todd and J.E. Giddens), American Society of Agronomy, Madison, Wisconsin, pp. 17-28.

Coleman, D.C., Ingham, E.R., Hunt, H.W. *et al.* (1990) Seasonal and faunal effects on decomposition in semiarid prairie, meadow and Lodgepole Pine forest. *Pedobiologia*, **34**, 207-19.

Collins, N.M. (1979) Observations on the foraging habits of *Hospitalitermes umbrinus* (Haviland), (Isoptera, Termitidae) in the Gunong Mulu National Park, Sarawak. *Ecological Entomology*, **4**, 231-8.

Collins, N.M. (1983) Termite populations and their role in litter removal in Malaysian rainforest., in *Tropical Rainforests: Ecology and Management* (eds S.L. Sutton, T.C. Whitmore and L.C. Chadwick), Blackwell Scientific Publications, Oxford, pp. 311-325.

Comeau, P.G. and Kimmins, J.P. (1989) Above- and below-ground biomass and production of Lodgepole Pine on sites with differing soil moisture regimes. *Canadian Journal of Forest Research*, **19**, 447-54.

Cooke, R.C. and Godfrey, B.E.S. (1964) A key to the nematode-destroying fungi. *Transactions of the British Mycological Society*, **47**, 61-74.

Cooke, A. and Luxton, M. (1980) Effect of microbes on food selection by *Lumbricus terrestris*. *Revue d'Ecologie et de Biologie du Sol*, **17** (3), 365-70.

Cooke, A. (1983) The effect of fungi on food selection by *Lumbricus terrestris* L., in *Earthworm Ecology* (ed. J. E. Satchell), Chapman & Hall, London, pp. 365-73.

Cookson, L.J. (1987) ^{14}C-lignin degradation by three Australian termites. *World Science and Technology*, **21**, 11-25.

Corliss, J.O. and Esser, S.C. (1974) Comments on the role of the cyst in the life cycle and survival of free-living Protozoa. *Transactions of the American Microscopy Society*, **93**, 578-93.

Cornell, R.M. and Schwertmann, U. (1996) *The Iron Oxides*, VCH, N.Y. and Weinheim.

Cornu, S. (1995) *Cycles biogéochimiques du silicium, du fer et de l'aluminium en forêt amazonienne*, thèse de l'Université Aix-Marseille III.

Cornu, S, Ambrosi, J.P., Lucas, Y. *et al* (1997) A comparative study of the soil solution chemistry of two Amazonian forest soils (Central Amazonia, Brazil). *Hydrology and Earth System Sciences*, **1(2)**, 313-24

Cornu, S., Luizáo, F., Rouiller, J. *et al.* (1997) Comparative study of litter decomposition and mineral element release in two Amazonian forest ecosystems: litter bag experiments. *Pedobiologia*, **41**, 456-71.

Cortez, J. and Schnitzer, M. (1979) Nucleic acid bases in soils and their association with organic and inorganic soil components. *Canadian Journal of Soil Science*, **59**, 277-86.

Cortez, J. and Bouché, M.B. (1987) Composition chimique du mucus cutané de *Allolobophora chaetophora chaetophora* (Oligochaeta: Lumbricidae). *Comptes-Rendus de l'Académie des Sciences, Paris*, **305** (III), 207-10.

Cortez, J. and Hameed, R. (1988) Effets de la maturation des litières de ray-gras (*Lolium perenne* L.) dans le sol sur leur consommation et leur assimilation par *Lumbricus terrestris* L. *Revue d'Ecologie et de Biologie du Sol*, **25** (4), 397-412.

Cortez, J., Hameed, R. and Bouché, M.B. (1989) C and N transfer in soil with or without earthworms fed with ^{14}C- and ^{15}N-labelled wheat straw. *Soil Biology & Biochemistry*, **21** (4), 491-7.

Cotrufo, M.E. and Ineson, P. (1995) Effects of enhanced atmospheric CO_2 and nutrient supply on the quality and subsequent decomposition of fine roots of *Betula pendula* Roth and *Picea sitchensis* (Bong) Carr. *Plant and Soil*, **170** (2), 267-77.

Cotton, D.C.F. and Curry, J.P. (1980) The effects of cattle and pig slurry fertilizers on earthworms (Oligochaeta, Lumbricidae) in grassland managed for silage production. *Pedobiologia*, **20**, 189-96.

Coûteaux, M.M. (1972) Distribution des Thécamoebiens de la litière et de l'humus de deux sols forestiers d'humus brut. *Pedobiologia*, **12**, 237-43.

Coûteaux, M.M. (1976) *Dynamisme de l'Équilibre des Thécamoebiens du Sol dans quelques Sols Climaciques*, Muséum National d'Histoire Naturelle, Paris.

Coûteaux, M.M. (1978) Etudes quantitative des thécamoebiens édaphiques dans une savane à *Loudetia* à Lamto (Côte d'Ivoire). *Revue d'Ecologie et de Biologie du Sol*, **15** (3), 401-12.

Coûteaux, M.M. and Chardez, D. (1981) Thécamoebiens édaphiques et muscicoles de Guyane Française. *Revue d'Ecologie et de Biologie du Sol*, **18** (2), 192-208.

Coûteaux, M.M. and Pussard, M. (1983) Nature du régime alimentaire des protozoaires du sol, in *New Trends in Soil Biology* (eds Ph. Lebrun, H.M. André, A. de Medts *et al.*), Dieu-Brichart, Ottignies, pp. 179-95.

Coûteaux, M.M. (1985) Relation entre la densité apparente d'un humus et l'aptitude à la croissance de ses Ciliés. *Pedobiologia*, **28**, 289-303.

Coûteaux, M.M. and Ogden, G. (1988) The growth of *Tracheleuglypha dentata* (Rhizopoda: Testacea) in clonal cultures under different trophic conditions. *Microbial Ecology*, **15**, 81-93.

Coûteaux, M.M., Faurie, G., Palka, L. *et al.* (1988) La relation prédateur-proie (protozoaires-bactéries) dans les sols: rôle dans la régulation des populations et conséquences sur les cycles du carbone et de l'azote. *Revue d'Ecologie et de Biologie du Sol*, **25** (1), 1-31.

Coûteaux, M.M., Mousseau, M., Célérier, M.L. *et al.* (1991) Increasing atmospheric CO_2 and litter quality: decomposition of Sweet Chestnut leaf litter with animal food webs of different complexities. *Oikos*, **61**, 54-64.

Coventry, R.J., Taylor, R.M. and Fitzpatrick, R.W. (1983) Pedological significance of the gravels in some red and grey earths of central north Queensland. *Australian Journal of Soil Research*, **21**, 219-40.

Cowan, J.A., Humphreys, G.S., Mitchell, P.B. *et al.* (1985) An assessment of pedoturbation by two species of mound-building ants, *Camponotus intrepidus* (Kirby) and *Iridomyrmex purpureus* (F. Smith). *Australian Journal of Soil Research*, **22**, 95-107.

Cowie, R.H. and Wood (1989) Damage to crops, forestry and rangeland by fungus-growing termites (Termitidae: Macrotermitinae) in Ethiopia. *Sociobiology*, **15** (2), 139-53.

Cowie, R.H., Logan, J.W.M. and Wood (1989) Termite (Isoptera) damage and control in tropical forestry with special reference to Africa and Indo-Malaysia: a review. *Bulletin of Entomological Research*, **79**, 173-84.

Cowling, E.B. and Merrill, W. (1966) Nitrogen in wood and its role in wood deterioration. *Canadian Journal of Botany*, **44**, 1539-54.

Cox, G.W., Mills, J.N. and Ellis, B.A. (1992) Fire ants (Hymenoptera: Formicidae) as major agents of landscape development. *Environmental Entomology*, **21**, 281-6.

Cragg, J.B. (1961) Some aspects of the ecology of moorland animals. *Journal of Ecology*, **49** (3), 477-506.

Crawford, R.L. (1981) *Lignin Biodegradation and Transformation*, John Wiley, New York.

Cresswell, H.P. and Kirkegaard, J.A. (1995) Subsoil amelioration by plant roots - the process and the evidence. *Australian Journal of Soil Research*, **33** (2), 221-39.

Cross, A.F. and Schlesinger, W.H. (1995) A literature review and evaluation of the Hedley fractionation: Applications to the biogeochemical cycle of soil phosphorus in natural ecosystems. *Geoderma*, **64** (3-4), 197-214.

Crossley, D.A.J., Reichle, D.E. and Edwards, C.A. (1971) Intake and turnover of radioactive cesium by earthworms (Lumbricidae). *Pedobiologia*, **11**, 71-6.

Crossley, D.A. (1977) The roles of terrestrial saprophagous arthropods in forest soils: current status of concepts, in *The Role of Arthropods in Forest Ecosystems* (ed. W.J. Mattson), Springer Verlag, New York, pp. 49-56.

C.S.T.C. (Chinese Soil Taxonomic Classification Research Group) (1990) *Chinese Soil Taxonomic System (1st Proposal)*, Science Press, Beijing.

Cuendet, G. (1983) Predation on earthworms by the Black-head Gull (*Larus ridibundus* L.), in *Earthworm Ecology: from Darwin to Vermiculture* (ed. J.E. Satchell), Chapman and Hall, London, pp. 415-24.

Cuevas, E. and Medina, E. (1988) Nutrient dynamics within Amazonian forests. II. Fine root growth, nutrient availibility and leaf litter decomposition. *Oecologia*, **76**, 222-35.

Culver, D.C. and Beattie, A.J. (1983) Effects of ant mounds on soil chemistry and vegetation patterns in a Colorado montane meadow. *Ecology*, **64**, 485-92.

Curl, E.A., Gudauskas, R.T., Harper, J.D. *et al.* (1983) Modification of the rhizosphere fungus flora by mycophagous insects, in *Proceedings of the 3rd International Mycological Congress*, The Mycology Society, Tokyo.

Curl, E.A. and Truelove, B. (1986) *The Rhizosphere*, Springer-Verlag, Berlin.

Curl, E.A. (1988) The role of soil microfauna in plant-disease suppression. *CRC Critical Reviews in Plant Sciences*, **7** (3), 175-96.

Cushman, J.H., Lawton, J.H. and Manly, B.F.J. (1993) Latitudinal patterns in European ant assemblages: variation in species richness and body size. *Oecologia*, **95**, 30-7.

Cutler, D.W., Crump, M. and Sandon, H. (1922) A quantitative investigation of the bacterial and protozoan population of the soil, with an account for the protozoan fauna. *Philosophical Transactions of the Royal Society of London (Series B)*, **211**, 317-50.

d'Aguilar, J. (1962) Superfamille des Dascilloidea, in *Entomologie Appliquée à l'Agriculture* (ed. A.S. Balachowsky), Masson, Paris, pp. 204-300.

Dangerfield, J.M. and Telford, S.R. (1989) Are millipedes important for soil fertility? *The Zimbabwe Science News*, **23** (7/9), 66-8.

Dangerfield, J.M. (1990) Abundance, biomass and diversity of soil macrofauna in savanna woodland and associated managed habitats. *Pedobiologia*, **34**, 141-50.

Daniel, O. (1991) Leaf-litter consumption and assimilation by juveniles of *Lumbricus terrestris* L. (Oligochaeta, Lumbricidae) under different environmental conditions. *Biology and Fertility of Soils*, **12**, 202-8.

Daniel, O. and Anderson, J.M. (1992) Microbial biomass and activity in contrasting soil materials after passage through the gut of the earthworm *Lumbricus rubellus* Hoffmeister. *Soil Biology & Biochemistry*, **24** (5), 465-70.

Darbyshire, J.F. and Greaves, M.P. (1967) Protozoa and bacteria in the rhizosphere of *Sinapis alba* L., *Trifolium repens* L. and *Lolium perenne* L. *Canadian Journal of Microbiology*, **13**, 1057-68.

Darbyshire, J.F. and Greaves, M.P. (1973) Bacteria and Protozoa in the rhizosphere. *Pesticide Science*, **4**, 349-60.

Darbyshire, J.F. (1976) Effect of water suction on the growth in soil of the ciliate *Colpoda steini* and the bacterium *Azotobacter chroococcum*. *Journal of Soil Science*, **27**, 369-76.

Darbyshire, J.F., Griffiths, B.S., Davidson, M.S. *et al.* (1989) Ciliate distribution amongst soil aggregates. *Revue d'Ecologie et de Biologie du Sol*, **26** (1), 47-56.

Darici, C., Schaefer, R. and Zaidi, Z. (1986) Effets du type d'argile sur les activités microbiennes de divers sols tropicaux, in *Comptes Rendus du 111ᵉ Congrès National des Sociétés Savantes*, Poitiers, pp. 105-29.

Darlington, J.P.E.C. (1982) The underground passages and storage pits used in foraging by a nest of the termite *Macrotermes michaelseni* in Kajiado, Kenya. *Journal of Zoology*, **198**, 237-47.

Darlington, J.P.E.C. (1984) A method for sampling the populations of large termite nests. *Annal of Applied*

Biology, **104**, 427-36.

Darlington, J.P.E.C. and Dransfield, R.D. (1987) Size relationships in nest populations and mound parameters in the termite *Macrotermes michaelseni* in Kenya. *Insectes Sociaux*, **34**, 165-80.

Darlington, J.P.E.C. (1994) Nutrition and evolution in fungus-growing termite., in *Nourishment and Evolution in Insect Societies* (eds J.H. Hunt and C.A. Nalepa), Westview Press, Boulder, pp. 105-30.

Dart, P.J. (1971) Scanning electron microscopy of plant roots. *Journal of Experimental Botany*, **22**, 163-8.

Dart, P.J. (1974) The infection process, in *The Biology of Nitrogen Fixation* (ed. A. Quispel), North Holland Publishing, Amsterdam, pp. 382-429.

Darwin, C. (1881) *The Formation of Vegetable Mould Through the Action of Worms with Observations on Their Habits*, John Murray, London.

Dash, M.C., Senapati, B.K. and Mishra, C.C. (1980) Nematode feeding by tropical earthworms. *Oikos*, **34**, 322-5.

Dash, M.C., Nanda, B. and Mishra, P.C. (1981) Digestive enzymes in three species of Enchytraeidae. *Oikos*, **36**, 316-8.

Dash, M.C., Behera, N., Satpathy, B. *et al.* (1985) Comparison of two methods for microbial biomass estimation in some tropical soils. *Revue d'Ecologie et de Biologie du Sol*, **22** (1), 13-20.

Dash, M.C. (1990) Oligochaeta: Enchytraeidae, in *Soil Biology Guide* (ed. D.L. Dindal), John Wiley & Sons, New York, pp. 311-40.

David, J.F. (1982) Variabilité dans l'espace et dans le temps des cycles de vie de deux populations de *Cylindroiulus nitidus* (Verhoeff) (Iulidae). *Revue d'Ecologie et de Biologie du Sol*, **19** (3), 411-25.

David, J.F. (1987) Consommation annuelle d'une litière de chêne par une population adulte du diplopode *Cylindroiulus nitidus*. *Pedobiologia*, **30**, 299-310.

David, J.F., Ponge, J.F. and Delecour, F. (1993) The saprophagous macrofauna of different types of humus in Beech forests of the Ardenne (Belgium). *Pedobiologia*, **3**, 49-56.

Davidson, E.A., Galloway, L.F. and Strand, M.K. (1987) Assessing available carbon: comparison of techniques across selected forest soils. *Communications in Soil Science and Plant Analysis*, **18**, 45-64.

Davidson, E.A., Eckert, R.W., Hart, S.C. *et al.* (1989) Direct extraction of microbial biomass nitrogen from forest and grassland soils of California. *Soil Biology & Biochemistry*, **21**(6), 773-8.

Davidson, E.A. and Ackerman, I.L. (1993) Changes in soil carbon inventories following cultivation of previously untilled soils. *Biogeochemistry*, **20**, 161-93.

Davidson, E.A. (1994) Climate change and soil microbial processes: secondary effects are hypothesised from better known interacting primary effects, in *Soil Response to Climate Change* (eds M.D.A. Rounsevell and P.J. Loveland), Springer-Verlag, Berlin, pp. 155-68.

de Faria, S.M., Lewis, G.P., Sprent, J.I. *et al.* (1989) Occurence of nodulation in the Leguminosae. *New Phytologist*, **111**, 607-19.

de Kimpe, C.R., Laverdiere, M.R. and Martel, Y.A. (1979) Surface area and exchange capacity of clay in relation to the mineralogical composition of gleysolic soils. *Canadian Journal of Soil Science*, **59**, 341-7.

de Laune, R.D., Reddy, C.N. and Patrick, W.H. (1981) Organic matter decomposition in soil as influenced by pH and redox conditions. *Soil Biology & Biochemistry*, **13**, 533-4.

de Leij, F.A.A., Whipps, J.M. and Lynch, J.M. (1994) The use of colony development for the characterization of bacterial communities in soil and on roots. *Microbial Ecology*, **27**, 81-97.

DeNiro, M.J. and Epstein, S. (1978) Influence of diet on the distribution of carbon isotopes of animals. *Geochimica and Cosmochimica Acta*, **42**, 495-506.

de Puytorac, P. (1954) Contribution à l'étude cytologique et taxonomique des Infusoires. *Annales des Sciences Naturelles, Zoologie*, **11**, 82-270.

de Ruiter, P.C., Vanveen, J.A., Moore, J.C. *et al.* (1993) Calculation of nitrogen mineralization in soil food webs. *Plant and Soil*, **157** (2), 263-73.

Decaëns, T., Lavelle, P., Jimenez Jaen, J.J. *et al.* (1994) Impact of land management on soil macrofauna in the Oriental Llanos of Colombia. *European Journal of Soil Biology*, **30** (4), 157-68.

Delecour, F. (1980) Essai de classification pratique des humus. *Pedologie*, **30** (2), 225-41.

Deleporte, S. (1981) Peuplement en Diptères Sciaridae d'une litière de chêne. *Revue d'Ecologie et de Biologie du Sol*, **18**, 231-42.

Deleporte, S. (1987) Rôle du Diptère Sciaridae *Bradysia confinis* (Winn., Frey) dans la dégradation d'une litière de feuillus. *Revue d'Ecologie et de Biologie du Sol*, **24**, 341-58.

Deleporte, S. and Rouland, C. (1991) Etude préliminaire de l'équipement digestif osidasique de *Bradysia confinis* (Diptera, Sciaridae): implications dans la dégradation de la matière organique. *Comptes Rendus de l'Académie des Sciences, Paris*, **312** (III), 165-70.

Delettre, S. and Baillot, S. (1977) Sur la résistance des larves de Chironomidae Orthocladiinae à l'assèchement du sol. *Comptes Rendus de l'Académie des Sciences, Paris*, **284** (17), 1717-9.

Deligne, J., Quennedey, A. and Blum, M.S. (1981) The enemies and defense mechanisms of termites, in *Social Insects* (ed. H.R. Hermann), Academic Press, New York, pp. 1-76.

Delvaux, B., Herbillon, A.J., Dufey, J.E. *et al.* (1990) Surface properties and mineralogy of hydrated halloysitic clays. I. Existence of interlayer K^+ specific sites. *Clay Minerals*, **25**, 129-39.

Demeure, Y., Freckman, D. and Van Gundy, S.D. (1979) Anhydrobiotic coiling of nematodes in soil. *Journal of Nematology*, **11** (2), 189-95.

Demeure, Y. and Freckman, D. (1981) Recent advances in the study of anhydrobiotic nematodes, in *Plant Parasitic Nematodes* (eds B.M. Zuckerman and R.A. Rohde), Academic Press, New York, pp. 205-26.

Dent, D.L. and Pons, L.J. (1995) A world perspective on acid sulphate soils. *Geoderma*, **67**, 263-76.

Devigne, J. and Jeuniaux, C. (1961) Origine tissulaire des enzymes chitinolytiques intestinaux des Lombrics. *Archives Internationales de Physiologie et de Biochimie*, **69** (2), 223-34.

Dexheimer, J., Marx, C., Gianinazzi-Pearson, V. *et al.* (1985) Ultrastructural Cytochemistry of the host-fungus interfaces in the endomycorrhizal association *Glomus mosseae/Allium cepa*. *Zeitung für Pflanzenphysiologie*, **92**, 191-206.

Dexter, A.R. (1986) Model experiments on the behaviour of roots at the interface between a tilled seed-bed and a compacted sub-soil. III. Entry of pea and wheat roots into cylindrical biopores. *Plant and Soil*, **95**, 149-61.

Dexter, A.R. (1988) Advances in the characterization of soil structure. *Soil and Tillage Research*, **11**, 199-238.

Dexter, A.R. (1991) Amelioration of soil by natural processes. *Soil and Tillage Research*, **20**, 87-100.

Di Castri, F. (1988) Enhancing the credibility of ecology: interacting along and across hierarchical scales. *Geojournal*, **17** (1), 5-35.

Dickinson, C.H. and Pugh, G.J.F. (eds) (1974) *Biology of Plant Litter Decomposition*, Academic Press, London.

Dickschen, F. and Topp, W. (1987) Feeding activities and assimilation efficiencies of *Lumbricus rubellus* (Lumbricidæ) on a plant-only diet. *Pedobiologia*, **30**, 31-7.

Didden, W.A.M. (1990) Involvement of Enchytraeidae (Oligochaeta) in soil structure evolution in agricultural fields. *Biology and Fertility of Soils*, **9**, 152-8.

Dietz, S. and Bottner, P. (1981) Etude par autoradiographie de l'enfouissement d'une litière marquée au ^{14}C en milieu herbacé. *Colloques Internationaux du CNRS*, **303**, 125-32.

Dighton, J. (1991) Acquisition of nutrients from organic resources by mycorrhizal autotrophic plants. *Experientia*, **47**, 362-69.

Dighton, J., Thomas, E.D. and Latter, P.M. (1987) Interactions between tree-roots, mycorrhizas, saprotrophic fungus, and the decomposition of organic substrate in a microcosm. *Biology and Fertility of Soils*, **4**, 145-50.

Dindal, D.L. (1990) *Soil Biology Guide*, John Wiley & Sons, New York.

Dix, N.J. (1985) Changes in relationship between water content and water potential after decay and its significance for fungal successions. *Transactions of the British Mycological Society*, **85**, 649-53.

Dixon, R.K., Wright, G.M., Behrns, G.T. *et al.* (1980) Water deficits and root growth of ectomycorrhizal White Oak seedlings. *Canadian Journal of Forest Research*, **10**, 545-8.

Dommergues, Y. and Mangenot, F. (1970) *Ecologie Microbienne du Sol*, Masson et C[ie], Paris.

Dommergues, Y., Dreyfus, B., Diem, H.G. *et al.* (1985) Fixation de l'azote et agriculture tropicale. *La Recherche*, **162**, 22-31.

Dommergues, Y. (1997) Contribution of actinorhizal plants to tropical soil productivity and rehabilitation. *Soil Biology & Biochemistry*, **29**, 931-41.

Domsch, K.H., Bech, T., Anderson, J.P.E. *et al.* (1979) A comparison of methods for microbial population and biomass studies. *Pflanzenernaehr Bodenkd*, **142**, 520-33.

Dorioz, J.M. and Robert, M. (1987) Aspects microscopiques des relations entre les microorganismes ou végétaux et les argiles. Conséquences sur les microorganisations et microstructurations des sols, in *Micromorphologie des Sols* (eds N. Fedoroff, L.M. Bresson and M.A. Courty), AFES, Paris, pp. 353-61.

Dorioz, J.M., Robert, M. and Chenu, C. (1993) The role of roots, fungi and bacteria on clay particle organization. An experimental approach. *Geoderma*, **56**, 179-94.

Dormaar, J.F. (1988) Effect of plant roots on chemical and biochemical properties of surrounding discrete soil zones. *Canadian Journal of Soil Science*, **68**, 233-42.

Douglas, J.T., Gross, M.J. and Hill, D. (1980) Measurements of pore characteristics in a clay soil under ploughing and direct drilling, including the use of a radioactive tracer (^{144}Ce) technique. *Soil Tillage Research*, **1**, 11-8.

Dowding, P. (1976) Allocation of resources, nutrient uptake and release by decomposer organisms, in *The Role of Terrestrial and Aquatic Organisms in Decomposition Processes. Proceedings of the 17th Symposium of the British Ecological Society* (eds J.M. Anderson and A. Macfadyen), Blackwell Scientific Publications, Oxford, pp. 169-83.

Dozsa-Farkas, K. (1978) Die ökologische bedeutung des mikrohabitates für das vorkommen einiger enchytraeiden-arten. *Pedobiologia*, **18**, 366-72.

Dozsa-Farkas, K. (1987) Effects of human treading on enchytaeid fauna of Hornbeam-Oak forests in Hungary. *Biology and Fertility of Soils*, **3**, 91-3.

Drees, L.R., Wilding, L.P., Smeck, N.E. *et al.* (1989) Silica in soils: quartz and disordered silica polymorphs, in *Minerals in Soil Environments* (eds J.B. Dixon and S.B. Weed), Soil Science Society of America, Madison, Wisconsin, pp. 913-74.

Drew, M.C. and Saker, L.R. (1980) Assessment of a rapid method, using soil cores, for estimating the amount and distribution of crop roots in the field. *Plant and Soil*, **55**, 297-305.

Drew, M.C. (1983) Plant injury and adaptation to oxygen deficiency in the root environment: a review. *Plant and Soil*, **75**, 179-99

Duboisset, A. (1995) *Caractérisation et quantification par analyse d'image des modifications structurales engendrées par* Pontoscolex corethrurus *(Glossoscolescidae, Oligochaeta) sur un sol ferrallitique cultivé (Pérou)*, Rapport de DEA, Université Nancy I.

Duchaufour, P. (ed.) Pedologie: Pedogenesis and Classification, Allen and Unwin, London.

Duchaufour, P. (1997) *Abrégéde Pédologie- Sol, Végétation, Environnement*, 5th edition, Masson, Paris.

Duddridge, J.A., Malibari, A. and Read, D.J. (1980) Structure and function of mycorrhizal rhizomorphs with special reference to their role in water transport. *Nature*, **287**, 834-6.

Duhoux, E., Prin, Y. and Dommergues, Y. (1993) Comparison of aerial nodulation of *Casuarina* sp. with legume stem nodulation, in *Symbioses in Nitrogen-fixing Trees* (eds N.S. Subba Rao and C. Rodriguez-

Barrueco), Oxford and IBH Publishing Co, New Delhi, pp.85-93.

Dunger, W. (1964) *Tiere im Bodem*, Ziemsen, Wittenberg Lutherstadt.

Dunger, W. (1975) On the delimitation of soil microathropod coenoses in time and space, in *Progress in Soil Zoology* (ed. J. Vanek), Academia, Prague, pp. 43-9.

Duvigneaud, P. (1971) Concepts sur la productivité primaire des écosystèmes forestiers. *Ecology and Conservation*, **4**, 111-40.

Duxbury, J.M., Smith, M.S. and Doran, J.W. (1989) Soil organic matter as a source and a sink of plant nutrients, in *Dynamics of Soil Organic Matter in Tropical Ecosystems* (eds D.C. Coleman, J.M. Oades and G. Uehara), Niftal Project, University of Hawaï, Honolulu, pp. 33-67.

Eady, R.R. (1992) The dinitrogen-fixing bacteria, in *The Prokaryotes*, Springer-Verlag, Berlin, pp. 534-53.

Easey, J.F. and Holt, J.A. (1989) Population estimation of some mound-building termites (Isoptera: Termitidae) using radioisotope methods. *Material and Organismen*, **24**, 81-91.

Edwards, C.A. (1974) Macroarthropods, in *Biology of Plant Litter Decomposition* (eds C.H. Dickinson and G.H.F. Pugh), Academic Press, London, pp. 533-54.

Edwards, C.A. and Lofty, J.R. (1982) Nitrogenous fertilizers and earthworm population in agricultural soils. *Soil Biology & Biochemistry*, **14**, 515-21.

Edwards, P.J. and Grubb, P.J. (1982) Studies of mineral cycling in a montane rain forest in New Guinea. IV. Soil characteristics and the division of mineral elements between the vegetation and soil. *Journal of Ecology*, **70**, 649-66.

Edwards, C.A. (1983) Earthworm ecology in cultivated soils, in *Earthworm Ecology: from Darwin to Vermiculture* (ed. J.E. Satchell), Chapman and Hall, London, pp. 123-38.

Edwards, R. and Mill, A.E. (1986) *Termites in Buildings*, Rentokil, England.

Edwards, C.A. and Fletcher, K.E. (1988) Interactions between earthworms and microorganisms in organic matter breakdown. *Agriculture, Ecosystems and Environment*, **24**, 235-47.

Edwards, C.A., Bohlen, P.J., Linden, D.R. *et al.* (1995) Earthworms in agroecosystems, in *Earthworm Ecology and Biogeography in North America* (ed. P.F. Hendrix), Lewis Publishers, Boca Raton, Florida, pp. 185-215.

Edwards, C.A. and Bohlen, P.J. (1996) *Biology and Ecology of Earthworms*, Chapman & Hall, London.

Eggleton, P. (1994) Termites live in a pear-shaped world: a response to Platnick. *Journal of Natural History*, **28**, 1209-12.

Eggleton, P. and Bignell, D.E. (1995) Monitoring the response of tropical insects to changes in the environment: troubles with termites, in *Insects in a Changing Environment* (eds R. Harrington and N. Stork), Academic Press, London, pp. 473-97.

Eggleton, P., Bignell, D.E., Sands, W.A. *et al.* (1995) The species richness of termites (Isoptera) under differing levels of forest disturbance in the Mbalmayo Forest Reserve, southern Cameroon. *Journal of Tropical Ecology*, **11**, 85-98.

Eggleton, P., Bignell, D.E., Wood, T.G. *et al.* (1996) The diversity, abundance and biomass of termites under differing levels of disturbance in the Mbalmayo Forest Reserve, Southern Cameroon. *Philosophical Transactions of the Royal Society of London*, **B 351**, 51-68.

Einhellig, F.A., Stille Muth, M. and Schon, M.K. (1985) Effects of allelochemicals on plant-water relationships, in *The Chemistry of Allelopathy. Biochemical Interactions among Plants* (ed. A.C. Thompson), American Chemical Society, Washington D.C., pp. 179-96.

Eissenstat, D.M. and Yanai, R.D. (1997) The ecology of root lifespan. *Advances in Ecological Research*, **27**, 2-59.

El Balkhi, M., Mangenot, F., Proth, J. *et al.* (1978) Influence de la percolation d'une solution de saccharose sur la composition qualitative et quantitative de la microflore bactérienne d'un sol. *Soil Science and Plant Nutrition*, **24**, 15-25.

Eldridge, D.J. (1994) Nests of ants and termites influence infiltration in a semi-arid woodland. *Pedobiologia*, **38** (6), 481-92.

Elkin, G.H. (1992) Biological nitrogen fixation systems in tropical ecosystems: an overview, in *Biological Nitrogen Fixation and Sustainability of Tropical Agriculture* (eds K. Mulongoy, M. Gueye and D.S.C. Spencer), John Wiley & Sons, New York, pp. 27-40.

Elkins, N.Z. and Whitford, W.G. (1982) The role of microarthropods and nematodes in decomposition in a semi-arid ecosystem. *Oecologia (Berlin)*, **55**, 303-10.

Elkins, C.B. (1985) Plant roots as tillage tools, in *Proceedings of the International Conference on Soil Dynamics*, Auburn University, Auburn USA, pp. 519-23.

Elkins, N.Z., Sabol, G.Z., Ward, T.J. *et al.* (1986) The influence of subterranean termites on the hydrological characteristics of a Chihuahuan desert ecosystem. *Oecologia*, **68**, 521-8.

Ellenberg, H., Mayer, R. and Schauermann, J. (1986) *Ökosystemforschung: Ergebnisse des Solling-Projekts*, Ulmer, Stuttgart.

Elliott, E.T., Coleman, D.C. and Cole, C.V. (1979) The influence of amoebae on the uptake of nitrogen by plants in gnotobiotic soil, in *The Soil-Root Interface* (eds J.L. Harley and R.S. Russell), Academic Press, London, pp. 221-30.

Elliott, E.T., Anderson, R.V., Coleman, D.C. *et al.* (1980) Habitable pore space and microbial trophic interactions. *Oikos*, **35**, 327-35.

Elliott, E.T. (1986) Aggregate structure and carbon, nitrogen and phosphorus in native and cultivated soils. *Soil Science Society of America, Journal*, **50** (3), 627-33.

Elliott, E.T., Hunt, H.W. and Walter, D.E. (1988) Detrital foodwebs interactions in north American grassland ecosystems. *Agriculture, Ecosystems and Environment*, **24**, 41-56.

Elliott, P.W., Knight, D. and Anderson, J.M. (1990) Denitrification in earthworm casts and soil from pasture under different fertilizer and drainage regimes. *Soil Biology & Biochemistry*, **22** (5), 601-5.

Ellsworth, D.S. and Reich, P.B. (1993) Canopy structure and vertical patterns of photosynthesis and related leaf traits in a deciduous forest. *Oecologia*, **96** (2), 169-78.

Elzein, A. and Balesdent, J. (1995) A mechanistic simulation of the vertical distribution of carbon concentrations and residence times in soils. *Soil Science Society of America, Journal*, **59** (5), 1328-35.

Emerson, A.E. (1955) Geographical origins and dispersions of termite genera. *Fieldiana, Zoology*, **37**, 465-521.

Emerson, W.W., Foster, R.C. and Oades, J.M. (1986) Organo-mineral complexes in relation to soil aggregation and structure, in *Interactions of Soil Minerals with Natural Organics and Microbes* (eds P.M. Huang and M. Schnitzer), Soil Science Society of America, Madison, Wisconsin, pp. 521-48.

Epstein, E. (1972) *Mineral Nutrition of Plants*, John Wiley, New York.

Eriksson, K.E., Grunewald, A. and Vallander, L. (1980) Studies of growth conditions in wood for three white-rot fungi and their cellulase-less mutants. *Biotechnology and Bioengineering*, **22**, 363-76.

Ernsting, G. and Joose, E.N.G. (1974) Predation on two species of surface dwelling Collembola. A study with radio-isotope labelled prey. *Pedobiologia*, **14**, 222-31.

Eschenbrenner, V. (1986) Contribution des termites à la micro-agrégation des sols tropicaux. *Cahiers ORSTOM, série Pédologie*, **XXII** (4), 397-408.

Eschenbrenner, V. (1994a) The influence of fungus-cultivating ants (Hymenoptera, Formicidae, Attini) on the morphology of Andosols in Martinique, in *Soil Micromorphology: Studies in Management and Genesis* (eds A.J. Ringrose-Voase and G.S. Humphreys), Elsevier, Amsterdam, pp. 405-10.

Eschenbrenner, V. (1994b) Pedoturbation by gardening ants in tropical Andosols (Martinique, F.W.I.).

Quantitative assessment, in *Proceedings of the 15th World Congress of Soil Science,* Vol. 4a (eds J.D. Etchevers, A. S. Aguilar, R.E. Núñez *et al.*), ISSS, Acapulco, pp. 102-15.

Eswaran, H., Van Den Berg, E. and Reich, P. (1993) Organic carbon in soils of the world. *Soil Science Society of America, Proceedings*, **57**, 192-4.

Ettershank, G. (1968) The three-dimensional gallery structure of the nest of the meat ant *Iridomyrmex purpureus* (Sm.) (Hymenoptera: Formicidae). *Australian Journal of Zoology*, **16**, 715-23.

Fairhurst, C. (1974) The adaptive significance of variations in the life-cycles of Schizophiline millipedes, in *Myriapoda* (ed. J.G. Blower), Zoological Society of London, London, pp. 575-87.

Fairley, R.I. and Alexander, I.J. (1982) Seasonal dynamics of fine roots and mycorrhizas in a Sitka Spruce plantation, in *Abstracts. Conference on Tree Root Systems, and their Mycorrhizas,* I.U.F.R.O., Edinburgh.

Fairley, R.I. (1985) Grass root production in restored soil following opencast mining, in *Ecological Interactions in Soil ; Plants, Microbes and Animals* (eds A.H. Fitter, D. Atkinson, D.J. Read *et al.*), Blackwell Scientific Publications, Oxford, pp. 81-5.

Fanning, D.S. and Fanning, P.C. (1989) *Soil Morphology, Genesis and Classification*, John Wiley & Sons, New York.

FAO (eds) (1984) *Legume Inoculants and Their Use*, FAO, Rome.

FAO-UNESCO (eds) (1990) *FAO-UNESCO Soil Map of the World: Revised Legend*, FAO, Rome.

Fardjah, M., Lemée, G. and Pontailled, J.Y. (1980) Dynamique comparée de l'eau sous hêtraie et dans des coupes nues ou à *Calamagrostis epigeios* en forêt de Fontainebleau. *Bulletin d'Ecologie*, **11**, 11-31.

Farrar, D.M. and Coleman, J.D. (1967) The correlation of surface area with other properties of nineteen British soils. *Journal of Soil Science*, **18**, 118-24.

Febvay, G. and Kermarrec, A. (1986) Digestive physiology of leaf-cutting ants, in *Fire Ants and Leaf-Cutting Ants* (eds C.S. Lofgren and R.K. Van der Meer), Westview Press, Boulder, Colorado, pp. 274-88.

Feller, C. (1979) Une méthode de fractionnement granulométrique de la matière organique des sols: application aux sols tropicaux à texture grossière, très pauvres en humus. *Cahiers ORSTOM, série Pédologie*, **XVII** (4), 339-46.

Feller, C., Albrecht, A. and Tessier, D. (1995) Aggregation and organic matter storage in kaolinitic and smectitic tropical soils, in *Structure and Organic Matter Storage in Agricultural Soils* (eds M.R. Carter and B.A. Stewart), Lewis Publishers, Boca Raton, Florida, pp. 309-59.

Fenchel, T. and Harrison, P. (1976) The significance of bacterial grazing and mineral cycling for decomposition of particulate detritus, in *The Role of Terrestrial and Aquatic Organisms in Decomposition Processes. Proceedings of the 17th symposium of the British Ecological Society* (eds J.M. Anderson and A. Macfadyen), Blackwell Scientific Publications, Oxford, pp. 285-99.

Fenwick, I.M. and Knapp, B.J. (1982) *Soils: Process and Response*, Duckworth, London.

Fernández, O.A., Montani, T. and Distel, R.A. (1988) El sistema radical de especies de zonas aridas y semi-aridas. Algunas estrategias de supervivencia. *Interciência*, **13** (1), 25-30.

Ferrar, P. (1982) Termites of a South African savana. II. Densities and populations of smaller mounds, and seasonality of breeding. *Oecologia*, **52**, 133-8.

Ferrier, R.C. and Alexander, I.J. (1985) Persistence under field conditions of excised fine roots and mycorrhizas of spruce, in *Ecological Interactions in Soil ; Plants, Microbes and Animals* (eds A.H. Fitter, D. Atkinson, D.J. Read *et al.*), Blackwell Scientific Publications, Oxford, pp. 175-9.

Ferrière, G. (1980) Fonctions des lombriciens. VII. Une méthode d'analyse de la matière organique végétale ingérée. *Pedobiologia*, **20**, 263-73.

Ferris, H. and McHenry, M. (1976) Nematode community structure in a vineyard soil. *Journal of Nematology*, **8**, 131-7.

Ferris, H. (1993) New frontiers in nematode ecology. *Journal of Nematology*, **25** (3), 374-82.

Ferry, B. (1994) *Les Humus Forestiers des Ghats Occidentaux en Inde du Sud. Facteurs climatiques, édaphiques et biologiques intervenant dans le stockage de la matière organique du sol*, Institut Français de Pondichery, Pondichery, India.

Fisher, M.J., Rao, I.M., Ayarza, M.A. *et al.* (1994) Carbon storage by introduced deep-rooted grasses in the South American savannas. *Nature*, **371**, 236-8.

Fitter, A.H., Atkinson, D., Read, D.J. *et al.* (eds) (1985) *Ecological Interactions in Soil. Plants, Microbes and Animals*, Blackwell Scientific Publications, Oxford.

Fitter, A.H. (1985) Functional significance of root morphology and root system architecture, in *Ecological Interactions in Soil. Plants, Microbes and Animals* (eds A.H. Fitter, D. Atkinson, D.J. Read *et al.*), Blackwell Scientific Publications, Oxford, pp.87-106.

Fitter, A.H. and Garbaye, J. (1994) Interactions between mycorrhizal fungi and other soil organisms. *Plant and Soil*, **159**, 123-32.

Fitzpatrick, E.A. (1967) Soil nomenclature and classification. *Geoderma*, **1**, 91-105.

Flach, W., Cady, J.G. and Nettleton, W.D. (1968) Pedogenetic alteration of highly weathered parent material. *Transactions of the Ninth International Congress of Soil Science*, **4**, 155-74.

Flogaitis, E. (1984) Le peuplement de macroarthropodes édaphiques d'une forêt tempérée mixte: composition, phénologie et organisation spatiale. *Pedobiologia*, **27**, 1-14.

Fogel, R. (1983) Root turnover and productivity of coniferous forests. *Plant and Soil*, **71**, 75-85.

Fogel, R. (1985) Roots as primary producers in below-ground ecosystems, in *Ecological Interactions in Soil ; Plants, Microbes and Animals* (eds A.H. Fitter, D. Atkinson, D.J. Read *et al.*), Blackwell Scientific Publications, Oxford, pp. 23-36.

Folgarait, P. (1998) Ant biodiversity and its relationship to ecosystem functioning: a review. *Biodiversity and Conservation*, **7**, 1221-44

Folkoff, M.E. and Meentemeyer, V. (1985) Climatic control of the assemblages of secondary clay minerals in the A-horizon of United States soils. *Earth Surface Processes and Landforms*, **10**, 621-33.

Foster, R.C. and Rovira, A.D. (1978) The rhizosphere - the ultrastructure of the rhizosphere of *Trifolium subterraneum* L., in *Microbial Ecology* (eds M.W. Loutit and J.A.R. Miles), Springer Verlag, Berlin, pp. 278-90.

Foster, R.C. (1981) The ultrastructure and histochemistry of the rhizosphere. *New Phytologist*, **89**, 263-73.

Foster, R.C. (1985) *In situ* localization of organic matter in soils. *Quæstiones Entomologicæ*, **21**, 609-33.

Foster, R.C. (1988) Microenvironments of soil microorganisms. *Biology and Fertility of Soils*, **6**, 189-203.

Foster, R.C. and Dormaar, J.F. (1991) Bacteria-grazing amoebae *in situ* in the rhizosphere. *Biology and Fertility of Soils*, **1**, 83-7.

Foster, R.C. (1994) The ultramicromorphology of soil biota *in situ* in natural soils: a review, in *Soil Micromorphology: Studies in Management and Genesis. Proceedings of the 11th International Working Meeting on Soil Micromorphology* (eds A.J. Ringrose-Voase and G.S. Humphreys), Elsevier, Amsterdam, pp. 381-93.

Fournier, A. (1991) *Phénologie, croissance et production végétales dans quelques savanes d'Afrique de l'ouest*, thèse de 3e cycle, Université Paris VI.

Fowler, H.G., Forti, L.C., Pereira-Da Silva, V. *et al.* (1986) *Economics of Grass-Cutting Ants*, in *Fire Ants and Leaf-Cutting Ants*, (eds C.S. Lofgren and R.K. Vander Meer) Westview Press, Boulder, Colorado.

Fox, I. (1957) The insect family of Japygidae (order Thysanura) in Puerto Rico. *Journal of Agriculture, University of Puerto Rico*, **41**, 35-7.

Fragoso, C. and Lavelle, P. (1987) The Earthworm community of a Mexican tropical rainforest (Chajul, Chiapas), in *On earthworms* (eds A.M. Bonvicini and O. Omodeo), Mucchi, Modena, pp. 281-97.

Fragoso, C. and Lavelle, P. (1992) Earthworm communities of tropical rainforests. *Soil Biology & Biochemistry*, **24** (12), 1397-408.

Fragoso, C., Lavelle, P., Blanchart, E. *et al.* (1999) Earthworm communities of tropical agroecosystems: origin, structure and influence of management practices, in *Earthworm Management in Tropical Agroecosystems* (eds P. Lavelle, L. Brussaard and P. Hendrix), CAB International, Wallingford, 27-56.

Fraile, J.M. (1989) *Poblaciones de lombrices de tierra (Oligochaeta: Annelida) en una pastura de* Cynodon plectostachyus *(pasto estrella) asociada con árboles de* Erythrina poeppigina *(puro), una pastura sin árboles de* Corida alliodora *(laurel), una pastura sin árboles y vegetacien a libre crecimiento en el Catie, Turrialba Costa Rica.* Tesis M. Sci., Universidad de Costa Rica.

Francis, R. and Read, D.J. (1994) The contributions of mycorrhizal fungi to the determination of plant community structure. *Plant and Soil*, **159**, 11-25.

François, C., Rafidison, Z., Villemin, G. *et al.* (1986) The accumulation and fate of brown pigments in leaves from a litter of *Fagus sylvatica* L.: a morphological and chemical study, in *Proceedings of the 7th Symposium of Environmental Biogeochemistry* (eds G. Giovannozzi-Sermanari and P. Nannipieri), Vitrbo, Italy.

Francl, L.J. (1986) Spatial analysis of *Heterodera glycines* populations in field plots. *Journal of Nematology*, **18** (2), 183-9.

Frank, A.B. (1885) über die auf Wurzelsymbiose beruhende Ernährung gewisser Baüme durch unterirdische Pilze. *Berichte Deutschen Botanische Gesellschaft*, **3**, 128-45.

Freckman, D.W., Mankau, R. and Sher, S.A. (1986) Nematode community structure in desert soils: nematode recovery. *Journal of Nematology*, **7**, 343-46.

Freckman, D.W. and Mankau, R. (1986) Abundance, distribution, biomass and energetics of soil nematodes in a northern Mojave Desert ecosystem. *Pedobiologia*, **29**, 129-42.

Freckman, D.W. (1988) Bacterivorous nematodes and organic-matter decomposition. *Agriculture, Ecosystems and Environment*, **24**, 195-217.

Freckman, D.W. and Womersley, C. (1983) Physiological adaptation of nematodes in Chihuahuan Desert soil, in *New Trends in Soil Biology* (eds Ph. Lebrun, H.M. André, A. de Medts *et al.*), Dieu-Brichart, Ottignies, pp. 395-403.

French, J.R.J., Robinson, P.J., Pahl, P.J. *et al.* (1987) Termite-fungi interactions. III. Response of *Coptotermes lacteus* in field mounds to fungus-decayed softwood blocks. *Material und Organismen*, **22**, 111-26.

Frenot, Y. (1985) Etude de l'introduction accidentelle de *Dendrobaena rubida* tenuis (Oligochaeta, Lumbricidae) à l'Île de la Possession. *Bulletin d'Ecologie*, **16** (1), 47-53.

Friese, C.F. and Allen, M.F. (1991) The spread of VA mycorrhizal fungal hyphae in the soil: inoculum types and external hyphal architecture. *Mycologia*, **83**, 409-18.

Friese, C.F. and Allen, M.F. (1993) The interaction of harvester ants and vesicular-arbuscular mycorrhizal fungi in a patchy semi-arid environment: the effects of mound structure on fungal dispersion and establishement. *Functional Ecology*, **7**, 13-20.

Fry, B. and Sherr, E.B. (1984) $\delta^{13}C$ measurement as indicators of carbon flow in marine and freshwater ecosystems. *Contributions in Marine Science*, **27**, 13-47.

Fusseder, A. (1987) The longevity and activity of the primary root of maize. *Plant and Soil*, **101**, 257-65.

Fyfe, W.S., Kronberg, B.I., Leonardos, O.H. *et al.* (1983) Global tectonics and agriculture: a geochemical perspective. *Agriculture, Ecosystems and Environment*, **9**, 383-99.

Fyles, I.H., Juma, N.G. and Robertson, J.A. (1988) Dynamics of microbial biomass and faunal populations in long-term plots on a gray luvisol. *Canadian Journal of Soil Science*, **68** (1), 91-100.

Gadd, G.M. (1993) Interactions of fungi with toxic metals. *New Phytologist*, **124**, 25-60.

Gadgil, R.L. and Gadgil, P.D. (1971) Mycorrhiza and litter decomposition. *Nature*, **233**, 133.

Gaff, D.F. (1981) The biology of resurrection plants, in *The Biology of Australian Plants* (eds J.S. Pate and A.J. McComb), University of Western Australia Press, Nedlands, pp. 114-46.

Gahoonia, T.S. and Nielsen, N.E. (1992) The effects of root-induced pH changes on the depletion of inorganic and organic phosphorus in the rhizosphere. *Plant and Soil*, **143**, 185-91.

Gallon, J.R. and Chaplin, A.E. (1987) The biochemistry of N_2 fixation: integration with metabolism, in *An Introduction to Nitrogen Fixation* (eds J.R. Gallon and A.E. Chaplin), Cassell Educational Ltd., London, pp. 128-78.

Garay, I. (1980) *Relation entre la dynamique des humus forestiers et la structure des peuplements de microarthropodes*, thèse de l'Université Paris VI.

Garay, I. (1981a) Etude d'un écosystème forestier mixte. IV. Données générales sur les Microarthropodes édaphiques et structure du peuplement d'Acariens Oribates. *Revue d'Ecologie et de Biologie du Sol*, **18** (3), 339-56.

Garay, I. (1981b) Le peuplement de microarthropodes dans la litière sur pied de *Espeletia lutescens* et *Espeletia timotensis*. *Revue d'Ecologie et de Biologie du Sol*, **18** (2), 209-19.

Garay, I., Nazoa, S. and Abbadie, L. (1986a) Etude d'une litière forestière mixte à charme et chêne. I. - Décomposition de la matière organique et dynamique des éléments biogènes. *Acta Œcologia, Œcologia Generalis*, **7** (2), 151-69.

Garay, I., Mollon, A. and Flogaitis, E. (1986b) Etude d'une litière forestière mixte à charme (*Carpinus betulus* L.) et chêne (*Quercus sessiliflora* Smith). II.- Succession des macroarthropodes au cours de la décomposition. *Acta Œcologia, Œcologia Generalis*, **7** (3), 263-88.

Garay, I. (1989) *Relations entre l'Hétérogénéité des Litières et l'Organisation des Peuplements d'Arthropodes Édaphiques*, Publication du Laboratoire de Zoologie no.35, Ecole Normale Supérieure, Paris.

Garbaye, J. (1991) Biological interactions in the mycorrhizosphere. *Experientia*, **47**, 370-75.

Garbaye, J. (1994) Helper bacteria: a new dimension to the mycorrhizal symbiosis. *New Phytologist*, **128**, 197-210.

Gardner, J.H. and Malajczuk, N. (1988) Recolonisation of rehabilitated bauxite mine sites in Western Australia by mycorrhizal fungi. *Forest Ecology and Management*, **24**, 27-42.

Garnier-Sillam, E., Villemin, G., Toutain, F. *et al.* (1985) Formation de micro-agrégats organo-minéraux dans les fèces de termites. *Comptes Rendus de l'Academie des Sciences, Paris*, **III** (301), 213-8.

Garnier-Sillam, E., Villemin, G., Toutain, F. *et al.* (1987) Contribution à l'étude du rôle des termites dans l'humification des sols forestiers tropicaux, in *Micromorphologie des Sols* (eds N. Fedoroff, L.M. Bresson and M.A. Courty), AFES, Paris, pp. 331-5.

Garnier-Sillam, E., Toutain, F., Villemin, G. *et al.* (1988a) Transformation de la matière organique végétale sous l'action du termite *Macrotermes mülleri* (Sjöstedt) et de son champignon symbiotique. *Canadian Journal of Microbiology*, **34** (11), 1247-55.

Garnier-Sillam, E., Toutain, F. and Renoux, J. (1988b) Comparaison de l'influence de deux termitières (humivore et champignonniste) sur la stabilité structurale des sols forestiers tropicaux. *Pedobiologia*, **32** (1/2), 89-97.

Garnier-Sillam, E. (1989) The pedological role of fungus-growing termites (Termitidae: Macrotermitinae) in tropical environments, with special reference to *Macrotermes muelleri*. *Sociobiology*, **15** (2), 181-96.

Garnier-Sillam, E., Toutain, F., Villemin, G. *et al.* (1989) Etudes préliminaires des meules originales du termite xylophage *Sphaerotermes sphaerothorax* (Sjöstedt). *Insectes Sociaux*, **36** (4), 293-312.

Garnier-Sillam, E. and Harry, M. (1995) Distribution of humic compounds in mounds of some soil-feeding termite species of tropical rainforests: its influence on soil structure stability. *Insectes Sociaux*, **42** (2), 167-85.

Garrett, S.D. (1951) Ecological groups of soil fungi and survey of substrate relationships. *New Phytologist*,

50, 149-66.

Garrett, S.D. (1963) *Soil Fungi and Soil Fertility*, Pergamon, Oxford.

Garten, C.T. (1976) Correlations between concentrations of elements in plants. *Nature*, **261**, 686-8.

Gates, G.E. (1970) On some exotic earthworms from Mexico. *Mexicana de Historia Natural*, **XXXI**, 281-7.

Gay, F.J. and Greaves, T. (1940) The population of a mound colony of *Coptotermes lacteus* (Frogg.). *Journal of the Council for Scientific and Industrial Research*, **13**, 145-9.

Gay, F.J. and Calaby, J.H. (1970) Termites from the Australian Region, in *Biology of Termites* (eds K. Krishna and F.M. Weesner), Academic Press, New York, pp. 393-448.

Gay, F.J. and Wetherly, A.H. (1970) The population of a large mound of *Nasutitermes exitiosus* (Hill) (Isoptera: Termitidae). *Journal of the Australian Entomological Society*, **9**, 27-30.

Geoffroy, J.J. (1981) Modalités de la coexistence de deux Diplopodes *Cylindroiulus punctatus* (Leach) et *Cylindroiulus nitidus* (Verhoeff) dans un écosystème forestier du Bassin Parisien. *Acta Œcologica, Œcologia Generalis*, **2**, 227-43.

Gerard, B.M. (1963) The activities of some species of Lumbricidae in pasture-land, in *Soil Organisms* (eds J. Doeksen and J. Van der Drift), North Holland Publishing Co., Amsterdam, pp. 49-52.

Ghilarov, M.S. and Chernov, Y.J. (1974) Some results of soil-zoological studies in the U.S.S.R. in frame of the International Biological Programme. *Academy of Sciences, USSR, Moscow*, **4**, 563-79.

Ghilarov, M.S. (1977) Why so many species and so many individuals can coexist in the soil, in *Soil Organisms as Components of Ecosystems* (eds U. Lohm and T. Persson), Ecological Bulletins/NFR, Stockholm, pp. 593-8.

Ghilarov, M.S. (1983) Some essential physiological characters of soil invertebrates, in *New Trends in Soil Biology* (eds Ph. Lebrun, H.M. André, A. de Medts *et al.*), Dieu-Brichart, Ottignies, pp. 315-20.

Ghosh, K. and Schnitzer, M. (1979) UV and visible absorption spectroscopic investigations in relation to macromolecular characteristics of humic substances. *Journal of Soil Science*, **30**, 735-45.

Gianinazzi-Pearson, V. and Gianinazzi, S. (1986) Connaissances actuelles des bases physiologiques et biochimiques des effets des endomycorhizes sur le comportement des plantes. *Physiologie Végétale*, **24** (2), 253-62.

Gibson, D.J., Colquhoun, I.A. and Greig-Smith, P. (1985) A new method of measuring nutrient supply rates in soils using ion-exchange resins, in *Ecological Interactions in Soil ; Plants, Microbes and Animals* (eds A.H. Fitter, D. Atkinson, D.J. Read *et al.*), Blackwell Scientific Publications, Oxford, pp. 73-9.

Gijsman, A.J. (1990) Rhizosphere pH along different root zones of Douglas-fir (*Pseudotsuga menziesii*), as affected by source of nitrogen, in *Plant Nutrition - Physiology and Applications* (ed. M.L. van Beusichem), Kluwer Academic Publishers, Dordrecht, pp. 45-51.

Giller, P.S. (1984) *Community Structure and the Niche*, Chapman & Hall, London.

Giller, K.E. and Wilson, K.J. (1991) *Nitrogen Fixation in Tropical Cropping Systems*, CAB International, Wallingford.

Gillman, L.R., Jefferies, M.K. and Richards, G.N. (1972) Non-soil constituents of termite (*Coptotermes acinaciformis*) mounds. *Australian Journal of Biological Sciences*, **25**, 1005-13.

Gilmour, J.T. (1984) The effects of soil properties on nitrification and nitrification inhibition. *Soil Science Society of America, Journal*, **48** (6), 1262-6.

Gilot, C. (1994) *Effets de l'introduction du ver géophage tropical* Millsonia anomala. *Omodeo en systèmes cultivés sur les caractéristiques des sols et la production végétale en moyenne Côte d'Ivoire*, thèse de l'Université Paris VI/INAPG.

Gilot, C., Lavelle, P., Blanchart, E. *et al.* (1995) Biological activity of soil under rubber plantations in Côte d'Ivoire. *Acta Zoologica Fennica*, **196**, 186-9.

Gisin, H. (1943) Ökologie und Lebensgemeinschaften der Collembolen im Schweizerischen Exkursionsgebiet Basels. *Revue Suisse de Zoologie*, **50** (4), 183-9.

Gogala, N. (1991) Regulation of mycorrhizal infection by hormonal factors produced by hosts and fungi. *Experimentia*, **47**, 331-40.

Goodey, T. (1963) *Soil and Freshwater Nematodes*, Methuen, London.

Goodland, R.J.A. (1965) On termitaria in a savanna ecosystem. *Canadian Journal of Zoology*, **43**, 641-50.

Gordon, D. (1991) Behavioural flexibility and the foraging ecology of seed-eating ants. *American Naturalist*, **138**, 379-411.

Gordon, D. (1992) How colony growth affects forager intrusion in neighbouring harvester ant colonies. *Behavioural Ecology and Sociobiology*, **31**, 417-27.

Gordon, D. (1995) The development of an ant colony's foraging range. *Animal Behaviour*, **49**, 649-59.

Gordon, D.M. and Kulig, A.W. (1996) Founding, foraging and fighting: colony size and the spatial distribution of harvester ant nests. *Ecology*, **77**, 2393-409.

Gorissen, A., Vanginkel, J.H., Keurentjes, J.J.B. *et al.* (1995) Grass root decomposition is retarded when grass has been grown under elevated CO_2. *Soil Biology & Biochemistry*, **27** (1), 117-20.

Goss, M.J., Ehlers, W., Boone, F.R. *et al.* (1984) Effects of soil management practice on soil physical conditions affecting root growth. *Journal for Agricultural Engineering Research*, **30**, 131-40.

Goss, M.J. (1991) Consequences of the activity of roots on soil, in *Plant Root Growth. An Ecological Perspective* (ed. D. Atkinson), Blackwell Scientific Publications, London, pp. 171-86.

Gotwald, W.H. (1974) Predatory behaviour and food preferences of driver ants in selected African habitats. *Annals of the Entomological Society of America*, **67**, 877-86.

Gotwald, W.H. (1986) The beneficial economic role of ants, in *Economic Impact and Control of Social Insects* (ed. S.B. Vinson), Praeger Special Studies, New York, pp. 290-313.

Gotwald, W.H. (1995) *Army Ants: The Biology of Social Predation*, Cornell University Press, Ithaca.

Gourbière, F. (1982) Pourriture blanche de la litière d'*Abies alba* Mill. I.- Evolution de la litière sous l'action des basidiomycètes du genre *Collybia*. *Revue d'Ecologie et de Biologie du Sol*, **19** (2), 163-75.

Gouzou, L., Burtin, G., Philippy, R. *et al.* (1993) Effect of inoculation with *Bacillus polymyxa* on soil aggregation in the wheat rhizosphere: preliminary examination. *Geoderma*, **56**, 479-91.

Graff, O. (1971) Stickstoff, Phosphor und Kalium in der Regenwurmlosung auf der Wiesenversuchsflächte des Sollingprojektes, in *IV Colloquium Pedobiologiae* (ed. J. d'Aguilar), INRA, Paris, pp. 503-11.

Graham, R.C., Tice, K.R. and Guertal, W.R. (1994) The pedologic nature of weathered rock, in *Whole Regolith Pedology* (eds R.B. Brown and J.H. Huddleston), Soil Science Society of America, Madison, Wisconsin, pp. 21-40.

Grandval, A. (1993) *Mise en évidence d'une structure spatiale du système de décomposition en relation avec la qualité de la litière en forêt guyanaise*, Rapport de DEA, Université Paris VI, Paris.

Grassé, P.P. (1984) *Termitologia. II*, Masson, Paris.

Grassé, P.P. (1986) *Termitologia. III*, Masson, Paris.

Graves, J.D., Satkins, N.K., Fitter, A.H. *et al.* (1997) Intraspecific transfer of carbon between plants linked by a common mycorrhizal network. *Plant and Soil*, **192**, 153-9.

Gray, T.R.G. and Postgate, J.R. (eds) (1976) *The Survival of Vegetative Microbes*, Cambridge University Press, Cambridge.

Greaves, T. (1962) Studies of foraging galleries and the invasion of living trees by *Coptotermes acinaciformis* and *C. brunneus* (Isoptera). *Australian Journal of Zoology*, **10** (4), 630-51.

Greaves, T. (1967) Experiments to determine the populations of tree-dwelling colonies of termites (*Coptotermes acinaciformis* (Froggart) and C. *frenchi* (Hill). *Division of Entomology, CSIRO Australia,*

Technical Paper, **7**, 19-33.

Greaves, M.P. and Darbyshire, J.F. (1972) The ultrastructure of the mucilaginous layer on plant roots. *Soil Biology & Biochemistry*, **4**, 443-9.

Green, R.D. and Askew, G.P. (1965) Observations on the biological development of macropores in soils of Romney Marsh. *Journal of Soil Science*, **16**, 342-9.

Green, R.N., Trowbridge, R.L. and Klinka, K. (1993) Towards a taxonomic classification of humus forms. *Forest Science*, **29** (1), 1-49.

Greenland, D.J. and Kowal, J.M.L. (1960) Nutrient content of moist tropical forest in Ghana. *Plant and Soil*, **12**, 154-74.

Greenland, D.J., Wild, A. and Adams, D. (1992) Organic matter dynamics in soils of the tropics-from myth to complex reality, in *Myths and Science of Soils of the Tropics* (ed. R. Lal and P. Sanchez), Soil Science Society of America, Madison, Wisconsin, pp. 17-35.

Greenslade, P.J.M. (1975) Dispersion and history of a population of the meat ant *Iridomyrmex purpureus* (Hymenoptera: Formicidae). *Australian Journal of Zoology*, **23**, 495-510.

Greenslade, P.J.M. (1983) Adversity selection and the habitat templet. *American Naturalist*, **122**, 352-65.

Gregorich, E.G., Monreal, C.M., Schnitzer, M. *et al.* (1996) Transformation of plant residues into soil organic matter: chemical characterization of plant tissue, isolated soil fractions, and whole soils. *Soil Science*, **161**, 680-93.

Grier, C.G., Vogt, K.A., Keyes, M.R. *et al.* (1981) Biomass distribution and above- and below-ground production in young and mature *Abies amabilis* zone ecosystems of the Washington Cascades. *Canadian Journal of Forest Research*, **11**, 155-67.

Griffin, D.M. (1972) *Ecology of Soil Fungi*, Chapman & Hall, London.

Griffin, D.M. (1981a) Water and microbial stress. *Advances in Microbial Ecology*, **5**, 91-136.

Griffin, D.M. (1981b) Water potential as a selective factor in the microbial ecology of soils, in *Water Potential Relations in Soil Microbiology* (eds J.F. Parr, W.R. Gardner and L.F. Elliot), Soil Science Society of America, Madison, Wisconsin, pp. 141-51.

Griffin, G.J., Hale, M.G. and Shay, F.J. (1976) Nature and quantity of sloughed organic matter produced by roots of axenic peanut plants. *Soil Biology & Biochemistry*, **8**, 29-32.

Griffin, G.J. and Roth, D.A. (1979) Nutritional aspects of soil mycostasis, in *Soil-Borne Plant Pathogens* (eds B. Schippers and W. Gams), Academic Press, London, pp. 79-96.

Griffiths, B.S. and Caul, S. (1993) Migration of bacterial-feeding nematodes, but not protozoa, to decomposing grass residues. *Biology and Fertility of Soils*, **15**, 201-7.

Griffiths, R.P., Bradshaw, G.A., Marks, B. *et al.* (1996) Spatial distribution of ectomycorrhizal mats in coniferous forests of the Pacific Northwest, USA. *Plant and Soil*, **180** (1), 147-58.

Griffiths, R.P., Baham, J.E. and Caldwell, B.A. (1994) Soil solution chemistry of ectomycorrhizal mats in a forest soil. *Soil Biology & Biochemistry*, **26**, 331-7.

Grimaldi, C. and Pedro, G. (1996) Importance de l'hydrolyse acide dans les systèmes pédologiques des régions tropicales humides. Rôle de la forêt et conséquences sur la genèse des sables blancs tropicaux. *Comptes Rendus de l'Académie des Sciences, Paris*, **323** (IIa), 483-92.

Grimaldi, M., Sarrazi, A. and Luizão, F. (1993) Effets de la déforestation et des cultures sur la structure des sols argileux d'Amazonie brésilienne. *Cahiers d'Agriculture*, **2**, 36-47.

Grime, J.P. (1979) *Plant Strategies and Vegetation Processes*, John Wiley, Chichester.

Grimont, P.A.D., Popoff, M.Y., Grimont, F. *et al.* (1980) Reproductibility and correlation study of three deoxyribonucleic acid hybridization procedures. *Current Microbiology*, **4**, 325-30.

Grobbelaar, N. (1993) The cycad-cyanobacterium symbiosis, in *Symbioses in Nitrogen-Fixing Trees* (eds N.S. Subba Rao and C. Rodriguez-Barrueco), Oxford & IBH Publishing Co, New-Dehli, pp. 95-140.

Grubb, P.J. and Edward, P.J. (1982) Studies of mineral cycling in a montaine rain forest in New Guinea. III. The distribution of mineral elements in the above-ground material. *Journal of Ecology*, **70**, 623-48.

Guckert, A., Breisch, H. and Reisinger, O. (1975) Interface sol-racine. I.- Etude au microscope électronique des relations mucigel-argile-microorganismes. *Soil Biology & Biochemistry*, **7**, 241-50.

Guckert, A. and Balandreau, J. (1981) Translocation of assimilates to the roots and its effect on nitrogen fixation, in *Agricultural Yield Potentials in Continental Climates, Proceedings of the International Colloquium of the Potash Institute*, Bern, pp. 127-36.

Guckert, A. (1985) Root exudation and microbial activity at the soil-root interface, in *Proceedings of the 2nd IPI-ISS Workshop on Potassium*, Academia Sinica, Nanjing, pp. 49-60.

Guillet, B., Vedy, J.C., Rouiller, J. *et al.* (1979) Migration de particules argileuses dans les milieux de géochimie organique très acide: Exemple des podzols sur colluvium de grès Vosgien, in *Phénomènes de Transport de Matière dans l'Ecorce Terrestre* (ed. G. Millot), CNRS, Strasbourg, pp. 13-8.

Gunnarsson, T. and Tunlid, A. (1986) Recycling of fecal pellets in isopods: microorganisms and nitrogen compounds as potential food for *Oniscus asellus* L. *Soil Biology & Biochemistry*, **18** (6), 595-600.

Guthrie, R.L. and Witty, J.E. (1982) New designations for soil horizons and layers and the new Soil Survey Manual. *Soil Science Society of America, Proceedings*, **46**, 443-4.

Haber, W.A., Frankie, G.W., Baker, H.G. *et al.* (1981) Ants like flower nectar. *Biotropica*, **13**, 211-4.

Hafidi, N. (1989) *Relation entre l'hétérogénéité de la structure des litières et le peuplement des macro-arthropodes édaphiques en forêt tempérée*, thèse de l'Université Paris Sud-Orsay.

Hagedorn, H.C. and Jensen, B.N. (1923) Zur mikrobestimmung des Blutzuckers mittels Ferricyanid. *Biochemische Zeitschrift*, **135**, 46-58.

Haider, K. and Martin, J.P. (1980) Biochemistry of humus formation and its interaction with clays. *Colloques Internationaux du CNRS*, **303**, 163-172.

Haimi, J., Huhta, V. and Boucelham, M. (1992) Growth increase of birch seedlings under the influence of earthworms. A laboratory study. *Soil Biology & Biochemistry*, **24** (12), 1525-8.

Haines, P.J. and Uren, N.C. (1990) Effects of conservation tillage farming on soil microbial biomass, organic matter and earthworm populations, in north-eastern Victoria. *Australian Journal of Experimental Agriculture*, **30**, 365-71.

Hale, W.G. (1965a) Post-embryonic development in some species of Collembola. I. *Pedobiologia*, **5** (1/2), 146-52.

Hale, W.G. (1965b) Post-embryonic development in some species of Collembola. II. *Pedobiologia*, **5** (3), 161-77.

Hale, M.G., Moore, L.D. and Griffin, G.J. (1981) Root exudates and exudation, in *Interactions between Non-Pathogenic Soil Microorganisms and Plants* (eds Y.R. Dommergues and S.V. Krupa), Elsevier, Amsterdam, pp. 163-203.

Halffter, G. and Edmunds, W.D. (1981) Evolución de la nidificación y de la cooperación bisexual en *Scarabaeinae* (Ins.: Col.). *Anales de la Escuela Nacional de Ciencias Biologicas*, **25**, 117-44.

Haller, T. and Stolp, H. (1985) Quantitative estimation of root exudation of maize plants. *Plant and Soil*, **86**, 207-16.

Halsall, D.M., Leigh, J.H., Gollasch, S.E. *et al.* (1995) The role of allelopathy in legume decline in pastures. 2. Comparative effects of pasture, crop and weed residues on germination, nodulation and root growth. *Australian Journal of Agricultural Research*, **46** (1), 189-207.

Hamblin, A.P. (1985) The influence of soil structure on water movement, crop root growth and water uptake. *Advances in Agronomy*, **38**, 95-158.

Hameed, R., Cortez, J. and Bouché, M.B. (1993) Influence de la qualité de la litière apportée à *Lumbricus terrestris* L. sur la dynamique de l'azote minéral et la production végétale. *Pedobiologia*, **37**, 178-92.

Hamilton, W.E. and Sillman, D.Y. (1989) Influence of earthworm middens on the distribution of soil microarthropods. *Biology and Fertility of Soils*, **8**, 279-84.

Handel, S.N. and Beattie, A.J. (1990) Seed dispersal by ants. *Scientific American*, **262**, 58-64.

Handley, W.R.C. (1954) Mull and mor formation in relation to forest Soils. *Bulletin of the Forestry Commission, London*, **23**, 115.

Handley, W.R.C. (1961) Further evidence for the importance of residual protein complexes on litter decomposition and the supply of N for plant growth. *Plant and soil,* **15**, 37-73.

Hanlon, R.D.G. and Anderson, J.M. (1980) Influence of macroarthropod feeding activities on microflora in decomposing oak leaves. *Soil Biology & Biochemistry*, **12**, 255-61.

Hanlon, R.D.G. (1981) Some factors influencing microbial growth on soil animal faeces. II. Bacterial and fungal growth on soil animal faeces. *Pedobiologia*, **21**, 264-70.

Hansson, A.C. and Steen, E. (1984) Methods of calculating root production and nitrogen uptake in an annual crop. *Swedish Journal of Agricultural Research*, **14**, 192-200.

Hansson, A.C., Pettersson, R. and Paustian, K. (1987) Shoot and root production and nitrogen uptake in Barley, with and without nitrogen fertilization. *Journal of Agronomy and Crop Science*, **158** (3), 163-71.

Hansson, A.C., Andrén, O., Boström, S. *et al.* (1990) Structure of the agroecosystem, in *Ecology of Arable Land-Organisms, Carbon and Nitrogen Cycling* (eds O. Andren, T. Lindberg, K. Paustian *et al.*), Ecological Bulletins/NFR, Stockholm, pp. 41-83.

Harley, J.L. (1989) The significance of mycorrhiza. *Mycological Research*, **92**, 129-39.

Harmand, J.M. (1997) Rôle des espèces ligneuses à croissance rapide dans le fonctionnement biogéochimique de la jachère. Effets sur la restauration de la fertilité des sols ferrigineux tropicaux (Bassin de la Benoue au Nord-Cameroun), thèse de l'Université Paris VI.

Harmer, R. and Alexander, I.J. (1985) Effects of root exclusion on nitrogen transformations and decomposition processes in Spruce humus, in *Ecological Interactions in Soil ; Plants, Microbes and Animals* (ed. A.H. Fitter, D. Atkinson, D.J. Read *et al.*), Blackwell Scientific Publications, Oxford, pp. 267-77.

Harper, J.L., Lovell, P.H. and Moore, K.G. (1970) The shapes and sizes of seeds. *Annual Review of Ecology and Systematics*, **1**, 327-56.

Harris, J.W. (1971) Microbiological studies at the Osage site, 1970. *US IBP Grassland Biome, Technical Report*, **102**.

Harris, W.V. (1970) Termites of the Palearctic region, in *Biology of Termites* (eds K. Krishna and F.M. Weesner), Academic Press, New York, pp. 295-313.

Harris, P.J. and Hartley, R.D. (1976) Detection of bound ferulic acid in cell walls of the Gramineae by ultraviolet fluorescence microscopy. *Nature*, **259**, 508-10.

Harris, W.F., Santantonio, D. and McGinty, D. (1980) The dynamic below-ground ecosystem, in *Forests: Fresh Perspectives from Ecosystem Analysis* (ed. R.H. Waring), Oregon State University, Corvallis, pp. 119-29.

Hartenstein, R., Neuhauser, E.F. and Kaplan, D.L. (1979) Reproductive potential of the earthworm *Eisenia fetida*. *Oecologia*, **43**, 329-40.

Hasebe, A., Kanazawa, S. and Takai, Y. (1984) Microbial biomass in paddy soil I. Microbial biomass calculated from direct count using fluorescence microscope. *Soil Science and Plant Nutrition*, **30** (2), 175-87.

Hasebe, A., Kanazawa, S. and Takai, Y. (1985) Microbial biomass in paddy soil II. "Microbial biomass carbon" measured by Jenkinson's fumigation method. *Soil Science and Plant Nutrition*, **31** (3), 349-59.

Hasegawa, M. and Takeda, H. (1996) Carbon and nutrient dynamics in decomposing pine needle litter in relation to fungal and faunal abundances. *Pedobiologia*, **40** (2), 171-84.

Haselwandter, K. and Bowen, G.D. (1996) Mycorrhizal relations in trees for agroforestry and land rehabilitation.

Forest Ecology and Management, **81**, 1-17.

Hassall, M. and Rushton, S.P. (1982) The role of coprophagy in the feeding strategies of terrestrial isopods. *Oecologia*, **53**, 374-81.

Hassall, M. and Jennings, J.B. (1975) Adaptative features of gut structure and digestive physiology in the terrestrial isopod *Philoscia muscorum* (Scop.). *Biological Bulletins*, **149**, 348-64.

Hassall, M., Visser, S. and Parkinson, D. (1986) Vertical migration of *Onychiurus subtenuis* (Collembola) in relation to rainfall and microbial activity. *Pedobiologia*, **29**, 175-82.

Hassink, J., Bouwman, L.A., Zwart, K.B. *et al.* (1993) Relationships between habitable pore space, soil biota and mineralization rates in grassland soils. *Soil Biology & Biochemistry*, **25** (1), 47-55.

Hattori, T. (1973) *Microbial Life in the Soil, an Introduction*, Marcel Dekker, New York.

Hattori, T. and Hattori, R. (1976) The physical environment in soil microbiology. An attempt to extend principles of microbiology to soil micro-organisms. *Critical Reviews in Microbiology*, **26**, 423-61.

Hawkes, G.E., Powlson, D.S., Randal, E.W. *et al.* (1984) A ^{31}P nuclear magnetic resonance study of the phosphorus species in the alkali extracts from long-term field experiments. *Journal of Soil Science*, **35**, 35-45.

Hawksworth, D.L. and Mound, L.A. (1991) Biodiversity databases: the crucial significance of collections, in *The Biodiversity of Microorganisms and Invertebrates: Its Role in Sustainable Agriculture* (ed. D.L.Hawksworth), CAB International, Wallingford, pp. 17-29.

Hayman, D.S. (1983) The physiology of vesicular-arbuscular endomycorrhizal symbiosis. *Canadian Journal of Botany*, **61**, 944-63.

Haynes, R.J. (1990) Active ion uptake and maintenance of cation-anion balance: a critical examination of their role in regulating rhizosphere pH. *Plant and Soil*, **126**, 247-64.

Heal, O.W. (1965) Observations on testate amoebae (Protozoa, Rhizopoda) from Signy Island, South Orkney Islands. *British Antarctic Survey Bulletin*, **6**, 43-7.

Heal, O.W., Flannagan, P.W., French, D.D. *et al.* (1981) Decomposition and accumulation of organic matter in tundra, in *Tundra Ecosystems: a Comparative Analysis* (eds L.C. Bliss, O.W. Heal and J.J. Moore), Cambridge University Press, Cambridge, pp. 587-633.

Heal, O.W. and Dighton, J. (1985) Resource quality and trophic structure in the soil system, in *Ecological Interactions in Soil ; Plants, Microbes and Animals* (eds A.H. Fitter, D. Atkinson, D.J. Read *et al.*), Blackwell Scientific Publications, Oxford, pp. 339-54.

Healy, I.N. and Russell-Smith, A. (1971) Abundance and feeding preferences of fly larvae in two woodland soils. *Annales de Zoologie, Ecologie Animale*, **Hors-série**, 177-92.

Healy, B. (1980) Distribution of terrestrial Enchytraeidae in Ireland. *Pedobiologia*, **20**, 159-75.

Heath, G.W., Arnold, M.K. and Edward, C.A. (1966) Studies in leaf litter breakdown. I. Breakdown rates of leaves of different species. *Pedobiologia*, **6**, 1-12.

Heath, B., Sollins, P., Perry, D.A. *et al.* (1988) Asymbiotic nitrogen fixation in litter from Pacific Northwest forests. *Canadian Journal of Forest Research*, **18**, 68-74

Hedley, M.J., Nye, P.H. and White, R.E. (1982) Plant-induced changes in the rhizosphere of rape (*Brassica napus* var. Emerald) seedlings. II. Origin of the pH change. *New Phytologist*, **91**, 31-44.

Helal, H.M. and Sauerbeck, D. (1986) Effect of plant roots on carbon metabolism of soil microbial biomass. *Zeitschrift für Pflanzenernährung und Bodenkunde*, **149**, 181-8.

Hendricks, C.W. and Pascoe, N. (1988) Soil microbial biomass estimated using 2450 MHz microwave irradiation. *Plant and Soil*, **110**, 39-47.

Hendrix, P.F., Parmelee, R.W., Crossley, D.A.J. *et al.* (1986) Detritus food webs in conventional and non-tillage agroecosystems. *BioScience*, **36** (6), 374-80.

Herbers, J.M. (1989) Community structure in north temperate ants: temporal and spatial variation. *Oecologia*,

81, 201-11.

Herrera, R., Jordan, C.F., Klinge, H. *et al.* (1978) Amazon ecosystems. Their structure and functioning with particular emphasis onnutrients. *Interciência*, **3** (4), 223-31.

Hesselman, H. (1926) Studies över barrskogen humusstäcke, des egenskape och beroende av skogsvarden. *Meddelanden frAn Statens Skogsförsöksanstalt*, **22**, 169-552.

Heulin, T., Guckert, A. and Balandreau, J. (1987) Stimulation of root exudation of rice seedlings by *Azospirillum* strains: carbon budget under gnotobiotic conditions. *Biology and Fertility of Soils*, **4**, 9-14.

Heulin, T. and Berge, O. (1992) Ecologie des bactéries fixatrices d'azote associées à la rhizosphère des céréales: mieux décrire pour mieux comprendre, in *Comptes Rendus de la 5e Conférence de l'Association Africaine pour la Fixation de l'Azote*, Rabat, Maroc.

Hewitt, A.E. (1992) *New Zealand Soil Classification*, DSIR Land Resources, Lower Hutt.

Higashi, S. and Yamaguchi, K. (1979) Influence of a supercolonial ant *Formica* (*Formica*) *yessensis* Forel on the distribution of other ants in Ishikari Coast. *Japanese Journal of Ecology*, **29** (3), 257-64.

Higashi, S. and Ito, F. (1989) Defense of termitaria by termitophilous ants. *Oecologia*, **80**, 145-7.

Higashi, M., Abe, T and Burns, T. (1992) Carbon-nitrogen balance and termite ecology. *Proceedings of the Royal Society of London*, **B 249**, 303-8.

Higashi, S. and Peeters, C.P. (1990) Worker polymorphism and nest structure in *Myrmecia brevinoda* Forel (Hymenoptera: Formicidae). *Journal of the Australian Entomological Society*, **29**, 327-31.

Higashida, S. and Takao, K. (1985) Seasonal changes in microbial activities in the surface soil of a grassland. *Soil Science and Plant Nutrition*, **31** (4), 647-51.

Hill, G.F. (1942) *Termites (Isoptera) from the Australian Region*, Council for Scientific and Industrial Research, Melbourne.

Hill, N.M. and Patriquin, D.G. (1990) Evidence for the involvement of *Azospirillum brasilense* and *Helicomyces roseus* in the aerobic nitrogen-fixing cellulolytic system from sugarcane litter. *Soil Biology & Biochemistry*, **22**, 313-9.

Hiltner, R. (1904) Uber neuere Erfahrungen und Probleme auf dem Gebiet der Bodenbakteriologie und unter besonderer Berücksichtigung der Gründüngung und Brache. *Arb. Deut. Landw. Ges.*, **98**, 59-78.

Hindwood, K.A. (1959) The nesting of birds in the nests of social insects. *Emu*, **59**, 1-36.

Hoagland, R.E. and Williams, R.D. (1985) The influence of secondary plant compounds on the associations of soil microorganisms and plant roots, in *The Chemistry of Allelopathy, Biochemical Interactions among Plants* (ed. A.C. Thompson), American Chemical Society, Washington D.C., pp. 301-25.

Hoeksema, K.J. and Jongerius, A. (1959) On the influence of earthworms on the soil structure in mulched orchards, in *Proceedings of the International Symposium on Soil Structure*, Rijkslandbouwhogeschool, Bodemkundig Laboratorium, Ghent, pp. 188-94.

Hofer, R.M. (1996) Root hairs, in *Plant Roots* (ed. Y. Waisel), Marcel Dekker, New York, pp. 111-26.

Hogg, B. and Hudson, H.J. (1966) Microfungi of the leaves of *Fagus sylvatica*. I. The microfungal succession. *Transactions of the British Mycology Society*, **49**, 185-92.

Holden, J. (1975) Use of nuclear staining to assess rates of cell death in cortices of cereal roots. *Soil Biology & Biochemistry*, **7**, 333-4.

Holdridge, L.R. (1947) Determination of world plant formations from simple climatic data. *Science*, **105**, 367-8.

Hole, F.D. (1961) A classification of pedoturbations and some other processes and factors of soil formation in relation to isotropism and anisotropism. *Soil Science*, **91**, 375-7.

Holland, E.A. and Lamarque, J.F. (1997) Modeling bio-atmospheric coupling of the nitrogen cycle through NO_x emissions and NO_y deposition. *Nutrient Cycling in Agroecosystems*, **48**, 7-24.

Holmgren, G.G.S. (1988) The point representation of soil. *Soil Science Society of America, Proceedings*,

52, 712-6.

Holt, J.A. and Easey, J.F. (1985) Polycalic colonies of some mound-building termites (Isoptera: Termitidae) in northeastern Australia. *Insectes Sociaux*, **32**, 61-9.

Holt, J.A. (1987) Carbon mineralization in semi-arid northeastern Australia: the role of termites. *Journal of Tropical Ecology*, **3**, 255-63.

Holt, J.A. (1995) Termites and mound dwelling microorganisms: a symbiosis promoting organic matter mineralization and nutrient cycling, in *The Termite-Symbionts System, Programme Report* (ed. T. Abe), Center for Ecological Research, Kyoto University.

Holt, J.A. (1996) Mound-building termites and soil microbial biomass. An interaction influencing termite abundance. *Insectes Sociaux*, **43**, 427-34.

Hoogerkamp, M., Rogaar, H. and Eijsackers, H.J.P. (1983) Effect of earthworms on grassland on recently reclaimed polder soils in the Netherlands, in *Earthworm Ecology: from Darwin to Vermiculture* (ed. J.E. Satchell), Chapman & Hall, London, pp. 85-105.

Hopp, H. and Hopkins, H.T. (1946) Earthworms as a factor in the formation of water-stable aggregates. *Journal of Soil and Water Conservation*, **1**, 11-3.

Hopp, H. (1973) *What Every Gardener Should Know about Earthworms*, Garden Way Pub. Co., Vermont.

Hota, P.K., Pani, S.C., Senapati, B.K. *et al.* (1988) Population dynamics and vertical distribution of soil nematodes in tropical irrigated upland and lowland rice fields. *Pedobiologia*, **32**, 213-9.

House, G.J. and Parmelee, R.W. (1985) Comparison of soil arthropods and earthworms from conventional and no-tillage agroecosystems. *Soil and Tillage Research*, **5**, 351-60.

Howe, H.F. (1984) Constraints on the evolution of mutualisms. *American Naturalist*, **123**, 764-777.

Hölldobler, B. (1983) Territorial behaviour in the green tree ant *(Oecophylla smaragdina)*. *Biotropica*, **15**, 241-50.

Hölldobler, B. and Wilson, E.O. (1990) *The Ants*, Harvard University Press, Harvard, Massachussets.

Huang, P.M. and Violante, A. (1986) Influence of organic acids on crystallization and surface properties of precipitation products of aluminium, in *Interactions of Soil Minerals with Natural Organics and Microbes* (eds P.M. Huang and M. Schnitzer), Soil Science Society of America, Madison, Wisconsin, pp. 429-51.

Hubac, C. (1978) Evolution des sucres solubles et de l'amidon au cours de l'assèchement, puis de la réhumidification, chez les plantes plus ou moins résistantes à la sécheresse (*Carex* et *Gossypium*). Comparaison entre les parties aériennes et souterraines, in *Proceedings of the Symposium on Root Physiology and Symbiosis* (eds A. Riedacker and J. Gagnaire-Michard), INRA, Nancy, pp.31-43.

Huhta, V., Karpinnen, E., Nurminen, M. *et al.* (1969) Effect of sylvicultural practices upon arthropod, annelid and nematode populations in a coniferous forest soil. *Acta Zoologica Fennici*, **4**, 87-143.

Huhta, V. (1979) Effects of liming and deciduous litter on earthworm (Lumbricidae) populations of a Spruce forest, with an inoculation experiment on *Allolobophora caliginosa*. *Pedobiologia*, **19**, 340-5.

Huhta, V., Hyvönen, R., Kaasalainen, P. *et al.* (1986) Soil fauna of Finnish coniferous forests. *Annales Zoologica Fennici*, **23**, 345-60.

Hulugalle, N.R. and Lal, R. (1986) Root growth of maize in a compacted gravelly tropical alfisol as affected by rotation with a woody perennial. *Field Crops Research*, **13**, 33-44.

Humbel, F.X. (1978) Caractérisation par des mesures physiques hydriques et d'enracinement de sols de Guyane française à dynamique de l'eau superficielle. *Science du Sol, Bulletins de l'AFES*, **2**, 83-4.

Humphreys, G.S. (1981) The rate of ant mounding and earthworm casting near Sydney, New South Wales. *Search*, **12** (5), 129-30.

Humphreys, G.S. (1994) Bioturbation, biofabrics and the biomantle: an example from the Sydney Basin. *Micromorphology in Soil: Studies in management and Genesis* (eds A.J. Ringrose-Voase and G. Humphreys), Amsterdam, pp. 787-798.

Hunt, J. (1974) Temporal activity patterns in two competing ant species (Hymenoptera: Formicidae). *Psyche*, **81**, 237-42.

Hunt, H.W., Coleman, D.C., Ingham, E.R. *et al.* (1987) The detrital food web in a shortgrass prairie. *Biology and Fertility of Soils*, **3**, 57-68.

Hurpin, B. (1962) Super-Famille des Scàrabeoidea, in *Entomologie appliquée à l'Agriculture. Tome I. Coléoptères* (ed. A.S. Balachowsky), Masson, Paris, pp. 24-204.

Huss-Danell, K. (1997) Actinorhizal symbioses and their N_2 fixation. *New Phytologist*, **136**, 375-405.

Huttel, C. (1975) Root distribution and biomass in three Ivory Coast rain forest plots, in *Tropical Ecological Systems. Trends in Terrestrial and Aquatic Research* (eds F.B. Golley and E. Medina), Springer-Verlag, New York, pp. 123-30.

Ingham, R.E. (1981) *The role of soil nematodes in ecosystem regulation*, Ph.D. thesis, Colorado State University.

Ingham, E.R. and Klein, D.A. (1984) Soil fungi: measurement of hyphal length. *Soil Biology & Biochemistry*, **16**, 279-80.

Ingham, E.R., Trofymow, J.A., Ames, R.N. *et al.* (1986a) Trophic interactions and nitrogen cycling in a semi-arid grassland soil. I. Seasonal dynamics of the natural populations, their interactions and effects on nitrogen cycling. *Journal of Applied Ecology*, **23** (2), 597-614.

Ingham, E.R., Cambardella, C. and Coleman, D.C. (1986b) Manipulation of bacteria, fungi and protozoa by biocides in Lodgepole Pine forest soil microcosms: effects on organism interactions and nitrogen mineralization. *Canadian Journal of Soil Science*, **66**, 261-72.

Ingham, E.R., Coleman, D.C. and Moore, J.C. (1989) An analysis of food-web structure and function in a shortgrass prairie, a mountain meadow, and a Lodgepole Pine forest. *Biology and Fertility of Soils*, **8**, 29-37.

Insam, H. (1990) Are the soil microbial biomass and basal respiration governed by the climatic regime? *Soil Biology & Biochemistry*, **22** (4), 525-32.

Irons, J.G., Oswood, M.W., Stout, R.J. *et al.* (1994) Latitudinal patterns in leaf litter breakdown: is Temperature really important? *Freshwater Biology*, **32** (2), 401-11.

ISSS Working group RB (1998a) *World Reference Base For Soil Resources*. FAO, Rome.

ISSS Working group RB (1998b) *World Reference Base For Soil Resources: Introduction*. (eds J.A. Deckers, F.O. Nachtergaele and O. Spaargaren), ISSS-ISRIC-FAO, Acco, Leuwen.

Isbell, R.F. and Smith, G.M. (1976) Some properties of red, yellow and grey massive earths in north Queensland. *CSIRO Australia, Division of Soils, Technical Paper*, **30**.

Isbell, R.F., Stephenson, P.J. and Murtha, G.G. (1976) Red basaltic soils in north Queensland. 1. Environment, morphology, particle size characteristics and clay mineralogy. *CSIRO Australia, Division of Soils, Technical Paper*, **28**, 1-22.

Isbell, R.F. (1990) Principles, needs and applications of soil classification systems, in *Transactions of the 14th International Congress of Soil Science*, ISSS, Kyoto, Japan, pp. 4-9.

Isbell, R.F. (1996) *The Australian Soil Classification*, CSIRO Publishing, Melbourne.

Jackson, R.B. and Caldwell, M.M. (1993) Geostatistical patterns of soil heterogeneity around individual perennial plants. *Journal of Ecology*, **81**, 683-92.

Jaillard, B. (1983) Pédologie - Mise en évidence de la calcitisation des cellules corticales de racines de Graminées en milieu carbonaté. *Comptes Rendus de l'Académie des Sciences, Paris*, **297**, 293-6.

Jaillard, B. and Hinsinger, P. (1993) Alimentation minérale des végétaux dans le sol. *Techniques Agricoles*, **1210**, 1-13.

Jaillard, B., Ruiz, L. and Arvieu, J.C. (1996) pH mapping in transparent gel using color indicator video-densitometry. *Plant and Soil*, **183**, 85-95.

Jalali, B.L. and Suryanaryana, D. (1971) Shift in the carbohydrate spectrum of root exudates of peas, cotton and barley. *Plant and Soil*, **34**, 261-7.

James, S.W. (1991) Soil nitrogen, phosphorus, and organic matter processing by earthworms in tallgrass prairie. *Ecology*, **72** (6), 2101-9.

Janeau, J.L. and Valentin, C. (1987) Relation entre les termitières de *Trinervitermes sp.* et la surface du sol: réorganisation, ruissellement et érosion. *Revue d'Ecologie et de Biologie du Sol*, **24**, 637-48.

Jansson, P.E. and Berg, B. (1985) Temporal variation of litter decomposition in relation to simulated soil climate. Long-term decomposition in a Scots Pine forest. V. *Canadian Journal of Botany*, **63** (6), 1008-16.

Janzen, D.H. (1974) Tropical blackwater rivers, animals, and mass fruiting by *Dipterocarpaceae*. *Biotropica*, **6** (2), 69-103.

Janzen, D.H. (1985) The natural history of mutualisms, in *The Biology of Mutualism* (ed. D.H. Boucher), Croom Helm, London, pp. 40-99.

Jeanson, C. (1979) Structuration du sol par la faune terricole, incidences sur les concentrations organo-minérales, in *Migrations Organo-Minérales dans les Sols Tempérés*, CNRS, Nancy, France, pp. 113-23.

Jenik, J. (1978) Roots and root systems in tropical tree morphologic and ecologic aspects, in *Tropical Trees as Living Systems* (eds P.B. Tomlinson and M.H. Zimmerman), Cambridge University Press, Cambridge, pp. 323-49.

Jenkinson, D.S. (1966) The priming action. *Journal of Applied Radiation Isotopes*, **Supplement**, 199-208.

Jenkinson, D.S. and Powlson, D.S. (1976) The effects of biocidal treatements on metabolism in soil. V. A method of measuring soil biomass. *Soil Biology & Biochemistry*, **8**, 209-13.

Jenkinson, D.S. and Ayanaba, A. (1977) Decomposition of ^{14}C labelled plant material under tropical conditions. *Soil Science Society of America, Journal*, **41**, 912-5.

Jenkinson, D.S. and Rayner, J.H. (1977) The turnover of soil organic matter in some of the Rothamsted classical experiments. *Soil Science*, **123**, 298-305.

Jenkinson, D.S. and Ladd, J.N. (1981) Microbial biomass in soil: measurement and turnover, in *Soil Biochemistry* (eds E.A. Paul and J.N. Ladd), Marcel Dekker, New York, pp. 415-71.

Jenkinson, D.S., Fox, R.H. and Rayner, J.H. (1985) Interactions between fertilizer nitrogen and soil nitrogen - the so-called "priming-effect". *Journal of Soil Science*, **36**, pp. 425-44.

Jenkinson, D.S., Hart, P.B.S., Rayner, J.H. *et al.* (1987) Modelling the turnover of organic matter in a long-term experiment of Rothamsted. *INTECOL Bulletin*, **15**, 1-8.

Jenkinson, D.S. (1988) Determination of microbial biomass carbon and nitrogen in soil, in *Advances in Nitrogen Cycling in Agricultural Ecosystems* (ed. J.R. Wilson), CAB International, Wallingford, pp. 368-86.

Jenkinson, D.S. (1990) Introduction to the global nitrogen cycle. *Soil Use and Management*, **6**, 56-63.

Jennings, T.J. and Barkham, J.P. (1979) Litter decomposition by slugs in mixed deciduous woodland. *Holarctic Ecology*, **2**, 21-9.

Jennings, D.H. (1989) Some perspectives on nitrogen and phosphorus metabolism in fungi, in *Nitrogen, Phosphorus and Sulphur Utilization by Fungi* (eds L. Boddy, R. Marchant and D.J. Read), Cambridge University Press, Cambridge, pp. 1-31.

Jennings, D.H. (1995) *The Physiology of Fungal Nutrition*, Cambridge University Press, Cambridge.

Jenny, H. (1941) *Factors of Soil Formation - A System of Quantitative Pedology*, McGraw-Hill, New York.

Jenny, H., Gessel, S.P. and Bingham, F.T. (1949) Comparative study of decomposition rates of organic matter in temperate and tropical regions. *Soil Science*, **68**, 419-23.

Jenny, H. (1980) *The Soil Resource*, Springer Verlag, New York.

Jimenez, J.J., Moreno, A.G., Lavelle, P. *et al.* (1998b) Population dynamics and adaptative strategies of *Martiodrilus carimaguensis* (Oligochaeta, Glossoscolecidae), a native species from the well-drained savannas of Colombia. *Applied Soil Ecology*., **9**: 153-60

Jimenez, J.J., Moreno, A.G., Decaëns, T. *et al.* (1998a) Earthworm communities in native savannas and man made pastures of the Eastern Plains of Colombia. *Biology and Fertility of Soils*, **28**, 101-110.

Johansson, M.B. (1994) Decomposition rates of Scots Pine needle litter related to site properties, litter quality, and climate. *Canadian Journal of Forest Research*, **24** (9), 1771-81.

Johnson, D.W., Cole, D.W., Gessel, S.P. *et al.* (1977) Carbonic acid leaching in a tropical, temperate, subalpine, and northern forest soil. *Arctic and Alpine Research*, **9**, 329-43.

Johnson, D.L., Watson-Stegner, D., Johnson, D.N. *et al.* (1987) Proisotropic and proanisotropic processes of pedoturbation. *Soil Science*, **143**, 278-92.

Johnson, D.L. (1990) Biomantle evolution and the redistribution of earth materials and artifacts. *Soil Science*, **149** (2), 84-102.

Johnson, C. (1994) Fruiting of hypogeous fungi in dry sclerophyll forest in Tasmania, Australia: seasonal variation and annual production. *Mycological Research*, **98**, 1173-82.

Johnson, D.L. and Hole, F.D. (1994) Soil formation theory: a summary of its principal impacts on geography, geomorphology, soil-geomorphology, quaternary geology and paleopedology, in *Factors of Soil Formation: A Fiftieth Anniversary Retrospective* (eds R.G. Admunson, J.F. Harden and M.J. Singer), Soil Science Society of America, Madison, Wisconsin, pp. 111-26.

Jones, P.C.T. and Mollison, J.E. (1948) A technique for the quantitative estimation of soil microorganisms. *Journal of General Microbiology*, **2**, 54-69.

Jones, F.G.W. (1959) Ecological relationships of nematodes, in *Plant Pathology, Problems and Progress* (eds C.S. Holton, G.W. Fischer, R.W. Fulton *et al.*), University Wisconsin Press, Madison, Wisconsin, pp. 395-411.

Jones, J.A. (1990) Termites, soil fertility and carbon cycling in dry tropical Africa: an hypothesis. *Journal of Tropical Ecology*, **6**, 291-305.

Jones, C.G., Lawton, J.H. and Shachak, M. (1994) Organisms as ecosystem engineers. *Oikos*, **69** (3), 373-86.

Jones, D.L. and Darrah, P.R. (1994) Simple method for ^{14}C-labelling root material for use in root decomposition studies. *Communications in Soil Science and Plant Analysis*, **25** (15/16), 2737-43.

Jones, D.T. and Gaythorne-Hardy, F. (1995) Foraging activity of the processional termite *Hospitalitermes hospitalis* (Termitidae: Nasutitermitinae) in the rainforest of Brunei, north-west Borneo. *Insectes Sociaux*, **42**, 359-69.

Jones, C.G., Lawton, J.H. and Shachak, M. (1997) Positive and negative effects of organisms as physical ecosystem engineers. *Ecology*, **78** (7), 1946-57.

Jonkman, J.C.M. (1980a) The external and internal structure and growth of nests of the leaf-cutting ant *Atta vollenweideri* Forel, 1983 (Hym.: Formicidae). Part I. *Zeitschrift für Angewandte Entomologie*, **89**, 158-73.

Jonkman, J.C.M. (1980b) The external and internal structure and growth of nests of the leaf-cutting ant *Atta vollenweideri* Forel, 1983 (Hym.: Formicidae). Part II. *Zeitschrift für Angewandte Entomologie*, **89**, 217-46.

Joose, N.G. (1970) The formation and biological significance of aggregations in the distribution of Collembola. *Netherlands Journal of Zoology*, **20** (3), 299-314.

Jordan, C.F. (1982) The nutrient balance of an Amazonian rain forest. *Ecology*, **63**, 647-57.

Jordan, C.F. (1989) *An Amazonian Rain Forest*, UNESCO/The Parthenon Publishing Group, Paris/Carnford.

Jordan, E.D., Thomas, E., Hsieh, C.Y. *et al.* (1993) Volatiles from litter and soil associated with *Ceratiola ericoides*. *Phytochemistry*, **33** (2), 299-302.

Joschko, M., Söchtig, W. and Larink, O. (1992) Functional relationships between earthworm burrows and soil water movement in column experiments. *Soil Biology & Biochemistry*, **24** (12), 1545-7.

Josens, G. (1983) The soil fauna of tropical savannas. III. The termites, in *Tropical Savannas* (ed. F. Bourlière), Elsevier, New York, pp. 505-524.

Judas, M. (1992) Gut content analysis of earthworms (Lumbricidae) in a Beechwood. *Soil Biology & Biochemistry*, **24** (12), 1413-17.

Juste, C. (1965) *Contribution à la dynamique de l'aluminium dans les sols acides du Sud-Ouest atlantique*, thèse de Docteur-Ingénieur, Université Nancy I.

Kabata-Pendias, A. and Pendias, H. (1992) *Trace Elements in Soils and Plants*, CRC Press, Boca Raton, Florida.

Kaczmarek (1984) A comparison of bacterial and fungal biomass in several cultivated soils. *Acta Microbiologica Polonica*, **33** (3/4), 239-47.

Kahn, J.F. (1984) *L'espace forestier racinaire*, thèse d'Etat, Université de Montpellier.

Kalela, E.K. (1957) Uber veränderungen in den wurzelverhältnissen der kifernbeständer in laufe der vegetationsperiode. *Acta Forestalia Fennica*, **65**, 5-42.

Kamill, B.W., Steinberg, Y. and Whitford, W.G. (1985) Soil microarthropods from the Chihuhuan Desert of New Mexico. *Journal of Zoology, London*, **205**, 273-86.

Kanyonyo ka Kajondo, J.B. (1984) *Ecologie alimentaire du ver de terre détritivore* Millsonia lamtoiana *(Acanthodrilidae, Oligochètes) dans la savane de Lamto (Côte d'Ivoire)*, thèse de 3e cycle, Université Paris VI.

Kellogg, C.E. (1974) Soil genesis, classification and cartography. *Geoderma*, **12**, 347-62.

Kemp, P.B. (1955) The termites of north-eastern Tanganyika: their distribution and biology. *Bulletin of Entomological Research*, **46**, 113-35.

Kendrick, W.B. (1992) *The Fifth Kingdom*, 2nd Edition, Mycologue Publications, Waterloo, Ontario.

Keogh, R.G. (1979) Lumbricid earthworm activities and nutrient cycling in pasture ecosystem, in *Proceedings of the 2nd Australasian Conference on Grasslands and Invertebrate Ecology* (eds T.K. Crosby and R.P. Pottinger), Government Printer, Wellington, pp. 49-51.

Kerry, B. (1980) Biocontrol: fungal parasites of female cyst nematodes. *Journal of Nematology*, **12**, 253-9.

Kethley, J. (1990) Acarina: Prostigmata (Actinedida), in *Soil Biology Guide* (ed. D.L. Dindal), John Wiley & Sons, New York, pp. 667-756.

Khalaf, F. and Al-Hashash, M. (1983) Aeolian sedimentation in the north-western part of the Arabian Gulf. *Journal of Arid Environments*, **6**, 319-32.

Kiffer, E., Kilbertus, G. and Proth, J. (1977) Rhizoplan et mycorrhizoplan de *Picea abies* (L.) Karst. *Bulletin de l'Académie et de la Société Lorraines des Sciences*, **16** (4), 125-36.

Kilbertus, G., Proth, J. and Mangenot, F. (1977) Sur la répartition et la survivance des microorganismes du sol. Etude électronique. *Bulletin de l'Académie et de la Société Lorraines des Sciences*, **14** (3), 93-104.

Kilbertus, G. and Vannier, G. (1979) Microbial analysis and weight estimation of feces produced by four sympatric Collembola species in forest litter. *Revue d'Ecologie et de Biologie du Sol*, **16** (2), 169-80.

Kilbertus, G. (1980) Etude des microhabitats contenus dans les agrégats du sol, leur relation avec la biomasse bactérienne et la taille des procaryotes présents. *Revue d'Ecologie et de Biologie du Sol*, **17** (4), 543-57.

Kilbertus, G. and Vannier, G. (1981) Relations microflore-microfaune dans la grotte de Ste Catherine (Pyrénées ariégeoises). II. Le régime alimentaire de *Tomocerus minor* (Lubbock) et *Tomocerus problematicus* Cassagnau (Insectes Collemboles). *Revue d'Ecologie et de Biologie du Sol*, **18** (3), 319-38.

Kilbertus, G. and Mangenot, F. (1981) Origine et composition des matières organiques impliquées dans la formation des agrégats organominéraux. *Colloques Internationaux du CNRS*, **303**, 103-11.

Kilbertus, G., Schwartz, R. and Alberti, G. (1982) La répartition quantitative des microorganismes dans les sols de forêt (chênes, pins). Indice d'activité microbiologique. *Revue d'Ecologie et de Biologie du Sol*, **19** (4), 513-23.

Kinden, D.A. and Brown, M.F. (1976) Electron microscopy of vesicular-arbuscular mycorrhizae of Yellow

Poplar. IV. Host-endophyte interactions during arbuscular deterioration. *Canadian Journal of Microbiology*, **22**, 64-75.

King, R.L. (1977) The plant ecology of ant-hills in calcareous grasslands. I. Patterns of species in relation to ant-hills in southern England. *Journal of Ecology*, **65**, 235-56.

Kirk, T.K. and Moore, W.E. (1972) Removing lignin in wood with white rot fungi and digestibility of resulting wood. *Wood Fiber*, **4**, 72-9.

Kirk, T.K. and Farrell, R.L. (1987) Enzymatic "combustion": the microbial degradation of lignin. *Annual Revue of Microbiology*, **41**, 465-505.

Kitazawa, T. (1971) Biological regionality of the soil fauna and its function in forest ecosystem types, in *UNESCO, Productivity of Forest Ecosystem, Proceedings of the Brussels Symposium 1969 (Ecology and Conservation 4)*, pp. 485-98.

Kitazawa, T. (ed.) (1977) *Ecosystem Analysis of the Subalpine Coniferous Forest of the Shiga Yama IBP area, Central Japan*, University of Tokyo Press, Tokyo.

Kjöller, A. and Struwe, S. (1982) Microfungi in ecosystems: fungal occurence and activity in litter and soil. *Oikos*, **39** (3), 389-422.

Kjöller, A. and Vestberg, K. (1985) Bacterial dynamics during decomposition of Alder litter. *Soil Biology & Biochemistry*, **12**, 425-31.

Kladivko, E.J., Mackay, A.D. and Bradford, J.M. (1986) Earthworms as a factor in the reduction of soil crusting. *Soil Science Society of America, Journal*, **50** (1), 191-6.

Klekowski, R.Z., Wasilewska, L. and Paplinska, E. (1972) Oxygen consumption by soil-inhabiting nematodes. *Nematologica*, **18**, 391-403.

Klinge, H. (1975) Bilanzierung von hauptnährstoffen im Ökosystem Tropischer Regenwald (Manaus) - vorlaüfige Daten. *Biogeographica*, **7**, 59-76.

Klironomos, J.N. and Kendrick, W.B. (1996) Palatibility of microfungi to soil arthropods in relation to the functioning of arbuscular mycorrhizae. *Biology and Fertility of Soils*, **21**, 43-52.

Knight, D., Elliot, P.W., Anderson, J.M. *et al.* (1992) The role of earthworms in managed, permanent pastures in Devon, England. *Soil Biology & Biochemistry*, **24**, 1511-7.

Knowles, R. (1978) Free-living bacteria, in *Limitations and Potentials for Biological Nitrogen Fixation in the Tropics* (eds J. Döbereiner, R.H. Burris, A. Hollaender *et al.*), Plenum Publishing Co., New York, pp. 25-40.

Kobus, J. and Pacewiczowa, T. (1966) Effect of clay minerals on the biological activity of light soils. *Roczniki Gleboznawcze*, **12** (1), 53-65.

Kollmannsperger, F. (1956) Lumbricidae of humid and arid regions and their effect on soil fertility, in Proceedings of the 6th International Congress of Soil Science, Paris, Report C, 293-7.

Kooyman, C. and Onck, R.F.M. (1987) Distribution of termite (Isoptera) species in southwestern Kenya in relation to land use and the morphology of their galleries. *Biology and Fertility of Soils*, **3**, 69-73.

Kotyk, A., Fischer-Schliebs, E. and Lüttge, U. (1991) Medium acidification by maize root tips and its inhibition by heavy water. *Botanica Acta*, **104** (6), 433-8.

Kozlovskaja, L.S. (1971) Der Einfluss der Wirbellosen auf die Tägigkeit der Mikroorganismen in Torfböden, in *Organismes du Sol et Production Primaire . IV. Colloquium Pedobiologiae* (eds J. d'Aguilar, C. Athias-Henriot, A. Bessard *et al.*), INRA, Paris, pp. 81-8.

Krantz, G.W. and Ainscough, B.D. (1990) Acarina: Mesostigmata (Gamasida), in *Soil Biology Guide* (ed. D.L. Dindal), John Wiley & Sons, New York, pp. 583-665.

Kretzschmar, A. (1982) Description des galeries de vers de terre et variations saisonnières des réseaux (observations en conditions naturelles). *Revue d'Ecologie et de Biologie du Sol*, **19**, 579-591.

Kretzschmar, A. (1987) Caractérisation microscopique de l'activité des lombriciens endogés, in *Micromorphologie des Sols* (eds N. Fedoroff, L.M. Bresson and M.A. Courty), AFES, Paris, pp. 325-31.

Kretzschmar, A. (1989) *Galeries de lombriciens en réseaux: structures fonctionnelles et signatures comportementales*, thèse de l'Université Paris Sud-Orsay.

Kretzschmar, A. and Monestiez, P. (1992) Physical control of soil biological activity due to endogeic earthworm behaviour. *Soil Biology & Biochemistry*, **24** (12), 1609-14.

Kreutzer, W.A. (1972) *Fusarium spp.* as colonists and potential pathogens in root zones of grassland plants. *Phytopathology*, **62**, 1066-70.

Kristufek, V., Ravasz, K. and Pizl, V. (1992) Changes in densities of bacteria and microfungi during gut transit in *Lumbricus terrestris* and *Aporrectodea caliginosa* (Oligochaeta: Lumbricidae). *Soil Biology & Biochemistry*, **24** (12), 1499-500.

Kristufek, V., Tajovsky, K. and Pizl, V. (1994) Ultrastructural analysis of the intestinal content of earthworm *Lumbricus rubellus* Hoffm. (Annelida, Lumbricidae). *Acta Microbiologica et Immunologica Hungarica*, **41** (3), 283-90.

Kubiena, W.L. (1953) *The Soils of Europe*, CSIC/Thomas Murby & Co, Madrid/London.

Kucera, C.L., Dahlman, R.C. and Koelling, M.R. (1967) Total net productivity and turnover on an energy basis for tallgrass prairie. *Ecology*, **48**, 536-41.

Kuikman, P.J. and van Veen, J.A. (1989) The impact of Protozoa on the availability of bacterial nitrogen to plants. *Biology and Fertility of Soils*, **8**, 13-8.

Kumada, K. (1987) *Chemistry of Soil Organic Matter*, Japan Scientific Societies Press, Tokyo.

Kunicki-Goldfinger, W., Drozanski, W. and Baszczak, D. (1967) Bactéries servant de nourriture aux amibes du sol. *Acta Microbiologica Polonica*, **6**, 331-4.

Kurcheva, G.F. (1967) Influence of soil invertebrates on litter decomposition rates in a forest near Koursk. Experimental study (in russian). *Pedobiologia*, **7** (2/3), 228-38.

Kusnezov, N. (1957) Numbers of species of ants in fauna of different latitudes. *Evolution*, **11**, 298-9.

Kuzmin, L.L. (1976) Free-living nematodes in the tundra of western Taimyr. *Oikos*, **27**, 501-5.

Kühnelt, W. (1961) *Soil Biology*, Faber and Faber, London.

l'Huillier, L. and Edighoffer, S. (1996) Extractability of nickel and its concentration in cultivated plants in nickel rich ultramafic soils of New Caledonia. *Plant and Soil*, **186**, 255-64.

Laffan, M.D. (1988) *Soils and Land Use on the Atherton Tableland, North Queensland*, CSIRO Division of Soils, Soils and Land Use Serie, No. 61.

Lagerlöf, J. (1987) *Dynamics and activity of microarthropods and enchytraeids in four cropping systems*, Ph.D. thesis, Swedish University of Agricultural Sciences.

Lagerlöf, J. and Andrén, O. (1989) Abundance and activity of soil mites (Acari) in four cropping systems. *Pedobiologia*, **32**, 129-45.

Lagerlöf, J. and Andrén, O. (1991) Abundance and activity of Collembola, Protura and Diplura (Insecta, Apterygota) in four cropping systems. *Pedobiologia*, **35**, 335-350.

Lai, P.Y., Tamashiro, M., Yates, J.R. *et al.* (1983) Living plants in Hawaii attacked by *Coptotermes formosanus*. *Proceedings of the Hawaiian Entomological Society*, **24**, 283-6.

Laishram, I.D. and Yadava, P.S. (1988) Lignin and nitrogen in the decomposition of leaf litter in a subtropical forest ecosystem at Shiroy hills in north-eastern India. *Plant and Soil*, **106**, 59-64.

Lakhani, K.H. and Satchell, J.E. (1970) Production by *Lumbricus terrestris* L. *Journal of Animal Ecology*, **39**, 473-92.

Lal, R. (1984) Soil erosion from tropical arable land and its control. *Advances in Agronomy*, **37**, 187-248.

Lamotte, M. (1989) Place des animaux détritivores et des microorganismes décomposeurs dans les flux d'énergie des savanes africaines. *Pedobiologia*, **33**, 17-35.

Lamotte, M. (1975) The structure and function of a tropical savannah ecosystem, in *Tropical Ecological Ecosystems: Trends in Terrestrial and Aquatic Research* (eds F.B. Golley and E. Medina), Springer Verlag, New York, pp. 179-222.

Lamparski, F., Kobel-Lamparski, A. and Kaffenberger, R. (1987) The burrows of *Lumbricus badensis* and *Lumbricus polyphemus*, in *On Earthworms* (eds A.M. Bonvicini Pagliai and P. Omodeo), Mucchi, Modena, pp. 131-40.

Lattaud, C., Locati, S., Mora, P. *et al.* (1997a) Origin and activities of glycolytic enzymes in the gut of the tropical geophagous earthworm *Millsonia anomala* from Lamto (Côte d'Ivoire). *Pedobiologia*, **41** (4), 242-51.

Lattaud, C., Zhang, B.G., Locati, S. *et al.* (1997b) Activities of the digestive enzymes in the gut and in tissue culture of a tropical geophagous earthworm, *Polypheretima elongata* (Megascolecidae). *Soil Biology & Biochemistry*, **29** (3/4), 335-9.

Latter, P.M. and Howson, G. (1978) Studies on the microfauna of blanket bog with particular reference to Enchytraeidae. II. Growth and survival of *Cognettia sphagnetorum* on various substrates. *Journal of Animal Ecology*, **47**, 425-48.

Laudelout, H. (1990) *La Jachère Forestière sous les Tropiques Humides*, UCL, Chimay.

Lavelle, P. (1971) Etude préliminaire de la nutrition d'un ver de terre africain *Millsonia anomala* (Acanthodrilidae, Oligochètes). *Annales de Zoologie et Ecologie Animale*, **Hors-série**, 133-45.

Lavelle, P. (1975) Consommation annuelle de terre par une population naturelle de vers de terre (*Millsonia anomala* Omodeo, Acanthodrilidae-Oligochetes) dans la savane de Lamto (Côte d'Ivoire). *Revue d'Ecologie et de Biologie du Sol*, **12** (1), 11-24.

Lavelle, P. (1978) *Les vers de terre de la savane de Lamto (Côte d'Ivoire): peuplements, populations et fonctions dans l'écosystème*, thèse d'Etat, Université Paris VI/ENS.

Lavelle, P., Maury, M.E. and Serrano, V. (1981) Estudio cuantitativo de la fauna del suelo en la region de Laguna Verde, Vera-Cruz. Epoca de Lluvias. *Publicación del Instituto de Ecologia (México)*, **6**, 75-105.

Lavelle, P. (1981) Stratégies de reproduction chez les vers de terre. *Acta Œcologica, Œcologia Generalis*, **2**, 117-33.

Lavelle, P. and Meyer, J.A. (1983) "Allez-les-Vers", a simulation model of dynamics and effect on soil of populations of *Millsonia anomala* (Oligochæta, Megascolecidæ), in *New Trends in Soil Biology* (eds Ph. Lebrun, A.M. André, A. de Medts *et al.*), Dieu-Brichart, Ottignies, pp. 503-19,.

Lavelle, P., Zaidi, Z. and Schaefer, R. (1983a) Interactions between earthworms, soil organic matter and microflora in an African savanna soil, in *New Trends in Soil Biology* (eds Ph. Lebrun, A.M. André, A. de Medts *et al.*), Dieu-Brichart, Louvain-la-Neuve, pp. 253-9.

Lavelle, P., Rangel, P. and Kanyonyo, J. (1983b) Intestinal mucus production by two species of tropical earthworms: *Millsonia lamtoiana* (Megascolecidae) and *Pontoscolex corethrurus* (Glossoscolecidae), in *New Trends in Soil Biology* (eds Ph. Lebrun, H.M. André, A. de Medts *et al.*), Dieu-Brichart, Ottignies, pp. 405-10.

Lavelle, P. (1983a) The soil fauna of tropical savannas. II-The earthworms, in *Tropical Savannas* (ed. F. Bourlière), Elsevier Scientific Publishing Co, Amsterdam, pp. 485-504.

Lavelle, P. (1983b) *Agastrodrilus* Omodeo and Vaillaud, a genus of carnivorous earthworms from the Ivory Coast, in *Earthworm Ecology: from Darwin to Vermiculture* (ed. J.E. Satchell), Chapman & Hall, London, pp. 425-9.

Lavelle, P. (1983c) The structure of earthworm communities, in *Earthworm Ecology: from Darwin to Vermiculture* (ed. J.E. Satchell), Chapman & Hall, London, pp. 449-66.

Lavelle, P. and Kohlmann, B. (1984) Etude quantitative de la macrofaune du sol dans une forêt tropicale humide du Mexique (Bonampak, Chiapas). *Pedobiologia*, **27**, 377-93.

Lavelle, P. (1984) The soil system in the humid tropics. *Biology International*, **9**, 2-15.

Lavelle, P. (1985) Les systèmes d'interaction ressources-populations-environnement. *Bulletin d'Ecologie*, **16** (1), 1-8.

Lavelle, P. (1986) Associations mutualistes avec la microflore du sol et richesse spécifique sous les tropiques: l'hypothèse du premier maillon. *Comptes Rendus de l'Académie des Sciences, Paris*, **III 302** (1), 11-4.

Lavelle, P., Barois, I., Fragoso, C. *et al.* (1987) Adaptative strategies of *Pontoscolex corethrurus* (Glossoscolecidæ, Oligochæta), a peregrine geophagous earthworm of the humid tropics. *Biology and Fertility of Soils*, **5**, 188-94.

Lavelle, P. (1987) The importance of biological processes in productivity of soils in the humid tropics, in *Geophysiology of the Amazon* (eds R.E. Dickinson and J. Lovelock), John Wiley & Sons, New York, pp. 175-214.

Lavelle, P. (1988) Assessing the abundance and role of invertebrate communities in tropical soils: aims and methods. *Journal of African Zoology*, **102**, 275-83.

Lavelle, P. and Pashanasi, B. (1989) Soil macrofauna and land management in Peruvian Amazonia (Yurimaguas, Loreto). *Pedobiologia*, **33**, 283-91.

Lavelle, P., Schaefer, R. and Zaidi, Z. (1989) Soil ingestion and growth in *Millsonia anomala*, a tropical earthworm, as influenced by the quality of the organic matter ingested. *Pedobiologia*, **33**, 379-88.

Lavelle, P., Spain, A.V., Blanchart, E. *et al.* (1992a) The impact of soil fauna on the properties of soils in the humid tropics, in *Myths and Science of Soils of the Tropics* (eds R. Lal and P.A. Sanchez), Soil Science Society of America, Madison, Wisconsin, pp. 157-85.

Lavelle, P., Melendez, G., Pashanasi, B. *et al.* (1992b) Nitrogen mineralization and reorganization in casts of the geophagous tropical earthworm *Pontoscolex corethurus* (Glossoscolecidae). *Biology and Fertility of Soils*, **14**, 49-53.

Lavelle, P., Blanchart, E., Martin, A. *et al.* (1993) A hierarchical model for decomposition in terrestrial ecosystems. Application to soils in the humid tropics. *Biotropica*, **25** (2), 130-50.

Lavelle, P. and Gilot, C. (1994) Priming effects of macroorganisms on microflora: a key process of soil function?, in *Beyond the Biomass* (eds K. Ritz, J. Dighton and K.E. Giller), John Wiley & Sons, Chichester, pp. 173-80.

Lavelle, P., Lattaud, C., Trigo, D. *et al.* (1994a) Mutualism and biodiversity in soils. *Plant and Soil*, **170** (1), 23-33.

Lavelle, P., Dangerfield, M., Fragoso, C. *et al.* (1994b) The Relationship between soil macrofauna and tropical soil fertility, in *Tropical Soil Biology and Fertility* (eds M.J. Swift and P. Woomer), John Wiley-Sayce, New York, pp. 137-69.

Lavelle, P., Chauvel, A. and Fragoso, C. (1995) Faunal activity in acid soils, in *Plant Soil Interactions at Low pH* (eds R. A. Date, N.J. Grundon, G.E. Rayment *et al.*), Kluwer Academic Publishers, Dordrecht, pp. 201-11.

Lavelle, P. (1997) Faunal activities and soil processes: adaptive strategies that determine ecosystem function. *Advances in Ecological Research*, **27**, 93-132.

Lavelle, P., Pashanasi, B., Charpentier, F. *et al.* (1998) Large-scale effects of earthworms on soil organic matter and nutrient dynamics, in *Earthworm Ecology* (ed. C.A. Edwards), St. Lucie Press, Boca Raton, Florida, pp. 103-22.

Laverack, M.S. (1963) *The Physiology of Earthworms*, Pergamon Press, Oxford.

Le Tacon, F. (1985) Les mycorhizes: une coopération entre plantes et champignons. *La Recherche*, **166**, 624-32.

Leclère-Sonrier, E. (1977) *Etude de l'action de microorganismes de la rhizosphère sur la perte de produits carbonés par les racines de maïs*, Rapport de DEA, INP, Nancy.

Lee, K.E. and Wood, T.G. (1971a) Physical and chemical effects on soils of some Australian termites and their pedological significance. *Pedobiologia*, **11**, 376-409.

Lee, K.E. and Wood, T.G. (1971b) *Termites and Soils*, Academic Press, London.

Lee, K.E. (1985) *Earthworms: their Ecology and Relationships with Soils and Land Use*, Academic Press, Sydney.

Lee, K.E. and Foster, R.C. (1991) Soil fauna and soil structure. *Australian Journal of Soil Research*, **29**, 745-75.

Leep, N.W. (1992) Uptake and accumulation of metals in bacteria and fungi, in *Biogeochemistry of Trace Metals* (ed. D.C. Adriano), Lewis Publishers, Boca Raton, Florida, pp. 277-98.

Legay, B. and Schaefer, R. (1980) Echange respiratoire de divers sols: sa dépendance du régime hydrique, in *Comptes Rendus du 105ᵉ Congrès National des Sociétés Savantes*, Caen, pp. 169-85.

Legros, J.P. (1992) Soil of alpine mountains, in *Weathering, Soils and Paleosols*, Elsevier, Amsterdam, pp. 153-81.

Leigh, J.H., Halsall, D.M. and Holgate, M.D. (1995) The role of allelopathy in legume decline in pastures .1. Effects of pasture and crop residues on germination and survival of Subterranean Clover in the field and nursery. *Australian Journal of Agricultural Research*, **46** (1), 179-88.

Lelong, F. and Souchier, B. (1972) Nature et genèse des argiles dans les profils vosgiens de la séquence sols bruns acides-podzols sur granite. *Science de la Terre*, **17**, 353-79.

Lemée, G. (1978) La hêtraie naturelle de Fontainebleau, in *Problèmes d'Ecologie: Ecosystèmes Terrestres* (eds M. Lamotte and F. Bourlière), Masson, Paris, pp. 75-128.

Leneuf, N. (1959) *L'altération des granites calco-alcalins et des granodiorites en Côte d'Ivoire forestière et des sols qui en sont dérivés*, Thèse d'Etat, Université Paris VI.

Leneuf, N. and Aubert, G. (1960) Essai d'évaluation de la vitesse de ferrallitisation, in *Transactions of the 7th International Congress of Soil Science*, ISSS, Madison, Wisconsin, pp. 225-8.

Lensi, R., Domenach, A.M. and Abbadie, L. (1992) Field study of nitrification and denitrification in a wet savanna of West Africa (Lamto, Côte d'Ivoire). *Plant and Soil*, **147** (1),107-113.

Lenz, M. (1985) Is interspecific variability of lower termite neotenic number due to adaptive thresholds for neotenic elimination? Considerations from studies on *Porotermes adamsoni* (Froggatt) (Isoptera: Termopsidae), in *Caste Differentiation in Social Insects, Current Themes in Tropical Science* (eds J.A.L. Watson, B.M. Okot-Kotber and C. Noirot), Pergamon Press, Oxford, pp. 125-45.

Lenz, M. (1994) Food resources, colony growth and caste development in wood-feeding termites, in *Nourishment and Evolution in Insect Societies* (eds J.H. Hunt and C.A. Nalepa), Westview Press, Boulder, Colorado, pp. 159-209.

Lepage, M. (1972) Recherches écologiques sur une savane sahélienne du Ferlo septentrional, Sénégal: données préliminaires sur l'écologie des termites. *La Terre et la Vie*, **26**, 383-409.

Lepage, M. (1974) *Les termites d'une savane sahélienne (Ferlo septentrional, Sénégal): peuplement, populations, consommation*, thèse d'Etat,Université de Dijon.

Lepage, M.G. (1981a) L'impact des populations récoltantes de *Macrotermes michaelseni* (Sjöstedt) (Isoptera, Macrotermitinae) dans un écosystème semi-aride (Kajiado-Kenya). I.-L'activité de récolte et son déterminisme. *Insectes Sociaux*, **28** (3), 297-308.

Lepage, M.G. (1981b) L'impact des populations récoltantes de *Macrotermes michaelseni* (Sjöstedt) (Isoptera, Macrotermitinae) dans un écosystème semi-aride (Kajiado-Kenya). II.-La nourriture récoltée, comparaison avec les grands herbivores. *Insectes Sociaux*, **28**(4), 309-19.

Lepage, M. (1983) Structure et dynamique des peuplements de termites tropicaux. *Acta Œcologica, Œcologia Generalis*, **4** (1), 65-87.

Lepage, M., Abbadie, L. and Mariotti, A. (1993) Food habits of sympatric termite species (Isoptera, Macrotermitinae) as determined by stable carbon isotope analysis in a Guinean savanna (Lamto, Côte d'Ivoire). *Journal of Tropical Ecology*, **9**, 303-11.

Lepp, N.W. (1992) Uptake and accumulation of metals in bacteria and fungi, in *Biogeochemistry of Trace Metals* (ed.by D.C. Adriano), Lewis Publishers, Boca Raton, Florida, pp. 277-98.

Leprun, J.C. and Roy-Noël, J. (1976) Minérologie des argiles et répartition des nids épigés de deux espèces du *Macrotermes* au Sénégal occidentale (Presqu'île du Cap Vert). *Insectes Sociaux*, **23**, 535-47.

Leroy, C., Toutain, F. and Lavelle, P. (1992) Variations des caractéristiques de l'humus forestier d'un sol ferrallitique (Guyane) selon l'essence arborée considérée. Résultats préliminaires. *Cahiers ORSTOM, série Pédologie*, **27** (1), 37-48.

Lespinat, P.A., Guckert, A. and Breisch, H. (1975) The incorporation of plant root exudates in soil organic matter, in *Transactions of the International Symposium "Humus et Planta" VI*, Prague, Czech Republic, pp. 169-74.

Lespinat, P.A. and Berlier, Y. (1975) Les facteurs externes agissant sur l'excrétion racinaire. *Société Botanique de France, Collection Rhizosphère*, **XX**, 21-30.

Levey, D.J. and Byrne, M.M. (1993) Complex ant-plant interactions: rain forest ants as secondary dispersers and post-dispersal seed predators. *Ecology*, **74**, 1812-902.

Levings, S.C. and Traniello, J.F.A. (1981) Territoriality, nest dispersion and community structure in ants. *Psyche*, **88** (3-4), 265-319.

Levings, S.C. and Adams, E.S. (1984) Intra- and interspecific territory in *Nasutitermes* (Isoptera: Termitidae) in a Panamanian mangrove forest. *Journal of Animal Ecology*, **53**, 705-14.

Leyval, C. and Berthelin, J. (1982) Rôle des microflores symbiotiques et non symbiotiques sur l'altération de la biotite et la croissance du mais (*Zea mays*); influence des conditions du milieu. *Science du Sol*, **1**, 3-14.

Lévesque, M.P. and Dinel, H. (1982) Some morphological and chemical aspects of peats applied to the characterisation of histosols. *Soil Science*, **133**, 324-32.

Lévieux, J. (1976) Deux aspects de l'action des fourmis (Hymenoptera, Formicidae) sur le sol d'une savane préforestière de Côte d'Ivoire. *Bulletin d'Ecologie*, **7** (3), 283-95.

Lévieux, J. and Aouti, A. (1978) Données démographiques obtenues en élevage sur une population du diplopode *Pachybolus laminatus* Cook de Basse Côte d'Ivoire. *Bulletin d'Ecologie*, **9** (2), 141-8.

Lévieux, J. (1983) The soil fauna of tropical savanna. IV. The ants, in *Tropical Savannas* (ed. F. Bourlière), Elsevier, Amsterdam, pp. 525-40.

Lewis, D.H. (1985) Symbiosis and mutualism: crisp concepts and soggy semantics, in *Biology of Mutualism* (ed. D.H. Boucher), Croom Helm, Beckenham, pp. 29-39.

Li, X.L., George, E. and Marschner, H. (1991) Extension of the phosphorus depletion zone in VA-mycorrhizal White Clover in a calcareous soil. *Plant and Soil*, **136**, 41-8.

Lie, T.A., Goktan, D., Engin, M. *et al.* (1987) Co-evolution of the legume-Rhizobium association. *Plant and Soil*, **100**, 171-81.

Ligthart, T.N. (1996) *Development of earthworm burrow systems and the influence of earthworms on soil hydrology*, Ph.D. thesis, University of Wageningen.

Likens, G.E., Driscoll, C.T., Uso, D.C. *et al.* (1994) The biogeochemistry of potassium at Hubbard Brook. *Biogeochemistry*, **25**, 61-125.

Ljungström, P.O. (1972) Introduced earthworms of South Africa. On their taxonomy, distribution, history of introduction and on the extermination of endemic earthworms. *Zoologischer Jährbucher Abteilungen Systematik, Ökologie und Geographie der Tiere*, **99**, 1-81.

Lobo, P.F., Flexor, J.M., Rapaire, J.L. *et al.* (1974) Essai de détermination du temps de résidence des fractions humiques de deux sols ferralitiques par l'utilisation du radiocarbone naturel et thermonucléaire. *Cahiers ORSTOM, série Pédologie*, **12** (1), 115-23.

Lobry de Bruyn, L.A. and Conacher, A.J. (1990) The role of termites and ants in soil modification: a review. *Australian Journal of Soil Research*, **28** (1), 55-93.

Lockaby, B.G. and Adams, J.C. (1985) Pedoturbation of a forest soil by fire ants. *Soil Science Society of America, Journal*, **49** (1), 220-3.

Lohm, U., Lundkvist, H., Persson, T. *et al.* (1977) Effects of nitrogen fertilization on the abundance of enchytraeids and microarthropods in Scots Pine forests. *Studia Forestalia Suecica*, **140**, 1-23.

Lohm, U. and Persson, T. (eds) (1977) *Soil Organisms as Components of Ecosystems*, Ecological Bulletins/NFR, Stockohlm.

Lofs-Holmin, A. (1983) Influence of agricultural practices on earthworms (Lumbricidae). *Acta Agricultura Scandinavica*, **33**, 225-34.

Lombardi Neto, F. and Bertoni, J. (1975) Tolerancia de perdas para solos do Estado de Sao Paulo. *Boletim Tecnico del Instituto de Agronomia*, **28**, 1-12.

Lopes-Assad, M.L. (1987) *Contribution à l'étude de la macroporosité lombricienne de différents types de sols de France*, thèse de l'Université de Montpellier/ENSAM.

Loquet, M. and Vinceslas, M. (1987) Cellulolyse et ligninolyse liées au tube digestif d'*Eisenia fetida andrei* Bouché. *Revue d'Ecologie et de Biologie du Sol*, **24** (4), 559-71.

Lossaint, P. (1959) *Etude expérimentale de la mobilisationn du fer des sols sous l'influence des litières forestières*, thèse d'Etat, Université de Strasbourg.

Lousier, J.D. and Bamforth, S.S. (1990) Soil Protozoa, in *Soil Biology Guide* (ed. D.L. Dindal), John Wiley & Sons, New York, pp. 97-136.

Loustau, D. (1984) *Morphologie et fonctionnement biodynamique de quelques humus hydromorphes de l'est de la France*, thèse de 3e cycle, Université Nancy I.

López-Hernandez, D., Niño, M., Nannipieri, P. *et al.* (1989) Phosphatase activity in *Nasutitermes ephratae* termite nests. *Biology and Fertility of Soils*, **7**, 134-7.

López-Hernández, D., Fardeau, J.C. and Lavelle, P. (1993) Phosphorus transformations in two P-sorption contrasting tropical soils during transit through *Pontoscolex corethrurus* (Glossoscolecidae, Oligochaeta). *Soil Biology & Biochemistry*, **25** (6), 789-92.

Lucas, R.E. and Davies, J.F. (1961) Relationship between pH values of organic soils and availabilities of 12 plant nutrients. *Soil Science*, **92**, 177-82.

Lucas, Y. and Chauvel, A. (1992) Soil formation in tropically weathered terrains, in *Regolith Exploration Geochemistry in Tropical and Subtropical Terrains* (eds C.R.M. Butt and H. Zeegers), Elsevier, Amsterdam, pp. 57-77.

Lucas, Y., Luizão, F.J., Chauvel, A. *et al.* (1993) The relation between biological activity of the rain forest and mineral composition of soils. *Science*, **260**, 521-3.

Lucas, Y., Nahon, S., Cornu, S. *et al.* (1996) Genèse et fonctionnement des sols en milieu équatorial. *Comptes Rendus de l'Académie des Sciences, Paris*, **322** (IIa), 1-16.

Luizão, R.C.C. (1994) *Soil biological studies in contrasting types of vegetation in central Amazonian rain forests*, Ph.D. thesis, University of Stirling.

Lumaret, J.P. and Kirk, A. (1987) Ecology of dung beetles in the French Mediterranean region (Coleoptera: Scarabaeidae). *Acta Zoologica Mexicana (nueva serie)*, **24**, 1-55.

Lundkvist, H. (1981) *Enchytraeidae (Oligochaeta) in pine forest soils: population dynamics and response to environmental changes*, Ph.D. thesis, Uppsala University.

Luxton, M. (1972) Studies on the oribatid mites of a Danish Beech forest. I. Nutritional biology. *Pedobiologia*,

12, 434-63.

Luxton, M. (1982) The biology of mites from Beech woodland soil. *Pedobiologia*, **23** (1), 1-8.

Lyford, W.H. (1963) Importance of ants to brown podzolic soil genesis in New England. *Harvard Forest Papers*, **7**, 18.

Lyford, W.H. and Wilson, B.F. (1964) Development of the root system of *Acer rubrum* L. *Harvard Forest Papers*, **10**, 1-17.

Lynch, J.M. (1982) Interactions between bacteria and plants in the root environment, in *Bacteria and Plants* (eds M.E. Rhodes-Roberts and F.A. Skinner), Academic Press, London, pp. 1-23.

Lynch, J.M. (1990) Beneficial interactions between microorganisms and roots. *Biotechnology Advances*, **8**, 335-46.

M'Ba, C. (1987) Vermicomposting and biological N-fixation, in *Proceedings of the Symposium on Soil Biology and Conservation of the Biosphere* (ed. J. Szegi), Akademiai Kiado, Budapest, pp. 547-52.

MBoukou-Kimbatsa, I.M.C., Bernhard-Reversat, F. and Loumeto, J.J. (1998) Change in soil macrofauna and vegetation when fast growing trees are planted on savanna soils. *Forest Ecology and Management*, **110**, 1-12.

Macias, F. and Chesworth, W. (1992) Weathering in humid regions, with emphasis on igneous rocks and their metamorphic equivalents, in *Weathering, Soils and Paleosols* (eds I.P. Martini and W. Chesworth), Elsevier, Amsterdam, pp. 283-306.

Maeda, T., Takenaka, H. and Warkentin, B.P. (1977) Physical properties of allophane soils. *Advances in Agronomy*, **29**, 229-64.

Maertens, C. and Clauzel, Y. (1980) Rôle du flux de sève brute dans les racines sur l'absorption du potassium et du calcium par les systèmes racinaires de jeunes plantes de *Zea mays*. *Comptes Rendus de l'Académie des Sciences, Paris*, **290** (D), 1285-7.

Maggenti, A.R. (1961) Revision of the genus *Plectus* (Nematoda: Plectidae). *Proceedings of the Helminthological Society of Washington*, **28**, 138-66.

Magnusson, T. (1994) Studies of the soil atmosphere and related physical characteristics in peat forest soils. *Forest Ecology and Management*, **67**, 203-24.

Malaisse, F. (1978) High termitaria, in *Biogeography and Ecology of Southern Africa* (ed. M.J.A. Werger), Dr. W. Junk Publisher, The Hague, pp. 1281-300.

Malcevski, S. (1978) Considérations quantitatives sur la nématofaune d'une savane herbeuse à Andropogonées de Lamto (Côte d'Ivoire). *Revue d'Ecologie et de Biologie du Sol*, **15**, 487-96.

Malcolm, R.L. (1993) Concentration and composition of dissolved organic carbon in soils, stream, and groundwaters, in *Organic Substances in Soil Water: Natural Constituents and Their Influences on Contaminant Behaviour* (eds A.J. Beck, K.C. Jones, M.H.B. Hayes *et al.*), Royal Society of Chemistry, Cambridge, pp. 19-30.

Maldague, M.E. (1961) Relations entre le couvert végétal et la microfaune. *Publications de l'Institut National d'Etude Agronomique, Congo, Série Sciences*, **90**

Maldague, M.E. (1964) Importance des populations de termites dans les sols équatoriaux, in *Proceedings of the 4th International Congress of Soil Science*, ISSS, Bucarest, Romania, pp. 743-52.

Mandimba, G., Heulin, T., Bally, R. *et al.* (1986) Chemotaxis of free-living nitrogen-fixing bacteria towards maize mucilage. *Plant and Soil*, **90**, 129-39.

Mando, A. (1997) Effects of termites and mulch on the physical rehabilitation of structurally crusted soils in the Sahel. *Land Degradation and Development*, **8**, 269-78.

Mangenot, F. and Toutain, F. (1980) Les litières, in *Actualités d'Ecologie Forestière* (ed. P. Pesson), Gauthier-Villars, Paris, pp. 3-59.

Mani, M.S. (1962) *Introduction to High Altitude Entomology*, Methuen, London.

Marcuzzi, G. (1970) Experimental observations on the role of *Glomeris spp.* in the process of humification of litter. *Pedobiologia*, **10** (6), 401-6.

Margulis, L. and Schwartz, K.V. (1998) *Five Kingdoms - An Illustrated Guide to the Phyla of Life on Earth*, Freeman, New York.

Marinissen, J.C.Y. (1995) *Earthworms, soil-aggregates and organic matter decomposition in agro-ecosystems in the Netherlands*, Ph.D. thesis, University of Wageningen.

Marrs, R.H., Thompson, J., Scott, D. *et al.* (1991) Nitrogen mineralization and nitrification in *terra firme* forest and savanna soils on Ihla de Maraca, Roraima, Brazil. *Journal of Tropical Ecology*, **7**, 123-37.

Marschner, H. and Römheld, V. (1983) *In vivo* measurement of root-induced pH changes at the soil-root interface: effect of plant species and nitrogen source. *Zeitschrift für Pflanzenphysiologie*, **111** (3), 241-51.

Marschner, H., Römheld, V., Horst, W.J. *et al.* (1986) Root-induced changes in the rhizosphere: importance for the mineral nutrition of plants. *Zeitschrift für Pflanzenernährung und Bodenkunde*, **149**, 441-56.

Marschner, H. and Dell, B. (1994) Nutrient uptake in mycorrhizal symbiosis. *Plant and Soil*, **159**, 89-102.

Marschner, H. (1995) *Mineral Nutrition of Higher Plants*, Academic Press, London.

Marshall, T.J. (1947) Mechanical composition of soil in relation to field description of texture. CSIR Australian Bulletin. No. 224.

Marshall, V.G. (1972) Comparison of two methods of estimating efficiency of funnel extractors for soil microarthropods. *Soil Biology & Biochemistry*, **4**, 417-26.

Martikainen, E. and Huhta, V. (1990) Interactions between nematodes and predatory mites in raw humus soil: a microcosm experiment. *Revue d'Ecologie et de Biologie du Sol*, **27** (1), 13-20.

Martin, J.A. (1977) Factors influencing the loss of organic carbon from wheat roots. *Soil Biology & Biochemistry*, **9**, 1-7.

Martin, M.M. and Martin, J.S. (1979) Distribution and origins of the cellulolytic enzymes of the higher termite *Macrotermes natalensis*. *Physiological Zoology*, **52**, 11-21.

Martin, M.M. (1984) The role of ingested enzymes in digestive processes of insects, in *Invertebrate-Microbial Interactions* (eds J.M. Anderson, A.D.M. Rayner and D.W.H. Walton), Cambridge University Press, Cambridge, pp. 155-72.

Martin, J.P. and Haider, K. (1986) Influence of mineral colloids on turnover rates of soil organic carbon, in *Interactions of Soil Minerals with Natural Organics and Microbes* (eds P.M. Huang and M. Schnitzer), Soil Science Society of America, Madison, Wisconsin, pp. 284-304.

Martin, A., Cortez, J., Barois, I. *et al.* (1987) Les mucus intestinaux de ver de Terre, moteur de leurs interactions avec la microflore. *Revue d'Ecologie et de Biologie du Sol*, **24** (4), 549-58.

Martin, A. (1989) *Effet des vers de terre tropicaux géophages sur la dynamique de la matière organique du sol dans les savanes humides*, thèse de l'Université Paris Sud-Orsay.

Martin, A., Mariotti, A., Balesdent, J. *et al.* (1990) Estimate of organic matter turnover rate in a savanna soil by ^{13}C natural abundance measurements. *Soil Biology & Biochemistry*, **22** (4), 517-23.

Martin, S. (1990) *Modélisation de la dynamique et du rôle d'une population de vers de terre* Millsonia anomala *dans les savanes humides de Côte d'Ivoire*, thèse de l'Université Paris VI.

Martin, A. (1991) Short- and long-term effects of the endogeic earthworm *Millsonia anomala* (Omodeo) (Megascolecidae, Oligochaeta) of tropical savannas, on soil organic matter. *Biology and Fertility of Soils*, **11**, 234-8.

Martin, A., Mariotti, A., Balesdent, J. *et al.* (1991) Soil organic matter assimilation by a geophagous tropical earthworm based on $\delta^{13}C$ measurements. *Ecology*, **73** (1), 118-28.

Martin, A., Balesdent, J. and Mariotti, A. (1992) Earthworm diet related to soil organic matter dynamics through ^{13}C measurements. *Oecologia*, **91**, 23-9.

Martinez, A.T., Montes, C., Toutain, F. *et al.* (1980) Influence de l'épaisseur de la litière et du type de sol sur les processus de biodégradation des feuilles de hêtre. *Revue d'Ecologie et de Biologie du Sol*, **17** (3), 307-25.

Martinez-Romero, E. (1994) Recent developments in *Rhizobium* taxonomy. *Plant and Soil*, **161**, 11-20.

Martius, C., Wassmann, R., Thein, U. *et al.* (1993) Methane emission from wood-feeding termites in Amazonia. *Chemosphere*, **26**, 63-632.

Martius, C. (1994) Diversity and ecology of termites in Amazonian forests. *Pedobiologia*, **38**, 407-28.

Mary, B., Mariotti, A. and Morel, J.L. (1992) Use of ^{13}C variation at natural abundance for studying the biodegradation of root mucilage, roots and glucose in soil. *Soil Biology & Biochemistry*, **24** (10), 1065-72.

Maser, C., Trappe, J.M. and Nussbaum, R.A. (1978) Fungal-small mammal interrelations with emphasis on Oregon coniferous forests. *Ecology*, **59**, 799-809.

Mason, P.A., Pelham, J. and Last, F. (1977) Stem anatomy and sheathing mycorrhizas in the *Betula verrucosa-Amanita muscaria* relationship. *Nature*, **26**, 675-98.

Massot, C. and Cancela da Fonseca, J.P. (1986) Etude, par l'analyse en composantes principales, de la méso-faune de la litière sous hêtre de quatre parcelles forestières exploitées différemment. *Revue d'Ecologie et de Biologie du Sol*, **23** (1), 19-27.

Massoud, Z. (1971) Un élément caractéristique de la pédofaune: les Collemboles, in *La Vie Dans les Sols* (ed. P. Pesson), Gauthier-Villars, Paris, pp. 33-388.

Massoud, Z. (1976) Essai de synthèse relatif à la phylogénie des Collemboles. *Revue d'Ecologie et de Biologie du Sol*, **13** (1), 35-42.

Massoud, Z., Betsch, J.M. and Thibaud, J.M. (1984) Expérience de piétinement contrôlé du sol d'une forêt périurbaine: effets sur le peuplement de Collemboles. *Revue d'Ecologie et de Biologie du Sol*, **21** (4), 507-18.

Matsumoto, T. (1976) The role of termites in an equatorial rain forest ecosystem of West Malaysia. I. Population density, biomass, carbon, nitrogen and calorific content and respiration rate. *Oecologia*, **22**, 153-78.

Matsumoto, T. and Abe, T. (1979) The role of termites in an equatorial rain forest ecosystem of west Malaysia. II. Leaf litter consumption on the forest floor. *Oecologia*, **38**, 261-74.

Mavingui, P., Laguerre, G., Berge, O. *et al.* (1992) Genetic and phenotypic diversity of *Bacillus polymyxa* in soil and in the wheat rhizosphere. *Environmental Microbiology*, **28** (6), 1894-903.

May, R.M. (1978) The dynamics and diversity of insect faunas, in *Symposium of the Royal Entomological Society of London* (eds L.A. Mound and N. Waloff), Blackwell Scientific Publications, Oxford, pp. 188-204.

May, R.M. (1986) How many species are there? *Nature*, **324**, 514-5.

Mazaud, D. and Bouché, M.B. (1980) Introduction en surpopulation et migrations de lombriciens marqués, in *Soil Biology as Related to Land Use Practices* (ed. D.L. Dindal), E.P.A., Washington D.C., pp. 667-701.

Macfadyen, A. (1961) Improved funnel-type extractors for soil arthropods. *Journal of Animal Ecology*, **30**, 171-184.

McArthur, R.H. and Wilson, E.O. (1967) *The Theory of Island Biogeography*, Princeton University Press, Princeton, N.J.

McBrayer, J.F., Ferris, J.M., Metw, L.J. *et al.* (1977) Decomposer invertebrate population in U.S. forest biomes. *US International Biological Program, Report*, **17**, 89-96.

McBride, M.B. (1994) *Environmental Chemistry of Soils*, Oxford University Press, New York.

McClaugherty, C. and Berg, B. (1987) Cellulose, lignin and nitrogen concentrations as rate regulating factors in late stages of forest litter decomposition. *Pedobiologia*, **30**, 101-12.

McDonald, P., Edwards, R.A. and Greenhalgh, J.F.D. (1988) *Animal Nutrition*, Longman, Harlow.

McDonald, R.C., Isbell, R.F., Speight, J.G. *et al.* (1990) *Australian Soil and Land Survey Field Handbook*,

Inkata Press, Melbourne.

McDougall, B.M. (1970) Sites of exudation of ^{14}C labelled compounds from wheat roots. *New Phytologist*, **69**, 999-1003.

McGinley, M.A., Dhillion, S.S. and Neumann, J.C. (1994) Environmental heterogeneity and seedling establishment: ant-plant-microbe interactions. *Functional Ecology*, **8**, 607-15.

McInerney, M. and Bolger, T. (1996) The effect of macrofaunal activity on soil organic matter decomposition in forest ecosystems, in *Soil Organisms and Soil Resource Management, XII International Colloquium on Soil Zoology* (eds T. Bolger and J. Curry), University College Dublin, Dublin, pp. 134.

McIvor, J.G., Gardener, C.J. and Spain, A.V. (1991) A Description of the ECCOSAT Experimental Site at Hillgrove near Charters Towers, north Queensland. *CSIRO Australia, Division of Tropical Crops and Pastures, Technical Memorandum* No. 70.

McKay, W.P. (1981) A comparison of the nest phenologies of three species of *Pogonomyrmex* harvester ants (Hymenoptera: Formicidea). *Psyche*, **88**, 25-74.

McKay, A.D. and Kladivko, E.J. (1985) Earthworms and rate of breakdown of soybean and maize residues in soil. *Soil Biology & Biochemistry*, **17** (6), 851-7.

McKay, W.P., Silva, S. and Whitford, W.G. (1987) Diurnal activity patterns and vertical migration in desert soil microarthropods. *Pedobiologia*, **30**, 65-71.

McKay, H.M. and Malcolm, D.C. (1988) A comparison of the fine root component of a pure and mixed coniferous stand. *Canadian Journal of Forest Research*, **18**, 1416-26.

McKeague, J.A., Cheshire, M.V., Andreux, F. *et al.* (1986) Organo-mineral complexes in relation to pedogenesis, in *Interactions of Soil Minerals with Natural Organics and Microbes* (eds P.M. Huang and M. Schnitzer), Soil Science Society of America, Madison, Wisconsin, pp. 549-92.

McLaren, J. (1975) Soil as a system of humus and clay immobilized enzymes. *Chemica Scripta*, **8**, 97-99.

McLean, S.F.J. (1975) Ecological adaptations of tundra invertebrates, in *Physiological Adaptation to the Environment* (ed. J. Vernber), Intext, New York, pp. 269-300.

McLean, S.F.J. (1980) The detritus-based trophic system, in *The Coastal Tundra at Barrow (Alaska)* (eds J. Brown, L.L. Tieszen and F.L. Bunnell), Dowden, Hutchinson and Ross, Stroudsburg, Pennsylvania, pp. 411-57.

McMahan, E.A. (1979) Temporal polyethism in termites. *Sociobiology*, **4**, 153-68.

McNeil, M. (1964) Lateritic soils. *Scientific American*, **211** (5), 521-3.

McTiernan, K.B., Ineson, P. and Coward, P.A. (1997) Respiration and nutrient release from tree leaf litter mixtures. *Oikos*, **78**(3), 527-38.

Meadows, P. and Meadow, A. (1991) *The Environmental Impact of Burrowing Animals and Animal Burrows*, Clarendon Press, Oxford.

Meek, B.D., de Tar, W.R., Rolph, D. *et al.* (1991) Infiltration rate as affected by an alfalfa and no-till cotton cropping system. *Soil Science Society of America, Journal*, **54**, 505-8.

Meentemeyer, V. (1978) Macroclimate and lignin control of litter decomposition rates. *Ecology*, **59**, 465-72.

Meentemeyer, V., Box, E.O. and Thompson, R. (1982) World patterns and amounts of terrestrial plant litter production. *BioScience*, **32**, 125-8.

Meetting, J.B.Jr. (1993) Structure and physiological ecology of soil microbial communities, in *Soil Microbial Ecology* (ed. J.B. Meeting Jr.), Marcel Dekker, New York, pp. 3-25.

Meiklejohn, J. (1962) Microbiology of the nitrogen cycle in some Ghana soils. *Empire Journal of Experimental Agriculture*, **30**, 115-26.

Mellilo, J.M., Aber, J.D. and Muratore, J.F. (1982) Nitrogen and lignin control of hardwood leaf litter decomposition dynamics. *Ecology*, **63**, 621-6.

Mengel, K. and Kirkby, E.A. (1987) *Principles of Plant Nutrition*, International Potash Institute, Worblaufen-Bern, Switzerland.

Mermut, A.R., Arshad, M.A. and Arnaud, R.J. (1984) Micropedological study of termite mounds of three species of *Macrotermes* in Kenya. *Soil Science Society of America, Proceedings*, **48**, 613-20.

Mesarovic, M.D., Macko, D. and Takahara, Y. (1970) *Theory of Hierarchical, Multilevel, Systems*, Academic Press, New York.

Mew, G. and Lee, R. (1981) Investigation of the properties and genesis of West Coast wet land soils, South Island, New Zealand. 1. Type localities, profile morphology, and soil chemistry. *New Zealand Journal of Science*, **24**, 1-24.

Meyer, J.A. (1960) Résultats agronomiques d'un essai de nivellement des termitières réalisé dans la cuvette centrale Congolaise. *Bulletin d'Agriculture du Congo Belge*, **51**, 1047-59.

Meyer, F.H. and Göttsche, D. (1971) Distribution of root tips and tender roots of Beech, in *Ergebnisse des Sollingprojekts der DFG (IBP), Mitt. Nr. 9* (ed. H. Ellenberg), Springer Verlag, Berlin, pp. 48-52.

Meyer, F.H. (1973) Distribution of ectomycorrhizae in native and man-made forests, in *Ectomycorrhizae* (eds G.C. Marks and T.T. Kozlowski), Academic Press, New York, pp. 79-105.

Mériaux, S. (1979) Le sol et l'eau, in *Constituants et Propriétés du Sol* (eds M. Bonneau and B. Souchier), Masson, Paris, pp. 266-312.

Michaelsen, W. (1903) *Die Geographische Verbreitung der Oligochaeten*, R. Friedlander und Sohn, Berlin.

Miedema, R., Brouer, J., Geiger, S.C. *et al.* (1994) Variability in the growth of *Faidherbia albida* near Niamey, Niger, Africa: micromorphological aspects of termite activity, in *Soil Micromorphology: Studies in Management and Genesis* (eds A.J. Ringrose-Voase and G.S. Humphreys), Elsevier, Amsterdam, pp. 411-9.

Miklos, A.A. (1992) *Biodynamique d'une couverture pédologique dans la région de Botucata (S.P., Brésil)*, thèse de l'Université Paris VI.

Milchunas, D.G., Lauenroth, W.K., Singh, J.S. *et al.* (1985) Root turnover and production by ^{14}C dilution: implications of carbon partitioning in plants. *Plant and Soil*, **88**, 353-65.

Mill, A.E. (1992) Termites as agricultural pests in Amazonia, Brazil. *Outlook on Agriculture*, **21**, 41-6.

Miller, R.M. and Jastrow, J.D. (1990) Hierarchy of root and mycorrhizal fungal interactions with soil aggregation. *Soil Biology & Biochemistry*, **22** (5), 579-84.

Miller, L.R. (1991) A revision of the *Termes-Capritermes* branch of the Termitinae in Australia (Isoptera: Termitidae). *Invertebrate Taxanomy*, **4**, 1147-1282.

Millot, G. (1977) Géochimie de la surface et formes du relief. Présentation. *Bulletin des Sciences Géologiques*, **30** (4), 229-33.

Millot, G. (1979) Clay. *Scientific American*, **240** (4), 76-87.

Milnes, A.R. (1992) Calcrete, in *Weathering, Soils and Paleosols* (eds I.P. Martini and W. Chesworth), Elsevier, Amsterdam, pp. 309-47.

Minderman, G. and Daniels, L. (1967) Colonisation of newly fallen leaves by microorganisms, in *Progress in Soil Biology* (eds O. Graff and J.E. Satchell), North Holland, Amsterdam, pp. 3-9.

Minderman, G. (1968) Addition, decomposition and accumulation of organic matter in forests. *Journal of Ecology*, **56**, 355-62.

Misra, R.K., Dexter, A.R. and Alston, A.M. (1986) Penetration of soil aggregates of finite size. II. Plant roots. *Plant and Soil*, **94**, 59-85.

Mitchell, M. (1977) Life history strategies of oribatid mites, in *Biology of Oribatid Mites* (ed. D.L. Dindal), SUNY College of Environmental Science and Forestry, Syracuse, New York, pp. 65-9.

Mitsui, Y. (1985) Distribution and ecology of nematode-trapping fungi in Japan. *Japan Agricultural Research Quaterly*, **18** (3), 182-93.

Mocquard, J.P., Juchault, P., Jambu, P. *et al.* (1987) Essai d'évaluation du rôle des crustacés oniscoides dans la transformation des litières végétales dans une forêt feuillue de l'ouest de la France. *Revue d'Ecologie et de Biologie du Sol*, **24** (3), 311-28.

Molfetas, S. (1982) Etude d'un écosystème forestier mixte. VIII.-Les Isopodes. *Revue d'Ecologie et de Biologie du Sol*, **19** (3), 427-738.

Molina, R., Massicote, H. and Trappe, J.M. (1992) Specificity phenomena in mycorrhizal symbiosis: community-ecological consequences and practical implications, in *Mycorrhizal Functioning, An Integrative Plant-Fungal Process* (ed. M.F. Allen), Chapman & Hall, New York, pp. 357-423.

Mollon, A. (1982) *Les Larves de Diptères d'une Chênaie-Charmaie: Structure spatio-temporelle du peuplement*, thèse de 3e cycle, Université Paris VI.

Molloy, L. (1988) *Soils in the New Zealand Landscape*, Mallinson Rendel, Wellington.

Monier, G. (1965) Action des matières organiques sur la stabilité structurale des sols. *Annales Agronomiques*, **16**, 471-534.

Montagnini, F., Ramstad, K. and Sancho, F. (1993) Litterfall, litter decomposition and the use of mulch of four indigenous tree species in the Atlantic lowlands of Costa-Rica. *Agroforestry Systems*, **23** (1), 39-61.

Moore, A.W., Russell, J.S. and Coldrake, J.E. (1967) Dry matter and nutrient content of a subtropical semiarid forest of *Acacia harpophylla* F. Muell. (Brigalow). *Australian Journal of Botany*, **15**, 11-24.

Moore, A.M. (1986) Temperature and moisture dependence of decomposition rates of hardwood and coniferous leaf litter. *Soil Biology & Biochemistry*, **18** (4), 427-35.

Moore, J.C. (1988) The influence of microarthropods on symbiotic and non-symbiotic mutualism in detrital-based below-ground food webs. *Agriculture, Ecosystems and Environment*, **24**, 147-59.

Moore, J.C., de Ruiter, P.C. and Hunt, H.W. (1993) Soil invertebrate/micro-invertebrate interactions: disproportionate effects of species on food web structure and function. *Veterinary Pathology*, **48**, 247-60.

Moore, P.D., Webb, J.A. and Collinson, M.E. (1991) *Pollen Analysis*, Blackwell Scientific Publications, Oxford.

Morré, D.J., Jones, D.D. and Mollenhauser, H.H. (1967) Golgi apparatus mediated polysaccharide secretion by outer root cap cells of *Zea mays*. I. Kinetics and secretory pathway. *Planta*, **74**, 286-301.

Moorhead, D.L. and Reynolds, J.F. (1989) The contribution of abiotic processes to buried litter decomposition in the northern Chihuahuan Desert. *Oecologia*, **79**, 133-5.

Mordelet, P., Barot, S. and Abbadie, L. (1996) Root foraging strategies and soil patchiness in a humid savanna. *Plant and Soil*, **182** (1), 171-6.

Morel, J.L., Habib, L., Plantureux, S. *et al.* (1991) Influence of maize root mucilage on soil aggregate stability. *Plant and Soil*, **136**, 111-9.

Morton, J.B. and Benny, G.L. (1990) Revised classification of arbuscular mycorrhizal fungi (Zygomycetes): a new order, Glomales, two new suborders, Glomineae and Gigasporineae, and two new families Acaulosporaceae and Gigasporaceae, with an emendation of Glomaceae. *Mycotaxon*, **37**, 471-91.

Moser, M. and Haselwandter, K. (1983) Ecophysiology of mycorrhizal symbioses, in *Physiological Plant Ecology III* (eds O.L. Lange, P.S. Nobel, C.B. Osmond *et al.*), Springer-Verlag, Berlin, pp. 392-411.

Mosse, B., Stribley, D.P. and Le Tacon, F. (1981) Ecology of mycorrhizae and mycorrhizal fungi, in *Advances in Microbial Ecology* (ed. M. Alexander), Plenum Press, New York, pp. 137-210.

Mousain, D. and Lamond, M. (1978) Influence de la concentration en phosphate (H_2PO_4-Y) de la solution nutritive sur la mycorhization du *Pinus pinaster* par le *Pisolithus tinctorius* et l'*Hebeloma cylindrosporum*, in *Proceedings of the Symposium on Root Physiology and Symbiosis* (eds A. Riedacker and J. Gagnaire-Michard), INRA, Nancy, pp.473-75.

Mousain, D. (1983) Aspects écologiques de la symbiose mycorhizienne. *Science du Sol*, **123** (3), 341-8.

Muckenhausen, E. (1965) The soil classification of the Federal Republic of Germany. *Pedologie, Special Series*, **3**, 57-74.

Munch, J.C. and Ottow, J.C.G. (1986) Nature des produits gazeux formés dans les sols à partir de différentes microflores dénitrifiantes. *Science du Sol*, **24**, 337-50.

Munns, D.N. (1978) Legume-rhizobium relations: soil acidity and nodulation, in *Mineral Nutrition of Legumes in Tropical and Subtropical Soils* (eds C.S. Andrew and E.J. Kamprath), Commonwealth Scientific and Industrial Research Organisation, Melbourne, pp. 247-63.

Munro, P.E. (1966) Inhibition of nitrifiers by grass root extracts. *Journal of Applied Ecology*, **3**, 231-8.

Murray, R.G.E., Brenner, D.J., Colwell, R.R. *et al.* (1990) Report of the *ad hoc* committee on approaches to taxonomy within the Proteobacteria. *International Journal of Systematic Bacteriology*, **40**, 213-5.

Murtha, G.G. (1982) *Soils and Land Use on the Southern Section of the Townsville Coastal Plain, North Queensland. Soils and Land Use Series No.59*, Division of Soils, CSIRO Australia.

Murtha, G.G. (1986) Soils of the Tully-Innisfail area, North Queensland. *CSIRO Australia, Division of Soils, Divisional Report*, No.**82**.

Müller, P.E. (1887) *Studien über die natürlichen Humusformen und deren Einwirkungen auf Vegetation und Boden*, Springer, Berlin.

Myers, R.J.K., Palm, C.A., Cuevas, E. *et al.* (1994) The synchronization of nutrient mineralisation and plant nutrient demand, in *The Biological Management of Tropical Fertility* (eds P.L. Woomer and M.J. Swift), Wiley-Sayce, Chichester, pp. 81-116.

Myrold, D.D., Elliott, L.F., Papendick, R.I. *et al.* (1981) Water potential-water content characteristics of wheat straw. *Soil Science Society of America, Proceedings*, **45**, 329-33.

Nadelhoffer, K.J. and Fry, B. (1994) Nitrogen isotope studies in forest ecosystems, in *Stable Isotopes* (eds K. Lajtha and R.H. Michener), Blackwell Scientific Publications, Oxford, pp. 22-44.

Nagel-de-Boois, H.M. (1971) Preliminary estimate of production of fungal mycelium in forest soil layers. *Annales de Zoologie, Ecologie Animale*, **28**, 447-53.

Nalepa, C.A. (1994) Nourishment and the origin of termite eusociality, in *Nourishment and Evolution in Insect Societies* (eds J.H. Hunt and C.A. Nalepa), Westview Press, Boulder Colorado, pp. 57-104.

Nannipieri, P., Johnson, R.L. and Paul, E.A. (1978) Criteria for measurement of microbial growth and activity in soil. *Soil Biology & Biochemistry*, **10**, 223-9.

Nel, J.J.C. (1968) Aggressive behaviour of the harvester termites *Hodotermes mossambicus* (Hagen) and *Trinervitermes trinervoides* (Sjöstedt). *Insectes Sociaux*, **15**, 145-56.

Needham, A.E. (1957) Components of nitrogenous excreta in the earthworms *Lumbricus terrestris* L. and *Eisenia fetida* (Savigny). *Journal of Experimental Biology*, **34**, 425-46.

Nemeth, A. (1981) *Estudio ecológico preliminar de las lombrices de tierra (Oligochaeta) en ecosistemas de bosque humedo tropical en San Carlos de Rio Negro, Territorio Federal Amazonas*, Thesis de Licenciatura, Universidad Central de Venezuela.

Neptune, A.M.L., Tabatabai, M.A. and Hanway, J.J. (1975) Sulfur fractions and carbon-nitrogen-phosphorus-sulphur relationships in some Brazilian and Iowa soils. *Soil Science Society of America, Proceedings*, **39**, 51-4.

Neuhauser, E.F. and Hartenstein, R. (1978) Phenolic content and palatability of leaves and wood to soil isopods and diplopods. *Pedobiologia*, **18**, 99-109.

Neves, M.C.P. (1982) Energy cost of biological nitrogen fixation, in *Biological Nitrogen Fixation Technology for Tropical Agriculture* (eds P.H. Graham and S.C. Harris), CIAT, Cali, pp. 77-92.

Newman, E.I. (1969) Resistance to water flow in soil and plant. 1. Soil resistance in relation to the amount of root: theoretical estimates. *Journal of Applied Ecology*, **6**, 1-12.

Newman, E.I. and Rovira, A.D. (1975) Allelopathy among some British grassland species. *Journal of Ecology*, **63**, 722-37.

Newman, E.I. and Reddell, P. (1987) The distribution of mycorrhizas among families of vascular plants. *New Phytologist*, **106**, 745-51.

Newman, E.I. (1995) Phosphorus inputs to terrestrial ecosystems. *Journal of Ecology*, **83**, 713-26.

Newton, A.F. (1990) Insecta: Coleoptera Staphylinidae (Adults and Larvae), in *Soil Biology Guide* (ed. D.L. Dindal), John Wiley & Sons, New York, pp. 1137-74.

Nicolai, V. (1988) Phenolic and mineral content of leaves influences decomposition in European forest ecosystems. *Oecologia*, **75**, 575-9.

Nicolardot, B. and Chaussod, R. (1982) Problèmes posés par l'estimation de la biomasse microbienne du sol, in *Comptes Rendus de la 6e Réunion des microbiologistes de l'INRA*, INRA, Murat-le-Quaire, pp. 69-73.

Nicolardot, B., Chaussod, R. and Catroux, G. (1982) Revue des principales méthodes disponibles pour mesurer la biomasse microbienne et ses activités. *Sciences du Sol*, **4**, 253-61.

Nielsen, C.O. (1961) Respiratory metabolism of some populations of enchytraeid worms and free-living nematodes. *Oikos*, **12**, 17-35.

Nielsen, G.A. and Hole, F.D. (1964) Earthworms and the development of coprogenous A1 horizons in forest soils of Wisconsin. *Soil Science Society of America, Proceedings*, **28**, 426-30.

Nkunika, P.O.Y. (1990) Field composition and size of the populations of the primitive damp wood termite, *Porotermes adamsoni* (Isoptera: Termopsidae) in South Australia. *Sociobiology*, **16**, 251-8.

Nogueira, S.B. and Martinho, M.R. (1983) Leaf-cutting ants (*Atta sp.*) damage to and distribution along Brazilian roads, in *Social Insects in the Tropics* (ed. P. Jaisson), Université Paris-Nord, Paris, pp. 181-6.

Noirot, C. (1970) The nests of termites, in *Biology of Termites* (ed. K. Krishna and F.M. Weesner), Academic Press, New York, pp. 73-125.

Noirot, C. (1989) Social structure in termite societies. *Ethology, Ecology and Evolution*, **1**, 1-17.

Noirot, C. (1990) Sexual castes and reproductive strategies in termites, in *Social Insects, An Evolutionary Approach to Castes and Reproduction* (ed. W. Engels), Springer-Verlag, Berlin, pp. 5-35.

Noirot, C. (1992) From wood-feeding to humus-feeding: an important trend in termite evolution, in *Biology and Evolution of Social Insects* (ed. J. Billen), Leuven University Press, Louvain, pp. 107-19.

Nooren, C.A.M., van Breemen, N., Stoorvogel, J.J. *et al.* (1995) The role of earthworms in the formation of sandy surface soils in a tropical forest in Ivory Coast. *Geoderma*, **65**, 135-48.

Nordström, S. and Rundgren, S. (1974) Environmental factors and lumbricid associations in southern Sweden. *Pedobiologia*, **14**, 1-27.

Norrish, K. and Rosser, H. (1983) Mineral phosphate, in *Soils, An Australian Viewpoint*, CSIRO/Academic Press, Melbourne/London, pp. 335-61.

Norton, D.C. (1978) *Ecology of Plant Parasitic Nematodes*, John Wiley & Sons, New York.

Norton, R.A. (1990) Acarina: Oribatida, in *Soil Biology Guide* (ed. D.L. Dindal), John Wiley & Sons, New York, pp. 779-804.

Nosek, J. (1973) *The European Protura: Their Taxonomy, Ecology and Distribution with Keys for Determination*, Museum d'Histoire Naturelle, Geneva.

Novack, R.O. and Whittingham, J. (1968) Soil and litter microfungi of the maple-elm-ash floodplain community. *Mycologia*, **60**, 776-87.

Nusbaum, C.J. and Ferris, H. (1973) The role of cropping systems in nematode population management. *Annual Review of Phytopathology*, **11**, 423-40.

Nye, P.H. (1954) Some soil-forming processes in the humid tropics. I. A field study of a catena in the West African forest. *Journal of Soil Science*, **5** (1), 7-21.

Nye, P.H. (1955a) Some soil-forming processes in the humid tropics. II. The development of the upper slope member of the catena. *Journal of Soil Science*, **6** (1), 51-62.

Nye, P.H. (1955b) Some soil-forming processes in the humid tropics. III. Laboratory studies on the development of a typical catena over a granitic gneiss. *Journal of Soil Science*, **6** (1), 63-72.

Nye, P.H. (1955c) Some soil-forming processes in the humid tropics. IV. The action of the soil fauna. *Journal of Soil Science*, **6** (1), 73-83.

Nye, P.H. and Tinker, P.B. (1978) *Solute Movement in the Soil-Root System*, Blackwell Scientific Publications, Oxford.

Nye, P.H. (1981) Changes of pH across the rhizosphere induced by roots. *Plant and Soil*, **61**, 7-26.

O'Connor, F.B. (1967) The Enchytraeidae, in *Soil Biology* (eds A. Burges and R. Raw), Academic Press, New York, pp. 213-56.

Oades, J.M. (1987) Aggregation in soils, in *Soil Structure and Aggregate Stability* (ed. P. Rengasamy), Institute for Irrigation and Salinity Research, Tatura, pp. 74-101.

Oades, J.M. (1993) The role of biology in the formation, stabilization and degradation of soil structure. *Geoderma*, **56**, 377-400.

Odum, H.T. (1970) Summary: an emerging view of the ecological system at El Verde, in *A Tropical Rain Forest* (eds H.T. Odum and R.F. Pigeon), Division of Technical Information, US Atomic Energy Commission, Oak Ridge, Tennessee, pp. I.191-I.289.

Odum, E.P. (1983) *Basic Ecology*, Saunders College Publishing, Philadelphia.

Odum, E.P. (1985) Trends expected in stressed ecosystems. *Bioscience*, **35**, 419-21.

Okello-Oloya, T., Spain, A.V. and John, R.D. (1985) Selected chemical characteristics of the mounds of two species of *Amitermes* (Isoptera, Termitinae) and their adjacent surface soils from northeastern Australia. *Revue d'Ecologie et de Biologie du Sol*, **22** (3), 291-311.

Okello-Oloya, T. and Spain, A.V. (1986) Comparative growth of two pasture plants from north eastern Australia on the mound materials of grass and litter-feeding termites (Isoptera: Termitidae) and on their associated surface soil. *Revue d'Ecologie et de Biologie du Sol*, **23**, 381-92.

Oldeman, R.A.A. (1990) *Forests: Elements of Silvology*, Spinger Verlag, Berlin.

Oliver, J.H. (1971) Parthenogenesis in mites and ticks (Arachnida: Acari). *American Zoology*, **11** (2), 283-99.

Olson, J.S. (1963) Energy storage and the balance of producers and decomposers in ecological systems. *Ecology*, **44**, 322-31.

Omodeo, P. (1952) Particolarita della zoogeografia dei lombrichi. *Bollettino di Zoologia*, **XIX** (4-5-6), 349-69.

Ouedraogo, P. (1997) *Rôle des termites dans la structure et la dynamique d'une brousse tigrée soudano-sahélienne*, thèse de 3ᵉ cycle, Université Paris VI.

Ovington, J.D. (1961) Some aspects of energy flow in plantations of *Pinus sylvestris* L. *Annals of Botany*, **25**, 12-20.

Paine, R.T. (1966) Food web complexity and species diversity. *American Naturalist*, **100**, 65-75.

Palm, C.A. and Sanchez, P.A. (1990) Decomposition and nutrient release patterns of the leaves of three tropical legumes. *Biotropica*, **22**, 330-8.

Palm, C.A. and Sanchez, P.A. (1991) Nitrogen release from the leaves of some tropical legumes as affected by their lignin and polyphenolic contents. *Soil Biology & Biochemistry*, **23** (1), 83-8.

Pan, W.L., Jackson, W.A. and Moll, R.H. (1985) Nitrate uptake and partitioning by Corn *(Zea mays* L.*)* root systems and associated morphological differences among genotypes and stages of root development. *Journal of Experimental Botany*, **36** (170), 1341-51.

Papendick, R.I., Cochran, V.I. and Woody, W.M. (1971) Soil water potential and water content profiles with wheat under low spring and summer rainfall. *Agronomy Journal*, **63**, 731-4.

Papendick, R.I. and Campbell, G.S. (1981) Theory and measurement of water potential, in *Water Potential Relations in Soil Microbiology* (eds J.F. Parr, W.R. Gardner and L.F. Elliot), Soil Science Society of America, Madison, Wisconsin, pp. 1-22.

Parker, L.W., Fowler, H.G., Ettershank, G. *et al.* (1982) The effects of subterranean termite removal on desert soil nitrogen and ephemeral flora. *Journal of Arid Environments*, **5**, 53-9.

Parkinson, D. (1982) Filamentous fungi, in *Methods of soil analysis, Part 2. Chemical and microbial properties* (eds A.L. Page, R.H. Miller and D.R. Keeney), American Society of Agronomy, Soil Science Society of America, Madison, Wisconsin, pp. 949-68.

Parle, J.N. (1963) Microorganisms in the intestines of earthworms. *Journal of General Microbiology*, **31**, 1-11.

Parmelee, R.W. and Alston, D.G. (1986) Nematode trophic structure in conventional and no-tillage agroecosystems. *Journal of Nematology*, **18** (3), 403-7.

Parmelee, R.W., Beare, M.H., Cheng, W. *et al.* (1990) Earthworms and enchytraeids in conventional and no-tillage agroecosystems: a biocide approach to assess their role in organic matter breakdown. *Biology and Fertility of Soils*, **10**, 1-10.

Parmelee, R.W., Ehrenfeld, J.G. and Tate, R.L.I. (1993) Effects of pine roots on microorganisms, fauna, and nitrogen availability in two soil horizons of a coniferous forest spodosol. *Biology and Fertility of Soils*, **15**, 113-9.

Parton, W.J., Anderson, D.W., Cole, C.V. *et al.* (1983) Simulation of soil organic matter formation and mineralization in semiarid agroecosystems, in *Nutrient Cycling in Agricultural Ecosystems* (eds R.R. Lowrance, R.L. Todd, L.E. Asmussen *et al.*), Georgia Experimental Station, Athens, Georgia, pp. 533-50.

Pashanasi, B., Lavelle, P., Alegre, J. *et al.* (1996) Effect of the endogeic earthworm *Pontoscolex corethrurus* on soil chemical characteristics and plant growth in a low-input tropical agroecosystem. *Soil Biology & Biochemistry*, **28** (6), 801 - 10.

Paton, T.R., Humphreys, G.S. and Mitchell, P.B. (1995) *Soils: A New Global View*, Yale University Press, New Haven.

Patrick, W.H. and Jugsujinda, A. (1992) Sequential reduction and oxidation of inorganic nitrogen, manganese, and iron in flooded soil. *Soil Science Society of America, Proceedings*, **56**, 1071-3.

Patriquin, D.G. (1982) Nitrogen fixation in sugarcane litter. *Biological Agriculture and Horticulture*, **1**, 39-64.

Patwardhan, A.S., Nieber, J.L. and Moore, I.D. (1988) Oxygen, carbon dioxide, and water transfer in soils: mechanisms and crop response. *Transactions of the American Society of Agricultural Engineers*, **31**, 1383-95.

Pau, R.N. (1989) Nitrogenases without molybdenum. *Trends in Biochemical Sciences*, **14**, 183-6.

Paul, E.A. (1984) Dynamics of organic matter in soils. *Plant and Soil*, **76**, 275-85.

Paul, E.A. and Clark, F.E. (1989) *Soil Microbiology and Biochemistry*, Academic Press, London.

Peachey, J.E. (1963) Studies on Enchytraeidae (Oligochaeta) of moorland soils, in *Progress in Soil Zoology* (ed. P.H. Murphy), Butterworths, London, pp. 286-93.

Peakall, R., Handel, S.M. and Beattie, A.J. (1991) The evidence for, and importance of, ant pollination, in *Ant-Plant Interactions* (eds C.R. Huxley and D.F. Cutler), Oxford University Press, Oxford, pp. 421-9.

Pearce, M.J. and Waite, B.S. (1994) A list of termite genera (Isoptera) with comments on taxonomic changes and regional distribution. *Sociobiology*, **23** (3), 247-63.

Pearson, R. and Parkinson, D. (1961) The sites of excretion of ninhydrin positive substances by Broad Bean seedlings. *Plant and Soil*, **13**, 391-6.

Pedro, G. (1964) Contribution à l'étude expérimentale de l'altération géochimique des roches cristallines. *Annales Agronomiques*, **15**, 85-191, 243-333, 9-456.

Pedro, G. (1979) Caractérisation générale des processus de l'altération hydrolytique. Base des méthodes géochimique et thermodynamique. *Science du Sol*, **2-3**, 93-105.

Pedro, G. and Siefferman, G. (1979) Weathering of rocks and formation of soils, in *Review of Research in Modern Problems of Geochemistry* (ed. F.R. Siegel), UNESCO, Paris, pp. 39-55.

Pedro, G. (1983) Les constituants amorphes des sols, éléments d'introduction. *Sciences du Sol,* **3-4**, 137-52.

Pedro, G. (1989) Geochemistry, mineralogy and microfabric of soils, in *Soils and their Management: a Sino-European Perspective* (eds E. Maltby and T. Wollersen), Elsevier Applied Sciences, Bruxelles, pp. 59-90.

Peoples, M.B., Freney, J.R. and Mosier, A.R. (1995) Minimizing gaseous losses of nitrogen, in *Nitrogen Fertilization in the Environment* (ed. P.E. Bacon), Marcel Dekker, N.Y., pp. 565-602.

Perrin, R. (1990) Interactions between mycorrhizae and diseases caused by soil-borne fungi. *Soil Use and Management*, **6**, 189-95.

Perry, D.H., Lenz, M. and Watson, J.A.L. (1985) Relationships between fire, fungal rots and termite damage in Australian forests. *Australian Forestry*, **48**, 46-53.

Persson, T. and Lohm, U. (1977) *Energetical Significance of the Annelids and Arthropods in a Swedish Grassland Soil*, NFR, Stockhom.

Persson, T. (1978) Root dynamics in a young Scots Pine stand in Central Sweden. *Oikos*, **30**, 508-19.

Persson, T. (1980) Death and replacement of fine roots in a mature Scots Pine stand. *Ecological Bulletins*, **32**, 252-60.

Persson, H. (1982) Changes in the tree and dwarf shrub fine roots in mature Scots Pine stand. *Swedish Coniferous Forest Project, Technical Report*, **31**.

Persson, H. (1985) The dynamics of fine roots of forest trees. *Ekologiya*, **4**, 33-9.

Persson, T. (1988) Effects of liming on the soil fauna in forest: a literature review. *Naturvardsverket, Stockholm, Report,* **3418**.

Persson, T. (1989) Role of soil animals in C and N mineralization, in *Ecology of Arable Land* (eds M. Clarholm and L. Bergström), Kluwer Academic Publishers, Dordrecht, pp. 185-9.

Persson, T., Bååth, E., Clarholm, M. *et al.* (1980) Trophic structure, biomass dynamics and carbon metabolism of soil organisms in a Scots Pine forest. *Ecological Bulletins*, **32**, 419-59.

Petal, J. (1978) The role of ants in ecosystems, in *Production Ecology of Ants and Termites* (ed. M. Brian), Cambridge University Press, Cambridge, pp. 292-325.

Petersen, H. (1980) Population dynamics and metabolic characterization of Collembola species in a Beech forest ecosystem, in *Proceedings of the 7th International Soil Zoology Colloquium* (ed. D.L. Dindal), E.P.A., Washington D.C., pp. 806-33.

Petersen, H. and Luxton, M. (1982) A comparative analysis of soil fauna populations and their role in decomposition processes. *Oikos*, **39** (3), 287-388.

Petersen, H. (1995) Temporal and spatial dynamics of soil Collembola during secondary succession in Danish heathland. *Acta Zoologica Fennica*, **196**, 190 - 4.

Petersen, L.W., Moldrup, P., Jacobson, O.H. *et al.* (1996) Relations between specific surface area and soil physical and chemical properties. *Soil Science*, **161**, 9-21.

Peterson, B.J. and Fry, B. (1987) Stable isotopes in ecosystem studies. *Annual Review of Ecology and Systematics*, **18**, 239-320.

Pettersson, R., Hansson, A.C., Andrén, O. *et al.* (1986) Above- and below-ground production and nitrogen uptake in Lucerne *(Medicago sativa)*. *Swedish Journal of Agricultural Research*, **16**, 167-77.

Petz, W., Foissner, W., Wirnsberger, E. *et al.* (1986) Mycophagy, a new feeding strategy in autochthonous soil ciliates. *Naturwissenschaften*, **73** (570).

Philips, J.R. (1990) Acarina: Astigmata (Acaridida), in *Soil Biology Guide* (ed. D.L. Dindal), John Wiley & Sons, New York, pp. 757-78.

Phillipson, J., Abel, R., Steel, J. *et al.* (1976) Earthworms and the factors governing their distribution in a Beech wood. *Oecologia*, **33**, 291-309.

Phillipson, J. (1983) Slug numbers, biomass and respiratory metabolism in a Beech woodland, Wytham, Oxford. *Oecologia*, **60**, 38-45.

Pianka, E.R. (1970) On *r*- and *K*-selection. *American Naturalist*, **104**, 592-7.

Piearce, T.G. (1978) Gut contents of some lumbricid earthworms. *Pedobiologia*, **18**, 3-157.

Piearce, T.G. and Phillips, M.J. (1980) The fate of ciliates in the earthworm gut: an *in vitro* study. *Microbial Ecology*, **5**, 313-9.

Pipe, A.E. and Shubert, L.E. (1984) The use of algae as indicators of soil fertility, in *Algae as Ecological Indicators*, (ed. L.E. Shubert), Academic Press, London, pp. 214-33.

Pirozynski, K.A. and Dalpé, Y. (1989) Geological history of the Glomaceae with particular reference to mycorrhizal symbiosis. *Symbiosis*, **7**, 1-36.

Pochon, J. and Tardieux, P. (1962) *Techniques d'Analyses en Microbiologie du Sol*, Editions La Tourelle, Paris.

Poesen, J. and Lavee, H. (1994) Rock fragments in top soils: significance and processes. *Catena*, **23**, 1-28.

Poinsot, N. (1968) Cas d'anhydrobiose chez le collembole *Subisotoma variabilis* Gisin. *Revue d'Ecologie et de Biologie du Sol*, **5**, 585-6.

Pojasok, T. and Kay, B.D. (1990) Effect of root exudates from corn and Bromegrass on soil structural stability. *Canadian Journal of Soil Science*, **70**, 351-62.

Polglase, P.J. and Attiwill, P.M. (1992) Nitrogen and phosphorus cycling in relation to stand age of *Eucalyptus regnans* F. Muell. 1. Return from plant to soil in litterfall. *Plant and Soil*, **142**, 157-66.

Ponge, J.F. (1980) Les biocénoses des collemboles de la Forêt de Sénart, in *Actualités d'Ecologie Forestière* (ed. P. Pesson), Gauthier-Villars, Paris, pp. 151-76.

Ponge, J.F. (1983) Les collemboles, indicateurs du type d'humus en milieu forestier. Résultats obtenus au sud de Paris. *Acta Œcologica, Œcologia Generalis*, **4**, 359-74.

Ponge, J.F. (1984) Etude écologique d'un humus forestier par observation d'un petit volume. I. Couche L1 d'un moder sous *Pinus sylvestris*. *Revue d'Ecologie et de Biologie du Sol*, **21**, 161-87.

Ponge, J.F. (1985) Etude écologique d'un humus forestier par l'observation d'un petit volume. II. La couche de L2 d'un moder sous *Pinus sylvestris*. *Pedobiologia*, **28**, 73-114.

Ponge, J.F. (1988) Etude écologique d'un humus forestier par l'observation d'un petit volume. III. La couche de F1 d'un moder sous *Pinus sylvestris*. *Pedobiologia*, **31**, 1-64.

Ponge, J.F. (1990) Ecological study of a forest humus by observing a small volume. I - Penetration of pine litter by mycorrhizal fungi. *European Journal of Forest Pathology*, **20**, 290-303.

Ponge, J.F. (1991) Food resources and diets of soil animals in a small area of Scots Pine litter. *Geoderma*, **49**, 33-62.

Pontin, A.J. (1963) Further considerations of competition and the ecology of the ants *Lasius flavius* (F.) and *L. niger* (L.). *Journal of Animal Ecology*, **32** (3), 565-74.

Pop, V.V. and Postolache, T. (1987) Giant earthworms build up vermic mountain rendzinas, in *On Earthworms* (eds A.M. Bonvicini Pagliai and P. Omodeo), Mucchi, Modena, pp. 141-50.

Pope, D.F., Thompson, A.C. and Cole, A.W. (1985) Phytotoxicity of root exudates and leaf extracts of nine plant species, in *The Chemistry of Allelopathy. Biochemical Interactions among Plants* (ed. A.C. Thompson), American Chemical Society, Washington D.C., pp. 219-34.

Post, W.M., Emmanuel, W.R., Zinke, P.J. *et al.* (1982) Soil carbon pools and world life zones. *Nature*, **298**, 156-9.

Post, W.M., Pastor, J., Zinke, P.J. *et al.* (1985) Global patterns of soil nitrogen storage. *Nature*, **317**, 613-6.

Post, W.M., Chávez, F., Mulholland, P.J. *et al.* (1992) Climatic feedbacks in the global carbon cycle, in

The Science of Global Change, The Impact of Human Activities on the Environment (eds D.A. Dunnette and R.J. O'Brien), American Chemical Society, Washington D.C., pp. 392-412.

Postma, J. and van Veen, J.A. (1990) Habitable pore space and survival of *Rhizobium leguminosarum* biovar *trifolii* introduced into soil. *Microbial Ecology*, **19**, 149-61.

Potter, D.A., Bridges, B.L. and Carter, G.F. (1985) Effect of N fertilization on earthworm and microarthropod populations in Kentucky bluegrass turf. *Journal of Agronomy*, **77**, 367-72.

Potts, R.C. and Hewitt, P.H. (1974) Some properties and reaction characteristics of the partially purified cellulase from the termite *Trinervitermes trinervoides* (Nasutitermitinae). *Comparative Biochemistry and Physiology*, **47 B**, 327-37.

Price, D.W. (1973) Abundance and vertical distribution of microarthropods in the surface layers of a Californian pine forest soil. *Hilgardia*, **42** (4), 121-48.

Price, P.W. (1984) Alternative paradigms in community ecology, in *A New Ecology: Novel Approaches to Interactive Systems* (eds P.W. Price, C.N. Slobodchikoff and W.S. Gaud), John Wiley, New York, pp. 1-14.

Prikryl, Z. and Vancura, V. (1980) Root exudates of plants. VI. Wheat root exudation as dependent on growth, concentration gradient of exudates and the presence of bacteria. *Plant and Soil*, **57**, 69-83.

Procter, D.L.C. (1977) Nematode densities and production on Truelove Lowland, in *Truelove Lowland, Devon Island, Canada: A High Arctic Ecosystem* (ed. L.C. Bliss), University of Alberta Press, pp. 347-61.

Procter, D.L.C. (1990) Global overview of the functional roles of soil-living nematodes in terrestrial communities and ecosystems. *Journal of Nematology*, **22** (1), 1-7.

Prot, J. (1980) Migration of plant parasitic nematodes towards plant roots. *Revue de Nématologie*, **3**, 305-18.

Proth, J. (1978) *Evolution de la microflore d'une rendzine forestière récemment privée de ses apports naturels en litière de Charme. Etude microbiologique et ultra structurale*, thèse de 3ᵉ cycle, Université Nancy I.

Pugh, G.J.F. (1980) Strategies in fungal ecology. *Transactions of the British Mycological Society*, **75**, 1-14.

Pye, K. (1987) *Aeolian Dust and Dust Deposits*, Academic Press, London.

Quinlan, R.J. and Cherret, J.M. (1979) The role of fungus in the diet of the leaf-cutting ant *Atta cephalotes* (L.). *Ecological Entomology*, **4**, 151-60.

Quispel, A., Rodriguez-Barrueco, C. and Subba Rao, N.S. (1993) Some general considerations on symbioses of nitrogen-fixing trees, in *Symbioses in Nitrogen-Fixing Trees* (eds N.S. Subba Rao and C. Rodriguez-Barrueco), Oxford and IBH Publishing Co., New Delhi, pp. 1-32.

Rademacher, W. (1994) Gibberellin formation in microorganisms. *Plant Growth Regulation*, **15** (3), 303-14.

Radulovich, R., Solorzano, E. and Sollins, P. (1989) Soil macropore size distribution from water breakthrough curves. *Soil Science Society of America, Proceedings*, **53**, 556-9.

Radulovich, R. and Sollins, P. (1991) Nitrogen and phosphorus leaching in zero-tension drainage from a humid tropical soil. *Biotropica*, **23**, 84-7.

Rafidison, G.Z. (1982) *Rôle de la faune dans l'humification: transformation des feuilles de hêtre par un ver anécique* (Nicodrilus velox), thèse de 3ᵉ cycle, Université Nancy I.

Rapaire, J.L. and Turenne, J.F. (1977) Mesures d'activité spécifique de fractions de matière organique appliquées à l'étude de l'évolution des sols de Guyane, in *Soil Organic Matter Studies*, IAEA, Vienne, pp. 179-86.

Rashid, G.H. and Schaefer, R. (1985) The seasonal pattern of carbon dioxide evolution from two temperate forest catena soils. *Revue d'Ecologie et de Biologie du Sol*, **22** (4), 419-31.

Ratcliffe, F.N., Gay, F.J. and Greaves, T. (1952) *Australian Termites*, CSIRO, Melbourne.

Raw, F. (1967) Arthropoda (except Acari and Collembola), in *Soil Biology* (eds A. Burges and F. Raw), Academic Press, London, pp.323-62.

Read, D.J., Malibari, A. and Whittingham, J. (1979) Nutrient and water transport between plants connected by mycorrhizal fungi, in *Proceedings of the 4th North American Conference on Mycorrhizae*, Fort Collins, Colorado.

Read, D.J., Francis, R. and Finlay, R.D. (1985) Mycorrhizal mycelia and nutrient cycling in plant communities, in *Ecological Interactions in Soil ; Plants, Microbes and Animals* (eds A.H. Fitter, D. Atkinson, D.J. Read *et al.*), Blackwell Scientific Publications, Oxford, pp. 193-217.

Read, D.J., Leake, J.R. and Langdale, A.R. (1989) The nitrogen nutrition of mycorrhizal fungi and their host plants, in *Nitrogen, Phosphorus and Sulphur Utilization by Fungi* (eds L. Boddy, R. Marchant and D.J. Read), Cambridge University Press, Cambridge, pp. 181-204.

Read, D.J. (1991) Mycorrhizas in ecosystems. *Experientia*, **47**, 376-91.

Reagan, T.E. (1986) Beneficial effects of the imported fire ant: a field ecology approach, in *Fire Ants and Leaf-Cutting Ants* (eds C.S. Lofgren and R.K. Van der Meer), Westview Press, Boulder, Colorado, pp. 58-71.

Reddell, P. and Spain, A.V. (1991a) Earthworms as vectors of viable propagules of mycorrhizal fungi. *Soil Biology & Biochemistry*, **23**, 767-74.

Reddell, P. and Spain, A.V. (1991b) Transmission of infective *Frankia* (Actinomycetales) propagules in casts of the endogeic earthworm *Pontoscolex corethrurus* (Oligochaeta: Glossoscolecidae). *Soil Biology & Biochemistry*, **23**, 775-78.

Reddell, P.R., Spain, A.V., Milnes, A.R. *et al.* (1992) Indicators of ecosystem recovery, in *Proceedings of the 7th Annual Environmental Workshop, Yeppoon Queensland*, Australian Mining Industry Council, Dickson, pp. 115-27.

Reeve, R., Isbell, R.F. and Hubble, G.D. (1963) Soil and climatic data for the Brigalow lands, eastern Australia. *CSIRO Australia, Division of Soils, Divisional Report*, **7/61**.

Reichle, D.E., McBrayer, J.F. and Ausmus, S. (1975) Ecological energetics of decomposer invertebrates in a deciduous forest and total respiration budget, in *Progress in Soil Zoology* (ed. J. Vanek), Academia, Prague, pp. 283-93.

Reichman, O.J. and Smith, S.C. (1990) Burrows and burrowing behaviour by mammals, in *Current Mammalogy* (ed. H.H. Genoways), Plenum Press, New York, pp. 197-244.

Reisinger, O., Toutain, F., Mangenot, F. *et al.* (1978) Etude ultrastructurale des processus de biodégradation. I. Pourriture blanche des feuilles de Hêtre (*Fagus sylvatica* L.). *Canadian Journal of Microbiology*, **24**, 725-33.

Reisinger, O. and Kilbertus, G. (1980) Mécanismes et facteurs de biodégradation en milieu forestier, in *Actualité d'Ecologie Forestière* (ed. P. Pesson), Gauthier-Villars, Paris, pp. 61-86.

Rempe, E.K.H. (1972) Significance of vitamins synthetized by the root microflora. *Fiziol Rast*, **19**, 663-8.

Renard, K.G., Foster, G., Yoder, D. *et al.* (1994) RUSLE revisited: status, questions answers and the future. *Journal of Soil and Water Conservation*, **49**, 213-20.

Reversat, G. (1981a) Effects of ageing and starvation on respiration and food reserve content in adult *Hirschmanniella spinicaudata*. *Revue de Nématologie*, **4** (1), 125-30.

Reversat, G. (1981b) Consumption of food reserves by starved second stage juveniles of *Meloidogyne javanica* under conditions inducing osmobiosis. *Nematologica*, **27**, 207-14.

Reynolds, J.W. and Cook, D.G. (1976) *Nomenclatura Oligochaetologica*, The University of New Brunswick, Fredericton.

Rice, E.L. (1974) *Allelopathy*, Academic Press, London.

Richter, M., Wilms, W. and Scheffer, F. (1968) Determination of root exudates in a sterile continuous flow-culture. II: Short term and long term variations of exudation intensity. *Plant Physiology*, **43**, 1743-54.

Ricou, G. (1978) La prairie permanente du nord-ouest français, in *Problèmes d'Ecologie: Ecosystèmes Terrestres* (eds M. Lamotte and F. Bourlière), Masson, Paris, pp. 17-74.

Rinaudo, G., Dreyfus, B. and Dommergues, Y. (1983) *Sesbania rostrata* green manure and the nitrogen content of rice crops and soil. *Soil Biology & Biochemistry*, **15** (1), 111-13.

Ringrose-Voase, A.J. (1987) A scheme for the quantitative description of soil macrostructure by image analysis. *Journal of Soil Science*, **38**, 343-56.

Ripley, E.A., Redmann, R.E. and Crowder, A.A. (1996) *Environmental Effects of Mining*, St Lucie Press, Delray Beach, Florida.

Rissing, S.W. and Wheeler, J. (1976) Foraging responses of *Veromessor pergandei* to changes in seed production (Hymenoptera: Formicidae). *Pan-Pacific Entomologist*, **52** (1), 63-72.

Rittenhouse, R.L. and Hale, M.G. (1971) Loss of organic compounds from roots. II. Effect of O_2 and CO_2 tension on release of sugars from peanut roots under axenic conditions. *Plant and Soil*, **35**, 311-21.

Robert, M. and Delmas (1984) La pédologie expérimentale et les aspects géochimiques et minéralogiques de l'évolution des sols, in *Livre Jubilaire du Cinquantenaire* (ed. A.F.E.S.), Maugein-Lachaise, Malermont, pp. 195-209.

Robert, M. and Berthelin, J. (1986) Role of biological and biochemical factors in soil mineral weathering, in *Interactions of Soil Minerals with Natural Organics and Microbes* (eds P.H. Huang and M. Schnitzer), Soil Science Society of America, Madison, Wisconsin, pp. 453-94.

Robert, M. and Tessier, D. (1992) Incipient weathering: some new concepts on weathering, clay formation and organisation, in *Weathering, Soils and Palaeosols* (eds I.P. Martini and W. Chesworth), Elsevier, Amsterdam, pp. 71-105.

Robertson, G.P. and Rosswall, T. (1986) Nitrogen in West Africa: the regional cycle. *Ecological Monographs*, **56** (1), 43-72.

Robertson, G.P. (1994) The impact of soil and crop management practices on soil spatial heterogeneity, in *Soil Biota. Management in Sustainable Farming Systems* (eds C.E. Pankhurst, B.M. Doube, V.V.S.R. Gupta *et al.*), CSIRO, Melbourne, pp. 156-61.

Robinson, J.B.D. (1958) Some chemical characteristics of "termite soils" in Kenya coffee fields. *Journal of Soil Science*, **9**, 58-65.

Robinson, D. and Rorison, I.H. (1983) A comparison of the responses of *Lolium perenne* L, *Holcus lanatus* L. and *Deschampsia flexuosa* (L.) Trin. to a localised supply of nitrogen. *New Phytologist*, **94**, 263-73.

Roces, F. and Nunez, J.A. (1989) Brood translocation and circulation variation in temperature preference in the ant *Camponotus mus*. *Oecologia*, **81**, 33-7.

Rodgers, G.A. and Stewart, W.D.P. (1977) The cyanophyte-hepatic symbiosis. 1. Nitrogen fixation and the interchange of nitrogen and carbon. *New Phytologist*, **78**, 459-71.

Rodin, L.E. and Bazilevich, N.I. (1967) *Production and Mineral Cycling in Terrestrial Vegetation*, Oliver & Boyd, London.

Rodriguez, C., Bettucci, L. and Roquebert, M.F. (1990) Fungal communities of volcanic ash soils along an altitudinal gradient in Mexico. I. Composition and organisation. *Pedobiologia*, **34**, 43-9.

Roessner, J. (1986) Untersuchungen zur Reduktion von Nematoden im Bodem durch Regenwuermer. *Mededelingen van de Rijksuniversiteit Landbouwwentenschappen Gent*, **51** (3b), 1311-8.

Roger, P.A. (1995) Biological N_2-fixation and its management in wetland rice cultivation. *Fertilizer Research*, **42**, 261-76.

Room, P.M. (1971) The relative distribution of ant species in Ghana's cocoa farms. *Journal of Animal Ecology*, **40** (3), 735-51.

Room, P.M. (1975) Diversity and organization of the ground foraging ant faunas of forest, grassland and tree crops in Papua New Guinea. *Australian Journal of Zoology*, **23**, 71-89.

Roonwal, M.L. (1970) Termites in the oriental region, in *Biology of Termites, vol. 2* (eds K. Krishna and

F.M. Weesner), Academic Press, New York, pp. 315-91.

Roose, E.J. and Lelong, F. (1976) Les facteurs de l'érosion hydrique en Afrique tropicale. Etude sur petites parcelles expérimentales de sol. *Revue de Géographie Physique et de Géologie Dynamique*, **4**, 365-74.

Roose, E. (1980) *Dynamique actuelle de sols ferrallitiques et ferrugineux tropicaux d'Afrique Occidentale*, thèse d'Etat, Université d'Orléans.

Roose, E., Kabore, V. and Guenat, C. (1996) Le zai: fonctionnement, limites et amélioration d'une pratique traditionnelle africaine de réhabilitation de la végétation et de la productivité des terres dégradées en région soudano-sahélienne (Burkina-Faso). *Cahiers ORSTOM, série Pédologie*, **28**, 159-79.

Rose, C.J. and Wood, A.W. (1980) Some environmental factors affecting earthworm populations and Sweet Potato production in the Tari Basin, Papua New Guinea highlands. *Papua New Guinea Agricultural Journal*, **31** (1-4), 1-13.

Ross, D.J., Tate, K.R., Cairns, A. *et al.* (1980) Microbial biomass estimations in soils from tussock grasslands by three biochemical procedures. *Soil Biology & Biochemistry*, **12**, 275-383.

Ross, P.J., Williams, J. and McCrown, R.L. (1985) Soil temperature and the energy balance of vegetative mulch in the semi-arid tropics. I. Static analysis of the radiation balance. *Australian Journal of Soil Research*, **23**, 493-514.

Ross, D.J. (1986) Soils of the Wycanna Woodland Experiment Centre, Talwood, South Maranoa Region, Queensland. *CSIRO Australia, Division of Soils, Divisional Report,* No.**83**.

Rossi, J.P., Lavelle, P. and Tondoh, J.E. (1995) Statistical tool for soil biology X. Geostatistical analysis. *European Journal of Soil Biology*, **31** (4), 173-81.

Rossi, J.P., Delaville, L. and Quénéhervé, P. (1996) Microspatial structure of a plant parasitic nematode community in a sugarcane field in Martinique. *Applied Soil Ecology*, **3**, 17-26.

Rossi, J.P., Lavelle, P. and Albrecht, A. (1997) Relationships between spatial pattern of the endogeic earthworm *Polypheretima elongata* and soil heterogeneity. *Soil Biology & Biochemistry*, **29** (3-4), 485-8.

Rossi, J.P. (1998) *Rôle fonctionnel de la distribution spatiale des vers de terre dans une savane humide de Côte d'Ivoire*, thèse de l'Université Paris VI.

Rossi, J.P. and Lavelle, P. (1998) Earthworm aggregation in the savannas of Lamto (Côte d'Ivoire). *Applied Soil Ecology*, **7** (2), 195-200.

Rouatt, J.W., Katznelson, H. and Payne, T.M.B. (1960) Statistical evaluation of the rhizosphere effect. *Soil Science Society of America, Proceedings*, **24**, 271-73.

Rouelle, J. (1977) *Ecologie du cycle de l'azote*, thèse de 3e cycle, Université de Lyon.

Rouelle, J. (1983) Introduction of amoebæ and *Rhizobium japonicum* into the gut of *Eisenia foetida* (Sav.) and *Lumbricus terrestris* L., in *Earthworm Ecology* (ed. J.E. Satchell), Chapman & Hall, London, pp. 375-81.

Rouelle, J., Pussard, M., Randriamamonjizaka, J.L. *et al.* (1985) Interactions microbiennes (Bactéries, Protozoaires), alimentation des vers de terre et minéralisation de la matière organique. *Bulletin d'Ecologie*, **16** (1), 83-8.

Rouland, C., Civas, A., Renoux, J. *et al.* (1988a) Synergistic activities of the enzymes involved in cellullose degradation, purified from *Macrotermes mülleri* and from its symbiotic fungus *Termitomyces* sp. *Comparative Biochemistry and Physiology*, **91 B** (3), 459-65.

Rouland, C., Civas, A., Renoux, J. *et al.* (1988b) Purification and properties of cellulases from the termite *Macrotermes mülleri* (Termitidae, Macrotermitinae) and its symbiotic fungus *Termitomyces* sp. *Comparative Biochemistry and Physiology*, **91 B** (3), 449-58.

Rouland, C., Brauman, A., Keleke, S. *et al.* (1990) Endosymbiosis and exosymbiosis in the fungus-growing termites, in *Microbiology in Poecilotherms* (ed. R. Lésel), Elsevier Science Publishers, Amsterdam, pp. 79-82.

Rouland, C., Brauman, A., Labat, M. *et al.* (1993) Nutritional factors affecting methane emission from termites. *Chemosphere*, **26** (1-4), 617-22.

Rovira, A.D. (1959) Root excretions in relation to the rhizosphere effects. IV. Influence of plant species, age of plant, light, temperature and calcium nutrition on exudation. *Plant and Soil*, **11**, 53-64.

Rovira, A.V. and Davey, C.B. (1974) Biology of the rhizosphere, in *The Plant Root and its Environment* (ed. E.W. Carson), University Press, Charlottesville, Virginia, pp. 153-204.

Rovira, A.D. (1979) Biology of the soil-root interface, in *The Soil-Root Interface* (eds J.L. Harley and R. Scott Russell), Academic Press, London, pp. 145-60.

Rovira, A.D., Foster, R.C. and Martin, J.K. (1979) Note on terminology: origin, nature and nomenclature of the organic materials in the rhizosphere, in *The Soil-Root Interface*, (eds J. L. Harley and R. Scott Russell), Academic Press, London, pp. 1-4.

Rowell, M.J., Ladd, J.N. and Paul E.A. (1973) Enzymically active complexes of proteases and humic analogues. *Soil Biology & Biochemistry*, **14**, 257-63.

Römbke, J. (1991) Estimates of the Enchytraeidae (Oligochaeta, Annelida) contribution to energy flow in the soil system of an acid Beech wood forest. *Biology and Fertility of Soils*, **11**, 255-60.

Rusek, J. (1985) Soil microstructures-contributions of specific soil organisms. *Quaestiones Entomologicae*, **21**, 497-514.

Rushton, S.P. and Hassal, M. (1983) Food and feeding rates of the terrestrial isopod *Armadillidium vulgare* (Latreille). *Oecologia*, **57**, 415-9.

Ryan, J.K. (1977) Synthesis of energy flow and population dynamics of Truelove Lowland invertebrates, in *Truelove Lowland, Devon Island, Canada: A High Arctic Ecosystem* (ed. L.C. Bliss), University of Alberta Press, Alberta, pp. 325-46.

Ryszkowski, L. (1975) Energy and matter economy of ecosystems, in *Unifying Concepts in Ecology* (eds W.H. van Dobben and R.H. Lowe-McConnell), Dr. W. Junk Publisher, The Hague, pp. 109-26.

Saari, L.L. and Ludden, P.W. (1986) The energetics and energy costs of symbiotic nitrogen fixation, in *Plant-Microbe Interactions. Molecular and Genetic Perspectives* (eds T. Kosuge and E. Nester), MacMillan, New York, pp. 147-93.

Sabatier, G. and Vartanian, N. (1983) Cinétique de la rhizogenèse adaptative à la sécheresse. Relation avec l'évolution des paramètres hydriques et morphologiques chez le *Sinapis alba*, à deux niveaux d'énergie lumineuse. *Physiology of Plants*, **59**, 501-7.

Sabatier, D. and Puig, H. (1986) Phénologie et saisonnalité de la floraison et de la fructification en forêt dense Guyanaise. *Mémoires du Muséum National d'Histoire Naturelle, Série B, Botanique*, **132**, 173-84.

Sadaka, N. and Poinsot-Balaguer, N. (1987) Relations trophiques entre le Collembole *Onychiurus zschokkei* Handschin et des feuilles de chêne vert *Quercus ilex* L. à divers stades de décomposition. *Revue d'Ecologie et de Biologie du Sol*, **24** (3), 329-40.

Safir, G.R., Boyer, J.S. and Gerdemann, J.W. (1972) Nutrient status and mycorrhizal enhancement of water transport on Soybean. *Plant Physiology*, **49**, 700-3.

Salati, E. and Vose, P.B. (1984) Amazon Basin: a system in equilibrium. *Science*, **225**, 129-38.

Salick, J., Herrer, R. and Jordan, C.F. (1983) Termitaria: nutrient patchiness in nutrient-deficient rain forests. *Biotropica*, **15**, 1-7.

Salick, J. and Tho, Y.P. (1984) An analysis of termite faunae in Malayan rainforests. *Journal of Applied Ecology*, **21**, 547-61.

Salisbury, F.B. and Ross, C.W. (1991) *Plant Physiology*, Wadsworth Publishing Co., Belmont, California.

Sallih, Z., Bottner, P., Billès, G. *et al.* (1987) Interaction racines-microorganismes: carbone et azote de la biomasse microbienne développée en présence de racines. *Revue d'Ecologie et de Biologie du Sol*, **24** (3), 459-71.

Sallih, Z. and Bottner, P. (1988) Effect of wheat *(Triticum aestivum)* roots on mineralization rates of soil organic matter. *Biology and Fertility of Soils*, **7**, 67-70.

Salt, G., Hollick, F.S.J., Raw, F. *et al.* (1948) The arthropod population of pasture soil. *Journal of Animal Ecology*, **17**, 139-50.

Samtsevich, S.A. (1971) Root excretions of plants. An important source of humus formation in the soil, in *Transactions of the International Symposium "Humus et Planta" V*, Prague, Czech Republic, pp. 147-54.

Sands, W. (1970) The association of termites and fungi, in *Biology of Termites* (eds K. Krishna and F.M. Weesner), Academic Press, New York, pp. 495-524.

Sarathchandra, S.U., Perrott, K.W. and Upsdell, M.P. (1984) Microbiological and biochemical characteristics of a range of New Zealand soils under established pasture. *Soil Biology & Biochemistry*, **16**, 177-83.

Sarkar, J.M., Leonowicz, A. and Bollag, J.M. (1989) Immobilization of enzymes on clays and soils. *Soil Biology & Biochemistry*, **21** (2), 223-30.

Satchell, J.E. (1971) Feasibility study of an energy budget for Meathop Wood, in *Productivité des Ecosystèmes Forestiers* (ed. P. Duvigneaud), UNESCO, Paris, pp. 619-30.

Satchell, J.E. (1974) Introduction: litter - interface of animate/inanimate matter, in *Biology of Plant Decomposition* (eds C.H. Dickinson and G.J.F. Pugh), Academic Press, New York, vol. 1, pp. 13-44.

Satchell, J.E., Martin, K. and Krishnamoorthy, R.V. (1984) Stimulation of microbial phosphatase production by earthworm activity. *Soil Biology & Biochemistry*, **16** (2), 195.

Sauerbeck, D. and Johnen, B.G. (1977) Root formation and decomposition during plant growth, in *Soil Organic Matter Studies*, International Atomic Energy Agency, Vienna, pp. 141-8.

Saulnier, L. and Athias-Binche, F. (1986) Modalités de la cicatrisation des écosystèmes méditerranéens après incendie: cas de certains arthropodes du sol. *Vie et Milieu*, **36** (3), 191-204.

Saur, E. and Ponge, J.F. (1988) Alimentary studies on the Collembolan *Paratullbergia callipygos* using transmission electron microscopy. *Pedobiologia*, **31**, 355-79.

Saussey, M. (1966) Contribution à l'étude des phénomènes de diapause et de régénération caudale chez *Allolobophora icterica* (Savigny). *Mémoires de la Société Linnéenne de Normandie*, **1**, 1-158.

Savant, N.K., Snyder, G.H. and Datnoff, L.E. (1997) Silicon management and sustainable rice production. *Advances in Agronomy*, **58**, 151-99.

Sanchez, P.A., Gichuru, M.P. and Katz, L.B. (1982) Organic matter in major soils of the tropical and temperate regions, in *Proceedings of the 11th International Congress of Soil Science*, ISSS, New Delhi, pp. 99-114.

Sanchez, P. and Salinas, J.G. (1983) Low input technology for managing oxisols and ultisols in tropical America. *Advances in Agronomy*, **34**, 279-405.

Schaefer, M. and Schauermann, J. (1990) The soil fauna of Beech forests: comparison between a mull and a moder soil. *Pedobiologia*, **34** (5), 299-314.

Schaller, F. (1953) Untersuchungen zur Fortpflanzungsbiologie arthropleoner Collembolen. *Zeitschrift für Morphologie und Ökologie der Tiere*, **41**, 265-77.

Schaller, F. (1970) Collembola (Springschwänze), in *Handbuch der Zoologie*, vol. 4 (II) (ed. W. Kukenthal), Walter de Gruyer, Berlin, pp. 1-72.

Scharpenseel, H.W., Wurzer, M., Freytag, J. *et al.* (1984) Biotisch und abiotisch gesteuerter abbau von organischer substanz im boden. *Zeitschrift für Pflanzenernährung und Bodenkunde*, **147** (4), 502-16.

Scharpenseel, H.W. (1988) Organic matter characteristics, in *Proceedings of the 2nd Regional Workshop on Land Development and Management of Acid Soils in Africa,* IBSRAM, 83-100.

Scharpenseel, H.W., Becker Heidmann, P., Neue, H.U. *et al.* (1989) Bomb-carbon, ^{14}C-dating and ^{13}C-measurements as tracers of organic matter dynamics as well as of morphogenetic and turbation process. *The Science of the Total Environment*, **81/82**, 99-110.

Schauermann, J. (1973) Zum Energieumsatz phytophager Insekten im Buchenwald. II Die produktionsbiologische Stellung der Rüsselkäfer (Curculionidae) mit rhizophagen Larvenstadien. *Oecologia*, **13**, 313-50.

Schauermann, J. (1977) Zur abundanz- und Biomassendynamik der Tiere in Buchenwäldern des Solling. *Verhandlungen der Gesellschaft für Ökologie*, 113-24.

Schäfer, D., Ziegler, F. and Zech, W. (1993) Einfluss der mineralischen Bodensubstanz auf den Abbau von Buchenstreu und den Aufbau organo-mineralischer komplexe. *Mitteilungen der Deutschen Bodenkundlichen Gesellschaft*, **71**, 213-6.

Scheu, S. (1987) The influence of earthworms (Lumbricidae) on the nitrogen dynamics in the soil litter system of a deciduous forest. *Oecologia*, **72**, 197-201.

Scheu, S. (1991) Mucus excretion and carbon turnover of endogeic earthworms. *Biology and Fertility of Soils*, **12**, 217-20.

Scheu, S. (1992) Changes in the lumbricid coenosis during secondary succession from a wheat field to a Beechwood on limestone. *Soil Biology & Biochemistry*, **24** (12), 1641-6.

Scheu, S., Maraun, M., Bonkowski, M. *et al.* (1996) Microbial biomass and respiratory activity in soil aggregates of different sizes from three Beechwood sites on a basalt hill. *Biology and Fertility of Soils*, **21**, 69-76.

Schimel, D.S., Braswell, B.H., Holland, E.A. *et al.* (1994) Climatic, edaphic, and biotic controls over storage and turnover of carbon in soils. *Global Biogeochemical Cycles*, **8**, 279-93.

Schippers, B. and Gams, W. (1979) *Soil Borne Plant Pathogens*, Academic Press, London.

Schlesinger, W.H. (1997) *Biogeochemistry*, 2nd edition, Academic Press, San Diego, California.

Schmitt, D.P. and Norton, D.C. (1972) Relationship of plant parasitic nematodes to sites in native Iowa prairies. *Journal of Nematology*, **4**, 200-6.

Schnitzer, M. (1986) Binding of humic substances by soil mineral colloids, in *Interactions of Soil Minerals with Natural Organics and Microbes* (eds P.M. Huang and M. Schnitzer), Soil Science Society of America, Madison, Wisconsin, pp. 77-101.

Schnürer, J. and Rosswall, T. (1982) Fluorescein diacetate hydrolysis as a measure of total microbial activity in soil and litter. *Applied and Environmental Microbiology*, **43**, 1256-61.

Schnürer, J., Clarholm, M. and Rosswall, T. (1985) Microbial biomass and activity in an agricultural soil with different organic matter contents. *Soil Biology & Biochemistry*, **17** (5), 611-8.

Schnürer, J. (1985) *Fungi in Arable Soil*, Sveriges lantbruksunivesitet, Uppsala.

Schnürer, J., Clarholm, M. and Rosswall, T. (1986) Fungi, bacteria and protozoa in soil from four arable cropping systems. *Biology and Fertility of Soils*, **2**, 119-26.

Scholes, M.C. and Sanchez, P.A. (1990) Low soil nitrogen mineralization rates in a humid tropical pasture. *Tropical Ecology*, **31** (1), 12-5.

Scholes, M., Scholes, R.J. and Retief, J. (1993) Nitrogen and phosphorus nutrition of *Digitaria eriantha*: is it limited by absorption capacity or rate of supply?, in *XVII International Grassland Congress*, Palmerston North, New Zealand.

Scholes, R.J. and Walker, B.H. (1993) *An African Savanna: Synthesis at the Nylsvley Study Site*, Cambridge University Press, Cambridge.

Schowalter, T.D., Sabin, T.E., Stafford, S.G. *et al.* (1991) Phytophage effects on primary production, nutrient turnover, and litter decomposition of young Douglas-fir in Western Oregon. *Forest Ecosystem and Management*, **42**, 229-43.

Schönborn, W. (1968) Taxonomish-faunistich Probleme bei der Bsiedlung des Bodens an Hand der beschalten

Amöben (Testacea). *Pedobiologia*, **8**, 23-9.

Schönborn, W. (1982) Estimation of annual production of Testacea (Protozoa) in mull and moder (II). *Pedobiologia*, **23**, 383-93.

Schulten, H.R. and Schnitzer, M. (1997) Chemical model structures for soil organic matter and soils. *Soil Science*, **162**, 115-30.

Schulten, H.R. and Schnitzer, M. (1998) The chemistry of soil organic nitrogen: a review. *Biology and Fertility of Soils*, **26**, 1-15.

Schultz, P.A. (1991) Grazing preferences of two collembolan species, *Folsomia candida* and *Proisotoma minuta*, for ectomycorrhizal fungi. *Pedobiologia*, **35**, 313-25.

Schulze, D.G. (1989) An introduction to soil mineralogy, in *Minerals in Soil Environments* (eds J.B. Dickson and S.B. Weed), Soil Science Society of America, Madison, Wisconsin, pp. 1-34.

Schumacher, A. and Whitford, W.G. (1974) The foraging ecology of two species of Chihuahuan desert ants: *Formica perpilosa* and *Trachymyrmex smithi neomexicanus* (Hymenoptera Formicidae). *Insectes Sociaux*, **21**, 317-30.

Schwartz, R. (1981) *Influence de la monoculture des résineux sur la microflore des sols et la décomposition des litières*, thèse de 3e cycle, Université Nancy I.

Schwartz, P.H. and Klassen, W. (1981) Estimate of losses caused by insects and mites to agricultural crops, in *CRC Handbook of Pest Management in Agriculture* (ed. D. Pimental), CRC Press, Boca Raton, Florida, pp. 1-15.

Schwertmann, U., Kodama, H. and Fischer, W.R. (1986) Mutual interactions between organics and iron oxides, in *Interactions of Soil Minerals with Natural Organics and Microbes* (eds P.H. Huang and M. Schnitzer), Soil Science Society of America, Madison, Wisconsin, pp. 453-95.

Schwertmann, U. (1988) Occurrence and formation of iron oxides in various pedoenvironments, in *Iron in Soils and Clay Minerals* (eds J.W. Stucki, B.A. Goodman and U. Schwertmann), Reidel Publishing Co., Dordrecht, pp. 267-308.

Schwertmann, U. and Taylor, R.M. (1989) Iron oxides, in *Minerals in Soil Environments* (eds J.B. Dickson and S.B. Weed), Soil Science Society of America, Madison, Wisconsin, pp. 379-438.

Schwertmann, U. and Herbillon, A.J. (1992) Some aspects of fertility associated with the mineralogy of highly weathered tropical soils, in *Myths and Science of Soils of the Tropics* (eds R. Lal and P.A. Sanchez), Soil Science Society of America, Madison, Wisconsin, pp. 47-59.

Scott, J.A., French, N.R. and Leetham, J.W. (1979) Patterns of consumption in grasslands, in *Perspectives in Grassland Ecology* (ed. N.R. French), Springer Verlag, New York, pp. 89-105.

Seastedt, T.R. (1984) The role of microarthropods in decomposition and mineralisation processes. *Annual Review of Entomology*, **29**, 25-46.

Seastedt, T.R., Todd, T.C. and James, S.W. (1987) Experimental manipulations of arthropod, nematode and earthworm communities in a north American tallgrass prairie. *Pedobiologia*, **30**, 9-17.

Seastedt, T.R., James, S.W. and Todd, T.C. (1988) Interactions between Protozoa and other organisms. *Agriculture, Ecosystems and Environment*, **24**, 219-28.

Sedlag, U. (1951) Untersuchungen über den Ventraltubus der Collembolen. *Wissenchaftliche Beitrage der Martin-Luther-Unoversität Halle-Wittenberg*, **1**, 93-127.

Seguin, G. (1970) *Les sols de vignobles du Haut Médoc. Influence sur l'alimentation en eau de la vigne et sur la maturation du raisin*, thèse d'Etat, Université de Bordeaux.

Seiler, W., Conrad, R. and Scharffe, D. (1984) Field studies of methane emission from termite nests into the atmosphere and measurements of methane uptake from tropical soils. *Journal of Atmospheric Chemistry*, **1**, 171-86.

Senapati, B.K. (1980) *Aspects of ecophysiological studies on tropical earthworms (distribution, population dynamics, production, energetics and their role in the decomposition process)*, Ph.D. thesis, University of Sambalpur.

Senapati, B.K., Panigrahi, P.K. and Lavelle, P. (1994) Macrofaunal status and restoration strategy in degraded soil under intensive tea cultivation in India, in *Proceedings of the 15th World Congress of Soil Science* (eds J.D. Etchevers, A. S. Aguilar, R.E. Núñez *et al.*), ISSS, Acapulco, pp. 64-75.

Sen-Sharma, P.K. (1986) Economically important termites and their management in the Oriental Region, in *Economic Impact and Control of Social Insects* (ed. S.B. Vinson), Praeger, New-york, pp. 69-102.

Setälä, H., Haimi, J. and Huhta, V. (1988) A microcosm study on the respiration and weight loss in Birch litter and raw humus as influenced by soil fauna. *Biology and Fertility of Soils*, **5**, 282-7.

Setälä, H. (1990) *Effects of soil fauna on decomposition and nutrient dynamics in coniferous forest soil*, University of Jyväskylä, Jyväskylä.

Sextone, A.J., Revsbech, N.P., Parkin, T.B. *et al.* (1985) Direct measurement of oxygen profiles and denitrification rates in soil aggregates. *Soil Science Society of America, Journal*, **49**, 645-51.

Shafer, S.R., Rhodes, L.H. and Riedel, R.M. (1981) In-vitro parasitism of endomycorrhizal fungi of Ericaceous plants by the mycophagous nematode *Aphelenchoides bicaudatus*. *Mycologia*, **73**, 141.

Sharma, M.L. and Uehara, G. (1968) Influence of soil structure and water relations in low humic latosols: I. Water retention. *Soil Science Society of America, Proceedings*, **32**, 765-70.

Sharma, M.L., Tunny, J. and Tongway, D.J. (1972) Seasonal changes in sodium and chloride concentration of saltbush *(Atriplex spp)* leaves as related to soil and plant water potential. *Australian Journal of Agricultural Research*, **23**, 1007-19.

Sharp, R.E. and Davies, W.J. (1985) Root growth and water uptake by maize plants in drying soil. *Journal of Experimental Botany*, **36** (170), 1441-56.

Sharpley, A.N. and Syers, J.K. (1977) Seasonal variation in casting activity and in the amounts and release to solution of phosphorus forms in earthworm casts. *Soil Biology & Biochemistry*, **9**, 227-31.

Shaw, C. and Pawluk, S. (1986) The development of soil structure by *Octolasion tyrtaeum, Aporrectodea turgida* and *Lumbricus terrestris* in parent materials belonging to different textural classes. *Pedobiologia*, **29**, 327-39.

Shay, F.J. and Hale, P.G. (1973) Effect of low levels of calcium on exudation of sugar acid derivatives from intact peanut roots under axenic conditions. *Plant Physiology*, **51**, 1061-3.

Shearer, G. and Kohl, D.H. (1989) Estimates of N_2 fixation in ecosystems: the need for and basis of the ^{15}N natural abundance method, in *Stable Isotopes in Ecological Research* (eds P.W. Runder, J.R. Ehleringer and K.A. Nagy), Springer-Verlag, Berlin, pp. 342-74.

Shelton, T.G. and Grace, K.G. (1996) Review of agonistic behaviours in the Isoptera. *Sociobiology*, **28**, 155-76.

Shields, J.A., Paul, E.A., Lowe, W.E. *et al.* (1973) Turnover of microbial tissue in soil under field conditions. *Soil Biology & Biochemistry*, **5**, 753-64.

Shoji, S., Ito, T., Nakamura, S. *et al.* (1987) Properties of humus of andisols from New Zealand, Chile and Ecuador. *Japanese Journal of Soil Science and Plant Nutrition*, **58**, 473-9.

Short, G.E. and Lacy, M.L. (1974) Germination of *Fusarium solani* f. sp. *pisi* chlamydospores in the spermosphere of pea. *Phytopathology*, **64**, 558-62.

Siepel, H. (1994) Life-history tactics of soil microarthropods. *Biology and Fertility of Soils*, **18**, 263-78.

Simard, S.W., Perry, D.A., Jones, M.D. *et al.* (1997) Net transfer of carbon between ectomycorrhizal tree species in the field. *Nature*, **388**, 579-82.

Simon, L., Bousquet, J., Lévesque, R.C. *et al.* (1993) Origin and diversification of endomycorrhizal fungi and coincidence with vascular land plants. *Nature*, **363**, 67-9.

Simons, W.R. (1973) Nematode survival in relation to soil moisture. *Medelingen van de Landbouwhogeschool te Wageningen*, **73** (3), 1-85.

Simonson, R.W. (1978) A multiple-process model of soil genesis, in *Quaternary Soils* (ed. W.C. Mahaney), University of East Anglia, Norwich, pp. 1-25.

Simonson, R.W. (1984) *Soil classification in the past - roots and philosophies*, Report, International Soil Reference and Information Centre.

Simonson, R.W. (1995) Airborne dust and its significance to soils. *Geoderma*, **65**, 1-43.

Sims, P.L. and Singh, J.S. (1971) Herbage dynamics and net primary production in certain ungrazed and grazed grasslands in North America, in *Preliminary Analysis of Structure and Function in Grasslands* (ed. N.R. French), Colorado State University, Fort Collins, Colorado, pp. 59-124.

Singh, B.N. (1941) Selectivity in bacterial food by soil amoebae in pure mixed culture and in sterilized soil. *Annals of Applied Biology*, **28**, 52-64.

Singh, K.P. and Srivatsava, S.K. (1984) Spatial distribution of fine root mass in young trees (*Tectona grandis*) of varying girth sizes. *Pedobiologia*, **27**, 161-70.

Singh, S., Raghubanshi, A.S., Singh, R.S. *et al.* (1989) Microbial biomass acts as a source of plant nutrients in dry tropical forest and savanna. *Nature*, **338** (6215), 499-500.

Singh, K.P. and Shekhar, C. (1986) Seasonal pattern of total soil respiration, its fractionation and soil carbon balance in a wheat-maize rotation cropland at Varanasi. *Pedobiologia*, **29**, 305-18.

Singh, R.S., Srivastava, S.C., Raghubanashi, A.S. *et al.* (1991) Microbial C, N and P in dry tropical savanna: effects of burning and grazing. *Journal of Applied Ecology*, **28**, 869-78.

Singh, S. and Singh, J.S. (1995) Microbial biomass associated with water-stable aggregates in forest, savanna and cropland soils of a seasonally dry tropical region, India. *Soil Biology & Biochemistry*, **27** (8), 1027 - 33.

Singleton, P.W. and Bohlool, B.B. (1983) Effect of salinity on the functional components of the soyabean-*Rhizobium japonicum* symbiosis. *Crop Science*, **73**, 815-8.

Sisworo, W.H., Mitrosuhardjo, M.M., Rasjid, H. *et al.* (1990) The relative roles of N fixation, fertiliser, crop residues and soil in supplying N in multiple cropping systems in the humid, tropical upland cropping system. *Plant and Soil*, **121**, 73-82.

Skjemstad, J.O., Clarke, P., Taylor, J.A. *et al.* (1996) The chemistry and nature of protected organic carbon in soil. *Autralian Journal of Soil Research*, **34**, 251-71.

Skjemstad, J.O., Clarke, P., Golchin, A. *et al.* (1997) Characterization of soil organic matter by solid-state ^{13}C NMR spectroscopy, in *Driven by Nature: Plant Litter Quality and Decomposition* (eds G. Cadish and K.E. Giller), CAB International, Wallingford, pp. 253-71.

Skuterad, L., Travnikova, I.G., Balonov, M.I. *et al.* (1997) Contribution of fungi to radiocaesium accumulation by rural populations in Russia. *The Science of the Total Environment*, **193**, 237-42.

Slankis, V. (1973) Hormonal relationships in mycorrhizal development, in *Ectomycorrhizae* (eds G.C. Marks and T.T. Kozlowski), Academic Press, London, pp. 231-98.

Slapokas, T. (1991) *Influence of Litter Quality and Fertilization on Microbial Nitrogen Transformation in Short-Rotation Forests*, Report No.49, Swedish University of Agricultural Sciences.

Slavikova, J. (1958) Einfluss der Buche (*Fagus sylvatica* L.) als Edifikatr auf die Entwicklung der Krautschiicht in den Buchenphytozönosen. *Preslia*, **30**, 19-42.

Slaytor, M. (1992) Cellulose digestion in termites and cockroaches: what role do symbionts play? *Comparative Biochemistry and Physiology*, **103B**, 775-84.

Slaytor, M. and Chappell, D.J. (1994) Nitrogen metabolism in termites. *Comparative Biochemistry and Physiology*, **107B**, 1-10.

Sleaford, F., Bignell, D.E. and Eggleton, P. (1996) A pilot analysis of gut contents in termites from

the Mbalmayo Forest Reserve, Cameroon. *Ecological Entomology*, **21** (3), 279-88.
Sleeman, J.R. and Brewer, R. (1972) Microstructures of some Australian termite nests. *Pedobiologia*, **12**, 347-73.
Smith, P.F. (1962) Mineral analysis of plant tissues. *Annual Review of Plant Physiology*, **13**, 81-108.
Smith, R.M. and Stamey, W.L. (1965) Determining the range of tolerable erosion. *Soil Science*, **100**, 414-24.
Smith, W.H. (1970) Root exudates of seedling and mature Sugar Maple. *Phytopathology*, **60**, 701-3.
Smith, B.N. and Epstein, S. (1971) Two categories of $^{13}C/^{12}C$ ratios for higher plants. *Plant Physiology*, **47**, 380-4.
Smith, W.H. (1972) Influence of artificial defoliation on exudates of Sugar Maple. *Soil Biology & Biochemistry*, **4**, 111-4.
Smith, A.D.M. and Wallace, H.R. (1976) Fluctuations in the distribution and numbers of *Helicotylenchus dihystera* in Kikuyu turf (*Pennisetum clandestinum*). *Nematologica*, **22**, 145-52.
Smith, S.E. and Read, D.J. (1997) *Mycorrhizal Symbiosis*, Academic Press, San Diego.
Sneath, P.H.A. (1962) Longevity of microorganisms. *Nature*, **195**, 643-6.
Soepadmo, E. (1977) *Ulmaceae, in Flora Malesiana* (ed. C.G.G.J. Steenis), Noordhoff, Jakarta, pp.31-76.
Sohlenius, B. (1980) Abundance, biomass and contribution to energy flow by soil nematodes in terrestrial ecosystems. *Oikos*, **34**, 186-94.
Sohlenius, B. and Sandor, A. (1987) Vertical distribution of nematodes in arable soil under grass *(Festuca pratensis)* and barley *(Hordeum distichum)*. *Biology and Fertility of Soils*, **3**, 19-25.
Sohlenius, B., Boström, S. and Sandor, A. (1988) Carbon and nitrogen budgets of nematodes in arable soil. *Biology and Fertility of Soils*, **6**, 1-8.
Soil Classification Working Group (ed.) (1991) *Soil Classification, A Taxonomic System for South Africa*, Department of Agricultural Development, Pretoria.
Soil Science Society of America (ed.) (1997) *Glossary of Soil Science Terms*, Soil Science Society of America, Madison, Wisconsin.
Soil Survey Staff (ed.) (1993) *Soil Survey Manual*, U.S. Department of Agriculture, Washington D.C.
Soil Survey Staff (ed.) (1999) *Soil Taxonomy: A Basic System of Soil Classification for Making and Interpreting Soil Surveys*, 2nd edition, U.S.Department of Agriculture, Soil Conservation Service, Washington D.C.
Solbrig, O.T. (1991) *From Genes to Ecosystems: A Research Agenda for Biodiversity*, IUBS-SCOPE-UNESCO, Paris.
Sollins, P., Spycher, G. and Glassman, C.A. (1984) Net nitrogen mineralisation from light and heavy-fractions of forest soil organic matter. *Soil Biology & Biochemistry*, **16** (1), 31-7.
Soma, K. and Saitô, T. (1979) Ecological studies of soil organisms with references to the decomposition of pine needles. I. Soil macrofaunal and mycofloral surveys in coastal pine plantations. *Revue d'Ecologie et de Biologie du Sol*, **16**, 337-54.
Sougoufara, B., Duhoux, E., Corbasson, M. *et al.* (1987) Improvement of nitrogen fixation by *Casuarina equisetifolia* through clonal selection. *Arid Soil Research and Rehabilitation*, **1**, 129-32.
Southwood, T.R.E. (1977) Habitat, the templet for ecological strategies? *Journal of Animal Ecology*, **46**, 337-65.
Sowden, F., Chen, Y. and Schnitzer, M. (1977) The nitrogen distribution in soils formed under widely differing climatic conditions. *Geochimica et Cosmochimica Acta*, **41**, 1524-1526.
Söderström, B.E. (1979) Seasonal fluctuations of active fungal biomass in horizons of a podzolized pine-forest soil in central Sweden. *Soil Biology & Biochemistry*, **11**, 149-54.
Spain, A.V. (1973) A preliminary study of spatial patterns in the accession of conifer litterfall, its physical components and selected of its plant nutrient elements. *Journal of Applied Ecology*, **10**, 561-71.
Spain, A.V., Isbell, R.F. and Probert, M.E. (1983a) Organic matter contents of Australian soils, in *Soils:*

An Australian Viewpoint, CSIRO/Academic Press, Melbourne/London, pp. 551-63.

Spain, A.V., Okello-Oloya, T. and Brown, A.J. (1983b) Abundance, above-ground masses and basal areas of termite mounds at six locations in tropical north-eastern Australia. *Revue d'Ecologie et de Biologie du Sol*, **20** (4), 547-66.

Spain, A.V. (1984) Litterfall and the standing crop of litter in three tropical Australian rainforests. *Journal of Ecology*, **72**, 947-61.

Spain, A.V., Sinclair, D.F. and Diggle, P.J. (1986) Spatial distributions of the mounds of harvester and forager termites (Isoptera: Termitidae) at four locations in tropical North-eastern Australia. *Acta Œcologia, Œcologia Generalis*, **7** (4), 335-52.

Spain, A.V. and Lefeuvre, R.P. (1987) Breakdown of four litters of contrasting quality in a tropical Australian rainforest. *Journal of Applied Ecology*, **24**, 279-88.

Spain, A.V. and McIvor, J.G. (1988) The nature of herbaceous vegetation associated with termitaria in north-eastern Australia. *Journal of Ecology*, **76**, 181-91.

Spain, A.V., Unwin, G. and Sinclair, D.F. (1989a) Soils and vegetation at three rainforest sites in tropical north-eastern Queensland. *CSIRO Australia, Division of Soils, Divisional Report*, **105**.

Spain, A.V., Hodgen, M.J. and Bailey, S.W. (1989b) Seasonal variation in selected chemical properties of litterfall and the litter layers in three Australian rainforests. *CSIRO Australia, Division of Soils, Technical Memorandum*, **14**.

Spain, A.V. (1990) Influence of environmental conditions and some chemical properties on the carbon and nitrogen contents of some tropical Australian rainforest soils. *Autralian Journal of Soil Research*, **28**, 825-39.

Spain, A.V., Saffigna, P.G. and Wood, A.W. (1990) Tissue carbon sources for *Pontoscolex corethrurus* (Oligochaeta, Glossoscolecidae) in a sugarcane ecosystem. *Soil Biology & Biochemistry*, **22** (5), 703-6.

Spain, A.V., Lavelle, P. and Mariotti, A. (1992) Stimulation of plant growth by tropical earthworms. *Soil Biology & Biochemistry*, **24** (12), 1629-34.

Spain, A.V. and Hodgen, M.J. (1994) Changes in the composition of sugarcane harvest residues during decomposition as a surface mulch. *Biology and Fertility of Soils*, **17** (3), 225-31.

Spain, A.V. and Reddell, P. (1996) $\delta\ ^{13}C$ values of selected termites (Isoptera) and termite-modified materials. *Soil Biology & Biochemistry*, **28** (12), 1585-93.

Spalding, R.F. and Exner, M.E. (1993) Occurrence of nitrate in groundwater - a review. *Journal of Environment Quality*, **22**, 392-402.

Sparling, G.P. (1992) Ratio of microbial biomass carbon to soil organic carbon as a sensitive indicator of changes in soil organic matter. *Australian Journal of Soil Research*, **30**, 195-207.

Spragg, W.T. and Paton, R. (1980) Tracing, trophallaxis and population measurement of subterranean termites (Isoptera) using a radioactive tracer. *Annals of the Entomological Society of America*, **73**, 708-14.

Sprent, J.I. (1984) Nitrogen fixation, in *Advanced Plant Physiology* (ed. M.B. Wilkins), Longman Scientific & Technical, New York, pp. 249-76.

Sprent, J.I. (1987) *The Ecology of the Nitrogen Cycle*, Cambridge University Press, London.

Sprent, J.I. and Sprent, P. (1990) *Nitrogen Fixing Organisms*, Chapman & Hall, London.

Springett, J.A., Brittain, D.E. and Springett, B.P. (1970) Vertical movement of Enchytraeidae (Oligochaeta) in moorland soils. *Oikos*, **21**, 16-21.

Springett, J.A. and Latter, P.M. (1977) Studies on the micro-fauna of blanket bog with particular reference to Enchytraeidae. I. Field and laboratory tests of microorganisms as food. *Journal of Animal Ecology*, **46**, 959-74.

Spycher, G., Sollins, P. and Rose, S. (1983) Carbon and nitrogen in the light fraction of a forest soil: vertical distribution and seasonal patterns. *Soil Science*, **135**, 79-87.

Srivastava, S.C. (1992) Microbial C, N and P in dry tropical soils: seasonal changes and influence of soil moisture. *Soil Biology & Biochemistry*, **24**, 711-4.

St. John, T.V., Coleman, D.C. and Reid, C.P.P. (1983) Growth and spatial distribution of nutrient-absorbing organs: selective exploitation of soil heterogeneity. *Plant and Soil*, **71**, 487-93.

St. John, T.V. and Uhl, C. (1983) Mycorrhizae in the rain forest at San Carlos de Rio Negro, Venezuela. *Acta Cientifica Venezolana*, **34**, 233-7.

Staff, H. (1988) Litter decomposition in Beech forests - effects of excluding tree roots. *Biology and Fertility of Soils*, **6**, 302-5.

Stahl, D.A. and Amann, R.I. (1991) Development and application of nucleic acid probes in bacterial systematics, in *Sequencing and Hybridization Techniques in Bacterial Systematics* (eds E. Stackebrandt and M. Goodfellow), John Wiley, New York, pp. 205-48.

Standen, V. (1973) The production and respiration of an enchytraeid population in blanket bog. *Journal of Animal Ecology*, **2**, 219-45.

Standen, V. (1984) Production and diversity of enchytraeids, earthworms, and plants in fertilized hay meadow plots. *Journal of Applied Ecology*, **21**, 293-312.

Stanghellini, M.E. and Hancock, J.G. (1971) Radial extent of the bean spermosphere and its relation to the behavior of *Pythium ultimum*. *Phytopathology*, **61**, 165-8.

Stanton, N. and Tepedino, V.J. (1977) Island habitats in soil communities. *Ecological Bulletins*, **25**, 511-4.

Stark, N. and Spratt, M. (1977) Root biomass and nutrient storage in rain forest oxisols near San Carlos de Rio Negro. *Tropical Ecology*, **18**, 1-9.

Stark, N. (1978) Man, tropical forests and the biological life of a soil. *Biotropica*, **10**, 1-10.

Stark, N.M. and Jordan, C.F. (1978) Nutrient retention by the root mat of an Amazonian rain forest. *Ecology*, **59** (3), 434-7.

Stafford-Smith, D.M. and Morton, S.R. (1990) A framework for the ecology of arid Australia. *Journal of Arid Environments*, **18**, 255-78.

Steinberg, Y. and Wallwork, J.A. (1985) Composition and vertical distribution patterns of the microarthropod fauna in a Negev desert soil. *Journal of Zoology*, **206**, 329-39.

Steinberger, Y., Orion, D. and Whitford, W.G. (1988) Population dynamics of nematodes in the Negev Desert soil. *Pedobiologia*, **31**, 223-8.

Stephens, P.M., Davoren, C.W., Ryder, M.H. *et al.* (1994) Enhanced growth of "take-all" affected plants associated with *Pseudomonas corrugata* 2140R and the earthworm *Aporrectodea trapezoides* (Lumbricidae) in *Increasing Plant Productivity with Rhizosphere Bacteria*, Proceedings of the 3rd International Workshop on Plant Growth Promoting Bacteria, Division of Soils, CSIRO, Adelaide, pp. 57-58

Stevenson, F.J. (1986) *Cycles of Soil: Carbon, Nitrogen, Phosphorus, Sulfur, Micronutrients*, John Wiley, New York.

Stewart, W.D.P. and Rodgers, G.A. (1977) The cyano-hepatic symbiosis. 2. Morphology and physiology. *New Phytologist*, **78**, 441-58.

Stienstra, A.W., Gunnewiek, P.K. and Laanbroek, H.J. (1994) Repression of nitrification in soils under a climax grassland vegetation. *FEMS Microbiology Ecology*, **14** (1), 45-52.

Stirling, G.R. (1984) Biological control of *Meloidogyne javanica* with *Bacillus penetrans*. *Phytopathology*, **74**, 55-60.

Stockdill, S.M.J. (1982) Effects of introduced earthworms on the productivity of New Zealand pastures. *Pedobiologia*, **24**, 29-35.

Stolzy, L.H., Focht, D.D. and Fluhler, H. (1981) Indicators of soil aeration status. *Flora*, **171**, 236-65.

Stone, E.L. and Kalisz, P.J. (1991) On the maximum extent of tree roots. *Forest Ecology and Management*,

46, 59-102.

Stoorvogel, J.J., Smaling, E.M.A. and Janssen, B.H. (1993) Calculating soil nutrient balances in Africa at different scales. *Fertilizer Research*, **35**, 227-35.

Stoorvogel, J.J., van Breemen, N. and Janssen, B.H. (1997) The nutrient input by Harmattan dust to a forest ecosystem in Ivory Coast, Africa. *Biogeochemistry*, **37** (2), 145-57.

Stork, N.E. and Eggleton, P. (1992) Invertebrates as determinants and indicators of soil quality. *American Journal of Alternative Agriculture*, **7** (1/2), 38-47.

Stork, N.E. and Brendell, M.J.D. (1993) Arthropod abundance in lowland rainforest of Seram, in *The Natural History of Seram, Mulaka, Indonesia* (eds I.D. Edwards, A.A. McDonald and J. Proctor), Intercept Ltd, Andover, pp. 115-130.

Stotzky, G. and Rem L.T. (1966) Influence of clay minerals on micro-organisms. 1. Montmorillonite and kaolinite on bacteria. *Canadian Journal of Microbiology*, **12**, 547-63.

Stout, J.D. and Heal, O.W. (1967) Protozoa, in *Soil Biology* (eds A. Burges and F. Raw), Academic Press, London, pp. 149-95.

Stout, J.D. and Goh, K.M. (1980) The use of radiocarbon to measure the effects of earthworms on soil development. *Radiocarbon*, **22** (3), 892-6.

Stout, J.D., Goh, K.M. and Rafter, T.A. (1981) Chemistry and turnover of naturally occurring resistant organic compounds in soil, in *Soil Biochemistry* (eds E.A. Paul and J.N. Ladd), Marcel Dekker, New York, pp. 1-73.

Stout, J.D., Bamforth, S.S. and Lousier, J.D. (1982) Protozoa, in *Methods of Soil Analysis, Part. 2 Chemical and Microbiological Properties* (eds A.L. Page, R.H. Miller and D.R. Keeney), American Society of Agronomy, Soil Science Society of America, Madison, Wisconsin, pp. 1103-20.

Stöckli, A. (1928) Studien über den Einfluss der Regenwürmer auf die Beschaffenheit des Bodens. *Landwirtschaftliches Jahrbuch der Schweiz*, **42**, 1-12.

Strading, D. (1987) Nutritional ecology of ants, in *Nutritional Ecology of Insects, Mites, Spiders and Related Invertebrates* (eds F. Slansky and J.G. Rodriguez), John Wiley, New York, pp. 927-69.

Strong, W.L. and La Roi, G.H. (1985) Root density-soil relationships in selected boreal forests of Central Alberta, Canada. *Forest Ecology and Management*, **12**, 233-51.

Stucki, J.W., Gan, H. and Wilkinson, H.T. (1992) Effects of microorganisms on phyllosilicate properties and behaviour, in *Interacting Processes in Soil Science* (eds R.J. Wagenet, P. Baveye and B.A. Stewart), Lewis Publishers, Boca Raton, Florida, pp. 227-54.

Sturm, H. (1983) Zur Bodenfauna der andinen Paramoregion. *Amazoniana*, **8**, 129-47.

Sumner, M.E. and Stewart, B.A. (eds) (1992) *The Electrical Double Layer and Clay Dispersion*, Lewis Publishers, Boca Raton, Florida.

Sumner, M.E. (1995) Sodic soils: new perspectives, in *Australian Sodic Soils: Distribution, Properties and Management* (eds R. Naidu, M.E. Sumner and P. Rengasamy), CSIRO Australia, East Melbourne, pp. 1-34.

Sundman, V. and Sivelä, S. (1978) A comment on the membrane filter technique for estimation of length of fungal hyphae in soil. *Soil Biology & Biochemistry*, **10**, 399-401.

Sutherland, J.R. and Fortin, J.A. (1968) Effect of the nematode *Aphelenchus avenae* on some ectotrophic, mycorrhizal fungi and on a Red Pine mycorrhizal relationship. *Phytopathology*, **58**, 519-23.

Sutton, S.L. (1972) *Woodlice*, Ginn, London.

Sutton, J.C. (1973) Development of vesicular-arbuscular mycorrhizae in crop plants. *Canadian Journal of Botany*, **51**, 2487-93.

Swift, M.J. (1973) The estimation of mycelial biomass by determination of the hexosamine content of wood tissue decayed by fungi. *Soil Biology & Biochemistry*, **5**, 321-32.

Swift, M.J. (1976) Species diversity and the structure of microbial communities in terrestrial habitats, in *The role of Terrestrial and Aquatic Organisms in Decomposition Processes* (eds J.M. Anderson and A. Macfadyen), Blackwell Scientific Publications, Oxford, pp. 185-222.

Swift, M.J., Heal, O.W. and Anderson, J.M. (1979) *Decomposition in Terrestrial Ecosystems*, Blackwell Scientific Publications, Oxford.

Swift, M.J., Russell-Smith, A. and Perfect, T.J. (1981) Decomposition and mineral-nutrient dynamics of plant litter in a regenerating bush-fallow in sub-humid tropical Nigeria. *Journal of Ecology*, **69**, 981-95.

Swift, M.J. (1982) Basidiomycetes as components of forest ecosystems, in *Decomposer Basidiomycetes, British Mycological Society Symposium 4* (eds J.C. Frankland, J.N. Hedger and M.J. Swift), Cambridge University Press, Cambridge, pp. 308-37.

Swift, M.J. and Boddy, L. (1984) Animal-microbial interactions in wood decomposition, in *Animal Microbial Interactions* (eds J.M. Anderson, A.D.M. Rayner and D.W.H. Walton), Cambridge University Press, Cambridge, pp. 89-131.

Swift, M.J. (1984) Microbial diversity and decomposer niches, in *Current Perspectives in Microbial Ecology* (eds M.J. Klug and C.A. Reddy), American Society for Microbiology, Washington D.C., pp. 8-16.

Swift, M.J. (1986) Tropical Soil Biology and Fertility (TSBF): inter-regional research planning workshop. Report of the Third Workshop of the Decade of the Tropics/TSBF Program. *Biology International*, **Special Issue 13**.

Syers, J.K., Sharpley, A.N. and Keeney, D.R. (1979) Cycling of nitrogen by surface-casting earthworms in a pasture ecosystem. *Soil Biology & Biochemistry*, **11**, 181-5.

Szábo, I.M. (1974) *Microbial Communities in a Forest Rendzina Ecosystem*, Akademiai Kiado, Budapest.

Szlavecz, K. (1985) The effect of microhabitats on the leaf litter decomposition and on the distribution of soil animals. *Holartic Ecology*, **8** (1), 33-8.

Tamm, C.O. (1991) *Nitrogen in Terrestrial Ecosystems*, Springer-Verlag, Berlin.

Tanner, E.V.J. (1977) Four montane rainforests of Jamaica: a quantitative characterisation of the floristics, the soils and the foliar mineral levels, and a discussion of the interrelationships. *Journal of Ecology*, **65**, 883-918.

Tardieu, F. (1988) Analysis of the spatial variability of maize root density. II. Distances between roots. *Plant and Soil*, **107**, 267-72.

Tardy, Y. and Roquin, C. (1992) Geochemistry and evolution of lateritic landscapes, in *Weathering, Soils & Paleosols* (eds I.P. Martini and W. Chesworth), Elsevier, Amsterdam, pp. 407-43.

Tashiro, H. (1990) Insecta: Coleoptera Scarabaidae larvae, in *Soil Biology Guide* (ed. D.L. Dindal), John Wiley & Sons, New York, pp. 1191-209.

Tate, R.L. (1979) Effect of flooding on microbial activity in organic soils: carbon metabolism. *Soil Science*, **128**, 267-72.

Taube, M. (1992) Evolution of matter and energy, in *The Natural Environment and the Biogeochemical Cycles* (ed. O. Hutzinger), Springer-Verlag, Berlin, pp. 65-182.

Tayasu, I., Sublmoto, A., Wada, E. *et al.* (1994) Xylophagous termites depending on atmospheric nitrogen. *Naturwissenschaften*, **81**, 229-31.

Tayasu, I., Abe, T., Eggleton, P. *et al.* (1997) Nitrogen and carbon isotope ratios in termites: an indicator of trophic habit along the gradient from wood-feeding to soil-feeding. *Ecological Entomology*, **22**, 343-51.

Tayasu, I., Inoue, T., Miller, L.R. *et al.* (1998) Confirmation of soil-feeding termites (Isoptera: Termitidae: Termitinae) in Australia using stable isotope ratios. *Functional Ecology*, **12**, 536-42.

Taylor, S.R. and McClennan, S.M. (1985) *The Continental Crust. Its Composition and Evolution*, Blackwell Scientific Publications, Oxford.

Tester, M., Smith, S.E. and Smith, F.A. (1987) The phenomenon of "non-mycorrhizal" plants. *Canadian Journal of Botany*, **65**, 419-31.

Theng, B.K.G. (1987) Clay-humic interactions and soil structural stability, in *Soil Structure and Aggregate Stability* (ed. P. Rengasamy), Institute for Irrigation and Salinity Research, Tatura, pp. 32-73.

Theng, B.K., Tate, K.R. and Sollins, P. (1989) Constituents of organic matter in temperate and tropical soils, in *Dynamics of Soil Organic Matter in Tropical Ecosystems* (eds D.C. Coleman, J.M. Oades and G. Uehara), University of Hawaii, Hawaii, pp. 5-31.

Thimm, T. and Larink, O. (1995) Grazing preferences of some Collembola for endomycorrhizal fungi. *Biology and Fertility of Soils*, **19** (2-3), 266-8.

Thomas, A., Nicholas, D.P. and Parkinson, D. (1965) Modifications of the agar film technique for assaying lengths of mycelium in soil. *Nature*, **175**, 953-4.

Thomas, R.S., Dakessian, S., Ames, R.N. *et al.* (1986) Aggregation of a silty clay loam soil by mycorrhizal onion roots. *Soil Science Society of America, Journal*, **50**, 1494-99.

Thomas, R.J. (1987) Distribution of *Termitomyces* Heim and other fungi in the nests and major workers of *Macrotermes bellicosus* (Smeathman) in Nigeria. *Soil Biology & Biochemistry*, **19**, 329-33.

Thomas, R.J. and Asakawa, N.M. (1993) Decomposition of leaf litter from tropical forage grasses and legumes. *Soil Biology & Biochemistry*, **25** (10), 1351-61.

Thompson, M.J., Zhang, H., Kazemi, M. *et al.* (1989) Contribution of organic matter to cation exchange capacity and specific surface area of fractionated soil materials. *Soil Science*, **148**, 250-7.

Thorne, B.L. and Kimsey, R.K. (1983) Attraction of Neotropical *Nasutitermes* to carrion. *Biotropica*, **15**, 295-6.

Tiedje, J.M., Sextone, A.J., Parkin, T.B. *et al.* (1984) Anaerobic processes in soil. *Plant and Soil*, **76**, 197-212.

Tiedje, J.M. (1995) Approaches to the comprehensive evaluation of Prokaryote diversity of a habitat, in *Microbial Diversity and Ecosystem Function* (eds D. Allsop, R.R. Colwell and D.L. Hawksworth), UNEP/CAB International, Wallingford, pp. 73-82.

Tiessen, H., Stewart, J.W.B. and Hunt, H.W. (1984) Concept of soil organic matter transformations in relation to organo-mineral particle-size fractions. *Plant and Soil*, **76**, 287-95.

Tieszen, L.L. (1991) Natural variations in the carbon isotope values of plants: implications for archaeology, ecology and paleoecology. *Journal of Archaeological Science*, **18**, 227-48.

Tilman, D. (1982) *Resource Competition and Community Structure*, Princeton University Press, Princeton..

Tisdall, J.M. and Oades, J.M. (1982) Organic matter and water-stable aggregates in soils. *Journal of Soil Science*, **33**, 141-63.

Tobin, J.E. (1994) Ants as primary consumers, diet and abundance in the Formicidae, in *Nourishment and Evolution in Insect Societies* (eds J.H. Hunt and C. Nalepa), Westview Press, Boulder, Colorado, pp. 279-307.

Tomati, U., Grappeli, A. and Galli, E. (1988) The hormone-like effect of earthworm casts on plant growth. *Biology and Fertility of Soils*, **5**, 288-94.

Toomsan, B., McDonagh, J.F., Limpinuntana, V. *et al.* (1995) Nitrogen fixation by groundnut and soya bean and residual nitrogen benefits to rice in farmers' fields in Northeast Thailand. *Plant and Soil*, **175**, 45-56.

Torsvik, V.L. and Goksoyr, J. (1978) Determination of bacterial DNA in soil. *Soil Biology & Biochemistry*, **10**, 7-12.

Torsvik, V., Salte, K., Sorheim, R. *et al.* (1990) Comparison of phenotypic diversity and DNA heterogeneity in a population of soil bacteria. *Applied Environmental Microbiology*, **56**, 776-81.

Touchot, F., Kilbertus, G. and Annie, G. (1983) Rôle d'un collembole (*Folsomia candida*) au cours de la dégradation des litières de charme et de chêne, en présence ou en absence d'argile, in *New Trends in Soil Biology* (eds Ph. Lebrun, H.M. André, A. de Medts *et al.*), Dieu-Brichart, Ottignies, pp. 269-80.

Touchot, F. (1984) *Relations microfaune-microflore dans le cadre de la biodégradation des écorces et de litières*, thèse de 3e cycle, Université Nancy I.

Toutain, F. (1974) *Etude ecologique de l'humification dans les hêtraies acidophiles*, thèse d'Etat, Université Nancy I.

Toutain, F. (1981) Les humus forestiers: structure et mode de fonctionnement. *Revue Forestière Française*, **XXXIII**, 449-77.

Toutain, F., Villemin, G., Albrecht, A. *et al.* (1982) Etude ultrastructurale des processus de biodégradation. II. Modèle enchytraeides-litière de feuillus. *Pedobiologia*, **23**, 145-56.

Toutain, F. (1984) Biologie des sols, in *Livre Jubilaire du Cinquantenaire* (ed. AFES), Maugein-Lachaise, Malermont, pp. 253-71.

Toutain, F. (1987a) Les litières: sièges de systèmes interactifs et moteurs de ces interactions. *Revue d'Ecologie et de Biologie du Sol*, **24**, 231-42.

Toutain, F. (1987b) Activité biologique des sols, modalités et lithodépendance. *Biology and Fertility of Soils*, **3**, 31-8.

Tôth, J.A. and Hammer, I. (1977) Quantitative microbiological studies on the soil of "Sikfokut Project". *Acta Biologica Debrecina*, **14**, 33-44.

Traniello, J.F.A. and Levings, S.C. (1986) Intra- and intercolony patterns of nest dispersion in the ant *Lasius neoniger*. Correlations with territoriality and foraging ecology. *Oecologia*, **69**, 413-9.

Trappe, J.M. (1987) Phylogenetic and ecological aspects of mycotrophy in the angiosperms from an evolutionary standpoint, in *Ecophysiology of VA Mycorrhizal Plants* (ed. G.R. Safir), CRC Press, Boca Raton, Florida, pp. 5-25.

Trehen, P. (1971) Répartition des Diptères à larves édaphiques en fonction de la texture et des états hydriques de divers types de sols. *Annales de Zoologie, Ecologie Animale*, **Hors-série**, 273-88.

Trigo, D. and Lavelle, P. (1993) Changes in respiration rate and some physicochemical properties of soil during gut transit through *Allolobophora molleri* (Lumbricidae, Oligochaeta). *Biology and Fertility of Soils*, **15**, 185-8.

Trigo, D. and Lavelle, P. (1995) Soil changes during gut transit through *Octolasion lacteum* Oerley (Lumbricidae, Oligochaeta). *Acta Zoologica Fennica*, **196**, 129-31.

Trinick, M.J. and Hadobas, P.A. (1988) Biology of the *Parasponia-Bradyrhizobium* symbiosis. *Plant and Soil*, **110**, 177-85.

Trofymow, J.A. and Coleman, D.C. (1982) The role of bacterivorous and fungivorous nematodes in cellulose and chitin decomposition, in *Nematodes in Soil Ecosystems* (ed. D.W. Freckman), University of Texas Press, Austin, pp. 117-38.

Trofymow, J.A., Coleman, D.C. and Cambardella, C. (1987) Rates of rhizodeposition and ammonium depletion in the rhizosphere of axenic oat roots. *Plant and Soil*, **97**, 333-44.

Turchenek, L.W. and Oades, J.M. (1979) Fractionation of organo-mineral complexes by sedimentation and density techniques. *Geoderma*, **21**, 311-43.

Turner, D.P. and Franz, E.H. (1985) The influence of Western Hemlock and Western Red Cedar on microbial numbers, nitrogen mineralization, and nitrification. *Plant and Soil*, **88**, 259-67.

Uehara, G. (1982) Soil science for the tropics, in *Engineering and Construction in Tropical and Residual Soils. Proceedings of the ASCE Geotechnical Engineering Division Specialty Conference*, American Society of Civil Engineers, Honolulu, Hawaii, pp. 13-26.

Ugolini, F.C. (1986a) Pedogenic zonation in the well-drained soils of the Arctic regions. *Quaternary Research*, **26**, 100-20.

Ugolini, F.C. (1986b) Processes and rates of weathering in cold and polar desert environments, in *Rates of Chemical Weathering of Rocks and Minerals* (eds S.M. Colman and D.P. Dethier), Academic Press,

Orlando, pp.193-235 .

Ulber, B. (1983) Einfluß von *Onychiurus fimatus* Gisin (Collembola, Onychiuridae) und *Folsomia fimetaria* (L.) (Collembola, Isotomidae) auf *Pythium ultimum* Trow., einen erreger des wurzelbrandes der zuckerrübe, in *New Trends in Soil Biology* (eds Ph. Lebrun, A.M. André, A. de Medts *et al.*), Dieu-Brichart, Louvain-la-Neuve, pp. 261-68.

Urbanek, J. and Dolezal, F. (1992) Review of some case studies on the abundance and on the hydraulic efficiency of earthworm channels in Czechoslovak soils, with reference to the subsurface pipe drainage. *Soil Biology & Biochemistry*, **24** (12), 1563-71.

Urbasek, F. (1990) Cellulase activity in the gut of some earthworms. *Revue d'Ecologie et de Biologie du Sol*, **27** (1), 21-8.

Usher, M.B. (1975) Some properties of the aggregations of soil arthropods: Cryptostigmata. *Pedobiologia*, **15**, 355-74.

Vachon, M. (1947) Contribution à l'étude du développement post-embryonnaire de *Pachybolus ligulatus* Voges (Myriapodes, Diplopodes, Spirobolides). Les étapes de la croissance. *Annales des Sciences Naturelles, Zoologie*, **11** (2), 109-21.

van Breemen, N. and Feijtel, T.C.J. (1990) Soil processes and properties involved in the production of greenhouse gases, with special relevance to soil taxonomic systems, in *Soils and the Greenhouse Effect* (ed. A.F. Bouwman), John Wiley, London, pp. 195-223.

van Breemen, N. (1993) Soils as biotic constructs favouring net primary productivity. *Geoderma*, **57** (3), 183-211.

van Cleemput, O. and El-Sebaay, A.S. (1985) Gaseous hydrocarbons in soil. *Advances in Agronomy*, **38**, 159-81.

van Gansen, P. (1962) Structures et fonctions du tube digestif du lombricien *Eisenia foetida* Savigny. *Journal of Microscopy*, **1**.

van Noordwijk, M. and de Willigen, P. (1986) Quantitative root ecology as element of soil fertility theory. *Netherlands Journal of Agricultural Science*, **34**, 273-81.

van Noordwijk, M., Brouwer, G. and Klaas, H. (1993) Concepts and methods for studying interactions of roots and soil structure. *Geoderma*, **56**, 351-75.

van Noordwijk, M. and Brouwer, G. (1993) Gas-filled root porosity in response to temporary low oxygen supply in different growth stages. *Plant and Soil*, **152** (2), 187-99.

van Praag, H.J., Sougnez-Remy, S., Weissen, F. *et al.* (1988) Root turnover in a Beech and a Spruce stand of the Belgian Ardennes. *Plant and Soil*, **105**, 87-103.

van Raij, B. and Peech, M. (1972) Electrochemical properties of some oxisols and alfisols of the tropics. *Soil Science Society of America, Proceedings*, **36**, 587-93.

van Vliet, P.C.J., West, L.T., Hendrix, P.F. *et al.* (1993) The influence of Enchytraeidae (Oligochaeta) on the soil porosity of small microcosms. *Geoderma*, **56** (1-4), 287-99.

Vance, E.D., Brookes, P.C. and Jenkinson, D.S. (1987) An extraction method for measuring soil microbial biomass C. *Soil Biology & Biochemistry*, **19** (6), 703-7.

Vancura, V. (1967) Root exudates of plants. III: Effect of temperature and "cold shock" on the exudations of various compounds from seeds and seedlings of maize and cucumber. *Plant and Soil*, **27**, 319-28.

Vancura, V. and Garcia, J.L. (1969) Root exudates of reversibly wilted millet plants (*Panicum milliaceum* L.). *Oecologia Plantarum*, **4**, 93-8.

Vancura, V. (1980) Fluorescent pseudomonads in the rhizosphere of plants and their relation to root exudates. *Folia Microbiologica*, **25**, 168-73.

Vannier, G. (1970) *Réaction des Microarthropodes aux variations de l'état hydrique du sol - Techniques relatives à l'extraction des arthropodes du sol*, CNRS, Paris.

Vannier, G. (1978) La résistance à la dessication chez les premiers arthropodes terrestres. *Bulletin de la Société d'Ecophysiologie*, **3** (1), 13-42.

Vannier, G. (1985) Modes d'exploitation et partage des ressources alimentaires dans le système saprophage par les microarthropodes du sol. *Bulletin d'Ecologie*, **16** (1), 19-34.

Vannier, G. (1987) The porosphere as an ecological medium emphasized in Professor Ghilarov's work on soil animal adaptations. *Biology and Fertility of Soils*, **3**, 39-44.

Vargas, R. and Hattori, T. (1986) Protozoan predation of bacterial cells in soil aggregates. *FEMS Microbiology, Ecology*, **38**, 233-42.

Vartanian, N. (1977) *Influence des facteurs hydriques de l'environnement sur le système racinaire: aspects morphologiques, histologiques et ecophysiologiques*, thèse d'Etat, Université Paris XI.

Vartanian, N. and Chauveau, M. (1986) Etude *in vitro* de la reprise de la respiration racinaire chez la Moutarde et le Colza après une période de sécheresse. *Plant and Soil*, **92**, 255-64.

Vasconcelos, H.L. and Cherrett, J.M. (1997) Leaf-cutting ants and early forest regeneration in central Amazonia: effects of herbivory on tree seedling establishment. *Journal of Tropical Ecology*, **13**, 357-70.

Vedy, J.C. and Bruckert, S. (1979) Les solutions du sol. Composition et signification pédogénétique, in *Constituants et Propriétés du Sol* (eds M. Bonneau and B. Souchier), Masson, Paris, pp. 161-81.

Vegter, J.J. and Huyer-Brugman, F.A. (1983) Comparative water relations in Collembola: transpiration, dessication tolerance and effects of body size, in *New Trends in Soil Biology* (eds Ph. Lebrun, H.M. André, A. de Medts *et al.*), Dieu-Brichart, Ottignies, pp. 411-6.

Veivers, P.C., Mühlemann, R., Slaytor, M. *et al.* (1991) Digestion, diet and polyethism in two fungus-growing termites: *Macrotermes subhyalinus* and *M. michaleseni* (Sjöstedt). *Journal of Insect Physiology*, **37**, 675-82.

Venter, J.M. and Reinecke, A.J. (1988) The life-cycle of the compost worm *Eisenia foetida* (Oligochaeta). *South African Journal of Zoology*, **23** (3), 161-5.

Vercoe, T.K. (1987) Fodder potential of selected Australian tree species, in *Australian Acacias in Developing Countries* (ed. J.W. Turnbull), ACIAR, Canberra, pp. 95-100.

Veresoglou, D.S. and Fitter, A.H. (1984) Spatial and temporal patterns of growth and nutrient uptake of five coexisting grasses. *Journal of Ecology*, **72**, 259-72.

Verhoef, H.A. and Nagelkenke, C.J. (1977) Formation and ecological significance of aggregations in Collembola. An experimental study. *Oecologia*, **31**, 215-26.

Verhoef, H.A. and van Selm, A.J. (1983) Distribution and population dynamics of Collembola in relation to soil moisture. *Holarctic Ecology*, **6**, 387-94.

Verhoef, H.A. and Brussaard, L. (1990) Decomposition and nitrogen mineralization in natural and agro-ecosystems: the contribution of soil animals. *Biogeochemistry*, **11**, 175-211.

Vernon, P. and Vannier, G. (1990) Adaptation écophysiologique des insectes arctiques et antarctiques. *Bulletin de la Société d'Ecophysiologie*, **15** (1), 7-21

Villalobos, F.J. and Lavelle, P. (1990) The soil Coleoptera community of a tropical grassland from Laguna Verde, Veracruz (Mexico). *Revue d'Ecologie et de Biologie du Sol*, **27** (1), 73-93.

Villalobos, F.J. (1991) The community structure of soil Coleoptera (Melolonthidae) from a tropical grassland in Veracruz, Mexico. *Pedobiologia*, **35**, 225-38.

Villenave, C., Charpentier, F., Lavelle, P., *et al* (1999) Effects of earthworms on soil organic matter and nutrient dynamics following earthworm incubation in field experimental situations, in *Earthworm Management in Tropical Agroecosystems* (eds. P. Lavelle, L. Brussaard and P. Hendrix), CAB International, Wallingford, pp. 173-98.

Vincent, W.F. (1988) *Microbial Ecosystems of Antarctica*, University Press, Cambridge.

Visser, S. and Parkinson, D. (1975) Fungal succession on Aspen Poplar leaf litter. *Canadian Journal of Botany*, **53**, 1640-51.

Vitousek, P.M. and Sanford, R.L.J. (1986) Nutrient cycling in moist tropical forest. *Annual Review of Ecology and Systematics*, **17**, 137-67.

Vitousek, P.M. and Howarth, R.W. (1991) Nitrogen limitation on land and in the sea: how can it occur? *Biogeochemistry*, **13**, 87-115.

Vitousek, P.M., Turner, D.R., Parton, W.J. *et al.* (1994) Litter decomposition on the Mauna Loa environmental matrix, Hawaii - patterns, mechanisms, and models. *Ecology*, **75** (2), 418-29.

Vitousek, P.M., Aber, J.D., Howarth, R.W. *et al.* (1997) Human alteration of the global nitrogen cycle: sources and consequences. *Ecological Applications*, **7** (3), 737-50.

Vogt, K.A., Edmonds, R.L., Grier, C.C. *et al.* (1980) Seasonal changes in mycorrhizal and fibrous-textured root biomass in 23 and 180 year old Pacific Silver Fir stands in Western Washington. *Canadian Journal of Forest Research*, **10**, 523-9.

Vogt, K.A., Edmonds, R.L. and Grier, C.C. (1981) Seasonal changes in biomass and vertical distribution of mycorrhizal and fibrous-textured conifer fine roots in 23 and 180 year old subalpine *Abies amabilis* stands. *Canadian Journal of Forest Research*, **11**, 223-9.

Volz, P. (1951) Untersuchungen über die mikrofauna des Waldbodens. *Zoologischer Jährbucher Abteilungen Systematik, Ökologie und Geographie der Tiere*, **79**, 514-66.

Volz, P. (1964) Uber die soziologizsche und physiognomische Forschungrichtung in der Bodenzoologie. *Verhandlungen der Deutschen Zoologischen Gesellschaft Keil*, **28**, 522-32.

Voroney, R.P. (1983) *Decomposition of crop residues*, Ph.D. thesis, University of Saskatchewan, Canada.

Wada, K. (1985) The distinctive properties of andosols. *Advances in Soil Science*, **2**, 174-229.

Wagner, D. (1997) The influence of ant nests on Acacia seed production, herbivory and soil nutrients. *Journal of Ecology*, **85** (1), 83-93.

Wainwright, M. (1988) Metabolic diversity of fungi in relation to growth and mineral cycling in soil - a review. *Transactions of the British Mycological Society*, **90**, 159-70.

Waksman, S.A. (1952) *Soil Microbiology*, John Wiley & Sons, New York.

Walker, T.W. and Adams, A.F.R. (1958) Studies on soil organic matter: 1. Influence of phosphorus content of parent materials on accumulations of carbon, nitrogen, sulfur, and organic phosphorus in grassland soils. *Soil Science*, **85**, 307-18.

Walker, T.W. and Syers, J.K. (1976) The fate of phosphorus during pedogenesis. *Geoderma*, **15**, 1-19.

Walker, C.M. and Trappe, J.M. (1993) Names and epithets in the Glomales and Endogonales. *Mycological Research*, **97**, 339-44.

Wallace, H. (1961) The orientation of *Ditylenchus dipsaci* to physical stimuli. *Nematologica*, **6**, 222-36.

Wallace, M.M.H. (1968) The ecology of *Sminthurus viridis*. II. Diapause in the aestivating egg. *Australian Journal of Zoology*, **16**, 871-83.

Waller, D.A. and La Fage, J.P. (1987) Nutritional ecology of termites, in *Nutritional Ecology of Insects, Mites and Spiders* (eds F.J. Slansky and J.G. Rodriguez), John Wiley & Sons, New York, pp. 487-532.

Wallwork, J.A. (1967) Acari, in *Soil Biology* (eds A. Burges and F. Raw), Academic Press, London, pp. 363-95.

Waloff, N. and Blackith, R.E. (1962) The growth and distribution of mounds of *Lasius flavus* (Fabricius) (Hymenoptera: Formicidae) in Silwood Park, Berkshire. *Journal of Animal Ecology*, **31**, 421-37.

Walsh, G.E. (1974) Mangroves: a review, in *Ecology of Halophytes* (eds R.J. Reimold and W.H. Queen), Academic Press, New York, pp. 51-174.

Wang, S.C., Kao, M. and Huang, P.M. (1980) The effect of pH on the catalytic synthesis of humic substances by illite. *Soil Science*, **129**, 333-8.

Wang, D., McSweeney, K., Lowery, B. *et al.* (1995) Nest structure of *Lasius neoniger* Emery and its implications to soil modification. *Geoderma*, **66**, 259-72.

Warburg, M.R. (1987) Isopods and their terrestrial environment. *Advances in Ecological Research*, **17**, 187-240.

Wardle, D.A. (1992) A comparative assessment of factors which influence microbial biomass carbon and nitrogen in soils. *Biological Reviews of the Cambridge Philsophical Society*, **67**, 321-58.

Wardle, D.A. and Lavelle, P. (1997) Linkages between soil biota, plant litter quality and decomposition, in *Driven by Nature: Plant Litter Quality and Decomposition* (eds G. Cadisch and K.E. Giller), CAB International, Wallingford, pp. 107-25.

Warembourg, F.R. and Roumet, C. (1989) Why and how to estimate the cost of symbiotic N_2 fixation? A progressive approach based on the use of ^{14}C and ^{15}N isotopes. *Plant and Soil*, **115**, 167-77.

Warneke,C.,Karl,T.,Judmaier,H.,Hansel,A.,Jordan,A.and Lindinger,W.(1999)Acetone,methanol,and other partially oxidized volatile organic emissions from dead plant matter by abiological processes:significance for atmospheric chemistry.Global Biogeochemical Cycles,13, 9-17.

Warner, G.S., Nieber, J.L., Moore, I.D. *et al.* (1989) Characterising macropores in soil by computed tomography. *Soil Science Society of America, Proceedings*, **53**, 653-60.

Warnock, A.J., Fitter, A.H. and Usher, M.B. (1982) The influence of a springtail *Folsomia candida* (Insecta, Collembola) on the mycorrhizal association of leek *Allium porrum* and the vesicular-arbuscular mycorrhizal endophyte *Glomus fascinatus*. *New Phytologist*, **90**, 285-.

Wasilewska, L. (1971) Trophic classification of soils and plant nematodes. *Wiadmosci Ekologie*, **17**, 379-88.

Wasilewska, L. (1979) The structure and function of soil nematode communities in natural ecosystems and agrocenoses. *Polish Ecological Studies*, **5** (2), 97-145.

Wasilewska, L. (1989) Impact of human activities on nematode communities in terrestrial ecosystems, in *Ecology of Arable Land, Perspectives and Challenges* (eds M. Clarholm and L. Bergström), Kluwer Academic Publishers, Dordrecht, pp. 123-32.

Waterman, P.G. (1983) Distribution of secondary metabolites in rain forest plants: towards an understanding of cause and effect, in *Tropical Rainforest: Ecology and Management* (eds S.L. Sutton, T.C. Whitmore and A.C. Chadwick), Blackwell Scientific Publications, Oxford, pp. 167-79.

Waterman, P.G. and McKey, D. (1989) Herbivory and secondary compounds in rain-forest plants, in *Tropical Rain Forest Ecosystems* (eds H. Lieth and M.J.A. Werger), Elsevier, Amsterdam, pp. 513-36.

Watson, J.A.L. (1969) *Schedorhinotermes derosus*, a harvester termite in northern Australia (Isoptera: Rhinotermitidae). *Insectes Sociaux*, **26**, 173-8.

Watson, J.A.L. and Gay, F.J. (1970) The role of grass-eating termites in the degradation of a mulga ecosystem. *Search*, **1**, 43.

Watson, J.P. (1972) The distribution of gold in termite mounds and soils at a gold anomaly in Kalahari sand. *Soil Science*, **113**, 317-321.

Watson, J.A.L., Barrett, R.A. and Green, J.P. (1988) Growth of mounds of the Australian harvester termite, *Drepanotermes perniger* (Froggatt) (Termitinae). *Sociobiology*, **14**, 217-44.

Watson, J.A.L. and Gay, F.J. (1991) Isoptera, in *The Insects of Australia* (eds I.D. Nauman, P.B.Carne, J.F. Lawrence *et al.*), Melbourne University Press, Carlton, Victoria, pp. 330-47.

Watson, A. (1992) Desert soils, in *Weathering, Soils & Paleosols* (eds I.P. Martini and W. Chesworth), Elsevier, Amsterdam, pp. 225-260.

Wauthy, G. (1982) Revue critique des relations entre la faune, la matière organique et les microorganismes dans les horizons organiques forestiers. *Agronomie*, **2**(7), 667-75.

Way, M.J. and Khoo, K.C. (1992) Role of ants in pest management. *Annual Review of Entomology*, **37**, 479-503.

Wayne, L.G., Brenner, D.J., Colwell, R.R. *et al.* (1987) Report of the *ad hoc* committee on reconciliation of approaches to bacterial systematics. *International Journal of Systematics and Bacteriology*, **37**, 463-4.

Weary, G.C. and Merriam, H.G. (1978) Litter decomposition in a Red Maple woodlot under natural conditions and under insecticide treatment. *Ecology*, **59**, 180-4.

Weber, N.A. (1972) Gardening ants, the attines. *Memories of the American Philosophical Society*, **92**, 1-146.

Weber, N.A. (1982) Fungus ants, in *Social Insects* (ed. H.R. Hermann), Academic Press, New York, pp. 255-363.

Webster, J. (1970) Coprophilous fungi. *Transactions of the British Mycological Society*, **54**, 161-80.

Wells, T.C.E., Sheail, J., Ball, D.F. *et al.* (1976) Ecological studies in the Porton Ranges. Relationship between vegetation, soils and land-use history. *Journal of Ecology*, **64**, 589-624.

Wetterer, J.K. (1994) Nourishment and evolution in fungus-growing ants and their fungi, in *Nourishment and Evolution in Insect Societies* (eds J.H. Hunt and C. Nalepa), Westview Press, Boulder Colorado, pp. 309-28.

Wheeler, W.M. (1907) The fungus-growing ants of North America. *Bulletin of the American Museum of Natural History*, **23**, 669-807.

Wheeler, W.M. (1910) *Ants, their Structure, Development and Behaviour*, Columbia University Press, New York.

Wheeler, W.M. (1936) Ecological relations of ponerine and other ants to termites. *Proceedings of the American Academy of Arts and Sciences*, **71**, 159-243.

Whelan, J. (1986) An analysis of acarine communities in the herbage and soil of three adjoining grassland sites in Co. Kildare, Ireland. *Pedobiologia*, **29**, 149-64.

White, D.C. (1988) Validation of quantitative analysis for microbial biomass, community structure, and metabolic activity. *Archiv für Hydrobiologie. Beiheft*, **31**, 1-18.

Whitford, W.G., Schaefer, D. and Wisdom, W. (1986) Soil movement by desert ants. *Southern Naturalist*, **31**, 273-4.

Whitford, W.G. (1989) Abiotic controls on the functional structure of soil food webs. *Biology and Fertility of Soils*, **8**, 1-6.

Whitford, W.G. (1991) Subterranean termites and long term productivity of desert rangelands. *Sociobiology*, **19**, 235-43.

Widden, P. (1986) Seasonality of forest soil microfungi in southern Quebec. *Canadian Journal of Botany*, **64**, 1413-23.

Widden, P. (1987) Fungal communities in soils along an elevation gradient in southern England. *Mycologia*, **79** (2), 298-309.

Wielemaker, W.G. (1984) *Soil formation by termites, a study in the Kisii area, Kenya*, Ph.D. thesis, University of Wageningen.

Wieser, W. (1966) Copper and the role of isopods in degradation of organic matter. *Science*, **153**, 67-9.

Wiken, E.B., Broersma, K., Lakulich, L.M. *et al.* (1976) Biosynthetic alteration in a British Columbia soil by ants (*Formica fusca* Linné). *Soil Science Society of America, Proceedings*, **40**, 422-6.

Wilding, L.P., Smeck, N.E. and Hall, G.F. (1983) *Pedogenesis and Soil Taxonomy II. The Soil Orders*, Elsevier, Amsterdam.

Williams, M.A.J. (1968) Termites and soil development near Brocks Creek, Northern Territory. *Australian Journal of Soil Science*, **31**, 153-4.

Wilson, E.O. (1959) Some ecological characteristics of ants in New Guinea rain forests. *Ecology*, **40**, 437-47.

Wilson, E.O. and Taylor, R.W. (1967) The ants of Polynesia. *Pacific Insects Monographs*, **14**, 1-109.

Wilson, E.O. (1971) *The Insect Societies*, Belknap Press, Cambridge Massachusetts.

Wilson, E.O. (1976) Behavioural discretization and the number of castes in ant species. *Behavioural Ecology and Sociobiology*, **1**, 141-54.

Wilson, E.O. (1988) The biogeography of the West Indian ants (Hymenoptera: Formicidae), in *Zoogeography of Caribean Insects* (ed. J. Liebherr), Cornell University Press, Ithaca, pp. 214-30.

Winter, J.P., Voroney, R.P. and Ainsworth, D.A. (1990) Soil microarthropods in long-term no-tillage and conventional tillage corn production. *Canadian Journal of Soil Science*, **70**, 641-53.

Wise, K.A.J. (1965) New records of Collembola and Acari in Antarctica. *Pacific Insects*, **6**, 522-3.

Wise, K.A.J. and Gressitt, J.L. (1965) Far southernmost animals and plants. *Nature*, **207**, 101-2.

Wise, K.A.J. and Spain, A.V. (1967) Entomological investigations in Antarctica, 1963-64. *Pacific Insects Monograph*, **9**, 271-93.

Wise, K.A.J. and Shoup, J. (1971) Entomological investigations in Antarctica, 1964-65 season. *Pacific Insects Monograph*, **25**, 27-56.

Wittich, W. (1952) Der heutige Stand unseres Wissens vom Humus une neue Wege zur Lösung des Rohhumusproblems im Walde. *Schriften aus der Forstlichen Fakultät der Universität Göttingen*, **4**, 1-106.

Witty, J.F. and Minchen, F.R. (1990) Rhizobium ecology in tropical pasture systems, in *Nitrogen Fixation: Achievements and Objectives* (eds P. M. Gresshoff, L. Evans Roth, A. G. Stacey *et al.*), Chapman & Hall, New York, pp. 285-92.

Woese, C.R. (1992) Prokaryote systematics: the evolution of a science, in *The Prokaryotes: A Handbook on the Biology of Bacteria: Ecophysiology, Isolation, Identification, Applications, Volume 1* (eds A. Balows, H.G. Truper, M. Dworkin *et al.*), Springer-Verlag, New York, pp. 3-18.

Wolters, V. (1988) Effects of *Mesenchytraeus glandulosus* (Oligochaeta, Enchytraeidae) on decomposition processes. *Pedobiologia*, **32**, 387-98.

Wolters, V. (1991) Soil invertebrates - effects on nutrient turnover and soil structure - a review. *Zeitschrift für Pflanzenernährung und Bodenkunde*, **154**, 389-402.

Wood, T.G. and Lee, K.E. (1971) Abundance of mounds and competition among colonies of some Australian termite species. *Pedobiologia*, **11**, 341-66.

Wood, T.G. and Sands, W.A. (1978) The role of termites in ecosystems, in *Production Ecology of Ants and Termites* (ed. M.V. Brian), Cambridge University Press, Cambridge, pp. 245-92.

Wood, T.G., Johnson, R.A. and Ohiagu, C.E. (1977) Populations of termites (Isoptera) in natural and agricultural ecosystems in southern Guinea savanna near Mokwa, Nigeria. *Geo-Eco-Trop*, **1**, 139-48.

Wood, T.G. (1978) Food and feeding habits of termites, in *Production Ecology of Ants and Termites* (ed. M.V. Brian), Cambridge University Press, Cambridge, pp. 55-80.

Wood, T.G. (1979) The termite (Isoptera) fauna of Malesian and other tropical rainforests, in *The Abundance of Animals in Malesian Rainforests. Transactions of the 6th Aberdeen-Hull Symposium on Malesian Ecology* (ed. A.G. Marshall), Aberdeen, pp. 113-32.

Wood, T.G., Johnson, R.A., Bacchus, S. *et al.* (1982) Abundance and distribution of termites (Isoptera) in a riparian forest of the Southern Guinea savanna vegetation zone of Nigeria. *Biotropica*, **14**, 25-39.

Wood, T.G., Johnson, R.A. and Anderson, J.M. (1983) Modification of soils in Nigerian savanna by soil-feeding *Cubitermes* (Isoptera, Termitidae). *Soil Biology & Biochemistry*, **15**, 575-79.

Wood, T.G. (1988) Termites and the soil environment. *Biology and Fertility of Soils*, **6**, 228-36.

Wood, T.G. (1991) Termites in Ethiopia: the environmental impact of their damage and resultant control measures. *Ambio*, **20**, 136-8.

Wood, T.G. and Pearce, M.J. (1991) Termites in Africa: the environmental impact of control measures and damage to crops, trees, rangeland and rural building. *Sociobiology*, **19**, 221-34.

Wood, T.G. (1996) The agricultural importance of termites in the tropics. *Agricultural Zoology Reviews*, **7**, 117-55.

Woodin, S., Graham, B. and Killick, A. (1992) Nutrient limitation of the long term response of heather *Calluna vulgaris* (L. Mull.) to CO_2 enrichment. *New Phytologist*, **122**, 635-42.

Woomer, P. and Bohlool, B.B. (1989) Rhizobium ecology in tropical pasture systems, in *Persistence of Forage Legumes* (ed. G.C. Marten), American Society of Agronomy, Madison, Wisconsin, pp. 233-45.

Woomer, P.L. and Swift, M.G. (1994) *The Biological Management of Tropical Soil Fertility*, John Wiley &

Sons, Chichester.

Wright, M.A. (1972) Factors governing ingestion by the earthworm *Lumbricus terrestris* with special reference to apple leaves. *Annals of Applied Biology*, **70**, 175-88.

Wu, J., O'Donnell, A.G., He, Z.L. *et al.* (1994) Fumigation-extraction method for the measurement of soil microbial S. *Soil Biology & Biochemistry*, **26** (1), 117-25.

Wynn-Williams, D.D. (1983) Microflora-microfauna interactions in Antarctic moss peat decomposition processes, in *New Trends in Soil Biology* (eds Ph. Lebrun, A.M. André, A. de Medts *et al.*), Dieu-Brichard, Louvain-la-Neuve, pp. 237-46.

Wyss, U. (1981) Ectoparasitic root nematodes: feeding behaviour and plant cell responses, in *Plant Parasitic Nematodes* (eds B.M. Zuckerman and R.A. Rohde), Academic Press, New York, pp. 325-51.

Yamin, M.A. (1981) Cellulose metabolism by the flagellate *Trichonympha* from a termite is independent of endosymbiotic bacteria. *Science*, **211**, 58-9.

Yang, W., Spencer, R.J. and Krouse, R.H. (1996) Stable sulfur isotope hydrogeochemical studies using desert shrubs and tree rings, Death Valley, California, USA. *Geochemica et Cosmochimica Acta*, **60** (16), 3015-22.

Ydrogo, H. (1994) *Efecto de la inoculación de lombrices de tierra* Pontoscolex corethrurus *(Glossoscolecidae) en las micorrizas vesiculo arbusculares y en la etapa de crecimiento de araza* (Eugenia stipitata), *achiote* (Bixa orellana) *y pijuayo* (Bactris gasipaes) *en suelos ultisoles de Yurimaguas (Peru)*, tesis de licenciatura, Universidad de Tarapoto.

Yeates, G.W. (1981) Soil nematode populations depressed in the presence of earthworms. *Pedobiologia*, **22**, 191-5.

Yeates, G.W. and Coleman, D.C. (1982) Role of nematodes in decomposition, in *Nematodes in Soil Ecosystems* (ed. D.W. Freckman), University of Texas Press, Austin, pp. 55-80.

Yeates, G.W. (1991) Nematode populations at three forest sites in New Caledonia. *Journal of Tropical Ecology*, **7**, 411-3.

Yong, R.N. and Warkentin, B.P. (1975) *Soil Properties and Behaviour*, Elsevier, Amsterdam.

Young, J.P.W. (1992) Phylogenetic classification of nitrogen-fixing organisms, in *Biological Nitrogen Fixation*, Chapman & Hall, New York, pp. 43-86.

Youssef, R.A., Kanazawa, S. and Chino, M. (1989) Distribution of microbial biomass across the rhizosphere of barley (*Hordeum vulgare* L.) in soils. *Biology and Fertility of Soils*, **7**, 341-5.

Yule, D.F., Reddy, M.V., Reddy, V.R. *et al.* (1991) Effects of soil management on population abundance and biomass of earthworm in a semi-arid tropical alfisol, in *Soil Tillage and Agricultual Sustainability. Proceedings of the 12th International Soil Tillage and Resource Organization*, Ibadan, Nigeria, pp. 63-71.

Zachariae, G. (1962) Zur Methodik der Geländeuntersuchugen in der Bodenzoologie. *Zeitschrift für Pflantzenernährung, Düngung, Bodenkunde*, **97**, 224-33.

Zaitseva, V.E. and Zviagintzev, D.G. (1978) Temporal and spatial variations in the number of soil bacteria by a luminescent technique. *Mikrobiologija*, **47**, 342-346.

Zech, W., Guggenberger, G., Haumaier, L. *et al.* (1996) Organic matter dyamics in forest soils of temperate and tropical ecosystems, in *Humic Substances in Terrestrial Ecosystems* (ed. A. Piccolo), Elsevier, Amsterdam, pp. 101-70.

Zelles, L., Rackwitz, R., Bai, Q.Y. *et al.* (1995) Discrimination of microbial diversity by fatty acid profiles of phospholipids and lipopolysaccharides in differently cultivated soils, in *The Significance and Regulation of Soil Biodiversity* (eds H.P. Collins, G.P. Robertson and M.J. Klug), Kluwer Academic Publishers, Dordrecht, pp.115-22.

Zhang, B.G., Rouland, C., Lattaud, C. *et al.* (1993) Activity and origin of digestive enzymes in gut of the tropical earthworm *Pontoscolex corethrurus*. *European Journal of Soil Biology*, **29** (1), 7 - 11.

Zinke, P.J. (1962) The pattern of influence of individual forest trees on soil properties. *Ecology*, **43**, 130-3.

Zinkler, D. (1983) Ecophysiological adaptations of litter-dwelling Collembola and tipulid larvae, in *New Trends in Soil Biology* (eds Ph. Lebrun, H.M. André, A. de Medts *et al.*), Dieu-Brichart, Ottignies, pp. 335-43.

INDEX

Key: Tables are indicated in italics; figures are indicated in bold text.

MIX
Papier aus verantwortungsvollen Quellen
Paper from responsible sources
FSC® C105338

If you have any concerns about our products,
you can contact us on
ProductSafety@springernature.com

In case Publisher is established outside the EU,
the EU authorized representative is:
Springer Nature Customer Service Center GmbH
Europaplatz 3, 69115 Heidelberg, Germany

Printed by Libri Plureos GmbH
in Hamburg, Germany